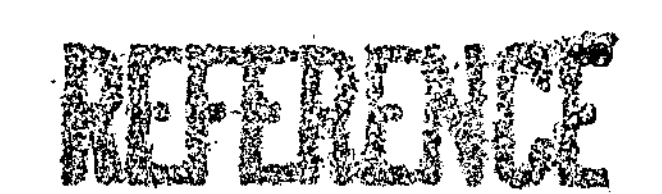

New Dictionary of Scientific Biography

Published by special arrangement with the American Council of Learned Societies

The American Council of Learned Societies, organized in 1919 for the purpose of advancing the study of the humanities and of the humanistic aspects of the social sciences, is a nonprofit federation comprising thirty-three national scholarly groups. The Council represents the humanities in the United States in the International Union of Academies, provides fellowships and grants-in-aid, supports research-and-planning conferences and symposia, and sponsers special projects and scholarly publications.

MEMBER ORGANIZATIONS

American Philosophical Society, 1743
American Academy of Arts and Sciences, 1780
American Antiquarian Society, 1812
American Oriental Society, 1842
American Numismatic Society, 1858
American Philological Association, 1869
Archaeological Institute of America, 1879
Society of Biblical Literature, 1880
Modern Language Association of America, 1883
American Historical Association, 1884
American Economic Association, 1885
American Folklore Society, 1888
American Society of Church History, 1888
American Dialect Society, 1889
American Psychological Association, 1892
Association of American Law Schools, 1900
American Philosophical Association, 1900
American Schools of Oriental Research, 1900
American Anthropological Association, 1902
American Political Science Association, 1903
Bibliographical Society of America, 1904
Association of American Geographers, 1904
Hispanic Society of America, 1904
American Sociological Association, 1905
American Society of International Law, 1906
Organization of American Historians, 1907
American Academy of Religion, 1909
College Forum of the National Council of Teachers of English, 1911
Society for the Advancement of Scandinavian Study, 1911
College Art Association, 1912
National Communication Association, 1914
History of Science Society, 1924
Linguistic Society of America, 1924
Medieval Academy of America, 1925
American Association for the History of Medicine, 1925
American Musicological Society, 1934
Economic History Association, 1940
Society of Architectural Historians, 1940
Association for Asian Studies, 1941
American Society for Aesthetics, 1942
American Association for the Advancement of Slavic Studies, 1948
American Studies Association, 1950
Metaphysical Society of America, 1950
North American Conference on British Studies, 1950
American Society of Comparative Law, 1951
Renaissance Society of America, 1954
Society for Ethnomusicology, 1955
Society for French Historical Studies, 1956
International Center of Medieval Art, 1956
American Society for Legal History, 1956
American Society for Theatre Research, 1956
African Studies Association, 1957
Society for the History of Technology, 1958
Society for Cinema and Media Studies, 1959
American Comparative Literature Association, 1960
Law and Society Association, 1964
Middle East Studies Association of North America, 1966
Latin American Studies Association, 1966
Association for the Advancement of Baltic Studies, 1968
American Society for Eighteenth Century Studies, 1969
Association for Jewish Studies, 1969
Sixteenth Century Society and Conference, 1970
Society for American Music, 1975
Dictionary Society of North America, 1975
German Studies Association, 1976
American Society for Environmental History, 1976
Society for Music Theory, 1977
National Council on Public History, 1979
Society of Dance History Scholars, 1979

New Dictionary of Scientific Biography

VOLUME 4

IBN AL-HAYTHAM–LURIA

Noretta Koertge
EDITOR IN CHIEF

CHARLES SCRIBNER'S SONS
An imprint of Thomson Gale, a part of The Thomson Corporation

Detroit • New York • San Francisco • New Haven, Conn. • Waterville, Maine • London

New Dictionary of Scientific Biography

Noretta Koertge

For more information, contact:
Gale Group
27500 Drake Rd.
Farmington Hills, MI 48331-3535
Or you can visit our Internet site at
http://www.gale.com

LIBRARY OF CONGRESS CATALOGING-IN-PUBLICATION DATA

New dictionary of scientific biography / Noretta Koertge, editor in chief.
p. cm.
Includes bibliographical references and index.
ISBN 978-0-684-31320-7 (set : alk. paper)—ISBN 978-0-684-31321-4 (vol. 1 : alk. paper)—ISBN 978-0-684-31322-1 (vol. 2 : alk. paper)—ISBN 978-0-684-31323-8 (vol. 3 : alk. paper)—ISBN 978-0-684-31324-5 (vol. 4 : alk. paper)—ISBN 978-0-684-31325-2 (vol. 5 : alk. paper)—ISBN 978-0-684-31326-9 (vol. 6 : alk. paper)—ISBN 978-0-684-31327-6 (vol. 7 : alk. paper)—ISBN 978-0-684-31328-3 (vol. 8 : alk. paper)
1. Scientists—Biography—Dictionaries. I. Koertge, Noretta.

Q141.N45 2008
509.2'2—dc22
[B]

2007031384

Editorial Board

I

IBN AL-HAYTHAM, ABU ʿALI AL-HASAN IBN AL-HASAN,

called **al-Baṣri** (of Baṣra, Iraq), **al-Miṣri** (of Egypt), also known as **Alhazen,** the Latinized form of first name, **al-Hasan** (*b.* 965; *d.* Cairo, c. 1040) *optics, astronomy, mathematics.* For the original article on Ibn al-Haytham see *DSB,* vol. 6.

There are still many questions relating to Ibn al-Haytham's biography, such as his origin, his education, the apparently high and demanding administrative position he reluctantly filled in the small Bûyid principality known as al-Baṣra-and-al-Ahwâz, the whens and wheres of the travels he found distracting, the reason and date of his immigration to Egypt, and the exact date of his death. Scholars have been rewarded, however, by the survival of a large number of his writings, some of which are quite long, in elementary and higher mathematics, physics, several aspects of astronomy and cosmology, all of which promise to receive scholarly attention for a long time.

Perspectives on Natural Philosophy. In his brief, tantalizing "Autobiography" (rescued by Ibn Abî Uṣaybiʿa in the thirteenth century), Ibn al-Haytham clearly states that early in his intellectual development he embraced Aristotelian empiricism which, as shown in his later mature *Optics,* he combined with the mathematism he also inherited from Euclid, Ptolemy, and Apollonius. The *Optics* in fact displays a deliberate synthesis (he called it *tarkîb*) which accepts the Aristotelian ontology of substances and qualities (*maʿânî*), like light and color, to which Ibn al-Haytham added the non-Aristotelian concept of point-forms (not his own term) that behave as the origin of light- and color-radiation from shining "points" (Sabra, 1980), and that naturally extend rectilinearly in all directions. Thus was established the foundation of his physical-mathematical, and experimental, theory of light and vision.

Ibn al-Haytham was not an atomist. But he subscribed to a natural-philosophical doctrine according to which the element earth is divisible down to a point where the earth turns to water, further to air, then to fire, and finally to ether, which alone is not divisible (Ibn al-Haytham III.60, *On light,* 2–19). He had this doctrine in mind when he argued, in *Optics,* that normally light shines from a luminous point in a transparent matter, say water or air, along every single straight line passing through that point, but when a minimal thickness of the matter is reached, the light vanishes. Ibn al-Haytham called the light extending along the thinnest possible width of matter "the least light" (*aqallu 'l-qalîl min al-ḍawʾ*)—a concept for which Isaac Newton found a role in his *Opticks,* namely as a suitable definition of "ray."

Ibn al-Haytham's position as physicist or natural philosopher is conveyed repeatedly in the *Optics,* that light does *not* behave in the way it does "for the sake of the eye/sight" (*li-ajl al-baṣar*), but, rather, the seeing eye just registers what it simply *happens* to receive from the passing light. His intention, as was later well-known, was to bury the "visual-ray theory" accepted by Euclid, Ptolemy, Galen, and al-Kindî. And yet, as shown elsewhere by this author, he continued to think in terms of a "single-ray theory" of vision, which he maintained up to Chapter 6 of Book Seven, where he came upon a simple, crucial experiment simply proving for the first time that "*all* that sight perceives, it perceives by refraction" of rays emanating from a single shining point in the shape of a cone. In the

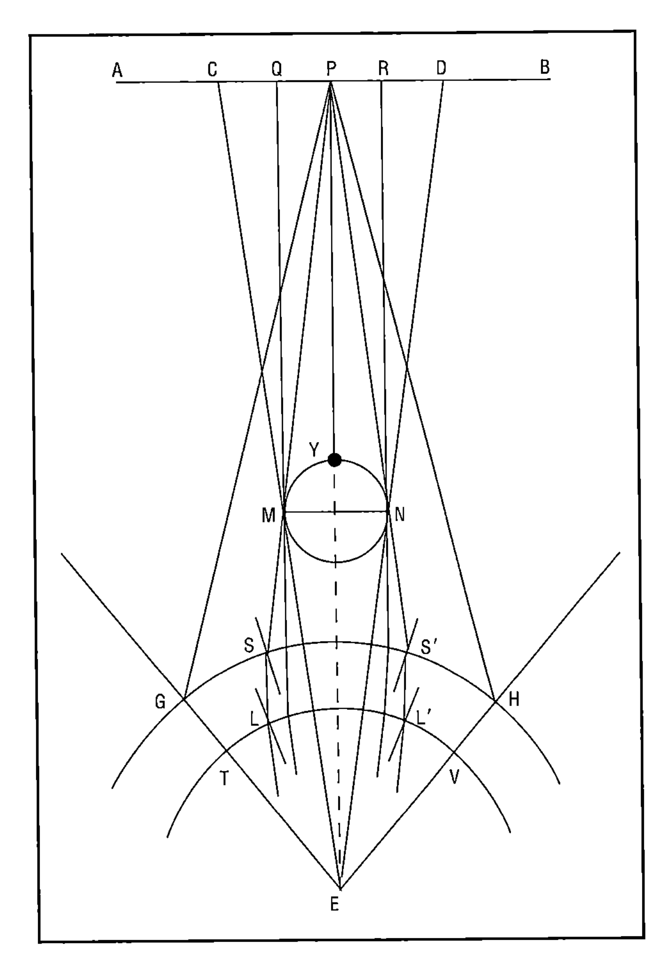

Figure 1. *GEH defines the visual cone. The light issuing from the luminous point P along the perpendicular PYE is obstructed by the needle MYN. Only rays such as PN and PM might reach the surface of the eye GH and the parallel surface TV of the crystalline humor, having been refracted first at the surface of the eye, to be refracted again at TV.*

experiment described in that location, the light proceeding from the shining point along the perpendicular to the surface of the eye cannot enter that surface or the parallel convex surface of the crystalline humor—the perpendicular ray being obstructed by a needle held close to the eye that appears as a shadow over the point seen only by the refracted rays (as in the diagram representing Ibn al-Haytham's clear and original argument) (Sabra, 2003, 99–102).

It was also as a physicist that Ibn al-Haytham searched, in Books Four and Seven, for explanations of optical reflection and refraction which he borrowed from the Muʿtazilite *mutakallimûn,* the dynamical concepts of *iʿtimâd*l (endeavour or effort) and *tawlîd*l (engendering), a substitute for Aristotelian causation. But he was not a practitioner of theological *kalâm,* as is clear from the titles of early essays of his that have not survived (Sabra, 1998, 10–11).

Other Primary Sources. The majority of the extant works making up List III are concerned with astronomy and mathematics. As of 2007 they demand still more attention than they have so far attracted. Some or perhaps many of the short treatises on these subjects may be routine, but Ibn al-Haytham appears to have looked deeper into the astronomical system proposed by the astronomer he referred to as "the excellent Ptolemy/*al-fâḍil Baṭlamyûs,*" while also judging the *Almagest* a book containing "more errors than can be enumerated." In 1962 Schlomo Pines drew attention to Ibn al-Haytham's work, III.64: *Aporias Against Ptolemy* / *al-Shukûk ʿalâ Baṭlamyûs,* which displayed some of "the more serious" of Ptolemy's errors and contradictions, including the well-known equant hypothesis. But an edition of the Arabic text was only published in 1971 (reprinted in 1996).

In mathematics, Ibn al-Haytham's interests obviously leaned more toward geometry and number theory than algebra and practical arithmetic. So, here again, he was, logically and philosophically, more involved with concepts of "the knowns" (III.54: *al-maʿlûmât*) than with "approximations," and with "analysis and synthesis" (III.53: *al-taʾlîl wa ʾl-tarkîb*) than with calculation—see examples, below.

It has been argued that "Muhammad ibn al-Ḥasan" and "al-Ḥasan ibn al-Ḥasan" were two individuals, the first being "a philosopher" and non-practicing physician and author of works constituting Lists I and II, which have been preserved, together with List III, by Ibn Abî Uṣaybiʾa; the other, a mathematician and author of works mostly contained in List III (Rashed 1993). This was an unfortunate proposition, ignoring facts discussed elsewhere (Sabra, 1998), and the old well-known Muslim custom of often naming a newborn male after the name of the prophet, "Muhammad," until a less common name was also adopted (a custom which has sometimes led bibliographers to drop the first name in their alphabetical ordering). Clearly, the author of the *Optics* presented himself as both a physicist or natural philosopher as well as a mathematician, a fact not uncommon in Arabic science.

SUPPLEMENTARY BIBLIOGRAPHY

Titles preceded by their numbers in IAU's version of List III, containing 92 titles (see above).

WORKS BY IBN AL-HAYTHAM

III.1. *M(aqâla). fî Hayʾat al-ʿâlam: On the Configuration of the World,* edited and translated by Y. Tzvi Langermann. New York: Garland Publishing, 1990.

III.3. *K(itâb). Al-Manâẓir – sabʿ Maqâlât: The Optics—Seven Books:* Bks I-II-III, *On Direct Vision,* the Arabic text, and Arabic-Latin glossaries, edited by A.I. Sabra. Kuwait: National Council for Culture, Arts and Letters (NCCAL), 1983 (repr. 2006). *The Optics of Ibn al-Haytham, Bks I–III,* translated, with Introduction and Commentary, by A. I. Sabra, 2 vols. London: The Warburg Institute, 1989. *Alhacen's Theory of Visual Perception, Bks I–II–III,* edited and translated by A. Mark Smith of the Medieval Latin version, 2 vols. Philadelphia: American Philosophical Society, 2001. The Arabic Text of Bks IV–V, *On Reflection, and Images Seen by Reflection,* 2 vols., edited by A. I. Sabra, Kuwait: National Council for Culture, Arts and Letters, 2002. *Alhacen on the Principles of Reflection,* Bks IV–V, edition and translation of the Latin text by A. Mark Smith, 2 vols. Philadelphia: American Philosophical Society, 2006.

III.4. *M. fî Kayfiyyat al-arṣâd: On the Method of [astronomical] Observations,* edited by A. I. Sabra, *Journal for the History of Arabic Sciences* 2 (1978): 3–37, 155.

III.7. *M. fî Samt al-qibla bi'l-'isâb: "Ibn al-Haytham's Universal Solution for Finding the Direction of the Qibla by Calculation,"* edited and translated by A'mad S. Dallâl, *Arabic Science and Philosophy* 5 (1995): 145–93.

III.10. *M. fî Ḥisâb al- mu'âmalât: "Der Mu'âmalât des Ibn al-Haitams," [On Business Arithmetic],* edited and German translation by Ulrich Rebstock, *Zeitschrift für Geschichte der Arabisch-Islamischen Wissenschaften* (*ZGAIW*) 10 (1995–1996): 61–121.

III.12. *M. fî Ru'yat al-kawâkib: On Seeing the Stars,* edited and translated by A. I. Sabra and Anton Heinen, *Zeitschrift für Geschichte der Arabisch-Islamischen Wissenschaften* 7 (1991–1992): 31–72. See below, III.38.

Rashed, Roshdi. *Les mathématiques infinitésimales du IXe au XIe siècles,* vol. II, *On Ibn al-Haytham.* London: Al-Furqân Islamic Heritage Foundation, 1993. This book includes a biographical study (see below) and editions and French translation of the following mathematical works:

III.16. *M. fî Misḥat al-kura: T(raité). sur la mesure de la sphère,* pp. 294–323.

III.17. *M. fî Misḥat al-mujassam al-mukâfî: T. sur la mesure du paraboloïd:* pp. 208–293.

III.20. *M. mukhtaṣarah fî l-Ashkâl al-hilâliyya: Q(awl). fî l-Hilâliyyât: T. sur les lunes,* pp. 70–81.

III.21. *M. mustaqṣâh fî l-ashkâl al-hilâliyya: T. sur les figures des lunules,* pp. 102–175.

III.26. *M. fî anna 'l-kura awsa'al-ashkâl al-mujassamah allatî i'âṭâtuhâ mutasâwiyah, wa anna 'l-dâ'ira awsa'al-ashkâl al-musaṭṭa'a allatî i'âṭâtuhâ mutasâwiyah: T. sur la sphère qui est la plus grande des figures solides ayant des périmètre égaux, et sur le cercle qui est la plus grandes des figures planes ayant des périmètres égaux,* pp. 384–459.

III.30. *M./Q(awl). fî Tarbî'al-dâ'ira: T. sur la quadrature du cercle,* pp. 82–101. See also: Tamara Albertini, "*La Quadrature du cercle d'Ibn al-Haytham: Solution philosophique ou mathématique?," Journal for the History of Arabic Sciences,* 9 (1991): 5–21.

III.38. *M. fî Ḥall shukûk al-maqâla 'l-'ûlâ min Kitâb al-Majisṭî yushakkiku f'hâ ba'ḍu ahl al-'ilm:* partially edited and translated by A.I. Sabra as "*On Seeing the Stars, II.* Ibn al-Haytham's 'Answers' to the 'Doubts' Raised by [Abû l-Qâsim] Ibn Ma'dân," *Zeitschrift für Geschichte der Arabisch-Islamischen Wissenschaften* 10 (1995–1996): 1–59. Contains detailed explanation by Ibn al-Haytham of the so-called moon-illusion problem, composed after the *Optics,* to which it refers.

III.40. *Q. fî Qismat al-miqdârayn al-mukhtalifayn al-madhkûrayn fî l—maqâla al-' âshira min Kitâb Uqlîdis: T. sur la division de deux grandeurs différente mentionnées dans la première proposition du dixième livre de l'ouvrage d'Euclide,* pp. 324–329.

III.49. *M fî 'l-Athar 'alladhî fî 'l-qamar: Treatise on the Mark [seen on the face] of the moon.* Edited by A. I. Sabra, *Journal for the History of Arabic Sciences* I (1977): 5–19.

III.53. *M. fî 'l-Ta'lîl wa 'l-Tarkîb: "L' analyse et la synthèse,"* edited and translated by Roshdi Rashed, *Mélanges de l'Institut Dominicain d'Etudes Orientales du Caire* 20 (1991): 31–231.

III.54. *M. fî 'l-Ma'lûmât:* "Les connus," edited and translated by Roshdi Rashed, *Mélanges de l'Institut Dominicain d Études Orientales du Caire,* 21 (1993): 87–275.

III.61, or III.63. *M. fî Ḥarakat al-iltifâf: "On the Motion of* iltifâf," or "*M. fî Ḥall shukûk arakat [ajrâm] al-iltifâf,*" edited by A. I. Sabra. *Journal for the History of Arabic Sciences* 3 (1979): 183–212. Arabic text and English summary.

III.74. *M. fî 'Amal al-musabba' fî 'l-dâ'ira:* "La construction de l'heptagone régulier par Ibn al-Haytham," edited and translated by Roshdi Rashed, *Journal for the History of Arabic Sciences* 3 (1979): 309–387.

III.77. *M. fî 'Amal al-kura 'l-mu'riqa: "T. sur la sphère ardente."* In *Géométrie et dioptrique au Xe siècle,* edited and translated by Roshdi Rashed, 111–132. Paris: Les Belles Letters, 1993.

Ibn al-Haytham's Completion of the Conics. Edited by Jan P. Hogendijk. New York: Springer Verlag, 1985. Based on a single surviving manuscript.

M. fî Thamarat al-'ikma: "On the Fruit of Wisdom," In *Dhakî Naguib Ma'mûd, Kitâb Tadhkârî,* the Arabic text edited and introduced by M. 'Abd-al-Hâdî Abû Rîda. Kuwait: Kuwait University, 1987.

OTHER SOURCES

Eastwood, Bruce S. "Alhazen, Leonardo, and Late-Medieval Speculation on the Inversion of Images in the Eye." *Annals of Science* 43 (1986): 413–446.

El-Bizri, Nader. "A Philosophical Perspective on Alhazen's *Optics.*" *Arabic Sciences and Philosophy* 15 (2005): 183–218.

Fennane, Khalid Bouzoubaâ. "Réflexions sur le principe de continuité à partir du Commentaire d'Ibn al-Haytham sur la Proposition I.7 des *Eléments* d'Euclide." *Arabic Sciences and Philosophy* 13 (2003): 101–136.

Hogendijk, Jan P. "al-Mu'taman's Simplified Lemmas for Solving 'Alhazen's Problem'." In *From Baghdad to Barcelona.* Studies in the Islamic Exact Sciences in Honour of Prof. Juan Vernet, edited by Josep Casulleras and Julio Samsó, vol. I. Barcelona: Instituto Millas Vallicrosa de Historia de la Ciencia Arabe, 1996.

Lindberg, David C. *Theories of Vision from al-Kindi to Kepler.* Chicago: University of Chicago Press, 1976. Chapter 4 is especially worth noting.

Omar, Saleh B. *Ibn al-Haytham's* Optics: *A Study of the Origins of Experimental Science.* Minneapolis, MN: Bibliotheca Islamica, 1977.

Raynaud, Dominique. "Ibn al-Haytham sur la vision binoculaire: Un précurseur de l'optique physiologique." *Arabic Sciences and Philosophy* 13 (2003) : 79–99.

Sabra, Abdelhamid I. "An Eleventh-Century Refutation [by Ibn al-Haytham] of Ptolemy's Planetary Theory." In *Science and History: Studies in Honor of Edward Rosen* (*Studia Copernicana XVI*), 117-31. Wroclaw, Poland: The Polish Academy of Sciences Press, 1978.

———. "Sensation and Inference in Alhazen's Theory of Visual Perception." In *Studies in Perception. Interrelations in History of Philosophy and Science,* edited by Peter Machamer and Robert G. Turnbull. Columbus: Ohio State University Press, 1978.

———. "'Form' in Ibn al-Haytham's Theory of Vision." *Zeitschrift für Geschichte der Arabisch-Islamischen Wissenschaften* 5 (1980): 115–140.

———. "Psychology *versus* Mathematics: Ptolemy and Alhazen on the 'Moon Illusion.' " In *Mathematics and Its Application to Science and Natural Philosophy in the Middle Ages: Essays in Honor of Marshall Clagett,* edited by Edward Grant and John E. Murdoch. Cambridge, U.K.: Cambridge University Press, 1987.

———. "The Physical and the Mathematical in Alhazen's Theory of Vision." In *Optics, Astronomy and Logic.* Aldershot, U.K.: Variorum, 1994.

———. "One Ibn al-Haytham or two? An Exercise in Reading the Bio-Bibliographical Sources." *Zeitschrift für Geschichte der Arabisch-Islamischen Wissenschaften* 12 (1998): 1–50.

———. "Conclusion." *Zeitschrift für Geschichte der Arabisch-Islamischen Wissenschaften* 15 (2002–2003): 95–108.

———. "Ibn al-Haytham's Revolutionry Project in Optics: The Achievement and the Obstacle." In *The Enterprise of Science in Islam, New Perspectives,* edited by Jan P. Hogendijk and Abdelhamid Sabra. Cambridge, MA: MIT Press, 2003.

———. "The 'Commentary,' That Saved the Text. The Hazardous Journey of Ibn al-Haytham's Arabic *Optics.*" *Early Science and Medicine* 12 (2007): 117–133.

———. "Alhazen's *Optics in Europe: Notes on What It Said and What it Did Not Say.*" Preprint: *Workshop on "Inside the Camera Obscura."* Berlin: The Max Planck Institute for the History of Science, forthcoming.

Sesiano, Jacques. "Un mémoire d'Ibn al-Haytham sur un problème arithmétique solide." *Centaurus* 20 (1976): 189–195.

Simon, Gérard. "L'*Optique* d'Ibn al-Haytham et la tradition ptolémeénne." *Arabic Sciences and Philosophy* 2 (1992): 203–235.

Smith, A. Mark. "Alhazen's Debt to Ptolemy's *Optics.*" In *Nature, Experiment, and the Sciences,* edited by Trevor H. Levere and William R. Shea. Dordrecht, Netherlands: Kluwer, 1990.

———. "Ptolemy, Alhazen, and Kepler and the Problem of Optical Images." *Arabic Sciences and Philosophy* 8 (1998): 9–44.

———. "The Latin Source of the Fourteenth-Century Italian Translation of Alhacen's *De aspectibus* (Vat. Lat. 4595)." *Arabic Sciences and Philosophy* 11 (2001): 27–43.

———. "The Alhacenian Account of Spatial Perception and Its Epistemological Implications." *Arabic Sciences and Philosophy* 15 (2005): 219–240.

Abdelhamid I. Sabra

IBRĀHĪM IBN SINĀN IBN THĀBIT IBN QURRA

(*b.* Baghdad [?], 908; *d.* Baghdad, 946), *mathematics, astronomy.* For the original article on Ibrāhīm see *DSB,* vol 7.

Ibrāhīm Ibn Sinān Ibn Thābit Ibn Qurra composed a survey of the whole field of geometrical analysis, contributed an original quadrature of the parabola, and gave the first general treatment of horizontal sundials. He was a member of a family that had produced distinguished scientists since the time of his grandfather, Thābit ibn Qurra, and Muḥammad al-Nadīm wrote that "During his time no one appeared who was more brilliant than he was" (al-Nadim 1970, vol. ii, 649). As he made clear in the preface to his work on the parabola, he saw it as his role to preserve and add to his family's reputation, and in this he was eminently successful.

Early Life. As the son of the personal physician to the Caliph al-Muqtadir, Ibrāhīm would have been raised in a well-to-do household and would have had the best education possible. Clearly precocious, he wrote in his autobiography that he began his mathematical researches at the age of fifteen and wrote a number of treatises by the time he was eighteen.

Both Arabic biographical sources and Ibrāhīm's autobiography indicate that at some point his family suffered persecution, probably when Ibrāhīm was in his mid-twenties, at the hand of the Caliph al-Qāhir, and were forced to take refuge in Khorasan, a large region, not closely defined, east of Iraq. Ibrāhīm's father, Sinān, converted from the Sabian sect to Islam under pressure from al-Qāhir, although it appears that his children may have remained Sabian. Ibrāhīm married and had a son named Isḥāq.

Dials. Ibrāhīm describes in his autobiography, composed sometime between the age of twenty-five and his death at the age of thirty-eight, the works he had written. Those he characterized as astronomical are *On Shadow Instruments*; *On the Motions of the Sun*; and *On Ptolemy's Approximate Methods for the Anomalies of Mars, Saturn, and Jupiter.* In the group he calls mathematical are *On Tangent Circles; On Analysis and Synthesis; Selected Problems* [in Geometry]; *Measurement of the Parabola*; and *Drawing the Three* [Conic] *Sections.*

Of *On Shadow Instruments,* there survive only Book 1 and a fragment of the beginning of Book 2 of his revision of an original that had been written in his teens. (He wrote later that he detested the prolixity of his teenage work.) *Of Shadow Instruments* was the first treatise on a general method for the design of plane sun dials, whatever the orientation of the dial's face. He based it on the idea that a plane dial with arbitrary orientation may be considered a horizontal dial for an appropriate place on the earth's surface.

The surviving fragment from Book 2 deals with the shape of hour lines on sundials, lines described by the tip of the gnomon's shadow at a fixed time of day over the course of a year. Ibrāhīm's grandfather, Thābit, had explicitly stated, but not proved, that some hour lines were not straight. Ibrāhīm proved this, but only part of the proof survives in the existing fragment of Book 2. (Paul Luckey's restoration of the proof implied that Ibrāhīm's proof only worked for certain hour lines, and this has been confirmed by a remark of Ibn al-Haytham in his *Treatise on the Hour Lines.*) Finally, although the surviving sections deal with plane dials, we learn from the autobiography that missing parts of Book 2 as well as the whole of Book 3 dealt with nonplanar dials—concave or convex spherical dials as well as concave conic dials.

In his autobiography, Ibrāhīm stated that after writing about the armillary sphere for his colleagues, he wrote a work on it "in different terms" for a craftsman who was making one for him. This adds further evidence to what is already known from Abū al-Wafā' al-Būzjānī in his *Geometrical Constructions for Artisans* about communications between mathematicians and craftsmen in medieval Islam.

Geometry. The first three of Ibrāhīm's mathematical works are all variations on a single theme: the geometrical method of analysis and synthesis. His complaint about what he called the "abbreviated" argument of his contemporaries in this regard is well-known, and his autobiography makes it plain that he wrote these three works to provide examples to students of three versions of this ancient method. The student would begin by progressing in stages through the (mostly) easy problems in *On Tangent Circles* (now lost), which contained a careful version of current practice. The student would, next, find in his *On Analysis and Synthesis* a careful discussion of the full method as found in Apollonius's *Cutting Lines in Ratios.* (Here the student would also find a classification of problems according to the number of solutions and the need for further conditions.) Finally, in *Selected Problems* the student would find (mostly analyses only, omitting the syntheses, of) forty-one geometrical problems to illustrate the current abbreviated practice.

In *On Drawing the Three Sections,* Ibrāhīm wrote that there was no instrument by which one could draw the conic sections. Consequently, he said, he wrote this treatise on how to find as many points as one wishes on a given section. (Abū al-Wafā' al-Būzjānī, in his treatment of burning mirrors in his *Geometrical Constructions,* reproduced Ibrāhīm's method for drawing a parabola.) However, Abū Sahl al-Qūhī wrote a late-tenth-century treatise, *On the Complete Compass,* which describes just the kind of instrument that Ibrāhīm said did not exist in his time. Thus, Ibrāhīm's remark provides a fairly narrow window, around the mid-tenth century, for the introduction of the complete compass in medieval Islam.

In the preface to his *Measurement of the Parabola,* Ibrāhīm referred to two earlier versions, both of which were lost. He also called special attention to the fact that he used only three theorems and was able to give a direct proof rather than—as in earlier treatments—a proof by contradiction.

Additionally, Ibrāhīm wrote a commentary on the *Conics* of Apollonius (lost), a *Description of the Notions/Theorems Used in Astronomy and Geometry,* and a "Letter of Ibrāhīm b. Sinān to Abū Yūsuf al-Hasan b. Isrā'īl on the Astrolabe." (This last work is, however, disputed.) Finally, he stated in his autobiography that he gathered his miscellaneous papers into a volume of about three hundred pages, of which he had the sole copy.

SUPPLEMENTARY BIBLIOGRAPHY

WORKS BY IBRĀHĪM IBN SINĀN

Die Schrift des Ibrāhīm b. Sinān b. Thābit über die Schatteninstrumente. Translation and commentary by Paul Luckey. Edited by Jan P. Hogendijk. Frankfurt am Main, Germany: Institute for the History of Arabic-Islamic Science at the Johann Wolfgang Goethe University, 1999.

The Works of Ibrāhīm Ibn Sinān. Edited by A. S. Saidan. Kuwait City: National Cultural Council, 1983.

OTHER SOURCES

Al-Nadim, Muḥammad. *The Fihrist.* Translated by Bayard Dodge. New York and London: Columbia University Press, 1970.

Rashed, Roshdi. *Les mathématiques infinitésimales du IXe au XIe siècle.* Vol. 1, *Fondateurs et commentateurs.* London: Al-Furqan Islamic Heritage Foundation, 1996.

———, and Hélène Bellosta. *Ibrāhīm b. Sinān: Logique et géométrie au Xe siècle.* Leiden, The Netherlands; Boston: Brill, 2000.

Rosenfeld, Boris A. "Geometrical Transformations in the Medieval East." XIIe Congrès International d'Histoire des Sciences 3, series A (1971): 129–131.

Sezgin, Fuat. *Geschichte des Arabischen Schrifttums.* Vols. 5–7. Leiden, The Netherlands, E. J. Brill, 1974, 1978–1979.

J. L. Berggren

IMES, ELMER SAMUEL (*b.* Memphis, Tennessee, 12 October 1883; *d.* New York, New York, 11 September 1941), *physics, vibrational-rotational spectra, education.*

Imes conducted the first high-precision experiments measuring the infrared spectrum of several diatomic molecules: HCl, HBr, and HF. This work provided experimental verification that both the rotational and vibrational energy levels of molecules are quantized. His detailed spectrum for HCl also provided the first direct evidence for the influence of different isotopes on the spectra of molecules.

Elmer Samuel Imes was the oldest of three male children born to Benjamin Albert Imes and Elizabeth Wallace. Elmer's grandparents, Samuel and Sarah Moore Imes, were free blacks who lived as farmers in south-central Pennsylvania. His parents met while they were attending Oberlin College and married in 1880. The father obtained degrees from Oberlin College (1877) and Oberlin Seminary (1880). Both parents then became home missionaries for the Congregational Church. One of Elmer's brothers, Albert Lovejoy, lived most of his adult life in Great Britain where he engaged in a variety of business ventures. The other, William Lloyd, became a very successful minister, theologian, and civil rights activist. William pastored integrated churches in the northeastern section of the country and was, at one time, the dean of the chapel at Fisk University in Nashville, Tennessee.

Education and Early Career. Elmer Imes attended grammar school (c. 1889–1895) in Oberlin, Ohio, and high school in (c. 1895–1899) in Normal, Alabama. He entered Fisk University circa 1899 and graduated in 1903 with a BA degree. From 1904 to circa 1909, he taught mathematics and physics at Albany Normal Institute in Albany, Georgia. Realizing both the need and desire to obtain an advanced degree, Imes enrolled in the graduate program at Fisk University circa 1909 and completed MA degree requirements in 1910. He remained at Fisk until 1914. During this period he was involved in teaching general science and mathematics classes.

By 1914, Imes, having reached the limits of what could be achieved educationally at a "Negro" southern university, clearly understood that advancement to the next level of training could occur only if he were accepted at and attended one of the northern state universities. In 1915, he enrolled at the University of Michigan as a graduate student in the Department of Physics.

University of Michigan. After a year of probational study, Imes was selected as a graduate fellow in 1916, a position that he held until his graduation in 1918. He began his research under the direction of Harrison M. Randall (University of Michigan, MS 1884, PhD 1902) who had recently returned from Germany where he studied infrared spectroscopy in the laboratory of Friedrich Paschen at Tübingen University. Imes's experimental work centered on the precise measurement of the rotational-vibrational spectra of a number of diatomic molecules. In 1919 he published essentially the full contents of his doctoral dissertation in a long paper in *Astrophysical Journal.* In 1920 Randall published a joint paper on additional work that he and Imes completed with regard to the three molecules HCl, HBr, and HF. These new results appeared in *Physical Review* and were presented at the annual meeting of the American Physical Society in New York in 1920.

Imes's stay at the University of Michigan was relatively calm, and available evidence shows that he earned the respect of both his teachers and fellow graduate students. William F. G. Swann of the Bartol Research Foundation of the Franklin Institute, in Swarthmore, Pennsylvania, states

> It was the writer's privilege to become acquainted with Professor Imes in his graduate student days at the University of Michigan, where his research laboratory was a mecca for those who sought an atmosphere of calm and contentment.… Imes could always be relied upon to bring to any discussion an atmosphere of philosophic soundness and levelheaded practicalness. (1941, p. 601)

In his last year at the university, Imes was initiated into Sigma Xi, the Scientific Research Society, and later was listed in the sixth and several subsequent editions of *American Men of Science.* Imes was among the first African Americans to receive these recognitions.

Reactions to Imes's Work. Diatomic molecules can both vibrate along the bond length and rotate about their center of gravity. Classical physics predicted that the spectra of these combined phenomena should consist of three rather sharp lines lying in the infrared region. However, early spectroscopic studies, starting in the nineteenth century and extending into the first decade and a half of the twentieth century, showed unexpected "band Spectra," that is, very broad regions of absorption spread over two frequency intervals roughly consistent with the classical predictions. A number of theoretical explanations were constructed with their physical bases continuously modified as the foundations of physics made the transformation from a classical to a quantum world view. Several prominent theorists (T) and experimentalists (E) involved in this transformation did at least a portion of their work exploring molecular band spectra. According to a detailed study of this issue by Chiyoko Fujisaki (1983a), prior to 1916 such individuals included Eva V. Bahr (E), Niels Bjerrum (T), W. Burmeister (E), Paul Ehrenfest (T),

Imes

Edwin C. Kemble (E, T), Walther Nernst (T), Lord Rayleigh (T), and Heinrich Rubens (E).

Some of these individuals attended the Solvay Conference held in Brussels from 30 October to 3 November 1911. This first in what would become a series of conferences was sponsored by industrialist Ernest Solvay at the behest of Nernst for the purpose of exploring quanta and radiation. During his final talk of the conference on "The Current Status of the Specific Heat Problem," Albert Einstein raised the issue of spectra and the rotational frequency of molecules. A discussion ensued concerning methods for quantizing this particular degree of freedom. Einstein's previous work on the specific heat of solids and its subsequent success had already paved the way toward acceptance that molecular vibrational energy was quantized.

During the period from 1911 to 1916, the data obtained from various experiments were confusing and in some cases contradictory. Some experiments showed broad bands, while others gave results with bands resolved into a series of separated lines. From a later point of view, the main reason for confusion and contradictions can be viewed as arising from differing resolutions of the spectrometers used by the various experimentalists. Whereas those who were utilizing what one would later call low resolution devices essentially observed two broad bands, those utilizing what one would later call high resolution devices observed structure to the absorption bands. Thus, the data obtained from the high resolution spectrometers gave direct, but not conclusive, evidence for quantization of rotational motion and additional confirmation for quantization of vibration behaviors for diatomic molecules. By 1914, these results were being applied to the calculation of the rotational contribution to the specific heats of solids and gases. A thorough discussion of this work and the major scientists involved is given in the review by Fujisaki (1983a,b). By 1916, there was general agreement that high-precision spectra were needed for diatomic molecules to both give the detailed structure of the band spectra and provide the data required for the formulation of an adequate theory of these phenomena. In 1916, under the direction of Randall at the University of Michigan, Imes began a course of studies that led to the design and construction of a series of infrared spectrometers of continuously increasing resolving powers. Imes used these instruments to measure the absorption bands for three hydrogen halides in the near infrared region: HCl, HBr, and HF. The final spectrometer, used for his most precise work, consisted of two spectrometers connected in tandem. The first was a prism spectrometer, while the second was a grating spectrometer. Imes's most detailed work was done for HCl. For this molecule, he found two bands near 1.76 and 3.46 microns. (The band at 1.76 microns corresponds to the first overtone of the band at 3.46 microns. The nonlinearity of the force between the H and Cl atoms is indicated by $1.76 \times 2 \neq 3.46$) His instrument was able to resolve the band at 3.46 microns into twelve pairs of peaks and the band at 1.76 microns into eight pairs of peaks. Imes used this design and construction of the spectrometer, and the measured data, as the basis for his dissertation. All these results appeared in *Astrophysical Journal* in an extended paper in 1919.

Imes's measurements were immediately recognized by the scientific community as providing accurate experimental proof that rotational energy was also quantized, as was the case for molecular vibrational energy. The clear interpretation of his spectra was that superimposed on the spectral lines coming from the vibrational transitions of these diatomic molecules were peaks corresponding to transitions between various quantized rotational energy states.

A detailed theoretical interpretation of Imes's spectra was soon completed by Adolf Kratzer (1920a) within the framework of the emerging quantum viewpoint. He used his theory and Imes's data to calculate the bond lengths between the hydrogen and halogen atoms for HCl, HBr, and HI and found values consistent with previous estimates.

Imes's data for HCl also indicated that each of the twelve peaks within the band located at 3.46 microns were themselves separated into two peaks (Imes and Randall, 1920). Both Kratzer (1920a) and F. Wheeler Loomis (1920) gave compelling theoretical arguments that this doublet substructure was a consequence of chlorine having isotopes of mass numbers 35 and 37. This means that HCl^{35} and HCl^{37} have slightly different locations for their absorption lines. An assessment of Imes's work is provided by Earle Plyler:

> Up until the work of Imes, there was doubt about the universal applicability of the quantum theory to radiation in all parts of the electromagnetic spectrum. Some held it was useful only for atomic spectra (electronic spectra); some held that it was applicable for all electromagnetic radiation. Ames's high resolution work on Hal, Her and HF was the first clear cut experimental verification of the latter hypothesis, namely, that the rotational energy levels of molecules are quintile as well as the vibration and electronic levels. Thus, Imes's work formed a turning point in the scientific thinking, making it clear that quantum theory was not just a novelty, useful in limiting fields of physics, but, of widespread and general application. (Plyler, 1974)

Although Plyler's assessment may be an exaggeration in giving sole credit to Imes, his work certainly made important contributions toward widespread acceptance of the new quantum theory.

The New York Years. After receiving his PhD in physics from the University of Michigan, Imes spent four years in New York City (1918–1922) where he was self-employed as a consulting physicist. A major reason for moving to New York was the lack of employment opportunities for an African American having advanced education in most regions of the country. Even locations such as Boston were generally hostile to blacks. Another reason was that New York was becoming the mecca for African Americans in various fields of the arts, literature, entertainment, and political theory and practice, and the center of the contemporary civil rights movement. The time period from about 1919 to 1930 is historically referred to as the Harlem Renaissance. It was in this environment that Imes met Nella Larsen. The two married on 3 May 1919.

Larsen was an active member of the Harlem Renaissance who would eventually write two novels, *Quicksand* (1928) and *Passing* (1929). Both Imes and Larsen had overlapping interests in various scholarly, literary, and political subjects. Thus, as members of the "Negro elite," they had ready access to and associated with many other members of the Harlem Renaissance movement such as W. E. B. Du Bois, Arna Bontemps, Aaron Douglas, Langston Hughes, Charles S. Johnson, and Walter White. When Imes later returned to Fisk University, many of these individuals would become connected to that institution either by joining the faculty or by exerting influence on its academic and teaching philosophy.

During the years 1922 to 1930, Imes was employed at three engineering-based firms in the New York metropolitan area: Federal Engineers Development Corporation (1922–1924), Burrows Magnetic Equipment Corporation (1924–1927), and E. A. Evert Signal Supplies (1927–1930). His official title was research physicist or research engineer. Imes's research and development work led to the issuance of four patents. These patents were for devices or techniques for improving the measurement of magnetic properties of various materials.

By 1929, Imes had grown restless. Although he had regular employment and participated in the social, cultural, and political activities of the Harlem Renaissance, all of this was not to his satisfaction. He wanted to return to a more academic setting. The opportunity to do so occurred when he was invited to organize a formal program of physics education and research at Fisk University. He accepted the offer and returned to his alma mater in 1930 as chair of the Physics Department. His wife did not accompany him to Nashville.

Fisk University. Imes had two major tasks to complete upon his return to Fisk. The first was to create a set of upper-level physics courses suitable for BA and MS degree programs. The second was to recruit students to enroll in these programs. He succeeded with both efforts. In the graduating class of 1935, he had three students who finished Fisk with BA degrees in physics and entered the graduate physics program at the University of Michigan. Of the three, one eventually received a doctorate (James Raymond Lawson), one an MS degree (Lewis Clark), and the third dropped out of the program.

Both undergraduate and graduate students were involved in research. The areas selected were topics related to x-rays and magnetics. These two fields were directly related to Imes's work as a research engineer/physicist in New York. To enhance the students' knowledge in these areas, he arranged for them to study at the University of Michigan during the summer to master the subject of x-rays and various techniques needed for their use in his research projects. He also returned to the University of Michigan on several occasions to carry out experimental research on the fine structure of the infrared rotational spectrum of acetylene (C_2H_2). Other work related to magnetic materials was done as summer research at New York University. None of these research efforts resulted in formal scientific publications.

Imes belonged to several professional research organizations: the American Physical Society, the American Institute of Electrical Engineers, and the American Society for Testing and Materials. Because of Jim Crow laws then in place, he was able to attend meetings of these societies only when they met in the northern region of the country or in Canada.

Imes knew many on the Fisk faculty from associations formed when he lived in New York. Also, the Imes family name was well known within the elite, educated African American population, and his wife's fame as a writer was of great value in forming other connections at the university, particularly among the nonscience faculty. In addition to constructing a curriculum for the physics program, he was also asked by the administration to help with the design of a new science building. This effort led to correspondence with other researchers, equipment designers, and manufacturers. Imes was also in charge of the film equipment and used this knowledge to aid various university clubs when a need for this equipment arose. One of the tasks that he especially enjoyed was the planning and execution of the Fisk University Annual Spring Arts Festival. This event brought to campus persons in the studio and performing arts, literature, and the sciences. Many of these individuals were of international renown and their presence and performances had great appeal to the wider Nashville community.

Imes had received a superior "classical" education as an undergraduate student at Fisk. His studies included four languages (French, Greek, Latin, and Spanish), along with separate courses in chemistry, biology, physics, and mathematics. Other courses covered English literature, ethics, sociology, and logic. Now as a professor at Fisk, he wanted to provide the full academic community a broad knowledge of the nature of science, how it could be applied, and its influence on general issues of importance to society. To this end, he developed a course, Cultural Physics, and wrote a book-length manuscript detailing his thoughts on these topics. He used these materials as a course text for one of the general science requirements. He also had another reason for creating this course, namely, continuing the longstanding debate over the issues as to what should be the primary focus of Negro higher education, that is, vocational training versus liberal arts. Fisk University had its genesis in the liberal arts tradition and all of Imes's close associates agreed with this direction for their efforts in education.

Nella Larsen divorced Imes in 1933. They had not lived together for any extended period since 1930. Larsen had no desire to live in the segregated south. She also wanted to remain in the New York metropolitan area where most of her friends and literary colleagues lived and where she felt her writing career would prosper. Her basis for divorce was on the legal grounds of cruelty. However, a much more dramatic reason was hinted at in a front-page story that appeared in the *Baltimore Afro-American* newspaper (7 October 1933). This article suggested that Imes was involved in an affair "with a white member of the Fisk University's administrative staff."

Scientific Legacy. Throughout his life and his career in science and education, Imes was held in high regard by his many friends and colleagues. His experimental work was immediately understood and interpreted as providing detailed experimental proof that rotational energy states were quantized just as was the case for the vibrational motions of molecules. Imes's *Astrophysics Journal* paper was highly cited in the research literature for more than a decade and reproductions of his measured vibrational-rotational spectrum for HCl were placed in textbooks and research monographs. While Imes did not carry out additional research in the field of infrared spectroscopy, he was clearly recognized as a major figure among the small group of spectroscopists who measured with high precision diatomic spectra during the period 1916–1923.

On 11 September 1941, after a long battle with cancer, Imes died in New York City. His longtime friend and colleague, William Francis Gray Swann, wrote an obituary for the journal *Science.* He stated:

> In the death of ... Imes science loses a valuable physicist, an inspiring personality and a man cultured in many fields.... In his passing, his many friends mourn the loss of a distinguished scholar and a fine gentleman. (pp. 600–601)

BIBLIOGRAPHY

Imes's papers are located at Fisk University Library Special Collections; the Carl Van Vechten Personal Collection, New York Public Library; and the James Weldon Johnson Collection, Beinecke Library, Yale University.

WORKS BY IMES

"Measurements on the Near Infra-Red Absorption of Some Diatomic Gases." *Astrophysical Journal* 50 (1919): 251–276.

With Harrison McAllister Randall. "The Fine-Structure of the Near Infra-red Absorption Bands of the Gases HCl, HB, and HF." *Physical Review* 15 (1920): 152–155.

"Cultural Physics." Book manuscript, Fisk University Library, Special Collections, c. 1935.

OTHER SOURCES

Barr, E. Scott. "Historical Survey of the Early Development of the Infrared Spectral Region." *American Journal of Physics* 28 (1960): 42–54.

Davis, Thadious M. *Nella Larsen: Novelist of the Harlem Renaissance.* Baton Rouge: Louisiana State University Press, 1994. This excellent biography of Larsen provides the best available assessment of Imes's life outside of his scientific activities.

Fujisaki, Chiyoko. "From Deslandres to Kratzer (I): Development of the Understanding of the Origin of Infrared Band Spectra (1880–1910)." *Historia Scientiarum* 24 (1983a): 53–75.

———. "From Deslandres to Kratzer (II): Development of the Understanding of the Origin of Infrared Band Spectra (1913–1920)." *Historia Scientiarum* 25 (1983b): 57–86. An excellent overview of Imes's work and its significance is given in Section IX.

Herzberg, Gerhard. *Molecular Spectra and Molecular Structure: I. Spectra of Diatomic Molecules.* 2nd ed. New York: D. van Nostrand, 1950.

Kratzer, Adolf. "Die Ultraroten Rotationsspektren der Halogen-Wasserstoff." *Zeitschrift für Physik* 3, no. 5 (1920a): 289–307.

———. "Eine Spektroskopische Bestatigung der Isotopen Chlor." *Zeitschrift für Physik* 3, no. 5 (1920b): 460–465.

Loomis, F. Wheeler. "Infrared Spectra of Isotopes." *Astrophysical Journal* 52 (1920): 248–256.

Plyler, Earle. "The Influence of Elmer S. Imes." Notes taken by Professor Nelson Fuson (16 August 1974) on a symposium talk by Dr. Earle Plyler (Professor Emeritus of Physics, Florida State University) at the Fisk Infrared Institute's 25th Anniversary Celebration. Copy in Fisk University Library, Special Collections.

Ruark, Arthur Edward, and Harold Clayton Urey. *Atoms, Molecules and Quanta.* New York: McGraw-Hill, 1930.

Swann, William F. G. "Elmer Samuel Imes." *Science* 94 (1941): 600–601. An obituary.

Ronald E. Mickens

INGLADA, VICENTE (*b.* Alicante, Spain, 9 January 1879; *d.* Madrid, 7 January 1949), *geophysics, mathematics, seismology, seismometry.*

Inglada was one of the founders of mathematical seismology. He developed and simplified the formulas for the calculation of the hypocenter in nearby earthquakes. His scientific contributions also included the formulation of the calculation of epicentral coordinates.

Early Life and Education. Inglada's rise from the placid Mediterranean littoral to the heart of the prestigious Geophysical Observatory of Toledo can be understood as a reflection of the social change experienced by Spain after the loss of the overseas colonies (Cuba, the Philippines, and Puerto Rico) in 1898. His hometown, Alicante, had sent numerous fellow citizens as soldiers to Spain's overseas dominions, and so Inglada fixed his gaze on the army. His father Rafael Inglada, a maritime shipping agent, and his mother Antonia Ors, readily agreed; Vicente, after a brilliant career at his home school, went to Toledo to study at the Infantry Academy. The seat of the legendary Alcázar fortification heightened not only Inglada's dedication to a technical-military vocation, but also his devotion to mathematics. The range of facilities, libraries, and tutorship opened his mind, and soon he was feeling as allured by military tactics and strategy as by the subject of topography—presented in the officer course instituted by the government as a result of the overseas confrontations. Inglada certainly faced the possibility of being sent to war: In Cuba, the war against the United States was at its zenith; in the Philippines, the Tagals had mutinied (1896). All this provoked preoccupation and uncertainty in him: on the one hand, it prompted him to continue his military career, but on the other, it made him doubt whether he could earn a livelihood in the alluring borderland between military art and the exact sciences.

The balance remained unaltered for a time. The colonel of Inglada's regiment appreciated his intelligence and recommended that he carry on his studies at the Escuela Superior de Guerra (military higher school) in Madrid. In 1898 Inglada joined this academy, where he encountered Joaquín Fanjul, Manuel Goded, and José Moscardo—all of whom were to have leading roles in the Spanish Civil War (1936–1939)—and Arturo Mifsut, a professor of astronomy and geodesy, who was to provide Inglada with both a model and an audience. This military science faculty instilled in Inglada a special character and preparation, partly as a consequence of the rigorous admission exams and partly because of the amalgam of *theoria* and *praxis* (four theoretical courses and three practical) that it espoused. However, Spain's military debacle in 1898, embodied by the losses of Cuba, Puerto Rico, and the Philippines, closed many doors. In this dark time, a post in the topographical commission for the island of Minorca appeared in 1906 as a glimmer of hope. The 480 days of topographical tasks probably decided Inglada's future, as this experience led to his being named an engineer in the geographical engineer corps in 1907.

Work at the Institute. From 1907 to 1910 Inglada worked at the Spanish Geographical and Statistical Institute, serving in diverse topographical commissions and cultivating studies of extraordinary range, including the techniques of geology and geodesy, the preparation of cartographical maps, meteorology, magnetism, gravimetry, and what was to become his specialty, seismology. Inglada, named a captain of general staff (and the best in his year), acquired a mass of technical expertise, up-to-date geophysical methods, and a virtuoso mastery of differential equations for the calculation of seismic foci from his fieldwork. He was appointed director of the Seismological Station of Toledo in 1910.

For Inglada, simplicity was always paramount. By 1916 the Russian Prince Boris Golitsyn had improved the precision of seismic records through a seismograph equipped with a galvanometric register, but the subsequent calculation of the depth of focus required the tortuous method of the Hungarian seismologist Radó Von Kövesligethy. In September 1919, however, Inglada inspected in person (as a governmental commissioner) the consequences of a devastating tremor: the Bajo Segura earthquake, in his native Alicante. The problem of the focal depth drew Inglada's attention both as a seismologist and as a mathematician; he flung himself into data collection and traced the curves of equal seismic intensity (or isosists), and emerged, two years later, with the elements of the so-called Inglada formulas for the calculation of the hypocenter in nearby earthquakes. The formulas are a blend of parts of earlier formulations (from Kövesligethy), mathematical ingenuity, and observational experience. From the use of coils, pendulums, and bands of smoked paper relating to a focal location, Inglada deduced, as was well known from the classical treatises of seismology, that isosists were the intersection of the seismic waves (coming from the focus) with Earth's surface. He visualized those curves as a succession of concentric circles around the epicenter, and, consequently, he inferred from geometrical considerations that the depth of the hypocenter was directly proportional to the difference of the radii of two consecutive isosists. Such lines, like isobars, allowed him to determine the "eye of hurricane." Later he introduced the use of abacuses (instead of regular tables) for the calculation of focal depth of distant earthquakes. Andrija Mohorovičić, Harold Jeffreys, and Beno Gutenberg, and even Kövesligethy himself expressed their appreciation for Inglada's skillful, sensitive, decisive simplification of these and the forgoing constructs.

Inglada's writings were always diffused with a subtle idea: the *harmonia mundi*—the harmony between the different parts of Earth; the planet as a "scientific unity." Between 1919 and 1925, while at the height of his reputation, Inglada opted to follow the theories, if not the methods, of early geophysicists: Golitsyn, Mohorovičić, and Alfred Wegener. But he found the contemporary treatises so compartmentalized that he felt compelled to pursue the unity of the sciences. "In dividing science into compartments: geology, geophysics, geodesy …, these being division of the first order, or volcanology, seismometry, gravimetry … (second order), and in cultivating each of them in isolation and exclusivity, a scientist does not even imitate the example of the outer layer of the Earth's crust, fragmented in blocks, it is true, but in such a way that its fractured surfaces, far from separating them, serve to put them in contact with each other" (Inglada, 1923b, p. 89). By 1925 Inglada had laid out in his books *El Interior de la Tierra* (1919), *La Corteza Terrestre* (1923a), and *La Sismología* (1923b) his particular conception, roughly equivalent to the twenty-first century notion of interdisciplinarity.

Return to the Academy. Just at the time of his greatest scientific reputation, Inglada, who had already published several didactic lessons on geometry, returned to academia. The internationally acclaimed mathematical seismologist became a professor at the Military Higher School in Madrid in 1923, a position he retained until 1928. Inglada was an excellent teacher, mainly because his lectures on algorithms, astronomy, geodesy, and meteorology were firmly based on his own field experiences, inquiries, and ideas. A flair for oratory (rather than for rhetoric), a prodigious memory, and an uncommon facility of speech fitted him for public lectures, and he successfully took part in erudite conferences from 1924 to 1930, informing his colleagues about "Inglada's formulas" and the science of earthquakes. In 1927 he dazzled the Prague Assembly of the International Union for Geodesy and Geophysics, and, one year later, he was elected to the Real Academia de Ciencias of Madrid.

Inglada's stance on geophysical questions was characterized by his receptiveness to new ideas; his attitude on sociocultural discussions, by his responsiveness to public activity. Hence it is somewhat surprising that, following the closure of the school in 1928 and his subsequent appointment as a technical secretary at the Geographical Institute, he opted for an intellectual retreat and for the total interruption of his hitherto fecund literary production. Although he is well known for his organizational works and improvements in geophysical instrumentation that occurred during the time when he was not publishing (1928–1940), Inglada displayed an equal, if not greater, talent for languages. In the 1930s he learned on his own Basque, Latin, Greek, and Hebrew, which, when added to those he had previously mastered (above all German, but also French, English, Italian, and Portuguese), completed a rich polyglot repertory. Earlier on he had demonstrated his linguistic gifts; his command of Esperanto, forged by preparing more than fifty articles and translations (including *Don Quixote's* first edition) and recognized by his appointment as academician of the Esperanto language in Paris, was formidable.

Research at the National Institute of Geophysics. Only after the civil war, when Inglada became deputy director in 1941 of the newly formed National Institute of Geophysics—a center mainly devoted to applied geophysics—did Inglada return to his research. In the 1940s, he focused recent seismometrical techniques—whose existence had passed practically unnoticed in Spain during the civil war—on a wide variety of problems: the relationship between cyclones and microseismic activity; the investigation of isostasy; and in particular, the study of bathyseisms (or earthquakes of deep origin) for the understanding of Earth's crust and core. Inglada returned to exhausting industriousness in the austere life of postwar Madrid. He took charge of the National Seismological Service, published long monographs, and scrutinized thousands of seismograms and bibliographical notes. Inglada's meticulousness is reflected in the extreme nearsightedness of his last years.

BIBLIOGRAPHY

WORKS BY INGLADA

El Interior de la Tierra según Resulta de las Recientes Investigaciones Sismométricas. Madrid: Talleres del Instituto Geográfico y Estadístico, 1919.

Nuevas Fórmulas para Abreviar el Cálculo de la Profundidad Aproximada del Foco Sísmico por el Método de Kövesligethy, y su Aplicación a Algunos Temblores de Tierra. Madrid: Talleres del Instituto Geográfico, 1921.

La Corteza Terrestre. Madrid: Talleres del Instituto Geográfico, 1923a.

La Sismología. Sus Métodos. El Estado Actual de sus Problemas Fundamentales. Madrid: Talleres del Instituto Geográfico, 1923b.

Las Observaciones Gravimétricas. Madrid: Talleres del Instituto Geográfico, 1923c.

Calcul des Coordonnées du Foyer Séismique au Moyen des Heures de P' ou P observées au Voisinage de l'Épicentre. Paris: Publications du Bureau Central Séismologique Internationale, 1927a, fasc. 5, 3–58.

"Estudio de Sismos Españoles. El Terremoto del Bajo Segura de 10 de Septiembre de 1919. Cálculo de su Profundidad Hipocentral y de la Hora Inicial de sus Sacudidas en el Foco y en el Epicentro." *Revista de la Real Academia de Ciencias Exactas, Físicas y Naturales* 23 (1927b): 1–72.

"Über die Berechnung der Herdtiefe auf Grund der Lage des Inflexions punktes der P-Laufzeitkurve." *Zeitschrift für Geophysik* 3 (1927c): 317–325.

"Die Berechnung der Herdkoordinaten eines Nahbebens aus den Eintrittszeiten der in einigen benachbarten Stationen aufgezeichneten P-order P-Wellen." *Beiträge zur Geophysik* 19 (1928): 73–98.

Trascendencia Científica del Fenómeno Sísmico. Madrid: Talleres del Instituto Geográfico y Catastral, 1929. Read at his reception ceremony as an academician at the Real Academia de Ciencias Exactas, Físicas y Naturales, and replied to by José Mª. de Madariaga.

La Prospección Sísmica en España. Madrid: Nuevas Gráficas, 1930. Read to the inaugural session of the 1930–1931 academic course at the Real Academia de Ciencias Exactas, Físicas y Naturales.

Nomogramas para la Determinación de los Ángulos de Incidencia, Distancias Epicentrales y Tiempos de Recorrido de los Rayos Sísmicos. Madrid: Talleres del Instituto Geográfico y Catastral, 1941.

"Tabelle degli Intervalli $t_{pP} - t_P$ et$_{sS} - t_S$ per l'Analisi dei Terremoti ad Ipocentro Profondo." *Rivista di Geofisica Pura et Applicata* 3, no. 1 (1941).

"Contribución al Estudio del Batisismo Sudamericano de 17 de Enero de 1922." *Revista de Geofísica* 2 (1943): 185–225; 3 (1944): 1–21, 197–227, 553–581. Also in *Memorias del Instituto Geográfico y Catastral.* Madrid: Instituto Geográfico y Catastral, 1943.

"Resultados de las Recientes Investigaciones Isostáticas." *Revista de Geofísica* 4 (1945): 163–207.

"Curso de Sismología: Los Ciclones y la Agitación Microsísmica." *Revista de Geofísica* 5 (1946): 413–432; 6 (1947): 1–30.

OTHER SOURCES

Anduaga, Aitor. "La Institucionalización y Enseñanza de la Meteorología y la Geofísica en España, 1800–1950." PhD diss., University of the Basque Country, Bilbao, 2001.

———. "Earthquakes, Damage, and Prediction: The Spanish Seismological Service, 1898–1930." *Earth Sciences History* 23, no. 2 (2004): 175–207.

Davison, Charles. *The Founders of Seismology.* Cambridge, U.K.: Cambridge University Press, 1927.

Fontserè, Eduard. "Sismología en España." In *Enciclopedia Universal Ilustrada Europeo Americana,* edited by José Espasa de Hijos, Vol. 70. Madrid: Espasa-Calpe, 1923. A general but valuable treatment on Spanish seismology written by one of the greatest authorities in the matter.

Galitzin, Boris. *Conferencias sobre Sismometría,* translated by V. Inglada, José García Siñeríz, and Wenceslao del Castillo from the German version of O. Hecker. Madrid: Instituto Geográfico y Estadístico, 1921.

Gómez de Llarena, Joaquín. "Vicente Inglada Ors (1879–1949)." *Boletín de la Real Sociedad Española de Historia Natural* 42 (1949): 555–561. An authoritative introduction to Inglada's work.

Inglada García-Serrano, Vicente. "Notas Necrológicas: Don Vicente Inglada Ors." *Revista de Geofísica* 8 (1949): 63–71. A valuable obituary notice.

Macelwane, James B. "Vicente Inglada Ors, 1879–1949." *Bulletin of the Seismological Society of America* 39 (1949): 219–220.

Rodríguez de la Torre, Fernando. "Vida y Obra de Vicente Inglada Ors (1879–1949)." *Revista del Instituto de Estudios Alicantinos* 32 (1982): 13–77. This is by far the best bibliography of Inglada's works and secondary literature.

Aitor Anduaga

INGOLD, CHRISTOPHER KELK

(*b.* Forest Gate, Essex, England, 28 October 1893; *d.* Edgware, London, 8 December 1970), *physical organic chemistry, structure of benzene, electronic theory of organic reactions, reaction mechanisms, systematizer of organic chemistry.*

A titan of twentieth-century chemistry, Ingold was a pioneer in the field that, from 1940, was called physical organic chemistry. Organic chemistry had previously focused on classifying the compounds of carbon in terms of their properties and structures. Using the principles and tools of physical chemistry, Ingold sought instead to understand organic chemistry in terms of the types of reactions these compounds underwent. Not only a prolific and original experimentalist, who waged with his co-workers research "campaigns" to develop his theories by means of veritable onslaughts of publications, he was also an important systematizer of the field, who was able to stand back from research detail, drawing it together to forge a new orientation for his subject. As had been the case with previous chemical systematizers, part of his success and enduring reputation lay in his ability to communicate and convince through the creation of evocative new terminology that encapsulated his ideas.

Early Life and Education. Though born to the east of London, Ingold moved with his family when very young to the Isle of Wight off the central south coast of England. His father had been a silk mercer in the metropolis, but set up as a confectioner after the move, a business that his wife carried on after his early death. In this environment, the young Ingold enjoyed considerable independence to explore the island and developed early what would be lifelong enthusiasm for outdoor pursuits, such as bird-watching, to which he would later add others such as rock climbing. He was educated in a recently established local, county funded secondary school, where he was by all accounts well taught in a range of subjects. In his final years at school, he studied mathematics, physics, and

Christopher Kelk Ingold. SPL / PHOTO RESEARCHERS, INC.

chemistry, and capped his school career by passing in 1911 the Intermediate BSc of the University of London, the most advanced level national examination for secondary school students at that time, thereby qualifying for university entrance.

As was usual at the time for a young man educated at a local secondary school who had the talent and the ambition to seek higher education, Ingold went on with a county scholarship to a local institution, the Hartley University College, which was on the mainland just across the water in Southampton. There students prepared for the external degree examinations of the University of London and Ingold gained in 1913 a second-class honors degree in chemistry. Reminiscing shortly after retirement, Ingold commented that he had chosen chemistry for his degree studies, rather than physics in which he was a stronger student, because of the inspiring teaching at Hartley of Professor David R. Boyd (1872–1955), who imbued his students with the "spirit of science." Boyd, on his own, had to cover the whole of the chemistry syllabus. By incorporating recent research findings, he taught the subject as a "living" science with exciting frontiers. Ingold followed this pedagogic strategy in his own teaching.

Ingold went on, with a set of scholarships, to the Royal College of Science in South Kensington, London, a constituent institution of Imperial College of Science and Technology, which had been formed in 1908, to work with the new professor of organic chemistry, Jocelyn Field Thorpe (1872–1940), a classical organic chemist, trained in Germany. Ingold obtained in 1914 the Associateship of the Royal College of Science with first-class honors, the equivalent of a second BSc degree, and immediately went on to undertake postgraduate research under Thorpe on a structural problem, with a view to achieving a DIC (Diploma of Imperial College). However, World War I intervened, and Ingold became engaged in emergency production at the college first of an analgesic and then of a war gas, a teargas code-named SK—for South Kensington. In 1915 he sought deferral of a prestigious research fellowship in order to move to the Cassel Cyanide Company in Glasgow to oversee their manufacture of SK, work for which he later received the British Empire Medal. After the war, Ingold remained at this industrial post although, having continued to collaborate with Thorpe throughout his time in Glasgow—achieving the DIC in 1916 and then a University of London MSc and the Associateship of the Royal Institute of Chemistry in 1919—he had already decided that he wanted to follow an academic career. He returned to academic life at Imperial College early in 1920 as a demonstrator in industrial organic chemistry, taking a considerable cut in salary to do so.

Early Career: Imperial College London and Leeds. On returning to London, Ingold rapidly made up for lost time and began what would become his standard practice of publishing concerted batches of papers on various aspects of a problem. In 1921 he submitted successfully a group of published papers together with a summary of the work to the University of London for the research degree of DSc. An early problem resulted in a revision of the accepted view of a fixed geometry with uniform bond angles for compounds of tetrahedral carbon. Another was the structure of benzene, which would be a recurring topic during his career. In 1922 he was awarded the first ever Meldola Medal, administered by the Royal Institute of Chemistry, for the most outstanding research during the year by a chemist under thirty years of age. He became a fellow of the institute in 1923 and was awarded the Meldola Medal for a second time in that year. It was also the year of his marriage to (Edith) Hilda Usherwood (1898–1988).

The couple met at Imperial College, where the future Mrs. Ingold arrived early in 1921 to do chemical research toward what was then a new higher degree in Britain, a

PhD. Hilda Ingold was born in London into a family that included distinguished schoolteachers on both sides. She attended the renowned North London Collegiate School for Girls, where she excelled, and then studied chemistry and botany at Royal Holloway College, graduating in 1920 with a first-class honors degree in chemistry from the University of London. She completed her PhD early in 1923. After marriage, Hilda Ingold moved with her husband's career, though she continued her own research, earning in 1926 a University of London DSc, which had been redefined as a higher doctorate after the institution of the PhD. Three children followed, but Hilda Ingold managed to continue publishing occasionally, sometimes jointly with her husband, until 1947, working as a researcher for many years in an unpaid capacity. However, from the onset of World War II, she took on a primarily administrative role in her husband's department, for which she proved to have a genuine ability, and did not retire until 1968.

By the end of 1924, Christopher Ingold's tally of publications was fifty-four, including a five hundred–page book on synthetic dyestuffs coauthored with Thorpe. In that year, at the very early age of thirty, on the basis of work during the previous ten years according to the citation, he was elected a Fellow of the Royal Society. The year 1924 also saw his appointment to the chair of organic chemistry at Leeds University, where he remained for six years and assembled a large research group. He was to comment in the 1950s that it was during this period that he developed a pedagogy that aimed to present organic chemistry to students more rationally and less empirically than had been customary. This systematizing impulse, which he suggested was already present in his teaching in the 1920s, would also come to characterize his research program from that time.

The Royal Society Fellowship citation mentioned not only Ingold's productive experimental work in organic chemistry and an innovative and promising mathematical approach to the subject, but also important papers published in the field of physical chemistry. At Leeds, he would expand his interest in physical chemistry, later crediting the then-professor of physical chemistry, Harry Medforth Dawson (1876–1939), with awakening his interest in kinetic studies. From this time, Ingold's research began to concentrate on the mechanisms of organic reactions. Over the period 1923–1927, he was involved in a protracted debate among a number of organic chemists, which has since been much analyzed by historians. In the event, the chief antagonists were Ingold himself and Robert Robinson (1886–1975), who was then at the University of Manchester. The debate was essentially about the positions of electrons in molecules and their consequent behavior in chemical reactions. Summarizing the debate, the historian of chemistry John Shorter has suggested that, during the course of his experimental work done in connection with it, Ingold shifted his position from supporting initially Bernard Flürscheim's (1874–1955) theory of alternating affinities along carbon chains, which were variable and not strictly quantitative, to favoring explicitly electronic explanations. Robinson had been developing an electronic explanation simultaneously (and previously) from the theories of his former mentor and, at the time of the debate, Manchester colleague, Arthur Lapworth (1872–1941). It was Ingold who went on to become identified with the electronic theory, an outcome that rankled with Robinson, who was primarily a natural products chemist, for the rest of his life.

Ingold would from this time onward devote his career to generalizing his study of reaction mechanisms in the light of the electronic theory using the methods of physical chemistry. This resulted in a new, powerful way of understanding organic chemistry that would come to form the basis of research and teaching in the field. His first general statement appeared in the American Chemical Society's prestigious *Chemical Reviews* in 1934 and served to consolidate Ingold's standing in the field. It had been prepared in 1932 during a leave of absence from his United Kingdom post, on health grounds, which he spent at Stanford University in California. "Principles of an Electronic Theory of Organic Reactions" ran to fifty pages and remained for many years the key treatment of the subject. Based, he suggested, on the initial part of a lecture course he delivered at Stanford, it had the explicitly generalizing aim of giving "a connected statement of principles" of the theory, which had "emerged piece-meal" in the literature in the course of various investigations of particular applications of it, such that the theory itself had been obscured. The paper set out a classification in physical terms of measurable electronic effects in molecules, introducing a broad distinction between permanent "polarization" effects (to do with the positions of charges in isolated molecules) and impermanent "polarizability" effects (to do with the mobility of charges in molecules in the presence of neighboring ions or molecules). An integral part of the argument was the development of a number of new terms, which would become standard in the field, thereby consolidating support for the theory itself.

University College London: The Early Years. In 1930 Ingold was appointed to the second chair of chemistry at University College London (UCL), perhaps ironically in succession to Robinson, who had held it as a chair of organic chemistry for two years en route from Manchester to Oxford. By this time, Ingold's publications totaled 135. UCL provided an ideal "physicalist" environment for the development of Ingold's work along the path on which he had set out in Leeds. The eminent physical chemist

Frederick George Donnan (1870–1956) was head of the department. Donnan believed strongly that the future of chemistry as a whole lay in interpreting it in the light of the methods and theories of physical chemistry. He argued successfully through the administration the retitling of the organic chemistry chair to adjective-free "chemistry" on the grounds that subject subdivisions impeded the proper development of the science. This conception of the unity of chemical science based on physical methods was one that Ingold shared and would develop at UCL.

Ingold's first task there was to be build up an independent research presence, as only one student had moved with him from Leeds. He was joined immediately by a postdoctoral fellow, Edward David Hughes (1906–1963), who had recently completed a PhD at University College Bangor for which Ingold had acted as external examiner. Thus began one of chemistry's most remarkable research partnerships that would end only with Hughes's early death in 1963. Apart from a brief spell from 1943 to 1948, when he held a chair at his alma mater, Hughes spent his entire career in the UCL department, albeit in various rather unsatisfactory junior capacities before his sojourn in Bangor, after which he returned to a UCL chair. The two men could not have been more contrasting in appearance, with Hughes on the short side and portly and Ingold rather tall and ascetic. They were also complementary in personality. Hughes was the one whom students felt more comfortable approaching, but had a quirky, less engaging lecture technique. By all accounts, Ingold was a masterly lecturer.

Structure and Mechanism in Organic Chemistry. The Bangor chemistry department, headed by Kennedy Joseph Previte Orton (1872–1930), was a pioneering British center for physical approaches to organic chemistry, known particularly for work on reaction mechanisms and kinetics. Hughes's expertise in the methodology of chemical kinetics proved the ideal complement to Ingold's interest in electronic theory. Indeed, most biographies that have considered the question are agreed that it is difficult to disentangle their mutual contributions. Together they published 138 papers, which comprised just under half of Ingold's outputs in the period and just under 60 percent of Hughes's. In his Royal Society memoir of Hughes, Ingold discussed with almost no mention of his own role their pathbreaking work in the 1930s. In this early collaboration, they developed a scheme for classifying a very large number of reactions of aliphatic compounds (consisting of nonring carbon chains) across what had previously been considered unrelated areas into two broad, interconnected families—*nucleophilic* (a term they coined to mean "nucleus seeking") substitution and elimination reactions. Nucleophilic substitution reactions they designated as either S_N1 or S_N2, according to whether one or two molecules were involved in the rate-determining step (the slow step) of the reaction. That is, whether the reaction was unimolecular or bimolecular. Elimination reactions were similarly designated E2 or E1. They also coined the term *electrophilic* to denote electron-seeking reactants/reactions.

In their hands and the hands of their students, this fundamental conceptual framework provided a powerful tool for understanding how varying conditions and structures affected and altered the way that reactions progressed. It also enabled Hughes and Ingold to explain and generalize certain puzzles in classical organic chemistry, such as the problem known as the Walden inversion, which had exercised chemists for some forty years. Their framework was not accepted without considerable opposition from more traditional chemical quarters and, according to Ingold, was known as the "British 'heresy'" in some places abroad. He suggested that this was because their work exemplified the view that physical chemistry was the basis of all chemistry, including organic chemistry and therefore required a considerable shift in orientation on the part of classically trained chemists. Securing its acceptance involved sustained campaigns in communications as well as in chemistry. Ingold's famous systematizing textbook aimed at university students, *Structure and Mechanism in Organic Chemistry,* was published in 1953 although finished by the end of 1951. Based on the Baker Lectures that he gave at Cornell in 1950–1951, it was a magnum opus in every sense—a highly structured eight hundred–page work that laid out in great detail his entire scheme. It quickly became, as Ingold's biographer Kenneth Leffek has called it, a "bible" for organic chemistry. In 1952 Ingold received the Royal Society's highest honor, its Royal Medal, for the experimental depth and theoretical breadth of his contributions across the field of organic reaction mechanisms.

The Structure of Benzene. In the same remarkably fruitful period in the mid-1930s, Ingold returned to the problem of the structure of the benzene molecule (C_6H_6—six carbon atoms joined in a ring, each with one atom of hydrogen attached). As was characteristic of all his work, he saw the potential of bringing together a number of new techniques to tackle this classic problem, which had engaged chemists from the mid-1860s. A new method for preparing deuterium, the isotope of hydrogen with twice its mass, for which Hughes set up a production plant, made possible a strategy of substituting deuterium atoms successively for the hydrogen atoms in the benzene molecule. Deploying the new Raman spectroscopy and new techniques in infrared spectroscopy provided a two-pronged approach to analyzing the molecular vibrations of the resulting compounds and comparing them with those of normal benzene. By mid-1935, Ingold and coworkers had synthesized fully deuterated (C_6D_6) and mono- and di-deuterated (C_6H_5D and 1:4–$C_6H_4D_2$) benzene and measured their infrared and Raman, as well as their ultraviolet, spectra. Responding to an earlier letter from a Copenhagen group reporting for certain deuterated benzenes only Raman frequencies, they published preliminary results in the form of a letter to the editor of *Nature* (135 [1935]: 1033–1034). Drawing also on quantum mechanical calculations by Edward Teller (1908–2003), who held a visiting post at UCL as a refugee during 1934–1935, this work led to the conclusion that the benzene molecule did have a plane regular hexagonal structure, which had recently been predicted theoretically by Edgar Bright Wilson (1908–1992) and Linus Pauling (1901–1994). The complete analysis of the structure required additionally the synthesis and spectroscopic analysis of the full set of deuterated benzene molecules. The experimental "campaign," which started with an eight-paper series in the *Journal of the Chemical Society* in 1936, was completed after the war with publication of part twenty-one of the series in 1946. In recognition of this work, Ingold was awarded the Bakerian lectureship of the Royal Society in 1938 and its Davy Medal in 1946.

UCL: Head of Department and Director of Laboratories, 1937–1961. Ingold succeeded Donnan as head of the department and director of laboratories in October 1937, posts that, formally, he held until his retirement in 1961. Particularly toward the end of his term, however, Ingold was much aided in these roles by Hughes, who formally became deputy head in 1957 and was appointed to the substantive posts on Ingold's retirement. Over the period of Ingold's headship, the number of academic staff in the department grew considerably; the establishment roughly doubled and there were numerous honorary members as well. There was also a constant stream of research visitors. In terms of funding, throughout Ingold's headship, the chemistry department had the highest internal grant from UCL, a long way ahead of all other departments. However, despite a steady stream of successful applications to industrial and government sources, especially for equipping the department with the modern instruments that were becoming so essential for chemical work, the level of external income in 1960–1961 was similar to that in the mid-1930s under Donnan. The decade of the 1950s was financially difficult for the department and for the college as a whole. Ingold's management of departmental finances was, however, so skillful that his staff were generally not aware of the difficulties. Student numbers, both undergraduate and postgraduate were similar at either end of Ingold's headship.

This apparent continuity belies the very considerable disruption of the war years, when the chemistry department was evacuated in two sections to Wales in 1939; Ingold accompanied the special honors degree students and the postgraduates to Aberystwyth on the west coast. Some research continued during the war, but the rate of publication declined, as did student numbers. On returning to London in 1944, Ingold faced a considerable task in rebuilding his department, particularly in staffing, since a number of senior staff did not return, but pursued opportunities that had arisen during the war. This circumstance was, however, also an opportunity to reshape the staffing complement, as a number of those who left had been Donnan's appointees. Ingold recruited heavily from his own research school. Adhering to his view of the unity of chemistry, his strategy was to appoint individuals with expertise in a range of techniques or fields who, he anticipated, would interact constructively to move chemistry based on physical principles forward across a broad front. For example, assessing crystallography as crucial to the development of chemistry, in 1947, he recruited Kathleen Lonsdale (1903–1971). Furthermore, he argued, without internal disciplinary boundaries, resources could be deployed much more efficiently.

This policy of mutual support was also evident in Ingold's continuing a practice that had been in place since Donnan's time. It was expected that all researchers would attend advanced lectures given by senior staff members, as well as the colloquia given by students and visitors at various stages in their research. Ingold's pointed and precise interventions across a wide range of topics at colloquia were keenly anticipated by research students, sometimes with trepidation. On the undergraduate side, Donnan and Ingold had been active in the 1930s in discussions of the reform of the requirements for the BSc degree of the University of London. Changes very much along the lines that they had advocated were finally implemented after the war. The special honors degree in chemistry was extended from a two-year to a three-year course, with students required to take at least one year of ancillary mathematics and two years of physics. Ingold's view of the nature of chemistry as a subject resting on physical principles that required mathematical understanding thus permeated the curriculum. Although he was very active on key university and college committees, especially when it was a case of arguing for his department or his subject, unlike Hughes, Ingold did not become heavily involved in chemical institutions. This was another aspect in which they were complementary. Ingold was, however, president of the Chemical Society in 1952 and 1953.

After the war, the research tempo picked up quickly and Ingold and his coworkers resumed their extraordinary productivity. Two further "incursions into molecular spectroscopy," as Ingold's Royal Society obituarist Charles W. Shoppee termed them, took place. Work using deuterated benzene continued with investigation of the excited states

Ingold

of benzene by means of ultraviolet spectroscopy, the first successful such analysis of a molecule consisting of more than two atoms. Work on the excited states of acetylene, using deuterated acetylene, followed, resulting in the first demonstration of a change of shape in the geometry of a molecule on excitation. In the same period, Ingold returned to the problem of aromatic nitration, which had been the focus of his studies of electrophilic reactions in the 1930s, proving in 1946 that the active agent in the nitration reaction of aromatic molecules by nitric acid was the nitronium ion, a species predicted in 1903. This was the subject of another of his famous publication campaigns: Ingold wrote twenty-two papers on aromatic nitration, which occupied nearly three hundred consecutive pages in the *Journal of the Chemical Society* in 1950. There was also a series of papers on the mechanisms of elimination reactions. During the 1950s and 1960s Ingold worked jointly with Robert S. Cahn (1899–1981) and Vladimir Prelog (1906–1998) on the sequence rule for specifying the positions of four different groups tetrahedrally bonded to a chiral carbon atom, that is, one that gives rise to optical activity. Their CIP (Cahn-Ingold-Prelog) sequence rule became the international standard to specify the R (rectus, or right) or S (sinister, or left) handedness of an optically active molecule.

Ingold retired formally in 1961, and became professor emeritus. He retained a role in the department as a special lecturer, and was reappointed annually. He greatly enjoyed having the freedom from departmental duties so that he could accept scientific invitations that took him on travels around the world. Ingold was knighted in 1958. In addition to recognition through awards from the Royal Society, he received the major awards and medals of the Chemical Society as well as a medal from the American Chemical Society and numerous honorary degrees from universities around the world. The one great honor that he did not receive was the Nobel Prize, which has occasioned speculation. Ingold was truly a titan in his field who not only made major, enduring discoveries, but also changed the very way that the field would be approached. Ingold's assessment of Hughes's significance in his Royal Society memoir applies equally well to the significance of Ingold's own career: "this work has changed the aspect of organic chemistry, by progressively replacing empiricism by rationality and understanding, to a degree which is now manifest in the terminology and teaching of the subject, and in the research activity all along its advancing frontier" (1964, p. 165).

BIBLIOGRAPHY

A list of 443 publications by Ingold dating from 1915 to 1969 appears in C. W. Shoppee, "Christopher Kelk Ingold (1893–1970)," Biographical Memoirs of Fellows of the Royal Society *18 (1972): 349–411 (on 385–411). There is no Ingold archive per se; however, documents and correspondence by and relating to Ingold are located in a number repositories of collections of other scientists, including that of Charles Coulson (the Bodleian Library, Oxford) and Robert Robinson (the Royal Society, London, http://www.nationalarchives.gov.uk/nra/), and in institutions with which he was connected, including, of course, University College London (http://www.ucl.ac.uk/efd/recordsoffice/ and http://www.chem.ucl.ac.uk/) and also the University of London (http://www.shl.lon.ac.uk/). For other archival materials in England and Wales, see "Access to Archives; The English Strand of the UK Archives Network" (http://www.a2a.org.uk). For a listing of cataloged archives of British scientists contemporary to Ingold, see "National Cataloguing Unit for the Archives of Contemporary Scientists" (http://www.bath.ac.uk/ncuacs/). Also see The Royal Society, "Ingold, Sir Christopher Kelk," Certificates of Election and Candidature, GB 117 The Royal Society, RefNo EC/1924/06" (http://www.royalsoc.ac.uk).*

WORKS BY INGOLD

"Principles of an Electronic Theory of Organic Reactions." *Chemical Reviews* 15 (1934): 225–274.

"The Structure of Benzene" (Bakerian Lecture). *Proceedings of the Royal Society of London* 169, ser. A, no. 937 (1938): 149–173.

Structure and Mechanism in Organic Chemistry. Ithaca, NY: Cornell University Press, 1953; 2nd ed., 1969.

"Organic Chemistry Begins to Grow Up." *Journal of the Chemical and Physical Society* (of UCL) 1 (1956): 147–151.

"The Education of a Scientist." *Nature* 196, no. 4859 (1962): 1030–1034.

"Edward David Hughes (1906–1963)." *Biographical Memoirs of Fellows of the Royal Society* 10 (1964): 147–182.

OTHER SOURCES

Brock, William H. *Fontana History of Chemistry.* London: Fontana, 1992. Especially chapters 13 and 14, which provide a good general, contextual introduction to the field of physical organic chemistry.

Davenport, Derek A., and Paul R. Jones, eds. "C. K. Ingold: Master and Mandarin of Physical Organic Chemistry." *Bulletin for the History of Chemistry* 19 (1996). A collection of papers from a symposium of the same title held at the American Chemical Society National Meeting in Chicago, 24–25 August 1993, in commemoration of the centennial of the birth of Professor Ingold.

Leffek, Kenneth T. *Sir Christopher Ingold: A Major Prophet of Organic Chemistry.* Victoria, BC, Canada: Nova Lion, 1996.

———. "Ingold, Sir Christopher Kelk (1893–1970)." In *Oxford Dictionary of National Biography,* edited by H. C. G. Matthew and Brian Harrison. Oxford: Oxford University Press, 2004.

Nye, Mary Jo. *From Chemical Philosophy to Theoretical Chemistry: Dynamics of Matter and Dynamics of Disciplines, 1800–1950.* Berkeley: University of California Press, 1993.

Ridd, John H., ed. *Studies on Chemical Structure and Reactivity, Presented to Sir Christopher Ingold.* London: Methuen, 1966.

Roberts, Gerrylynn K. "C. K. Ingold at University College London: Educator and Department Head." *British Journal for the History of Science* 29 (1996): 65–82.

Schofield, Kenneth. "The Development of Ingold's System of Organic Chemistry." *Ambix* 41 (1994): 87–107.

———. "The Growth of Physical Organic Chemistry." N.p., 1996. Copies are deposited with the Royal Society in London and with the History of Chemistry Research Group, The Open University in Milton Keynes, England. This is a more technical treatment of the subject.

Shoppee, C. W. "Christopher Kelk Ingold (1893–1970)." *Biographical Memoirs of Fellows of the Royal Society* 18 (1972): 349–411.

Shorter, John. "Electronic Theories of Organic Chemistry." *Natural Product Reports* 4 (1987): 61–66. This has become the standard analysis of the famous Robinson-Ingold controversy of the 1920s.

———. "Physical Organic Chemistry." Chap. 7 in *Chemical History: Reviews of the Recent Literature,* edited by Colin A. Russell and Gerrylynn K. Roberts. Cambridge, U.K.: Royal Society of Chemistry, 2005. A bibliographic essay, which charts the history of the subject, this contains numerous references to studies of various aspects of Ingold's work, including an important series of papers by Martin Saltzman, as well as related topics.

Gerrylynn K. Roberts

J

JĀBIR IBN ḤAYYĀN (*fl.* late eighth and early ninth centuries), *alchemy.* For the original article on Jābir see *DSB*, vol. 7.

Several studies have been published since the 1970s bringing new documents, evidence, and arguments on the question of the biography of Jābir and the authenticity of the treatises attributed to him. Major contributions come from Fuat Sezgin in volume 4 of his *Geschichte des arabischen Schrifttums* (1971, pp. 132–269). Sezgin adds thirty new titles of treatises to the 1943 catalog by Paul Kraus. Sezgin develops the idea that alchemy began very early in the Islamic world, in the first century AH (seventh century CE). According to his conclusions, Jābir had existed, indeed lived at the time of Jaʿfar Muḥammad al-Ṣādiq and actually composed a large number of the texts attributed to him (Jābir). This thesis consequently implies that the Greek philosophers and scientists quoted by Jābir were translated into Arabic one or two centuries earlier than is usually believed. (For this last point, see also Alexander of Aphrodisias, *On Aristotle's "On Coming-to-Be and Perishing" 2.2–5,* 2004.) Thus the whole of Islamic philosophy and science would have commenced in conditions completely different from what is generally thought. Similar ideas were advocated in a more moderate way by Seyyed Hossein Nasr (1968), Toufic Fahd (1970), Henry Corbin (1986b), and Syed Nomanul Haq (1994).

However, it seems difficult to overcome the arguments put forward by Kraus. Sezgin asserts a number of arguments and hypotheses about the ancientness of the Jābirian treatises, but these mere suppositions do not constitute evidence or certainty; the numerous anachronisms and the use of later translations for Greek terms in the Jābirian corpus confirm Kraus's conclusions. The very existence of Jābir—unknown in all the ancient bio-bibliographical works—still seems very doubtful, as Manfred Ullmann pointed out in his concise 1972 account. By contrast, the idea that the corpus was composed gradually seems very likely. This had already been observed by Ernst Darmstaedter in his 1925 study of the text of Jābir's *Book of Mercy.* Similar observations were made by Pierre Lory in 1996 on the basis of a comparative study of the manuscripts of the *Book of Seventy.* To a primitive kernel of alchemical assertions were added several technical developments and a third layer of more theoretical additions. On the basis of the study of six manuscripts of the *Book of Seventy,* some of them being recent and complete, such as Hüseyin Celebi 743 or Carullah 1554, one being more ancient and incomplete, Dār al-Kutub, Tabīʿiyyāt, 731, with others containing several lacks, such as Dār al-Kutub, Taymûr, Tabīʿiyyāt 67, one can draw the hypothesis of two successive additions by authors belonging to the same alchemical school.

The doctrinal dimensions of the Jābirian corpus—that is to say, his Shiite ideas—are another controversial issue. Kraus was convinced that "Jābir" was the name of a group of Ismaili propagandists working for the Fatimid dynasty (1943, Introduction). He had prepared a whole monograph on this question but died before finishing it. This point was studied again by Yves Marquet (1988). He followed Kraus's views but was inclined to identify the Jābirian school with a trend belonging to the revolutionary and millenarian movement of the Carmatians who established a powerful state in Northern Arabia during the tenth century CE. Lory also dealt with this question (2000, 2003). He deems that the Jābirian school belonged to an original nonsectarian and millenarian Shiite movement that could in no way belong to Ismailism because of its viewpoint on the succession of the Shiite imam Ja'far Muhammad al-Sādiq. Ismailis confess indeed that the successor of Ja'far as Imam was his son Ismā'īl and after him Muhammad son of Ismā'īl; whereas Jâbir explicitly argues in his *Book of the Fifty* that the successor of Ja'far was Mūsā son of Ja'far.

The scientific side of Jābir's treatises was further explained by several scholars, among them Friedemann Rex (1975), Lory (1988), and Haq (1994), who shed some light in presenting and commenting on certain texts of Jābir while translating or editing them.

SUPPLEMENTARY BIBLIOGRAPHY

Alexander of Aphrodisias. *On Aristotle's "On Coming-To-Be and Perishing 2.2-5."* Translated by Emma Gannagé and edited by Peter Adamson. Ithaca, NY: Cornell University Press, 2005.

Corbin, Henry. "Le livre du glorieux de Jābir ibn Ḥayyān," 1950. In *L'Alchimie comme Art hiératique.* Paris: Editions de l'Herne, 1986a.

———. *Histoire de la philosophie islamique.* New ed. Paris: Gallimard, 1986b.

Darmstaedter, Ernst. "Liber misericordiae Geber, eine lateinische Übersetzung des grosseren *kitāb al-rahma.*" *Archiv für Geschichte der Medizin* 18 (1925): 181–197.

Fahd, Toufic. "Ja'far al-Sâdiq et la tradition scientifique arabe." In *Le chiisme imamat.* Paris: Presses Universitaires de France, 1970.

Haq, Syed Nomanul. *Names, Natures, and Things: The Alchemist Jābir ibn Hayyān and his Kitāb al-Ahjār (Book of Stones).* Dordrecht and Boston: Kluwer Academic, 1994.

Kraus, Paul. "Studien zu Jābir ibn Ḥayyān," 1931. In *Alchemie, Ketzerei, Apokryphen im frühen Islam,* edited by Rémi Brague. Hildesheim: Georg Olms Verlag, 1994.

———. *Jābir ibn Ḥayyān, contribution a l'histoire des idees scientifiques dans l'Islam.* Vol. 1, *Le corpus des ecrits jabiriens*; vol. 2, *Jābir et la science grecque.* 1942, 1943. Republished, Paris: Les Belles Lettres, 1986.

Lory, Pierre. *L'élaboration de l'Elixir suprême: Quatorze traités de Jâbir ibn Ḥayyân sur le Grand Oeuvre alchimique.* Damascus: Institut Français de Damas, 1988.

———. *Dix traités d'alchimie de Jâbir ibn Ḥayyân.* Paris: Actes Sud/Sindbad, 1996.

———."Eschatologie alchimique chez Jâbir ibn Ḥayyân." In "Mahdisme et millénarisme en Islam," *Revue des Mondes Musulmans et de la Méditerranée* 91–92–93–94 (2000): 73–91.

———. *Alchimie et mystique en terre d'Islam.* Paris: Verdier, 2003.

Marquet, Yves. *La philosophie des alchimistes et l'alchimie des philosophes: Jâbir ibn Ḥayyân et les "Frères de la Pureté."* Paris: Maisonneuve et Larose, 1988.

Nasr, Seyyed Hossein. *Science and Civilization in Islam.* Cambridge, MA: Harvard University Press, 1968. Republished, Lahore, Pakistan: Suhail Academy, 1983.

Rex, Friedemann. *Zur Theorie der Naturprozesse in der früharabischen Wissenschaft: Das Kitâb al-ikhâj, übersetzt und erklärt, ein Beitrag zum alchemistischen Weltbild der Jâbit-Schriften (8./10. Jahrhundert n. Chr.* Wiesbaden: Stein, 1975.

Sezgin, Fuat. *Geschichte des arabischen Schrifttums.* Vol. 4, pp. 132–269. Leiden: Brill, 1971.

Ullmann, Manfred. *Die Natur- und Geheimwissenschaften im Islam.* Leiden: Brill, 1972.

Pierre Lory

JAGHMĪNĪ, SHARAF AL-DĪN MAḤMŪD IBN MUḤAMMAD IBN 'UMAR AL-

(*fl.* Khwārizm [now northwest Uzbekistan], first half of thirteenth century), *astronomy.*

Jaghmīnī is best known as the author of the elementary astronomical text *al-Mulakhkhaṣ fī al-hay'a al-basīṭa* (Epitome of simplified theoretical astronomy). This extremely popular work inspired the writing of at least twenty commentaries and supercommentaries; thus Jaghmīnī's *Mulakhkhaṣ* and commentaries provide an example of an active, continuous tradition of astronomical learning within Islamic societies, one that spanned a period of at least five centuries.

There has been some confusion about Jaghmīnī's dates, but he can safely be situated in the early thirteenth century. Several sources give the date of composition of the *Mulakhkhaṣ* as 618 H/1221–1222 CE, and an extant Istanbul manuscript (Laleli, 2141.3, ff. 61b–81a) contains a copy dated 644 H/1246–1247 CE. Claims that Jaghmīnī lived circa 1344–1345 CE were due in part to mistaking him for another Jaghmīnī, a physician who lived at that time. Also contributing to this dating error is the fact that the *Mulakhkhaṣ* was modified over time, by copyists in such a way that it incorporated material from later sources, thus making it appear that it was dependent on these later sources. In one case, for example, Jaghmīnī's original list of Ptolemaic values for maximum daylight and latitudes of climes was replaced with a list of values derived later from Naṣīr al-Dīn al-Ṭūsī's *Tadhkira fī 'ilm al-hay'a.*

Jaghmīnī's *Mulakhkhaṣ* is a simplified textbook on astronomy written in Arabic. It is part of a genre of elementary theoretical astronomical literature within Islam explaining the physical structure (*hay'a*) of the universe for nonspecialists. The *Mulakhkhaṣ* contains a summary of the configuration of both the celestial and terrestrial worlds, providing the arrangement of the Ptolemaic celestial orbs as well as the sublunar levels. It is composed of an introduction and two sections. The introduction explains the divisions of the bodies in general; section one (divided into five parts) presents an explanation of the celestial orbs and what pertains to them; and section two (divided into three parts) provides an explanation of Earth and what pertains to it. The *Mulakhkhaṣ* contains no mathematical proofs, for which one would presumably look in Ptolemy's *Almagest.* And unlike some of the other existing elementary and intermediate theoretical astronomical textbooks, the *Mulakhkhaṣ* does not contain any treatment of distances between Earth and the celestial bodies or their sizes, thus distinguishing it from works by Shams al-Dīn al-Kharaqī (d. 1138–1139), Mu'ayyad al-Dīn al-ʿUrḍī (d. 1266), and Naṣīr al-Dīn al-Ṭūsī (d. 1274). Jaghmīnī apparently wrote a separate treatise on this subject in a unique manuscript (Cairo, Dār al-kutub MS Ṭalʿat majāmīʿ 429.2, f. 4a–4b).

Among the eleven known commentators on the *Mulakhkhaṣ* are Faḍlallāh al-ʿUbaydī (d. 1350), Kamāl al-Dīn al-Turkmānī (fl. 1354), the theologian al-Sayyid al-Sharīf al-Jurjānī (d. 1413), and Qāḍīzāde al-Rūmī (d. c. 1440). Qāḍīzāde's commentary deserves particular mention since it became the subject of numerous supercommentaries that were widely circulated. Qāḍīzāde completed the work in 1412 and dedicated it to Ulugh Beg, the Timurid sultan whose madrassa (school) in Samarqand was a place where astronomy and mathematics were taught. It was probably used as an intermediate astronomy textbook to be read with Jaghmīnī's original text (often the commentary contained the text within it). Qāḍīzāde's commentary eventually became part of the madrassa curriculum in Ottoman lands. This transmission was facilitated by ʿAlī Qushjī, who had been a key scientist at Samarqand but spent his last years in Istanbul, and by Qāḍīzāde's student Fatḥallāh al-Shirwānī (d. 1486), who wrote a supercommentary on it (which he presented in 1473 to Mehmed II, the conqueror of Constantinople). Qāḍīzāde's commentary was used in eighteenth-century Cairo as well as in Muslim India (along with its Persian translation), where it was lithographed and where Indian scholars wrote glosses on it.

In addition to the eleven commentators, eight supercommentators have been identified. Fatḥallāh al-Shirwānī's supercommentary contains references to other *Mulakhkhaṣ* commentaries as well as notes from his lectures at Samarqand. Two other authors are Bahā' al-Dīn al-ʿĀmilī (d. 1622) and ʿAbd al-ʿAlī al-Birjandī (d. 1525–1526), both attendees of Isfahan courts and whose works also surface in India. The supercommentary of Birjandī, who was a second-generation scholar of the Samarqand school, became an advanced-level textbook in the Ottoman madrassas.

Jaghmīnī's *Mulakhkhaṣ* exists in thousands of copies, either independently or as part of commentaries, supercommentaries, and glosses some of these were translated into Persian and Turkish. Needless to say, these numbers are overwhelming; unfortunately the vast majority of these works and their exemplars, many containing significant marginalia, have yet to be seriously examined. In addition to providing information on teachers and students of astronomy, as well as the institutions in which this subject was taught during the latter period (1200–1900) of Islamic science, these notes and annotations give an indication of how early modern European science was assimilated in the Islamic world. Jaghmīnī's *Mulakhkhaṣ*, its transformation, and its wide dissemination clearly demonstrate that scientific texts were actively being taught in the madrassas. As such the *Mulakhkhaṣ* and its commentary tradition provide an important means to understand some of the dynamics of learning within Islamic societies.

BIBLIOGRAPHY

WORKS BY JAGHMĪNĪ

al-Mulakhkhaṣ fī al-hay'a al-basīṭa. Istanbul, Laleli MS, 2141.3, ff. 61b–81a.

Rudloff, G., and A. Hochheim. "Die Astronomie des Maḥmûd ibn Muḥammed ibn ʿOmar al-Ǵagmînî." *Zeitschrift der Deutschen Morganländischen Gesellschaft* 47 (1893): 213–275. Reprinted in *Islamic Mathematics and Astronomy,* vol. 77, *Miscellaneous Texts and Studies on Islamic Mathematics and Astronomy* II, edited by Fuat Sezgin. Frankfurt, Germany: 1998. A German translation based upon a later version of the text.

OTHER SOURCES

Fazlıoğlu, İhsan. "Osmanlı felsefe-biliminin arkaplanı: Semerkand matematik-astronomi okulu." *Dîvân İlmî Araştırmalar* 14 (2003): 1–66.

İhsanoğlu, Ekmeleddin, et al. *Osmanlı astronomi literatürü tarihi* (*OALT*) [History of astronomy literature during the Ottoman period]. 2 vols. Istanbul: IRCICA, 1997. For a listing of *Mulakhkhaṣ* manuscripts by Ottoman commentators, see vol. 1, pp. 8–21, 23, 24, 27, 40–41, 47, 54, 135, 141, 249, 291, 330, and vol. 2, pp. 463–464, 744–745, 786.

———, ed. *History of the Ottoman State, Society, and Civilisation.* 2 vols. Istanbul: IRCICA, 2002. For a discussion of the tradition of the *Mulakhkhaṣ* in the Ottoman Empire see vol. 2, pp. 522–523, 525, 533, 535, 545–546, 548, 586–587.

İzgi, Cevat. *Osmanlı Medreselerinde İlim.* Vol. 1, pp. 370–392. Istanbul: İz, 1997. On the teaching of the *Mulakhkhaṣ* in the Ottoman madrassas.

Kātib Čelebī. *Kashf al-ẓunūn ʿan asāmī al-kutub wa-'l-funūn.* Vol. 2, collections 1819–1820. Istanbul, 1943.

King, David A. *A Survey of the Scientific Manuscripts in the Egyptian National Library*, p. 150. Winona Lake, IN: Eisenbrauns, 1986.

Ragep, F. Jamil. *Naṣīr al-Dīn al-Ṭūsī's* Memoir on Astronomy (*al-Tadhkira fī' ilm al-hay'a*). 2 vols. New York: Springer, 1993.

———. "On Dating Jaghmīnī and His *Mulakhkhaṣ*." In *Essays in Honour of Ekmeleddin Ihsanoğlu,* edited by Mustafa Kaçar and Zeynep Durukal, 461–466. Istanbul: IRCICA, 2006. An overview of the controversy on dating Jaghmīnī.

Ragep, Sally P. "Jaghmīnī." In *Biographical Encyclopedia of Astronomers,* general editor Thomas Hockey. New York: Springer, forthcoming.

Storey, Charles A. *Persian Literature: A Bio-Bibliographical Survey.* Vol. 2, pt. 1, pp. 50–51. London: Luzac, 1972.

Suter, Heinrich. "Die Mathematiker und Astronomen der Araber und ihre Werke." *Abhandlungen zur Geschichte der mathematischen Wissenschaften* 10 (1900): 164–165.

———; revised by Juan Vernet. "al-Djaghmīnī." In *Encyclopaedia of Islam.* 2nd ed. Vol. 2, p. 378. Leiden, Netherlands: E.J. Brill, 1965.

Ṭāshkubrīzade, Aḥmad b. Muṣṭafā. *Miftāḥ al-sa'āda wa-miṣbāḥ al-siyāda.* Vol. 1, p. 349. Beirut: Dār al-kutub al-'ilmiyya, 1985.

Sally P. Ragep

JAMES, WILLIAM (*b.* New York, New York, 11 January 1842; *d.* Chocorua, New Hampshire, 26 August 1910), *psychology, philosophy.* For the original article on James see *DSB,* vol. 7.

James is widely known as the father of American psychology, and he is fondly remembered as a public intellectual with the clarity and insight to create usable popularizations of complex academic and scientific work. His major professional contribution to psychology was to remind his fellows in the field of the worlds of mind and behavior that lie beyond the grasp of even the most elegant theories. His chief philosophical contribution was his unblinking attention to the concreteness of experience in psychological events and feelings, religious beliefs, scientific inquiry, and the philosophy of empiricism itself. Although he had often been dismissed as a mere popularizer in the decades after his death in 1910, recent work on James has stimulated a steadily growing appreciation for his substantial contributions. Clarity has been no vice as he helped to shape humanistic orientations, process thinking, and phenomenology in psychology, philosophy, and religious studies. In fact, in the early twenty-first century there is wide recognition of his transdisciplinary significance for his pioneering critique of overreliance on scientific authority, for his anticipations of neuroscience, for his rhetorical gifts, for his spirituality, and for his prophetic warnings about a culture dominated by corporate institutions and commodified values.

Background and Upbringing. William James's upbringing, in the household of a Swedenborgian radical and in the context of growing scientific authority in the middle-to-late nineteenth century, shaped his life, temperament, and work. He taught, wrote, and lectured on the borderland of science and religion, gravitating toward mediation of these fields, with all their kindred associations, in his psychology, philosophy, and religious thought.

Because the spirituality of his father, the elder Henry James (to distinguish him from his second son, the novelist), included endorsement of the empirical spirituality of Emanuel Swedenborg, an eighteenth-century mystic popular in Romantic cultural circles, the religious training that the oldest son, William, received was open to the study of nature. By age nineteen, with his father's encouragement, he went to the Lawrence Scientific School at Harvard University, where he met a brand of science shaped by the professional demands for rigorous empiricism. He studied chemistry, anatomy, and physiology before transferring to the Harvard Medical School; this succession of studies reflected his temperamental ambivalence and contributed to his vocational indecision. These and related troubles with depression, eye and back ailments, philosophical uncertainty, and hesitancy to marry—precisely because he dreaded passing on his troubles to a next generation—plunged him into a period of indecision and personal crisis, which prolonged his formative years and postponed the beginning of his working life.

Crisis and Construction of Worldview. James is perhaps as famous for this period of crisis as for his mature theories, in part because many of those ideas first began to emerge in his youthful drama of pain and recovery. The very paths he took while struggling in his youth showed not only the first stirrings of his mature ideas, but also their roots in the scientific and religious commitments of his early adulthood. While he found it difficult to accept his father's ideas directly, he gravitated toward theories and cultural experiences that, similarly, expressed the immaterial factors of life (including spirituality and consciousness) within the natural world rather than in orthodox references to incomprehensible factors or another world. In particular, he was attracted to the pre-Christian ancients, sectarian medicine, humanistic psychology, and voluntaristic philosophy. Just as Henry Sr. approached nature "as if it had some life in it" (in William's words), so the son, now on a professional path in his thirties, inquired into the experiences of the natural world and human nature, noticing the role of immaterial factors to

William James. HULTON ARCHIVE/GETTY IMAGES.

complement the emerging understanding of nature in terms of physics and chemistry as assumed by most modern professional science (*Correspondence,* vol. 4, p. 227). William endorsed the work of modern science, benefiting deeply from the latest research in his own profession, most notably in his landmark *Principles of Psychology* (1890). However, he maintained an impulse to pull his profession toward a "program of the future of science" that would be less wedded to a materialistic philosophy, such as the automaton theory of consciousness, which he noticed to be a frequent accompaniment of professional science but an unnecessary stowaway on the path of inquiry. He offered wry praise for the scientific enthusiasm of his times as "a temporarily useful excentricity" (1902, p. 395).

A Psychology of Philosophizing. Despite materialist and positivist enthusiasms for Darwinism swiftly spreading in influence from the 1860s, when James first studied science, he noticed its hypothetical and probabilistic qualities. This type of thinking shaped his theory of "The Sentiment of Rationality" (1879), in which he proposed that the motivation to philosophize stemmed from the craving for explanatory sufficiency in clarity or simplicity that shaped human thinking in general, even in science.

Openness to experience unconstrained by the abstractions of theory—whether scientific, philosophical, or religious—took his work out of the mainstream profession of scientific psychology and into philosophical and religious studies. He was also concerned about the overreliance on experimental method in the psychological laboratory. In *Pragmatism* (1907) he emphasized the usefulness of theories as instruments of inquiry, but he also insisted that we should not confuse these directional tools for the whole forest of experience. In "The Will to Believe" (1895) he proposed that when evaluating a belief—in science or religion—it is more important to pursue the prospective possibilities in experience itself rather than remain in fear of crossing abstract boundaries of prior theory formation. This essay, and his many accompanying "essays in popular philosophy," never called for rejection of scientific inquiry, but for a continuation of its methods into religion and other spheres conventionally segregated from science. He even called his research into *Varieties of Religious Experience* (1902) a "science of religions" with which he explored personal religious experiences that generally underlie institutional structures; finding evidence of "the more" in human psychology, he proposed that it served as the basis for orthodox religious affiliations. Toward the end of his life, James called his emphasis on pure experience unadorned by abstraction "radical empiricism," and his essays in elaboration of this philosophy were collected posthumously.

James was not particularly concerned with the specialized discourses of psychology, philosophy, religious studies, and science studies that today claim pieces of his corpus. He contributed to each of these fields as they now stand, but his own personal and intellectual motivations were more broadly based. Of course, he lived in a time when educational institutions were only beginning to specialize; in fact, he remarked that the first lecture he ever heard in psychology was his own. In addition, the very indecisiveness that troubled him, and that has encouraged much of the psychological scrutiny of James himself, actually contributed to his appetite to work in all these fields. He knew the pain of choice, and its necessity when selecting paths forward, but he was also attentive to the tyranny of choices that ruled out other important factors in life, hence his insistence on recognizing the "ever not quite" factors in our knowledge of all fields.

Meliorism, a Hope in the Making. His hopes were hard won, and this lent his philosophy an authentic and experience-chastened quality that catapulted his work onto a public stage, where he became popular beyond the appeal of most academics. Perhaps his most alluring mediation was between pessimism and optimism. Why endorse an outlook that is explicitly against one's own good? Yet at the same time, how turn away from the truly tragic and burdensome facts of life? His answer was to adopt neither, but instead to endorse what he called "meliorism," the commitment to hoping and working for the best. This gritty buoyancy is at the heart of James's commitment and a reason for the enduring appeal of his words even decades after his death in 1910. Characteristically, he was scientific enough to doubt the orthodox claims for an afterlife, but he was religious enough to notice their power to motivate and their plausibility in the depths of "psychic experiences." Besides, after decades of committed search into the natural world and human psyche, and with a sharp sense that human experience involved both bodily and mental factors, he faced death with the wistful realization that "I'm just getting fit to live" (*Letters*, vol. 2, p. 214).

SUPPLEMENTARY BIBLIOGRAPHY

WORKS BY JAMES

The Varieties of Religious Experience: A Study in Human Nature. New York: Longman, Green, 1902.

The Works of William James. Edited by Fredson Bowers and Frederick H. Burkhardt. 19 vols. Cambridge, MA: Harvard University Press, 1975–1988. There are many editions of James's writings; this collection and the *Correspondence* below are the most authoritative. *Works* includes all of James's published books and collections of many of his unpublished notes, lectures, and essays thoroughly annotated and contextualized.

The Correspondence of William James. Edited by Ignas K. Skrupskelis and Elizabeth M. Berkeley. 12 vols. Charlottesville: University Press of Virginia, 1992–2004. A collection of correspondence to and from James, including 70 percent of the 9,300 known letters that he wrote in his lifetime, with the remaining calendared for ease of reference.

OTHER SOURCES

Bjork, Daniel. *William James: The Center of His Vision.* New York: Columbia University Press, 1988. Provides a detailed account of James's life, especially in his relations with his wife, Alice Howe Gibbens James, and emphasizes William James's exuberant, distinctive genius in dealing with knotty psychological and philosophical questions.

Cotkin, George. *William James, Public Philosopher.* Baltimore, MD: Johns Hopkins University Press, 1990. A historical account of the cultural and political position of James, especially in his adult life.

Croce, Paul Jerome. *Science and Religion in the Era of William James.* 2 vols. Chapel Hill: University of North Carolina Press, 1995 (Vol. 1) and forthcoming (Vol. 2). Sets James in his personal and cultural contexts and emphasizes the intermingling of mind and body in his work.

Donnelly, Margaret E., ed. *Reinterpreting the Legacy of William James.* Washington, DC: American Psychological Association, 1992. A collection of essays by psychologists in evaluation of the meaning and legacy of James's psychological theories, written from the perspective of pluralism, phenomenology, evolutionary biology, and history and with attention to theories of the self, emotions, clinical work, free will, and parapsychology.

Feinstein, Howard M. *Becoming William James.* Ithaca, NY: Cornell University Press, 1984. A psychobiography that evaluates the young man in relation to family dynamics.

Levinson, Henry Samuel. *The Religious Investigations of William James.* Chapel Hill: University of North Carolina Press, 1981. A religious studies evaluation of his development of a science of religions.

Menand, Louis. *The Metaphysical Club.* New York: Farrar, Straus and Giroux, 2001. A highly readable narrative synthesis of recent scholarship on James in relation to other founders of pragmatism—Oliver Wendell Holmes Jr., Charles Sanders Peirce, and John Dewey—and in the context of countless stories of American social and cultural history.

Myers, Gerald E. *William James: His Life and Thought.* New Haven, CT: Yale University Press, 1986. Covers much more thought than life, with an emphasis on his philosophical psychology.

Seigfried, Charlene Haddock. *William James's Radical Reconstruction of Philosophy.* Albany: State University of New York Press, 1990. Presents James's philosophical development toward a postmodern view of the constructed character of knowledge.

Simon, Linda. *Genuine Reality: A Life of William James.* Chicago: University of Chicago Press, 1999. Offers a biographical overview with more attention to social context and personal issues than to theory formation.

Taylor, Eugene. *William James on Consciousness beyond the Margin.* Princeton, NJ: Princeton University Press, 1996. Challenges the conventions in the history of psychology that James abandoned psychology after 1890 with evidence that his experimental psychopathology was directed toward an alternative psychology.

Townsend, Kim. *Manhood at Harvard: William James and Others.* New York: Norton, 1996. Examines James's ambivalence in terms of gender and finds the popular philosopher both subject to a strong and brittle masculinity and intelligently rising above it.

Paul Jerome Croce

JANET, PIERRE (*b.* Paris, France, 30 May 1859; *d.* Paris, 24 February 1947), *psychology, philosophy, psychotherapy, hypnosis.*

In his day Janet enjoyed a worldwide reputation as a psychologist and psychotherapist. A specialist in psychopathology, he proposed a theory of the psyche that was widely seen as competing with Sigmund Freud's. After World War I, he worked out a general theory of the evolution of the mind based on a notion of "conduct" that he contrasted with the idea of behavior promoted by John B. Watson and the American behaviorist school.

Life and Work. Janet's background was middle-class. His father, Jules Janet, was a Paris lawyer, and his mother, Fanny Hummel, was a devout Catholic from Alsace. In his "Autobiography," he tells how in his youth his interests were divided between an enthusiasm for the natural sciences, particularly botany, and mystical inclinations, and how he hoped to resolve this contradiction by taking up philosophy (1930). This choice was also much affected by his uncle Paul Janet (1823–1899), an influential philosopher of the spiritualist school. Eclectic spiritualism was in fact the predominant philosophical doctrine in France for the greater part of the nineteenth century. It traced its origins to Maine de Biran and its chief exponent was Victor Cousin. Typically, thanks to Cousin, senior students in French secondary schools learned philosophy with a spiritualist bent. Paul Janet described himself as a spiritualist philosopher who, as such, drew a distinction between two irreducible realities, the material and the spiritual; at the same time he declared himself a "liberal" open to positive science. From 1870 on, under the Third Republic, Paul Janet exercised great institutional power. He was a professor at the Sorbonne and reformed the French secondary-school philosophy curriculum. In 1879, Paul Janet published a celebrated philosophy textbook that went through twelve editions before 1919.

Following in his uncle's footsteps, Pierre Janet became a student of philosophy, passing the competitive entrance examination to the elite École Normale Supérieure in 1879. Among his fellow students were the future sociologist Émile Durkheim, the future philosopher Henri Bergson, and the future socialist politician Jean Jaurès. In 1882 he passed the *agrégation* exam, which qualified him to teach philosophy in *lycées.* He did so for several months in Châteauroux, in central France, and then at Le Havre, in Normandy, where he remained until 1889. Attracted by medicine and scientific psychology, he persuaded local hospital doctors to allow him to observe patients, notably hysterics. Another Le Havre physician, Joseph Gibert, drew his attention to the work of magnetizers (or mesmerists) and introduced him to Léonie, an uneducated woman who seemed to possess a remarkable ability to be hypnotized mentally from a distance and to read the thoughts of others. With Gibert, Janet made various experiments regarding mental suggestion, or with what would now be called telepathy, and published the findings in 1885. These experiments aroused widespread interest, and helped make Janet's name known in medical and scholarly circles.

It was not long, however, before Janet realized that twenty years earlier Léonie had in fact been coached by magnetizers and that she was by no means the naïve subject that she appeared to be. He promptly abandoned his "parapsychological" research and directed his attention to the scientific psychology of Théodule Ribot (1839–1916). Ribot was a philosopher, not a physician, and his ambition was to found a "positive" psychology in France that was not beholden to the metaphysics of the spiritualists. This "evolutionist" psychology was inspired rather by the philosopher Herbert Spencer and the neurologist John Hughlings Jackson, and sought a basis in physiology and medicine. According to Ribot, it was by studying mental disturbances, conceived of as regressions relative to the development of both the individual and the species, that the psychologist could understand the normal, more highly evolved development of mental functions. Ribot accordingly gave pride of place to psychopathology as the best line of approach, and he steered the new French psychology in that direction.

Ribot had played an important institutional role, for in 1876 he had founded the *Revue Philosophique,* which published not only properly philosophical work but also contributions from physicians and philosophers looking to lay the groundwork for a physiological psychology. It was in this journal that Pierre Janet published his first articles. In 1888, with the support of Paul Janet, Ribot became the first person to occupy the Chair of Experimental and Comparative Psychology at the celebrated Collège de France.

In 1889, Janet defended a thesis in philosophy entitled "Psychological Automatism: An Essay in Experimental Psychology Concerning the Inferior Forms of Human Activity." His uncle Paul was a member of the jury. This thesis, which adumbrated all the main themes of Janet's future work, was hailed at once as an event in the philosophical and medical world. After delivering it, Janet taught philosophy in a Parisian *lycée* and embarked on medical studies. He began frequenting the wards of the neurologist Jean-Martin Charcot at the Salpêtrière hospital in Paris. At that time Charcot was at the pinnacle of his fame, but his work on hysteria and hypnosis was being challenged by Hippolyte Bernheim, a professor at the Faculty of Medicine of Nancy in eastern France. Bernheim claimed that the neurological symptoms associated with hysteria and hypnosis, as identified by Charcot, were the result of a kind of training: hysteria and hypnosis were not pathological states but psychological phenomena for which suggestion was largely responsible. It was in the context of this debate that Charcot encouraged the work of the young psychologist Janet, which he hoped would erode Bernheim's case. In 1890 he set up a clinical psychology laboratory *(Laboratoire de psychologie dela clinique)* for his protégé at the Salpêtrière. In 1893, under Charcot's supervision, Janet presented his medical thesis,

"Contribution to the Study of the Mental Accidents of Hysterics."

In 1894 Janet married Marguerite Duchesne. From 1895 to 1897, standing in for Ribot, he was responsible for the teaching of experimental psychology at the Collège de France. From 1898 to 1902, he held the post of *chargé de cours* at the Sorbonne, still in the field of experimental psychology. When Ribot retired from the Collège de France in 1902, he and Bergson put Janet's name forward as a successor. Despite the competing candidacy of the psychologist Alfred Binet, Janet won the appointment. After long years as a secondary-school teacher, he now took up a respected, tenured position that he was to occupy until his retirement in 1935. Janet's lectures at the Collège de France always attracted a large audience. In 1901 Janet had founded the Société de Psychologie, and in 1904, with Georges Dumas, he launched the *Journal de Psychologie Normale et Pathologique*, the second French journal devoted to scientific psychology (the first being *L'Année Psychologique*, initiated by Henri Beaunis and Alfred Binet in 1895). At the start of the twentieth century, a strong claim could be made that Janet was the most eminent representative of French psychology.

In parallel with his university career, Janet practiced medicine and psychotherapy both in the hospital setting and in a private capacity. He amassed many thousands of clinical records that he filed carefully in a room in his apartment, but which were destroyed after his death in accordance with his wishes. Janet's scientific and therapeutic work continued up until his death soon after World War II.

The Subconscious, Hysteria, and Psychasthenia. In his work *L'Automatisme psychologique* (1889), Janet followed Charcot in promoting hypnosis as an experimental method capable of eliciting and reproducing the hysterical phenomena in which he was interested. He endorsed Ribot's hypothesis, according to which a continuity existed between the normal and the pathological such that the laws governing illness applied equally to health. His perspective, like Ribot's, was distinctly evolutionist in character, and underpinned by a model borrowed from Jackson. Janet's aim was to reveal lower, elementary and automatic activity that was generally inhibited in normal individuals by higher functions.

Janet parted ways with Ribot, however, on the question of the relationship between psychology and physiology. Whereas Ribot subordinated the first to the second, Janet made psychology responsible for the study of such automatic phenomena as somnambulism, hallucinations, automatic writing, and dual personality, which, because they appeared to occur in the absence of consciousness, had been left to the scrutiny of the physiologists. Physiology, he felt, was unreliable when mental phenomena needed to be explained. Automatic phenomena were always accompanied, in his view, by a particular kind of consciousness that lay outside and below normal awareness. The consciousness of hysterics, because of their psychological weakness, could contain and synthesize only a small number of phenomena. This "shrinking of the field of consciousness," as Janet called it, accounted for a "dissociation" of the personality: psychological phenomena outside normal consciousness governed the automatic forms of behavior that gave rise to a second consciousness and a dual personality.

For Janet, then, the object of psychology was still consciousness. He distinguished between a normal or personal consciousness and a pathological consciousness, which might appear to be unconscious but in fact was not. It was in this connection that he popularized the term "subconscious," which he preferred to "unconscious." He did not therefore challenge the postulate of the spiritualists, according to which there was no psychology apart from the psychology of consciousness. Indeed, he referred often to Maine de Biran and drew a clear distinction between automatic activity on the one hand (based on habit, and conservative and repetitive), and voluntary (creative, synthesizing, organizing) activity on the other. In contrast to Maine de Biran, though, Janet assigned a psychological status to subconscious phenomena. In common with Bergson and Freud, at that moment, Janet held that phenomena that were unconscious or devoid of a certain kind of consciousness fell nevertheless within the province of psychology and ought not to be assigned to physiology.

In *L'Automatisme psychologique* Janet showed himself to a genuine heir to Ribot and Charcot, but at the same time he sought to reconcile this positivist heritage with his uncle's spiritualism.

In his medical thesis, published in 1893 as *L'État mental des hystériques* (The mental state of hystericals), Janet treated hysteria as a mental illness. Like Freud and Breuer at the same period, he adopted Charcot's idea that hysteria was a phenomenon linked to a traumatic event. It might seem, therefore, that Janet was close to the two Viennese authors at that time; in reality, however, despite some common ground, especially the quest for a psychological explanation of hysteria, there was a wide gap between the respective approaches of Janet and the Freudians.

Janet ascribed what he called "hysterical accidents" connected to patients' histories to the abovementioned shrinking of the field of consciousness. Because of their "psychological impoverishment," hysterics were able (to use today's terminology) to process and synthesize no more than a few pieces of information at a time, so that a certain number of items—notably those which underlay subconscious fixations—remained outside the field of personal consciousness. As a corollary, such excluded fixed ideas tended only to exacerbate psychological disintegration.

Character traits traditionally attributed to hysterics, such as instability, impulsiveness, and volatility, as well as long-lasting hysterical stigmata (anesthesias, paralyses, etc.), were also viewed by Janet as effects of a "weakness of psychological synthesis." The main defining characteristic of hysteria was the tendency to develop a dual personality.

While Freud and Joseph Breuer tended to see the traumatic event as the root cause of dissociation, Janet looked upon it rather as a consequence of the shrinking of the field of consciousness, and hence of the psychological weakness of the hysteric. Freud and Breuer thus drew a picture very different from Janet's, highlighting the intelligence, liveliness of mind, and psychological fortitude of their hysterical patients.

Beginning in 1901, Janet framed a psychological theory that allowed him to group various illnesses together in a single clinical category, *psychasthenia* (whose etymological meaning is "lack of mental strength"). This condition was said to begin with a weakening of nervous functions that occasioned a state described at the time, following the American physician George M. Beard, as "neurasthenic." In Janet's account, a "primitive neurasthenia" underlay deteriorations of various bodily functions and various mental disturbances such as an absence of will (abulia), indifference, and apathy. Subsequently there arose what Janet called "feelings of inadequacy" (*sentiments d'incomplétude*), which in turn produced a constellation of symptoms, including depersonalization, shame, anxiety, impressions of automatism, and so on. The feelings of inadequacy could signal the development of tics, phobias, and eventually, in many cases, obsessive ideas (ideas of illness, crime, sexual transgression, etc.). Such obsessions, as distinct from the *idées fixes* of hysteria, were conscious rather than subconscious, for the shrinking of the field of consciousness was not very marked in psychasthenics, who were quite aware of the contradictions between their obsessive ideas and their "normal tendencies."

Janet's clinical description of psychasthenia was buttressed by an energetic conception of mental functioning: psychasthenics' low psychological energy level made complex, well-adapted and voluntary actions too costly for them, so causing the drop in mental level that generated their seemingly bizarre actions and symptoms.

Whatever the mental disturbance at issue, Janet tended to invoke psychological impoverishment and weakness, and in his clinical notes he paid special attention to the complaints actually articulated by patients. Behind this approach was a vision of the human being to which voluntary action and the unity of the self were cardinal. The term depression appeared frequently in Janet's writings, and in any case the omnipresence therein of the theme of psychological weakness arguably foreshadowed the modern preoccupation with depressive illness.

Janet as Psychotherapist. Unlike other psychotherapists of his era, who were content merely to suppress symptoms by means of suggestion, Janet sought to understand their genesis. The basis of his clinical approach was "psychological analysis," which consisted fundamentally in the identification of the root causes of the condition and, usually, in the "dissociation of fixed ideas" by using suggestion to suppress them. Alternatively, should suggestion fail, other ideas might be substituted: for example, as he recounts in "Histoire d'une idée fixe" (1898), Janet gradually was able to change his patient Justine's terrifying hallucinations of dead cholera victims into amusing visions, so that over the course of her treatment hysterical crises were replaced by fits of laughter. Janet also—though much more rarely—succeeded, like Freud and Breuer, in effecting a cure through the retrieval of traumatic memories. This was true in the case of Irène, who had forgotten things about her mother's agony and death and was reliving them in a hallucinatory way; since she refused to be hypnotized, Janet stimulated her memory until her amnesia was dispelled and she was able to recall her memories at will (Janet, 1904).

Janet's psychological analysis was inseparable from a synthetic endeavor to restore mental integrity by teaching patients how to manage their psychological economy in a better way. Very often his therapeutic approach incorporated an educational or re-educational attempt to improve the exercise of voluntary attention. In "L'Influence somnambulique et le besoin de direction" (1897), he compared his therapeutic intervention to that of a teacher or a spiritual guide (*directeur de conscience*) and underscored the relational aspect of treatment. According to Janet, the physician's task was a paradoxical one in that the patient had to be coaxed into a curative relationship of dependence—even of "somnambulistic passion" —with the therapist, while at the same time being led toward emancipation from that dependence. Later, while emphasizing the social aspect of the therapeutic relationship, Janet (1919) wrote that an "adoptive" attitude of the patient toward the therapist was a desirable goal.

Janet was an eclectic therapist who borrowed from the old "magnetic" techniques and continued, if needed, to use hypnotic suggestion long after it had lost the respect of many of his colleagues; he had no fear of playing the pedagogue, the spiritual guide, or even the exorcist. Indeed he adopted practices associated with Catholicism, as for instance in the case of Achille (1898b), a patient who presented all the classical features of

Janet

Pierre Janet. © BETTMANN/CORBIS.

demonic possession. After failing to hypnotize Achille, Janet reports, he had the idea of acting like a "modern exorcist" and addressing himself to the Devil. He discovered by this means that Achille had had an extramarital affair, and, suffering the effects of remorse, had been harboring a "dream," which was subconscious, in which he felt he was damned and possessed by the Devil. Janet conducted the treatment in such a way that Achille forgot both his transgression and his remorse. In this case history, Janet noted, "Knowing how to forget is sometimes as much a quality as knowing how to learn, because forgetting is prerequisite to moving forward, to progress, to life itself.... One of the most valuable contributions that pathological psychology could make would be to discover a reliable way to precipitate the forgetting of specific psychological phenomena" (Janet, 1898b, p. 404). Janet's chief therapeutic concern here was apparently not a cathartic retrieval of memories, as promoted at the same period by Freud and Breuer, but rather a process of learning how to forget.

Janet's Psychology of Conduct. In 1919, Janet published *Psychological Healing* (*Les Médications psychologiques*), a three-volume work in which he presented himself as a seasoned practitioner. The first volume dealt with the history of psychotherapy, and as such constituted the first notable and well-documented book on the origins of French psychology. In the second two volumes, Janet reviewed contemporary psychotherapeutic approaches. He included a polemical text against psychoanalysis that he had first presented in 1913 at the International Congress of Medicine in London and first published in 1914. It criticized the claims of psychoanalysis to originality, as well as its systematic and sectarian character. Despite these criticisms, however, Janet's approach was more measured than might have been expected with respect to the Freudians, and indeed he later defended psychoanalysis on occasion. It was not after all impossible, seeing that he had not a little in common with his Viennese colleague, for Janet to see himself as in some sense the "French Freud." His contemporaries, especially the early French psychoanalysts, reinforced this image, which was not truly discarded until after his death.

In *Psychological Healing*, Janet restated theories about hysteria and psychasthenia that he had first framed before World War I. He reiterated the need to distinguish between psychological strength and psychological tension: the first meant the quantity of energy available and the second the subject's capacity to mobilize that energy at a particular level relative to an evolutionary scale. Mental disturbances were specific to either psychological strength or psychological tension. Janet's aim was to construct a pathological and therapeutic psychology of the entirety of mental illnesses on the basis of this opposition.

Almost all of the first volume of Janet's work *De l'angoisse à l'extase* (From anxiety to ecstasy), published in 1926, was devoted to a mystically inclined woman whom he called Madeleine and whom he had observed at the Salpêtrière over a few years early in the century. Madeleine presented contractures, stigmata evocative of the wounds of Christ, and ecstatic states. Janet had been able to observe her at leisure in his laboratory, and in her case too he had assumed the role of a lay spiritual guide. It was on the basis of this case, in the second volume of *De l'angoisse à l'extase* (1928), that he constructed a schema, conceived from an evolutionist perspective, of the development of feelings and the corresponding ways of regulating action. It was in this volume that Janet presented a first outline of a vast psychological system that he elaborated upon subsequently in his many publications and in his courses at the Collège de France. Yet he never produced a grand synthesis: the closest he came was a short encyclopedia article (1938) in which he stated the basic principles and sketched the main features of his projected theory.

The "psychology of conduct" that Janet defined in this article was founded on the unifying hypothesis according to which all psychological phenomena were actions, thus obviating the need for a distinction to be drawn in manuals of psychology between action and thought. Consciousness itself became action thanks to language. Janet's psychology of conduct implied that the study of consciousness was lacking in the behaviorism then triumphant in the United States, which Janet considered inadequate to the study of man. He based his thinking on the idea of *tendency,* defined as "a disposition of the living organism to perform a specific action." Tendencies in Janet's system were classified and ordered in a hierarchical manner, from the primordial and automatic to the most recently evolved, which were also the most fragile, and hence the first to be lost in mental illness. Clearly Janet was borrowing here from Ribot's evolutionism and applying it to all "tendencies." Drawing on psychopathology, he extended his objects of study to the child, to primitive humans, and to animals. Forms of conduct were underpinned by tendencies common to man and animals ranging from the reflex to the social. Intellectual tendencies were incipient in lower animals and fully developed only in man. For Janet, this category included language and symbolic thought. Medium-range tendencies underlay belief systems, which were founded on speech as a substitute for action. Lastly, the highest-level tendencies were the basis of the kind of conduct adopted by individuals in order to conform to the moral rules and logical laws developed over time by social groups; such higher tendencies also supported those individual and particular kinds of conduct which allowed humanity to evolve in a progressive way. Prior to this synthetic sketch of 1938, Janet's work had been increasingly focused on conduct of the social kind.

As for how Janet himself evaluated the import of his system of psychology, it is striking that he seemed at times to distance himself somewhat from his grand evolutionist narratives. Was this perhaps his way of acknowledging criticism? In the France of the late 1930s, certainly, the growing influence on psychopathology of psychoanalysis and phenomenology had begun to make Janet's approach seem old fashioned, and in the 1950s the overwhelming success of Freudianism marginalized it almost completely. This situation endured until Janet and his work were rediscovered by Henri F. Ellenberger (1970).

BIBLIOGRAPHY

WORKS BY JANET

L'Automatisme psychologique: Essai de psychologie expérimentale sur les formes inférieures de l'activité humaine. Paris: Alcan, 1889.

The Mental State of Hystericals: A Study of Mental Stigmata and Mental Accidents, translated by Caroline Rollin Corsin. Bristol: Thoemmes, 1998 [1893].

"L'influence somnambulique et le besoin de direction." *Revue Philosophique* 43 (1897): 128–143.

"Histoire d'une idée fixe." In *Névroses et idées fixes,* Vol 1. Paris: Alcan, 1898a.

"Un cas de possession et l'exorcisme moderne." In *Névroses et idées fixes,* Vol. 1. Paris: Alcan, 1898b.

Les Obsessions et la psychasthénie. Paris: Alcan, 1903.

"L'amnésie et la dissociation des souvenirs." *Journal de psychologie normale et pathologique* 1 (1904): 28–37.

Psychological Healing, 2 vols. Translated by Eden and Cedar Paul. London: Allen & Unwin, 1925 [1919].

De l'angoisse à l'extase. 2 vols. Paris: Société Pierre Janet, 1975 [1926, 1928].

"Autobiography." In *A History of Psychology in Autobiography,* Vol. 1, edited by Carl Murchison. Stanford, CA: Stanford University Press, 1930.

"La psychologie de la conduite." In *Encyclopédie Française,* Vol. 8: *La Vie mentale.* Paris: Société de Gestion de l'Encyclopédie Française, 1938.

OTHER SOURCES

Brooks, John, III. *The Eclectic Legacy. Academic Philosophy and the Human Sciences in Nineteenth-Century France.* London: Associated University Press, 1998.

Carroy, Jacqueline, and Régine Plas. "How Pierre Janet Used Pathological Psychology to Save the Philosophical Self." *Journal of the History of the Behavioral Sciences* 36, no. 3 (2000): 231–240.

———, Annick Ohayon, and Régine Plas. *Histoire de la psychologie en France XIXe–XXe siècles.* Paris: La Découverte, 2006.

Ellenberger, Henri F. *The Discovery of the Unconscious: The History and Evolution of Dynamic Psychiatry.* New York: Basic Books, 1970.

Ohayon, Annick. *L'impossible rencontre: psychologie et psychanalyse en France 1919–1969.* Paris: La Découverte, 1999.

Prévost, Claude M. *La psycho-philosophie de Pierre Janet: Économies mentales et progrès humain.* Paris: Payot, 1973.

Jacqueline Carroy
Régine Plas

JANSSEN, PAUL ADRIAAN JAN

(*b.* Turnhout, Belgium, 12 September 1926; *d.* Rome, Italy, 11 November 2003), *medicine, pharmacology, chemistry.*

Janssen spearheaded research that resulted in more than eighty medicines, of which five were put on the World Health Organization's list of essential drugs. These drugs span a broad range of therapeutic areas, including psychopharmacology, neuropharmacology, gastroenterology, cardiology, parasitology, mycology, virology, immunology, anesthesiology and analgesia.

Paul Janssen's father, Constant, was a general practitioner. In 1934, he founded N.V. Producten Richter, a company that initially imported and sold products from the Hungarian company Richter. During and after World War II, Constant successfully developed his own products. At that time, Paul Janssen was studying classical humanities at the Jesuit college in Turnhout. Not yet seventeen years old, he enrolled at the Facultés Notre-Dame de la Paix in Namur, studying sciences at this Jesuit institution for two years. This quality, combined with the strict discipline and intensive pace of the curriculum, laid the foundations for his subsequent scientific research. At the end of 1945, he enrolled at the Faculty of Medicine at Catholic University in Leuven. By then he was already convinced that there was a relationship between the chemical structure of a molecule and its pharmacological effect. This would be the underlying principle of his research activities at Janssen Pharmceutica, the company he would one day found: the synthesis of chemical molecules to explore the relationship between their pharmacological structure and activity.

A Pharmaceutical Entrepreneur. While at the university he traveled to the United States and over a period of a few months attended summer courses in pharmacology and biochemistry at Cornell Medical School (given by Harry Gold), Harvard University (Edwin Cohn), and the University of Chicago (Carl Pfeiffer). He also visited a number of private research laboratories that were not represented in Europe at that time. A final important factor in his training as a research scientist was a stint as a part-time assistant to Nobel Prize winner Corneel Heymans at the State University of Ghent, where Janssen received his medical degree in 1951 (and a postdoctoral degree in pharmacology in 1956). After that he had a number of internships in Paris, Vienna, Heidelberg, and Ghent, and during his military service, he was a part-time assistant to Jozef Schuller at the Pharmacological Institute of Cologne University. In 1953, Dr. Paul, as he was familiarly called by his staff, started up his own research facility at his father's company, which focused increasingly during the postwar years on the production and sale of products that the research unit had developed itself (AD-Vitan, Perdolan, Rubalgan, Bronchosedal, and others). N.V. Laboratoria Pharmaceutica was founded on April 5, 1956, and Paul Janssen became the company's managing director and research director. The expansion of the company led to the purchase of a new industrial site in Beerse, near Turnhout. Paul Janssen started there with thirty employees in new research laboratories brought under an independent company in 1958; he called them N.V. Researchlaboratoria Dr. C. Janssen in honor of his father. The company changed the location of its registered office from Turnhout to Beerse in October 1961.

Janssen's research activities were so successful that the future of his company became a matter of concern to him in the event that anything should happen to him. Johnson & Johnson (J&J) knew about Janssen's achievements, and William Lycan, then director of the Research Division of J&J and a member of its board of directors, paid a visit to the Beerse company. Through a share swap, the family company, with all its departments, was taken over by J&J on 24 October 24 1961. On 10 February 1964, the name of the company in Beerse became Janssen Pharmaceutica N.V.

It was a decisive step in the expansion of the international pharmaceutical strategy of J&J, and the American parent company saw the merger as an investment in Janssen's innovative genius, an investment that subsequently proved highly lucrative; under his inspiring leadership, Janssen Pharmaceutica grew in the space of a few decades into an international company with affiliates in over forty countries.

Janssen established a production plant in China. As far back as 1978, he set up contacts with Ma Haide (whose real name was George Hatem), a doctor of Lebanese origin and one of Mao Zedong's personal physicians, who asked Janssen to help modernize the Chinese pharmaceutical industry. In July 1981, a pharmaceutical countertrade agreement was concluded with China, the subject of which was Janssen's anthelmintic compound, mebendazole, a product for the control of worm infections in livestock. In 1983, the Chinese launched a production plant in Hanzhong with the help of specialists from Janssen. Janssen Pharmaceutica was the first Western pharmaceutical company in the world to carry out such a large project in China. Intensive cooperation was rewarded in 1985 with a joint venture, which led to the setting up of a pharmaceutical production plant in Xian under the name Xian-Janssen Pharmaceutical Ltd. After only ten years, the turnover of this company amounted to 253 million euros, and it was acclaimed on a number of occasions as the best joint venture in the country by *Fortune* magazine. The most important drug produced there was Motilium, which soon passed the one billion tablet mark. As a token of the high regard in which Janssen was held, the Dr. Paul Janssen Faculty for Pharmacy was established at Xi -Jing University in China in 2002.

Between 1965 and 2000, the annual turnover of Janssen Pharmaceutica N.V. grew spectacularly. Of course, the secret of this remarkable result was due, first and foremost, to the free-ranging creativity of its exceptional founder and scientific leader. Apart from that, his success was also based on the principle of building research around people, who were inspired by his example and rewarded his trust with dedication and hard work. His research concept was also his management philosophy. It rested on four pillars: teamwork by a cross-disciplinary team within the flattest possible organizational structure; bold basic research that sought out the challenges of the unknown; feedback from practitioners; and an uncompromising commitment to freedom of thought.

In the eyes of Paul Janssen people and not investment were the key; one had to trust one's scientists.

Research Results. As a medicinal chemist, Janssen paid particular attention to the presumed relationship—and even more so to the gaps in this relationship— between a chemical structure and the pharmacological activity of organic molecules. Discrepancies between *in vitro* and *in vivo* experiments were also a direct impetus to further investigation. This alertness to the unexpected contributed to many important discoveries. For him, it was the observable effect of a substance in a functional test model that counted. The general pharmacology with *in vivo* tests was the key focus of his research. The introduction of rigorously validated and standardized functional animal models is one of his most important contributions to medicinal research. Janssen's research efforts were driven by scientific concepts. For example, as a medical student Janssen had learned about pethidine, a new drug that had been introduced into medicine in 1939. Very early on this medication seemed to cast a spell over him, perhaps because he had observed that its structure recurred in a broad class of morphinomimetic compounds, the diphenylpropylamines, and also because it had multiple medicinal properties; pethidine—it was called meperidine in the United States—had been characterized as an atropine-like antispasmodic/antidiarrheal agent, but clinically had been found to be an addictive morphinomimetic analgesic. What also interested him was the fact that pethidine was a piperidine derivative and so it would be fairly easy to iteratively synthesize analogs from it. The guiding concept was his curiosity about whether he could separate the two properties of this compound from one another, in particular its analgesic action and its antidiarrheal/antispasmodic effects. Following Dr. Paul Ehrlich he had a chemical starting point or lead compound and two bioassays for efficacy evaluation: the Straub mouse tail test for opiate activity and the electrically driven guinea pig ileum in vitro for antispasmodic activity. After synthesizing 500 analogs, he came up with eight marketable drugs.

Nonmorphine-like Analgesics. The right combination of neurolepsy and analgesia offered new possibilities for anesthetics, giving rise to fentanyl (R 4263) and a series of fentanyl derivatives, which were developed between 1974 and 1976. These drugs included sufentanil (R 33800), which in 1984, under the trade name Sufenta, made possible the first artificial human heart implant; alfentanil (R 39209) for short surgical operations; carfentanil (R 33799) for the immobilization of large animals such as elephants; and lofentanil (R 34995), which has seen stalwart service as a potent painkiller in the fight against cancer. The last two of these substances are the most powerful analgesics in existence, exerting ten thousand times the effect of morphine. The efforts to relieve chronic pain led to the use of transdermal patches and intravenous delivery systems, thereby lowering the risk of opioid dependence. Durogesic was launched in the 1990s in the form of a patch that controls pain for seventy-two hours, without the side effects of morphine.

Paul Adriaan Jan Janssen. *Paul Janssen in his laboratory.* COURTESY OF JANSSEN PHARMACEUTICA NV, BELGIUM.

Neuroleptics. Until midway through the twentieth century, the treatment of psychoses left much to be desired. Then Rhône-Poulenc synthesized chloropromazine, but it had numerous unwanted effects. In 1958, however, the laboratories at Janssen developed a new chemical series of butyrophenones. The most potent molecule in this series, without any analgesic effect, but with a potent neuroleptic activity, was haloperidol (R 1625). This drug, under the trade name Haldol, eventually brought about a veritable sea change in psychiatric practice, such that straitjackets and electroshock became things of the past. Its use opened the way for psychotherapy and various types of psychosocial support, and the number of psychiatric patients in specialized hospitals in the United States fell by

80 percent from 1965 to 1988. Into the early twenty-first century, haloperidol remained on the World Health Organization's (WHO) List of Essential Drugs.

In the 1960s, more variants were created on the same theme. Further research revealed that the antipsychotic effect of haloperidol was related to blockade of the central dopaminergic receptors. It was observed that certain analogs of haloperidol had an additional effect on the serotonergic receptors in the brain. These investigations used radioactively marked haloperidol and spiperone. More and more receptors and variants were described, and the research at Janssen made a fundamental contribution (including the identification of the serotonin-S_2 receptor). In 1982, chemists at Janssen found the first molecule that bonded specifically with the serotonin receptor; ketanserin (R41468) was a highly specific antagonist of the S_2 receptors, a potential anti-hypertensive agent, but also suitable for obtaining greater insight into the system that regulates blood pressure. Two years later the neuroleptic risperidone (R64766) came along, which combined the best qualities of haloperidol with the beneficial characteristics of ritanserin (R55667). In addition, this compound stood out for its ability to prolong slow-wave sleep and yielded good results as an add-on to Haldol therapy. It was given the trade name Risperdal and became a trailblazer for second-generation antipsychotics, improving treatment for a substantial number of patients with schizophrenia or associated symptoms. It was also found to be effective in relieving conduct disorders and psychological symptoms (depression, anxiety), so that, in particular, more elderly patients could live independently for a longer time. Next, research scientists at Janssen discovered nebivolol (R 67555). This substance, the result of a chemical application that made totally symmetrical molecules, lowers blood pressure and improves cardiac function, which was not the case with conventional β-blockers, so that the end product was better than ketanserin. It was licensed out to another company, and since the 1990s it has been marketed worldwide. In this way, the laboratories at Janssen have made an invaluable contribution to the development of neuropharmacology, which went hand in hand with discoveries in other areas of disease control.

Gastrointestinal Disorders. In 1956, diphenoxylate was discovered as an antidiarrheal agent at Janssen, but the search continued for a molecule whereby the antidiarrheal effect could be separated from the action on the central nervous system. In 1968, an attempt was made to synthesize variants of the familiar haloperidol; that led to loperamide (R 18553), which was free of morphine-like effects. Under the trade name Imodium, it became the most widely sold product worldwide for the symptomatic treatment of diarrhea, thanks to its combined mucosal and myenteric action. After a series of about one hundred molecules were synthesized in routine tests for potential neuroleptics, domperidone (R33812) was introduced to control vomiting. Due to its structure, domperidone could not cross the blood-brain barrier, but it blockaded the dopamine receptors in the stomach and intestines. The molecule was marketed under the trade name Motilium, with the indication of restoring the motility of the esophagus and the stomach. This motility is regulated by neurotransmitters that work in tandem; dopamine is the inhibitor and acetylcholine the motor. The chemists at Janssen found a starting point in clebopride, a synthetic product that acted on the dopamine receptor, and set out to make a variant that would have an action on acetylcholine. This quest led to the discovery of cisapride (R51619). Under the trade name Prepulsid, this product brought relief to millions of patients with esophageal reflux and dyspepsia. Gastrointestinal motility is therefore a major specialization at Janssen.

Worm and Fungal Infections. Health problems in the developing world played a prominent part in Janssen's research activities. A number of veterinarians and microbiologists returned to Belgium following Congolese independence in 1960 and were recruited by Janssen. This out-of-Africa group would play an important role in various research programs for two decades. For worm control, Janssen used an aminothiazole derivative that he had developed himself. This medicine could expel three types of nematodes from chickens but was not active in other animal species. Janssen proceeded from the hypothesis that this finding was attributable to a difference in the biotransformation of the substance by the different species of animals. Consequently, he decided to separate and finally to test all the metabolites. One of these, tetramisole, displayed a broad activity spectrum, and on this basis levamisole was developed. This drug, however, did not eliminate all parasitic worms. In 1968, researchers at Janssen Pharmaceutica discovered mebendazole (R 17635), and somewhat later flubendazole (R 17889). These benzimidazole derivatives would conquer the world under the trade names Vermox, for treatment of humans, and Telmin and Mebenvet in veterinary medicine. Levamisole and mebendazole are included in the WHO List of Essential Drugs, and are still helping to cure millions of people, in both rich and poor countries. In developing countries worm infections, as a result of the enormous protein loss that they cause, play an important role in the spiral of poverty and hunger. From research into antimycotics, researchers knew of the imidazole ring, which is present in tetramisole and levamisole. These compounds were active against (worm) parasites. Was this not a trail that might lead to a medicine effective against (fungal) parasites?

In the course of this quest, researchers found metomidate (R 7315) and later etomidate (R 16659). Hypnomidate remained in the early 2000s the most potent hypnotic in anesthesia. Subsequent research came up with molecules that had an antimycotic activity. Miconazole (R 14889), known by the trade name Daktarin, is used to treat fungal infections of the skin. In 1977, an elaborated variant of miconazole, namely ketoconazole (R 41400), which had a broad activity spectrum, was administered orally for the first time. Ketoconozale is the human variant of micoazole and the first antimycotic for oral use by humans. The trade name is Nizoral, a drug capable of acting on fungi in the body itself, and as such, constituted a breakthrough in antimycotic research. The most recent azole is itraconazole (R 51211), known under the trade name Sporanox, and it too is active orally. It is a safe, broad-spectrum antimycotic, with a potent action on the cytochrome P 450-enzyme in fungi and yeasts. It is effective against Aspergillus, a fungal strain often found in AIDS patients. In this case, too, the discovery resulted from skillfully varying chemical structures with the aim of obtaining a specific pharmacological effect.

An innovation in plant protection came in 1979 with propiconazole (R 49362), which was ideally suited for the protection of crops against rust. Licensed out as Tilt, this product was responsible for a 30 percent increase in grain production worldwide. Janssen also brought about a substantial improvement in the protection of materials (wood, textiles, etc.). Since 2001 the spectacular result of Janssen's efforts in material protection is the rescue of the famous Chinese terracotta warriors in Xian from deterioration by fungi, whereby Janssen played its part in conserving a world heritage. In total, thirteen antimycotics were discovered by Janssen and his team. Levamisole, mebendazole, miconazole, and ketoconazole appear on the WHO List of Essential Drugs.

Antihistaminics for Allergies. In the 1970s, Janssen's laboratories were turning out two to three thousand new molecules each year, usually the fruits of both a prototypal combinatorial synthesis and high throughput screening, methods that had to wait until the 1990s before coming on line in the pharmaceutical industry. In this way, researchers at Janssen came up with a variant of the cinnarizine molecule, which offered possibilities for the treatment of allergies. Oxatomide (R 35443) proved to be an effective treatment for asthma. Even more specific was the discovery in 1977 of astemizole (R 43512). This substance had a long-lasting effect on hay fever, had hardly any side effects, and in particular did not induce drowsiness. A little later levocabastine (R 50547), the world's most potent antihistaminic, was developed. Because this medicine did cause drowsiness, Livostin (the trade name of Levocabastine) was used to treat allergic conjunctivitis.

The Human Immunodeficiency Virus. Already in 1990, Janssen stated that together with a number of colleagues, he had discovered a new class of highly effective, nontoxic substances that might be effective against AIDS. He was referring to TIBO derivatives. These were the first non-nucleoside inhibitors of the HIV-1RT (NNRTIs). However, he found them unsatisfactory and therefore started up the Center for Molecular Design (CMD) in 1996 with a view to deepening his investigations. With the aid of molecular modeling, organic chemistry, and a supercomputer, a cross-disciplinary team inspired by Janssen designed a whole series of new molecules. The affiliate Virco lent a hand in his project and focused on the development of diagnostic tests. The crystallographic studies were carried out at the Center for Advanced Biotechnology and Medicine at Rutgers University in New Brunswick, New Jersey, in the United States. This approach resulted in the identification of many promising potential drugs, including the dianilinopyrimidine (DAPY) analogs. Dapivirine is the generic name of R 147681/TMC120. It has been donated by J&J to the International Partnership for Nicrobicides (IPM) for development into an HIV-prophylactic drug. Dapivirine and Etravirine (R 165335 or TMC 125) were derived from these and were shown to be highly active in Phase I and Phase II trials. The simplicity of production of this drug makes it particularly suitable for developing countries, where the need is greatest. Janssen was the driving force behind the discovery of a new generation of anti-AIDS drugs, including rilpivirine (R278474 or TMC278), a substance characterized by an exceptional profile with respect to antiviral activity and biological availability. In Janssen's own words, this compound is "an absolute world champion."

Janssen was a medical and medicinal chemistry research genius, part of whose talent was his ability to gather around him the best brains in Flanders throughout the decades in which his career flourished. With more than one hundred patents to his name, he was named "the most successful drug discoverer of all time" in 2002 by *Nature Reviews.* Thanks to his cross-disciplinary approach, Janssen carried out research in a broad variety of therapeutic areas: mycology, psychiatry, parasitology, allergology, gastroenterology, pain control and anesthesia, veterinary medicine, and plant and material protection. His way of working was based on the relationship between the chemical structure and the pharmacological action of molecules, and the use of the appropriate experimental models to test these relationship. Janssen authored or co-authored more than 850 scientific publications and received a large number of accolades that included twenty-two honorary doctorates, five honorary professorships, and honorary memberships in more than thirty scientific institutions and organizations in his own country and beyond.

BIBLIOGRAPHY

Janssen Pharmaceutica, Beerse (Belgium) is the owner of an extensive collection of archives, a series of interviews, and company publications. A complete list of the publications of Dr. Paul Janssen is included in Niemegeers, C. J. E., ed, Historical Record of Janssen Research Publications (1952–1990). *4 Vols. Beerse: Janssen Pharmaceutica, 1992.*

WORKS BY JANSSEN

With A. H. M. Jageneau, P. J. A. Demoen, C. Van de Westeringh, et al. "Compounds Related to Pethidine. I: Mannich Bases Derived from Norpethidine and Acetophenones." *Journal of Medicinal and Pharmaceutical Chemistry* 1, no. 1 (1959): 105–120.

With C. Van de Westeringh, A. H. M. Jageneau, P. J. A. Demoen, et al. "Chemistry and Pharmacology of CNS Depressants Related to 4-(4-Hydroxy-4-Phenylpiperidino) Butyrophenone. Part I: Symthesis and Screening Data in Mice." *Journal of Medicinal and Pharmaceutical Chemistry* 1, no. 3 (1959): 281–297.

With J. M. Van Nueten. "Difenoxine (R15403), the Active Metabolite of Diphenoxylate (R01132). Part 3: Inhibition of the Peristaltic Activity of the Guinea Pig Ileum in Vitro." *Arzneimitiel Forschung* 22, no. 3 (1972): 518–520.

With R. A. Stokbroekx, J. Vandenberk, A. H. M. T. Van Heertum, et al. "Synthetic Antidiarrheal Agents. 2,2-Diphenyl-4-(4'-Aryl-4'-Hydroxypiperidino) Butyramides." *Journal of Medicinal Chemistry* 16 (1973): 782–786.

With C. I. F. Niemegeers and J. L. McGuire. "Domperidone, a Novel Gastrokinetic Drug." *The Pharmacologist* 20, no. 3 (1978): 209.

With D. Thienpont, J. Van Cutsem, F. Van Gerven, and J. Jeeres. "Ketoconazole: A New Broad Spectrum Orally Active Antimycotic." *Experientia* 35, no. 5 (1979): 606–607.

With F. de Clerck and J. L. David. "Inhibition of 5-Hydroxytryptamine-Induced and -Amplified Human Platelet Aggregation by Ketanserin (R41468), a Selective 5-HT2-Receptor Antagonist." *Agents and Actions* 12, no. 3 (1982): 388–397.

With A. Reyntjens, M. Verlinden, J. Schuurkes, and J. Van Nueten. "New Approach to Gastrointestinal Motor Dysfunction: Non-Antidopaminergic, Non-Cholinergic Stimulation with Cisapride." *Current Therapeutic Research* 36, no. 5 (1984): 1029–1037.

With J. E. Leysen, W. Gommeren, A. Eens, et al. "Biochemical Profile of Risperidone, a New Antipsychotic." *Journal of Pharmacology and Experimental Therapeutics* 247, no. 2 (1988): 661–670.

With L. Matthieu, P. de Doncker, G. Cauwenbergh, et al. "Itraconazole Penetrates the Nail via the Nail Matrix and the Nail Bed: An Investigation in Onychomycosis." *Clinical and Experimental Dermatology* 16 (1991): 374–376.

OTHER SOURCES

Ayd, Frank J., ed. *30 Years of Janssen Research in Psychiatry.* Baltimore, MD: Ayd Medical Publications, 1989.

Healy, David, ed. *The Psychopharmacologists.* 2 Vols. London: Chapman and Hall, 1996–1998.

Leysen, J., M. Borgers, R. Reneman, et al. "Belangrijkste bijdragen van Dr. Paul tot de geneeskunde en de medicinale chemie" [The Most Important Contributions of Dr. Paul to Medicine and Medicinal Chemistry]. In *Dr. Paul Janssen, 1926–2003: Een portret in woorden* [A Portrait in Words]. Beerse, Belgium: Janssen Pharmaceuticals, 2004.

Magiels, G. *Paul Janssen: Pionier in FARMA and in China* [Pioneer in the Pharmaceutical Industry and in China]. Antwerp, Belgium: Houtekiet, 2004.

The Paul Janssen Concept of Research Management. Beerse, Belgium: Janssen Research Foundation, 1995.

Schwartz, Harry. *Breakthrough: The Discovery of Modern Medicines at Janssen.* Plains, NJ: Skyline Publishing Group, 1989.

Theunissen, G. *Dr. Paul: De zoektocht naar betere geneesmiddelen eindigt nooit* [The Quest for Better Medicines Never Ends]. Beerse, Belgium, 1992.

Roland Baetens

JEFFREY, RICHARD CARL (*b.* Boston, Massachusetts, 5 August 1926; *d.* Princeton, New Jersey, 9 November 2002), *rational choice theory, probabilism, epistemology.*

Jeffrey was perhaps the leading philosophical proponent of Bayesian methods in decision theory, epistemology, and philosophy of science.

Early Years and Career Path. Richard Carl Jeffrey was born in Boston, Massachusetts, on 5 August 1926. He entered Boston University at age sixteen, where he became interested in philosophy and logic as a result of reading Rudolf Carnap's *Philosophy and Logical Syntax.* As Jeffrey would later declare, the exposure to Carnap's work made him into "a teenage logical positivist" (1998). During World War II Jeffrey was drafted out of college and into the U.S. Navy, where he served from 1944 to 1946. After the war, Jeffrey went to the University of Chicago, where he worked closely with Carnap on the foundations of probability and statistics, topics that remained central to his research throughout his career.

After earning his MA in philosophy at Chicago in 1952, he went to work on the design of computers at the Massachusetts Institute of Technology's (MIT) Digital Computer Laboratory and the Lincoln Laboratory. While in Boston he met Edie Kelman. They married in 1955. In that same year Jeffrey went to Princeton to study under the great philosopher of science Carl Gustav Hempel. Upon completing the PhD under Hempel in 1957, Jeffrey spent a year as a Fulbright Scholar at Oxford, after which he returned to MIT for a year as an assistant professor of electrical engineering. He then held a position as an assistant professor of philosophy at Stanford from 1959 to 1963. At the invitation of Kurt Gödel, Jeffrey spent 1963–1964 at the Institute for Advanced Study at Princeton. During this period he completed *The Logic of Decision,* his first book. Jeffrey went on to hold appointments in the philosophy departments at the City College of New York (1964–1967), the University of Pennsylvania (1967–1974), and finally Princeton (1974–1999), where he spent the bulk of his career. He wrote two books during this period—*Formal Logic: Its Scope and Limits* (1967) and *Computability and Logic* (1974, with George Boolos)—and many important papers, the best known of which are collected in *Probability and the Art of Judgment* (1992). After retiring from Princeton, Jeffrey held a visiting distinguished professorship of social science at the University of California, Irvine.

Contributions to Rational Choice Theory. Most of Jeffrey's research focused on rational choice theory, and on the application of probabilistic thinking to epistemology and the philosophy of science. Jeffrey's main contribution to rational choice theory, *The Logic of Decision,* was written while he was at the Institute for Advanced Study in Princeton. Combining new insights about decision making, and benefiting from the mathematics of Ethan Bolker (1966), Jeffrey developed a version of expected utility theory that differs from earlier versions, such as that of Leonard J. Savage (1954), in three important respects. First, whereas Savage distinguished among acts, states of the world, and consequences, Jeffrey formulated his theory using a single underlying algebra of propositions, leaving distinctions between acts, states, and consequences to be drawn at the level of application rather than theory.

Second, whereas most decision theories require the maximization of unconditional expected utility, Jeffrey advocated maximizing *conditional* expected utility, thereby allowing for the possibility that acts might influence the states used to frame decisions. This is achieved by replacing the standard formula for expected utility, $Exp(A) = \Sigma_S P(S) \times u(A, S)$ (where the probabilities $P(S)$ do not depend on the choice of an act), by a new formula, $des(A) = \Sigma_S P(S \mid A) \times des(A \,\&\, S)$. The resulting function is such that:

1. the choice of state partition does not matter, because $\Sigma_S P(S \mid A) \times des(A \,\&\, S) = \Sigma_{S^*} P(S^* \mid A) \times des(A \,\&\, S^*)$ whatever partitions S and S^* range over;
2. no "state-dependent" utilities are required, because, in any application, consequences are act/state conjunctions;
3. potential influences of choices on acts are reflected by the conditional probabilities $P(S \mid A)$; and
4. no explicitly or implicitly causal notions are invoked anywhere in the theory.

Jeffrey's theory is often referred to as "evidentialist," because it is natural to interpret $des(A)$ as expressing the value of A as an indicator or sign of desirable or undesirable results.

Third, the framework prevents preferences from uniquely determining subjective probabilities. In Jeffrey's theory, even if one fixes a zero and a unit for utility, a complete set of preferences can always be represented as maximizing expected utility relative to any of a family of probability/utility pairs (P_λ, des), where λ satisfies $-1/\sup\{des(X)\} \leq \lambda \leq -1/\inf\{des(X)\}$. Except in the rare case where the preference ranking requires an unbounded utility, which makes $\lambda = 0$ the only option, this family will contain an infinity of members defined by the identities $P_\lambda(X) = P_0(X)[1 + \lambda \times des_0(X)]$ and $des_\lambda(X) = des_0(X)[(1 + \lambda)/(1 + \lambda \times des_0(X))]$. Jeffrey saw this indeterminacy as an advantage of his theory, and argued that the uniqueness of the representation offered by other decision theories was spurious. Beliefs and desires, he maintained, are mostly imprecise and incomplete, and should be represented, in just the way his theory suggests, by sets of probability and utility functions. This was "Bayesianism with a Human Face" (1983). Jeffrey, together with Isaac Levi, Henry Kyburg, Bas van Fraassen, Teddy Seidenfeld, and others, did much to illuminate such "imprecise attitudes." However, unlike other defenders of this position, Jeffrey thought it impossible to detach probabilities from utilities. A complete representation of a rational agent's mental state will be a set of probability/utility pairs that typically cannot be decomposed into a Cartesian product of a set of probabilities with a set of utilities.

Ratificationism. An additional contribution to decision theory, which appeared with the second edition of *The Logic of Decision,* is the doctrine of ratificationism. This arose as a result of challenges raised by "causal" decision theorists. As noted, Jeffrey's utilities measure the degree to which news of an act indicates that desirable results will ensue. This can lead to flawed recommendations in decisions whose acts indicate desirable outcomes without doing anything to causally promote those outcomes. Causal decision theorists see such "Newcomb problems" as counterexamples to Jeffrey's theory, and use them as a rationale for including explicitly causal or counterfactual notions in the formulation of decision theory. This was anathema to Jeffrey, who saw the absence of causal notions as one of the main virtues of his theory. His response, which owes much to Ellery Eells (1982), was to argue that, when properly amended, his system does not recommend auspicious but inefficacious acts. The required amendment is the *maxim of ratification*: *one should choose* *for the person one expects to be once one has chosen.* Decision making is then a two-stage process in which one first eliminates all "unratifiable" acts, those for which *des*($A \mid A$ is chosen) < *des*($B \mid A$ is chosen) for some B, and then maximizes expected utility among the remaining "ratifiable" options. Jeffrey assumed that an agent's knowledge that she will choose A is sufficient to screen off any spurious evidentiary relationships that hold between A and states of the world. When combined with the requirement to choose only ratifiable options, this assumption eliminates auspicious but inefficacious acts from consideration in Newcomb problems.

Ratificationism has had a complicated history. In the face of strong criticism of the screening-off assumption, Jeffrey was forced to repudiate ratificationism as a solution to Newcomb problems. Yet, many causal decision theorists found ratificationism plausible and incorporated it into their theories. It also became clear that the ratificationist idea is built into the concept of a game-theoretic equilibrium. Recently, the idea of a ratifiable decision has reappeared in the work of the economists Michael Rabin and Botond Koszegi (2006) under the title of "personal equilibrium." All this suggests that the importance of ratifiability to decision theory goes far beyond the role for which it was initially introduced.

After giving up on ratificationism as a solution to Newcomb problems, Jeffrey shifted tactics in *Subjective Probability* (2004) and argued that these problems are not genuine decisions because agents who face them possess so much evidence about correlations between their acts and states of the world that they cannot properly see their choices as causes of outcomes. Jeffrey based this conclusion on a model of rational deliberation that requires conditional probabilities of the form *P*(*outcome* | *act*) to remain constant during deliberation. Some, for example, James M. Joyce (2007), have rejected this claim, and so remain convinced that Newcomb problems cause trouble for Jeffrey's theory.

Contributions to Epistemology and Philosophy of Science. Jeffrey's central contribution to epistemology was his thoroughgoing defense of radical probabilism, the idea that, "in principle, all knowledge might be merely probable" (1988, p. 135). Radical probabilism opposes "dogmatic" epistemologies that make states of "full belief" the core ingredients of the theory of knowledge, and that presuppose foundationalist conceptions of learning. For a radical probabilist, all information acquisition and processing is described probabilistically.

In a famous exchange with Isaac Levi, Jeffrey (1970) argued that the concept of full belief is largely irrelevant to epistemology because choices among actions are governed by subjective probabilities, and such probabilistic attitudes are appropriate in light of the ambiguous and inherently uncertain data we typically receive. The correct way to model epistemic states, according to Jeffrey, is not with a set of known propositions, but by a family of probability functions that best reflect uncertainty in light of evidence.

Jeffrey was particularly concerned to show that an adequate theory of learning did not require a concept of full belief. He often quoted Clarence Irving Lewis's remark "if anything is to be probable, then something must be certain" (1946, p. 186), using it as a foil. On Lewis's dogmatist picture, learning involves becoming certain of some proposition. This is represented probabilistically as simple Bayesian conditioning in which a person who undergoes a learning experience becomes certain of some proposition e and updates her "prior" subjective probability function P by moving to the "posterior" $P_e(\cdot) = P(\cdot \mid e)$. Jeffrey's rejection of this picture is based on a model that allows experiences to increase probabilities without raising them to certainty. Consider a beer drinker who starts out confident that he is being served lager but who, after taking a sip, suspects that he has mistakenly been given ale. The gustatory experience might have moved his degrees of confidence from the priors *P*(*lager*) = 0.95 and *P*(*ale*) = 0.05 to the posteriors Q(*lager*) = 0.45 and *Q*(*ale*) = 0.55. More generally, if $\langle e_j \rangle = \langle e_1, e_2, \ldots, e_n \rangle$ is a set of mutually exclusive, collectively exhaustive propositions, Jeffrey allows for experiences that fix posteriors $P_q(e_i) = q_i$. If this is the experience's only immediate effect, then ratios of probabilities within each cell of the partition remain static, so that $P_q(\cdot \mid e_i) = P(\cdot \mid e_i)$ for all i. This "sufficiency condition" ensures that the posterior is given by $P_q(\cdot) = \Sigma_i \, q_i \times P(\cdot \mid e_i)$. Jeffrey called this form of belief revision "probability kinematics," though most philosophers refer to it as "Jeffrey conditioning." It describes the sort of nonfoundationalist learning that underwrites radical probabilism.

Unlike standard conditioning, Jeffrey's operation is not commutative. If experience $\boldsymbol{q}$ sets probabilities $\langle q_i \rangle$ for $\langle e_i \rangle$ and experience $\boldsymbol{r}$ sets probabilities $\langle r_j \rangle$ for $\langle f_j \rangle$, and if the sufficiency condition holds at each stage, then it can happen that $P_{q,r} \neq P_{r,q}$, so that $\boldsymbol{q}$ followed by $\boldsymbol{r}$ is not evidentially equivalent to $\boldsymbol{r}$ followed by $\boldsymbol{q}$. Some have objected to this on the grounds that the order in which data is received should not affect conclusions drawn from it. This principle, however, is dubious in just those cases where Jeffrey conditioning fails to commute. Jeffrey shifts can occur in two ways, depending on whether experience fixes posterior probabilities for events in a way that depends on their priors. If the posterior probability that experience $\boldsymbol{q}$ sets for each e_i is independent of its prior probability, then $\boldsymbol{q}$ obliterates information about the prior probabilities of the e_i themselves: any two priors with $P(\cdot \mid e_i) = P^*(\cdot \mid e_i)$ for all e_i are mapped to same posterior $P_q = P^*_q$. The "order makes no difference" principle is bogus for experiences of this type. For, if $\boldsymbol{r}$ conveys information about the e_i, so that $P_r(e_i) \neq P(e_i)$ for some i, then it *should* matter whether $\boldsymbol{r}$ precedes or follows $\boldsymbol{q}$. When $\boldsymbol{r}$ precedes $\boldsymbol{q}$, the information it provides about the e_i is destroyed by $\boldsymbol{q}$ and so does not show up in $P_{r,q}$. This information is, of course, reflected in $P_{q,r}$. Thus, when the posterior probabilities for the e_i (or f_j) do not depend on their prior probabilities, one should expect Jeffrey conditioning to commute only when the experiences that generate them are "Jeffrey independent" in the sense that $P_r(e_i) = P(e_i)$ and $P_q(f_j) = P(f_j)$ for all i and j. As shown in Persi Diaconois and Sandy Zabell (1982), this is necessary and sufficient for $P_{q,r} = P_{r,q}$.

As Hartry Field (1978) showed, the "order makes no difference" principle is plausible in some cases where experiences preserve information from earlier shifts. Imagine that, instead of directly determining posteriors, an experience $\boldsymbol{q}$ only fixes ratios of posterior to prior probabilities for cells in a partition, so that $\boldsymbol{q}$'s immediate effect is to fix a Bayes factor $\alpha_i = P_q(e_i)/P(e_i)$ for each e_i. When sufficiency holds, the shift from P to P_q is then fully characterized by Bayes factors and priors: $P_q(\cdot) = P(\cdot) \times \Sigma_i \alpha_i \times P(e_i \mid \cdot)$. When such shifts occur in succession—one fixing Bayes factors α_i for the e_i, the other fixing Bayes factors B_j for the f_j—the result is $P_q(\cdot) = P(\cdot) \times \Sigma_{i,j} \alpha_i \times B_j \times P(e_i \,\&\, f_i \mid \cdot)$. Since this equation is symmetric in i and j, Jeffrey conditioning can commute when subsequent experiences preserve the information from previous experiences, which is exactly as it should be.

Lasting Influence. Jeffrey made contributions in many other areas of philosophy, among them an early and influential account of higher-order preferences (1974), a theory of conditionals (with Robert Stalnaker) that assigned indicative conditionals partially probabilistic semantic values (1994), an important paper on preference aggregation (with Matthias Hild and Mathias Risse), and a number of influential papers on the general philosophy of science (1956, 1969, 1975, and 1993).

Jeffrey had great influence within and outside of philosophy, both for the depth and inventiveness of his philosophical work and for the kindness and generosity with which he dealt with colleagues, students, and friends. Among his many academic honors, Jeffrey served as president of the Philosophy of Science Association and was a member of the American Academy of Arts and Sciences. The impact of his contributions can be seen directly in the work of the philosophers Abner Shimony, Brian Skyrms, William Harper, Stalnaker, van Fraassen, Alan Hájek, Brad Armendt, Joyce, Cristina Bicchieri, and Risse (his last PhD student), as well as the statisticians Diaconis and Zabell, the physician Michael Hendrickson, and the mathematician Carl G. Wagner.

Richard Jeffrey died on 9 November 2002 in Princeton at the age of seventy-six. He spent his last days putting the finishing touches on his last book, *Subjective Probability: The Real Thing!*, an elegant final expression of his thinking.

BIBLIOGRAPHY

WORKS BY JEFFREY

"Valuation and Acceptance of Scientific Hypotheses." *Philosophy of Science* 23 (1956): 237–246.

The Logic of Decision. New York: McGraw-Hill, 1965. Rev. 2nd ed., Chicago: University of Chicago Press, 1983.

Formal Logic: Its Scope and Limits. New York: McGraw-Hill, 1967.

"Statistical Explanation vs. Statistical Inference." In *Essays in Honor of Carl G. Hempel: A Tribute on the Occasion of His Sixty-Fifth Birthday*, edited by Nicholas Rescher. Dordrecht: Reidel, 1969.

"Dracula Meets Wolfman: Acceptance vs. Partial Belief." In *Induction, Acceptance, and Rational Belief*, edited by Marshall Swain. Dordrecht: Reidel, 1970.

"On Interpersonal Utility Theory." *Journal of Philosophy* 68 (1971): 647–657.

With George Boolos. *Computability and Logic.* Cambridge, U.K.: Cambridge University Press, 1974. Rev. 2nd ed., 1980.

"Preference among Preferences." *Journal of Philosophy* 71 (1974): 377–391.

"Probability and Falsification: Critique of the Popper Program." *Synthese* 30 (1975): 95–117 and 149–157.

"Bayesianism with a Human Face." In *Testing Scientific Theories*, edited by John Earman. Minnesota Studies in the Philosophy of Science 10. Minneapolis: University of Minnesota Press, 1983.

"Conditioning, Kinematics, and Exchangeability." In *Causation, Chance, and Credence*, vol. 1, edited by Brian Skyrms and William Harper. Boston: Kluwer Academic, 1988.

Probability and the Art of Judgment. Cambridge, U.K.: Cambridge University Press, 1992.

"Take Back the Day! Jon Dorling's Bayesian Solution to the Duhem Problem." In *Science and Knowledge*, edited by Enrique Villanueva. Atascadero, CA: Ridgeview, 1993.

With Robert Stalnaker. "Conditionals as Random Variables." In *Probability and Conditionals: Belief Revision and Rational Decision*, edited by Ellery Eells and Brian Skyrms. Cambridge, U.K.: Cambridge University Press, 1994.

"I was a Teenage Logical Positivist (Now a Septuagenarian Radical Probabilist)." PSA Presidential Address. Delivered Kansas City, 23 October 1998. Revised 10 April 2000. Available online from http://www.princeton.edu/~bayesway/KC.tex.pdf.

With Matthias Hild and Mathias Risse. "Preference Aggregation after Harsanyi." In *Justice, Political Liberalism, and Utilitarianism: Themes from Harsanyi and Rawls*, edited by

Maurice Salles and John A. Weymark. Cambridge, U.K.: Cambridge University Press, 2004.

Subjective Probability: The Real Thing. Cambridge, U.K.: Cambridge University Press, 2008.

OTHER SOURCES

Bolker, Ethan. "Functions Resembling Quotients of Measures." *Transactions of the American Mathematical Society* 124 (1966): 292–312.

Diaconis, Persi, and Sandy Zabell. "Updating Subjective Probability." *Journal of the American Statistical Association* 77 (1982): 822–830.

Eells, Ellery. *Rational Decision and Causality.* Cambridge, U.K.: Cambridge University Press, 1982.

Field, Hartry. "A Note on Jeffrey Conditionalization." *Philosophy of Science* 45 (1978): 361–367.

Joyce, James M. "Are Newcomb Problems Really Decisions." *Synthese* 156, no. 3 (2007): 537–562.

Koszegi, Botond, and Matthew Rabin. "A Model of Reference-Dependent Preferences." *Quarterly Journal of Economics* 121, no. 4 (2006): 1133–1166.

Levi, Isaac. "Probability and Evidence." In *Induction, Acceptance, and Rational Belief,* edited by Marshall Swain. Dordrecht: Reidel, 1970.

Lewis, Clarence Irving. *An Analysis of Knowledge and Valuation.* La Salle, IL: Open Court, 1946.

Savage, Leonard J. *The Foundations of Statistics.* New York: Wiley, 1954.

James M. Joyce

JEFFREYS, HAROLD (*b.* Fatfield, County Durham, England, 22 April 1891; *d.* Cambridge, England, 18 March 1989), *applied mathematics, statistics, theoretical seismology, astrophysics.*

Jeffreys's main achievements were in seismology, planetary geodynamics, and applied statistics: he provided proof of the liquid nature of Earth's metallic core (1926). He developed, from 1931 onward, improved travel-time tables for the propagation of seismic waves from earthquakes to enable better determinations of the location of earthquake foci and of the velocity structure (and hence inferred composition) of Earth's mantle and argued for regional differences in velocity structures beneath the crust (1954, 1962). He showed the presence of large-scale anomalies in Earth's gravitational field (1941) and the importance of cyclones as part of general atmospheric circulation (1927). In 1931 he also pioneered the field now known as Bayesian statistics.

Childhood and Education. Jeffreys was the only child of Robert Hall Jeffreys and his wife Elizabeth Mary Sharpe, both schoolteachers. As a child he was interested in natural history, botany, astronomy, and photography. His lifelong interest in celestial mechanics sprang from reading popular works by the astronomers Sir Robert Ball and Sir George Howard Darwin. In 1903 he gained a scholarship to Rutherford College school, Newcastle-upon-Tyne, and in 1907 he was awarded a scholarship to Armstrong College of Durham University, Newcastle (University of Newcastle from 1963), where he read mathematics with ancillary physics and chemistry, plus a year of geology, graduating first class in 1910 with distinction in mathematics. While there, he read Charles Darwin's work on the equilibrium of rotating, self-gravitating fluid masses, the origin of the solar system, and how Earth's tides influenced the evolution of the Earth-Moon system.

At that time, the Cambridge mathematical tripos was widely regarded as the best training for a mathematical career, and graduates from other universities commonly took it as though it were postgraduate study. Accordingly, Jeffreys was awarded an entrance scholarship to read mathematics at St. John's College, Cambridge. Although he found it considerably harder than that which he had studied at Armstrong College, he became a wrangler (i.e., a student who obtained first-class marks) in part 2 of the tripos in 1913 and was jointly awarded the Hughes Prize. Influenced by lectures of the mathematician Ebenezer Cunningham and the astronomer and mathematician Sir Arthur Eddington, Jeffreys became committed to astrophysics. In 1912 he was awarded the Adams Memorial Prize for an essay on precession (the path around the circumference of the cone traced by the obliquity of the rotation axis of a planet about the pole to the plane in which the planets appear to orbit the Sun) and nutation (the regular high-frequency wobble imposed on this circular path), subjects to which he would return throughout his career. His college scholarship was extended for a fourth year, and Jeffreys embarked on research. He was elected to fellowship of St. John's College in November 1914 (a position he held until his death) and was awarded the Isaac Newton Studentship (1914–1917), working mainly on cosmology and planetary geodynamics.

Personality. Slim and moderately tall, Jeffreys wore glasses, sported a small mustache, and was a heavy smoker. He was a keen cyclist and photographer, extremely knowledgeable about botany, a great reader of detective and other fiction and poetry, and an admirer of Henrik Ibsen's *Peer Gynt* (quirky quotations from all these sources enlivened chapter headings of his books). For many years he sang tenor in the Cambridge Philharmonic Choir, but despite being sociable he tended to be withdrawn, working almost entirely on his own and, in contrast to the clarity of his writing, was inarticulate and halting in conversation and capable of reducing an undergraduate class to one or two attendees with his stumbling delivery. His college tutorials could proceed in silence, but the paleomagnetist Raymond Hide recalled that "his grunts and murmurs should be taken seriously, because they were likely to contain pearls of wisdom" (Hide, 1997). In correspondence (e.g., with the paleomagnetist Stanley Keith Runcorn) Jeffreys could be acerbic.

Marriage. On 6 September 1940 Jeffreys married Bertha Swirles (1903–1999), daughter of William Alexander Swirles, commercial traveler, and his wife Harriet Blaxley, a primary school teacher. Having obtained first-class honors in the mathematical tripos (1924), Bertha became Hertha Ayrton Research Fellow at Girton College, Cambridge (1927–1928), supervised by Ralph Fowler and Douglas Hartree, spending the winter semester of 1927–1928 at Göttingen University, Germany, where she was supervised by Max Born and Werner Heisenberg. She was awarded her PhD in 1929 with a thesis on quantum mechanics. Following appointments at the universities of Manchester, Bristol, and Imperial College London, she returned to Cambridge as official fellow and lecturer in mathematics at Girton in 1938, later becoming its director of studies in mathematics and mechanical sciences (1949–1969) and vice-mistress (1966–1969). Following the marriage, she continued to publish in her own field but collaborated with Jeffreys on *Methods of Mathematical Physics* (1946) and in editing the volumes of his *Collected Papers*. She also aided him by translating, from the Russian, Evgenii Fedorov's book *Nutation and Forced Motion of the Earth's Pole* (1963) and an article and correspondence by the Croatian physicist Stjepan Mohorovičić. There were no children of the marriage.

Career. Following the outbreak of World War I, Jeffreys worked part time at the Cavendish Laboratory (1915–1917) on war-related problems. He then joined the Meteorological Office in South Kensington, London, as personal assistant to its director, Sir William Napier Shaw, but returned to Cambridge on weekends. In 1920 Shaw was preparing to retire and Jeffreys was appointed librarian, but the position eventually proved so unsatisfactory that he returned to St. John's in 1922 as college lecturer in mathematics. In 1926 he was appointed university lecturer, then reader in geophysics (1931) and Plumian Professor of Astronomy and Experimental Philosophy (1946). He retired in 1958; the last of his technical papers was published in 1987 when he was ninety-six.

Cosmology. Jeffreys's first papers on Earth (1915) resulted from reading *Dynamics of a System of Rigid Bodies* by the mathematician Edward John Routh. The "planetesimal hypothesis," proposed by the American astronomers Thomas Chrowder Chamberlin and Forest Ray Moulton in 1905, envisaged the origin of the solar system as attributable to a catastrophic close encounter between the Sun and a passing star, giving rise to a number of gaseous jets from the Sun, the inner parts of which condensed into the initial cores of the planets while the outer cooled to form a swarm of solid bodies rotating about the Sun. The planets then grew by gradual accretion. In an Adams Prize essay in 1917 the mathematician and astronomer Sir James Hopwood Jeans had proposed that a single large cigar-shaped streamer of hot gas torn from the Sun condensed into the planets. Jeffreys (1917–1918) concurred, except for postulating a much smaller primitive Sun than had Jeans. Despite severe objections being raised to the Jeans-Jeffreys theory by the American astronomer Henry Norris Russell in 1935, Jeffreys never accepted the alternative view. (Both these "catastrophic" hypotheses have since been replaced by the "nebular hypothesis," which envisages planetary growth by gradual accretion and collision within the dust cloud of a flat rotating protoplanetary disk surrounding the primitive Sun.)

Meteorology and Planetary Geodynamics. While at the Meteorological Office, Jeffreys worked on wartime problems in fluid dynamics having to do with the effects of tide and winds on sea conditions off the French coast, gunnery trajectories, and the atmosphere. With time to continue his own research, he showed (1920) that the worldwide effect of tidal friction on the seabed accounted for secular slowing of Earth's rotation. His legacy from this period includes a number of papers (1917–1933) on atmospheric convection, tides, and the interaction between wind and the sea, but after the 1930s he wrote relatively little on meteorology and fluid dynamics.

Statistics. Underpinning all of his research in seismology, general geophysics and related geological issues, celestial mechanics, and planetary geodynamics was his approach to scientific method and the application of mathematical and probabilistic techniques to the solution of practical problems in the treatment of gravity and seismic data. Examples include his development of methods for the numerical solution of differential equations; fitting a travel-time curve to observations of arrival times of seismic waves; and the lessening of the effect of unusual or erroneous values in estimating the mean and standard deviation of a frequency distribution in order to improve the accuracy of seismic travel-time tables (summarized in Jeffreys, 1939, ch. 4).

While a student at Cambridge, Jeffreys had been greatly influenced by *The Grammar of Science* (1892) by the statistician and philosopher of science Karl Pearson. He subsequently met the mathematician Dorothy Maud

Wrinch during his time in London. He and Wrinch now found common cause in the philosophy of science. They began collaborative work, initially writing on the methods of scientific inquiry and related probability theory and subsequently on seismology. Their collaboration continued following her return to Cambridge, as a lecturer in mathematics at Girton College (1921–1923). Much of Jeffreys's later work was influenced by the "simplicity postulate," which they first proposed in 1921: the belief that if a set of data might be accounted for by two or more alternative laws (models), all of which had a reasonable a priori probability, then the simplest model capable of making precise predictions that could be tested should be given the greatest prior probability and would therefore be preferable. Thus, fitting a simple equation might be preferable to a higher-order polynomial or differential equation, even if the latter fitted the data more exactly.

By 1939 Jeffreys had generalized this view, so that the complexity of a "law" was simply the number of adjustable parameters in it, the significance of each of which could be statistically tested. Although he and Wrinch introduced a notation for conditional probability in the 1920s, it was Jeffreys's book *Scientific Inference* (1931) in which the now-standard notation $P(q|h)$, to denote the probability P of a proposition q, given prior knowledge (initial data) h, first appeared. In this work he argued strongly for the statistical principle of "inverse probability" (first postulated by the eighteenth-century British mathematician Thomas Bayes) and developed methods to enable the utilization of different kinds of evidence with different reliabilities. His largely independent development of probability theory brought him into prolonged dispute (1932–1934) with the agricultural statistician Sir Ronald Aylmer Fisher. This arose mainly as the result of a mutual misunderstanding of the other's views, as the practical outcomes of their different approaches proved very similar. Jeffreys's *Theory of Probability* (1939), which introduced the methods of what came to be called Bayesian statistics, was cooly received at the time, being largely ignored by statisticians, but it was promoted by others, such as the New Zealand mathematician Keith Edward Bullen, a former PhD student of Jeffreys (1931–1934). Jeffreys also wrote several textbooks on mathematical methods: *Operational Methods in Mathematical Physics* (1927), *Cartesian Tensors* (1931), and *Asymptotic Approximations* (1962).

Geophysics. While in London, Jeffreys also met the geologist Arthur Holmes. Holmes was then beginning his work on radioactivity and the age of Earth. As a result, Jeffreys wrote several papers on the thermal history of Earth, the origin of the solar system, and the dynamics of the inner planets. In the years 1918–1929, based on estimates of the time taken for the eccentric orbit of Mercury to develop, Jeffreys made several estimates of the age of Earth of about 2.5 billion to 3 billion years. Although still too young (4.54 billion years is accepted in the early twenty-first century) these were, correctly, much older than the ages then being obtained by Holmes and others from studies of uranium and lead isotopes in minerals. In 1921 Jeffreys and Wrinch analyzed the seismic waves resulting from the accidental detonation of four thousand tons of explosives at Oppau, Germany, and this led him to suggest (1924) intentionally setting off large explosions in order to generate artificial seismic waves as a source for seismic studies (he would write in 1962 on travel times of seismic waves from thermonuclear weapons tests in the Pacific). Data from the Oppau explosion enabled him to confirm a hypothesis first mentioned to him by Holmes, that a granitic layer within the crust rested on one of more basic composition. The results of these and other investigations were brought together in Jeffreys's masterly book *The Earth: Its Origin, History, and Physical Constitution* (1924), which would run to six editions. Bullen coauthored with him two important works in early seismology: *Times of Transmission of Earthquake Waves* (1935) and *Seismological Tables* (1940).

Jeffreys's earlier work in meteorology and computational fluid dynamics made him realize that convection might occur within the body of Earth, but he believed that any such effect could only have existed early in its history. One of the consequences of his theoretical analyses of the cooling of Earth since its formation, and the effects of radioactivity, was his conclusion that contraction of Earth's crust as a result of cooling would produce sufficient compression to cause the development of mountain chains. He first postulated this in 1916 and persisted with it (e.g., in *Earthquakes and Mountains,* 1935), despite criticism from geologists. This theory also partly underpinned his lifelong opposition to the theory of continental drift. Another reason he was so unyielding on this topic was that in 1957 a German seismologist, Cinna Lomnitz, published an empirical relationship showing that the creep (slow but continuous deformation) observed in igneous rocks under long-continued stress in the laboratory increased logarithmically with time. Jeffreys in 1958 generalized this relationship to explain the damping of Earth's Chandler wobble (its fourteen-month nutation), the sharpness of transverse seismic waves at great distances, and the Moon's rotation and the persistence of its dynamic ellipticities, although (together with Stuart Crampin) in 1960 he revised values of some of the constants used in this last work. Jeffreys's modified Lomnitz law also implied that convection currents in Earth could not be maintained. This last, together with his apparent lack of knowledge of developments in paleomagnetism, led him to discount the mounting evidence from the 1950s onward that supported continental drift. Jeffreys

(1974, p. 401) still maintained that his theory showed that "continental drift—by convection, sea-floor spreading, and/or plate tectonics—cannot occur."

Surprisingly, he made no contributions to geomagnetism and was extremely skeptical of paleomagnetic results (which he always distrusted: having been warned in 1905 against mishandling magnets, he assumed rock specimens would prove equally unreliable). While admitting that some meteoritic craters existed on Earth, encouraged by experimental results obtained by Mohorovičić in 1928, he believed that lunar craters could equally well have been the result of explosive emission of internal gases or steam from a semifluid crust.

Honors. Jeffreys was awarded the DSc of Durham University in 1917 for his publications on geodynamics (the PhD at Cambridge did not exist until 1921). He was elected Fellow of the Royal Society in 1925. He served as secretary (1920–1927) and president (1955–1957) of the Geophysical Committee of the Royal Astronomical Society and president of the International Association of Seismology (1957–1960). He was awarded the Buchan Prize of the Royal Meteorological Society (1929); Gold Medal of the Royal Astronomical Society (1937); Murchison and Wollaston Medals of the Geological Society (1939, 1964); Royal and Copley Medals of the Royal Society (1948, 1960); the Prix Lagrange of the Académie Royale des Sciences, des Lettres et des Beaux-Arts de Belgique (1948); and the Vetlesen Prize of the Lamont Geological Observatory (subsequently the Lamont-Doherty Earth Observatory) of Columbia University, New York (1962), as well as five honorary degrees and honorary memberships of many societies. The importance of his statistical work was finally recognized by the statistical community in the 1960s, and he was awarded the Guy Medal in gold of the Royal Statistical Society in 1963. He was created knight bachelor in 1953.

BIBLIOGRAPHY

St. John's College, Cambridge, England, holds biographical papers, research, lectures, publications, societies and organizations, visits and conferences, correspondence, photographs, and sound and video recordings (206 files; catalog is available from http://janus.lib.cam.ac.uk/db/). The National Meteorological Library, Bracknall, England, contains an audiotaped interview, also available at American Institute of Physics, Center for Physics and Niels Bohr Library, College Park, Maryland.

WORKS BY JEFFREYS

The Earth: Its Origin, History, and Physical Constitution. Cambridge, U.K., and New York: Cambridge University Press, 1924.

Operational Methods in Mathematical Physics. Cambridge, U.K: Cambridge University Press, 1927.

The Future of the Earth. London: Kegan Paul, Trench, Trubner, 1929; New York: Norton, 1929.

Cartesian Tensors. Cambridge, U.K.: Cambridge University Press, 1931.

Scientific Inference. Cambridge, U.K.: Cambridge University Press, 1931.

Earthquakes and Mountains. London: Methuen, 1935.

With Keith E. Bullen. *Times of Transmission of Earthquake Waves.* Publications du Bureau Central Séismologique International, Serie A. Tarlouse: Bureau Séismologique International, 1935.

Theory of Probability. Oxford: Clarendon Press, 1939.

With Keith E. Bullen. *Seismological Tables.* Edinburgh: Neill & Company for British Association Seismological Investigations Committee, London, 1940.

With Bertha Swirles (Lady Jeffreys). *Methods of Mathematical Physics.* Cambridge, U.K.: Cambridge University Press, 1946.

Asymptotic Approximations. Oxford: Clarendon Press, 1962.

With Bertha Swirles (Lady Jeffreys), eds. *Collected Papers of Sir Harold Jeffreys on Geophysics and Other Sciences.* 6 vols. London, Paris, and New York: Gordon and Breach, 1971–1977. Contains the majority of Jeffreys's more than 440 papers, excepting work later incorporated in his books.

"Theoretical Aspects of Continental Drift." In *Plate Tectonics—Assessments and Reassessments,* Memoir 23, edited by Charles F. Kahle, 395-405. Tulsa, OK: American Association of Petroleum Geologists, 1974.

OTHER SOURCES

Aldrich, John. "The Statistical Education of Harold Jeffreys." *International Statistical Review* 73 (2005): 289–307. An earlier version is available from http://eprints.soton.ac.uk/33497.

Bolt, Bruce A. "Jeffreys and the Earth." In *Relating Geophysical Structures and Processes: The Jeffreys Volume,* edited by Keiiti Aki and Renata Dmowska, 1–10. Geophysical Monograph 76. Washington, DC: American Geophysical Union, 1993. A good survey of Jeffreys's work in geophysics and geodynamics.

Cook, Alan. "Sir Harold Jeffreys." *Biographical Memoirs of Fellows of the Royal Society* 36 (1990): 303–333. A good overall biographical account with a comprehensive bibliography on microfiche.

Galavotti, Maria C. "Harold Jeffreys' Probabilistic Epistemology: Between Logicism and Subjectivism." *British Journal for the Philosophy of Science* 54 (2003): 43–57.

Hide, Raymond. "Hide Receives the 1997 Bowie Medal." American Geophysical Union, 1997. Available from http://www.agu.org/inside/awards/bios/hide_raymond.html.

Howie, David. *Interpreting Probability: Controversies and Developments in the Early Twentieth Century.* Cambridge, U.K., and New York: Cambridge University Press, 2002. A detailed account of Jeffreys's work on probability and the Jeffreys-Fisher dispute.

Lapwood, E. Ralph. "Contributions of Sir Harold Jeffreys to Theoretical Geophysics." *Mathematical Scientist* 7 (1982): 69–84.

Lindley, Dennis V. "Jeffreys' Contribution to Modern Statistical Thought." In *Bayesian Analysis in Econometrics and Statistics: Essays in Honor of Harold Jeffreys,* edited by Arnold Zellner, 35–40. Amsterdam and New York: North-Holland, 1980.

M.A.K. and D.H.G. "Sir Harold Jeffreys (1891–1989)." *Annual Report of the Geological Society, London* (1989): 34–35.

Runcorn, Keith. "Sir Harold Jeffreys 1891–1989." *Nature* 339 (1989): 102.

Seidenfeld, Teddy. "Jeffreys, Fisher and Keynes: Predicting the Third Observation Given the First Two." In *New Perspectives on Keynes,* edited by Allin F. Cottrell and Michael S. Lawlor. Durham, N.C.: Duke University Press, 1995.

Spall, Henry. "Sir Harold Jeffreys—An Interview." [abridged from *Earthquake Information Bulletin,* 12 (1980)] Available from neic.usgs.gov/neis.

Swirles, Bertha (Lady Jeffreys). "Harold Jeffreys: Some Reminscences." *Chance* 4 (1991): 22–26.

Zellner, Arnold. "Jeffreys, Sir Harold (1891–1989)." In *International Encyclopaedia of the Social and Behavioural Sciences,* edited by Neil J. Smelser and Paul B. Bates, 7960–7963. Kidlington, Oxford: Pergamon Press. 2001.

Richard J. Howarth

JELÍNEK, JAN (*b.* Brno, Czechoslovakia, 6 February 1926; *d.* Brno, Czech Republic, 3 October 2004), *human paleontology, museology.*

Jelínek made important contributions to the human paleontology of Central Europe, especially the Moravian Middle and Late Pleistocene and in the promotion of the study of human evolution and paleoanthropology through museology. In paleoanthropology, Jelínek was recognized for his descriptions of Moravian Neandertal and upper Paleolithic fossils and his argument that modern humans traced at least part of their ancestry to Neandertals. In museology he founded in 1963 one of the first training programs in museum studies at Masaryk University in Brno, was president of the United Nations's international council of museums for five years, and was renowned for his innovative museum exhibits at the Anthropos Pavillion in Brno, Czech Republic.

Jelínek was born in Brno with one sister, Zdena; his mother, Marie Nevolova, was a housewife and his father, Jan, was a clerk. Through scouting, the young Jelínek developed an interest in Moravian natural history and prehistory, which he followed up in his academic work. He married Květa Rejzkova, a medical doctor, in 1952, and they had one son, Jan.

Jelínek studied with Karel Absolon and Vojtěch Suk at Masaryk University-Brno. After completing his degree, he joined the Moravian Museum in Brno as a researcher in 1947 and in 1958 was promoted to director. He immediately reorganized the museum, hiring young scholars specializing in prehistoric archaeology, such as Karel Valoch, and in zooarchaeology, including Rudolf Musil. With them, Jelínek initiated interdisciplinary, prehistoric research at key Paleolithic sites in Moravia including Mladeč; Předmostí; Stranska Skalá and, most extensively, Kůlna. He also participated in excavations at numerous Neolithic (e.g., Vedrovice) and more recent sites in Moravia. Overall, he wrote more than two hundred mostly single-authored articles in journals; book chapters; and books dealing with wide-ranging topics from New Guinea art to Neandertals to museology.

In 1962, Jelínek restarted the scientific journal *Anthropologie,* which was originally founded in 1923 by Jindřich Matiegka from Prague's Charles University but was suspended at his death in 1941. Jelínek also revived the *Anthropos* monographic series and published twenty volumes, including important site descriptions (e.g., Brno II, Kůlna, Šipka); conference proceedings of the 2nd Congress of the European Anthropology Association in 1982; and a special volume in *Anthropos* in 1986 honoring his sixtieth birthday. *Anthropologie* and the *Anthropos* monographic series are important resources covering many aspects of paleoanthropology and initially functioned to promote research in archaeology and physical anthropology in what was then Czechoslovakia. They were used through exchange to acquire foreign journals and books in the cash-strapped socialist economy of the time.

His *The Great Art of the Early Australians,* an exhaustive inventory and interpretation of prehistoric art of Arnhem Land in northern Australia, appeared in 1989. The volume, based on fieldwork from 1969 to 1973, contains 412 figures and photographs of what now is lost or endangered cave art. His *Das Grosse Bilderlexikon des Menschen in der Vorzeit* (*Great Pictorial Atlas of Prehistoric Man,* London: Hamlyn, 1976), first issued in German in 1972, has been translated into fourteen languages and is a comprehensive review of the paleontological and archaeological evidence for human evolution. In the early twenty-first century many of the images are readily available on the Internet, but when released, the book was the most conveniently accessible warehouse of photos and drawings documenting two million years of human evolution. Jelínek published in 2004 a detailed review of Saharan prehistoric art (*Sahara: Histoire de l'Art Rupestre Libyen*) based on his 1980s Libyan fieldwork and wrote a massive compendium of prehistoric and ethnographic-present living structures, titled *Střecha nad Hlavou: Kořeny Nejstarší Architektury* (A house to live in), that was published posthumously in 2006.

In the early 1960s Jelínek instituted a museology degree program at Masaryk University in Brno. In line with the practice of the Moravian Museum, he stressed an interdisciplinary focus and included many different programs at Masaryk. In the late 1960s he became involved with UNESCO's International Council of Museums (ICOM), serving on various committees and then as its president from 1971 to 1977. His creative ability and skill in the area of museum design resulted in the prescient Anthropos Pavilion in Brno, which opened in 1962. Although its founder was Absolon, it was Jelínek who convinced the central government to commit funds for a new building and his innovative presentation of prehistory that was responsible for its life-size reconstructions and original paintings by Zdeněk Burian.

Jelínek's theoretical contributions focused especially on morphological variation in fossil populations and its relevance to questions of evolutionary relationships and issues involving speciation and species identification in the genus *Homo*. From his earliest publications, Jelínek stressed the importance of variation in human populations, whether recent or fossil, and how this impacted evolutionary questions. These points were discussed in his classic paper on the Central European fossil record, "Neanderthal Man and *Homo sapiens* in Central and Eastern Europe," published in *Current Anthropology* in 1969. Jelínek maintained that the early Upper Paleolithic people were related to, not completely replaced by, Neandertals. He stressed the appearance of modern features in the Moravian European Neandertal finds (such Kůlna and Šipka) and retention of Neandertal features in specimens from the great Moravian sites of Brno, Dolní Věstonice, Mladeč, and Předmostí. These ideas were developed in detail in his colloborative work with David W. Frayer, Martin Oliva, and Milford H. Wolpoff on the Mladeč skeletal remains in 2006.

BIBLIOGRAPHY

A complete bibliography of Jelínek's work does not exist, but the author's tribute to him (David W. Frayer, "Some Parting Words for Jan Jelínek (February 6, 1926–October 3, 2004)," Journal of Human Evolution *49 (2005): 270–278) contains many of the most important references.*

WORKS BY JELÍNEK

"Neanderthal Man and *Homo sapiens* in Central and Eastern Europe." *Current Anthropology* 10 (1969): 475–503.

Das Grosse Bilderlexikon des Menschen in der Vorzeit. Prague: Artia, 1972.

"The Fields of Knowledge and Museums." *Journal of World History* 14 (1972): 13–23.

The Great Art of the Early Australians: The Studies of the Evolution and Rock Art in the Society of Australian Hunters and Gatherers. Anthropos Study in Anthropology, Palaeoethnology, and Quaternary Geology 25. Brno, Czechoslovakia: Moravian Museum-Anthropos Institute, 1989.

Sahara: Histoire de l'Art Rupestre Libyen. Grenoble, France: Jérôme Millon, 2004.

Střecha nad Hlavou: Kořeny Nejstarší Architektur. [A house to live in]. Brno, Czech Republic, 2006.

OTHER WORKS

Frayer, David W., Martin Oliva, and Milford H. Wolpoff. "Aurignacian Males from the Mladeč Caves, Moravia, Czech Republic." In *Modern Humans at the Moravian Gate: Mladeč Cave and Its Remains,* edited by Maria Teschler-Nicola. New York: Springer, 2006.

Novotny, Vladimir V., and Alena Mizerová. *Fossil Man, New Facts—New Ideas: Papers in Honor of Jan Jelínek's Life Anniversary.* Brno, Czechoslovakia: Anthropos Institute–Moravian Museum, 1986.

Wolpoff, Milford H., and David W. Frayer. "Aurignacian Female Crania and Teeth from the Mladeč Caves, Moravia, Czech Republic." In *Modern Humans at the Moravian Gate: Mladeč Cave and Its Remains,* edited by Maria Teschler-Nicola. New York: Springer, 2006.

David W. Frayer

JENNINGS, HERBERT SPENCER

(*b.* Tonica, Illinois, 8 April 1868; *d.* Santa Monica, California, 14 April 1947), *animal behavior, physiology, genetics, protozoology, zoology, philosophy of science.* For the original article on Jennings see *DSB,* vol. 7.

Since the original *DSB* article was published, new light has been shed on Jennings's approach to science, especially his epistemology and experimental method; his views on inheritance; and his political views, especially concerning eugenics.

Herbert Spencer Jennings was a leader in the experimentalization of zoological research in the United States in the first decades of the twentieth century. Beginning with his postdoctoral research on the behavior of lower organisms carried out between 1896 and 1906, and continuing with his studies of heredity and variation from 1907 through the remainder of his career, Jennings championed the use of experiment for biological study. Jennings's methodological orientation was deeply informed by his adherence to pragmatism, as articulated primarily by the philosophers John Dewey and William James. By 1913 Jennings formulated his own philosophy of biology, termed "radically experimental analysis," which bore many of the hallmarks of pragmatism, including a relentless commitment to action and to empiricism for the resolution of scientific problems. As a pragmatist, Jennings directed these same priorities to the social arena as well.

Herbert Spencer Jennings. © BETTMANN/CORBIS.

Jennings's pragmatism informed all aspects of his career, from his behavioral and genetics research programs with protozoa, to his substantial philosophical writings, to his public criticisms of eugenics during the 1920s.

Jennings's contact with pragmatism began early in his education, when he pursued two years of philosophical study with the young Dewey between 1890 and 1892, while an undergraduate at the University of Michigan. Jennings had already had considerable contact with literature and philosophy as a young child who passed much of his time reading widely from the large library of his father, George Nelson Jennings. The senior Jennings was a physician and enthusiastic evolutionist, who named his first-born son for the influential British philosopher, Herbert Spencer, while naming his second son George Darwin. The young Herbert Spencer Jennings came to reject much of the Spencerian thought that he had adopted at an early age when he was exposed to Dewey's experimentalism. Dewey claimed that ethical behavior was grounded not in natural law, as had been held by Spencer, but in scientific method. While ultimately pursuing a course of study in zoology, culminating in earning his PhD in zoology at Harvard University in 1896, Jennings never lost interest in philosophy and its relevance for political and scientific conduct.

In his dissertation research, Jennings had been deeply interested in the experimental goals and methods but largely critical of the mechanistic orientation of the new experimental program of development mechanics, or *Entwicklungsmechanik*, championed by Wilhelm Roux and others. Developmental mechanics imported experimental methods from the field of physiology to zoology, which until that time was largely characterized by the observational practices of natural history. Inspired by the new experimentalism, Jennings pursued postdoctoral study in the laboratory of the German physiologist Max Verworn in Jena in 1896–97. There, Verworn introduced Jennings to many of the experimental methods of physiology as well as to the microscopic single-celled organisms, the protozoa, for experimental study. Jennings was also further immersed in the evolutionism that suffused Jena, the epicenter of German evolutionary thought. Throughout his career, Jennings, like Verworn and his Jena colleague, Ernst Haeckel, embraced the protozoa for their biological importance as organisms possessing complex physiological processes and behaviors, yet were at the same time single cells.

Relatively early in his physiological studies of the behavior of protozoa, Jennings emphasized the rudimentary psychological capabilities of the organisms. In the course of this research he characterized the essential features of the avoiding reaction in protozoa, the process by which the microscopic organism stops its forward motion, swims backwards, and then starts off in a new direction when it confronts a new external stimulus. Jennings rejected the idea that external, physicochemical forces causally directed the organism's motions, as others such as Jacques Loeb claimed. He maintained that the avoiding reaction was driven by a mechanism internal to the organism. During a year of research at the Zoological Station in Naples, Italy, in 1903–04, however, he adopted a more radical position: Jennings now argued that the avoiding reaction in protozoa was an early manifestation of the method of "trial and error" deemed by the evolutionary psychologist C. Lloyd Morgan to be fundamental to the development of intelligence in higher organisms. This research culminated in the publication of his influential text *Behavior of the Lower Organisms* in 1906. During the same year Jennings accepted an appointment at the Zoological Laboratory at The Johns Hopkins University, which he directed from 1910 until he retired in 1938. (For discussion of the various academic positions held by Jennings prior to moving to Johns Hopkins, please see the original *DSB* article.)

Behavior of the Lower Organisms attracted considerable criticism both for its critical stance toward mechanistic biology and for its attribution of a developmental-evolutionary continuum "from the lowest organisms up to man" (p. 335) in behavioral features such as perception, choice, attention, and even consciousness. In the years that followed, Jennings dedicated considerable effort to epistemological writings informed by these earlier criticisms. Throughout his philosophical reflections he championed both a natural history-like attention to the individual organism and experiment in biology. Central to his concern was the development of a middle position in biology between vitalism on the one hand and mechanism on the other. Jennings was particularly concerned about the vitalism articulated by the German zoologist Hans Driesch, who maintained that two biological systems that are completely identical could behave differently under identical conditions. Jennings rejected Driesch's position of experimental indeterminism, because it threatened to halt experimental inquiry in biology altogether. Arising from these concerns was Jennings' formulation of his methodology of radically experimental analysis, whereby all biological questions are reduced to an experimental situation; in a given experimental situation, a preceding perceptual difference is sought for every existing perceptual difference.

In 1907 Jennings turned his attention from behavior to problems of heredity and variation. Following the rediscovery of Mendel's laws in 1900, biological interest in heredity quickly blossomed as experimentation grew in prominence among academic biologists. Jennings was in the vanguard of American zoologists to publish experimental research in heredity and genetics in the first decade of the twentieth century. As researchers experimented with a variety of organisms, Jennings initiated experimental studies with the same organisms that had been central to his behavior research: the protozoa, and primarily the ciliate Paramecium. Conceptually, Jennings likewise presented his research program in heredity as a continuation of his behavioral research, focusing on the inheritance of adaptive qualities, a problem that he viewed as intimately related to the development of adaptive behaviors in an individual organism. There was methodological continuity between the two research programs as well; Jennings aimed to maintain analytical focus on the individual organism and on relationships between individual organisms in his hereditary studies. This was quite a distinctive undertaking, as the developing protocols in genetics research also demanded statistical analysis of data derived from large quantities of organisms, a methodological domain in which Jennings was also a leader.

Whereas Jennings engaged in important statistical and theoretical work in the 1910s and 1920s that lent credence to Mendelism and chromosome theory, his own experimental research program was not Mendelian. Like those of other leading zoologists attracted to the study of heredity during the period, the broader intent of Jennings's research program was to demonstrate evolution experimentally. Furthermore, he aimed to uncover the most fundamental mechanisms of heredity and to advance a generalized conception of heredity. Unlike many, however, Jennings adopted a relatively pluralistic attitude toward the study of heredity. During this period he evenhandedly explored both selection and the inheritance of acquired characteristics as possible mechanisms of heredity and evolution. Jennings placed a great emphasis on the utility of the protozoa for such research, because their single cellularity made their exposure to environmental influences more feasible and hereditary effects in them more readily observable. Likewise, their uniparental form of reproduction made it possible to explore minute sources of heritable variation unobscured by biparental inheritance. Jennings came to view lineages of asexually reproducing protozoa as material embodiments of pure lines. Between 1909 and 1911 he facilitated discussions that helped to solidify the pure line and genotype concepts introduced by Wilhelm Johannsen.

The experience of World War I and the international political upheaval in the years that immediately followed was a transformative experience for Jennings, as it was for many scientists of his generation. Distress over growing American nativism in the postwar years and the associated rise of the eugenics movement prompted Jennings to largely abandon his experimental research program and dedicate his efforts throughout the 1920s to public criticism of eugenics. Jennings employed diverse strategies in this undertaking: He criticized fallacious reasoning and faulty assumptions underlying many eugenic arguments; in concert with social workers and educators, he countered genetic determinism by underscoring the role of environment and education in the development of human potential; and he worked to keep the field of genetics institutionally separate from the broader eugenics movement. At the end of the decade Jennings returned to experimental research with a narrowed research agenda. In a bid to generate new evidence further undercutting the eugenics movement, he reestablished a research program with Paramecium aimed at conclusively demonstrating the inheritance of environmental effects. The research planned by Jennings was largely carried out by his former student and research associate Tracy Sonneborn, and longtime research assistant Ruth Stocking Lynch. This work led to Sonneborn's discovery of mating types in *Paramecium aurelia* in 1937.

Jennings served in many leadership capacities and received many honors in the course of his career. He served as president of the American Society of Zoologists (1909) and the American Society of Naturalists (1910),

First Chairman of the Genetics Society of America (1922), Vice-President of Section F (Zoology) of the American Association for the Advancement of Science (1925), and member of the Council of the National Academy of Sciences (1934–1940). He served as a founding member of the editorial boards of *The Journal of Experimental Zoology* and *Genetics* and on the editorial boards of several other journals. He was a Terry Lecturer at Yale University (1933), Vanuxem Lecturer at Princeton University (1934), Eastman Visiting Professor at Oxford University (1935–1936), Leidy Lecturer at the University of Pennsylvania (1940), and Patten Lecturer at Indiana University (1943).

SUPPLEMENTARY BIBLIOGRAPHY

WORKS BY JENNINGS

Herbert Spencer Jennings Papers. Philadelphia, PA: American Philosophical Society Library. Contains extensive correspondence, notebooks, journals, manuscripts and lectures.

"Diverse Ideals and Divergent Conclusions in the Study of Behavior in Lower Organisms." *The American Journal of Psychology* 21 (1910): 349–370.

"Doctrines Held as Vitalism." *The American Naturalist* 47 (1913): 385–417.

"Causes and Determiners in Radically Experimental Analysis." *American Naturalist* 47 (1913): 349–60.

"Heredity and Environment." *Scientific Monthly* 19 (1924): 225–38.

"Diverse Doctrines of Evolution, Their Relation to the Practice of Science and of Life." *Science* 65 (1927): 19–25.

OTHER SOURCES

Kingsland, Sharon. "A Man Out of Place: Herbert Spencer Jennings at Johns Hopkins, 1906–1938." *American Zoologist* 27 (1987): 807–817.

Pauly, Philip J. "The Loeb-Jennings Debate and the Science of Animal Behavior." *Journal of the History of the Behavioral Sciences* 17 (1981): 504–515.

Schloegel, Judith Johns. "Intimate Biology: Herbert Spencer Jennings, Tracy Sonneborn, and the Career of American Protozoan Genetics." PhD dissertation. Bloomington: Indiana University, 2006.

———, and Henning Schmidgen. "General Physiology, Experimental Psychology and Evolutionism: Unicellular Organisms as Objects of Psychophysiological Research, 1877–1918." *Isis* 93 (2002): 614–645.

Sonneborn, Tracy M. "Herbert Spencer Jennings, April 8, 1868–April 14, 1947." *Biographical Memoirs of the National Academy of Sciences* 47 (1975): 143–223. A detailed biographical resource including a complete bibliography of Jennings's publications and listings of positions, honors, and distinctions.

Judith Johns Schloegel

JERNE, NIELS KAJ (*b.* London, England, 23 December 1911; *d.* Castillon-du-Gard, France, 7 October 1994), *immunology, molecular biology, serology, biosemiotics, theoretical biology.*

Jerne was one of the leading immunologists of the post–World War II period. He received the Nobel Prize in 1984 for three theoretical contributions: the natural selection theory of antibody formation, the somatic generation theory of antibody diversity, and the idiotypic network theory of the immune system. In addition, he invented a method for the detection of single-antibody-producing cells that became one of the most cited techniques in cellular immunology.

Latecomer to Science. Niels Jerne was the fourth of five children of the inventor and industrialist Hans Jessen Jerne (1877–1950) and Else Marie (née Lindberg, 1874–1956). His parents were Danish but had emigrated to London in 1910 to start a celluloid factory. Right after the outbreak of World War I the family moved to the Netherlands, where his father acquired a refrigerated storehouse in Rotterdam and later became reasonably wealthy. Later in life Jerne never felt exclusive ties to any single nation and often described himself as a citizen of the North Sea: Denmark provided his family roots, Dutch was the language in which he was most fluent, and England represented his cultural preferences and political sympathies. In addition to carrying a Danish passport he remained a British subject throughout his life.

Jerne's road to an academic career took a number of twists and turns. In his own words he was a *Spätzünder* (one who is late to ignite). After leaving high school in 1928 with fairly average grades he was employed as a junior clerk in the Elders and Fyffes banana company in Rotterdam. The diaries and correspondence of his late adolescent years bear witness to an arrogant and ironic young man who loved to play language games, immersed himself in romantic and modernist literature, and was drawn to Kierkegaard and Nietzsche. Accordingly he wanted to study philosophy but, at his father's request, began with chemistry at the University of Leiden in 1931. He preferred a fairly hectic schedule of extracurricular student union activities, to serious academic work, however, and dropped out of university after two years.

In 1934 his father supported a renewed attempt at a higher education, this time in medicine at the University of Copenhagen. Soon after his arrival in Denmark, however, Jerne met a Czech painter, Ilse ("Tjek") Wahl (1910–1945), whom he married and with whom he had a son in 1936. Thus he quit university again, began working for his father, who was experimenting with new bacteriological methods for bacon curing, and only resumed his medical studies in 1939. For most of the war years in

German-occupied Copenhagen he divided in his time among medical textbooks, basic clinical training, and a bohemian night life with his wife's artist friends.

In order to support his growing family (another son was born in 1941), Jerne took a part-time position as a secretary in the Department of Standardization at the Danish State Serum Institute in 1943. The department had been set up by the Health Committee of the League of Nations in the 1920s for the establishment of international serum standards. Here Jerne discovered his aptitude for mathematical analysis and statistical thinking ("chance governs all" became one of his favorite expressions), and he began to immerse himself in problems of biometrics and methods of biological standardization of antigens and antisera. Significantly, prime numbers and chess were among his favorite pastimes besides literature and language.

Antibody Avidity and the Selection Theory. His wife's suicide in the autumn of 1945 was a turning point in Jerne's life. He succeeded in finishing his medical degree in 1947 and decided, after internship and marriage with his former mistress, Adda Sundsig-Hansen (1914–1993), to go into research. The head of the standardization department, Ole Maaløe, who later became the pioneer of Danish molecular biology, encouraged him to follow up on his earlier observation that the measure of the strength of an antiserum depends on its concentration, indicating that the chemical reaction between antibody and antigen is reversible. Using a rabbit skin assay system to measure the amount of surplus toxin, Jerne set out to investigate the kinetics of the reaction between diphtheria toxin and diphtheria antitoxin to get a quantitative measure of the binding strength of the antibody (its "avidity"). His published dissertation, *A Study in Avidity* (1951), was received by the international serological community as an example of experimental rigor, statistical prowess, and conceptual clarity.

One of the unexpected findings of the dissertation was that avidity increases in the course of immunization. Stimulated by James D. Watson and Gunther Stent's visit to Maaløe's department in 1951–1952, Jerne began to orient himself to the problems and methods of phage research and created a new, and more sensitive, bacteriophage-antiphage assay system for studying the details of early avidity increase. After a couple of years of inconclusive experimentation, interrupted by an inspection tour he undertook for the World Health Organization (WHO) to standardization centers in Asia in 1953, he was struck, in the summer of 1954, by an experiment that, in his eyes, demonstrated the existence of specific antibodies without the prior presence of a corresponding antigen. A few weeks later he formulated a theory of antibody formation that could explain, he thought, a whole array of serological and immunological phenomena, including the avidity increase. All possible kinds of specific antibody molecules already exist in normal serum, he stated, and one of these specific molecules fits, "by chance," any given antigen, not necessarily exactly but approximately. The intruding antigen selects such specific fitting antibody molecules and transports them to antibody-producing cells, where they are further multiplied and released into the bloodstream.

The natural selection theory of antibody formation flew in the face of the ruling dogma of immunology. The existence of preformed antibodies had been highly disputed in immunology for more than half a century. In 1897 Paul Ehrlich had speculated that all cells in the body carry preformed molecular groups, so-called side chains (*Seitenketten*, what would be called receptors in the early twenty-first century), which were assumed to play a part in the absorption of specific nutritional substances. Intruding antigens—toxins, for example—would also be recognized by such side chains, which were supposed to be released into the bloodstream as specific antibodies.

After Karl Landsteiner's experiments, beginning in the 1910s, with synthetic haptens (small molecules that are coupled to the antigen and change its antigenic properties) Ehrlich's theory fell into disrepute. Whatever particular antigen was synthesized in the chemist's laboratory, it gave rise to a specific antibody response. It was considered impossible for organisms to have all these specific molecules in stock waiting for whatever might intrude. Instead, from the middle of the 1930s varieties of template theories became dominant among serologists and immunologists who claimed that the specificity of antibodies was determined de novo by the antigen. The chemically most sophisticated of these template theories was put forward by Linus Pauling, who suggested that specific antibodies were formed when normal, nonspecific globulin molecules that had not yet received their final tertiary configuration wrapped around antigen molecules and so took on a specific, complementary structure.

Accordingly, preformed antibodies were considered an anomaly, and Jerne's demonstration of specific antibody activity without prior antigen presence could easily be dismissed as an experimental error. Jerne was immediately convinced of the anti-template alternative interpretation, however. The writings of his intellectual hero, the British biometrician and evolutionist Ronald A. Fisher, were an important cognitive resource for thinking about the match between antigen and antibody in terms of random fitness and Darwinian selection. But there is also much to suggest that Jerne's dogged anti-template position drew on his personal and emotional proclivities. He repeatedly referred to himself as a person with a repertory of preformed states of mind that could be elicited in interactions with other people. The creation of the natural

Niels Kaj Jerne. AFP/GETTY IMAGES.

selection theory of antibody formation can thus be seen as an act of metaphorical projection: His experience of his own self became an emotional charged cognitive resource for the new immunological theory.

Jerne worked out the details of the natural selection theory during a fellowship at the California Institute of Technology in 1954–1955, and the paper was sent to the *Proceedings of the National Academy of Sciences* by his mentor, Max Delbrück. It received a positive response from molecular biologists inclined toward neo-Darwinism—for example, Salvador Luria—but most immunologists were lukewarm. The doyen of American immunology, Alwin Pappenheimer, thought the theory was "all baloney … all hocus pocus" (cited in Söderqvist, 2003, p. 200). Upon his return to Copenhagen, Jerne felt discouraged, and a year later, in 1956, he left his family and a permanent research job at the Serum Institute to take up an administrative position at the Section of Biological Standardization at the WHO in Geneva.

Immunology as a Theoretical Discipline. At the WHO, Jerne's main responsibility was to create new guidelines, so-called minimum requirements, for the standardization of vaccines and sera, including vaccines against smallpox, poliomyelitis, yellow fever, and cholera. Diplomatic skills and extensive committee work were needed to coax the expert recommendations through the internal bureaucracy, and Jerne's achievements soon earned praise in the organization. In 1960 he was assigned the task of organizing a new program to foster international cooperation and training in immunology, especially in the developing countries. Another advantage of working in Geneva was, of course, the tax-free salary at the WHO, the international atmosphere, and the city's culinary offerings. Jerne worked hard, but he also appreciated the material pleasures of life.

Meanwhile the natural selection theory began to win the attention of a growing number of immunologists, especially after Sir Frank Macfarlane Burnet (Nobel Prize winner in 1959 for his work on acquired immunity) revised it by suggesting that selection works at the level of cells rather than on circulating antibody molecules. Within a few years the clonal selection theory acquired the status of a central dogma in immunology, and as a consequence Jerne's star rose. He began considering a scientific comeback. The opportunity came in 1962 when the medical faculty at the University of Pittsburgh asked him to become professor and chairman of its Department of Microbiology.

The vigorous intellectual activity in Pittsburgh put Jerne under pressure to achieve. The opportunity came only a few months after his arrival when, together with his first and only postdoc, Albert Nordin, he constructed a powerful method for the quantification of antibody-producing cells in vitro. The idea was to combine the plaque assay technique he had used in Copenhagen a decade earlier with the well-known principle of complement-mediated immune hemolysis, discovered by Jules Bordet in 1901 and further developed by August von Wassermann for the diagnosis of syphilis. Each antibody-producing cell produced a zone of hemolyzed red blood cells around it, like small, bright stars in a dark sky. These aesthetic qualities, together with the ingeniousness and simplicity of the assay, rapidly made it one of the most frequently used quantitative methods in cellular immunology; the short paper in *Science* in March 1963 was one of the most cited papers in the biomedical literature in the 1960s through 1980s.

Together with a small group of coworkers Jerne spent the next couple of years refining the plaque assay method to get a quantitative picture of the kinetics of the early immune response, but without any conclusive results. He also cultivated other interests. In Geneva he had been interested in the analogies between language and the immune system; now he mused about the analogies between cognitive and immune learning—ideas that would later be taken up by Gerald Edelman in his Darwinian notion of the nervous system.

But these were sidetracks; it was Jerne's immunological star that rose. These were the years when the number of immunological journals, textbooks, and societies proliferated and when immunology became established as an independent scientific discipline with its own departments and chairs. Jerne received offers from the University of Copenhagen and Harvard Medical School, but chose instead, partly for financial, partly for nostalgic reasons (he loved to fashion himself as a European intellectual, while disdaining American culture) to take up a position as director of the venerable Paul-Ehrlich-Institut in Frankfurt, West Germany. A year later, in 1967, his reputation as the leading theoretician in the discipline was cemented at the Cold Spring Harbor symposium on immunology, where his final report of the meeting, "Waiting for the End," was considered a brilliant summary of the state of the field, especially in his problematization of the entrenched division between chemically and biologically oriented immunologists.

Antibody Diversity and Idiotypic Networks. Jerne's explicitly stated goal for going to Frankfurt was that he wanted to build up a European counterpart to the strong U.S. domination in the field. He traveled widely giving lectures and organized the first course in immunology for the newly founded European Molecular Biology Organization in 1967. However, he soon became frustrated by the German research bureaucracy and the relaxed work habits at the Paul-Ehrlich-Institut, so when the multinational pharmaceutical company Hoffmann-La Roche invited him, in 1968, to become the first director of a new research institute for immunology in Basel, Switzerland, Jerne quickly accepted.

Roche's funds were seemingly unlimited. Jerne traveled worldwide to recruit the best available scientists. His ideal was to establish a new kind of research institute based on the idea of a nonhierarchical, communicative network where both the architecture and the organization were designed to encourage maximum interaction between its members. When it opened in June 1971 the Basel Institut für Immunologie was the largest of its kind in the world, with a staff of around 150. It remained the world's leading center for immunological research throughout the 1970s and 1980s, housing, among others, three future Nobel laureates (Jerne, Susumu Tonegawa, and Georges Köhler). According to the immunological grapevine, Jerne spent most of his time in perpetual discussions and never visited the laboratories, because, as he allegedly once said, "the reality would confuse me" (Söderqvist, p. 269.)

In the late 1960s and early 1970s Jerne's immunological thinking developed along two lines. One basic question concerned the genetic basis for the origin of the antibody repertoire. Did the information for the coding of all specific antibodies exist in the germ line? Or did the final repertoire arise through somatic mutations during the differentiation of the antibody-producing cells? Jerne's contribution to the problem was to invoke the well-known phenomenon of self-tolerance. In 1969 he proposed that self-tolerance and antibody diversity were generated in the same process during embryonic development, and a year later he assigned a central role to the thymus, both as a breeder of mutant, specific lymphocytes and as a generator of the organism's tolerance against its own antigens. The theory was contested by many of his colleagues but was partly confirmed by Tonegawa in the mid-1970s, who received the Nobel Prize in 1987 for the discovery of the mechanism of somatic recombination which allows for a comparatively small number of genes to code for millions of specific kinds of antibodies.

The other question that incessantly occupied Jerne's mind was how the immune system is regulated. Even though he had rapidly accepted Burnet's clonal selection theory—"I hit the nail but Burnet hit it on its head" (Söderqvist, p. 222), he never gave up the idea that circulating antibodies might after all play a significant role in immune regulation. Throughout the 1960s he obtained experimental results pointing in this direction and consequently began to think of circulating antibodies as "receptors" for antigens. The findings of Henry Kunkel and Jacques Oudin that specific antibodies have unique antigenic determinants (idiotypes) furthered this line of thinking.

Further inspired by his reading of information theory and systems theory, Jerne began to think of the immune system as a self-regulating cybernetic system of idiotypic relations. In an article in *Scientific American* in 1973, Jerne proposed the idiotypic network theory of the immune system, according to which all antibodies and lymphocyte receptors are seen as mutually independent parts of a steady state system. In the following years Jerne published a series of theoretical articles to support the idea that every variable region of an antibody molecule (paratope) functions as a specific antigenic determinant (epitope) which is recognized by still another number of paratopes, which in turn function as specific epitopes, and so on ad infinitum. In Jerne's vision, the immune system was a kind of "hall of mirrors," a dynamic equilibrium system of mutual molecular recognition which in principle did not need any responses from the outside to work. The external world with its bacteria, viruses, and other antigenic determinants were of subordinate importance to the regulation; the system had an eigen behavior, that is, it was self-referential.

Several experimental studies seemed to confirm the regulatory effect of anti-idiotypic antibodies, and around 1980 Jerne's network theory was immensely popular among immunologists. Not everyone was convinced, however. The most vociferous of Jerne's critics, Melvin Cohn, asserted that the theory was unscientific, even absurd, because it was incapable of being tested empirically. Jerne was not indifferent to such criticisms but nevertheless felt that they missed the point. He held that science is not just an accumulating series of falsifiable propositions that experimentalists in true Popperian fashion try to falsify, but also (and mainly) an arena for the imaginative mind. Indeed, the notion of imagination was a recurrent one in Jerne's thinking; in one of the first international meetings on the formation of antibodies, in Prague 1958, he had asked the participants to "fly into the realm of imagination" and repeatedly stressed the importance of giving free rein to scientific fantasy.

Last Years. After his retirement from the Basel Institute in 1980, Jerne withdrew to his country home near Castillon-du-Gard, Languedoc, France, together with his third wife, Ursula (Alexandra) Kohl (b. 1936), whom he had married in 1964. There he continued to develop his ideas, first expressed in the early 1960s, about the analogies between linguistics and immunology. For example, drawing on American linguist Noam Chomsky's theory of generative grammar, Jerne suggested that the immune system was governed by a set of grammatical rules that allowed an open-ended number of sentences (antibody specificities). He increasingly viewed immunology as a philosophical subject and questioned the reductionism of experimental immunology. In his view the immune network was not built up by interacting material factors—it was an abstract principle, a pure recognition system. These ideas attracted only few adherents among the new generations of immunologists trained in more pedestrian sciences, such as biochemistry and molecular biology.

Although his influence on immunological practice waned rapidly in the 1980s, Jerne was nevertheless duly remembered and honored for his lifelong contributions to immunology. He received honorary doctorates from the Erasmus University in Rotterdam and from the Weizmann Institute of Science in Rehovot, Israel, and was awarded several prestigious scientific prizes, including the Paul Ehrlich Prize in 1982 and the Nobel Prize for Physiology or Medicine in 1984. Jerne shared the Nobel with Georges Köhler and César Milstein, who had invented the hybridoma technique for producing monoclonal antibodies a decade earlier. Jerne was cited for his "visionary theories," which had enabled modern immunology to make "major leaps of progress"; in fact he was the first medical Nobel Prize winner to be cited primarily for his theoretical contributions, and he did not shy away from emphasizing that he was not particularly fascinated by monoclonal antibodies which, in his mind, were mainly of practical interest. Neither did he see anything of immunological interest in the AIDS epidemic.

The expression of such politically incorrect views added to the web of myths and legends that were spun around Jerne in the immunological culture. Anecdotes circulated about his power of imagination, his working capacity, and his analytical keenness, but also about the aura of aloofness, elitism, and *Bildung* that surrounded him. Jerne was keen to present himself as an urbane European intellectual who would rather read Shakespeare and Proust than the *Journal of Immunology* and who preferred to sit in a wine bar and talk about art, politics, and life in general than work in the laboratory; conversation was his life's breath. There is also a Faustian theme in Jerne's life story: He wanted terribly to be unusual; he strove for perfection in science and eventually won the recognition of being "one of the most intelligent biologists of this century" (Burnet, 1968, p. 249). At the same time, he paid for his efforts by becoming a wanderer who continually tried to evade responsibility for his own life and the care of others, and he characterized himself as "a kind of misfit" (Söderqvist, p. 14). His life story epitomizes Alastair MacIntyre's claim that biography is "neither hagiography nor saga, but tragedy" (p. 213).

BIBLIOGRAPHY

WORKS BY JERNE

A Study of Avidity Based on Rabbit Skin Responses to Diphtheria Toxin-Antitoxin Mixtures. Copenhagen: Munksgaard, 1951.

"The Natural-Selection Theory of Antibody Formation." *Proceedings of the National Academy of Sciences* 41 (1955): 849–857.

With Albert A. Nordin. "Plaque Formation in Agar by Single Antibody-Producing Cells." *Science* 140 (1963): 405.

"Summary: Waiting for the End." *Cold Spring Harbor Symposia on Quantitative Biology* 32 (1967): 591–603.

"Generation of Antibody Diversity and Self-Tolerance: A New Theory." In *Immune Surveillance,* edited by Richard T. Smith and Maurice Landy. New York: Academic Press, 1970.

"The Immune System." *Scientific American* 229 (July 1973): 52–60.

"The Generative Grammar of the Immune System." In *Les Prix Nobel 1984: Nobel Prizes, Presentations, Biographies and Lectures.* Stockholm: Almqvist and Wicksell International, 1985.

OTHER SOURCES

Burnet, F. Macfarlane. "A Modification of Jerne's Theory of Antibody Production Using the Concept of Clonal Selection." *Australian Journal of Science* 20 (1957): 67–68.

———. *Changing Patterns: An Atypical Autobiography.* Melbourne: Heinemann, 1968.

MacIntyre, Alasdair. *After Virtue: A Study in Moral Theory.* 2nd ed. Notre Dame, IN: University of Notre Dame Press, 1984.

Söderqvist, Thomas. *Science as Autobiography: The Troubled Life of Niels Jerne.* New Haven, CT: Yale University Press, 2003.

Thomas Söderqvist

JOERGENSEN, AXEL CHRISTIAN KLIXBÜLL

SEE **Jørgensen, Axel Christian Klixbüll.**

JOHN OF ALEXANDRIA

SEE **John Philoponus.**

JOHN OF DUMBLETON

SEE **Dumbleton, John.**

JOHN PHILOPONUS (*b.* Ceaserea [?] c. 490; *d.* Alexandria, *c.* 570), *natural philosophy, theology.* For the original article on Philoponus see *DSB,* vol. 7.

Since the original publication of the *DSB* there have been large-scale changes in Philoponus's studies in particular and late antique philosophy generally. Thanks in great part to the work of Richard Sorabji, Philoponus and other late antique thinkers have been the subject of numerous monographs and articles. And Sorabji's large translation project of the *Commentaria in aristotelem graeca* and associated texts has led to the widespread availability of Philoponus's writings for English-language scholars.

Known also as John the Grammarian or John of Alexandria, John Philoponus was one of the most astute philosophical thinkers of the sixth century. A Christian Neoplatonist adhering to the Monophysite sect, he spent his career articulating a reformed Aristotelian physics that he hoped would both be amenable to Christians yet would retain the marked rationalism of pagan philosophy. He was educated at the Academy in Alexandria under Ammonius Hermeiou and probably taught there, although he likely did not hold the chair of philosophy. His earliest works were traditional commentaries on Aristotle and other elementary texts, such as a treatise on the astrolabe and an introduction to Nikomachus's *Arithmetic.* Philoponus's name is attached to five commentaries (on the *Categories,* the *Prior* and *Posterior Analytics, De anima,* and *De generatione et corruptione*), which in fact were lectures given by Ammonius that Philoponus edited and to which he added some of his own ideas. Few works can be dated with precision, but his commentary on the *Physics* (which includes the important *Corollary on Place* and *Corollary on Void*) was published in 517. An unfinished commentary on the *Meteorologica* was written shortly thereafter.

Beginning in the late 520s, Philoponus wrote a number of important treatises aimed at refuting arguments for the eternity of the world. The two most important of these are *Contra Proclum de aeternitate mundi* and the now-fragmentary *Contra Aristotelem de aeternitate mundi.* His final natural philosophical work, *De opificio mundi*—an exegesis of the opening book of Genesis in an attempt to reconcile Christianity with pagan natural philosophy—was most likely written in the 540s, although other evidence points to a date ranging from 557 to 560. The remainder of his career was dedicated primarily to specifically Christian theological issues, yet he retained a belief in the possibility of solving questions of the trinity and the nature of Christ with the same pagan rationality that he applied to natural philosophy. Indeed, by rigorously applying the Aristotelian definition of substance to the problem of the trinity, Philoponus developed a theology that his opponents would brand *Tritheism* (the Godhead contains three individual substances) for which he was anathematized from the church about a century after his death in 681. Although his anathema all but eliminated his influence in the Latin Christian thought, he had an impact on Islamic philosophy (particularly his arguments against eternity), which then later contributed to Latin Christian thought through the translation movement of the twelfth and thirteenth centuries.

Although scholars used to celebrate Philoponus for his apparent anticipation of many ideas developed during the scientific revolution of the seventeenth century (for example Galileo's impetus theory or Descartes's theory of matter as extension), later scholars placed him squarely in the intellectual context of the sixth century. It cannot be denied that Philoponus did in fact take some strikingly modern positions, but viewed as a whole he was a man of late antiquity, albeit with exceptional intellectual talent. Moreover, although often portrayed as a potent warrior in the battle to overcome Aristotle's towering influence on the history of Latin Christian thought, he nonetheless retained a deep admiration for and indebtedness to Aristotle and an even deeper admiration of the philosopher's commitment to rationality. That is, one should not be deceived by Philoponus's rejection of eternity or his attempt to unify heavenly and terrestrial physics, for example, into thereby believing that he rejected Aristotelianism altogether as would happen in the modern era. Basic Aristotelian

John Philoponus

principles—the four causes, the role of potentiality and actuality, substance and accident—were not only retained by Philoponus but were his central resources for developing a new Christian natural philosophy.

Nonetheless, John Philoponus was a reformer. He was inspired primarily by the desire to bring natural philosophy into line with Christian dogma. And the most sweeping and decisive reform was the polemical and philosophical attack against pagan philosophers' arguments for eternity. The attacks were systematic and comprehensive, taking each of the many arguments for eternity and drawing out contradictions, absurdities, and philosophical difficulties. Some of the arguments Philoponus deployed were not original, but he also presented many novel and fascinating ones. Aristotle rejected actual infinites as a general rule (e.g. infinite extension). This is one of the major tenets which Philoponus exploited to show that eternity entails actual infinites (e.g. an infinite number of past generations) and thus violates Aristotle's own criteria. He asked how there could be different numbers of infinites: Different planets rotate around the earth at different rates and so there has been one actual infinite number of rotations for Jupiter and another for Mars, an early instance of cardinal numeration of infinites. Philoponus also challenged eternity from the opposite perspective, pointing out the unactualizable character of eternities. If the future is infinite, then its potential will never be exhausted and thus an eternal future violates the principle of plenitude: A potentiality which is never actualized is not a true potentiality at all. Philoponus also articulated another problem with the supposed eternity of the cosmos, namely, how could a finite body like the heavens continue eternally without an infinite force to keep it in motion? But a finite body cannot contain an infinite force, so its existence must be limited. These brief examples by no means represent the full extent of Philoponus's powerful anti-eternity polemics, but they do give particular insight into the way in which he turns the pagans' own philosophical resources against them.

Philoponus also rejected the existence of Aristotle's fifth element, the aether. As a Christian, he denied the heavens were divine or endowed with intelligence and argued that they were composed of the traditional terrestrial elements. The heavens manifestly have some terrestrial qualities—like the sun's heat—regardless of the protestations of pagans to the contrary. He dissented from Aristotle's problematic theory of forced motion, attempting instead to develop an impetus theory. Further he suggested that the heaven's circular motion was not the natural motion of the aether but a forced motion imparted to them by the Creator at the beginning of the world, just as the motion of the sphere of fire (directly below the Moon and in which meteorological phenomena occur) is forced, according to the pagans, because its circular motion is not natural but imparted to it by the celestial spheres. Consequently, because all forced motions eventually end, so too will the rotation of the heavens. He also defined place, again rejecting Aristotle, as the empty three-dimensional space which a body occupies. This theoretical void space never exists without body because of nature's *horror vacui*, but qua empty can be subjected to thought experiments about how it might behave if it were in fact void space.

Philoponus's reformed natural philosophy was deeply controversial. Simplicius, his pagan counterpart at the Athenian Neoplatonic Academy, was scathing in his response to Philoponus's *Contra Aristotelem.* Simplicius saw it not only as philosophically disingenuous (to be fair, some arguments are) but also as a betrayal of pagan religion and its sacred race. And some Christians, particularly those from the diphysite Nestorian sect such as Cosmas Indicopleutes, objected just as vociferously. Cosmas argued that Christians should reject paganism entirely, including the tradition of Greek philosophical rationalism. For him, Philoponus's main fault lay in his failure to refer to scripture as a guide to forming a Christian natural philosophy. Philoponus responded with his *De opificio mundi,* an exegesis of the narrative of creation and fierce pro-rationalist polemic against Cosmas' provincialism and ignorance. Despite his controversial role, he steadfastly considered himself a champion of reasonable Christianity (he entitled his proposed solution to the doctrinal controversies surrounding the Second Council of Constantinople [553] the *Arbiter*) and created a comprehensive and coherent, if ultimately unsuccessful, alternative to the dominant Aristotelian natural philosophy of Late Antiquity.

BIBLIOGRAPHY

WORKS BY JOHN PHILOPONUS

Ioannis Alexandrini cognomine Philoponi De usu astrolabii eiusque constructione libellus. Edited by Heinrich Hase. Bonnae: E. Weberi, 1839.

Ioannou Grammatikou Alexandreos (tou Philoponou) eis to proton tes Nikomachou Arithmetikes eisagoges. Edited by Richard Hoche. Lipsiae: Teubner,1864.

Carte archéologique de la Gaule. Vols. 13–17 (1887–1909).

De aeternitate mundi contra Proclum. Edited by H. Rabe. Lipsiae: Teubner, 1899.

Against Aristotle, on the Eternity of the World. Translated by Christian Wildberg. Ithaca, NY: Cornell University Press, 1987.

Ancient Commentators on Aristotle. Edited by Richard Sorabji. Ithaca, NY: Cornell University Press, and London: Duckworth, 1987.

De opificio mundi. Edited by Clemens Scholten. Frieburg, Germany: Herder, 1997.

John Philoponus and the Controversies over Chalcedon in the Sixth Century: A Study and Translation of the Arbiter. Translated by Uwe Michael Lang. Leuven: Peeters, 2001.

Against Proclus' "On the Eternity of the World." Translated by Michael Share. Ithaca, NY: Cornell University Press, 2005–2006.

OTHER SOURCES

De Haas, Frans. *John Philoponus' New Definition of Prime Matter.* Leiden, Germany: Brill, 1997.

"Ioannes Philoponus." In *Paul's Real-Encyclopadie der Classischen Altertumswissenschaft,* 9. Stuttgart: Drunckenmüller, 1764–1795.

McKenna, John. *The Setting in Life of the* Arbiter *of John Philoponus.* Eugene, OR: Wipf and Stock Publishers, 1998.

Pines, Shlomo. "An Arabic Summary of a Lost Work of John Philoponus." *Israel Oriental Studies* 2 (1979): 320–352.

Scholten, Clemens. *Antike Naturphilosophie und Christlich Kosmologie in der Schrift* De opificio mundi *des Johannes Philoponos.* Berlin: Gruyter, 1996.

Sorabji, Richard, ed. *Philoponus and the Rejection of Aristotelian Science.* London: Duckworth, 1987.

———. *Aristotle Transformed: The Ancient Commentators and Their Influence.* London: Duckworth, 1990.

Wildberg, Christian. *John Philoponus' Criticism of Aristotle's Theory of Aether.* Berlin: De Gruyter, 1998.

Carl Pearson

JOHN THE GRAMMARIAN

SEE **John Philoponus.**

JOHNSTON, JAMES FINLAY WEIR

(*b.* Paisley, Scotland, 13 September 1796; *d.* Durham, England, 18 September 1855), *chemistry, agriculture, teaching, and popularization.*

Johnston was a founder of the British Association for the Advancement of Science, compiling a report on isomerism and isomorphism for its 1837 meeting. He lectured at the newly founded Durham University, became an authority on applied chemistry, agriculture, and geology, and wrote very successful, widely translated popular works.

Johnston was the eldest son of James Johnston, a merchant of Kilmarnock. At Glasgow University, paying his way by tutoring, he won prizes and medals in both science and humanities. Graduating with an MA in 1826, he opened a school in Durham; in 1829 he married a Northumberland heiress, Susan Ridley, nineteen years older than him. That year he visited the Swedish chemist Jöns Jakob Berzelius, and Johnston convinced him that the solid substance paracyanogen (which Johnston had analyzed) had the same components in the same proportions as cyanogen. He went on to attend that year's annual meeting of German men of science. Much impressed by this open forum, he published an account of it; and with his friend and patron David Brewster he proclaimed the decline of science and the need to promote it, especially in the provinces. A founder of the British Association for the Advancement of Science (BAAS), which met first in York in 1831, he was never in its inner circle of "gentlemen of science."

In 1832 another visit to Berzelius's prestigious laboratory fueled his interest in the way atoms might be arranged in compounds. In 1837 he presented his commissioned report to the BAAS meeting at Newcastle-upon-Tyne, on the puzzling relationships between chemical composition, chemical and physical properties, and crystalline form. Some different substances, such as paracyanogen and cyanogen, had the same composition: they were called isomers. Some elements (like carbon or sulfur) or compounds existed in two different physical forms: they were dimorphic. Some compounds, like those of ammonium and potassium, formed exactly similar crystals: they were isomorphic, and often isodimorphic, each existing in two different forms. And yet ammonium (NH_4) is composed of five atoms, while potassium has a single atom in its corresponding compounds: so crystalline form and atomic constitution could not be connected in any simple way. Johnston's achievement was thus in clearly describing and classifying phenomena that still eluded satisfactory explanation.

In 1833 he was appointed reader in chemistry (at a salary of £50 a year, plus fees from students attending lectures; laboratory instruction was extra) at the newly founded and staunchly Anglican University of Durham, despite his belonging to the Church of Scotland. The little university, only the third to be chartered in England, was lucky to get a man with his reputation and connections. In the absence of a medical school in Durham, then the normal locus for chemistry, he strenuously promoted courses in engineering, involving practical work as well as advanced chemistry, geology, and mathematics. This pioneering venture, at first popular, did not ultimately succeed: engineering employers refused to recognize the new paper qualification as worth anything, making graduates begin on the shop floor and pay the full premium demanded of apprentices.

Holding forth at the Literary and Philosophical Society in Newcastle and at meetings of similar groups in the north of England, and in Mechanics' Institutes (which he avidly promoted), Johnston became a very successful public lecturer. Such activity enhanced his professional reputation, and in 1837 he was elected a Fellow of the Royal Society. In 1841 the Chemical Society of London (in which expatriate Scots were prominent) was set up, and Johnston served on its council from 1842 to 1845. The 1840s were the "hungry forties," and Johnston turned toward agricultural chemistry. He pressed in vain for agricultural teaching in the university, but established a course for trainee teachers at the diocesan institution in Durham. In 1842 he published *Elements of Agricultural Chemistry and Geology* (superseding Humphry Davy's *Elements of Agricultural Chemistry*, 1813); its nineteenth edition was published in 1895. Johnston's brief *Catechism of Agricultural Chemistry and Geology* (1844) went through over thirty editions in his lifetime and was recommended by Leo Tolstoy, among others. Johnston's English style was agreeable, and his books were also widely successful in translation; in Germany and Scandinavia he thus became known for his accessible writings, his fame being second only to that of Justus von Liebig. Believing that the Dutch chemist G. J. Mulder had been wronged in a dispute with Liebig, he provided an introduction and notes for the translation of Mulder's *Chemistry of Vegetable and Animal Physiology* (1845). In 1849–1850 Johnston visited North America, where he had family, and in 1851 published *Notes on North America*, much concerned with agriculture (American wheat, after all, was a possible answer to European hunger). The book was noted for its lack of prejudice and exaggeration; visitors from Britain were customarily snobbish about democratic culture, but Johnston was not.

Johnston then settled down to write *The Chemistry of Common Life*, dedicated to Brewster and published in 1855, which became a classic of popular science writing. Beginning with food and drink, and going on to poisons and smells, it led the reader into thinking chemically, and it was up to date. It sold well, kept chemistry in the public eye, and can still be read with pleasure—evoking a time when chemistry was the fundamental, exciting, and popular science, and fertilizers and explosives were unquestionably benevolent. Johnston skillfully presented the science to anyone interested in how the world works; the book became in effect his memorial, because by the time it was published he had died from a lung infection caught abroad.

He had kept up Scottish connections, and had become professor to the Highland and Agricultural Society there. His volumes of the Royal Society's *Philosophical Transactions*, duly marked "Professor Johnston" in a clerk's hand, remained unopened and unbound; the run (now in Durham) was bequeathed to New College, Edinburgh, indicating that at the Disruption of 1843 his sympathies were with Thomas Chalmers, Hugh Miller, and the Free Church.

Johnston felt strongly about sanitary reform, Durham being a dirty little city in urgent need of cleaning up in cholera epidemics. He saw urban churchyards as a major source of infection; and was duly interred at Croxdale, a village south of Durham. He bequeathed the residue of his estate for educational purposes; these included the Johnston Laboratory in Newcastle when in 1870 Durham University established a College of Science there, and in 1899 the coeducational Johnston School in Durham, now a comprehensive high school. Capable of research at the frontier of science, Johnston was most successful when communicating interest in and enthusiasm for useful knowledge.

BIBLIOGRAPHY

The Royal Society of London has articles and letters; the National Library of Scotland has correspondence with Blackwood, his publisher.

WORKS BY JOHNSTON

The Economy of a Coal-field: An Exposition of the Objects of the Geological and Polytechnic Society of the West Riding of Yorkshire, and of the Best Means of Attaining Them. Durham, U.K.: Andrews, 1838.

Elements of Agricultural Chemistry and Geology. Edinburgh: Blackwood, 1842.

Catechism of Agricultural Chemistry and Geology. Edinburgh: Blackwood, 1844.

Lectures on Agricultural Chemistry and Geology. Edinburgh: Blackwood, 1844.

With T. Watkin. *Report on the Sanitary Condition of the City of Durham,* Durham, U.K.: Durham City Sanitary Association, 1847.

Contributions to Scientific Agriculture. Edinburgh: Blackwood, 1849.

Experimental Agriculture: Being the Results of Past, and Suggestions for Future Experiments in Scientific and Practical Agriculture. Edinburgh: Blackwood, 1849.

Notes on North America, Agricultural, Economical and Social. Edinburgh: Blackwood, 1851.

The Chemistry of Common Life. Edinburgh: Blackwood, 1855.

OTHER SOURCES

Anonymous obituary in *Quarterly Journal of the Chemical Society* 9 (1856): 157–159.

Knight, David. *The Transcendental Part of Chemistry.* Folkestone, U.K.: Dawson, 1978.

Lundgren, Anders, and Bernadette Bensaude-Vincent, eds. *Communicating Chemistry: Textbooks and Their Audiences.* Canton, MA: Science History Publications, 2000. Standard work.

Morrell, Jack, and Arnold Thackray. *Gentlemen of Science: Early Years of the British Association for the Advancement of Science.* Oxford: Oxford University Press, 1981. The classic work on the early BAAS.

———, and Arnold Thackeray, eds. *Gentlemen of Science: Early Correspondence of the British Association for the Advancement of Science.* Camden, 4th series London: Royal Historical Society, 1984. Includes letters.

Preece, Clive. "The Durham Engineer Students of 1838." *Transactions of the Architectural and Archaeological Society of Durham and Northumberland,* n.s. 6 (1982): 71–74.

David Knight

JORDAN, ERNST PASCUAL (*b.* Hannover, Germany, 18 October 1902; *d.* Hamburg, Germany, 31 July 1980), *theoretical physics, biophysics, cosmology.* For the original article on Jordan see *DSB,* vol. 17, Supplement II.

Pascual Jordan has remained an intriguing figure for historians of physics and for historians of the political relations of science in twentieth-century Germany. In studies of Jordan's scientific work, he frequently appears as a physicist gifted in generating provocative, novel approaches, not always fully appreciated at the time but perhaps even more impressive in retrospect. Reaching a definitive interpretation of Jordan's controversial political stances, especially vis-à-vis national socialism, remains problematic. Studies since 1990 have offered depth and complexity to scholars' picture of Jordan's relationship to Nazi ideology; they have also highlighted continuities between his actions during the National Socialist period and his radically conservative thought before 1933 and after 1945.

Jordan as Quantum Theorist. New assessments of Jordan's role in the development of quantum theory have arisen from particular details about his interactions with contemporaries and from reinterpretations of the history of quantum theory as a whole. The prior consensus was that in constructing matrix mechanics and transformation theory, Jordan was largely responsible for the mathematical structures of quantum theory; several historians, however, also ascribe to him a more conceptually central role than that of a mathematical technician. Mara Beller, for example, in arguing that the "Copenhagen interpretation" of quantum theory originated from a flux of dialogues among various theorists, calls attention to Jordan's role in formulating the indeterminacy principle. In her reading, Werner Heisenberg's famous paper on the subject, "Über den anschaulichen Inhalt der quantentheoretischen Kinematik und Mechanik," published in 1927, was a direct response to Jordan's *habilitation* lecture of earlier that year, "Philosophical Foundations of Quantum Theory." Jordan's lecture asserted that quantum theory was still incomplete, because one could imagine apparently deterministic experimental situations (e.g., tracing the trajectory of a particle in a cloud chamber) that were not described by the quantum formalism. The theory was statistical but had not yet been reduced to independent, elementary probabilities. Heisenberg was thereby inspired, according to Beller, to focus on the problem of measurement: necessarily incomplete knowledge of initial conditions produced quantum indeterminacy (Beller, 1999).These new emphases on the limitations of measurement and on essential quantum indeterminacy became fundamental to Jordan's own work after 1927.

Although Jordan's place in the development of matrix mechanics and transformation theory has received the most attention, Jordan himself considered his work on the "second quantization" of fields to be "his most important contribution to theoretical physics" (Schweber, 1994, p. 33). Jordan's role was, once again, highly dialogic. A series of papers—partly by himself and partly in collaboration with Oskar Klein, Wolfgang Pauli, and Eugene Wigner, inspired other physicists, notably Heisenberg and Paul Dirac—to examine the possibilities for field quantization more closely, albeit sometimes with skepticism toward Jordan's approach.

Jordan as Biophysicist. In the 1930s Jordan turned his attention increasingly to biophysics; indeed, based on his plans for the immediate postwar period, this was arguably his main field of interest as of 1945. Jordan's "amplifier theory," introduced in 1932, was an attempt to apply Niels Bohr's complementarity idea to biology; in turn, Bohr's subsequent statements on biology, as well as the initial work in this direction by Max Delbrück, can be understood as rejoinders to Jordan. Whereas Bohr's application of the complementarity idea to biology was primarily analogical, Jordan attempted to locate actual quantum phenomena in organic reactions. Meanwhile, biologists such as Erwin Bünning and Max Hartmann and biochemists such as Otto Meyerhof objected that Jordan ignored the causal methodologies of the life sciences, while Jordan's fellow positivists such as Edgar Zilsel and Otto Neurath attacked his ideas as thinly veiled neovitalism.

Undeterred, Jordan's biophysics took on an increasingly technical character from 1936 onward. He primarily deployed the "target theory," which had been developed since the early 1920s by radiation biologists such as Friedrich Dessauer in Germany, Fernand Holweck in France, and J. A. Crowther in Great Britain. Target theory sought to identify submicroscopic features through a statistical analysis of the effects of radiation on the organism. A target-theoretical analysis of x-ray mutagenesis formed the basis of a 1935 paper by Nikolai W.

Timoféeff-Ressovsky, Karl G. Zimmer, and Delbrück, which proposed that genes were single (large) molecules. For Jordan, this signally confirmed the amplifier theory: genetic phenomena such as mutations were essentially quantum-physical in character and therefore indeterministic. Jordan sought further confirmation of the amplifier concept in immunology, protein chemistry, sense physiology, and various psychoanalytical and parapsychological phenomena. He also worked on refining the statistical methodology of target theory—for example, by mathematically accounting for the "saturation" effect of densely ionizing radiation. (From 1938 onward Zimmer and Timoféeff-Ressovsky experimented with neutron radiation to test these theoretical refinements.) Once again Jordan's approach had its detractors; Timoféeff-Ressovsky and Zimmer, for example, while appreciating Jordan's more technical contributions, were generally reserved about any broader philosophical implications.

After World War II Jordan was in negotiation with the erstwhile Prussian Academy of Sciences, which had taken control of a number of research facilities in the Soviet Zone, including Timoféeff-Ressovsky's at the Kaiser Wilhelm Institute for Brain Research in the Berlin suburb of Buch. Timoféeff-Ressovsky and Zimmer having both been taken to the Soviet Union, the institute needed new leadership. The academy envisioned making it into a center for biophysics and contemplated appointing Jordan as its director or rather (once this appeared barred by political complications) as its leading scientific member, with target-theoretical radiation biology as a major research emphasis. Jordan remained in West Germany, however, after several years of career uncertainty. Nevertheless, the ambitious scope of the Buch plans indicates the importance Jordan attached to his biophysical work as of the late 1940s, as well as the seriousness with which it was regarded by (at least some) of his German colleagues.

Jordan as Mathematician and Cosmologist. Jordan's main scientific activity thereafter moved into two other fields, both of which, however, had roots in earlier work. One was algebraic theory, especially the theory of semigroups and the theory of nonassociative algebras—an interest that can be traced back to his introduction of matrices into quantum physics. Many of Jordan's publications in this field appeared in the proceedings of the Academy of Sciences and Literature in Mainz, of which Jordan was the founding vice president (1949) and later president (1963–1967).

The second field was cosmology. Jordan made Hamburg into the leading German center for general relativity theory and its cosmological applications. Once again his approach was idiosyncratic: in the late 1930s Jordan had become interested in a suggestion by Dirac that the supposed gravitational "constant" was, rather, decreasing over time. Working out the consequences of this hypothesis, Jordan developed a cosmological model in which the universe expanded from a single atomic starting point. Matter was created continuously over time, with embryonic stars appearing in sudden, explosive bursts. Applying the idea to geophysics, Jordan concluded that Earth was expanding, as manifest in phenomena such as rift valleys. Apart from this specific hypothesis, Jordan (in collaboration with a series of students) also undertook a broader program of research into the theory of the gravitational field.

Years later Jordan's cosmology would be commonly cited as an analogue to the 1961 theory of Carl Brans and Robert Dicke—originally articulated independently of Jordan's work—of a varying scalar field that would account for objects' mass. The "Jordan-Brans-Dicke field" has become one of the central theoretical concepts of particle cosmology, a burgeoning subfield of physics since the 1970s.

Jordan as Political Figure and Public Intellectual. Doubtless the most controversial aspect of Jordan's career remains his affiliation with national socialism. Jordan's joining the Nazi Party and the SA (*Sturmabteilung*, or storm troopers) in 1933 was not in itself unusual among German scientists; however, a series of publications suggested that Jordan's affinity for the Nazi cause was more than nominal. Articles in cultural-political journals argued that academia had to reorient itself toward service to the new state following the political revolution. *Physikalisches Denken in der neuen Zeit* (1935) pointed out ways in which modern physics contribute to building a strong military—for example, atomic energy. In several articles and, at greater length in *Physik und das Geheimnis des organischen Lebens* (1941), Jordan analogized between "steering centers" in the organism, such as genes, and the dictatorial state; parliamentary democracy was, conversely, the analogue of lifeless inorganic matter.

After 1945 Jordan argued that he had joined the Nazi Party out of a sense of career pressure or obligation to work for moderation within the Nazi power system. Jordan did campaign actively against "Aryan physicists" who saw much of modern physics as Jewish influenced and hence undesirable. *Anschauliche Quantentheorie* (1936) was in part a sustained argument that quantum physics was not mere mathematical formalism of the sort deplored by the Aryan physicists but was, rather, based solidly on observational encounters with nature. Likewise, Jordan engaged in a running literary battle in defense of his positivistic interpretation of quantum theory against Hugo Dingler, a philosopher associated with the Aryan physicists.

Jordan

However, in historical perspective, the claim that opposition to Aryan physics in itself constituted opposition to Nazism is difficult to sustain. Despite post-1945 self-representations, the mainstream physics community also made its accommodations with the Nazi regime. The circumstances behind Jordan's receipt of the 1942 Planck Medal, for example, indicate that his colleagues were well aware of the symbolic power of having a scientist who was conspicuously friendly to the regime as the bearer of the German Physical Society's top honor.

Moreover, Jordan's political interests were not unique to the Nazi period and so cannot be dismissed as mere opportunism. As early as 1930 Jordan (under a pseudonym) was publishing conservative-nationalist journal articles; he continued to do so under his own name after 1933. (On the connection to the "Domeier" articles see Beyler, 1994, pp. 207–224.) After World War II, Jordan's activism for right-wing causes if anything increased. He thus may not have been committed to all aspects of Nazi ideology but apparently found in Nazism some resonances with a lifelong political philosophy: opposition to communism and suspicion of liberalism as manifestations of materialism; enthusiasm for technology; and elitism in his understanding of history and society.

In the postwar period Jordan's activity as a public intellectual and popularizer of science flourished in the form of hundreds of books, articles, and public lectures. Above all, he sought to convey the message that modern science, and the positivist philosophy he associated with it, had brought about the end of materialism. This meant, in turn, a denial of materialism's denial of the possibility of religious faith. *Der Naturwissenschaftler vor der religiösen Frage: Abbruch einer Mauer* (1963 and subsequent editions) described the breaching of the "wall" that materialism had erected between religiosity and science.

The wall metaphor had obvious political overtones. According to Jordan, the overthrow of philosophical materialism also meant its demise as a political philosophy, and he spoke and wrote vociferously against communism and socialism, presented as misguided attempts to impose philosophical dogmas on society, and in favor of conservatism, seen as an attachment to empirically grounded realities. This ideological commitment was combined with a frank endorsement of a strong military posture in the West. In *Der gescheiterte Aufstand* (1956), for example, Jordan portrayed a future nuclear conflict as nearly inevitable but insisted that its worst effects could be overcome by the preparation of underground cities and similar measures. The following year, an imbroglio erupted in the wake of Chancellor Konrad Adenauer's suggestion that the German army might become equipped with nuclear weapons. The "Göttingen Eighteen"—a group of atomic physicists including Max Born and Heisenberg—issued a manifesto stating their refusal to undertake weapons research. Jordan publicly criticized his colleagues as, at best, politically naive. Results included public rejoinders and, privately, a bitter exchange of letters between Jordan and Max and Hedwig Born. Largely as a result of his advocacy for nuclear armament, Jordan was put forward as a Christian Democratic candidate for the German federal parliament in the 1957 election and won a seat that he held until 1961. This brief spell was the extent of Jordan's activity as a politician in the strict sense, but in many publications and lectures until his death in 1980 he continued to promote various conservative positions in politics and theology.

SUPPLEMENTARY BIBLIOGRAPHY

WORKS BY JORDAN

"Philosophical Foundations of Quantum Theory." *Nature* 119 (1927): 566–569, 779. Originally published as "Kausalität und Statistik in der modernen Physik." *Naturwissenschaften* 15 (1927): 105–110.

Physikalisches Denken in der neuen Zeit. Hamburg, Germany: Hanseatische Verlagsanstalt, 1935.

Anschauliche Quantentheorie: Eine Einführung in die moderne Auffassung der Quantenerscheinungen. Berlin: J. Springer, 1936.

Die Physik und das Geheimnis des organischen Lebens. Braunschweig, Germany: F. Vieweg, 1941.

Der gescheiterte Aufstand: Betrachtungen zur Gegenwart. Frankfurt, Germany: Vittorio Klostermann, 1956.

Der Naturwissenschaftler vor der religiösen Frage: Abbruch einer Mauer. Oldenburg, Germany: G. Stalling, 1963.

OTHER SOURCES

Overviews

Beyler, Richard H. "From Positivism to Organicism: Pascual Jordan's Interpretations of Modern Physics in Cultural Context." PhD diss., Harvard University, 1994. Surveys Jordan's career, concentrating particularly on his biophysical work and postwar popularizations of science.

Ehlers, Jürgen, and Engelbert Schücking. "'Aber Jordan war der Erste': Zur Erinnerung an Pascual Jordan." *Physik Journal* 11 (November 2002): 71–74. Memoir by two of Jordan's former students.

Ehlers, Jürgen, Dieter Hoffmann, and Jürgen Renn, eds. *Pascual Jordan (1902–1980): Mainzer Symposium 100.* Max-Planck Institut für Wissenschaftgeschichte preprint no. 329. Berlin, 2007. Includes memoirs and scientific studies by former students of Jordan, essays by historians, and a bibliography.

Hoffmann, Dieter, ed. *Pascual Jordan (1902–1980): Symposium zum 100. Geburtstag des Physikers.* Max-Planck-Institut für Wissenschaftsgeschichte preprint. Berlin, forthcoming 2007. Includes memoirs and scientific studies by former students of Jordan, studies by historians of physics, and an extensive bibliography.

Schücking, Engelbert L. "Jordan, Pauli, Politics, Brecht, and a Variable Gravitational Constant." *Physics Today* 52 (October 1999): 26–31. Memoir by a former student; provides insight on ideology, activity as popularizer of science, and cosmological theories.

Jordan and Quantum Physics

Beller, Mara. *Quantum Dialogue: The Making of a Revolution.* Chicago: University of Chicago Press, 1999. Attempts a historically contingent interpretation of the rise of the "Copenhagen interpretation" of quantum theory; considers Jordan in dialogue with other figures.

Darrigol, Olivier. *From c-Numbers to q-Numbers: The Classical Analogy in the History of Quantum Theory.* Berkeley: University of California Press, 1992. Examines the use of correspondence principles in the development of quantum theory; considers role of Jordan among other figures.

Greenspan, Nancy Thorndike. *The End of the Certain World: The Life and Science of Max Born.* New York: Basic Books, 2004. Includes discussion of Jordan's interactions with Born as a student and during the nuclear weapons controversy of the 1950s.

Schweber, Silvan S. *QED and the Men Who Made It: Dyson, Feynman, Schwinger, and Tomonaga.* Princeton, NJ: Princeton University Press, 1994. Includes a section on Jordan as progenitor of quantum field theory.

Jordan's Work in Other Scientific Fields

Aaserud, Finn. *Redirecting Science: Niels Bohr, Philanthropy, and the Rise of Nuclear Physics.* Cambridge, U.K., and New York: Cambridge University Press, 1990. Describes the origin of Jordan's biological theories and disagreements with Bohr thereon.

Beyler, Richard H. "Targeting the Organism: The Scientific and Cultural Context of Pascual Jordan's Quantum Biology, 1932–1947." *Isis* 87 (1996): 248–273. Analyzes Jordan's contributions to target theory in the context of his ideological stances before, during, and after the National Socialist era.

———. "Evolution als Problem für Quantenphysiker," translated by Rainer Brömer. In *Evolutionsbiologie von Darwin bis heute,* edited by Rainer Brömer, Uwe Hossfeld, and Nicolaas Rupke, 137–160. Berlin: VNB, 1999. Examines Jordan's biological theories in contrast to competing views presented by Erwin Schrödinger.

Kragh, Helge. *Cosmology and Controversy: The Historical Development of Two Theories of the Universe.* Princeton, NJ: Princeton University Press, 1994. Includes discussion of Jordan's cosmological models.

Jordan's Political and Philosophical Commitments

Beyler, Richard H. "The Demon of Technology, Mass Society, and Atomic Physics in West Germany, 1945–1957." *History and Technology* 19 (2003): 227–239. Discusses pessimism about and enthusiasm for technology, in particular nuclear technology, in postwar Germany, with Jordan as major example.

———. "Pascual Jordan: Freedom vs. Materialism." In *Eminent Lives in Twentieth-Century Science and Religion,* edited by Nicolaas A. Rupke, 157–176. Frankfurt, Germany: Peter Lang, 2007. Considers Jordan's religious views in the context of Cold War Germany.

———, Michael Eckert, and Dieter Hoffmann. "Die Planck-Medaille." In *Physiker zwischen Autonomie und Anpassung: Die Deutsche Physikalische Gesellschaft im Dritten Reich,* 217–235. Weinheim, Germany: Wiley-VCH, 2006. Analyzes the politics behind the granting of the Planck Medal of the German Physical Society, including Jordan's award in 1942.

Danneberg, Lutz. "Logischer Empirismus in Deutschland." In *Wien-Berlin-Prag: Der Aufstieg der wissenschaftlichen Philosophie,* edited by Rudolf Haller and Friedrich Stadler, 320–361. Vienna: Hölder-Pichler-Tempsky, 1993. Surveys controversies over positivism in Germany in the 1920s and 1930s, including Jordan's involvement.

Hentschel, Klaus, ed. *Physics and National Socialism.* Basel, Switzerland: Birkhäuser, 1996. Extensive documentation, background, and interpretation of primary texts, with Jordan among many other authors.

Hoffmann, Dieter. *Pascual Jordan im Dritten Reich: Schlaglichter.* Max-Planck-Institut für Wissenschaftsgeschichte preprint no. 248. Berlin, 2003. Includes reproductions of several articles and letters by Jordan from the National Socialist era.

Schirrmacher, Arne. *Dreier Männer Arbeit in der frühen Bundesrepublik: Max Born, Werner Heisenberg and Pascual Jordan als politische Grenzgänger.* Max-Planck-Institut für Wissenschaftsgeschichte preprint no. 296. Berlin, 2005. Examines Jordan as public and political figure in the 1950s, in comparison with Born and Heisenberg.

Wise, M. Norton. "Pascual Jordan: Quantum Mechanics, Psychology, and National Socialism." In *Science, Technology, and National Socialism,* edited by Monika Renneberg and Mark Walker, 224–254. Cambridge, U.K., and New York: Cambridge University Press, 1994. Seminal article analyzing Jordan's political thought in juxtaposition to his psychological and biological theories.

Richard H. Beyler

JØRGENSEN, AXEL CHRISTIAN KLIXBÜLL (*b.* Ålborg [Aalborg], Jutland, Denmark, 18 April 1931; *d.* Paris, France, 9 January 2001), *inorganic chemistry, coordination chemistry, physical chemistry, spectroscopy.*

Jørgensen was a major contributor to the renaissance of transition metal and *f*-element chemistry that took place after World War II. The theory of bonding in transition metal chemistry in 1950 was elementary and unsatisfactory. Linus Pauling had tried to adapt his hybridization theory, which had worked so well for main group compounds and had explained the change from paramagnetism to diamagnetism in hemoglobin upon oxygenation. However, one of the most striking features of transition metal chemistry, the varying colors of the compounds, was not really understood. Jørgensen's first major contribution was to carry out a wide survey of the electronic (ultraviolet and visible) spectra of complexes and to establish the general applicability of ligand field theory to this subject. His early work in Copenhagen was mainly concerned with *d-d* and *f-f* transitions, and in a later period in Geneva he turned to a systematic study of electron transfer bands. In 1965 he proposed the angular overlap approach to calculating *d*-orbital splittings, which remains the most useful theory to this day. The great volume of experimental data that he produced and, perhaps more importantly, organized, was of vital importance for the theoretical work of Leslie E. Orgel, John S. Griffith, and others. The understanding of the electronic structure of transition metal compounds that resulted from the close collaboration between Jørgensen and the theoreticians was essential for the development of synthetic inorganic chemistry by chemists such as Joseph Chatt, F. Albert Cotton, Sir Ronald S. Nyholm, and Sir Geoffrey Wilkinson and the mechanistic studies of Fred Basolo, Henry Taube, and Harry B. Gray.

Early Years. Axel Christian Klixbüll Jørgensen (Axel to his parents; Christian or Klixbüll to his friends) was the son of Sven Jørgensen, an officer on a Danish training ship, and his wife Ingrid (née Sørensen). When Jørgensen was one year old, the family moved to Copenhagen, where he was raised and educated. Because the Danish training ship was at sea at the time of the German invasion of Denmark on 9 April 1940, Christian grew up without his father's presence for more than five years. After the Japanese attack on Pearl Harbor the vessel served as a training ship for the U.S. Coast Guard with Christian's father as its engineer.

Education. Jørgensen was a child prodigy and an avid reader of books. By his teens he had already shown an interest in two of his subjects of lifelong study—spectroscopy and the lanthanides or rare earths (the inner transition elements with atomic numbers 57 to 71). Jørgensen attended the Vestre Borgerdydskole gymnasium (high school) from 1947 to 1950, receiving his *Abitur* degree. In the autumn of 1950 he matriculated at the University of Copenhagen to pursue a *Candidatus magisterii* degree so that he could teach at a secondary school. The curriculum consisted of astronomy, chemistry, mathematics, and physics and was almost identical for those who intended to major in each of these four fields. He studied with Jannik Bjerrum, one of the world's leading authorities on the stabilities of complex ions. In 1954 Jørgensen received his *Candidatus magisterii* degree in chemistry, mathematics, astronomy, and physics, with chemistry as his main subject, and he decided to continue his research on optical spectra of transition metal ions, then a neglected topic, in Bjerrum's laboratory, one of the world's leading centers for coordination chemistry.

The Copenhagen Period. While preparing his doctorate on electronic spectroscopy Jørgensen served as an instructor in Chemistry Department A (Inorganic Chemistry) of the Technical University of Denmark from 1953 to 1958. Although students had learned the color of transition metal compounds for years, their origins were not well understood in 1954. The physicists Hans A. Bethe and John H. Van Vleck had introduced the ideas of crystal field theory in the 1930s, mainly to describe the magnetic properties of transition metal compounds.

The electrons in an atom are distributed in orbitals, of four types (*s, p, d,* and *f*). The periodic table of the elements corresponds to the progressive filling of the different types of orbitals with an increasing number of electrons. The transition metals are the family of elements in which the *d* orbitals are partially occupied. Similarly the lanthanides and actinides form the family in which the *f* orbitals are partially occupied. The partial occupation of these orbitals confers a number of interesting physical and chemical properties on the complexes.

The metallic elements (i.e., those which form metals in the pure state) are usually found as positively charged ions (or cations) in their compounds. These positive ions, which are formed by the loss of one or more electrons, attract negatively charged ions (or anions) or the negative parts of a neutral molecule (such as the oxygen atom of water), to form a complex. The species, which are bound to the cation, are called ligands. The ligands modify the behavior of the cation to which they are bound, and the properties of the resulting complex may be controlled by a suitable combination of metal cation and ligands.

The principle of the theory is that the *d*-orbitals in a transition metal complex are unchanged except by the electrostatic field of the surrounding ligands, which leads to a loss in degeneracy or "splitting" of the *d*-orbitals. It had been applied to the electronic spectrum of the titanium(III) ion by Friedrich Ernst Ilse and Hermann Hartmann in 1951, but it remained largely neglected in chemistry. Ryutaro Tsuchida had proposed the existence of the spectrochemical series to explain the systematic shift in the absorption bands of complexes in 1938, but this also had had little impact on inorganic chemistry. At the same time as Jørgensen was carrying out his practical measurements, the theoreticians Leslie E. Orgel, and Yukito Tanabe and Satoru Sugano were developing the treatment of the crystal field splitting combined with electronic repulsion inside the *d*-orbital subshell.

Jørgensen undertook a broad survey of the spectra of inorganic complexes. Apart from the readily accessible *3d* complexes, he studied the *4d* and *5d* series and the lanthanide and actinide elements, as well as the influence on the ligand field splitting observed in the spectra of *d*-metal complexes of the ligands, and of the metal oxidation state. His doctoral thesis, *Energy Levels of Complexes and Gaseous Ions* (1957), summarizes his work, which had already appeared in more than thirty papers, mainly with Jørgensen as the sole author. The thesis treats the energy levels of trivalent lanthanides, shows that the spectra of actinide complexes confirm Glenn T. Seaborg's actinide hypothesis that the elements following actinium showed partial occupation of *f*-orbitals (it is perhaps worth pointing out that uranium and thorium were generally considered to be *6d* elements during the first half of the twentieth century), discusses the values of the electron repulsion parameters, and in particular shows the decrease of these parameters upon complexation, for which he coined (with Claus E. Schäffer) the expression *nephelauxetic effect.* From his survey of *d*-element complexes he established the systematics of the ligand field splitting and the application of Orgel and Tanabe-Sugano diagrams to interpret the spectra. From this and the nephelauxetic effect he argued that complex formation must involve an element of covalency, contrary to the essentially ionic model of the crystal field theory.

This period of his career was remarkable not only for his own productivity but also for his encyclopedic knowledge of the literature. This gave him a unique vision of the experimental data and allowed him to participate actively in the theoretical debate. He was a central figure in the establishment of ligand field theory as essential to inorganic chemistry.

From Copenhagen to Geneva. On 26 June 1957, shortly after finishing his doctorate, Jørgensen married Micheline Prouvez. The couple had a son, Philippe (born in 1963), and a daughter, Estelle (born in 1969). Micheline, who had been recovering from breast cancer, died of pleurisy on 3 November 1978 in Paris, and Jørgensen never remarried. Jørgensen became director of the Office of Science Adviser to the North Atlantic Treaty Organization (NATO) in Paris (1959–1960), where he helped to organize the summer schools that later became the NATO workshops and Advanced Study Institutes. In 1961 he became director of the Theoretical Inorganic Chemistry group at the Cyanamid European Research Institute (CERI), housed in an old mansion in Cologny, facing Lac Léman (Lake Geneva) near Geneva, Switzerland, where he worked from 1961 to 1968. Many of the workers here were eminent scientists who later occupied professorial chairs throughout the world. His tenure here was a second highly productive period in his career.

Jørgensen's work on absorption spectra continued during this period. He published extensively on electron transfer bands of transition metal and *f*-shell complexes, and in 1964 he reported the hypersensitive pseudo-quadrupole transitions of lanthanide complexes. In 1965 with Schäffer he proposed the angular overlap model to calculate the splitting of *d*-orbitals. This model, which has stood the test of time, considers the *d*-orbitals as acquiring antibonding character proportional to the square of the overlap integral between ligand and *d*-orbital. Somewhat paradoxically, it was initially conceived for lanthanide complexes but was found to be generally applicable to *d*-metal complexes.

The University of Geneva. In 1969 CERI was closed, and Jørgensen became invited professor (1969–1970) and professor (1969–1974) in the Department of Physical Chemistry at the University of Geneva, where he then served as professor in the Department of Inorganic, Applied, and Analytical Chemistry from 1974 until his retirement in 1997. This move was a watershed in his career. He abandoned experimental work on electronic spectroscopy, although he continued to act as an external consultant with other groups. He never built up a large research group. At the beginning of his time at the university he studied x-ray photoelectron spectroscopy, but the technique failed to become the instrument of investigation of chemical bonding that electronic spectroscopy had been earlier. He was the only inorganic chemist in the department and undoubtedly missed the active research atmosphere that he had known in Copenhagen and Cologny.

Jørgensen's major contribution to the literature in the final half of his career was the study of lanthanide spectra in solid-state matrices carried out with Renata Reisfeld of the Hebrew University of Jerusalem, who performed the experimental work. Apart from this work he published many general papers on chemical bonding, including some rather speculative articles on superheavy elements and the chemistry of quarks. Although more than half of his four hundred publications appeared during this period, they did not have the general impact of his earlier work.

On 29 April 1983 Jørgensen was awarded an honorary doctorate from the Philosophical Faculty II of the Universität Zürich "as a pioneer in the bridging of quantum mechanics and chemistry, which has explained in a prominent manner the electronic spectra of metal complexes and has proved fruitful for chemistry."

In addition to his membership in such scientific societies as the Royal Danish Academy of Sciences and Letters (Det Kongelige Danske Videnskabernes Selskab), the Chemical Society (London), and the Société de Chimie

Physique (Paris), Jørgensen belonged to nonscientific organizations such as the Association for Symbolic Logic, the International Banknote Society, and the Explorers Club. Although raised as a Lutheran, in 1959 he converted to Roman Catholicism. However, philosophically he felt close to the Eastern Orthodox churches. His hobbies included astrophysics, formal logic, science fiction, and banknote collection.

In 1997 Jørgensen began to suffer from frequent lapses of memory and exhibited erratic behavior. After a bookcase fell on him, he was hospitalized and diagnosed with dementia, forcing him to retire in 1997. He moved to France to be close to his children, and he died in a nursing home near Paris. Considering the major role that intellectual pursuits played throughout his entire life, his final years were especially poignant to those who knew or worked with him.

Contribution and Character. Although Jørgensen is often regarded as a theoretical chemist, his major contribution was undoubtedly experimental. His experimental work on electronic spectra was essential to validate or criticize the theories of transition metal compounds. He had an encyclopedic memory and could recall references or misprints in papers after twenty years or more. A second great strength was in organizing this material. For example, he showed that the ligand field splitting, generally designated as Δ, could be expressed as the product of two factors f and g, the first dependent on the metal and the second on the ligand. The analysis of the variation of these factors was essential to understanding the bonding. He developed similar schemes, in which experimental data could be reproduced from a limited number of metal and ligand parameters, for the nephelauxetic effect, and the frequency of electron transfer bands, where he introduced the notion of optical electronegativity. He equally analyzed the variation in electron repulsion and spin-pairing energies in d and f orbitals. He possessed a remarkable capacity for bringing together observations from apparently unrelated areas.

Jørgensen was an amiable, unassuming, and witty man with no trace of guile or self-aggrandizement. His reading was extremely broad, and he could talk for hours on a wide variety of subjects. Although he loved discussion, he did not actively seek conversation, least of all "small talk," and had few close friends. He spoke several languages, albeit with a strong Danish accent, and language was important to him. He liked to introduce new terms such as the *nephelauxetic effect, non-innocent ligands, optical electronegativity,* and *inorganic symbiosis,* although all were not always widely accepted. His style of writing was unique, generally directed to the reader who was at least partially familiar with the subject. He delighted in presenting subjects from a different viewpoint, as evidenced by his remark that phlogiston could be equated with the electron: oxidation corresponds to a loss of phlogiston (electrons); reduction corresponds to a gain of phlogiston (electrons). This style gives his writings at best a somewhat whimsical air but at worst makes his papers rather inaccessible. Ultimately, he suffered from this style, since the contributions that he made are no longer associated with him. All those working in transition metal chemistry or lanthanide and actinide chemistry owe a debt to the pioneering studies made by Jørgensen in the 1950s and 1960s.

BIBLIOGRAPHY

WORKS BY JØRGENSEN

Energy Levels of Complexes and Gaseous Ions. Copenhagen: Gjellerup, 1957.

With Claus E. Schäffer. "The Nephelauxetic Series of Ligands Corresponding to Increasing Tendency of Partly Covalent Bonding." *Journal of Inorganic & Nuclear Chemistry* 8 (1958): 143–148.

Absorption Spectra and Chemical Bonding in Complexes. Oxford: Pergamon Press, 1962.

With Claus E. Schäffer. "The Angular Overlap Model, an Attempt to Revive the Ligand Field Approaches." *Molecular Physics* 9 (1965): 401–412.

"Recent Progress in Ligand Field Theory." *Structure and Bonding* 1 (1966): 3–31. The first of Jørgensen's numerous articles in this journal.

"Electric Polarizability, Innocent Ligands, and Spectroscopic Oxidation States." *Structure and Bonding* 1 (1966): 234–248.

Modern Aspects of Ligand Field Theory. Amsterdam: North Holland, 1971.

With Renata Reisfeld. *Lasers and Excited States of Rare Earths.* Berlin: Springer-Verlag, 1977.

With William H. Brock, Kai A. Jensen, and George B. Kauffman. "The Origin and Dissemination of the Term 'Ligand' in Chemistry." *Ambix* 28 (1981): 171–183.

"Coordination Based on Known Free Ligands, Moderate Dissociation Rate, Weaker Electron Affinity of Central Atom Than Ionization Energy of Ligand, and Quantum Paradoxes." In *Coordination Chemistry: A Century of Progress,* edited by George B. Kauffman, 226–239. Washington, DC: American Chemical Society, 1994.

With Renata Reisfeld. "Coordination Compounds of Metal Ions in Sol-Gel Glasses." In *Coordination Chemistry: A Century of Progress,* edited by George B. Kauffman, 439–443. Washington, DC: American Chemical Society, 1994.

OTHER SOURCES

Basolo, Fred. *From Coello to Inorganic Chemistry: A Lifetime of Reactions.* New York: Kluwer Academic/Plenum, 2002.

Bjerrum, Jannik. *Metal Ammine Formation in Aqueous Solution: Theory of the Reversible Step Reactions.* Copenhagen: P. Haase and Son: 1941. One of the most influential books on coordination chemistry, regarded as a classic.

Brock, William H. *The Fontana History of Chemistry.* London: Fontana Press, 1992.

Day, Peter. "Christian Klixbüll Jørgensen (1931–2001): Inorganic Spectroscopist Extraordinaire." *Coordination Chemistry Reviews* 238–239 (2003): 3–8.

Kauffman, George B. "Impact: Interview with Jannik Bjerrum and Christian Klixbüll Jørgensen." *Journal of Chemical Education* 62 (1985): 1002–1005.

Schönherr, Thomas, ed. *Optical Spectra and Chemical Bonding in Inorganic Compounds: Special Volume I Dedicated to Professor Jørgensen.* Structure and Bonding, vol. 106. Berlin and New York: Springer, 2004. This volume contains a biographical article by Claus E. Schäffer (pp. 1–5), an article by Peter Day comparing Jørgensen with Michael Faraday in their approaches to naming new phenomena (pp. 7–18), twelve technical articles, and ten personal notes about Jørgensen by Fausto Calderazzo (pp. 237–239), Baldassare Di Bartolo (p. 240), Brian R. Judd (pp. 241–242), George B. Kauffman (pp. 242–245), Edwin Anthony C. Lucken (p. 245), Makota Morita (pp. 245–246), Dirk Reinen (pp. 246–247), Renata Reisfeld (pp. 247–249), H.-H. Schmidtke (pp. 249–250), and Alan F. Williams (pp. 250–253).

———. *Optical Spectra and Chemical Bonding in Transition Metal Complexes: Special Volume II Dedicated to Professor Jørgensen.* Structure and Bonding, vol. 107. Berlin and New York: Springer, 2004.

Williams, Alan F. "Obituary: Christian Klixbüll Jørgensen 1931–2001." *Chimia* 55 (2001): 472–473.

George B. Kauffman
Alan F. Williams

JÖRGENSEN, CHRISTIAN KLIXBÜLL

SEE **Jørgensen, Axel Christian Klixbüll.**

JUNG, CARL GUSTAV

(*b.* Kesswil, Switzerland, 26 July 1875; *d.* Küsnacht, Switzerland, 6 June 1961), *analytical psychology, psychiatry, comparative religion.* For the original article on Jung see *DSB,* vol. 7.

This article gives an account of the main shifts in the historical understanding of Jung's work since the 1970s, and in particular highlights the dismantling of the Freudocentric legend of Jung, the authorship of *Memories, Dreams, Reflections,* the contextualization of his work, and the commencement of research into his unpublished papers.

The Jungian Legend. For much of the twentieth century, Jung's name was considered to be so closely bound with that of Sigmund Freud that biographical, historical, and evaluative works gave pride of place to his relation with Freud. Such a framing of Jung's work forms part of the "Jungian legend." The key components of this may be summarized as follows: that Freud was the founder of psychotherapy; that Jung was a disciple of Freud and derived his ideas from him; that the two most important figures for Jung in the genesis of his work were Freud and Jung's Russian patient and associate Sabina Spielrein; that after his break with Freud, Jung had a breakdown; that during this "confrontation with the unconscious" he discovered (or invented) his ideas of the collective unconscious, archetypes, and individuation; that analytical psychology represents a revision of psychoanalysis; that Jung wrote an autobiography, which has been taken as the main source of information about his life and work; and that analytical psychology in the early twenty-first century directly descends from Jung, and indeed, was founded by him. Followers and critics have generally only differed in whether they have considered Jung's deviations from psychoanalysis in a positive or negative light.

In this form, the Jungian legend is in part a tributary to what has been called the "Freudian legend." The main elements of this interpretative narrative are the claims that psychoanalysis has had a major impact on twentieth-century society, and has led to wide-scale transformations in social life; that Freud discovered the unconscious; that Freud was the first to study dreams and discover their meaning; that Freud was the first to study sexuality scientifically and discovered infantile sexuality and its role; that his discoveries provoked a storm of disapproval due to Victorian repression; that he invented modern psychotherapy, and that psychoanalysis was the most advanced form of psychotherapy; and that these discoveries were based on his self-analysis and observation of patients.

The Freudocentric view of the origins of Jung's analytical psychology impeded the development of scholarly work on Jung, as it implied that the historical contextualization of Freud's work together with a close study of the Freud-Jung correspondence would be sufficient to grasp the genesis of analytical psychology, as these were seen to provide the theoretical backdrop that, together with the personal and affective impetus, would lead to Jung's later work. This perspective eventually came to be seen as seriously wanting, and a major realignment of the understanding of Jung's work has taken place. This has been brought about by a series of interlinked developments.

First, until the last quarter of the twentieth century, biographical and "historical" works on psychoanalysis and analytical psychology had been dominated by the work of disciples. Professional scholars started to work on these fields, and an extensive literature on Freud developed. It came to be seen that the epochal significance that had widely been attributed to Freud's work was actually the result of an active process of legend making on the part of Freud and his disciples. Freud's work began to be contextualized within the development of neurology, psychopathology, hypnosis, and evolutionary biology in the late nineteenth century. Second, the broader fields of the formation of late-nineteenth- and early-twentieth-century psychology, psychotherapy, and psychiatry began to attract scholarly attention after decades of neglect. Third, primary historical research on Jung commenced.

Memories, Dreams, Reflections. In particular, the best-selling *Memories, Dreams, Reflections,* which had been taken to be Jung's autobiography and hence the definitive firsthand statement on his life and work was shown to be nothing of the kind. Its history may be briefly narrated. For many years, the publisher Kurt Wolff had unsuccessfully tried to get Jung to write an autobiography. In the summer of 1956, he suggested a new project to Jung, along the lines of Johann Peter Eckermann's *Conversations with Goethe.* It was to be presented in the first person, and Jung's secretary, Aniela Jaffé, was proposed to take the role of Eckermann. Jaffé undertook a series of regular interviews with Jung, in which he spoke about a wide range of subjects. Jaffé, with the close involvement of Wolff, selected material from these interviews and arranged it thematically. This was then organized into a series of approximately chronological chapters.

During this process, Jung wrote a manuscript at the beginning of 1958 titled "From the Earliest Experiences of My Life." With Jung's permission, Jaffé incorporated this manuscript into *Memories.* His request to have this clearly demarcated from the rest of the book was not followed through. Passages were deleted and added by Jaffé, and further changes were made by others involved in the project. Jaffé also incorporated excerpted versions of some other unpublished manuscripts of Jung, such as material from a seminar he presented in 1925 seminar and accounts of some of his travels.

During the composition of the work, there were many disagreements between the parties involved concerning what the book should contain, its structure, the relative weighting of Jung's and Jaffé's contributions, the title, and the question of authorship. It was clear that for the publishers, an autobiography of Jung—or something that could be made to look as much like one as possible—held far greater sales potential than a biography by the then as yet unknown Jaffé. In 1960 a resolution was drawn up between the participants in which Jung formally stated that the work should not be regarded as an autobiography, but rather as a work by Jaffé to which he had made contributions. Jung's attitude toward the project fluctuated. After reading the early manuscript, he criticized Jaffé's handling of the text, complaining that he had been turned into an old maid. He died, however, before he could rectify the situation. He never saw the final manuscript, and the drafts that he did see went through considerable editing after his death.

Not incidentally, the published text represented the apotheosis of the Freudocentric legend of Jung. In the published text, Freud was the sole figure with a chapter dedicated to him. Jung's comments on contemporary figures such as Pierre Janet, Eugen Bleuler, William McDougall, William James, Alfred Binet, and others were simply omitted from the manuscript, which had the effect of dissociating Jung from much of his psychological network. The work was taken to be Jung's autobiography, and became the preeminent source of his life and work.

From a historiographical perspective, the Freudocentric legend of Jung had its genesis in Freud's *On the History of the Psycho-Analytical Movement.* In this work, Freud annexed Jung's early work on the associations experiment (in which subjects were asked to respond with one word in turn to a list of one hundred words) and on the psychology of dementia praecox (later called schizophrenia), claiming that what Jung had done was simply to apply psychoanalytic theory to interpret associations and the psychogenesis of psychotic symptoms. For Freud, Jung's work simply consisted in the application of the theory and procedures of psychoanalysis into areas in which they had not yet been utilized: experimental psychology and psychiatry. In actuality, Jung's early work represented a complex fusion of the French psychology of the subconscious together with German experimental psychopathology. Jung's alliance with Freud came to be seen within the context of a broader psychotherapeutic and psychogenic movement that was sweeping across America and Europe. Freud's assessment of Jung's work was that what was valuable in it lay in its application and extension of his own discoveries, and Jung's supposedly new innovations represented a secession from psychoanalysis and a descent into mystical obscurantism. This judgment has subsequently generally been subscribed to by the psychoanalytic community. A number of Jung's subsequent writings took the form of commencing with a critique of a number of Freud's positions before presenting his own. Jung's rhetorical mode of presentation came to be mistaken as indicating the genesis and source of his ideas.

Contextualizing Jung. The critique of the Freudian rewriting of history and the manner in which it cast Jung in a subordinate position to Freud together with the opening of the field of the history of late-nineteenth-century psychology created the possibility for the relocation of the genesis of Jung's work within a complex series of networks of debate and interactions. In particular, Jung's relations

Carl Gustav Jung. © BETTMANN/CORBIS.

with figures such as Bleuler, Janet, Théodore Flournoy, and James came in for extended study, together with his readings of figures such as Friedrich Nietzsche, Arthur Schopenhauer, Friedrich von Schiller, Henri Bergson, and Lucien Lévy-Bruhl.

Furthermore, Jung's psychology came to be resituated within the myriad attempts to form a science of psychology at the turn of the twentieth century, in the competing terrain of the human sciences. Much that had been attributed solely to Jung came to be seen as representing the confluence of networks of debates and discussions in European and American psychology, psychiatry, philosophy, and the human sciences spanning several decades. Jung's work came to be seen as drawing from quite distinct and counterposed intellectual traditions than those from which Freud drew. At the same time, historical work began on Jung's patients and his relations with some of his followers. In place of the conception of Jung as a solitary figure, the social and institutional networks between him and his followers started to be reconstructed, and the extent to which analytical psychology owed its existence to a collective endeavor began to be grasped.

The contextualisation of Jung's ideas has enabled the development of his work to be grasped, as well more clearly outlining what was particular to Jung. Much that was hitherto seen as idiosyncratic has come to be seen as embedded within intersecting networks of debate in late nineteenth and early twentieth century intellectual history. This recontextualization of Jung's works has gone hand in hand with the commencement of research in his unpublished papers. It is still not widely realized that Jung's *Collected Works* are far from complete, and that much of the existing secondary literature has been based on an incomplete textual corpus, together with unscholarly and not wholly reliable editions and deficient translations. In 1993 a catalog of Jung's papers at the Swiss Federal Institute of Technology was compiled, and they were made available for study. The scale of the unpublished materials proved to be quite unexpected. The holdings revealed sufficient unpublished manuscripts, notes, and drafts to fill ten volumes, together with an equal amount of unpublished seminars, and more than thirty thousand unpublished letters. In terms of bulk, the extent of the unpublished materials exceeds that of what has already been published. An initiative to publish this material is currently underway; see http://www.philemonfoundation.org.

Paradoxically, one of the main achievements of recent historical scholarship on Jung has been the demonstration of the unreliability of much of what had been taken as fact, and the growing realization of how much basic primary research still remains to be undertaken, and the necessity for a complete critical historical edition of his writings before his work can be adequately assessed.

SUPPLEMENTARY BIBLIOGRAPHY

WORK BY JUNG

Analytical Psychology: Notes of the Seminar Given in 1925. Edited by William McGuire. Bollingen series 99. Princeton, NJ: Princeton University Press, 1989.

OTHER SOURCES

Bishop, Paul, ed. *Jung in Contexts: A Reader.* London: Routledge, 1999.

Elms, Alan. "The Auntification of Jung." In *Uncovering Lives: The Uneasy Alliance of Biography and Psychology.* New York: Oxford University Press, 1994.

Shamdasani, Sonu. "Memories, Dreams, Omissions." *Spring: A Journal of Archetype and Culture* 57 (1995): 115–137.

———. *Jung and the Making of Modern Psychology: The Dream of a Science.* Cambridge, U.K.: Cambridge University Press, 2003.

Taylor, Eugene. "The New Jung Scholarship." *Psychoanalytic Review* 83 (1996): 547–568.

Sonu Shamdasani

JUNGIUS, JOACHIM (*b.* Lübeck, Germany, 22 October 1587; *d.* Hamburg, Germany, 23 September 1657) *natural philosophy, mathematics, logic.* For the original article on Jungius see *DSB,* vol. 7.

For several reasons, Jungius's systematic position in the history of science and knowledge remains difficult to establish: First, his published works were eclectic or subject to teaching requirements and therefore not necessarily representative for his own point of view; second, the vast majority of Jungius's manuscripts, including the two-thirds destroyed by fire in 1691, were scrap or reading notes and drafts rather than systematic treatises; third, Jungius considered his foremost task the critical examination and confutation of established knowledge aimed at creating the *tabula rasa* of an intellect undisturbed by preconceived opinions, upon which future generations should build an empirically founded system of knowledge.

Jungius is a typical representative of the early seventeenth-century crisis in natural philosophy. His work aims at overcoming the shortcomings of late Aristotelianism by combining an antimetaphysical attitude with intense empiricism and the idea of knowledge as a system, built *more geometrico* upon simple entities. Headmaster of a town school during the Thirty Years' War, Jungius was at the fringe of the academic communications network; he published little and mainly for local audiences. As a consequence, and despite the posthumous publications edited by devoted pupils, Jungius's highly idiosyncratic approach remained largely unknown at the time. There are exceptions, however: Gottfried Wilhelm Leibniz took an interest in Jungius's methodology, John Ray adopted parts of his botanical nomenclature, and Johann Wolfgang Goethe studied Jungius's plant morphology in connection with his own idea of metamorphosis.

During the nineteenth century the work of Jungius was rediscovered by German historians and was seen as sort of a counterpart to Francis Bacon's empiricism. In the 1930s this "national" portrayal culminated in seeing in Jungius the representative of a holistic approach to nature allegedly favored by the Germanic race. From the 1950s onward, the Hamburg-based Joachim Jungius Gesellschaft encouraged scholarly research on the philosophical and scientific work of its patron. Matter theory in relation to the intellectual foundation of early modern chemistry was the topic of Hans Kangro's monumental, if somewhat positivistic, study. In it, Jungius almost appears to be one of the founders of experimental science. More recent research has placed him more in the context of late humanist disputes on the value of metaphysics for the foundation of true knowledge, and within a framework of approaches and notions implicit to late Aristotelian modes of thought. In addition, the practice of teaching, arguing, and disputing within a school context have become clear as delimiting factors for Jungius in pursuing a program of his own.

Basically, Jungius's approach was critical, empirical, anti-metaphysical, and it aimed at a general reform of knowledge. As a philosopher, he summarized and, at the same time, transcended the Aristotelian logic and its metaphysical heritage. At the core of his *Logica Hamburgensis* of 1638 was a theory of knowledge meant to ground philosophical truth upon empirical demonstrations or at least upon probability (*logica engistica*). Consequently, metaphysics was replaced by what Jungius termed *protonoëtica,* a method to generate the finite number of simple notions (*protonoëmata*) upon which reasoning was to be founded.

This idea was by no means confined to philosophical reasoning alone, but expressed a much more general principle underlying both reality and knowledge. Nature, in Jungius's view, consists of hierarchies of increasing complexity, which can be split up into less complex and finally basic entities, out of which higher complexities can be constructed. Using analysis and synthesis as the primary operations (both mentally and practically) and starting from immediate sense perception, Jungius's method was meant to yield, by means of analysis, those simple, indivisible empirical entities out of which a new and true system of knowledge should in turn be built up by means of subsequent synthesis. In each of these steps, the mental and notional operations were thought to correspond to the respective entities and operations in nature.

This principle of decomposition and composition, modeled according to the mathematical method, was thought to rule everything: chemical transformations, plant morphology, teaching methods and the ways of reasoning. In the material world the required basic entities were supposed to be found in what Jungius called "hypostatical principles," ultimate parts of matter that could not decomposed further experimentally. This was the core argument of the *Praelectiones physicae,* Jungius's chief work in natural philosophy, begun as a course of lectures in 1629 and supplemented by a series of academic disputations on methodological and epistemological issues. However, Jungius never achieved a clear determination of these ultimate entities, nor did he ever give more than a vague idea of what his new system of knowledge would have been like. In fact, he considered his main task to be *doxoscopia,* that is, the falsification of established, and above all of empirically untenable, knowledge. Jungius's strict and somewhat naive empiricism, the lack of a proper experimental methodology, and his still largely Aristotelian terminology doomed this approach to failure. Toward the end of his life, he had collected some 150,000 sheets of scrap notes from almost every branch of knowledge, but this mass of information did not restructure itself to form a new system, as Jungius might have expected.

Taken together, Jungius's work appears as a symptom of rather than as a solution to the early seventeenth-century crisis of knowledge. With René Descartes, whose work Jungius seems to have known only selectively, he shared the conviction that the new system of knowledge had to be built upon simple notions that correspond to simple entities; with Francis Bacon, whom he admired, he shared the empiricist approach, but not the inductive methodology. But unlike these two scholars, Jungius remained much more closely bound to the context of late humanist and late Aristotelian school teaching within the Lutheran tradition.

SUPPLEMENTARY BIBLIOGRAPHY

The bibliographic survey in DSB, *vol. 7, pp. 195–196, and the bibliography in Hans Kangro,* Joachim Jungius' Experimente und Gedanken zur Begründung der Chemie als Wissenschaft: Ein Beitrag zur Geistesgeschichte des 17. Jahrhunderts, *Wiesbaden: Steiner, 1968, are still indispensable. Subsequent research is listed in Christoph Meinel, "Joachim Jungius," in* Grundriss der Geschichte der Philosophie: Die Philosophie des 17. Jahrhunderts, vol. 4: Das Heilige Römische Reich Deutscher Nation, Nord- und Ostmitteleuropa, *edited by Helmuth Holzhey and Wilhelm Schmidt-Biggemann, Basel: Schwabe, 2001, pp. 920–926, 983–984. The ca. 45,000 folios of Jungius's preserved papers, including drafts of manuscripts and collections of notes, are catalogued in Christoph Meinel,* Der handschriftliche Nachlass von Joachim Jungius in der Staats- und Universitätsbibliothek Hamburg, *Stuttgart: Hauswedell, 1984. The following sections contain later published original works by Jungius not mentioned in the* DSB *and select historical research published since 1984.*

WORKS BY JUNGIUS

Logicae Hamburgensis Additamenta, edited by Wilhelm Risse. Göttingen: Vandenhoeck & Ruprecht, 1977.

Praelectiones Physicae, edited by Christoph Meinel. Göttingen: Vandenhoeck and Ruprecht, 1982.

Disputationes Hamburgenses, edited by Clemens Müller-Glauser. Göttingen: Vandenhoeck and Ruprecht, 1988.

Elsner, Bernd, ed. *'Apollonius Saxonicus': Die Restitution eines verlorenen Werkes des Apollonius von Perga durch Joachim Jungius, Woldeck Weland und Johannes Müller.* Göttingen: Vandenhoeck and Ruprecht, 1988.

Aus dem literarischen Nachlaß von Joachim Jungius: Edition der Tragödie, Lucretia' und der Schul- und Universitätsreden, edited by Gaby Hübner. Göttingen: Vandenhoeck and Ruprecht, 1995.

Geometria empirica und Reiß-Kunst, edited by Bern Elsner. Göttingen: Vandenhoeck and Ruprecht, 2004.

Der Briefwechsels des Joachim Jungius, edited by Martin Rothkegel. Göttingen: Vandenhoeck and Ruprecht, 2005.

OTHER SOURCES

Meinel, Christoph. *In physicis futurum saeculum respicio: Joachim Jungius und die Naturwissenschaftliche Revolution des 17. Jahrhunderts.* Göttingen: Vandenhoeck and Ruprecht, 1984.

———. "Empirisme et réforme scientifique au seuil de l'époque moderne." In *Archives Internationales d'Histoire des Sciences* 37 (1987): 297–315.

———. *Die Bibliothek des Joachim Jungius: Ein Beitrag zur Historia litteraria der frühen Neuzeit.* Göttingen: Vandenhoeck and Ruprecht, 1992.

———. "Enzyklopädie der Welt und Verzettelung des Wissens: Aporien der Empirie bei Joachim Jungius." In *Enzyklopädien der Frühen Neuzeit: Beiträge zu ihrer Erforschung*, edited by Franz M. Eybl, Wolfgang Harms, Hans-Henrik Krummacher and Werner Welzig. Tübingen: Niemeyer, 1995.

Christoph Meinel

JUST, ERNEST EVERETT

(*b.* Charleston, South Carolina, 14 August 1883; *d.* Washington, D.C., 27 October 1941), *embryology, developmental biology, cell biology.*

Best known for his discovery of the "wave of negativity" that sweeps over the surface of the marine invertebrate egg upon fertilization, a wave that correlates with what has become known as the "fast block to polyspermy," Just more generally showed that the egg cell surface plays an important role in fertilization and development. He was the first to associate cell surface changes with stages of embryonic development experimentally.

As an African American in early-to-mid-twentieth-century America, Just lived and worked in a social and cultural milieu that was often hostile to his research efforts. He succeeded in obtaining a prestigious research fellowship from philanthropist Julius Rosenwald through the National Research Council in the middle of his career (1920–1930). Later on, however, he had considerable difficulty in obtaining financial support and spent much of his time asking for help from potential donors, including various American foundations, wealthy individuals—and even Italian dictator Benito Mussolini. Despite these potentially crippling challenges, Just made lasting contributions to biology.

Formative Years. Just was the son Charles Frazier Just Jr. and Mary Matthews Just. In 1887, after Charles's death, the family moved to James Island, just off the coast of

Ernest Everett Just. SCIENCE PHOTO LIBRARY

Charleston, South Carolina. In his early years, young Ernest was educated mainly by his mother, a strong-willed, independent woman who established the first school and church on the island. At age twelve, he left James Island to attend the Colored Normal Industrial Agricultural and Mechanics College at Orangeburg (now South Carolina State College), and at fifteen he left the South for New England. He entered Kimball Union Academy in Meriden, New Hampshire, at age seventeen and then went to Dartmouth College in Hanover, New Hampshire, where he studied biology, history, literature, and the classics. He graduated magna cum laude from Dartmouth in 1907 as an esteemed Rufus Choate scholar.

In the autumn of 1907, Just accepted a faculty position at Howard University, an African American school in Washington, D.C. His initial appointment was in English, but in 1910 he moved to the Biology Department and soon became the first head of the Department of Zoology. With funding from the Rosenwald Fund, he established a master's program in zoology. In 1915 Just was chosen from among a group of distinguished nominees to receive the first NAACP (National Association for the Advancement of Colored People) Spingarn Medal for his research excellence and his promotion of medical education at Howard. Despite a sometimes rocky relationship with the university administration and many trips to Europe, he remained a Howard faculty member until his death in 1941.

Woods Hole. In 1909 Just began making summer excursions to the Marine Biological Laboratory (MBL) in Woods Hole, Massachusetts, where he worked as an assistant to its director, the eminent embryologist Frank R. Lillie of the University of Chicago. Under Lillie's supervision, Just received his PhD in experimental embryology in 1916. His early studies at Woods Hole on the fertilization of the marine annelid *Platynereis megalops* formed the basis of his PhD thesis.

At Woods Hole, using a light microscope, Just recorded in detail the changes that take place within the egg during fertilization. In one particularly innovative study, Just exposed eggs of the sand dollar *Echinarachnius parma* to dilute seawater at various precisely timed intervals after insemination, and he carefully measured the position of membrane elevation relative to the point of sperm contact along with the time it took for each egg to rupture at this position. He found that the fertilization envelope forms as the result of a wave of structural instability that moves from the point of sperm entry to the opposite side. It has since become known that this wave of instability is actually a wave of cortical granule exocytosis that forms the fertilization envelope. As early as 1919, Just had observed that a "wave of negativity" moves from the point of sperm contact around the surface to the egg's opposite side. He noticed that the wave was associated with an immediate block to polyspermy (aberrant fertilization by multiple sperm), a block that preceded the liftoff of the fertilization membrane. He also observed the slow, mechanical block that occurred after membrane separation. Together, these two phenomena constitute what became known as the fast and slow blocks to polyspermy.

While at Woods Hole, Just studied the effects of a range of conditions—hypotonic and hypertonic seawater, temperature, degree of hydration, and ultraviolet irradiation, for example—on both normal development and parthenogenesis (egg activation without sperm) in a number of marine animals, including *E. parma,* the parchment worm *Chaetopterus pergamentaceus,* the clam worm *Nereis limbata,* and the sea urchin *Arbacia punctulata.* He also performed experiments that demonstrated the validity of Lillie's "fertilizin" hypothesis of fertilization. As early as 1906, Lillie had proposed that eggs release a diffusible substance (fertilizin) that, when it contacts spermatozoa, causes them to agglutinate. Fertilizin was hypothesized to have two ends, one that interacted with a receptor on the egg, and another that interacted with a receptor on the sperm. In this way, the fertilizin molecule allowed sperm and egg to come together. Lillie's hypothesis was in conflict with a rival idea of Jacques Loeb's known as the lysin hypothesis. Based on Loeb's observations of experimental parthenogenesis (see below), the hypothesis proposed that a cytolytic factor (lysin) in the sperm activates the egg, thereby initiating development. Throughout his career, Just sought experimental support for Lillie's hypothesis, proving that it was valid for a number of marine animals. From 1919 onward, in published papers and at scientific conferences, Just mounted a sustained attack on Loeb's rival hypothesis as well as on his work on experimental parthenogenesis. In 1930 Just wrote a long defense of the fertilizin hypothesis.

In the course of his work, Just discovered the important effects that environmental factors have on embryonic development. Based on these discoveries, he came to believe strongly that in order for laboratory experiments to be valid, the conditions in the laboratory must match as closely as possible those in nature. In marine settings at the MBL and, later on, at marine stations in Italy and France, he observed and carefully recorded the breeding behavior of the animals whose eggs he studied. Through his intimate knowledge of marine invertebrate natural history, and by testing the effects of different variables on development, he was able to formulate specific indices of normal development for a range of species. These indices, based for the most part on when and under what conditions fertilization envelope separation occurs, allowed him to predict with a high degree of certainty whether or not a particular egg, upon fertilization, would develop normally. So deep was Just's knowledge of marine invertebrate natural history, and so great was his ability to coax marine embryos to develop normally when other researchers failed, that he was often consulted on matters related to the proper handling of marine eggs. This led him to write and publish his first book, *Basic Methods for Experiments on Eggs of Marine Animals* (1939). He was widely regarded at Woods Hole and beyond as a brilliant experimental embryologist and a leading expert on embryo handling and culture.

Europe. Just's first European trip was to the Stazione Zoologica in Naples, Italy, in 1929, where for six months he studied cortical reactions in sea urchin eggs in order to test Lillie's fertilizin hypothesis further. Following up on an idea he had conceived earlier, he also showed that *P. megalops* of Woods Hole was not the same species as the Mediterranean annelid *Nereis dumerilii.* Just's second trip came only a year after his first. He received an invitation, exceedingly rare for an American at that time, to visit the Kaiser Wilhelm Institut für Biologie in Berlin for six months beginning in January 1930. While there, Just met and became friends with such eminent German embryologists as Max Hartmann (who had invited Just), Otto Mangold, and Richard Goldschmidt. He became particularly close to Johannes Holtfreter, Mangold's assistant, who in subsequent years published groundbreaking papers on the role of the cell surface in adhesion during amphibian embryo morphogenesis. Holtfreter's papers heavily cited Just's second book *The Biology of the Cell Surface* (1939) but, curiously, Just's work, though groundbreaking in its own right, was not cited much after that. After World War II, his work was almost completely forgotten.

Altogether, from his first trip in 1929 to his last in 1938, Just made nine visits to Europe to pursue research interests. Some of these were to Berlin, some were to Naples, and some were to Paris and Roscoff, a French village on the English Channel that had a small marine biological station. Although he had conceived of it earlier (around 1931), while at the Stazione Zoologica in Naples in 1934 Just began to work in earnest on *The Biology of the Cell Surface,* which was to be a fusion of his scientific work and his philosophical views on biology.

When at Woods Hole, Just had gained international recognition for the quality of his experimental technique and for his significant discoveries regarding the fertilization process, but he had not had the opportunity to explore more general theories about biology. At the Kaiser Wilhelm Institut für Biologie, however, he was encouraged to develop his ideas about the importance of the cell surface (the ectoplasm) and extend them to species other than marine invertebrates, such as amoebae. Emboldened by his European experience, Just underwent a transformation: he began to display increasing confidence in tackling the larger problems of biology. After 1936, except for one paper of medical significance and three papers on his work on fertilization of marine animals at Roscoff, Just's writings turned strongly in a philosophical direction. His papers had bold titles such as "A Single Theory for the Physiology of Development and Genetics" (1936) and "Phenomena of Embryogenesis and Their Significance for a Theory of Development and Heredity" (1937).

Philosophical Leanings. It is likely that Just's first paper, "The Relation of the First Cleavage Plane to the Entrance Point of the Sperm" (1912), in which he showed that the first cleavage plane of the fertilized *Nereis* egg is determined by the point of sperm entry, planted the seed of his later view that the organism is a holistic system having properties that emerge out of its complexity and organization. Just reasoned that if the plane of first cleavage is not preformed, but is contingent on the sperm entry point, then development must be an epigenetic rather than a preformationistic process. This holistic view, which considers living systems to be more than the sum of their parts, is becoming more prevalent at the beginning of the twenty-first century as biologists increasingly embrace systems biology. It was also common among embryologists at the beginning of the twentieth century. What set Just apart from his contemporaries, however, was his willingness to articulate openly the embryologist's view in the face of the opposing reductionistic views of such prominent biologists as Loeb and Thomas Hunt Morgan.

By 1900 Loeb had discovered that he could parthenogenetically activate sea urchin and marine annelid eggs through an experimental two-step method involving treatment with butyric acid followed by hypertonic seawater. This became known as Loeb's "superficial-cytolysis-corrective-factor" method. In papers, at scientific conferences, and in *The Biology of the Cell Surface,* Just challenged Loeb's method. He demonstrated not only that the order of treatment—butyric acid, then hypertonic seawater—was inconsequential, but that only the butyric acid was required to activate the egg. He showed that Loeb's conclusions were based on a mishandling of the eggs and that the cytolytic effect of the butyric acid was due to overexposure of the eggs to the acid. In sharp contrast to Loeb, a staunch mechanist who believed that he had created life through chemical means, Just believed that experimental parthenogenesis (and fertilization) showed the egg to be a dynamic system poised to respond to external agents of various kinds. It was not a passive physicochemical object upon which external agents acted to "cause" the development that ensued. Parthenogenesis and fertilization revealed what he called the independent irritability of the egg cell surface, its ability to respond to external stimuli. Thus, for Just, it was the cell surface and the structured layer just below it, the ectoplasm, that mediated fertilization and all subsequent developmental events. Likewise, it was the cytoplasmic surface, which was in contact with the environment, that was critical in the evolution of the first living thing, the ancestor of all life.

Just's focus on the cell cytoplasm (ectoplasm), as opposed to genes in the nucleus, led him to challenge Nobel laureate Morgan at the 1935 meeting of the American Society of Zoologists in Princeton, New Jersey. Morgan had proposed the gene theory, which postulated that genes located in linear arrays on chromosomes in the nucleus were the units of inheritance. Morgan believed that the purpose of the cytoplasm was only to execute the orders of genes in the nucleus. In contrast, Just believed that genes and chromosomes played secondary roles, and that the units of inheritance were located in the cytoplasm, not the nucleus.

At the Princeton meeting, Just presented his own theory of genetic restriction in opposition to Morgan's theory. The theory was founded on Just's detailed experiments showing that the cell nuclei increased in mass during the cleavage process that follows fertilization. It was an attempt to explain how a single-celled fertilized egg could differentiate into a multicellular organism; it proposed that the nuclei, and the genes within them, act only to remove selectively obstacles from the cytoplasm so that the "activity of the cytoplasm" can be released in one direction as opposed to another. This would lead to differentiation of cells into different types. Thus, although Morgan's purely nucleocentric view was at one extreme, Just's purely cytoplasmic view was clearly at the other. Morgan's gene theory was incorrect in attributing almost unlimited power to genes, but Just's genetic restriction theory was even more off the mark. Both were too one-sided, and neither succeeded in bridging the divide that separated the new geneticists (represented by Morgan) and the traditional embryologists (such as Just). Indeed, it is only at the beginning of the twenty-first century, with the rise of epigenetics and systems biology, that these two perspectives on development—the nucleocentric and the cytoplasmic—are becoming reconciled. The genecentric view, which has dominated biology in the latter half of the twentieth century, is giving way to a recognition of the importance of the gene in context.

Endings. In 1938 Just initiated a self-imposed exile in Europe and began working at the Station Biologique at Roscoff. In May 1940, however, Nazi Germany invaded France. All foreigners were ordered to leave the country, but Just decided to stay behind to finish writing his paper, "Unsolved Problems in General Biology" (1940). In June the Nazis seized control of Paris and the surrounding countryside, including Roscoff. Just was briefly imprisoned, but he was released with the help of friends. Although he had expected to reside permanently in Europe following his exile attempt, he was forced to return to the United States. In October 1941, having become gravely ill with pancreatic cancer, he died. He was fifty-eight years old.

Just was, by all accounts, a brilliant experimental embryologist who made important contributions to our understanding of fertilization and development. His experiments on the importance of the cell surface during development influenced others such as Holtfreter, who extended Just's work. His holistic view of the embryo is increasingly relevant as the organism is more and more viewed as an integrated system with properties that emerge from its complexity and organization. Just made these contributions despite the racist social and cultural milieu of his time, a milieu that influenced how his work was perceived as well as his ability to secure research support. The words of Just's friend and mentor, Lillie, expressed in his obituary of Just in the journal *Science,* are particularly poignant:

> An element of tragedy ran through all Just's scientific career due to the limitations imposed by being a Negro in America, to which he could

make no lasting psychological adjustment in spite of earnest efforts on his part.... In Europe he was received with universal kindness, and made to feel at home in every way; he did not experience social discrimination on account of his race, and this contributed greatly to his happiness there. Hence, at least in part, his prolonged self-imposed exile on many occasions. That a man of his ability, scientific devotion, and of such strong personal loyalties as he gave and received, should have been warped in the land of his birth must remain a matter for regret. (1942, p. 11)

BIBLIOGRAPHY

WORKS BY JUST

"The Relation of the First Cleavage Plane to the Entrance Point of the Sperm." *Biological Bulletin* 22 (1912): 239–252.

With Frank R. Lillie. "Breeding Habits of the Heteronereis Form of *Nereis limbata* at Woods Hole, Mass." *Biological Bulletin* 27 (1913): 147–169.

"The Present Status of the Fertilizin Theory of Fertilization." *Protoplasma* 10 (1930): 300–342.

"A Single Theory for the Physiology of Development and Genetics." *American Naturalist Supplement* 70 (1936): 267–312.

"Phenomena of Embryogenesis and Their Significance for a Theory of Development and Heredity." *American Naturalist* 71 (1937): 97–112.

Basic Methods for Experiments on Eggs of Marine Animals. Philadelphia: Blakiston's Son, 1939.

The Biology of the Cell Surface. New York: Garland Publishing, 1939.

"Unsolved Problems in General Biology." *Physiological Zoology* 13 (1940): 23–142.

OTHER SOURCES

Byrnes, W. Malcolm, and William R. Eckberg. "Ernest Everett Just (1883–1941)—An Early Ecological Developmental Biologist." *Developmental Biology* 296 (2006): 1–11. Highlights Just's appreciation of the role of environmental factors in development and ties this in with the emerging field of ecological developmental biology.

Gilbert, Scott F. "Cellular Politics: Ernest Everett Just, Richard B. Goldschmidt, and the Attempt to Reconcile Embryology and Genetics." In *The American Development of Biology,* edited by Ronald Rainger, Keith R. Benson, and Jane Maienschein. Philadelphia: University of Pennsylvania Press, 1988. Compares and contrasts Just with Goldschmidt, arguing that Just had a more egalitarian, decentralized view of the cell, while Goldschmidt had a more imperialistic view, with the nucleus controlling everything.

Gould, Stephen J. "Just in the Middle: A Solution to the Mechanist-Vitalist Controversy." In *The Flamingo's Smile: Reflections in Natural History.* New York: Norton, 1985. Makes the argument that Just's organicist views were "in the middle" between those of vitalism on the one hand and mechanism on the other.

Lillie, Frank R. "Obituary of Ernest Everett Just." *Science* 95 (1942): 11.

Manning, Kenneth R. *Black Apollo of Science: The Life of Ernest Everett Just.* New York: Oxford University Press, 1983. An excellent, well-researched biography of Just.

W. Malcolm Byrnes

K

KALCKAR, HERMAN MORITZ (*b.* Copenhagen, Denmark, 26 March 1908; *d.* Cambridge, Massachusetts, 17 May 1991), *biochemistry, enzymology, molecular biology.*

Kalckar's biochemical contributions were multifaceted. His initial work opened an investigative pathway to what has come to be called oxidative phosphorylation. As a mature investigator, his utilized his laboratory to play a significant role in the rapidly evolving field of enzymology, and Kalckar introduced innovative ways of measuring enzymatic activity. A significant part of his enzymological work focused on galactose metabolism, and he was one of the first investigators to clarify the nature of the genetic disorder galactosemia. Like many twentieth-century biochemists, Kalckar did not restrict his work to any single organism or tissue, and he contributed equally to the understanding of microbial physiology and human diseases.

Early Life and Education. Kalckar was the middle of three sons and described his early family life as "a middle-class, Jewish-Danish family—Danish for several generations." While not financially wealthy, his family life was intellectually rich and allowed Kalckar's "interest in the humanistic disciplines" to develop and thrive. His businessman father, Ludvig, had an avid interest in the theater and described seeing the 1879 world premiere of Henrik Ibsen's *A Doll's House.* His mother, Bertha Rosalie (née Melchior), was fluent in both German and French and introduced Kalckar to writers such as Gustave Flaubert, Marcel Proust, Johann Wolfgang von Goethe, Heinrich Heine, and Gotthold Ephraim Lessing. Although the family members were "mainly free-thinkers" and his father was primarily secular, his mother was more observant and encouraged some religious education. Thus, Kalckar learned "a minimum of Hebrew" at the Copenhagen main synagogue (Kalckar, 1991, p. 2).

Kalckar attended the Østre Borgerdyd Skole, located a short distance from home, which he described as "interesting and rewarding" and having an "Athenian flavor," a characteristic that arose from the headmaster, a world-renowned Greek scholar. His early scientific education also benefited from "a formidable and passionately devoted teacher in mathematical physics," who significantly influenced Kalckar's decision to pursue a career in research and teaching. The Danish Nobel Prize–winning (Physiology or Medicine, 1920) physiologist August Krogh shaped Kalckar's biological interests by some human physiology demonstrations. Kalckar commented that Krogh was "the only physiologist in the 1920s who took an active interest in introducing the principles of human physiology to Danish high school boys." The demonstrations were apparently elaborate, as Krogh used a number of research instruments from his own research program (Kalckar, 1991, pp. 2–3).

Kalckar's younger brother, Fritz, was a physicist and a student/colleague of Niels Bohr. Herman Kalckar also knew Bohr and was close friends with the Bohr family (his son Niels Kalckar was named after Bohr). In 1937 Fritz Kalckar died from a status epilepticus attack while in California. The Bohr family provided personal comfort to the Kalckar family, and Niels Bohr gave a eulogy at Fritz's cremation service (Kalckar, 1991, p. 8). The close Bohr friendship may explain the intense chemical focus that Kalckar brought to biological problems.

In 1933 Kalckar completed medical studies at the University of Copenhagen and the following year began graduate studies in the Department of Physiology. His research mentor was Ejnar Lundsgaard. In the early 1930s, Lundsgaard revolutionized the way physiologists understood cellular energetics by demonstrating that phosphocreatine was involved in muscle contraction. Several simultaneous events happened when Kalckar joined Lundsgaard's lab. The biochemist Fritz Lipmann was forced to leave Germany, and Lundsgaard offered him a position in Copenhagen. Also, Lundsgaard became physiology department chair, and thus his time was more restricted. Consequently, Lipmann became Kalckar's primary mentor and encouraged him to read "the newer literature on carbohydrate and phosphate metabolism in isolated phosphorylation tissue extracts or tissue particle preparations" (Kalckar, 1991, p. 5).

Kalckar began his research career at an important period in biochemistry's history. Eugene Kennedy observed that the central question facing many biologists at the time was: "How is energy captured by the oxidation of sugars and other foodstuffs linked to the reduction of molecular oxygen?" (1996, p. 151). Kalckar very indirectly set about to address this question and decided to study phosphate and carbohydrate metabolism in kidney cortex extracts.

His use of kidney cortex tissue seems a bit surprising because many workers at the time (including Hans Krebs) were working with minced pigeon breast muscle. Kalckar stated that one reason for his choice of tissue was his mentor's interest in carbohydrate transport in the kidney, a process thought "to be driven by phosphorylation-dephosphorylation cycle in the membrane." Thus, Kalckar "set out to scout for a really vigorous phosphorylation system that could 'drive'" carbohydrate transport in the kidney cortex (Kalckar, 1969, p. 171). Later, Kalckar explained that in order to study pigeon tissue he would have to train himself to decapitate pigeons, however he "greatly preferred to avoid killing animals, probably from a lack of courage." He was fortunate to obtain fresh cat and rabbit kidneys from the weekly physiology department perfusion experiments (Kalckar, 1991, p. 6).

Using Otto Warburg's manometric technique and apparatus, Kalckar began to measure relationships between oxygen utilization and phosphorylation in minced kidney tissue. Few modern investigators can appreciate the multitude of difficulties associated with this experimental approach, which required measuring changes in extremely small volumes of gases. The apparatus was standard biochemical equipment for much of the twentieth century; nevertheless, Efraim Racker half jokingly referred to the apparatus as "nystagmus inducing" (1965, p. 19). Regardless, Kalckar's choice of this experimental approach was important, because in the Warburg apparatus the minced tissue was continuously exposed to high oxygen levels.

Kalckar noted that his "studies on phosphorylation and respiration in kidney cortex extracts turned out to be rewarding" (1991, p. 6). Prior to his work, researchers were investigating various pathways whereby carbohydrates are oxidized to products such as lactic acid. These oxidations involved a number of phosphorylation reactions and produced a phosphorylated compound, ultimately identified as adenosine triphosphate (ATP). Referred to as "glycolysis," these reactions occurred in the absence of oxygen. While a variety of phosphorylated compounds were produced in addition to ATP, the role of phosphorylation was not understood; most biochemists believed that the phosphate group somehow made the compound more "fit" for reaction.

In a series of papers published between 1937 and 1939, Kalckar established that in kidney cortex tissue slices these phosphorylation reactions were "coupled" with oxygen consumption. In these experiments, Kalckar suspended minced kidney cortex in phosphate buffer, which was then incubated with various carbohydrates in a Warburg apparatus. At various times O_2 consumption was manometrically determined and inorganic phosphate (P_i) was chemically measured. In the presence of O_2, P_i levels were significantly reduced, whereas under anaerobic conditions no P_i was consumed. Carbohydrate (e.g., glucose) was also required for P_i consumption. The process, which Kalckar referred to as "aerobic phosphorylation," was the first direct demonstration that carbohydrate oxidation was directly "coupled" to carbohydrate phosphorylation (Kalckar, 1937; 1939; 1969, pp. 171–172); the process, later referred to as oxidative phosphorylation, is fundamental to life, and Kalckar's work helped establish the basic phenomenon and opened the way to its systematic exploration (Kennedy, 1996).

Early Career. Kalckar completed work for his PhD in 1939 and in January received a Rockefeller research fellowship to spend a year at the California Institute of Technology (Caltech). After a brief visit in London, Kalckar spent much of February in St. Louis visiting Carl and Gerty Cori's Washington University laboratory, which was then doing some of the most innovative biochemical work in the United States. The Cori lab had, unsuccessfully, tried to reproduce some of Kalckar's published work. Kalckar noted that in their technique, which involved "the old-fashioned Meyerhof extract technique, using test tubes," the tissue extracts were static. In Kalckar's technique, using the Warburg manometer, the tissue extract was vigorously shaken and thus highly aerated. Kalckar worked with the Coris' graduate student, Sidney

Colowick, and quickly demonstrated that when the cortex extracts were shaken, oxidative phosphorylation was easily observed (Kalckar, 1991, p. 10).

In early March 1939, Kalckar arrived in Pasadena, where he and his wife Vibeke Meyer rapidly became involved in the Caltech intellectual and social community, which included Max Delbrück, Linus Pauling, and two Bonner brothers, James and David. The Delbrück friendship, which had initially begun with a brief Copenhagen meeting, became lifelong. The Bonner brothers introduced Kalckar to various members of the Caltech faculty, and helped Herman and Vibeke settle into an American life by arranging housing and the purchase of a car.

Most likely at Delbrück's encouragement, Kalckar attended Cornelius B. van Niel's popular microbiology course taught at Pacific Grove. Van Niel was responsible for introducing numerous American scientists to the complexity of the microbial world. As a student of Albert Jan Kluyver, van Niel spread the gospel of biochemical unity, which Kluyver expressed in the aphorism, "From the elephant to the butyric acid bacterium it is all the same!" (Kamp et al., 1959, p. 20). The Pacific Grove experience "may have planted a seed that led later to Kalckar's interest in microbial molecular biology" (Kennedy, 1996, p. 153). Throughout his career, Kalckar, like many of his contemporaries, frequently turned to bacterial systems to address questions unresolvable in more complex biological systems.

The Pauling connection proved especially valuable. Most likely from his familiarity with Bohr's atomic concepts, Kalckar was influenced by the potential impact of Pauling's chemical ideas on biological problems. At Pauling's encouragement, Kalckar wrote an extensive review in which he developed many of our modern concepts of bioenergetics, including the notion that oxidation reactions are "coupled" to phosphorylation reactions via ATP (Kalckar, 1941).

Although not widely appreciated at the turn of the twenty-first century, Kalckar's 1941 *Chemical Reviews* paper influenced the way many scientists thought about the energetics of life processes. One reason for this lack of appreciation perhaps lies in the fact that Fritz Lipmann also published a similar paper, in which he developed the "high energy bond" concept, in 1941. While the Kalckar paper was influential in shaping scientists' views about bioenergetics, the Lipmann paper was more frequently cited. Nevertheless, Kalckar viewed his relationship with Lipmann as collaborative, stating: "While Lipmann was preaching the gospel of phosphorylation and the 'high energy phosphate bond' on the eastern seaboard ... I was beginning my missionary work at the same time in the 'Pacific triangle'—Caltech, Berkeley, and Stanford" (Kalckar, 1992, p. 43).

The outbreak of World War II, and German occupation of Denmark, stranded the Kalckars in the United States. Because of his earlier visit in the Cori lab, in 1940 Kalckar received an appointment as research fellow in the Pharmacology Department at Washington University, and he and Vibeke moved to St. Louis. Kalckar resumed work with Colowick studying phosphorylation reactions in muscle tissue. The work was important in terms of Kalckar's scientific maturation for two reasons: first, he began to focus more on individual enzymatic reactions, even partially purifying the responsible protein, than he had previously; second, he and Colowick discovered the enzyme myokinase (now called adenylate kinase), which catalyzed the reversible reaction:

$$\text{ATP} + \text{AMP} \rightleftarrows 2\ \text{ADP} \quad \text{(Reaction 1)}$$

Because many biochemical processes led to adenosine monophosphate (AMP) production, the enzyme was important in helping replenish cellular levels of ATP. Shortly after the collaboration began, Colowick was drafted to do war research, and Kalckar finished the myokinase characterization by himself.

In 1943 Oliver Howe Lowry invited Kalckar to join his lab at the New York Public Health Institute as a research associate. The appointment helped advance Kalckar's enzymology research in part because, as Paul Berg commented, Kalckar "developed a whole new approach to being able to use enzymes in a novel way" (2000, p. 26). The lab had a new, and relatively rare, Beckman DU ultraviolet spectrophotometer, and Kalckar rapidly became a virtuoso of the instrument, using it to develop a number of novel enzyme assay techniques. The work culminated in a series of three papers on purine metabolism enzymes (Kalckar, 1947a, b, c). All three papers were highly cited (3,200 citations in 2006), and the third paper was recognized as a "Citation Classic" in 1984. In a brief commentary on the paper, Kalckar noted that the paper's popularity most likely arose from its potential clinical applications such as diagnosing diseases such as arthritic urica or Lesch-Nyhan syndrome, a serious inborn metabolic error in infants (Kalckar, 1984).

Return to Denmark. When the war ended, Lundsgaard arranged for Kalckar to return to Copenhagen in 1946 where he ran the new Cytofysiologisk Institute, funded by the Rockefeller Foundation, Lederle Laboratories, as well as the Danish Carlsberg Foundation. His research focused primarily on the enzymology of nucleoside and nucleotide metabolism, and the lab rapidly became a magnet attracting numerous bright young biochemists from around the world, including Americans such as Paul Berg, Morris Friedkin, Walter McNutt, Günter Stent, and James Watson.

Kalckar's "Cytofysiologisk Institut" 1950. Back row, from left: Gunther Stent, Niels Ole Kjeldgaard, Hans Klenow, Jim Watson, and Vincent Price. Front row, from left: Kalckar, Audrey Jarnum, Jytte Heisel, Eugene Goldwasser, Walter McNutt, and E. Hoff-Jorgensen. COURTESY OF STEEN GAMMELTOFT, M.D.

Shortly after his return to Copenhagen, Kalckar's marriage to Vibeke dissolved. He married Barbara Wright, a developmental biologist and American postdoctoral fellow in his lab. Three children, Sonja, Nina, and Niels, were born of this marriage (Kennedy, 1996).

In early 1950, Kalckar suggested that such nucleotides as uridine diphosphate (UDP), UDP-glucose, or UDP-galactose were involved in the conversion of glucose-1-P to galactose-1-P, and his laboratory began to focus on galactose metabolism in microbial and animal tissues, a research program that dominated the rest of his career.

Return to the United States. Bernard Horecker, at the National Institutes of Health (NIH), invited Kalckar to come to Bethesda, Maryland, as a visiting scientist in 1952; later the appointment was made permanent in the National Institute of Arthritis and Metabolic Diseases. At the NIH, Kalckar expanded his research program on galactose metabolism and developed an interest in the genetic disease, galactosemia. The disease, which is characterized by liver enlargement, kidney failure, and mental retardation, occurs because of galactose (formed from the milk component lactose) accumulation. In untreated infants, mortality is as high as 75 percent. The prevailing view on galactosemia in 1952 is summarized in Reactions (2) through (4).

$$\text{Lactose (in milk)} \rightarrow \text{Glucose} + \text{Galactose} \quad \text{(Reaction 2)}$$

$$\text{Galactose} + \text{ATP} \rightleftarrows \text{Galactose-1-P (Gal-1-P)} \quad \text{(Reaction 3)}$$

$$\text{Gal-1-P} \rightleftarrows \text{Glucose-1-P} \quad \text{(Reaction 4)}$$

Reaction 3 is catalyzed by an enzyme called an "epimerase," and investigators speculated that galactosemia arose from a defect in this enzyme (which would lead to galactose accumulation via reversal of Reaction 3).

For a variety of reasons, Kalckar suspected another enzyme was defective; he thought the enzyme blocked was Galactose-1-P uridyl transferase (GALT), which catalyzes Reaction 5:

Gal-1-P + UDP-glucose ⇄ UDP-galactose + Glucose-1-P (Reaction 5)

If this enzyme was not functioning, Gal-1-P would accumulate leading to galactose build up through reversal of Reaction 3. Although galactosemia is relatively rare, Kalckar's associates were able to identify several afflicted individuals and obtain blood samples from them. If Kalckar's hypothesis was correct, the analysis should show two features. First, individuals with the disease should have decreased levels of GALT. Second, epimerase levels in afflicted individuals should be normal. Both features were observed in blood samples from galactosemia patients when compared with normal subjects (Kalckar, et al., 1956a and b).

In addition to a rich scientific environment, the NIH community offered personal advantages. Unlike many institutions at the time, the NIH had no official prohibitions regarding employing scientific couples; thus Kalckar's wife, Barbara Wright, also had a staff appointment. The environment was both intellectually and socially rewarding. Other married couples working at the NIH at the same time included: Thressa and Earl Stadtman; Marjorie and Evan Horning; and Martha Vaughan and Jack Orloff (Park, 2002).

In 1958 William McElroy offered Kalckar a full professorship in biology at Johns Hopkins University, where his biochemical interests in galactose metabolism continued. Perhaps reflecting the long-term influence of van Niel's course, his research focus shifted to bacterial systems.

While at Hopkins, Kalckar suggested (in a *Nature* paper) that the effects of isotope fallout from nuclear weapons testing could be measured by analysis of strontium-90 levels in children's deciduous teeth (Kalckar, 1958). The proposal was both important and original and elicited numerous state and local organizations to collect teeth for analysis; the data did indeed show a correlation between isotope levels and nuclear testing. Arguably the resulting program helped encourage public sentiment to ban atmospheric nuclear weapons testing.

In 1961 Kalckar moved again, to the Harvard Medical School where he was Henry S. Wellcome Research Biochemist and headed the Massachusetts General Hospital Biochemical Research Laboratory, a position previously held by Fritz Lipmann. His research program continued to focus on bacterial galactose utilization, however it tended to become more physiologically oriented (Wellcome Trust, 1963, p. 16). He became interested in cellular sensory and signaling processes. For example, in collaboration with Winfried Boos, he isolated a galactose binding protein in *Escherichia coli* that played an important role in cellular galactose transport. Later work demonstrated that the same protein played an important role in *E. coli*'s chemotactic response to galactose.

The Boston move also had significant personal ramifications for Kalckar: his marriage to Barbara Wright dissolved; he renewed a friendship with Agnete Fridericia, whom he had known as a student in Copenhagen. Kalckar commented that their "cheerful conversations during the 1960s, most of it in Danish, changed [his] life"; in 1968, he and Agnete were married (Kalckar, 1991, p. 27).

Kalckar retired as head of the Biochemical Research Laboratory in 1974 but remained at Massachusetts General as visiting professor. In 1979 he was appointed as a distinguished research professor in the Boston University chemistry department, a position that permitted him to continue his research work for the rest of his life (Kennedy, 1996, p. 159).

In a scientific career that spanned almost six decades of the twentieth century, Herman Kalckar witnessed many fundamental changes in biochemistry. Although urease had been crystallized in 1926, in 1934—when Kalckar began his graduate studies—some biochemists were still debating the nature of enzymes. At his death, after having been a founder of modern bioenergetics and enzymology, he was studying the genes responsible for enzyme action. These achievements received numerous recognitions: he was elected to the National Academy of Sciences, the Royal Danish Academy, and the American Academy of Arts and Sciences; he received honorary degrees from Washington University, the University of Chicago, and the University of Copenhagen.

On a personal level, Kalckar often appears enigmatic. On the one hand, Eugene Kennedy noted "The sweep of his intellect was very broad, his spirit was open and generous, and he had a wonderful sense of humor" (Kennedy, 1996, p. 160). Former associates speak fondly of exciting scientific discussions, which might be interrupted to talk about art, music, even Nordic mythology. As his work on strontium-90 levels in children's teeth suggests, Kalckar was passionately concerned about social issues like the spread of nuclear weapons and weapons testing.

However, Kalckar was equally known for his peculiar way of talking, which some hearers found unintelligible. He would, for example, often begin an argument with a conclusion and work backward to the premises. Boos, his former associate, referred to this speech mode as a "language" he called "Kalckarian": a language that Kalckar occasionally used to avoid conversations he found uninteresting.

Boos further observed that Kalckar divided the scientific world into two types. The Apollonian was a capable technician, who accumulated data to tell clear, logical stories, a perspective Kalckar found boring and lacking any sense of humor. The Dionysian, however, was more interested in the answers, even if they come in a dream, than in the logic and proof of the answer. For Boos, Kalckar "was the archetype of a Dionysian" (Boos, 1991, p. 8).

The term *Dionysian* can be used in another sense, related to the Roman incarnation of Bacchus, the God of wine and other pleasures. In this view, an individual is open to *all* that life offers and is not bound in a single worldview; Kalckar seems to fit this image as well. Science was one of life's rich rewards; good music, good friends, a cornucopia of human experiences were equally valuable.

BIBLIOGRAPHY

WORKS BY KALCKAR

"Phosphorylation in Kidney Tissue." *Enzymologia* 2 (1937): 47–53.

"The Nature of Phosphoric Esters Formed in Kidney Extracts." *Biochemical Journal* 33, no. 5 (1939): 631–641.

"The Nature of Energetic Coupling in Biological Syntheses." *Chemical Reviews* 28 (1941): 71–178.

With Sidney P. Colowick and Carl F. Cori. "Glucose Phosphorylation and Oxidation in Cell-Free Tissue Extracts." *Journal of Biological Chemistry* 137 (1941): 343–356.

"The Enzymatic Action of Myokinase." *Journal of Biological Chemistry* 143 (1942): 299–300.

"The Role of Myokinase in Transphosphorylations: II. The Enzymatic Action of Myokinase on Adenine Nucleotides." *Journal of Biological Chemistry* 148 (1943): 127–137.

With Manya Shafran. "Differential Spectrophotometry of Purine Compounds by Means of Specific Enzymes: I. Determination of Hydroxypurine Compounds." *Journal of Biological Chemistry* 167 (1947a): 429–443.

With Alice Neuman Bessmann. "Differential Spectrophotometry of Purine Compounds by Means of Specific Enzymes: II. Determination of Adenine Compounds." *Journal of Biological Chemistry* 167 (1947b): 445–459.

With Manya Shafran. "Differential Spectrophotometry of Purine Compounds by Means of Specific Enzymes: III. Studies of the Enzymes of Purine Metabolism." *Journal of Biological Chemistry* 167 (1947c): 461–475.

With Elizabeth P. Anderson and Kurt J. Isselbacher. "Galactosemia, a Congenital Defect in a Nucleotide Transferase." *Biochimica et Biophysica Acta* 20 (1956a): 262–268.

With Elizabeth P. Anderson, Kurt J. Isselbacher, and Kiyoshi Kurahashit. "Congenital Galactosemia, a Single Enzymatic Block in Galactose Metabolism." *Science* 123 (1956b): 635–636.

"An International Milk Teeth Radiation Census." *Nature* 182 (1958): 283–284.

Editor. *Biological Phosphorylations: Development of Concepts.* Englewood Cliffs, NJ: Prentice-Hall, 1969.

"This Week's Citation Classic." *Current Contents* 26 (1984): 148.

"50 Years of Biological Research: From Oxidative Phosphorylation to Energy Requiring Transport Regulation." *Annual Review of Biochemistry* 60 (1991): 1–37.

"High Energy Phosphate Bonds: Optional or Obligatory?" In *Phage and the Origins of Molecular Biology,* expanded ed., edited by John Cairns, Gunther S. Stent, and James D. Watson. Cold Spring Harbor, NY: Cold Spring Harbor Laboratory Press, 1992.

OTHER SOURCES

Berg, Paul. "A Stanford Professor's Career in Biochemistry, Science Politics, and the Biotechnology Industry." Berkeley: Regional Oral History Office, Bancroft Library, University of California, 2000. Available from http://www.calisphere.universityofcalifornia.edu. An oral history conducted in 1997 by Sally Smith Hughes.

Boos, Winfried. "Farewell to a Dionysian." *Biological Chemistry Hoppe-Seyler* 372, no. 11 (1991): 8–9. Original in German; translation kindly provided by Dr. Boos.

Garfield, Eugene. "Highly Cited Articles: 35. Biochemistry Papers Published in the 1940s." *Current Contents* 8 (1977): 5–11. Reprinted in: *Essays of an Information Scientist,* Vol. 3. Philadelphia, PA: ISI Press. Available from http://www.garfield.library.upenn.edu/essays.html.

Kamp, A. F., J. W. M. La Rivière, and W. Verhoeven. "Kluyver as Professor: Chronicles of the Laboratory." In their *Albert Jan Kluyver: His Life and Work,* pp. 14–48. New York: Interscience Publishers, 1959.

Kennedy, Eugene P. "Herman Moritz Kalckar, March 26, 1908–May 17, 1991." *Biographical Memoirs, National Academy of Sciences* (U.S.) 69 (1996): 148–165.

Lipmann, F. "Metabolic Generation and Utilization of Phosphate Bond Energy." *Advances in Enzymology* 1 (1941): 99–162.

Park, Buhm Soon. "Historian's Perspective: NIH Offered Haven from Antinepotism Rules." *Newsletter of the NIH Alumni Association* 14, no. 2 (2002): 18–19. Available from http://www.fnih.org/nihaa/NIHAA-Update.pdf.

Racker, Efraim. *Mechanisms in Bioenergetics.* New York: Academic Press, 1965.

Wellcome Trust. *Fourth Report, Covering the Period 1960–1962.* London: Author, 1963.

Rivers Singleton Jr.

KAMMERER, PAUL (*b.* Vienna, Austria-Hungary, 17 August 1880; *d.* Puchberg am Schneeberg, Lower Austria, 23 September 1926), *zoology, heredity, evolution.*

The main goal of Kammerer's research was to demonstrate the modifying power of the environment and the heritability of acquired characteristics, using the experimental methodology of twentieth-century biology. Conceptually, he aimed to reconcile this Lamarckian mode of heredity with the new science of genetics and an older interpretation of Darwinism that posited the evolutionary significance of acquired characteristics. Along with many other opponents of neo-Darwinism, Kammerer argued that Darwin had intended natural selection only as a means of eliminating unfavorable characteristics. New and favorable ones had to be acquired in response to environmental changes and to changed habits. Individual improvement and inheritance of acquired skills and behaviors, in addition to physical traits, were also the key to human evolution and cultural progress, according to Kammerer.

Kammerer was a prolific writer and popularizer. His books, articles, and lectures reached beyond Vienna to a broad public, as did his ideas about how to apply his experimental findings to human cultural evolution. Among professional scientists, his work was also well known, and it stimulated heated discussion about the nature of heredity and the causes of variation. But Kammerer's arguments for the inheritance of acquired characteristics were not widely accepted by scientists at the time, and they are not considered valid today. Moreover, he is suspected of tampering with, or perhaps even fabricating, his experimental evidence.

Family Background and Education. Paul Kammerer grew up in the secure, prosperous, and creative world of upper-middle-class Vienna in the last decades of the Habsburg Empire. His father Karl was a manufacturer whose family came from Siebenbürgen, a German-speaking enclave in what later became Romania. His mother Sophie was from a converted Jewish family on at least one side. His parents had both been married before, and Paul had four much-older half siblings.

Like many others of his generation and social class, Kammerer was intensely devoted to the arts, and even though he chose a scientific career, he always valued creativity and thought of himself as an artist among scientists. Kammerer studied zoology at the Vienna University and music theory at the conservatory, beginning in 1899. He wrote his dissertation on the evolutionary relationships between two species of salamander, under the direction of Berthold Hatschek, a protégé of Ernst Haeckel, who taught a comparative approach to morphology and evolution. Kammerer got his introduction to experimental methodology outside the university, under the tutelage of zoologist Hans Przibram, another student of Hatschek's who was rebelling against the comparative approach.

Kammerer taught biology courses at a variety of places, from the university, to the *Volkshochschule* (a system of adult-education centers), and a high school for girls. He also traveled to give talks throughout the Austrian Empire and Germany, often to chapters of the Monist League, which was associated with Ernst Haeckel. Kammerer shared the Monists' goals of popularizing science and especially scientific approaches to social and political problems.

Kammerer was said to be handsome, charismatic, and vain, and stories abound about his attractiveness to the Viennese ladies and his requited and unrequited loves. Kammerer was married in 1906 to Felicitas Maria Theodora ("Dora") von Wiedersperg, the daughter of Bohemian aristocrats, through whom Kammerer gained entry into Viennese high society. Kammerer left her to marry the painter Anna Walt around 1912 or 1913, but returned to Dora after a short, stormy marriage and a suicide attempt. He and Dora had a daughter named Lacerta, after the genus of mural lizards.

Kammerer also had some success in the arts. He published a book of lieder and had them performed in Vienna. He made the acquaintance of many of the eminent artistic figures of turn-of-the-century Vienna, such as Gustav Mahler, whom he idolized, and Mahler's wife Alma, whom he courted after the composer's death and to whom he dedicated a small book on the inheritance of musical talent.

Environmental Effects and Their Inheritance. In 1902, before finishing his dissertation, Kammerer went to work for Przibram at the newly founded Biologische Versuchsanstalt (Institute for Experimental Biology) in Vienna. This laboratory, also known as the Vivarium, after the zoological exhibition hall that previously occupied the same premises, was owned by Przibram and devoted to the new experimentalism, which Przibram wanted to apply to the widest possible range of organisms and scientific problems. As Przibram's assistant, Kammerer helped acquire the animal stocks, design and build aquaria and terraria, and develop methods of rearing the animals and controlling their environments. He also began doing experiments of his own and added an experimental portion to his dissertation project.

After defending his dissertation in 1904, Kammerer remained at the Vivarium to pursue a multifaceted career, combining experimental work on heredity, evolution, development, regeneration, and symbiosis with forays into evolutionary theory, and sidelines as a composer, teacher, and popular science writer and lecturer.

The themes of his later work were already discernable in Kammerer's dissertation on the evolutionary relationship between the spotted salamander, *Salamandra maculosa* and its alpine cousin *S. atra.* By manipulating their environments in the laboratory, Kammerer induced individuals of each species to take on characteristics of the other: for example, instead of the usual pattern of producing a large brood of eggs that hatched into tadpoles, *S. maculosa* was induced to give birth ovoviviparously on land to fewer, but fully metamorphosed young. *S. atra* was induced to make the opposite shift. In a follow-up study, Kammerer bred these salamanders and found that their offspring retained the modified reproductive strategies, even when reared in their original environments. That work earned him the Sömmering Prize from the Senckenberg Society of Naturalists in Frankfurt in 1909 as well as his habilitation, or qualification to teach courses at the university at the rank of privatdozent (lecturer), in 1911.

Kammerer made further demonstrations of the modifying power of the environment, using many of the other species and environments available at the Vivarium. For example, he produced color-variants in the mural lizard *Lacerta* by varying temperature, and he used an artificial lighting regimen to induce the blind cave salamander *Proteus* to develop large, functional eyes. More controversial were the experiments in which he also bred his modified individuals and reported inheritance of the new characteristics. In addition to his dissertation project, these breeding experiments included several on the effects of background colors on the patterns of spots on *S. maculosa*; several on the life cycle and breeding behavior of the midwife toad *Alytes obstetricans*; the effects of amputation on siphon length in the sea squirt or tunicate *Ciona intestinalis*; and more besides.

Kammerer argued that in such cases the environmentally induced modification had been communicated to the chromosomes, where incipient genes were forming that would perpetuate the modification, at least partially. Crosses between modified and unmodified stocks sometimes yielded hybrid progeny in Mendelian ratios, apparently supporting the idea that new genes were at least partially formed. Kammerer speculated that hormones might be the medium of communication that allowed the chromosomes to express themselves in the body and carried information about the body back to the chromosomes.

In any case, he argued that his experiments had vindicated the older interpretation of Darwinism that had allowed for inheritance of acquired characteristics, and undercut the neo-Darwinism of August Weismann that ruled it out. In particular, they had falsified Weismann's conception of an isolated germ plasm or hereditary material that could receive no communication back from the body and was mostly shielded from environmental influences as well. Kammerer felt he had demonstrated that there was no Weismannian barrier to communication with the hereditary material.

Most geneticists, neo-Darwinian evolutionists, and even some proponents of the inheritance of acquired char-

Paul Kammerer. © BETTMANN/CORBIS.

acteristics received these results with skepticism, if not hostility. They tried to explain away Kammerer's results as effects of selection in highly variable organisms, atavisms, lucky mutations, poor environmental controls, and other errors. In retrospect, inadvertent selection probably accounts for most of the results. Recording errors and poor environmental controls are also likely.

Opponents also complained that Kammerer's published documentation was not up to professional standards. His verbal descriptions of modified animals were vague, his drawings sketchy, his photographs fuzzy or retouched, and he had to defend himself repeatedly against accusations of unreliability and insinuations of fraud. Przibram and others who saw Kammerer's specimens vouched for their authenticity, however.

Human Cultural Evolution. Kammerer's public lectures and popular writings applied his evolutionary ideas and experimental results to human affairs. Kammerer believed that every individual had the potential to be improved and to contribute to biological or cultural progress. Inherited musical talent, for example, could be improved by practice and passed on in slightly enhanced form to the next generation. Such effects on posterity were what made the individual and his or her upbringing and education most valuable and effective. Education, training, and practice were not Sisyphean chores that had to be repeated entirely from scratch in every generation, as the Weismannian alternative implied.

During World War I Kammerer was called away from the Vivarium to work for the military censor. His experimental work came to a halt and his stocks of modified laboratory animals died out. In his spare time Kammerer wrote antiwar essays, mostly using evolutionary arguments about the benefits of cooperation and the purely eliminative effects of struggle and selection. He also published a textbook of general biology (1915) and a theoretical analysis of coincidences titled *Das Gesetz der Serie* (The law of series, 1919). The latter book was rather speculative, if not pseudoscientific, and it cost him a promotion at the university to *ausserordentlicher Professor* (extraordinary professor—just a rank, not a position on the payroll). Kammerer's rejection for this promotion has also been blamed on anti-Semitism, Kammerer's socialism and wartime pacifism, and rumors about his unreliability as a scientific observer, but Przibram, who was on the committee, said they were ready to overlook everything else before the book appeared.

After the war, Kammerer did little further experimental work. His writings and lectures shifted their focus to the prospects for getting humanity back on track towards evolutionary progress through the controlled acquisition of desirable characteristics. He now drew on the work of his Vivarium colleague, the endocrinologist Eugen Steinach, on the power of glandular secretions, especially the testicular secretion, to modify both body and behavior. In his lectures and popular writings he touted the "Steinach operation," a form of vasectomy, as a means of rejuvenating aging men by stimulating testicular secretion. He also favored other kinds of interventions, such as injections with glandular extracts, radiation treatments, and testicle transplants, as means of improving unfit individuals instead of selecting against them. He argued, based on his prewar experimental results, that the hormonal effects would become hereditary and offer a much more effective and humane alternative to negative eugenic measures, such as sterilization or restrictions on marriage.

Scandal and Suicide. Kammerer resigned from the Vivarium in 1921 and lived from lecturing and writing. A grand tour took him to Britain and the United States in 1923 and 1924, where he promoted the Steinach operation and his proposals for an alternative eugenics, and where he also exhibited his one remaining specimen of a midwife toad that had inherited an acquired characteristic: a nuptial pad. Normally absent in midwife toads, which mate on land, these dark and rough patches of skin on the front legs of the male are found in water-breeding frogs as adaptations for clasping the female. Kammerer's experimental midwife toads had acquired the habit of mating in the water and subsequently developed the pads.

Questions had been raised before about whether the dark patches were nuptial pads, and his tour revived them. The geneticist William Bateson was most aggressive in his questioning, and a testy exchange of letters to *Nature* ensued after Kammerer's visit to Cambridge. The British embryologist Ernest William MacBride also joined in on Kammerer's side.

Kammerer got a much warmer reception in the Soviet Union in May and June 1926. There, Marxist intellectuals at the Communist Academy in Moscow were suspicious of Western genetics and neo-Darwinism, and sympathetic to Kammerer's alternative. They also regarded him as a fellow socialist, and offered him a job as director of a small institute to be built in Moscow under the academy's auspices. Kammerer accepted the offer, but never returned to Moscow to take up the position.

Earlier in 1926, the American herpetologist G. Kingsley Noble had come to the Vivarium to examine the midwife toad and found that the dark color of the purported nuptial pad came from an injection of India ink. Separate accounts and interpretations of the finding, by Noble and Przibram, appeared in *Nature* on 7 August. Kammerer asserted his innocence but committed suicide six weeks later, leaving himself with a posthumous reputation as an ideologically motivated opponent of modern Darwinism and a fraud.

Kammerer's defenders, among them the novelist Arthur Koestler, claim that he was either framed by an unscrupulous Darwinian, eager to discredit unwelcome evidence for Lamarckism, or that a well-meaning assistant had merely touched up a fading specimen that really did once have a distinct nuptial pad. It is also possible that Kammerer inked it himself in order to reduce the glare when taking a photograph of the wet specimen.

BIBLIOGRAPHY

WORKS BY KAMMERER

"Beitrag zur Erkenntnis der Verwandtschaftsverhältnisse von Salamandra atra und maculosa. Experimentelle und statistische Studie." *Archiv für Entwicklungsmechanik der Organismen* 17 (1904): 165–264.

"Experimentelle Veränderung der Fortpflanzungstätigkeit bei Geburtshelferkröte (*Alytes obstetricans*) und Laubfrosch (Hyla arborea)." *Archiv für Entwicklungsmechanik der Organismen* 22 (1906): 48–140.

"Vererbung erzwungener Fortpflanzungsanpassungen. I. und II. Mitteilung: Die Nachkommen der spätgeborenen Salamandra maculosa und der frühgeborenen Salamandra atra." *Archiv für Entwicklungsmechanik der Organismen* 25 (1908): 7–51.

"Allgemeine Symbiose und Kampf ums Dasein als gleichberechtigte Triebkräfte der Evolution." *Archiv für Rassen- und Gesellschafts-Biologie* 6 (1909): 585–608.

"Vererbung erzwungener Fortpflanzungsanpassungen. III. Mitteilung: Die Nachkommen der nicht brutpflegenden Alytes obstetricans." *Archiv für Entwicklungsmechanik der Organismen* 28 (1909): 447–545.

"Mendelsche Regeln und Vererbung erworbener Eigenschaften." *Verhandlungen des naturforschenden Vereines in Brünn* 49 (1911): 72–110.

"Experimente über Fortpflanzung, Farbe, Augen, und Körperreduktion bei *Proteus anguineus* Laur. (Zugleich: Vererbung erzwungener Farbveränderungen. III. Mitteilung)." *Archiv für Entwicklungsmechanik der Organismen* 33 (1911–1912): 349–461.

"Adaptation and Inheritance in the Light of Modern Experimental Investigation." *Annual Report of the Board of Regents of the Smithsonian Institution* (1912): 421–441.

Über Erwerbung und Vererbung des musikalischen Talentes. Leipzig: Theodor Thomas, 1912.

"Vererbung erzwungener Farbveränderungen. IV. Mitteilung: Das Farbkleid des Feuersalamanders (*Salamandra maculosa* Laurenti) in seiner Abhängigkeit von der Umwelt." *Archiv für Entwicklungsmechanik der Organismen* 36 (1913): 4–193.

Allgemeine Biologie. Stuttgart: Deutsche Verlags-Anstalt, 1915.

Einzeltod, Völkertod, biologische Unsterblichkeit, und andere Mahnworte aus schwerer Zeit. Vienna: Anzengruber-Verlag Brüder Suschinsky, 1918.

Das Gesetz der Serie: Eine Lehre von den Wiederholungen im Lebens- und im Weltgeschehen. Stuttgart: Deutsche Verlags-Anstalt, 1919.

Menschheitswende. Wanderungen im Grenzgebiet von Politik und Wissenschaft. Zeit- und Streitschriften des "Friede," 2. Vienna: Verlag "der "Friede," 1919.

"Vererbung erzwungener Formveränderungen. I. Mitteilung: Brunftschwiele der Alytes-Männchen aus 'Wassereiern' (Zugleich Vererbung erzwungener Fortpflanzungsanpassungen. V. Mitteilung)." *Archiv für Entwicklungsmechanik der Organismen* 45 (1919): 323–370.

Sind wir Sklaven der Vergangenheit oder Werkmeister der Zukunft? Vienna, Anzengruber, 1921.

Über Verjüngung und Verlängerung des persönlichen Lebens: Die Versuche an Pflanze, Tier und Mensch gemeinverständlich dargestellt. Stuttgart: Deutsche Verlags-Anstalt, 1921.

"Breeding Experiments on the Inheritance of Acquired Characteristics." *Nature* 111, no. 2793 (1923): 637–640. See subsequent issues of *Nature* for responses from Bateson and others.

Rejuvenation and the Prolongation of Human Efficiency: Experiences with the Steinach-Operation on Man and Animals. Translated by A. Paul Maerker-Branden. New York: Boni and Liveright, 1923.

The Inheritance of Acquired Characteristics. Translated by A. Paul Maerker-Branden. New York: Boni and Liveright, 1924.

Neuvererbung, oder Vererbung erworbener Eigenschaften. Erbliche Belastung und erbliche Entlastung. Stuttgart-Heilbronn: Walter Seyfert, 1925.

OTHER SOURCES

"Eyes: Newt, Rat, Human." *Time,* 18 June 1923, p. 19.

"Famous European Biologist Visits Us: Dr. Paul Kammerer, Associate of Steinach, Driven from Vienna by Poverty." *New York Times,* 25 November 1923.

Bateson, William. "Dr. Kammerer's Testimony to the Inheritance of Acquired Characteristics." *Nature* 103, no. 3 (July 1919): 344–345.

Gliboff, Sander. "'Protoplasm … Is Soft Wax in Our Hands': Paul Kammerer and the Art of Animal Transformation." *Endeavour* 29 (2005): 162–165.

———."The Case of Paul Kammerer: Evolution and Experimentation in the Early Twentieth Century." *Journal of the History of Biology* 39, no. 3 (2006): 525–563.

Hirschmüller, Albrecht. "Paul Kammerer und die Vererbung erworbener Eigenschaften." *Medizinhistorisches Journal* 26 (1991): 26–77.

Hofer, Veronika. "Rudolf Goldscheid, Paul Kammerer und die Biologen des Prater-Vivariums in der liberalen Volksbildung der Wiener Moderne." In *Wissenschaft, Politik und Öffentlichkeit: Von der Wiener Moderne bis zur Gegenwart,* edited by Mitchel G. Ash and Christian H. Stifter. Vienna: WUV-Universitätsverlag, 2002.

Koestler, Arthur. *The Case of the Midwife Toad.* London: Hutchinson, 1971.

Mahler-Werfel, Alma. *Mein Leben.* Frankfurt am Main: Fischer, 1960. Reprint: Frankfurt: Fischer Taschenbuch Verlag, 1980. English: *And the Bridge Is Love.* New York: Harcourt Brace, 1958.

Noble, G. Kingsley. "Kammerer's Alytes (1)." *Nature* 118, no. 2962 (1926): 209–210.

Przibram, Hans. "Kammerer's Alytes (2)." *Nature* 118, no. 2962 (1926): 210–211.

Sander Gliboff

KAPITSA (OR KAPITZA), PETR LEONIDOVICH

(*b.* Kronstadt, Russia, 8 July 1894; *d.* Moscow, U.S.S.R, 8 April 1984), *physics of low temperatures, solid-state physics, engineering.*

Kapitsa contributed to the development of low-temperature physics. His 1930s studies on liquid helium earned a Nobel Prize (1978). An enigmatic figure, he served as a symbol of science in the Soviet Union during the Stalin era and beyond. He had an international reputation, living much of his early career in England, yet was not permitted in 1934 to return to his laboratory in Cambridge where he had worked with Ernest Rutherford for a dozen years, and nearly abandoned his career. He rose to the top of the physics establishment, yet fell under house arrest in Moscow in the late 1940s. He protected such leading Soviet physicists as Vladimir Fock and Lev Landau from almost certain death during the Great Terror in the 1930s, risking his own career to do so, and became a central figure of the conservative scientific establishment in the 1960s. Over the years, Kapitsa wrote dozens of letters to Joseph Stalin, Vyacheslav Molotov, and other Soviet leaders, and to his wife when separated from her. These letters, many of which have now been published in Russian and English, provide an excellent insight into the career and mindset of Kapitsa. In many ways, Kapitsa's career serves as a microcosm of Soviet physics, where gifted scientists had to maneuver through a number of pitfalls, political perils, international isolation, and a low level of industrial support for research.

Early Years. Kapitsa was born to a well-to-do family. His father was a military engineer, his mother a specialist in children's literature and folklore. He studied in a classical gymnasium without great success, then a *Realschule* from which he graduated with honors, and entered St. Petersburg Polytechnical Institute in 1912. The war found him in Scotland, and when he returned to Russia he served as an ambulance driver. Before the war's end he returned to the institute, where he met Abram Ioffe. Ioffe, a specialist in x-ray crystallography and physics of the solid state, was a leading organizer of the nascent discipline of physics in the czarist empire. He had recently established a seminar in which a number of future luminaries of Soviet physics including future Nobel laureate Nikolai Semenov and the theoretician Yakov Frenkel, and others participated.

Petr Leonidovich Kapitsa. *Kapitsa in ceremonial attire at the opening of a new laboratory in Cambridge.* HULTON ARCHIVE/ GETTY IMAGES.

Just after the Russian Revolution, Kapitsa joined Ioffe's newly founded Leningrad Physical Technical Institute, located across the street from the Polytechnical Institute. The Ioffe institute became known as the "cradle" of Soviet physics. A number of physicists who later established their own research centers and leading schools of research, won state prizes, and several who became Nobel laureates began their careers at Ioffe's institute. Revolution and civil war put heavy burdens on Petrograd, science, and especially on the middle and upper classes. The city lost hundreds of thousands of people to death, disease, and emigration. Kapitsa, his father, wife, and two children struggled to survive. Food and fuel were in short supply, starvation and epidemics broke out, and Kapitsa lost his father, wife, and children. In spite of broad social and personal turmoil, Kapitsa managed to conduct a few original investigations during this time on angular momentum associated with magnetization and x-ray crystallography, a technique in which a pattern produced using diffusion of x-rays through a crystal is recorded to reveal the crystal's structure.

Ioffe was both an institute builder and an ambassador of science who helped reestablish contacts with western scientists after World War I and the revolution. He journeyed abroad several times with scarce hard currency acquired from the Bolshevik government in order to obtain various apparatuses and reagents and renew subscriptions to foreign journals. Ioffe invited Kapitsa on a trip in 1921, in part to help Kapitsa recover from his great personal losses. During the first stages of the trip, while Ioffe was in Berlin, he secured a visa for Kapitsa to travel to England, other nations having refused to grant a visa to someone from the young revolutionary Soviet regime. There were few places where someone of Kapitsa's interests in the physics of crystals and of low temperatures might study abroad: Leiden, Netherlands (where the theoretician Paul Ehrenfest, a longtime acquaintance who had studied in St. Petersburg, had settled), Berlin, and Cambridge, England.

Rutherford's Cavendish Laboratory. Kapitsa was intrigued by the possibility of visiting Rutherford's Cavendish Laboratory in Cambridge. His first meetings with the formal Rutherford were difficult, but they became close over the years. According to an apocryphal story, Rutherford initially turned down Kapitsa's request to work with him, saying his laboratory was already too crowded with thirty scientists. Kapitsa then asked what accuracy Rutherford aimed at in his experiments. The startled Rutherford replied within 2 or 3 percent. Kapitsa responded that adding one more researcher to the laboratory would not be noticed since he would fall within experimental error. Kapitsa stayed in Cambridge from July 1921 until 1934; during this time he traveled widely in Europe and was a part of the international physics community.

From the start, Kapitsa impressed Rutherford with his ability to select research problems and his experimental acumen. Kapitsa's first research concerned "The Loss of Energy of an Alpha-Ray Beam in Its Passage through Matter," which involved the study of the velocities of alpha particles by measuring the curvature of their tracks in a magnetic field. The findings were published in the *Proceedings of the Royal Society.* Existing magnets were unable to produce large enough fields to curve the tracks. Kapitsa's idea was to use fields that were much larger but lasted a very short time. Kapitsa frequently used impulsive magnetic fields in the study of the solid state and low temperature and employed specially designed cloud chambers to track the particles. Kapitsa became more and more adept at producing higher magnetic fields with his impulsive field technique. The idea was to design a large dynamo that could generate very high power for a short time through a coil and thus produce a very powerful field. Toward this end, Kapitsa had to design a switch that opened and closed synchronously with the dynamo cycle, while the coil had to be strong enough to withstand very powerful fields.

Kapitsa and Rutherford had a formal relationship built on mutual respect and eventually deep friendship. Kapitsa may have been one of the few people who could joke around Rutherford, although initially Rutherford's insistence on doing things only his way, and Kapitsa's independence, left them acting coldly toward one another. Kapitsa awarded Rutherford the nickname "Crocodile." Contemporary observers and Kapitsa's own letters suggest a variety of reasons why Rutherford earned the name, but it was most likely owing to the initial fear that young scientists had for the imposing Rutherford. Eventually, however, Crocodile became a term of endearment. The Mond Laboratory, established with a grant in 1930 to house Kapitsa's high-field and cryogenic equipment, carries a crocodile etched on its outside wall.

In January 1923 Cambridge admitted Kapitsa as a PhD student. He earned the degree in summer 1923 (in part based on the already completed research) and was also admitted as a Fellow at Trinity College. He later became assistant director of Magnetic Research at Cavendish and then was elected a Fellow of the Royal Society in 1929. In Cambridge, perhaps missing the engagement of Ioffe's Polytechnical seminar, Kapitsa organized the Kapitsa Club, which ran for over forty years. At the weekly seminar participants discussed papers in an informal setting. One could attend and remain a member of the club only by giving a talk from time to time. The minutes and guest book of the club are a fascinating source of the history of twentieth-century physics. Dozens of leading scientists presented at the club on central issues of modern physics, including Paul Dirac, John Chadwick, and Landau. When Kapitsa was forced to remain in the U.S.S.R. in 1934, his British colleagues continued the meetings in his honor, although with some interruptions. When, in May 1966, Kapitsa was permitted to visit Cambridge again, the Kapitsa Club met one last time with Dirac and Kapitsa speaking. Kapitsa's Moscow seminar continued the tradition for Soviet physicists.

Kapitsa was an excellent experimentalist, a skilled engineer, and a handy inventor who produced most of the apparatus needed for his research. Unlike in the U.S.S.R., the Cavendish laboratory had the material that Kapitsa needed. He demonstrated his organizational and engineering skills when in 1925 he earned a large, British government grant (with Rutherford's support) to build a dynamo for the production of high magnetic fields. He studied how the electrical resistance of metals increases with magnetic field, a phenomenon known as Kapitsa's law of magnetoresistance, magnetization of various substances, and magnetostriction. He turned to the construction of a device to liquefy helium to pursue this study, and had great plans to expand his research program and facilities.

Although Kapitsa did not supervise many students directly, either in Cambridge or in Moscow, while working at the Mond Laboratory Kapitsa succeeded in finding short-term fellowships for such Soviet colleagues as nuclear specialists Aleksandr Leipunskii, Kirill Sinel'nikov, Soviet hydrogen bomb specialist and one of the designers of the atomic bomb Yuli Khariton, and theoretician Lev Landau. Some of the funds for the fellowships came from the Rockefeller Foundation's International Educational Board. Kapitsa always remained convinced that science was an international endeavor.

Superfluidity and Stalinist Rigidity. In summer 1934, during his annual summer trip home to Russia, the authorities refused to permit Kapitsa to return to England, although they permitted his Russian wife, Anna Alekseevna Krylova (daughter of the engineer Aleksei Krylov), to leave the country with their two sons. (He had remarried in 1927.) Unlike George Gamow, who managed to get a visa for a foreign conference and never returned to the U.S.S.R., and Vladimir Ipatiev, a member of the nobility and leading chemist who served the Bolshevik regime for over a decade before quitting the U.S.S.R. for life in America, Kapitsa would not be permitted to escape.

The Soviet government claimed that Kapitsa was needed for its industrial programs, alternately suggested he was involved in inappropriate military work in England, and also pressured him to remain by suggesting that he was not a patriot. Some Communist Party officials may also have detested Kapitsa over his enjoyment of British academic life, its formal teas, and the bourgeois lifestyle generally, even though Kapitsa was not a natty dresser. Kapitsa relaxed by motorcycling and driving a Triumph and a Vauxhall automobile. Lev Kamenev, a Bolshevik administrator, and perhaps Ioffe and Nikolai Bukharin (who was head of the Scientific-Technical Department of the Supreme Economic Council), had already approached Kapitsa about working as a consultant at the newly established Kharkov Physical Technical Institute in Kharkov. In any event, Kapitsa would be forced to stay.

Over the next two years, Dirac, Niels Bohr, Rutherford, and several other scientists attempted to secure Kapitsa's freedom to resume his work in Cambridge. They hesitated to do so publicly, preferring to work behind the scenes. Kapitsa's mood deteriorated. He grew isolated from his colleagues, some of whom in fact resented his English sojourn, and he was followed by the secret police. Initially, Kapitsa refused to do research for the Soviet state, and considered leaving physics, perhaps for physiology and biophysics working with Ivan Pavlov. Ultimately, however, Kapitsa began to think of doing research again, and he seems to have convinced Soviet leaders that his studies could not resume unless he acquired the equipment he had specially designed for the Mond Laboratory.

Eventually the Soviet and British authorities, with some assistance from the physics community, agreed to transfer Kapitsa's Mond Laboratory to Moscow for thirty thousand pounds. The equipment arrived from England in 1936 and was installed in the new institute created by and for Kapitsa, the Institute of Physical Problems. The authorities relaxed their surveillance of him and also assigned him a nice apartment, provided him with access to good stores and theater tickets, and even gave him an automobile. Kapitsa was able to produce liquid helium for experiments again early in 1937. With the opening of the institute, Kapitsa returned to full-time research on low-temperature physics, although he was often frustrated by the stark change in his situation and the heavy-handedness of the authorities. The Institute of Physical Problems was small by Soviet standards with only seven scientists in 1937, but its stellar researchers included Landau, among others. It grew to around fifty leading researchers and some two hundred personnel. Anna Kapitsa and the children came to Moscow in January 1936.

The Institute of Physical Problems itself is a lovely oasis in the center of Moscow, just 8 kilometers from the Kremlin, located behind tall, thick walls just off what is now Gagarin Square. The institute's comfortable buildings include an auditorium, dining hall, buffet, and well-equipped library. The buildings for the various departments, laboratories, and machine and glass-blowing shops sit on quiet, tree-lined streets. Kapitsa insisted on creating space for a garden, tennis court, and small park that dropped down toward the Moscow River. The institute became known for weekly seminars, built on the model of the Kapitsa Club in Cambridge, which became a scientific and cultural institution of the physics community and drew leading scholars from around Moscow to discuss important issues.

The entire Kapitsa affair indicates one of the trademarks of science in the Stalin period: the drive to create autarky. Few foreign scientists managed to visit the U.S.S.R. until the Khrushchev period, and fewer Soviet scientists were permitted to travel abroad. Foreign journals were censored, reaching the libraries of institutes after delays of nine months and longer. Scientists were denied the right to send reprints abroad or to publish in foreign journals without getting permission from an increasingly bureaucratic administration.

In 1937 Kapitsa discovered the superfluidity of liquid helium for which he won his Nobel prize. Kapitsa built on the work of a number of specialists studying low temperatures. Heike Kamerlingh Onnes of the Netherlands sought to liquefy gases at low temperatures. He used various apparatuses to produce liquid helium in 1908 and establish its boiling point at 4.2 K. Onnes received the Nobel prize in 1913 for this and other low temperature researches. Subsequently, Onnes and others discovered that liquid helium exhibits odd behavior. By 1924 Onnes had measured liquid helium's density, and determined that as the temperature lowers, the density goes through a sharp maximum at about 2.2 K. In 1927 Willem Keesom and Mieczyslaw Wolfke concluded that that liquid helium undergoes a phase transition at that temperature of roughly 2.2 K (now established at 2.18 K or –270.97 C). This temperature is called the lambda point because the graph of specific heat versus temperature resembles the Greek letter lambda. The two phases are called helium I and helium II. Kapitsa's contribution was to show that helium II was a superfluid.

While investigating the thermal conductivity of liquid helium, Kapitsa measured the flow as the fluid flows through a gap between two discs into a surrounding bath. Above the lambda point, there was little flow, but below the lambda temperature, the liquid flowed with such great ease that Kapitsa drew an analogy with superconductors. It was a liquid of zero viscosity. He discovered the phenomenon in 1937 and published a paper about it in *Nature* in January 1938. He wrote: "The helium below the lambda point enters a special state that might be called a 'superfluid.'"

Superfluids have the unique quality that all their atoms are in the same quantum state; they all have the same momentum, and if one moves, they all move. The best known superfluids are the two isotopes of helium, ^{3}He and ^{4}He. Superfluids move without friction through the tiniest of cracks, apparently defying gravity. If placed in an open container, the superfluid will rise up the sides and flow over the top. If the container is rotated from stationary, the fluid inside will never move. If light or other energy is introduced into a container of a superfluid, and there is an aperture at the top, the fluid will shoot out of the aperture.

At the same time, Kapitsa worked closely with industrial leaders to develop and manufacture devices for the production of liquid oxygen, in part for industrial uses. Kapitsa wanted to cheapen and simplify the large-scale production of liquid oxygen. Kapitsa proposed using an expansion turbine to cool air to its liquefaction temperature. His first machine delivered roughly 30 kilograms per hour with a short startup time and lower expenditure of energy than conventional machines. He patented the design of a turbine motor that overcame instabilities of very high rotation. Liquid oxygen is a powerful oxidizing agent, which means that organic materials will burn rapidly in liquid oxygen. It has therefore found use in rocket fuel. It also has medical applications, for example, in portable oxygen tanks, as well as on jet airplanes and in submarines. His relations with industry and political leaders grew strained during these activities, even though, while in evacuation in Kazan during World War II, he managed to set up a large-scale pilot plant.

Kapitsa wrote and saved a voluminous number of letters to his mother, to his wife, and to Soviet leaders. He continued writing them during both the initial period of his detention and isolation in the U.S.S.R. and after he had returned to work at the Institute of Physical Problems. His letters to officials in the commissariats of Education and of Heavy Industry, the Supreme Economic Council, and other bureaucracies, and to such leaders as Stalin and Molotov, described at length his view of how science ought to be funded and organized and criticized the way that the Soviet government saw things. One theme concerned international isolation. Another concerned the slow progress on opening his institute. Kapitsa also intervened personally through his letters, and perhaps at risk to himself, to save two theoreticians, Fock and Landau, who had been arrested during the Great Terror. Landau, who had escaped arrest in Kharkov, Ukraine, in 1937 by taking a position at Kapitsa's institute, was arrested in Moscow in 1938 and spent over a year in jail before being released to Kapitsa's custody. Landau nearly died in prison; Fock's 1937 ordeal ended after a few days.

A last theme was the government's heavy-handed administration of science to foster more rapid assimilation of scientific achievements into the economy. Kapitsa often invoked metaphors drawn from classical music to describe the treatment of Soviet science by know-nothing administrators. He likened science to a symphony, something that could be written with inspiration, but never according to orders from above. Kapitsa characterized the efforts of industrial ministers to force the pace of science as the equivalent of playing a violin with a hammer. Soviet science was too highly bureaucratized, Kapitsa believed, to compete with Western science.

During World War II scores of institutes, factories, and other strategic facilities were broken down, put into cases and boxes, and placed on railway cars for evacuation just ahead of rapidly advancing German armies. Kapitsa's institute was evacuated to Kazan from 1941 to 1943.

Kapitsa became closely involved with the Soviet atomic bomb project. He was critical of its administration, in particular its direction under secret police chief Lavrenty Beria, and the path chosen by the project leaders: to follow the U.S. lead rather than seek another, perhaps more original solution. He continued his extensive letter-writing campaign concerning what he perceived as organizational and directional problems, and subjected even Beria to hostile comments. His discontent with industry concerning oxygen-production equipment also flared. After these letters, Kapitsa was removed not only from any involvement in the project but also as director of his institute.

Again Kapitsa fell essentially under house arrest. He spent from 1946 until 1954 at his summer home in the Nikolina Hills, or *Nikolina Gora* in Russian, trying to get some research done. From time to time he gave talks in Moscow, but he mostly spent his efforts on putting together a small personal laboratory at what some called the *Izba* (for peasant hut, not *Institute*) of Physical Problems. Kapitsa worked on ball lightning and on the development of powerful high-frequency microwave generators (magnetrons). Anatolii Aleksandrov, another physicist who began his career in the Ioffe institute, who later became the president of the U.S.S.R. Academy of Sciences, was appointed director of the Institute of Physical Problems during this interregnum in Kapitsa's career. A number of physicists criticized Aleksandrov for accepting the directorship, but not Kapitsa, and certainly there was little choice in the matter in Stalin's Russia.

Statesman of Science. After Stalin's death in 1953, Kapitsa resumed direction of the Institute of Physical Problems after Landau led a campaign involving the nation's leading physicists to appeal to Georgii Malenkov, who succeeded Stalin for two years as premier, though only briefly wearing the additional mantle of party chief. (The latter role was assumed almost immediately by Nikita Khrushchev.) During this time, the weekly scientific seminar (the "Kapichnik") grew in importance, and Kapitsa became a senior statesman of Soviet science. Even when other institutes began to outdistance Kapitsa's in terms of equipment and research possibilities, the discussions at the Institute of Physical Problems remained a major institution of Soviet physics. During the reformist Khrushchev years (1955–1964) and the conservative Brezhnev era (1964–1983), Kapitsa contributed to the rapid growth of the scientific establishment as a member of the elite Soviet Academy of Sciences. Roughly speaking, the academy was the bastion of basic research, while universities provided young scientists to staff academy institutes, and industrial research was the province of ministerial institutes.

A bureaucratic separation between research, education, and industry plagued Soviet science. Kapitsa was one of the organizers of the prestigious new Moscow Physical Technical Institute (founded in 1946 as a department of Moscow University), which was formed to train and ensure a steady stream of promising young scientists into physics and engineering. In 1950 or 1951, Kapitsa was dismissed from teaching at the institute for his lack of loyalty to the regime.

Kapitsa's major achievements in the postwar years concerned the reestablishment of scientific contacts with the West, expansion of publication, and the popularization of science. In the 1950s, Kapitsa worked actively to permit Soviet scholars to travel abroad for conferences, while inviting foreign scientists to the U.S.S.R., and cooperated with such centers as Bohr's Institute of Theoretical Physics toward this end. But only in 1965 was Kapitsa himself allowed abroad again to visit Copenhagen to receive the Niels Bohr Gold Medal, and in only 1966 did the authorities permit him to return to Cambridge to receive the Rutherford Medal. He was long-term editor of the *Journal of Experimental and Theoretical Physics.* Kapitsa addressed the issue of how to improve the organization of increasingly bureaucratized and moribund science in a series of papers, but little came of his ideas or those of other senior statesmen. In addition to his Nobel Prize, Kapitsa won Stalin Prizes in 1941 and 1943, Orders of Lenin in 1943 and 1944 for his oxygen work, and the title Hero of Socialist Labor, and was elected corresponding member of the Soviet Academy of Sciences in 1929 and full member in 1939.

In his Nobel Prize speech, Kapitsa chose not to address his work on the physics of low temperatures, a field he acknowledged having left thirty years earlier, but his work and that of his institute on fusion (controlled thermonuclear synthesis). Kapitsa saw nuclear power as a solution to eventual shortages of fossil fuels, and believed that wind, solar, and hydroelectric power were also possibilities. He worried about waste and proliferation problems associated with nuclear fission, and thus pursued fusion. He mentioned pioneering Soviet efforts in the tokamak design for a fusion reactor. He then described a novel approach for heating the plasma to a sufficiently high temperature in order to fuse nuclei. This was to use a high-power continuous wave (CW) microwave generator called a "Nigotron." Kapitsa noted that the CW microwave generator was not invented with thermonuclear reactors in mind, but that, in its development, he and his colleagues discovered the hot plasma phenomenon.

Absent-minded, a joke teller, unconcerned about personal appearance, charming, and cheerful, at times moody, boastful, and self-confident, as his letters to the Kremlin leadership reveal, Kapitsa was an imposing figure. As laboratory leader, he argued that work must stop at 6 p.m., except by special permission, to allow time for reflection, and he insisted upon open disputation of results.

Kapitsa was deeply involved in important social issues for the last two decades of his life. He publicly denounced endemic anti-Semitism in the U.S.S.R.; he criticized the objectionable science and methods of Trofim Lysenko, who dominated Soviet biology after rejecting genetics; and he defended physicist and dissident Andrei Sakharov in discussions of whether to expel the latter from the Academy of Sciences. According to several sources, he pointed out at a Presidium meeting that Albert Einstein had left the Prussian Academy under duress after the rise of the Nazis. This put an end to suggestions that Sakharov be expelled. Kapitsa contributed regularly to Pugwash discussions over arms control. In March 1984, Kapitsa suffered a severe stroke, and he died on 8 April 1984.

BIBLIOGRAPHY

Archival material can be found in the archives of the Institute of Physical Problems, the Ioffe Physical Technical Institute, and the Russian Academy of Sciences (personal fund of P. L. Kapitsa).

WORK BY KAPITSA

"Viscosity of Liquid Helium Below the Λ-point." *Nature* 141 (1938): 74.

"Expansion Turbine Producing Low Temperatures Applied to Air Liquefaction." *Journal of Physics of the USSR* 1 (1939): 7.

Collected Papers of P. L. Kapitza, 4 vols. Edited by D. ter Haar. New York: Macmillan, 1964–1986.

OTHER SOURCES

Badash, Lawrence. *Kapitsza, Rutherford and the Kremlin.* New Haven, CT: Yale University Press, 1985.

Boag, J. W., P. E. Rubinin, and D. Shoenberg, comps. and eds. *Kapitza in Cambridge and Moscow.* Amsterdam: North Holland, 1990.

Esakov, Vladimir Dmitrievich. *Kapitsa, Kreml i nauka: v dvukh tomakh.* Moscow: Nauka, 2003.

Kapitsa, E. L., and P. E. Rubinin, comps. *Petr Leonidovich Kapitsa. Vospominaniia, Pis'ma, Dokumenty.* Moscow: Nauka, 1994.

Rubinin, P. E. *Pis'ma o Nauke.* Moscow: Moskovskii Rabochii, 1989.

Paul Josephson

KAPITZA, PETR

SEE **Kapitsa, Petr Leonidovich.**

KATZ, BERNARD

(*b.* Leipzig, Germany, 26 March 1911; *d.* London, United Kingdom, 20 April 2003), *neurophysiology, action potential, neurotransmitter, synaptic transmission.*

Katz shared a Nobel Prize with Alan Hodgkin in 1963 for their contribution to the development of the Goldman, Hodgkin, Katz (GHK) equations. The equations describe relations among electrical and chemical factors that influence electrical activity in nerve cell axons. He helped demonstrate that synaptic signaling between nerve and muscle or other nerve cells is driven by a chemical mechanism. In 1970 he received a second Nobel Prize for research that helped establish the role of the neurotransmitter acetylcholine (ACh) and the quantal nature of the mechanism that motor neurons use to influence activity at the end plates of postsynaptic muscle cells. He also helped make it plausible that nerve cells use similar chemical mechanisms to signal one other.

One of Hitler's Gifts to British Science. Bernard Katz was born of Jewish parents in Leipzig, Germany. Because his Russian father never became a German citizen, Katz was born stateless as well as Jewish, and he remained a stateless alien until 1941, when he became a British subject. He lived in Leipzig until he finished his medical degree at the University of Leipzig in 1934. In 1935 he migrated to London to work with Archibald V. Hill at University College. Katz was attracted by Hill's work in physiology, as well as his outspoken public opposition to anti-Semitism and Nazism. Hill's personality and attitude toward science inspired him. The motto Katz chose for an autobiographical essay—"To tell you the truth, sir, we do it because it's amusing"—is Hill's reply to an indignant crank who challenged Hill to say what practical use could be made of his highly technical work on muscle physiology. Throughout his career, Hill kept a picture of Adolf Hitler in his office as an expression of his gratitude for Katz and the other gifts the *Führer* gave him.

The disadvantages of being a Jew in Germany were more than obvious to Katz by the age of sixteen, when his reading of Theodor Herzl, the founder of the Zionist movement, and his conversations with like-minded friends inspired him to become a Zionist. Zionism helped sustain him in the face of German anti-Semitism, and eventually it led to an interview with Chaim Weizmann, who arranged some financial support for his move to England. By the end of Katz's time at the University of Leipzig, Hitler had assumed power. (After arriving in England Katz changed his name from Bernhard to Bernard in solidarity with his new home.)

Early Work. While Katz was in Leipzig, his teacher, Martin Gildenmeister, and other physiologists were working to construct precise mathematical descriptions of electrical activity in nerves. Although he later expressed reservations about the value of this enterprise, Katz wrote some successful manuscripts on related research topics. Several were published in *Pfluger's Archiv,* and one was awarded a prize for physiological research. Katz submitted his prize paper under the name Thomas Müller, partly in honor of Hermann Ludwig Ferdinand von Helmholtz's teacher, but mainly because, unlike "Katz," "Müller" is a good Aryan name.

Katz had more trouble with his name in England. Hill arranged for the German publishers of *Ergebnisse der Physiologie* to invite him to write a review of recent literature on nerve fiber responses to electric stimulation. When Katz submitted his manuscript, the editor informed him that *Ergebnisse* could not publish it without an Aryan coauthor. Katz relates in his autobiographical essay that someone suggested approaching Winston Churchill for this purpose! Katz withdrew the manuscript and submitted it to the Oxford University Press, which published it in 1939 under the title *Electric Excitation of Nerve.* In his preface Katz acknowledges the editors of *Ergebnisse,* saying that his appreciation for their encouragement is "in no way diminished by the fact that the ms was refused on 'racial' grounds by the prospective publishers in Berlin."

Katz's book discusses equations developed by physiologists to capture patterns in data from artificially stimulated nerve preparations. To obtain the data, they immersed bits of nerve fiber in solutions whose temperature and chemical compositions they could control. They then inserted microelectrodes into the nerve tissue through which to send direct and alternating current pulses; recording microelectrodes a short distance away were used to monitor responses to the pulses. Stimulus currents of different signs, durations, and strengths were administered in different patterns and time intervals. Investigators manipulated the temperature and chemical composition of the bath to see how these factors influenced electrical activity, and also to damp confounding effects, and to speed up or slow down reactions to make it easier to record them. In some experiments, chemicals were introduced into the nerve itself through tiny pipettes. Stimulating and recording electrodes were arranged in different positions relative to each other so that investigators could measure current development, spread, and decay at different locations along the nerve. These were the experimental methods on which Katz would rely throughout his career.

Paul Fatt, one of Katz's collaborators, recalls how difficult the experiments could be. In 1948 Katz worked in a rickety laboratory near a stairwell. When people tramped up and down the stairs, the floor vibrated so much that it was hard to implant the microelectrodes properly and keep them in position. Fatt recalls

> many times that BK went storming out of the room to remonstrate with someone walking along the corridor or climbing the spiral staircase. … [One] time BK charged out of the room, only to come back silently with a sheepish look—his lovely grin. It was AV [Hill] on the staircase and he had a particularly heavy tread. (D. Katz, Huxley, Fatt, et al., 2003)

Many of the equations Katz reviewed in his book describe "passive" electrical effects, that is, effects that investigators could predict from features of the stimulating current and background conditions in accordance with standard electrical theory. But under some conditions, electrical activity in a nerve fiber spikes nonlinearly in a way that suggests that, instead of responding passively to a stimulus pulse, the nerve fiber itself is doing something to influence current flow. Waves of uniformly elevated electrical activity that travel rapidly without decay down the axon of an excited nerve are called action potentials for this reason. Katz concludes his review with an account of Hodgkin's early investigations of the action potential propagation.

Electric Excitation in Nerve provides a vivid picture of state of the art of neurophysiology during Katz's first years in England. Throughout the book he emphasizes that until a great deal more was known about the anatomy and physiology of the nerve cell, no one would be able to say enough about "the intimate mechanism of the 'nerve membrane' and its functional changes" to explain how and why electrical stimulation produces its effects, or to describe "the physico-chemical nature of 'excitation.'" Katz wrote quotation marks around the terms nerve membrane and excitation to warn us that whatever they signified was poorly understood. Katz himself was more interested in investigating mechanisms than in describing regularities among their effects.

The question of whether chemical mechanisms figure in neuronal signaling, especially with regard to transmission across the synapse, was so far from settled that at a public lecture he attended in 1935 Katz was surprised to observe a rousing verbal battle on this subject between Henry H. Dale and John C. Eccles. Katz reports that as Edgar Adrian served as "a most uncomfortable and reluctant referee," Dale argued that sympathetic nerves use acetylcholine to transmit signals across the synapse while Eccles objected vehemently on the basis of pharmacological evidence.

By this time Hill had given up the search for physical-chemical transmission mechanisms to devote himself to the development of some of the quantitative descriptions that Katz had reviewed in *Electric Excitation.* In his 1966 autobiographical essay, Katz says he thinks this was "in some respects … a retrograde step" back to a research program that exhibited a naive pride in mathematical formulations. But even so, he says it was not "entirely unfashionable" at a time when "some of the most eminent neurologists" had yet to accept "even the basic concept of the membrane potential being involved in the process of electric excitation."

In 1939 Katz moved to Australia. He worked there with Eccles and Stephen Kuffler from time to time until he returned to Hill's department in London in 1946. In 1941 they published the results of frog muscle experiments to argue that nerve cells initiate electrical activity in muscle end plates through chemical rather than purely electrical means. When they stimulated a nerve in a neuromuscular synapse sufficiently to produce action potentials, measurable electrical responses occurred in the end plates of postsynaptic muscle cells. These muscle end plate potentials set up currents that spread passively over a short distance. On repeated stimulation, these potentials sum to magnitudes sufficient to produce action potentials and contractions in the muscle fiber. End plate responses to presynaptic stimulation varied with temperature in ways suggestive of chemical rather than purely electrical interactions. When curare (a drug known to inhibit skeletal muscular responses to ACh) was added gradually to the preparation, the magnitudes and frequencies of end plate potential responses to presynaptic stimulation decreased to extinction. When they added eserine (this drug blocks inhibitors to facilitate ACh interactions), end plate

Bernard Katz. © BETTMANN/CORBIS.

potentials occurred more often and their magnitudes increased. That made it plausible not only that neuromuscular signaling is chemical, but that ACh is the chemical that transmits impulses over the synapse. But the evidence that chemical mechanisms are involved in neuromuscular transmission did not convince the scientific community that nerve cells use neurotransmitters to signal one another. According to Eccles, the notion that central nervous system synaptic transmission is a purely electrical process lingered on until the early 1950s, when newly developed recording techniques enabled investigators to produce enough evidence to kill it.

The Goldman-Hodgkin-Huxley Equations. Katz enlisted in the Royal Australian Air Force as soon as he became a British subject. He served as a radar operator for three years, making use of electrical tricks he had learned from Otto Schmidt's physiology laboratory. During the next year he finished his military career as a liaison officer assigned to the University of Sydney radiophysics laboratory. In Sydney he met and married Marguerite Penly. On his return to England he began the work that led to a paper on the GHK equations that he and Hodgkin published in 1949.

Nerve axons are membrane-lined tubes filled with and bathed in solutions containing K^+, Na^+, Cl^-, and other charged ions. When a nerve is at rest, the charge on the inner surface of the axon membrane is negative relative to the charge on the outer surface. By convention, the membrane potential (the voltage difference between the charges on the inner and outer membrane surfaces) is said to be negative in this condition. From time to time the membrane depolarizes, which is to say that the inner surface becomes less negative relative to the outer surface. During depolarization, action potentials are generated. Action potential propagation stops during hyperpolarization, a redistribution of charges that moves the membrane potential back toward its resting value. According to what has become the standard account of these processes, the charges on either side of the axon membrane are carried by ions in solution at each surface.

Hodgkin and Katz conducted a series of experiments on squid giant axons immersed in salt solutions. They observed how membrane potentials and other electrical quantities changed at different temperatures during such manipulations as holding potassium and chlorine ion concentrations fixed while varying the amount of sodium in the solution. Their data indicated that membrane potential is sensitive to changes in potassium and chlorine concentrations, but that action potentials varied far more directly with sodium concentration. Sodium was the only ion whose addition to the bath produced action potentials, and action potentials did not occur when sodium content fell below a certain level. In keeping with David Goldman's 1943 discussion of electrical activity in an artificial membrane, Hodgkin and Katz proposed that changes in membrane potential result from ion flows across the membrane and, furthermore, that the depolarization that initiates the action potential results from an inward flow of sodium ions.

The reversal potential is the membrane potential at which there is no net ion flow across the membrane in either direction, as happens when the nerve is at rest and at the instant when an ion current changes direction during depolarization or hyperpolarization. Hodgkin and Katz (like Goldman) accepted Walther Nernst's assumption that charged ions in solution tend to flow toward regions of lower concentration and opposite charge. They also assumed that the membrane is not equally permeable to ions of different species. Accordingly, they proposed that at any given temperature the value of the reversal potential is a function of the ratios of inner to outer surface Na^+, K^+, and Cl^- concentrations, weighted by permeability coefficients. This is the GHK voltage equation. (The equation is written $E_{rev} = (RT/F) \ln \left(\frac{P_K[K]_o + P_{Na}[Na]_o + P_{Cl}[Cl]_i}{P_k[K]_i + P_{Na}[Na]_i + P_{Cl}[Cl]_o}\right)$. For each ion species, S, P_S is the permeability constant for S, $[S]_i$ is its concentration inside, and $[S]_o$, its concentration outside the membrane. E_{rev} is the reversal potential for the combined ion currents. T is temperature, F is Faraday's constant, and R is the ideal gas constant).

Using the equation that Hodgkin and Andrew Huxley had developed to describe how ion currents vary with membrane potentials, Hodgkin and Katz transformed the voltage equation into several current equations. Each of these describes the cross membrane current for one ion species as a function of membrane potential together with the ion's valence, the membrane's permeability to the ion, and the ion's concentrations on the inner and outer membrane surfaces. (The GHK voltage equations have the form $I_S = (P_S z_S{}^2)(EF^2/RT) \left(\frac{[S]_i - [S]_o) \exp(-z_S RE/RT}{1 - \exp(-z_S FE/RT)}\right)$. I_S is the current carried by ions of kind S. P_s is the membrane permeability for S ions. z is the valence of S ions. E is the membrane potential.) The GHK equations are accurate only to a rough approximation. Their importance derives from what neurophysiologists have learned about the action potential by investigating and finding factors to account for experimentally detectable discrepancies between GHK predictions and experimentally established values of the relevant quantities.

End Plate Potentials and Neurotransmitters. In 1946 Katz returned to University College in London. In 1948 he and Fatt began the investigations of signaling over neuromuscular junctions that earned him his second Nobel Prize. In 1952 he became professor of biophysics at University College. He chaired the department for twenty-six years.

With Fatt, and later with José del Castillo, Ricardo Miledi, and other distinguished collaborators, Katz measured electrical activity in neuromuscular preparations bathed in salt solutions. They found that when presynaptic nerves are at rest in the absence of any artificial stimulation, very small bursts of electrical activity occur at random intervals in muscle cell end plates on the far side of the neuromuscular junction. Katz and his associates called them miniature end plate potentials. Larger bursts of end plate electrical activity occurred in response to artificially induced presynaptic action potentials. End plate potentials came in different sizes, but their magnitudes appeared to be integral multiples of the magnitudes of the miniature potentials.

Katz and his associates established the chemical nature of neuromuscular signaling by demonstrating that they could both facilitate and damp end plate responses to presynaptic stimulation through manipulations that should have no such effects on the operation of a purely electrical mechanism. Neuromuscular interactions responded to temperature manipulations too small to facilitate or damp the flow of an electric current. Although calcium is not required for electrical transmission, presynaptic nerve cells did not excite muscle end plates in solutions that contained no calcium. Electricity can be transmitted through solutions containing magnesium irons, but presynaptic action potentials do not excite muscle end plates when calcium is replaced by magnesium. Furthermore, the relative magnitudes of electrical quantities in the axon, the synaptic cleft, and the muscle end plate were not related to one another, as would be expected for purely electrical transmission.

These and other experimental results suggested that end plate potentials are produced by neurotransmitter chemicals released in small quantities at random from resting nerves cells, and more regularly and in larger amounts in response to action potentials in excited cells. Such results suggest that miniature end plate potentials are caused by small amounts of a neurotransmitter chemical that leaks out resting presynaptic axons, and that full-fledged end plate potentials are caused by larger amounts of the same chemical released from excited axons in response to action potentials.

Assuming that the magnitudes of end plate potentials depend on how much neurotransmitter the axon releases, the fact that end plate potential magnitudes are multiple integrals of miniature end plate potential magnitudes suggests that neurotransmitters are not released molecule by molecule in continuously larger or small amounts. Accordingly, Katz and his collaborators proposed that neurotransmitter molecules are released in parcels, each one of which contains just enough of the chemical to produce a single miniature end plate potential. The size of the parcels may vary from case to case, but each one contains the smallest amount of neurotransmitter that can produce an end plate response under the circumstances. Similarly, miniature end plate potential magnitudes may vary, but whatever its magnitude may be, no smaller electrical response is possible in any given case. The reason that full-fledged end plate potentials differ from miniature end plate potentials as integral multiples is that the former are caused by the release of two or more parcels of neurotransmitter. Thus Katz could call the process "quantal" even though he recognized that the units of released neurotransmitter and of electrical response could vary in magnitude. By investigating synaptic transmission in squid stellate ganglia, Katz and his colleagues produced evidence that communication between nerves resembles neuromuscular signaling, in that it too depends upon a quantal chemical mechanism that requires calcium for its operation.

Katz and his colleagues continued to argue that the neurotransmitter for neuromuscular synaptic transmission is ACh. ACh introduced into the neuromuscular junction without presynaptic (or any other electrical) stimulation produced the same end plate potentials as presynaptic action potentials. Drugs that damp or enhance end plate responses to artificially introduced ACh have the same effect on responses to presynaptic stimulation in the absence of artificially introduced ACh.

In his 1969 Sherrington Lecture, Katz recapitulated a working hypothesis from a publication he published with del Castillo in 1956. Their idea was that each unit parcel of neurotransmitter is "pre-formed within a synaptic vesicle in the nerve terminal.... The transmitter substance parceled up inside a vesicular bag is separated from its postsynaptic target by ... the vesicular membrane ... and the membrane of the axon terminal." Katz supposed that axon potentials influence the membrane at the end of the axon to raise the probability that the vesicle will pass through it and release transmitter into the synaptic cleft. Some electron microscope evidence for the existence of neurotransmitter vesicles and the discharge of neurotransmitters from them was available to del Castillo and Katz in 1956. More visual evidence appeared during the next two years.

Katz completed his work on synaptic transmission by investigating the locations at which neurotransmitters are released, the postsynaptic locations at which they do their work, and the molecular biology of postsynaptic responses. His subsequent research included investigations of the biochemistry of the pineal gland and its role in melatonin production.

Katz retired in 1978 but continued for some years to referee papers for publication, and discuss ongoing research with erstwhile colleagues and students. In 1999 his wife died of a prolonged illness during which Katz visited her bedside every day to hold her hand and read to her. His son David says that after her death life lost much of its savor for Katz, and he gradually gave up his work. In addition to David, Katz was survived by another son, Jonathan, and three grandchildren.

BIBLIOGRAPHY

WORKS BY KATZ

Electric Excitation of Nerve: A Review. London: Oxford University Press, 1939.

With Alan L. Hodgkin. "The Effect of Sodium Ions on the Electrical Activity of the Giant Axon of the Squid." *Journal of Physiology* (London) 108 (1949): 37–77.

"Depolarization of Sensory Terminals and the Initiation of Impulses in the Muscle Spindle." *Journal of Physiology* (London) 111 (1950): 261–283.

With Paul Fatt. "Some Observations on Biological Noise." *Nature* 106 (1950): 597–598.

"An Analysis of the End-Plate Potential Recorded with an Intracellular Electrode." *Journal of Physiology* (London) 115 (1951): 320–350.

"Spontaneous Subthreshold Activity at Motor Nerve Endings." *Journal of Physiology* (London) 117 (1952): 109–128.

"The Membrane Change Produced by the Neuromuscular Transmitter." *Journal of Physiology* (London) 125 (1954): 546–565.

With José del Castillo. "Changes in Endplate Activity Produced by Pre-synaptic Polarization." *Journal of Physiology* (London) 124 (1954): 586–604.

———. "Quantal Components of the Endplate Potential." *Journal of Physiology* (London) 124 (1954): 560–573.

"On the Localization of Acetylcholine Receptors." *Journal of Physiology* (London) 128 (1955): 157–181.

"The Croonian Lecture: The Transmission of Impulses from Nerve to Muscle and the Subcellular Unit of Synaptic Action." *Proceedings of the Royal Society (Series B)* 155 (1962): 455–477.

With R. Miledi. "A Study of Spontaneous Miniature Potentials in Spinal Motoneurons." *Journal of Physiology* (London) 168 (1963): 389–422.

"The Measurement of Synaptic Delay and the Time Course of Acetylcholine Release at the Neuromuscular Junction." *Proceedings of the Royal Society (Series B)* 161 (1965): 483–495.

"Propagation of Electric Activity in Motor Nerve Terminals." *Proceedings of the Royal Society (Series B)* 161 (1965): 453–482.

Nerve, Muscle, and Synapse. New York: McGraw-Hill, 1966.

The Release of Neural Transmitter Substances. Sherrington Lectures X. Liverpool: Liverpool University Press, 1969.

"Sir Bernard Katz." In *The History of Science in Autobiography,* vol. 1, edited by Larry R. Squire, 348–381. Washington, DC: Society for Neuroscience, 1996. Autobiographical essay.

OTHER SOURCES

Bennett, Max R. "Sir Bernard Katz." *Journal of Neurocytology* 32 (June 2003): 431–436. Obituary.

Eccles, John C. *The Physiology of Synapses.* New York: Academic Press, 1964.

Hille, Bertil. *Ion Channels of Excitable Membranes.* 3rd ed. Sunderland, MA: Sinauer Associates, 2001. See pp. 449–470.

Katz, David, Sir Andrew Huxley, Paul Fatt, et al. "Memories of Bernard Katz: An Afternoon at University College, London, 8th October 2003." Available from http://www.physiol.ucl.ac.uk/Bernard_Katz/.

Valenstein, Elliot S. *The War of the Soups and the Sparks.* New York: Columbia University Press, 2005.

Jim Bogen

KEELING, CHARLES DAVID (*b.* Scranton, Pennsylvania, 20 April 1928; *d.* Hamilton, Montana, 20 June 2005), *atmospheric chemistry, geochemistry, carbon cycle, global climate change.*

Keeling proved that carbon dioxide (CO_2) was increasing in the atmosphere due to human burning of fossil fuels. Scientists had long been interested in the possibility that such a rise might be occurring and causing global warming, but before Keeling's systematic measurements, beginning in the 1950s, CO_2 observations were not accurate enough to reliably detect a global trend. Keeling's time series record of atmospheric CO_2 at Mauna Loa, commonly known as the "Keeling Curve," is generally regarded as being among the most solid evidence of a human impact on the planet as a whole.

Origins. Charles David Keeling was born in Pennsylvania and grew up in the Chicago area. His father, Ralph Franklin Keeling, was a financial analyst; his mother, Grace Sherburne Keeling, was an English teacher. They encouraged his aptitude and interest in music: he won numerous piano competitions as a boy, and often performed for social events, sometimes accompanying his younger sister, Lyla, a singer. He studied chemistry as an undergraduate at the University of Illinois, where he earned a BA in 1948, but his interest in science as career developed when Malcolm Dole of the Northwestern University Chemistry Department, and a friend of his parents, then offered him a graduate fellowship at Northwestern. He completed his PhD there in 1954. While at Northwestern, he developed lasting scientific interests outside of his doctoral field: in geology, meteorology and astronomy.

Keeling's doctoral thesis indicated that double bonds between carbon atoms in polyethylene moved to the end of polymer chains when subjected to a high-energy beam of neutrons. Although this research led to job offers from large chemical manufacturers, especially those involved in developing new plastics, and to promising career prospects in nuclear chemistry, Keeling instead sought opportunities that would allow for research in remote mountainous areas of western North America where he had vacationed as a boy and enjoyed hiking during summers in college. In the fall of 1953, he moved to the California Institute of Technology in Pasadena to become the first postdoctoral fellow of Harrison Brown, head of the geochemistry department there, who was then busy working on his influential book, *The Challenge of Man's Future.* Brown encouraged Keeling's first studies of CO_2.

Detecting and Explaining the CO_2 Rise. Keeling's CO_2 measurements were initially motivated by the question of whether carbonate in rivers and ground waters was in equilibrium with CO_2 in the nearby air, but he soon turned his principal attention to the air measurements themselves. Keeling's air sampling was more precise than that of his predecessors, and this led to the awareness that global trends in atmospheric CO_2 concentrations could be determined by careful and continuous monitoring at a few representative sites.

The notion that humans can influence climate has deep historical roots. The modern hypothesis that CO_2 from industrial emissions might produce a sustained rise in global temperatures emerged out of turn-of-the-twentieth century theorizing by Svante Arrhenius, his associate Nils Eckholm, and his contemporary Thomas Chrowder Chamberlin. It was then developed further in the studies of Guy Stewart Callendar, starting in the 1930s, and by Gilbert Plass in the 1950s. By the mid-1950s, oceanographer Roger Revelle, nuclear physicist Hans Suess, and geochemist Harmon Craig at the Scripps Institution of Oceanography in San Diego were shedding light on the questions of how much and how quickly anthropogenic CO_2 might be absorbed by the oceans, although significant uncertainties remained.

In 1954 planners for the upcoming International Geophysical Year (IGY) in 1957–1958 decided to use that funding opportunity to establish a worldwide network of stations for monitoring atmospheric carbon dioxide. Because CO_2 measurements from the techniques used previously (based on "wet chemical" methods) suggested that CO_2 levels varied widely depending on local conditions, a wide observation network was thought necessary in order to properly detect any overall average trend. When Keeling applied the process developed for his air-water comparisons to systematic measurements of CO_2 in air, over time, and in multiple locations, however, he came up with results that led him to question the accuracy of these prior approaches.

Rather than replicate the existing procedures, Keeling designed and constructed a "manometer" that could measure CO_2 from air—captured in air flasks and extracted by a liquid nitrogen "cold trap"—by means of the pressure it exerted on a column of mercury. With more repeatable gas handling, his approach was much less prone to error than prior wet chemical methods had been. Keeling applied his slow-to-operate but very precise methods to the analysis of samples he took in remote forest locations, where there was little risk of disturbance from local human sources of CO_2. There he consistently found that the concentrations varied systematically with the time of day. The maximum values occurred in the early morning and the minimum values in the early afternoon. He also measured the isotopic ratio of carbon-13 to carbon-12 in his CO_2 samples, which proved that these variations

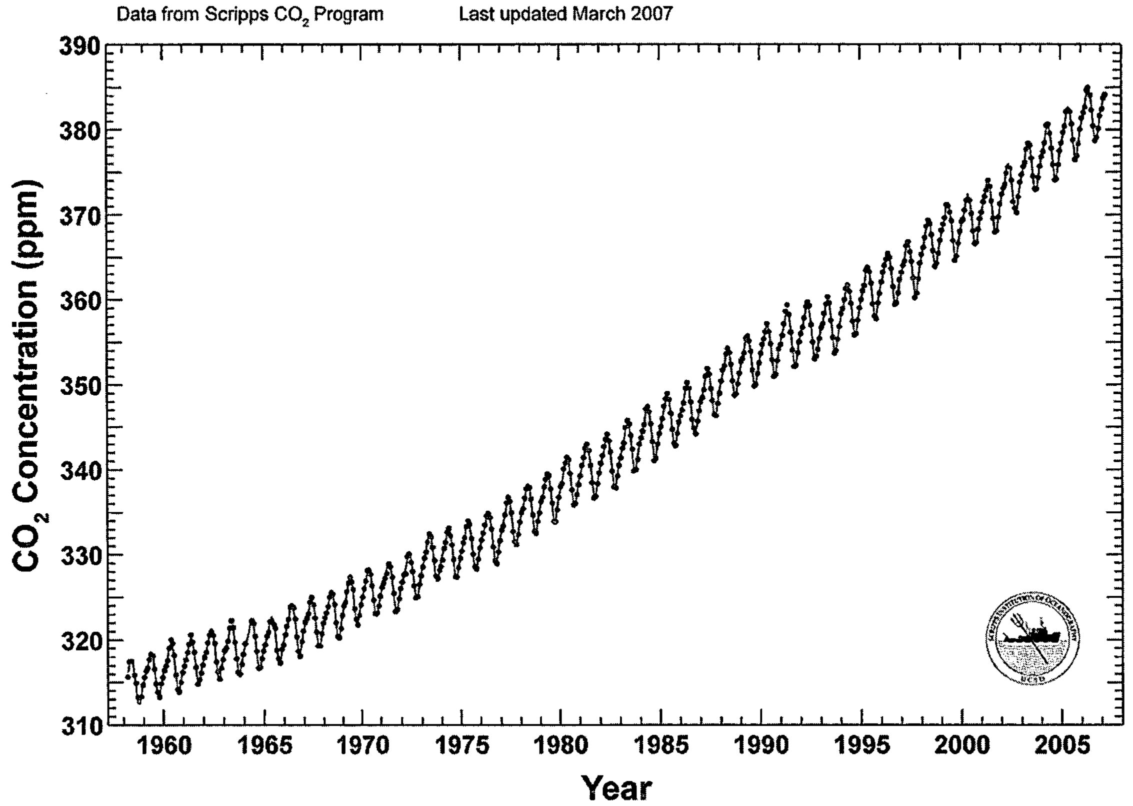

The Keeling Curve. The "Keeling Curve" of CO_2 measurement begun in 1958. SCRIPPS INSTITUTION OF OCEANOGRAPHY.

were the result of natural daily cycles of photosynthesis, respiration, and atmospheric mixing.

During the summer of 1955, as Keeling tracked these "diurnal" fluctuations in parks and wilderness areas up and down the Pacific Coast from Big Sur to the Cascades, the afternoon minimum concentrations of his samples remained surprisingly constant at about 315 parts per million (ppm) of CO_2. This finding was unexpectedly contrary to prior studies, and suggested to Keeling that he was detecting some kind of regional or even global "background level." Additional samples collected from barren mountain tops, in deserts, and at sea further supported the existence of this background level.

Keeling came to realize that the existence of a measurable, uniform average CO_2 concentration, characteristic of a large portion of Earth's atmosphere, meant that the global trend of CO_2 could be determined more readily than IGY planners had previously believed possible. By late 1955, Keeling's CO_2 research had come to the attention of Harry Wexler, director of meteorological research at the U.S. Weather Bureau, and Roger Revelle, the head of the Scripps Institution of Oceanography. Keeling had meanwhile formulated an approach for augmenting manometric methods with much faster infrared analyzers capable of continuous CO_2 measurement. He was invited to interview at both Scripps and the Weather Bureau in Washington, D.C., and was offered jobs by both organizations.

The Weather Bureau was completing construction of an observation station atop Mauna Loa in Hawaii, which Wexler thought would be ideal for the sort of continuous long-term measurement program that Keeling proposed. Revelle, still believing that CO_2 might be too variable to obtain meaningful trends from only a few locations (despite Keeling's 1955–1956 data suggesting a stable "background" level), favored the original IGY scheme of taking CO_2 "snapshots" from a large number of locations during the eighteen months of the IGY, in order to establish a "baseline" value. Any persisting growth in the atmospheric average of CO_2 could be gauged by repeating

Charles David Keeling. *Charles David Keeling in his Scripps Institution of Oceanography laboratory, 1988.* SCRIPPS INSTITUTION OF OCEANOGRAPHY.

the survey in perhaps ten or twenty years. Keeling preferred the continuous approach endorsed by Wexler but was more attracted by the working environment at Scripps. An agreement was therefore reached to take advantage of the considerable funding under IGY, and pursue both continuous long-term CO_2 measurements and observations to set a "baseline." Wexler's Weather Bureau underwrote the establishment of continuous CO_2 measurement at Mauna Loa and Antarctica, while Scripps hired Keeling and helped to fund the geographically widespread sampling from ships and airplanes designed to obtain a "baseline" average. The twenty-eight-year-old scientist, appointed to run the IGY CO_2 program, came to Scripps in the summer of 1956 to take up this time-consuming double job. The baseline sampling was completed by 1962, and the Hawaii and Antarctica measurements continued thereafter.

Keeling's systematic use of carefully calibrated infrared analyzers and extensive flask sampling, meticulously double-checked, and in locations remote from human activity, yielded data over ten times more accurate than previous measurements. In 1960, after just two years of observations, he reported that the continuous Mauna Loa and Antarctic data (confirmed by the samplings conducted from airplanes and ships) showed that CO_2 fluctuated seasonally in the Northern Hemisphere, due to plant growth (confirmed by concurrent measurement of isotopic carbon), and suggested that global concentrations of CO_2 were rising in line with increasing fossil fuel consumption.

The difference between the seasonal peak in May and the seasonal trough in October was only about 5 ppm. This was small relative to the diurnal swing of as much as 100 ppm in the forested areas where Keeling's first samples were taken in 1955 and 1956, and thus went unnoticed at first. At Mauna Loa, however, the seasonal fluctuations were clearly apparent by 1959, the second year of data gathering, because the diurnal fluctuations there were only about 1 ppm. The annual CO_2 rise at Mauna Loa averaged slightly under 1 ppm in the late 1950s and early 1960s. Apart from a number of small but interesting temporary variations, the seasonal cycle and the rate of annual increase changed only slowly over the subsequent decades of Keeling's continuous measurements. By 2000 the seasonal amplitude was about 6 ppm,

Charles David Keeling. *Keeling and the graphs posted in the hallway outside his Scripps office showing CO_2 concentrations at Mauna Loa and other sampling stations.* SCRIPPS INSTITUTION OF OCEANOGRAPHY.

the year-to-year rise about 2 ppm, and the average overall level had grown to 370 ppm.

Once Keeling's initial results were publicized in 1960, interest in "snapshots" at ten- or twenty-year intervals quickly dwindled. The reliability and value of continuous, careful measurements from remote locations was undisputed and has remained so. By the early mid-1960s, Keeling had shown incontrovertibly that CO_2 was increasing in the atmosphere, and his subsequent research became the focal point for growing scientific interest in the implications of rising atmospheric CO_2 for the future climate of the planet.

After five years of measurements, Keeling also had enough observations to credibly estimate that approximately half of all industrially emitted CO_2 was remaining in the atmosphere on a consistent basis. This result was supported by figures showing that industrial production and atmospheric accumulation of the gas were increasing in tandem. In the early 1970s, using well over a decade of data revealing the long-term rate of CO_2 increase more precisely, he also refined estimates of emission levels by making adjustments for the lower carbon content of soft coals, and he developed global carbon cycle models to help track and predict the relative portions going into the air, oceans, and biosphere. Further insights into the carbon cycle also came from the pioneering surveys that Keeling performed in the 1960s concerning CO_2 partial pressure in seawater.

Variations in CO_2 and Changes in Climate. Having identified the global average trend of CO_2 and related it to fossil fuel consumption, Keeling turned his attention in the 1970s and 1980s to investigations of the causes and consequences of variations and anomalies in that average across space and time. Early in his continuous CO_2 tracking, for instance, it became evident that the characteristic seasonal swings due to biotic growth and decay were more pronounced in the Northern Hemisphere because of the preponderance of plant life there. The CO_2 observations also revealed systematic variations in the annual mean concentration with latitude. These variations were featured already in a model of atmospheric mixing developed with Bert Bolin in the early 1960s, and Keeling worked further on such modeling with Martin Heimann and other collaborators in the 1980s.

As the Mauna Loa record was extended, decade by decade, further variations around the general seesawing upward trend also became evident. There were ups and downs on both a three-to-four-year and a ten-year time scale. Unlike the rise of the Keeling Curve itself, these fluctuations were determined to have natural origins. In the mid-1970s Keeling's associate Robert Bacastow established that the three-to-four-year oscillations were tied to the El Niño weather phenomenon, and this synchronicity was incorporated into the carbon cycle modeling collaborations that followed in the 1980s. Isotopic measurements, made later on Keeling's flask samples in a collaboration with Wilhelm Mook, showed that these El Niño oscillations were mainly caused by surges in plant growth and decay and wildfires. In the 1990s Keeling also worked on explaining the ten-year swings in CO_2 by correlating them to cycles in ocean temperatures.

By this point, the longevity and precision of Keeling's CO_2 time series, and his accumulating experience in interpreting its nuances, were enabling him to examine possible interactions between CO_2 levels and changes in climate. In collaboration with biologists and geographers in the mid-1990s, he showed that a long-term increase in the magnitude and duration of seasonal declines in CO_2 was associated with increasing temperatures, vegetation, and longer growing seasons in higher northern latitudes. Seeking to also explain the ten-year ocean temperature cycles correlated with ten-year fluctuations in seasonal CO_2 amplitude, Keeling and his associate Tim Whorf hypothesized that long-term tidal cycles had been causing episodic cooling of ocean surface waters. Their preliminary findings, published in 1997, showed a close match between tides and the periodicity of air temperatures since 1855.

Charles Keeling. *Charles Keeling, 1969.* KEELING FAMILY PHOTOS.

Recognition, Influence, and Nonprofessional Life. Charles David Keeling was appointed as a tenured professor at the newly created University of California (UC), San Diego, in 1964. He retired in 2003 but remained active in science until his unexpected death of a heart attack in 2005. His collaborations with European scientists were furthered by sabbatical leaves in Stockholm in 1961–1962 (with Bert Bolin), Heidelberg in 1969–1970 (with Karl Otto Münnich), and Bern in 1979–1980 (with Hans Oeschger). Keeling wrote more than one hundred published scientific papers, many of them coauthored with scientific collaborators from around the world, or with associates working in the Scripps CO_2 Program that he established at UC San Diego and that continues in the early twenty-first century, directed by his son Ralph Keeling. Charles David Keeling co-convened international conferences on oceanic and atmospheric carbon dioxide, served on the Commission on Atmospheric Chemistry and Global Pollution of the International Association of Meteorology and Atmospheric Physics, and was the scientific director of the Central CO_2 Calibration Laboratory of the World Meteorological Organization from 1975 to 1995. The inexorable upward trend of the Keeling Curve underscored the long-term importance of research into the effects of increasing CO_2 upon climate, influenced early pioneers of global climate modeling such as Syukuro Manabe, and helped inspire the global measurement and study of other atmospheric gases

Over the course of his career, Keeling faced considerable and repeated difficulty in securing financial support for his CO_2 research due to shifting priorities and budgets within funding agencies, and, after the 1960s, by agency administrators who maintained that CO_2 monitoring had become "routine" and was duplicating newer government-run measuring programs. Except for a hiatus from February to April 1964, when Mauna Loa measurements were temporarily shut down, Keeling always managed to obtain funding and keep the measurements going, by assiduously persuading agencies and officials of the significance and unmatched reliability, consistency, and long-lived continuity of the data series being generated by his program.

Keeling received the Second Half Century Award of the American Meteorology Society in 1981, the Maurice

Charles Keeling. *Charles Keeling sampling in Montana, 2003.* KEELING FAMILY PHOTOS.

Ewing Medal of the American Geophysical Union in 1991, the Blue Planet Prize in 1993, the National Medal of Science in 2002, and the Tyler Prize for Environmental Achievement in 2005. He was a Fellow of the American Academy of Arts and Sciences, the American Geophysical Union, and the American Association for the Advancement of Science, and a member of the National Academy of Sciences and the American Philosophical Society.

Keeling married Louise Barthold in 1954, and they had five children. In addition to his scientific career, he was an avid hiker and conservationist, and in 1974–1975 chaired the citizens' General Plan revision committee of Del Mar, California, where he lived from the late 1950s. He enjoyed hiking near the summer home he established in Montana in the 1980s, sometimes taking air samples in the forest there. Keeling was an accomplished nonprofessional musical performer, teacher, and composer, and, throughout his life, often played and performed chamber music with family and friends. From 1964 to 1969 he directed the UC San Diego Madrigal Singers.

The author gratefully acknowledges the careful and extensive scrutiny and assistance of Professor Ralph Keeling, Scripps Institution of Oceanography.

BIBLIOGRAPHY

Archival Sources: C. D. Keeling Collection, Scripps Institution of Oceanography Archives, University of California, San Diego. A full list of Keeling's publications can be found on the Scripps CO_2 Program Web site, which is referenced under "Other Sources" below.

WORKS BY KEELING

"Variations in Concentration and Isotope Abundances of Atmospheric Carbon Dioxide." In *Proceedings of the Conference on Recent Research in Climatology,* edited by H. Craig, 43–49. San Diego: Committee on Research in Water Resources and University of California, Scripps Institution of Oceanography, 1957.

"The Concentration and Isotopic Abundances of Carbon Dioxide in Rural Areas." *Geochimica et Cosmochimica Acta* 13 (1958): 322–334.

"The Concentration and Isotopic Abundances of Carbon Dioxide in the Atmosphere." *Tellus* 12 (1960): 200–203.

With Bert Bolin. "Large-Scale Atmospheric Mixing as Deduced from the Seasonal and Meridional Variations of Carbon Dioxide." *Journal of Geophysical Research* 68 (1963): 3899–3920.

With Jack C. Pales. "The Concentration of Atmospheric Carbon Dioxide in Hawaii." *Journal of Geophysical Research* 70, no. 24 (1965): 6053–6076.

With Craig W. Brown, "The Concentration of Atmospheric Carbon Dioxide in Antarctica." *Journal of Geophysical Research* 70, no. 24 (1965): 6077–6085.

With Norris W. Rakestraw and Lee S. Waterman. "Carbon Dioxide in Surface Waters of the Pacific Ocean. 1. Measurements of the Distribution." *Journal of Geophysical Research* 70 (1965): 6087–6097.

"Carbon Dioxide in Surface Waters of the Pacific Ocean. 2. Calculation of the Exchange with the Atmosphere." *Journal of Geophysical Research* 70 (1965): 6099–6102.

"Is Carbon Dioxide from Fossil Fuel Changing Man's Environment?" *Proceedings of the American Philosophical Society* 114 (1970): 10–17.

"The Carbon Dioxide Cycle: Reservoir Models to Depict the Exchange of Atmospheric Carbon Dioxide with the Oceans and Land Plants." In *Chemistry of the Lower Atmosphere,* edited by S. I. Rasool, 251–329. New York: Plenum Press, 1973.

"Industrial Production of Carbon Dioxide from Fossil Fuels and Limestone." *Tellus* 25 (1973): 174–198.

"The Influence of Mauna Loa Observatory on the Development of Atmospheric CO Research." In *Mauna Loa Observatory 20th Anniversary Report,* edited by J. Miller, 36–54. Washington, DC: National Oceanographic and Atmospheric Administration, 1978.

With Roger Revelle. "Effects of El Niño/Southern Oscillation on the Atmospheric Content of Carbon Dioxide." *Meteoritics* 20 (1985): 437–450.

With Eric From. "Reassessment of Late 19th-Century Atmospheric Carbon Dioxide Variations." *Tellus* 38B (1986): 87–105.

With Martin Heimann, et al. "A Three Dimensional Model of Atmospheric CO_2 Transport Based on Observed Winds." In *Aspects of Climate Variability in the Pacific and the Western Americas,* edited by David H. Peterson, 165–363. Washington, DC: American Geophysical Union, 1989.

With Timothy P. Whorf, Martin Wahlen, and Johannes van der Plicht. "Interannual Extremes in the Rate of Rise of Atmospheric Carbon Dioxide since 1980." *Nature* 375 (1995): 666–670.

With John F. S. Chin and Timothy P. Whorf. "Increased Activity of Northern Vegetation Inferred from Atmospheric CO_2 Measurements." *Nature* 382 (1996): 146–149.

With Ranga B. Myneni, Compton J. Tucker, Ghassem Asrar, and Ramakrishna R. Nemani. "Increased Plant Growth in the Northern High Latitudes Due to Enhanced Spring Time Warming." *Nature* 386 (1997): 698–702.

With Timothy P. Whorf. "Possible Forcing of Global Temperature by the Oceanic Tides." *Proceedings of the National Academy of Sciences of the United States of America* 94 (1997): 8321–8328.

"Rewards and Penalties of Monitoring the Earth." *Annual Review of Energy and the Environment* 23 (1998): 25–82.

With Timothy P. Whorf. "The 1,800-Year Oceanic Tidal Cycle: A Possible Cause of Rapid Climate Change." *Proceedings of the National Academy of Sciences of the United States of America* 97 (2000): 3814–3819.

With Nicolas Gruber and Nicholas R. Bates. "Interannual Variability in the North Atlantic Ocean Carbon Sink." *Science* 298 (2002): 2374–2378.

OTHER SOURCES

Bolin, Bert, and Pieter Tans. "In Memory of Charles David Keeling." *Tellus* 58B (2006): 328–329.

Bowen, Mark. *Thin Ice: Unlocking the Secrets of Climate Change in the World's Highest Mountains.* New York: Henry Holt, 2005.

Fleming, James Rodger. *Historical Perspectives on Climate Change.* New York: Oxford University Press, 1998.

———. *The Callendar Effect.* Boston: AMS Books, 2007.

Heimann, Martin. "Charles David Keeling, 1928–2005." *Nature* 437 (15 September 2005): 331.

Scripps Institution of Oceanography. "CO_2 Program." Available from http://scrippsco2.ucsd.edu/.

Weart, Spencer R. *The Discovery of Global Warming.* Cambridge, MA: Harvard University Press, 2003.

Weiner, Jonathan. *The Next One Hundred Years: Shaping the Fate of Our Living Earth.* New York: Bantam, 1990.

Drew Keeling

KEELING, DAVE

SEE **Keeling, Charles David.**

KEKULÉ VON STRADONITZ (KEKULÉ), (FRIEDRICH) AUGUST

(*b.* Darmstadt, Germany, 7 September 1829; *d.* Bonn, 13 July 1896), *chemistry, benzene ring.* For the original article on Kekulé see *DSB,* vol. 7.

Kekulé's theoretical insights into the classification of organic compounds, valence and structural chemistry, and his suggestion of the hexagonal formula for benzene transformed the subject of organic chemistry. In his great paper of 1858, "On the Constitution and Metamorphoses of Chemical Compounds and on the Chemical Nature of Carbon," in which he extended the quadrivalence of carbon in methane to all of its compounds, Kekulé stressed how indebted he was to the English and French schools of chemistry for his interpretation. With hindsight it can be seen that at one stroke organic chemistry had been unified: chemists no longer needed to separate "types" for paraffins, ethers, and amines; all organic compounds were now embraced within the idea of carbon chains (catenation) and the notion of carbon's tetravalence.

Pathway to Insight. Historians have debated why this insight occurred to Kekulé. Alan J. Rocke has suggested that the concept of structure (application of valence rules to the supposed construction of molecules) probably owed much to the work of Adolphe Wurtz. In his influential *Méthode de chimie* (1854), Auguste Laurent had speculated that atoms might be divisible in order to explain why, for example, iron had both odd and even powers of combination to form ferrous and ferric salts. At the same time, Alexander Williamson and William Odling were developing the idea of double, triple, and mixed "types" in which one dibasic (diatomic) molecule linked together two monobasic radicals. In 1855, following Laurent's hint, Wurtz wondered whether oxygen was dibasic and

August Kekulé. *Friedrich August Kekulé von Stradonitz.* SPL / PHOTO RESEARCHERS, INC.

nitrogen tribasic because these elements were formed from two or three juxtaposed subatoms. According to this concept, polyvalent atoms were really aggregates of monovalent subatoms. This, as Rocke has pointed out, conforms to the later textbook rule that an element's chemical equivalent is "atomic weight divided by valence" and is still reflected in the definition of atomicity as the number of atoms in a molecule (rather than of subatoms in an atom).

Kekulé's reverie on a London omnibus (recounted in 1890 and occurring, if it was a real event, probably in 1855) therefore may have involved segmented wormlike entities made up from subatoms. The sausage-shaped graphic formulas that he first used in lectures in Heidelberg in 1857 followed from this model and seem confirmed by Kekulé's statement in 1867 that "polyvalent atoms, with respect to their chemical value [valence], can be viewed in a sense as a conglomeration of several monovalent atoms" (Kekulé, 1867, p. 217). The graphic formulas that Kekulé used in his *Lehrbuch der organischen Chemie* (1859–1887) were visualizations of Wurtz's speculation, though he continued to use type formulas as a means of classification. It was Alexander Crum Brown and Edward Frankland who slowly directed the chemical community toward graphic formulas in which the "carbon chain" property was made explicit. The heuristic nature of Kekulé's formulas may have been responsible for his refusal to accept that valence could vary. Although he was proved wrong, nevertheless, it was his conviction that carbon's valence was invariable that logically led him to posit the carbon chain, double and triple bonding, and the benzene ring (1865).

Benzene Ring and "Kekulé's Dreams." No one can be certain exactly when Kekulé hit upon the structure of benzene as a closed chain of six carbon atoms, though it probably occurred to him around the time of his first marriage in 1862. Toward the end of his life, in typical self-deprecatory fashion, he said that, in a dream, he had imagined a chain of dancing carbon atoms forming a closed circle, like a snake eating its own tail. Since the 1970s, provoked principally by the Czech American chemist and historian John H. Wotiz (1919–2001), historians have debated the validity and nature of Kekulé's two "dreams" on historical and chemical grounds and whether the stories they told were a cloak to disguise the fact that Kekulé was less original than he claimed. These reinterpretations of Kekulé's character and reputation led to resentments between chemist-historians and historians of chemistry that culminated in 1995 in a complicated legal action that affected a national society and brought ad hominem restrictions on a few historians' freedom of action (Wotiz, 1993, pp. 108–110).

Historians of chemistry have tended to stress the slow evolutionary continuity of Kekulé's work in organic chemistry rather than the sudden emergence of particular insights. They have also stressed the importance of the careful examination of "context of use" when considering whether or not Kekulé was indebted to Laurent, Archibald Scott Couper, and Frankland for ideas concerning constitutional formulas, or Aleksandr Butlerov for the concept of chemical structure, or to Laurent, Albert Ladenburg, and others for the hexagonal formula of benzene. This aspect of historiography has not always been appreciated by chemist-historians. Another debated issue has been whether Kekulé's benzene formula was indebted to the graphic circle (or ball) formula that Josef Loschmidt privately published in his *Chemische Studien* in 1861. Rocke and Schiemenz have argued plausibly that Loschmidt was portraying benzene as an indeterminate hexavalent superatom using Kekulé's previously published concept of quadrivalent carbon catenation, rather than trying to represent six individual carbon atoms connected in a hexagon. It must also be recognized that the "Benzolfest" of 1890 at which Kekulé recounted his reveries was designed less to honor Kekulé than to impress participants and readers of the central economic significance of chemistry for the Reich. Consequently, accounts of the speeches have to be used with caution as historical texts.

Later Years. Despite his declining powers, lethargy, and fixation with family history and ennoblement, which Hermann Kolbe exploited in inexorable criticism of Kekulé's structure theory, Kekulé did make some significant achievements at Bonn. This included, in 1872, the daring dynamic oscillation formula for benzene that explained the embarrassing lack of isomeric disubstituted derivatives in benzene that seemed otherwise possible. In the early 1880s, when several alternative possibilities had been touted, including the prism formula proposed by Ladenburg, Kekulé demonstrated a return of his old powers when he showed that a series of experimental transformations of pyrocatechol and quinone into dioxytartaric acid and trichloracetoacrylic acid were best and most simply explained if benzene was hexagonal.

While it is convenient to suppose that it was Kekulé's architectural training that helped him conceive molecular structure and to play with molecular models, what is more striking is the view he acquired from Williamson of the dynamic nature of molecules. Architecture is essentially static, whereas Kekulé's conception of structure was much more fluid and imprecise. To that end, the visionary giddy molecular dances of his Benzolfest address in 1890 ring true.

SUPPLEMENTARY BIBLIOGRAPHY

Kekulé's principal archives are housed in the Kekulé Sammlung, Institut für Organische Chemie, Technische Hochschule Darmstadt, Petersenstrasse 22, and the University of Ghent, Belgium. The Museum of Science and Technology at Ghent, Korte Meer 9, displays Kekulé's apparatus and models.

WORKS BY KEKULÉ

"Über die Constitution des Mesitylens." *Zeitschrift für Chemie* 10 (1867): 214–218.

With Justus von Liebig. *Liebigs Experimentalvorlesung: Vorlesungsbuch und Kekulés Mitschrift.* Edited by Otto Paul Krätz and Claus Priesner. Weinheim, Germany: Verlag Chemie, 1983. Facsimiles and transcriptions of Liebig's Giessen lectures on organic chemistry, together with Kekulé's student notes of 1848 and valuable commentaries.

OTHER SOURCES

Brock, William H. "August Kekulé (1829–96): Theoretical Chemist." *Endeavour* 20, no. 3 (1996): 121–125.

Brooke, John H. "Doing Down the Frenchies: How Much Credit Should Kekulé Have Given?" In *The Kekulé Riddle: A Challenge for Chemists and Psychologists,* edited by John H. Wotiz, 59–76. Clearwater, FL: Cache River Press, 1993.

Brush, Stephen G. "Dynamics of Theory Change in Chemistry, Part 1: The Benzene Problem, 1865–1945" and "Part 2: Benzene and Molecular Orbitals, 1945–1980." *Studies in*

History and Philosophy of Science 30, no. 1 (1999): 21–79; 30, no. 2 (1999): 263–302.

Fisher, Nicholas W. "Kekulé and Organic Classification." *Ambix* 21 (1974): 29–52.

Göbel, Wolfgang. *Friedrich August Kekulé*. Leipzig, Germany: Teubner, 1984.

Kauffman, George B. "Werner, Kekulé, and the Demise of the Doctrine of Constant Valency." *Journal of Chemical Education* 49 (1972): 813–817.

Meinel, Christoph. "Molecules and Croquet Balls." In *Models: The Third Dimension of Science,* edited by Soraya de Chadarevian and Nick Hopwood, 242–275. Stanford, CA: Stanford University Press, 2004.

Noe, Christian R., and Alfred Bader. "Facts Are Better Than Dreams." *Chemistry in Britain* 29 (February 1993): 126–128.

Paoloni, Leonello. "Stereochemical Models of Benzene, 1869–1875." *Bulletin for the History of Chemistry* 12 (1992): 10–23.

Ramsay, O. Bertrand, and Alan J. Rocke. "Kekulé's Dreams: Separating the Fiction from the Fact." *Chemistry in Britain* 20 (December 1984): 1093–1094.

Rocke, Alan J. "Kekulé, Butlerov, and the Historiography of the Theory of Chemical Structure." *British Journal for the History of Science* 14 (1981): 27–57.

———. "Subatomic Speculations and the Origin of Structure Theory." *Ambix* 30 (1983): 1–18.

———. "Hypothesis and Experiment in the Early Development of Kekulè's Benzene Theory." *Annals of Science* 42 (1985): 355–381.

———. "Kekulé's Benzene Theory and the Appraisal of Scientific Theories." In *Scrutinizing Science: Empirical Studies of Scientific Change,* edited by Arthur Donovan, Larry Laudan, and Rachel Laudan, 145–161. Dordrecht, Netherlands, and Boston: Kluwer, 1988.

———. "Waking Up to the Facts?" *Chemistry in Britain* (May 1993): 401–402. A reply to Noe and Bader.

Rudofsky, Susanna F., and John H. Wotiz. "Psychologists and the Dream Accounts of August Kekulé." *Ambix* 35 (1988): 31–38.

Russell, Colin A. "Kekulé and Frankland: A Psychological Puzzle?" In *The Kekulé Riddle: A Challenge for Chemists and Psychologists,* edited by John H. Wotiz, 78–101. Clearwater, FL: Cache River Press, 1993.

Schiemenz, Günter P. "Albert Ladenburg und die Kekuléformel des Benzols." *Mitteilungen der GDCh Fachgruppe Geschichte der Chemie* 1 (1988): 51–69. A revised English version is in *The Kekulé Riddle: A Challenge for Chemists and Psychologists,* edited by John H. Wotiz, 104–122. Clearwater, FL: Cache River Press, 1993.

———. "Goodbye, Kekulé? Josef Loschmidt und die monocyclische Struktur des Benzols." *Naturwissenschaftliche Rundschau* 46 (1993): 85–88.

———. "A Heretical Look at the Benzolfest." *British Journal for the History of Science* 26 (1993): 195–205. Article suppressed from Wotiz (1993) because of views on Loschmidt.

Seltzer, Richard J. "Influence of Kekulé Dream on Benzene Structure Disputed." *Chemistry & Engineering News,* 4 November 1985, pp. 22–23. A good account of historians' dispute followed by spirited correspondence lasting three months.

Wotiz, John H. Interview by Herbert T. Pratt at Newcastle, Delaware, and Washington, DC, 7, 8, and 10 August 2000. Philadelphia: Chemical Heritage Foundation, Oral History Transcript # 0197.

———, ed. *The Kekulé Riddle: A Challenge for Chemists and Psychologists.* Clearwater, FL: Cache River Press, 1993. Contains seventeen essays by chemists, historians, and psychologists.

Wotiz, John H., and Susanna Rudofsky. "Kekulé's Dream: Fact or Fiction? *Chemistry in Britain* 20 (August 1984): 720–723.

———. "The Unknown Kekulé." In *Essays on the History of Organic Chemistry,* edited by James G. Traynham, 21–34. Baton Rouge: Louisiana State University Press, 1987. Preliminary claim of Kekulé's fraudulence.

———. "Louis Pasteur, August Kekulé, and the Franco-Prussian War." *Journal of Chemical Education* 66 (1989): 34–36.

W. H. Brock

KELDYSH, MSTISLAV VSEVOLODOVICH

(*b.* Riga, Latvia, Russia, 10 February 1911; *d.* Moscow, U.S.S.R., 24 June 1978), *applied mathematics, mechanics, aircraft design, rocketry, space science.*

A famous Russian mathematician closely identified with the Russian space research program, Keldysh is ranked among the most prominent founders of the Soviet missile production centers serving both military and civil aims. He served as president of the Soviet Academy of Sciences between 1961 and 1975.

Origin and Educational Background. Mstislav was the fifth child of Vsevolod Mikhaylovich Keldysh (1878–1965), a prominent civil engineer, who was a professor at the Riga Polytechnic Institute. In 1915, after the beginning of World War I and the German invasion of Latvia, Keldysh's family moved to Moscow, following the evacuated Polytechnic Institute. After the war, in 1918, upon the return of the institute to Riga, Keldysh's father decided to stay in Moscow together with a larger part of the other Russian professors. After holding various university and engineering ranks he became a member of the Military Engineering College in 1932, and, subsequently vice president of the Soviet Academy of Building and Architecture. Keldysh's mother, Maria Alexandrovna Kovzan (1879–1957), was a general's daughter. She gave her seven children (three daughters and four sons) a good primary education and encouraged their interest in literature, music, and painting.

Keldysh

In 1927, after finishing secondary school, Mstislav, willing to follow his father's footsteps, decided to enter the prestigious Moscow Civil Engineering Institute. However, his application was rejected because he was too young. Keldysh was only sixteen, while enrollment age into Higher Technical Schools commonly began from eighteen. Regular universities did not have any age limits, so Keldysh entered the Mechanics and Mathematics Faculty of Moscow University. He did not side with Nikolay Luzin's "pure" mathematical school (the so-called Luzitania), which dominated at that time, and began to specialize in the theory of functions of the complex variable, which had a lot of practical applications. His main interest was applied mathematics with various contributions to engineering and technological issues. Keldysh's first research, presented in 1932, dealt with aerodynamics and was devoted to a more precise definition of Nikolai Zhukovski's formula (lift force of the wing) with regard to air compressibility.

From Aviation to Rocketry. In 1931, after his graduation, Keldysh joined the General Theoretical Group of the Central Institute of Aero-Hydrodynamics (CAHI). Simultaneously, he was appointed professor of mathematics at the Moscow Electro-Machine-Building Institute (1930) and then at Moscow University (1932–1953). At CAHI Keldysh began to work on different problems concerning aircraft construction. One of them was research on self-oscillation of a wing at high flying speed—the so called flutter phenomenon. At the early stage of aviation this unknown phenomenon had caused a few unexpected accidents. At the beginning of the 1930s, a special "flutter group" was created at the CAHI Experimental-Aerodynamic Department. During two years of work the group could not produce any satisfactory result. Solving the problem needed a combination of high engineering skills and advanced mathematical qualification. Keldysh united both these qualities in the same person. Soon after he became the head of the flutter group, the problem was successfully resolved. In coauthorship with Evgeny P. Grossman, Keldysh developed a general theory of the phenomenon and worked out ways to eliminate it for airplanes of various design.

Keldysh's success in solving the most complicated scientific and engineering problems was combined with great organizational abilities and high personal qualities. For example, in 1939, when Keldysh realized that flutter required radical changes in the technology of airplane design and a sharp increase of mathematical calculations, he organized the All-Union Flutter Seminar. Regular participants of the seminar, represented by employees from different design bureaus, were obliged not only to attend theoretical courses, but also to conduct detailed flutter analyses of the airplane designs currently developed at their laboratories and institutes. From the very beginning he commenced to establish new forms of scientific communication, which helped him unite efforts of the most prominent Soviet specialists. In the course of work on the flutter problem Keldysh met illustrious and already well-known airplane designers, such as Semyon A. Lavochkin and Andrey N. Tupolev. In 1942 Keldysh and Grossman were honored with the Stalin Prize for scientific work on prevention of break-up of airplanes.

The next step in Keldysh's scientific carrier was closely connected with the beginning of two major Soviet scientific projects—the space research program and the development of nuclear weapons. In 1944 he became the head of a newly established Mechanics Department at the Mathematical Institute of the Soviet Academy of Sciences. By that time Keldysh already had a doctoral degree in mathematics (1938) and was a corresponding member of the Academy of Sciences (1943). During this period he began to shift to the fields of rocket dynamics and applied celestial mechanics. In 1946, immediately after his election to full Soviet Academy membership, he was appointed a director of the secret Scientific Research Institute of Rocket Aviation (so-called NII-1). As in the case of the flutter group, Keldysh came to the institute when it was on the brink of collapse. Because of the conflict with the Ministry of Aircraft Construction that arose during the postwar industrial reorganization, the institute lost almost half of its staff and a significant part of its working area. Keldysh undertook a number of immediate measures (including a personal appeal to Joseph Stalin) to adjust the institute to the new political and professional situation. He raised standards for theoretical background for employees, broadened laboratories, and equipped the institute with the latest test stands. The institute began to study air-breathing and liquid-propellant rocket engines, and develop their fuel and strength characteristics. The first achievement of the institute was the creation and successful in-flight testing of the ramjet engine.

The scientific and technological goals of Keldysh's institute clearly meshed with the military ones. In 1948 he received an invitation to participate, as an expert, in the final testing of the ballistic missile R-1, developed at the secret Scientific Research Institute of Missile Armament. Here he met the legendary Soviet designer Sergey P. Korolev. Their friendship lasted for almost twenty years, till Korolev's death in 1966. Very soon Keldysh met another powerful figure in the Soviet scientific establishment: the director of the Soviet nuclear project, Igor V. Kurchatov. Already in 1946 Keldysh began to participate in numeral calculations of atomic problems. Soon after that he became head of the Atomic Problems Department, established in the Mathematical Institute. In 1953 this department broke away into an autonomous body, the Institute of Applied Mathematics. Since the very

Keldysh

beginning Keldysh continuously was director. The alliance of these three prominent specialists—Korolev, Kurchatov, and Keldysh—was informally named the "Three Ks."

Regime: Restrictions and Opportunities. Biographies of Keldysh written during the Soviet period do not reflect the reverse side of his brilliant carrier. During the Stalinist time he and all his family (his wife Stanislava Valerianovna, daughter Svetlana, and son Piotr) lived in constant fear for their lives. His "dubious" social origin (Keldysh's parents belonged to the gentry) was a problem not only for him, but also for his senior colleagues, at least those who recommended him for leading academic posts. In 1935 the KGB threw his mother in the notorious Lubianka prison. She was freed in a few weeks, but soon, in 1936, Keldysh's brother Mikhail (a postgraduate student at the Historical Faculty of Moscow University) was arrested and executed. The eldest brother Alexandr also was arrested on the pretext of being a French agent. Fortunately, before he could be sent to a confinement camp, the chief of Lubianka was arrested too. His successor started reviewing current sentences and found that Keldysh's brother had been framed. There is no primary evidence how Keldysh responded to these dramatic events. Some of his former colleagues mentioned that he preferred to employ people with social status similar to his own. The reason was not only sympathy or the wish to render support. Different treatment of family members was common during Stalin's rule (the most striking example is the case of two brothers, Sergey and Nikolay Vavilov: one was executed, but the other appointed president of the Academy of Sciences). Relatives of punished people usually worked with more diligence and could be subordinated with relative ease. Keldysh could simply reactivate that steady homology of Stalin's regime.

Keldysh never associated himself with so-called pure science. The university supervisors reproached him for his passion for practical problems, thinking it might ruin his mathematical talent. However, the "practical instinct" of Keldysh utterly fitted the ideal image of the Stalinist science, which demanded a "definite materialistic basis and practical orientation." This necessity to serve demands of both academic and administrative elites left its mark on a number of Keldysh's activities. For example, the title of his first scientific paper, devoted to Zhukovski's formula and concerning particular technical problems, sounded rather "academic": "Exterior Neumann's Problem for Non-Linear Differential Equation of Elliptical Type." As a mathematician, he was elected a full member of the academy in the Engineering Sciences Division of the Soviet Academy of Sciences. Such a practice of combining "technical" and "abstract" orientations provided more opportunities for certain specialists because, in Marxist terms, it demonstrated "the unity of science and praxis"—one of the most categorical slogans of the Stalinist ideology.

Both the image and related practice of science were transformed after Stalin's death in 1953. The changes in political leadership obviously inspired some reevaluations of the existing policies, not only on the part of the top Soviet officials, but also on the part of various interest groups, including scientists. Nuclear physicists represented one of these groups. In the mid-1950s they used their increased political capital and social status as politically important scientists to push for a major change in the political organization and management of science. One of the most important results of their activities was a reversal in the relationship between science and practice. Stalinist science was expected to serve the goals of industrial and economic development of the country. The post-Stalin reformers succeeded in designing and establishing a new concept of scientific organization, which justified institutional separation of the most advanced academic research from the process of technological modernization.

Keldysh, as well as Korolev, did not participate, at least openly, in the political maneuvering undertaken by nuclear physicists. For example, he did not send appeals to the top officials (as many prominent Soviet physicists did, such as Petr L. Kapitsa, Abram I. Alikhanov, Lev A. Artsimovich, and Igor E. Tamm); he did not explicitly express his opinion at the so-called *aktivy* meetings—businesslike discussion of necessary improvements in Soviet science, undertaken by the academy leadership in 1956 after Nikita Khrushchev's well-known secret report on Stalin's "personality cult" presented at the Twentieth Congress of the Communist Party (1956); and finally, he did not publish articles in the Soviet media during the several months of heated public debate at the end of 1959. In the period between 1956 and 1959 Keldysh tried with all his might to solve the next pressing engineering problem: the creation of a stratospheric intercontinental winged missile (Burya). This work led him to pioneering studies in what is now known as astronavigation and applied celestial mechanics. Simultaneously, he participated in Korolev's project on the creation of the intercontinental ballistic missile. Research undertaken under Keldysh's leadership during that period is difficult to summarize briefly, for it included a number of studies in the theory of rocket engines, the theory of combustion, gas dynamics, hypersonic aerodynamics, heat transfer and thermal protection during atmosphere entry with cosmic speed, calculating the mobility of propellant in tanks, and many others. These studies helped Korolev construct his famous *semiorka* (seven)—the two-stage ballistic rocket R-7, used for military aims and, simultaneously, for placing the first artificial satellite *Sputnik* into Earth orbit.

Mstislav Vsevolodovich Keldysh. *Mstislav Vsevolodovich Keldysh (center) with John Bardeen (right) and Walter Franz (left).*
PHYSICS TODAY COLLECTION/AIP/PHOTO RESEARCHERS, INC.

A triumphant solution of these complex engineering problems became possible owing to, at least, two of Keldysh's personal contributions: the application of new methods of calculation to aerodynamic and design analysis, and the organization of close cooperation between different engineering and scientific groups. While director of the intercontinental winged missile project, Keldysh established an unprecedented state-sponsored enterprise, which unified the leading design bureaus with well-equipped scientific institutes. For example, under his general leadership, the CAHI developed aerodynamic designs of aircraft; Alexei M. Isayev's design bureau constructed a liquid-propellant rocket engine for the first (accelerating) stage of the missile; Mikhail M. Bondaryuk's design bureau created a supersonic ramjet engine for the second (sustainer) stage; and the system of missile orientation was developed at the department of the Institute of Rocket Aviation. In 1959 the missile was successfully tested in flight. By 1960, Keldysh occupied several leading positions both in the Soviet space program and official science leadership. In 1958 he was appointed the chair of the Interdepartmental Council on Space Exploration. In 1959 he became vice president of the Soviet Academy of Sciences.

President of the Academy. Meanwhile, the process of academy reform reached its final stage. Discussion on the future structure of the academy split academicians in two confronting groups: representatives of so-called fundamental science, headed by nuclear physicists, on the one hand, and "practical" engineers, consolidated at the Engineering Sciences Division, on the other. The president of the academy (from 1951) Aleksandr N. Nesmejanov had taken a compromised position and tried to ease tension between engineers and fundamental scientists. That tactic provoked the most prominent and still very powerful nuclear physicists against him. At the elections in May 1961 Nesmejanov was forced to withdraw his candidacy. A few days before, he had recommended M. V. Keldysh to become his successor at that post. This decision seemed to be quite logical because Keldysh's hands-off policy toward academy reform kept him on friendly terms with both physicists and engineers. In other respects, it was rather a contradictory move, because Keldysh's promotion resulted in the disbanding of the academy's engineering science division—the scientific body of which Keldysh had been a member until he became president. Besides, Keldysh was an obvious "military scientist." All prizes, medals, and orders he had been awarded before 1961 were for military achievements. Nominally, Keldysh hardly fitted expectations of the nuclear physicists, who tried to establish a scientific organization not allied to war purposes. Finally, Keldysh himself was not pleased with such a radical change in his career. He saw that administrative duties of a chief Soviet scientist would not permit him to develop his own scientific ideas.

Still, in May 1961, soon after issuing a joint governmental and party decree "On Measures for Better Coordination of Scientific Research in the Country and the Work of the Academy of Sciences" (*Pravda*, 12 April 1961), which legalized the institutional division between fundamental and applied sciences, Keldysh was unanimously elected president of the academy. After the election Keldysh put in practice the changes proposed by the reform proponents: increasing the number of divisions, each of which was responsible for a certain important problem; liquidation of the Engineering Science Division; international opening of Soviet science; transition from hostility to peaceful competition in science; and promoting the propaganda of Soviet scientific achievements in the West. Keldysh put his fondest hopes in the development of space research.

Indeed, his election coincided with the greatest triumph of Soviet science and engineering, the first human space flight by Yuri Gagarin. Keldysh's first presidential speech dealt totally with the Soviet space program. During the next decades the Soviet media kept singing praises for conquering the cosmos, referring to a number of Soviet victories. S. P. Korolev was formally named the "chief astronautic designer," and Keldysh "chief astronautic theorist." However, the real advance in spacecraft technologies and relevant theoretical fields was not as fast as it seemed. Soviet specialists paid dearly for their assurance of successful realization of space projects. At the early stage of astronautics almost every second Soviet interplanetary launch was ineffective. Keldysh was exactly the person to take responsibility for these unavoidable mistakes and develop ways to eliminate them. For example, in the mid-1960s the shortage of precise data such as planetary ephemerides and the gravitational field of the Moon, among others, caused failures of a few Venus expeditions. Traditional astronomical ways of calculating these events gave standard error of about 500 miles, which was not enough for effective realization of interplanetary flights. Keldysh initiated the foundation of radar sets with the express purpose to provide computation centers with precise astronomical data. There were a lot of such divergences between the habitual and the required accuracy in other scientific and engineering fields, concerning new problems of space exploration. One of the most significant of Keldysh's results as president was overcoming these partly organizational, partly cognitive misapprehensions between both scientists and governmental leaders.

Engineering and scientific troubles were redoubled with aircraft designers' squabbles over leading positions in the space program. In the beginning of the 1960s other design bureaus began to compete with the original Soviet space exploration enterprise, constructed by Korolev. Designer Vladimir N. Chelomey became one of these challengers. Being a brother-in-law of the General Secretary of the Communist Party Khrushchev, and an employer of Khrushchev's son, he used his proximity to the Soviet political elite to deprive Korolev of his leadership. In designers' circles this conflict was called the "small civil war." Keldysh tried to be an impartial arbitrator in this struggle. After Khrushchev's resignation, when Korolev's proponents were ready to deal with Chelomey, Keldysh supported his project of the so-called light launch vehicle (Proton), which could place into Earth orbit a weight up to 20 tons. Successful realization of this project played the decisive part in the development of Soviet astronautics, in contrast to Korolev's project of a "heavy" vehicle (with payload capacity about 100 tons), which was closed in the beginning of the 1970s after a series of accidents during in-flight testing.

Ten years of strenuous work undermined Keldysh's health. In 1973 he went through a complicated operation on his legs (for thrombosis), after which he resigned from the post of president. Deep depression, caused by exhaustion, a serious disease, and an exaggerated sense of personal responsibility for the lag in the so-called space race, led him to an untimely death in 1978. Keldysh was buried with all due honors at the Kremlin wall.

BIBLIOGRAPHY

WORKS BY KELDYSH

"Nachalo Kosmicheskoy Ery" [On the eve of space era]. *Vestnik Akademii Nauk SSSR* 6 (1961): 16–18.

Kosmicheskie issledovaniya [Space research]. Moscow: Nauka, 1981.

Matematika: Izbrannye trudy [Mathematics: Selected papers]. Edited by K. I. Babenko. Moscow: Nauka, 1985.

Raketnaya tekhnika i kosmonavtika: Izbrannye trudy [Rocketry and astronautics: Selected papers]. Edited by V. S. Avduevsky and T. M. Eneev. Moscow: Nauka, 1988.

OTHER SOURCES

Bashilova, E. Yu. "'Dvizhenie v period tekhniki poletov shlo gorazdo bystree, chem eto mozhno bylo ozhidat'" ["Advance in spacecraft techniques was going forward much faster than one could expect"]. *Istorichesky Arkhiv* 1 (2001): 14–18.

Grigor'yan, A. T. "Vydayushchi'sya uchenyi i organizator nauki" [The prominent scientist and organizer of science]. *Voprosy istorii estestvoznaniya i tekhniki* 1 (1981): 77–79.

Siddiqi, Asif A. *Challenge to Apollo: The Soviet Union and the Space Race, 1945–1974.* Washington, DC: National Aeronautics and Space Administration, 2000. Systematic review and analysis of the Soviet side of the space race.

Vestnik Akademii Nauk SSSR 6 (1961): 3–15. Fragments of discussion at the General Meeting of the Academy during the presidential elections.

Zabrodin, A. V., ed. *M. V. Keldysh. Tvorcheskiy portret po vospominaniyam sovremennikov* [M. V. Keldysh in memoirs of contemporaries]. Moscow: Nauka, 2001.

Konstantin V. Ivanov

KEPLER, JOHANNES

(*b.* Weil der Stadt, Germany, 27 December 1571; *d.* Regensburg, Germany, 15 November 1630), *astronomy, optics, mathematics.* For the original article on Kepler, see *DSB* vol. 7.

Kepler reissued only one of his works during his lifetime: twenty-five years after the publication of his first book, the *Mysterium cosmographicum* (1596), he published a second edition. The heir of a rich humanist tradition, and mindful of his own place in history, Kepler was careful to treat his own work in its historical context (Grafton 1992). He reproduced the text of the original edition verbatim, and added a series of long notes detailing where he had changed his mind or how his ideas had developed in the meantime. In one of these he remarked on the significance of the *Mysterium* to his later work, noting "Almost every book on astronomy which I have published since that time could be referred to one or another of the important chapters set out in this little book." Scholarship published since the original *DSB* article in 1973 has only underscored Kepler's conclusion, both in the specific sense in which the *Mysterium* foreshadowed his later work, and in the general understanding of the coherent unity of Kepler's thought.

Teacher at Graz. Scholars have long recognized the importance of Kepler's education for his accomplishments in astronomy for two specific reasons. First, he was groomed from an early age for service in the Lutheran church in Würtemberg, and his education both in school and at the University of Tübingen was thus thoroughly theological. Second, he was unusual in that he was taught astronomy by Michael Maestlin, one of a very few convinced Copernicans teaching anywhere in Europe at the time. The confluence of these factors came when Kepler was assigned to teach mathematics at the Protestant school in Graz, Styria. Initially deeply disappointed by his unexpected career change, he found solace when he resolved that by vigorously defending the reality of the Copernican system he could "glorify God also in astronomy." The fruit of this resolution was Kepler's first book, the *Mysterium cosmographicum* (1596), which argued for the reality of Copernican heliocentrism on the basis of "physical, or if you prefer, metaphysical reasons." Late twentieth- and early twenty-first-century scholarship has broadened our appreciation of the role of Kepler's education in his scientific work by highlighting the importance not only of a Lutheran education but of one deeply influenced by the reforms of Philipp Melanchthon (Methuen 1998; Barker and Goldstein 2001).

The continuity of the research program of the *Mysterium cosmographicum* with that of the *Astronomia nova* (1609) is now also more fully appreciated. The original *DSB* article on Kepler mentions that in addition to the well-known polyhedral hypothesis, the *Mysterium* also contains a lesser-known physical hypothesis that attempts to derive the relationship between the planets' distances and periods based on the power of a planet-moving force coming forth from the Sun. What it does not mention is the attempt in Chapter 22 of the *Mysterium* to extend that analysis to the motion of a planet around its own circular eccentric. In that case, Kepler argued, when the planet was at aphelion or perihelion, it should move as though rotating uniformly around Ptolemy's equant point, or as though being moved by Copernicus's equivalent construction. In either case, he concluded, in the absence of solid celestial spheres—newly disproven by the comet of 1577—his own physical account of a planet-moving force could explain Ptolemy's and Copernicus's mathematical models.

The difference in character between the "chimerical" *Mysterium cosmographicum* and the formidably technical *Astronomia nova* is considerably less baffling when viewed in the context of Kepler's work under Tycho Brahe. Kepler's research program in the *Mysterium* was not abandoned but redirected. Confronting Brahe's suspicion and secretiveness, Kepler found it impossible to obtain data on all the planets with which to develop the main system-wide arguments of the *Mysterium cosmographicum.* However, there were outstanding issues in the planet-moving force hypothesis that he could investigate and develop. Specifically, there was the orbit of the Earth, which did not have a bisected eccentricity with an equant such as Kepler had proposed a physical account of, but a plain eccentric. In a heliocentric system, should the orbit of the Earth not be of the same general form as the other planets? Also, there was the question of whether it would not make more sense for the orbits of the planets to be referred to the actual Sun rather than the center of the Earth's orbit. These two issues were the first he addressed when he began his work under Brahe, and were both swiftly resolved in Kepler's favor.

Imperial Mathematician. At the turn of the twenty-first century, scholarship on Brahe and his research establishment (Thoren 1990; Christianson, 2000) also clarified much about Kepler's role when working for Brahe and the circumstances of his succession upon Brahe's death (Voelkel 2001). The uncanny convenience of the transfer of observations from the accomplished but conservative observer to the bold, young theorist had previously left this transition more or less unquestioned. Closer examination of the historical moment, informed by a modern appreciation of Brahe's social status, has called into question first—contrary to the long-held assumption—whether Brahe ever held the position of imperial mathematician at all, and second, the inevitability of Kepler's appointment to that post. It seems highly unlikely that anyone of Brahe's aristocratic social status would demean themselves by accepting a position as low as the emperor's "mathematician." And Brahe does seem to have promised his long-time assistant Christian Sørenson Longomontanus the former imperial mathematician Nicholas Reimers Ursus's job (Thoren 1990, p. 408).

In fact, it seems that Kepler's appointment as imperial mathematician after Brahe's death had far more to do with the *Rudolphine Tables,* and that that project in turn owed its existence to Brahe's persistent failure to wring any money out of the imperial treasury with which to pay his assistants. When Kepler had first arrived at Benatky outside of Prague early in 1600, he found Brahe's household crowded with longstanding associates and assistants who easily outranked him. It was only upon his return to Prague in 1601, by which time the others had left seeking greener pastures and Kepler and Brahe were both more desperate, that Brahe proposed the composition of a set of astronomical tables bearing Emperor Rudolf II's name, which project would require the funding of his remaining assistant, Kepler. Kepler's intimate association with the project from its inception undoubtedly accounts for the alacrity with which he was appointed imperial mathematician after Brahe's death, with responsibility for completing Brahe's unpublished works.

But the legal and financial circumstances surrounding Kepler's appointment had far greater influence on his subsequent work than was appreciated at the time of the original *DSB* article, especially with regard to Kepler's masterpiece, the *Astronomia nova.* Emperor Rudolf II promised Brahe's heirs—foremost among them was Brahe's son-in-law, the nobleman Franz Gansneb Tengnagel van Kamp—12,000 florins for Brahe's instruments and observations, but they were never paid more than a fraction of that amount. In their efforts to extract the balance of what was owed them, they first sought to withhold the observations from Kepler, and then resorted to having responsibility for the *Rudolphine Tables* taken away from Kepler and assigned to Tengnagel. In the process, they accused Kepler of sloth, and at the same time Kepler was ordered to name the works he would publish to justify his continuing employment as imperial mathematician. Grasping among his half-finished projects for works he could publish quickly, Kepler named two works, the *Astronomniae pars optica* (1603), the foundational work of seventeenth-century optics, and the *Astronomia nova.* Up until this moment, he had never discussed publishing his epoch-making researches on the orbit of Mars separately (Voelkel 1999b, 2001).

As mentioned above, the research Kepler would publish in the *Astronomia nova* represented the successful conclusion of one line of research coming from the *Mysterium cosmographicum.* Consider that when Kepler resolved to publish the work, he was just beginning to play with a form of the area law and was still years away from the discovery of the elliptical form of the orbit. What he had worthy of calling at that time "the key to a more penetrating astronomy" were the discoveries about the planets' orbits being best referred to the Sun, and the Earth's orbit being generally like the planets' orbits and thus amenable to the same physical explanation given in the *Mysterium.* But the circumstances of the *Astronomia nova*'s publication further influenced its presentation in a way that had been remarked on prior to the original *DSB* article but not explained until the turn of the twenty-first century.

The *Astronomia nova* has a literary form unique among works of science: it is a confessional narrative, in which Kepler recounts at length a series of attempts he made to solve the problem of Mars's true orbit. In it, he does not hesitate to include his failures and setbacks. And it is this feature of the book that has made it possible for so many historians to describe the sequence of steps he took in discovering his first two "laws" of planetary motion. However, Owen Gingerich, the author of the original *DSB* article, discovered around the time of its publication in 1973 that all was not well with the assumption that the *Astronomia nova* was a faithful account of Kepler's research. Comparing it with the sequence of investigations in Kepler's manuscript Mars notebook, he concluded "the book… represents a much more coherent plan of organization than a mere serial recital of his investigations would allow" (Gingerich 1993, p. 370). Despite the strong statement in Bruce Stephenson's masterful account of Kepler's physical astronomy that the *Astronomia nova* represented argument, not history (1987, pp. 2–3), the question as to the exact nature of the account Kepler offers did not become acute until Donahue's discovery of "patent fraud" and "cover-up" in it during the course of translating the *Astronomia nova* (Donahue 1988).

The convoluted style of the *Astronomia nova* has since been shown to arise not from some inherent quirk of Kepler's personality but from a conscious rhetorical strategy (Voelkel 2001). Faced with great skepticism from the astronomical community both to his program of physical astronomy in general and to the specific changes in planetary theory that he was proposing, Kepler resorted to a pseudo-historical narrative in the hopes of presenting himself as blameless for the changes he was proposing. Rather than being seen as a "novelty seeker," he sought to depict himself as forced along the path that led to the area law and the ellipse. This strategy had the additional benefit of insulating Kepler somewhat from Brahe's heirs, who had gained the right to censor any work based on Brahe's observations: embedding his physical reasoning so deeply in his account made it impossible to suppress. Finally, and most specifically, Kepler used his correspondence with David Fabricius to map out points of contention that he systematically addressed in the structure of the *Astronomia nova*. The nature and purpose of the *Astronomia nova* are thus understood in ways unsuspected at the time of the original *DSB* article.

Work at Linz. The *Astronomia nova* was the most brilliant and idiosyncratic of Kepler's works. The pressures that shaped it abated not long after its publication, and after the death of his patron Rudolf II and Kepler's relocation to Linz, Upper Austria, Kepler was able to devote himself to a busy publication schedule within the more conventional genres of astronomical texts, such as his *Ephemerides novae* (1617–1620) (Bialas and Papadimitriou 1980). The most important publication, however, for which successive Hapsburg emperors supported Kepler nearly until the end of his life, was the *Rudolphine Tables*, responsibility for which was returned to Kepler after Tengnagel's fall in 1611, but which were not finally published until 1627. In the meantime, Kepler tried another tack in finding support for his new astronomy, a comprehensive multivolume textbook aimed at the "school benches" called the *Epitome astronomiae Copernicanae* (Epitome of Copernican astronomy, 1618–1621), in conscious imitation of his teacher Maestlin's textbook, the *Epitome astronomiae.* He also returned to the very questions first addressed in the *Mysterium cosmographicum* in his *Harmonices mundi* (Harmony of the world, 1619) in far greater detail, though perhaps with less pure satisfaction (Stephenson 1994; Martens 2000).

But it would be the *Rudolphine Tables* by which astronomers would judge Kepler's new astronomy and come to recognize the 100-fold increase in accuracy over previous astronomical theory it represented. This was the project that had long bound Kepler to the Hapsburgs, and which astronomers had avidly awaited for decades. And despite two excellent review articles (Wilson 1989; Applebaum 1996), the story of the appropriation of Keplerian astronomy in the seventeenth century remains far from understood. It is clear now, however, that Kepler's influence must be sought not in a post-Newtonian framework, by searching for discussion of his "laws" (which were not known as such until the eighteenth century), but in the context of subtle but fundamental Keplerian reforms in the way planetary positions were calculated (Voelkel and Gingerich 2001). Just as continued and careful attention to the content and context of Kepler's work since the original *DSB* article has yielded new insights into his life and work, we can expect that a fuller appreciation of his influence will follow.

SUPPLEMENTARY BIBLIOGRAPHY

WORKS BY KEPLER

Gesammelte Werke, 22 vols. Munich: C.H. Beck, 1937–. Since the original *DSB* article, the publication of Kepler's collected works has progressed greatly, though it has not quite been brought to conclusion. The plan of publication remains 22 volumes, but due to reorganization and expansion of the content, Volumes 11, 20, and 21 have been split into two parts each, for a total of 25 volumes. For example, the sound decision to reproduce all—rather than a sample—of the *Ephemerides novae* (1617–1619) in facsimile, required Volume 11 to be split in two, with the *Ephemerides novae* in Volume 11, Part 1 (1983). This change left some room with the calendars and prognostica in Volume 11, Part 2 (1993), into which the *Somnium* was moved from Volume 12. The resulting room in Volume 12 (1990) was given to the manuscript conclusion of the defense from Kepler's mother's witch trial. The most significant change introduced by the reorganization has been in the amount of manuscript material presented in the *Gesammelte Werke,* which has essentially doubled. Volume 20, Part 1 (1988) and Part 2 (1998), and part of Volume 21, Part 1 (2002) together are devoted to various astronomical manuscripts, and the balance of Volume 21, Part 1 to mathematical manuscripts and documents on the Gregorian calendar reform. A last volume of manuscripts and the index are the last outstanding volumes.

"On the More Certain Fundamentals of Astrology." Edited by J. Bruce Brackenridge and translated by Mary Ann Rossi. *Proceedings of the American Philosophical Society* 123 (1979): 85–116.

Bialas, Volker, and Elli Papadimitriou. "Materialen zu den Ephemeriden von Johannes Kepler." *Nova Kepleriana, Neue Folge* 7 (1980). An essential companion volume to *Gesammelte Werke,* Vol. 11, Part 1 containing manuscript sources for the *Ephemerides novae.*

The Secret of the Universe—Mysterium cosmographicum. Translated by A. M. Duncan. Introduction and Commentary by E. J. Aiton. New York: Abaris Books, 1981.

Le secret du monde. Translation, introduction, and notes by Alain Segonds. Paris: Les Belles Lettres, 1984. Preferable to the Duncan/Aiton translation, especially with regard to notes.

Kepler

New Astronomy. Translated by William H. Donahue. Cambridge, U.K.: Cambridge University Press, 1992.

Discussion avec le messager celeste. Translation, introduction, and notes by Isabelle Pantin. Paris: Les Belles Lettres, 1993. Much preferable to Rosen's 1965 English translation.

The Harmony of the World. Translated by E.J. Aiton, A.M. Duncan, and J.V. Field. Philadelphia: American Philosophical Society, 1997.

Optics: Paralipomena to Witelo & Optical Part of Astronomy. Translated by William H. Donahue. Sante Fe, NM: Green Lion Press, 2000.

OTHER SOURCES

Aiton, Eric J. "Johannes Kepler and the Astronomy without Hypotheses." *Japanese Studies in the History of Science* 14 (1975): 49–71.

Applebaum, Wilbur. "Keplerian Astronomy after Kepler: Researches and Problems." *History of Science* 14 (1996): 451–504. Splendid review article on the fate of Keplerian astronomy in the seventeenth century.

Barker, Peter, and Bernard R. Goldstein. "Theological Foundations of Kepler's Astronomy." *Osiris* 16 (2001): 88–113.

Caspar, Max. *Kepler.* Translated and edited by C. Doris Hellman. New introduction and references by Owen Gingerich, with bibliographical citations by Gingerich and Alain Segonds. New York: Dover, 1993. A reproduction of the original 1959 translation made infinitely more useful with painstakingly supplied citations (which the original mysteriously lacked).

Christianson, John Robert. *On Tycho's Island: Tycho Brahe and His Assistants, 1570–1601.* Cambridge, U.K.: Cambridge University Press, 2000. A compelling study of Tycho Brahe's research institute, with biographical studies of his many known assistants and associates.

Davis, A. E. L. "Kepler's Resolution of Individual Planetary Motion," "Kepler's 'Distance Law'—Myth Not Reality," "Grading the Eggs (Kepler's Sizing Procedure for the Planetary Orbit)," "Kepler's Road to Damascus," and "Kepler's Physical Framework for Planetary Motion." *Centaurus* 35 (1992): 97–191. An entire number of *Centaurus* devoted to articles stemming from Davis's 1981 PhD Thesis.

Donahue, William H. "Kepler's Fabricated Figures: Covering up the Mess in the *New Astronomy.*" *Journal for the History of Astronomy* 19 (1988): 217–237.

———. "Kepler's First Thoughts on Oval Orbits: Text, Translation, and Commentary." *Journal for the History of Astronomy* 24 (1993): 71–100.

Field, J. V. *Kepler's Geometrical Cosmology.* Chicago: University of Chicago Press, 1988.

Gingerich, Owen. "Johannes Kepler." In *Planetary Astronomy from the Renaissance to the Rise of Astrophysics,* Vol. 2, Part A: *Tycho Brahe to Newton,* edited by René Taton and Curtis Wilson. Cambridge, U.K.: Cambridge University Press, 1989.

———. *The Eye of Heaven: Ptolemy, Copernicus, Kepler.* New York: American Institute of Physics, 1993. Reprints Gingerich's most important articles on Kepler from around the time of his original *DSB* article and shortly thereafter.

Grafton, Anthony. "Humanism and Science in Rudolphine Prague: Kepler in Context." In *Literary Culture in the Holy Roman Empire, 1555–1720,* ed. James A. Parente Jr., Richard Erich Schade, and George C. Schoolfield. Chapel Hill: University of North Carolina Press, 1991.

———. "Kepler as a Reader." *Journal of the History of Ideas* 53 (4, 1992): 561–572.

Hamel, Jürgen. *Bibliographia Kepleriana: Verzeichnis der gedruckten Schriften von und über Johannes Kepler.* Munich: C.H. Beck, 1998. Indispensable supplement to Caspar's *Bibliographia Kepleriana* (2nd ed., 1968); includes listings of locations for all of Kepler's surviving works.

Hübner, Jürgen. *Die Theologie Johannes Keplers zwischen Orthodoxie und Naturwissenschaft.* Tübingen: Mohr, 1975. The definitive treatment of Kepler's theology.

Jardine, N. *The Birth of History and Philosophy of Science: Kepler's* A Defence of Tycho against Ursus, *with Essays on its Provenance and Significance.* Cambridge, U.K.: Cambridge University Press, 1984.

Kothmann, Hella. "Die Reisen des Johannes Kepler. Ein Chronologie—Ein Itinerarium." In *Miscellanea Kepleriana,* edited by Friedericke Boockmann, Daniel A. Di Liscia, and Hella Kothmann. Augsburg: E. Rauner, 2005.

———. "Die Reisen des Johannes Kepler. Eine Chronologie — Ein Itinerarium. Teil 2: 1614–1630." *Berichte der Kepler-Kommission* 15 (2005): 43–67.

Kozhamthadam, Job. *The Discovery of Kepler's Laws: The Interaction of Science, Philosophy, and Religion.* Notre Dame: The University of Notre Dame Press, 1994.

Martens, Rhonda. *Kepler's Philosophy and the New Astronomy.* Princeton, NJ: Princeton University Press, 2000. The finest treatment to date of Kepler's philosophy.

Methuen, Charlotte. *Kepler's Tübingen: Stimulus to a Theological Mathematics.* Brookfield, VT: Ashgate, 1998.

Rosen, Edward. *Three Imperial Mathematicians: Kepler Trapped between Tycho Brahe and Ursus.* New York: Abaris Books, 1986.

Stephenson, Bruce. *Kepler's Physical Astronomy.* New York: Springer-Verlag, 1987. The finest treatment of the entirety of Kepler's program of physical astronomy.

———. *The Music of the Heavens: Kepler's Harmonic Astronomy.* Princeton: Princeton University Press, 1994. Easily the best account of Kepler's cosmology from the *Harmonices mundi* (1619).

Sutter, Berthold. *Der Hexenprozess gegen Katharina Kepler.* Weil der Stadt: Kepler-Gesellschaft, Heimatverein, 1979. The definitive treatment of Kepler's mother's witch trial.

Thoren, Victor E. *The Lord of Uraniborg: A Biography of Tycho Brahe.* Cambridge, U.K.: Cambridge University Press, 1990. The definitive twentieth-century biography of Tycho Brahe.

Voelkel, James R. *Johannes Kepler and the New Astronomy.* New York: Oxford University Press, 1999a.

———. "Publish or Perish: Legal Contingencies and the Publication of Kepler's *Astronomia nova.*" *Science in Context* 12 (1, 1999b): 33–59.

———. *The Composition of Kepler's* Astronomia Nova. Princeton, NJ: Princeton University Press, 2001.

———, and Owen Gingerich. "Giovanni Antonio Magini's 'Keplerian' Tables of 1614 and Their Implications for the Reception of Keplerian Astronomy in the Seventeenth Century." *Journal for the History of Astronomy* 32 (2001): 237–262.

Wilson, Curtis. *Astronomy from Kepler to Newton.* London: Variorum, 1989.

———. "Predictive Astronomy in the Century after Kepler." In *Planetary Astronomy from the Renaissance to the Rise of Astrophysics,* Vol. 2, Part A: *Tycho Brahe to Newton,* edited by René Taton and Curtis Wilson. Cambridge, U.K.: Cambridge University Press, 1989.

James R. Voelkel

KETTLEWELL, BERNARD

SEE **Kettlewell, Henry Bernard Davis.**

KETTLEWELL, HENRY BERNARD DAVIS

(*b.* Howden, Yorkshire [now Humberside], England, 24 February 1907; *d.* Steeple Barton, near Oxford, England, 11 May 1979), *evolution, ecological genetics, entomology.* For the original article on Kettlewell see *DSB,* vol. 17.

In the years since John Turner's original *DSB* entry on him, Kettlewell and his pioneering work on industrial melanism have attracted historical interest that extends well beyond its scientific merits. This interest stems from the prominent role his investigations and the phenomenon of industrial melanism have played in the teaching of biology, and the centrality of industrial melanism as an example of natural selection in creationism/intelligent design versus evolution debates. This update summarizes contemporary scientific appraisals of Kettlewell's classic work on industrial melanism. It then discusses how legitimate scientific questions surrounding Kettlewell and his research have led popular writers and others with a fairly obvious agenda astray. A concluding section summarizes work on Kettlewell by historians and philosophers and outlines outstanding questions for further research.

In textbooks and the popular press, Kettlewell's field experiments in the 1950s are widely portrayed as providing the first definitive evidence that industrial melanism is indeed an example of natural selection. This is inaccurate for two reasons. First, the evidence that industrial melanism in the peppered moth is an example of natural selection came from the work of geneticists who had established that the color of the peppered moth has a genetic basis (Bowater, 1914), and the observations of literally hundreds of amateur and professional collectors who documented the spread of the dark form in the vicinity of manufacturing areas during the heyday of the industrial revolution and its predictable decline owing to the passage of Clean Air legislation (Doncaster, 1906; Kettlewell, 1958; Cook, 2003; Majerus, 2005). The directional rise and fall of the dark form, not only in the peppered moth but also in literally hundreds of other species where the phenomenon is known to have occurred, definitively establishes that this is an example of natural selection (Majerus, 1998; Grant, 1999). This is true even in the absence of knowing precisely why the dark form is at an advantage in polluted environments. Kettlewell's field experiments are considered important because they provided the first experimental demonstration that bird predation was indeed the selective mechanism at work at a time when many lepidopterists publicly doubted that bird predation was a significant factor on moth populations (e.g., Allen, 1955).

Second, although they are often depicted in the popular press as definitive, scientists have long recognized several fundamental problems in the conduct of Kettlewell's initial field experiments (e.g., Clarke & Sheppard, 1966; Majerus, 1998; Grant, 1999). Some of these problems center around the design of his experiment, such as whether the elevated densities of moths he used to ensure his results were statistically significant might have led endemic birds to eat moths when they ordinarily do not. Other problems have been associated with the central assumptions of his investigations, such as where the moth rests by day. While industrial melanism has proven to be a more complicated phenomenon than simplistic textbook accounts would have us believe, at least eight studies since using variously modified experimental designs to get around these problems have confirmed the basic conclusions of Kettlewell's initial experiments, namely that birds prey upon peppered moths and further that they do so differentially, depending upon how well the moth matches its background (Majerus, 2005). This being said, it seems clear further research needs to be done on the role of other nonvisual components of selection, for example, prelarval survival differences (Cook, 2000, 2003).

Sympathizers with intelligent design have misinterpreted these valid concerns about how best to interpret Kettlewell's investigations as somehow calling into doubt whether industrial melanism is indeed an example of natural selection at all. Several have drawn attention to discrepancies between introductory textbook accounts on industrial melanism written for children and journal articles by scientists who work on the phenomenon as

Kettlewell

evidence of a conspiracy to promote evolutionary theory (e.g., Wells, 2000, but see Rudge, 2002).

Following their lead, Judith Hooper, a popular science writer, in the first book-length biography of Kettlewell, all but explicitly charges Kettlewell with committing fraud and suggests Edmund Briscoe Ford and his colleagues engaged in an elaborate coverup to prevent the truth from being widely known (Hooper, 2002). While the book captures a sense of the colorful personalities associated with what has come to be known as the "Oxford School of Ecological Genetics" (c.f. Turner, 1985, 1988, but see Cain, 1988), Hooper's interpretation of the historical events is fundamentally flawed and rests on shoddy historical research (Rudge, 2005). Scientists familiar with research on industrial melanism have unanimously condemned the book as revealing a woeful misunderstanding of fieldwork in biology (e.g., Coyne, 2002; Grant, 2002). Much of the book is devoted to portraying Kettlewell and his associates as the sort of people who would resort to committing fraud, primarily by drawing attention to the vested interest they had in being able to provide experimental evidence for natural selection in nature.

Her implied allegation that Kettlewell committed fraud centers around a reported increase in recapture rates during his original 1953 investigation, an increase she interprets as evidence Kettlewell "fudged" his data in order to appease his boss Ford. M. Young and I. Musgrave (2005) have shown that the discrepancy upon which Hooper's specific allegation that Kettlewell fudged his data is made is entirely explicable in the context of field studies. Numerous other interpretive problems in Hooper's account appear to stem from a lack of field experience (Grant, 2002), poor understanding of the process of natural selection (Majerus, 2005), and a fundamentally flawed understanding of issues associated with the nature of science (Rudge, 2005). It is also unclear that Ford was as intimately involved in Kettlewell's work as Hooper suggests (Berry, 1990).

Owen (1997) suggests Kettlewell's published work on industrial melanism inadequately acknowledges his debt to James W. Tutt, who is generally acknowledged as the first to publish the opinion that the reason why the dark form was becoming more common near industrial sites was because of the cryptic advantage of dark coloration in soot-darkened environments (Tutt, 1890). Joel B. Hagen (1999) and David Rudge (1999) have debated whether Kettlewell's experiment in the unpolluted wood should be interpreted as a control for the earlier experiment in a polluted setting. These papers also raise questions about how this episode has and should be depicted in science textbooks. Rudge (2003) emphasizes the role of visual imagery in accounting for why the example and Kettlewell's work on it has become so ubiquitous. Rudge (2000) and Douglas Allchin (2001) have also drawn attention to how this episode can and should be used to teach issues associated with the nature of science.

H. B. D. Kettlewell. *Kettlewell using a 120 watt infra red bulb to attract insects for study in trap room. Brazil, 1958.* DMITRI KESSEL/TIME LIFE PICTURES/GETTY IMAGES.

Several outstanding questions surround Kettlewell's life and work that will make him the object of continued interest by historians of science. As hinted at in Turner's original entry, Kettlewell is a fascinating individual in his own right as someone struggling to negotiate between two worlds: the world of the amateur entomologist and the world of the professional scientist. Additional work should be done to clarify the connections between his research on industrial melanism and previous research by amateur and professional scientists in Britain and other countries (briefly mentioned in Kettlewell, 1973, pp. 53–54, and Majerus, 1998). There are also outstanding questions to consider regarding whether and how his research on industrial melanism relates to his other research interests, such as his pioneering use of radioactive isotopes to track insect populations (Rudge, 2005, pp. 257–258). Further historical research on Kettlewell is also of instrumental value in making sense of the complex relationships amongst individuals associated with the Oxford School of Ecological Genetics. (The latter is a particularly important consideration given the relatively complete set of correspondence Kettlewell and Philip Sheppard left behind [archived at the Bodleian Library at Oxford University and the American Philosophical Society Library in Philadelphia respectively] compared to other individuals associated with this group, including Ford and Arthur J. Cain).

SUPPLEMENTARY BIBLIOGRAPHY

WORKS BY KETTLEWELL

"A Survey of the Frequencies of *Biston betularia* (L.) (Lep.) and Its Melanic Forms in Great Britain." *Heredity* 12 (1958): 51–72.

The Evolution of Melanism: The Study of a Recurring Necessity. Oxford: Clarendon Press, 1973.

OTHER SOURCES

Allchin, Douglas. "Kettlewell's Missing Evidence, a Study in Black and White." *Journal of College Science Teaching* 31 (2001): 240–245.

———. "Scientific Myth-Conceptions." *Science Education* 87 (2003): 329–351.

Allen, P. M. Review of E. B. Ford's *Moths. Entomologist's Record and Journal of Variation* 67 (1955): 103–104.

Berry, R. J. "Industrial Melanism and Peppered Moths (*Biston betularia* (L))." *Biological Journal of the Linnean Society* 39 (1990): 301–322.

Bowater, W. "Heredity of Melanism in the Lepidoptera." *Journal of Genetics* 3 (1914): 299–315.

Cain, Arthur J. "A Criticism of J. R. G. Turner's Article 'Fisher's Evolutionary Faith and the Challenge of Mimicry.'" In *Oxford Surveys in Evolutionary Biology,* vol. 5, edited by P. H. Harvey and L. Partridge. Oxford: Oxford University Press, 1988.

Clarke, C. A., and P. M. Sheppard. "A Local Survey of the Distribution of Industrial Melanic Forms in the Moth *Biston betularia* and Estimates of the Selective Values of These in an Industrial Environment." *Proceedings of the Royal Society of London* B, *Biological Science* 165 (1966): 424–439.

Cook, Laurence M. "Changing Views on Melanic Moths." *Biological Journal of the Linnean Society* 69 (2000): 431–441.

———. "The Rise and Fall of the Carbonaria Form of the Peppered Moth." *Quarterly Review of Biology* 78 (2003): 399–417.

Coyne, J. A. "Evolution under Pressure: A Look at the Controversy about Industrial Melanism in the Peppered Moth." *Nature* 418 (2002): 19–20.

Doncaster, L. "Collective Inquiry as to Progressive Melanism in Lepidoptera: Summary of Evidence." *Entomologist's Record and Journal of Variation* 18 (1906): 165–170, 206–208, 222–226, 248–264.

Grant, Bruce S. "Fine Tuning the Peppered Moth Paradigm." *Evolution* 5 (1999): 980–984.

———. "Sour Grapes of Wrath." *Science* 297 (2002): 940–941.

Hagen, Joel B. "Retelling Experiments: H. B. D. Kettlewell's Studies of Industrial Melanism in Peppered Moths." *Biology and Philosophy* 14 (1999): 39–54.

Hooper, Judith. *Of Moths and Men: An Evolutionary Tale.* New York: W.W. Norton, 2002.

Majerus, Michael E. N. *Melanism: Evolution in Action.* Oxford: Oxford University Press, 1998.

———. "The Peppered Moth: Decline of a Darwinian Disciple." In *Insect Evolutionary Ecology: Proceedings of the Royal Society's 22nd Symposium,* edited by M. D. E. Fellowes, G. J. Holloway, and J. Rolff. Cambridge, MA: CABI Publishing, 2005.

Owen, Denis F. "Natural Selection and Evolution in Moths: Homage to J. W. Tutt." *Oikos* 78 (1997): 177–181.

Rudge, David. "Taking the Peppered Moth with a Grain of Salt." *Biology and Philosophy* 14 (1999): 9–37.

———. "Does Being Wrong Make Kettlewell Wrong for Science Teaching?" *Journal of Biological Education* 35, no. 1 (2000): 5–11.

———. "Cryptic Designs on the Peppered Moth." *International Journal of Tropical Biology and Conservation (Revista de Biología Tropical)* 50, no. 1 (2002): 1–7.

———. "The Role of Photographs and Films in Kettlewell's Popularizations of the Phenomenon of Industrial Melanism." *Science & Education* 12 (2003): 261–287.

———. "Did Kettlewell Commit Fraud? Re-examining the Evidence." *Public Understanding of Science* 14, no. 3 (2005): 249–268.

Turner, J. R. G. "Fisher's Evolutionary Faith and the Challenge of Mimicry." In *Oxford Surveys in Evolutionary Biology,* vol. 2, edited by R. Dawkins and M. Ridley. Oxford: Oxford University Press, 1985.

———. "Reply: Men of Fisher's?" In *Oxford Surveys in Evolutionary Biology,* vol. 5, edited by P. H. Harvey and L. Partridge. Oxford: Oxford University Press, 1988.

Tutt, James W. "Melanism and Melanochroism in British Lepidoptera." *Entomologist's Record and Journal of Variation* 1 (1890): 5–7, 49–56, 84–90, 121–125, 169–172, 228–234, 293–300, 317–325.

Wells, Jonathan. *Icons of Evolution: Science or Myth? Why Much of What We Teach about Evolution Is Wrong.* Washington, DC: Regnery Publishing, 2000.

Young, M., and I. Musgrave. "Moonshine: Why the Peppered Moth Remains an Icon of Evolution." *Skeptical Inquirer* 29, no. 2 (2005): 23–28.

David Rudge

KEYNES, JOHN MAYNARD

(*b.* Cambridge, United Kingdom, 5 June 1883; *d.* Firle, Sussex, United Kingdom, 21 April 1946), *economics, macroeconomics, unemployment, inflation, probability, rationality, politics.* For the original article on Keynes see *DSB,* vol. 7.

Keynes was one of the greatest economists of the twentieth century, theoretically and practically. He also made a pioneering contribution to the philosophy of probability, and advanced political ideas relevant to modern societies. His energies were often focused on problems on a world scale, including World War I, the inflation of the 1920s, the Great Depression of the 1930s, World War II, and the post-1945 global trade and financial system. He was also a patron of the arts, an eloquent writer, a prolific correspondent, a longtime editor of the *Economic Journal,* an unflagging journalist, and a member of the Bloomsbury group of writers and artists.

Keynes was born into a middle-class family, with an academic father and a social activist mother. After Eton College, he graduated from Cambridge University with a degree in mathematics, not economics. He received his primary economics education from Alfred Marshall in preparation for the civil service examination, two years after which he began lecturing in economics at Cambridge. His first intellectual love at Cambridge, however, was philosophy. He became a follower of George Edward Moore, the Cambridge ethical philosopher, and was also influenced by Bertrand Russell.

Philosophy. It was one of Keynes's criticisms of Moore's practical ethics that led him to philosophical work on probability. Not published until 1921 because of World War I, Keynes's *Treatise on Probability* laid the foundations for the logical theory of probability as distinct from the relative frequency and subjective theories. The question Keynes sought to resolve was how to theorize rational but nonconclusive arguments. His solution, which conceived of probability as a logical relation between two sets of propositions (the premises and the conclusion of an argument), cast probability theory as the general logic of argument in which deductive logic was a special case. Such probabilities, known by logical intuition, express the degree of belief that it is rational to have in the conclusion, given the information supplied by the premises. Keynes developed these ideas into a distinctive theory of rational belief and action under uncertainty. On this theory, the rational is not necessarily identical with the true, and probabilities fall into heterogeneous noncomparable classes that limit their mathematical manipulation. His theory of rationality under uncertainty departs significantly from the theory of rationality deployed in mainstream economics.

Economics. Keynes first made his name internationally with his 1919 book, *The Economic Consequences of the Peace,* a trenchant critique of the rationality and morality of the Versailles Treaty at the end of World War I. His *Tract on Monetary Reform* of 1923 then explored the dele-

John Maynard Keynes. © HULTON-DEUTSCH COLLECTION/CORBIS.

terious effects of inflation and deflation, and proposed remedies to enhance price stability, including central bank regulation of the interest rate. In 1930 he produced *A Treatise on Money,* intended as his magnum opus on monetary theory. This remained within the quantity theory of money framework of his earlier work, but analyzed the price level in terms of efficiency wages and the gap between saving and investment with a view to embracing price level dynamics. Although original and insightful, the work came under considerable criticism, and Keynes set to work to remedy its inadequacies.

To this point, Keynes was essentially an orthodox economist in the Marshallian tradition, though always interested in criticism and innovation to improve theory and policy. The mass unemployment of the Great Depression, however, impelled him toward a new economic theory that rejected much orthodox thinking. This new approach, published in 1936 as *The General Theory of Employment, Interest, and Money,* inaugurated a revolution in economic theory and policy, put macroeconomics on a sounder footing as the study of the economic system as a whole, and encouraged the collection of aggregate economic statistics. Two main ideas informed the new theory.

First was the concept of unemployment equilibrium, which posited that deficiencies in aggregate demand could cause the economy to settle into equilibria with unemployed labor. The second was radical or nonprobabilistic uncertainty, which meant that much rational behavior was actually based on factors other than calculable forecasts, thus leading to suboptimal levels of private investment and aggregate demand. These ideas led to interdependency between real and monetary factors, which previous theory had kept separate. In policy terms, given that capitalism had no automatic tendency to full employment, state action would be required to achieve this goal, chiefly (but not exclusively) via investment in public works.

During World War II, Keynes vigorously assisted the British government's war effort and postwar planning. In *How to Pay for the War* (1940), he outlined a plan of deferred pay to manage civilian demand so as to avoid inflation and strengthen social justice. In 1941 he proposed a scheme for an International Clearing Union with adjustment requirements on both creditor and debtor nations. With its demise, he contributed to the discussions that led to the 1944 Bretton Woods system for international finance and trade (a system reflecting American more than British views), and negotiated the American loan of 1945, which saved Britain from financial disaster. He fought hard to retain Britain's independence within the Anglo-American alliance, but by war's end a marked transfer of economic and financial power from Britain to the United States had occurred. Throughout the 1940s he also promoted policies to generate high levels of postwar employment. He died in 1946 of a heart attack, mainly brought on by overwork.

In the twenty-first century, Keynesian ideas still have considerable influence and powers of rejuvenation, although they are less widely accepted relative to the 1950s and 1960s, when they formed part of the economic mainstream.

Politics. Keynes sought a particular middle way between laissez-faire liberalism and state socialism. He favored planning but not central planning, individual liberty but not unfettered economic freedom, and greater social justice but not complete equality of outcomes. His vision of a better world was driven not by acquisitive materialism but by ethical ends for which a well-functioning economy was a prerequisite. For Keynes, economic prosperity was not to be pursued for its own sake, but only as a means to noneconomic activities promoting greater goodness and civilization.

Keynes was a marvelous (but not always clear) writer who created numerous memorable passages. His most famous saying is probably "*In the long run* we are all dead," by which he meant that long-run thinking on its own is inadequate, and that rational policy making needs to be based on both short- and long-run considerations.

Keynes left many legacies, but the most enduring of all is an undying commitment to reason and persuasion, and to the capacity of humans to make the world a better place.

SUPPLEMENTARY BIBLIOGRAPHY

Keynes's extensive writings (not all of which have been published) have given rise to numerous interpretations and a huge secondary literature. While some of this literature is suitable for the general reader, much of it requires more specialized knowledge.

WORKS BY KEYNES

The Collected Writings of John Maynard Keynes. 30 vols. London: Macmillan; New York: St. Martin's Press, for the Royal Economic Society, 1971–1989. Contains all his writings published in his lifetime and much previously unpublished material. General readers could start with *Essays in Persuasion* (vol. 9), *Essays in Biography* (vol. 10), or *The Economic Consequences of the Peace* (vol. 2), before turning to other volumes.

OTHER SOURCES

Intellectual Biographies

Moggridge, D. E. *Maynard Keynes: An Economist's Biography.* London: Routledge, 1992. A lengthy, single volume biography.

Skidelsky, Robert. *John Maynard Keynes: A Biography.* Vol. 1, *Hopes Betrayed, 1883–1920.* London: Macmillan, 1983. The first of a long three-volume biography; see below for the subsequent volumes.

———. *John Maynard Keynes: A Biography.* Vol. 2, *The Economist as Saviour, 1920–1937.* London: Macmillan, 1992.

———. *John Maynard Keynes: A Biography.* Vol. 3, *Fighting for Britain, 1937–1946.* London: Macmillan, 2000.

———. *John Maynard Keynes, 1883–1946: Economist, Philosopher, Statesman.* New York: Penguin, 2005. A single volume condensation of the above three volumes.

Philosophy

Carabelli, Anna M. *On Keynes's Method.* Basingstoke, U.K.: Macmillan, 1988.

O'Donnell, R. M. *Keynes: Philosophy, Economics, and Politics: The Philosophical Foundations of Keynes's Thought and Their Influence on His Economics and Politics.* Houndmills, U.K.: Macmillan, 1989.

Economics

Cate, Thomas, ed. *An Encyclopedia of Keynesian Economics.* Cheltenham, U.K., and Brookfield, VT: Elgar, 1997. Technical work.

Clarke, Peter. *The Keynesian Revolution in the Making, 1924–1936.* Oxford: Clarendon Press; New York: Oxford University Press, 1988. General work.

Harcourt, G. C., and P. A. Riach, eds. *A "Second Edition" of the General Theory.* 2 vols. New York: Routledge, 1997. Technical work.

Moggridge, D. E. *Keynes.* London: Macmillan, 1976. General work.

Stewart, Michael. *Keynes and After.* 3rd ed. Harmondsworth, U.K.: Penguin, 1986. General work.

Trevithick, J. A. *Involuntary Unemployment: Macroeconomics from a Keynesian Perspective.* London and New York: Harvester Wheatsheaf, 1992. Technical work.

Rod O'Donnell

KIMURA, MOTOO (*b.* Okazaki, Japan, 13 November 1924; *d.* Mishima, Japan, 13 November 1994), *molecular evolution, neutral theory, population genetics, evolutionary genetics, diffusion equations.*

Kimura was the chief proponent of the neutral theory of molecular evolution. In addition, he was a noted and influential mathematical population geneticist who developed the use of diffusion equation approximations for problems in population biology.

Early Years and Education. Kimura was born in 1924 in Okazaki, Japan. His father was a businessman and Motoo was the first son. His father's interests in flowers and ornamental plants, along with his middle school teacher's encouragement, led Kimura to an early conviction that he would become a botanist. During school, Kimura also developed an interest in mathematics, but could see no connection to botany and did not pursue it with the same interest. In 1942 Kimura entered the National High School in Nagoya, where he eagerly studied plant cytogenetics under M. Kumazawa. Of special significance for Kimura's future, Kumazawa also taught a course in biometry. For the first time, Kimura realized that his mathematical skills could find a place in biology.

The urgent circumstances of World War II shortened Kimura's time spent in high school from three years to two-and-one-half, thus allowing him to enter the Kyoto Imperial University in 1944. Kimura entered as a student of botany under the Faculty of Sciences, but his main influence was Hitoshi Kihara, a geneticist in the Faculty of Agriculture. By the end of his first year at Kyoto, the atomic bombs had been dropped on Hiroshima and Nagasaki and Japan had surrendered. Wartime shortages worsened after the surrender, however, and Kimura—being a student away from home—was hit especially hard. Fortunately, Kimura had a cousin in Kyoto that he could call on occasionally for better food.

Kimura's cousin, Matsuhei Tamura, was an associate professor under Hideki Yukawa, the theoretical physicist who had predicted the existence of the meson and was considered by Kimura to be Japan's scientific hero. Tamura was a mathematical physicist and surely had an influence on Kimura, who had developed an ambition "to do something in genetics like what theoretical physicists were doing in physics" (Kimura, 1985, pp. 463–464). Although Tamura was not impressed with the idea of theoretical biology, he had a better understanding of what mathematical biology entailed than most biologists in Japan at the time.

Kimura's growing interest in mathematical treatments of genetics and biology flourished after he graduated and moved into Kihara's laboratory at Kyoto. Kihara's attitude was remarkable; it was just right for Kimura, and for the future of population genetics. Recognizing Kimura's talent, Kihara assigned him no specific duties, leaving him free to study. Kimura threw himself into the technical literature of mathematical genetics, then dominated by Sewall Wright, John B. S. Haldane, and Ronald Aylmer Fisher. Kimura took mathematics courses where he could, but he was largely self-taught with occasional help from Tamura. While a student, Kimura had read voraciously whatever genetics literature he could get. Pirated editions of Conrad H. Waddington's *An Introduction to Modern Genetics* (1939) and Theodosius Dobzhansky's *Genetics and the Origin of Species* (1937) led him to the work of Wright in particular. By graduation, he had begun to devote more and more of his time to studying Wright's mathematical papers. Indeed, Kimura spent a full year on Wright's 1931 paper, "Evolution in Mendelian Populations," alone, learning the math as he went.

When the National Institute of Genetics was founded in 1949 in Mishima, Kimura was hired as a research associate with Kihara's recommendation. He remained associated with the institute for the rest of his life. The institute was located in a wooden building that had been a wartime aircraft factory. It was hot in summer and cold in winter. Furthermore, at that time Mishima was a small, provincial city, lacking the cultural and intellectual attractions of Kyoto and Tokyo. No one in Mishima understood or cared about Kimura's work, which increased his sense of isolation. He made frequent trips to Kyoto and Tokyo for library facilities, and undoubtedly for intellectual refreshment. Undeterred, he began writing papers and the first annual report of the Genetics Institute contained five of his reports, some startlingly original. It is interesting to read these early reports as they foreshadowed some of the later work for which Kimura was to become famous.

One Japanese scientist who did show an interest in Kimura's work was Taku Komai, who had studied with Thomas Hunt Morgan at Columbia University. Duncan

Motoo Kimura. *Kimura with A. H. Sturtevant (left).* COURTESY OF THE ARCHIVES, CALIFORNIA INSTITUTE OF TECHNOLOGY.

McDonald and Newton Morton, two American geneticists at the Atomic Bomb Casualty Commission, also recognized Kimura's work. Together with Komai, they were able to find enough funding for Kimura to come to the United States. Kimura wanted to work with Wright, but by this time Wright was getting ready to retire from the University of Chicago and was not taking students. Instead Kimura went to Iowa State College (now Iowa State University) in 1953, where worked with America's best-known animal breeder and Wright acolyte, Jay L. Lush.

After entering Iowa State College, Kimura became dissatisfied with the research program, which was concerned with quantitative traits and emphasized subdivision of epistatic variance (that is, the variance component caused by gene interaction). Kimura understood this, but he really wanted to work on stochastic processes. Furthermore, he developed a strong dislike of Lush. When Kimura had first arrived in the United States, he had attended the Genetics Society of America meeting in Madison, Wisconsin, where he met James F. Crow. Crow was a population geneticist and one of the few who were acquainted with Kimura's work. Indeed, on the voyage from Japan Kimura had written a paper demonstrating how fluctuating selection could mimic the stochastic effects of random genetic drift. Kimura had cleverly found a transformation that converted a complicated partial differential equation into a simple heat-diffusion formula, known to every physics student. He gave the paper to Crow for comments, who suggested its publication in *Genetics.* Wright reviewed the paper with unusual enthusiasm, and it was soon published (Kimura, 1954). As Kimura's dissatisfaction grew at Iowa State, he decided to transfer to the University of Wisconsin and study with Crow. Crow was reasonably sure that Wright would soon be moving to Wisconsin, and so accepted Kimura as a student.

Kimura spent two years, 1954–1956, getting his PhD in Wisconsin. Before coming to the United States, Kimura had discovered the two Kolmogorov equations. These are partial differential equations, one known as "forward" and other "backward," used to describe random processes, such as Brownian motion and more general diffusion processes. Wright had used the forward equation—in fact he rediscovered it himself—but Kimura was the first geneticist to employ the backward equation. He realized while still in Wisconsin that this equation was especially useful for some previously unsolved problems. Later, for example, he used this to study the age of a mutant allele in a population.

Soon after arriving in Wisconsin, Kimura obtained the complete distribution of allele frequencies under neutral random drift, at any time from any arbitrary starting frequency. He soon extended this to three alleles, then to an indefinite number. He then included the effects of mutation, migration, and selection. These results were published in the *Cold Spring Symposium* (Kimura, 1955). By the time Kimura received his PhD, he was already a recognized leader in theoretical population genetics. Kimura then returned to Japan. Except for occasional stays abroad, usually a year or less, he spent the rest of his life in Mishima.

In Japan, Kimura continued to develop equations for stochastic genetic models of greater generality. He introduced the "infinite allele" and "infinite site" models, widely used for evolutionary studies many years later after the coming of molecular techniques. With his colleague Takeo Maruyama, he found a method for investigating several problems, such as the number of individuals in the path to fixation or loss, or the number of heterozygotes. Kimura also developed the "stepping-stone" model of population structure, which has served as a foundation for the study of migration by many other scientists. In addition, Kimura also did a number of studies of genetic load and on inbreeding theory. His interest in mathematical models also led him to pioneering uses of computer simulations in population genetics (Crow, 1995).

The Neutral Theory of Molecular Evolution. In the 1960s population genetics had the most beautiful theory in biology, but there were few opportunities to apply it. Molecular biology changed everything. Data on the rates of molecular evolution were appearing and were awaiting analysis. In 1968 Kimura proposed what would become known as the neutral theory of molecular evolution. Using protein sequence data generated by biochemists such as Emile Zuckerkandl and Emmanuel Margoliash, Kimura and his colleague Tomoko Ohta compared mammalian protein sequences and used the number of detected differences across species to calculate a rate of molecular evolution. Kimura then reasoned that if most mutations were in fact selected, then the rate of evolution calculated for mammals would create an intolerable genetic load (the amount of differential mortality and fertility required for such a rate was more than the population could sustain). Because mammals were not extinct or staggering under an enormous genetic load, Kimura concluded that most detected molecular variants were in fact selectively neutral, meaning that they produced no change in survival or fertility for their possessors (Kimura, 1968; also see Dietrich, 1994, and Suarez & Barahona, 1996).

Kimura's conclusion and argument were controversial, but the dispute between neutralists and selectionists was guaranteed in 1969 when Tom Jukes and Jack King wrote their neutralist manifesto under the provocative title of "Non-Darwinian Evolution." King and Jukes brought a large variety of evidence to bear in favor of large numbers of neutral mutations (1969). By using evidence from the growing field of molecular evolution to support the idea of neutral mutations and the importance of random drift, they spelled out the molecular consequences of the neutral hypothesis more clearly than Kimura had. King and Jukes built their case using phenomena such as synonymous mutations, the Treffers mutator, the relation between amino acid frequencies and the genetic code, and the growing body of data on specific proteins such as cytochrome c.

The neutral theory directly challenged the power of natural selection in evolutionary biology. Because the neutral theory claimed that random drift was more significant than natural selection at the molecular level, it helped drive a wedge between the way evolution was understood at the organismal and molecular levels—at the organismal level, natural selection predominated, while at the molecular level, random genetic drift was an important factor. By articulating a different set of evolutionary mechanisms for the molecular level, the neutral theory provided a theoretical foundation for the development of molecular evolution as a new field of biological inquiry.

Many biologists were extremely skeptical of the neutral theory. Classical geneticists believed, reasonably, that hardly any observable changes were completely neutral, in part, because they were thinking about morphological changes, not changes in nucleotides or amino acids. Kimura and Ohta pursued the neutral theory vigorously. One of the most attractive features of the approach was that it provided a basis for the molecular clock, described in 1964 by Zuckerkandl and Linus Pauling. Zuckerkandl and Pauling had observed that molecules collected substitutions at a remarkably constant rate. Hence, the number of changes between two species could be used to infer the time since they split from a common ancestor—a great boon to systematic biologists. The neutral theory proposed a mechanism for this constancy, because neutral evolution is driven by the mutation rate; meaning that when random drift is taken into account, the longtime average rate of nucleotide substitution becomes equivalent to the mutation rate, which was believed to be roughly constant.

In 1969 Kimura used the constancy of the rate of amino acid substitutions to argue powerfully for the importance of neutral mutations and random drift in molecular evolution. At the same time, Kimura was also calling on his earlier work on stochastic processes in population genetics to forge a solid theoretical foundation for the neutral theory. Kimura's diffusion equation method provided the theoretical techniques he needed to formulate specific models, which in turn enabled him to address issues such as the probability and time to fixation of a mutant substitution as well as the rate of mutant substitutions in evolution. Working in collaboration with

Tomoko Ohta, Kimura also extended the neutral theory to encompass the problem of explaining protein polymorphisms. This was a central concern of population genetics, and Kimura and Ohta were able to argue that protein polymorphisms were a phase in mutations' long journey to fixation (Kimura & Ohta, 1971).

Kimura found many other arguments in favor of the neutral theory over the next few years. For instance, amino acids in regions of less importance for the function of a polypeptide evolved faster than those important for the function. Particularly revealing was the insulin molecule, which has three regions, one of which is discarded and not used. The unused part evolved fastest. Within codons, synonymous changes were faster than nonsynonymous. To Kimura, slow evolution of some nucleotides was caused by "selective constraint": these regions already functioned well, and therefore most mutations were harmful. One of Kimura's most striking arguments came from the fact that the number of amino acid differences between the alpha and beta hemoglobins in humans was about the same as that between human beta and carp alpha. The first two have been in the same cell for some 400 million years, while the latter two have been in fish and the line leading to humans. The difference in selective forces could hardly be greater. If the amino acid changes were due to selection, the two sequences should be enormously different from each other; but they were not. Kimura summarized his views in a widely quoted book, *The Neutral Theory of Molecular Evolution,* published in 1983. He devoted much of his energy for the rest of his active life to finding more evidence and arguing his case.

DNA Enters the Debate. The availability of DNA sequence data in the mid-1980s transformed the debate over the neutral theory of molecular evolution. While earlier techniques, such as electrophoresis, allowed evolutionary biologists to estimate variability at the molecular level, DNA sequencing promised more direct measurements of genetic variability. More importantly, DNA sequence data made it possible to better distinguish drift from selection.

In the 1960s Kimura and King and Jukes proposed that synonymous changes, changes in DNA that do not produce a corresponding change in the amino acid of the protein coded for, should be neutral because they have no observable effect. Because these changes presumably have no selective effect, they should evolve more quickly than most of the nonsynonymous changes because most of these are harmful and are eliminated rather than contributing to evolutionary change. The rare advantageous change would evolve more quickly than the neutral change as positive selection pushed it to fixation in the population. Kimura proposed that the differences in synonymous and nonsynonymous substitution rates could form the basis for detecting positive selection (Kimura, 1983). In 1984 Martin Kreitman introduced DNA sequencing to evolutionary genetics and extended Kimura's idea of comparing synonymous and nonsynonymous substitutions. The McDonald-Kreitman test compares the ratio of nonsynonymous to synonymous changes within a species and between two species. If the sequences are neutral, the ratios should remain the same. If there is positive selection, then nonsynonymous changes should have accumulated over time, so there would be more nonsynonymous changes between species than within a species. This test and many other statistical tests that followed enabled evolutionary biologists to detect balancing selection, adaptive protein evolution, and population subdivision (McDonald & Kreitman, 1991; Kreitman, 2000).

During the last decade of Kimura's life, the debate surrounding the neutral theory died down. While the neutral theory may not have been accepted in full, it became a standard part of evolutionary theory. As sequence data accumulated, biologists realized that many organisms possess a great deal of noncoding DNA, which would then have been subject to neutral evolution. At the same time, the neutral theory forms a natural null hypothesis for studies of selection and for statistical tests of selection.

Despite the controversy surrounding the neutral theory, Kimura received numerous honors, including honorary degrees from the University of Chicago and the University of Wisconsin, the Japan Academy Prize in 1968, the Japanese Order of Culture (Emperor's Medal) in 1976, the Chevalier de L'Ordre National du Mérite in 1986, the Asahi Shimbun Prize in 1987, the John J. Carty Award from the (U.S.) National Academy of Sciences in 1987, and the Darwin Medal from the Royal Society in 1992. He is particularly honored in his hometown of Okazaki, thanks largely to efforts of his brother. In addition to a museum, Kimura is honored with a statue in the city (Crow, 1995).

Soon after his return to Japan from Wisconsin, Kimura married. He and Hiroko Kimura had one child, a son, Akio. He had one important hobby, orchid breeding. Every Sunday was devoted to this, and he produced several prize-winning clones. Throughout his life he also enjoyed philosophy, especially the writings of Bertrand Russell, and science fiction, where he was particularly fond of the writing of Arthur Clarke. Kimura's main interest in life was his work, especially after the neutral theory, for which he became a passionate advocate. Kimura's advocacy continued up to a short time before his death. In his late sixties, Kimura developed amyotrophic lateral sclerosis, and deteriorated very rapidly. His death came on his seventieth birthday, the result of an accidental fall.

BIBLIOGRAPHY

WORKS BY KIMURA

"Process Leading to Quasi-Fixation of Genes in Natural Populations Due to Random Fluctuation of Selection Intensities." *Genetics* 39 (1954): 280–295.

"Stochastic Processes and Distribution of Gene Frequencies under Natural Selection." *Cold Spring Harbor Symposium on Quantitative Biology* 20 (1955): 33–53.

Diffusion Models in Population Genetics. London: Methuen, 1964.

With James Crow. "The Number of Alleles that Can Be Maintained in a Finite Population." *Genetics* 49 (1964): 725–738.

"Evolutionary Rate at the Molecular Level." *Nature* 217 (1968): 624–626. Kimura's initial argument for neutral molecular evolution.

"The Rate of Molecular Evolution Considered from the Standpoint of Population Genetics." *Proceedings of the National Academy of Sciences of the United States of America*. 63, no. 4 (1969): 1181–1188.

With Tomoko Ohta. "Protein Polymorphism as a Phase in Molecular Evolution." *Nature* 229 (1971): 467–469.

The Neutral Theory of Molecular Evolution. Cambridge, U.K.: Cambridge University Press, 1983. Kimura's most extensive treatment of the neutral theory.

"Genes, Populations, and Molecules: A Memoir." In *Population Genetics and Molecular Evolution*. Edited by Tomoko Ohta and Kenichi Aoki. Tokyo: Japan Scientific Society Press, 1985.

"Molecular Evolutionary Clock and the Neutral Theory." *Journal of Molecular Evolution* 26 (1987): 24–33.

Population Genetics, Molecular Evolution, and the Neutral Theory: Selected Papers. Edited by Naoyuki Takahata. Chicago: University of Chicago Press, 1994. A collection of Kimura's most influential papers and a complete bibliography of his publications.

OTHER SOURCES

Crow, James. "Motoo Kimura (1924–1994)." *Genetics* 140 (1995): 1–5.

Dietrich, Michael R. "The Origins of the Neutral Theory of Molecular Evolution." *Journal of the History of Biology* 27 (1994): 21–59.

King, Jack L., and Thomas H. Jukes. "Non-Darwinian Evolution." *Science* 164 (1969): 788–798.

Kreitman, Martin. "Methods to Detect Selection in Populations with Application to the Human." *Annual Review of Genomics and Human Genetics* 1 (2000): 539–559.

McDonald, John H., and Martin Kreitman. "Adaptive Protein Evolution at the Adh Locus in *Drosophila*." *Nature* 351 (1991): 652–654.

Ohta, Tomoko, and John Gillespie. "Development of Neutral and Nearly Neutral Theories." *Theoretical Population Biology* 49 (1996): 128–142.

Provine, William. "The Neutral Theory of Molecular Evolution in Historical Perspective." In *Population Biology of Genes and Molecules*, edited by Naoyuki Takahata and James Crow. Tokyo: Baifukan, 1990.

Suarez, Edna, and Anna Barahona. "The Experimental Roots of the Neutral Theory of Molecular Evolution." *History and Philosophy of the Life Sciences* 18 (1996): 55–81.

Zuckerkandl, Emile, and Linus Pauling. "Evolutionary Divergence and Convergence in Proteins." In *Evolving Genes and Proteins*, edited by Vernon Bryson and Henry J. Vogel. New York: Academic Press, 1965.

Michael R. Dietrich
James F. Crow

KING, ADA AUGUSTA, COUNTESS OF LOVELACE

(*b.* London, England, 10 December 1815; *d.* London, 27 November 1852), *mathematics, computing.*

Ada Augusta King, Countess of Lovelace, was an early-nineteenth-century mathematician and scientist who is generally remembered for her work with Charles Babbage. She translated an early description of Babbage's machines, added an extensive note of her own to the description, and prepared a set of instructions for the machine. However, she had interests in science beyond calculation and illustrates how science engaged women in early Victorian Britain

Ada Lovelace, as she is commonly known, was the legitimate daughter of Lord Byron (George Gordon Byron) and Annabelle Milbanke. She was born eleven months after her parents married and five weeks before they separated. Her father left England shortly after her birth, never to return.

Lovelace was raised entirely by her mother, a strong-willed and tempestuous individual. She was schooled at home, as most girls were, and was taught the usual set of topics that were considered acceptable for young women: reading, grammar and spelling, arithmetic, music, geography, drawing, and French.

By the time she was an adolescent, Lovelace was starting to show signs that she could be as emotional as her father or as impetuous as her mother. After turning seventeen, she attempted to elope with her tutor. Friends intervened in the plot and returned Lovelace to her home. "There was, I hope, no real misconduct at the time," wrote an acquaintance, "and an open scandal was prevented" (Sophia De Morgan, quoted in Stein, 1985, p. 36).

In the spring of 1834, Lovelace decided that she would undertake a serious study of science. She seems to have been motivated, at least in part, by her remorse over the attempted elopement. "I find that nothing but a very close and intense application of subjects of a scientific nature," she wrote, "now seems at all to keep my imagination from

Ada Augusta King, Countess of Lovelace. SPL PHOTO RESEARCHERS, INC.

running wild" (to William King, March 9, 1834; Toole, 1998, pp. 57–58).

Lovelace began studying trigonometry and mathematical astronomy. She occasionally signed her letters, "Ever Yours Mathematically." Her interest in science was encouraged by two important new friends: Charles Babbage and Mary Somerville. She had met Babbage at a party, held the previous June, where he had demonstrated a model of his first computing device, the Difference Engine. Lovelace was impressed with the engine and referred to it as "the gem of all mechanism" (to William King, 1 September 1834; Toole, 1998, p. 60).

Somerville, who was thirty-five years Lovelace's senior, was England's most prominent woman scientist. Supported by wealth from an early marriage and encouraged by a sympathetic husband, she had pursued the study of mathematics and astronomy. In 1831 she had published an English translation of Pierre-Simon de Laplace's masterwork, *Traité de mécanique céleste* (*Celestial Mechanics*). To Lovelace, she played the role of mentor, correspondent, and occasional tutor.

In 1835 Lovelace married William King. During the first four years of her marriage, she gave birth to three children: Byron Noel King (1836), Anna Isabella King (1837), and Ralph Gordon Noel King (1839). Following the birth of Ralph, she decided to return to her study of science. She took instruction from Augustus De Morgan, a family friend and a professor at University College London. De Morgan led Lovelace through the basic concepts of calculus. Lovelace made steady progress in her study, but was impatient with the work. "I wish I went on quicker," she wrote. "I wish a human head, or my head at all events, could take in a great deal more and a great deal more rapidly than is the case" (to Augustus De Morgan, 13 September 1840, Toole, pp. 112–123).

In 1840 Lovelace's occasional correspondent, Babbage, traveled to Turin in order to give a series of talks on his computing machines. In addition to his Difference Engine, Babbage was working on the design of a more flexible device, which he called the Analytical Engine. Unlike his first machine, the Analytical Engine could be programmed, by encoding instructions on punched cards. His talks at Turin were summarized in an article by a military engineer named Luigi Menabrea, which was published in the journal *Bibliothèque Universelle de Genève.*

Lovelace learned about Menabrea's article from Charles Wheatstone, a member of the Lovelace social circle. She began the translation without telling Babbage. When he finally learned of the effort, he asked why Lovelace had "not written an original paper on a subject with which she was so intimately acquainted?" According to Babbage, Lovelace replied that the idea "had not occurred to her." Babbage then suggested that she write some notes that would illustrate the operation of the machine (Babbage, 1994, p. 102).

Working with guidance from Babbage, Lovelace prepared her manuscript. "We will terminate these notes," she wrote, "by following, in detail, the steps through which the engine could compute the Numbers of Bernoulli, this being (as we shall deduce it) a rather complicated example of its powers" (notes to Menabrea translation as reprinted in Morrison and Morrison, 1961, p. 286). This detailed example is her principal scientific legacy. In modern terms, these steps are a computing program.

Lovelace's partnership with Babbage ended badly. The two of them disagreed about their dealings with the publisher. They soon resolved their differences but never worked together again. Leaving Babbage behind, Lovelace was attracted to a wide array of subjects, not all of which could be considered scientific pursuits, even by the standards of the day. For the next eight years, Lovelace wandered through the various fields of science, reading German books, and corresponding with prominent

English scientists. Her only substantial scientific contribution of this period was the review of a French book on meteorology and agriculture, which she wrote jointly with her husband.

In the last years of her life, Lovelace lived an increasingly unstable existence. She suffered from ill health, became emotionally attached to a man who was not her husband, and amassed large gambling debts.

Lovelace's reputation has always been tied to Babbage. Her work was rediscovered in the 1940s, when the electronic computer brought a new assessment of Babbage's work. In 1979 the U.S. Defense Department named its new computer language Ada, in her honor. Like her father, she had an unconventional personality, far more unconventional than her mentor, Somerville. That personality made it difficult for her to accept the role of translator or expositor, the two positions in science that were accessible to women at the time.

BIBLIOGRAPHY

WORK BY KING

Translator (with notes). Menabrea, Luigi F. "Sketch of the Analytical Engine Invented by Charles Babbage." *Bibliothèque Universelle de Genève,* no. 82. Geneva: A. Cherbuliez, 1842. Available from http://www.fourmilab.ch/babbage/sketch.html. Reprinted in *Charles Babbage on the Principles and Development of the Calculator and Other Seminal Writings,* by Charles Babbage, edited by Philip Morrison and Emily Morrison, pp. 225–297. New York: Dover, 1961.

OTHER SOURCES

Babbage, Charles. *Passages from the Life of a Philosopher.* New Brunswick, NJ: Rutgers University Press, 1994.

Baum, Joan. *The Calculating Passion of Ada Byron.* Hamden, CT: Archon, 1986.

Fuegi, John, and Jo Francis. "Lovelace & Babbage and the Creation of the 1843 'Notes.'" *IEEE Annals of the History of Computing* 25, no. 4 (2003): 16–26.

Huskey, Velma, and Harry Huskey. "Lady Lovelace and Charles Babbage." *IEEE Annals of the History of Computing* 2 (1980): 299–329.

Lee, J. A. N. "Ada Augusta, Countess of Lovelace, 1815–1852: What Was Her Family Name?" *IEEE Annals of the History of Computing* 22, no. 4 (2000): 72–73.

Moore, Doris Langley-Levy. *Ada, Countess of Lovelace: Byron's Legitimate Daughter.* London: J. Murray, 1977.

Stein, Dorothy. *Ada: A Life and a Legacy.* Cambridge, MA: MIT Press, 1985.

Toole, Betty A. "Ada Byron, Lady Lovelace, An Analyst and Metaphysician." *IEEE Annals of the History of Computing* 18, no. 3 (1996): 4–12.

———. *Ada, the Enchantress of Numbers: Prophet of the Computer Age, a Pathway to the 21st Century.* Mill Valley, CA: Strawberry; Sausalito, CA: Orders to Critical Connection, 1998.

Wade, Mary Dodson. *Ada Byron Lovelace: The Lady and the Computer.* New York: Dillon, 1994.

Woolley, Benjamin. *The Bride of Science: Romance, Reason, and Byron's Daughter.* New York: McGraw-Hill, 1999.

David A. Grier

KING, CLARENCE RIVERS

(*b.* Newport, Rhode Island, 6 January 1842; *d.* Phoenix, Arizona, 24 December 1901), *economic and regional geology, federal science administration, mining consulting.* For the original article on King see *DSB,* vol. 7.

In August 1887 Clarence King decided that "The one important fact … of my twenty years [*sic*] Washington career and the most important contribution I ever made to science … was the crushing of the old system of personal surveys" (Marcus Benjamin Papers, Smithsonian Institution Archives). King thus ranked as his best achievement his role in discontinuing in 1879 the competing federal mapping and science surveys of the postbellum American West, led by Ferdinand Hayden and John Powell for the Interior Department and by Lieutenant George Wheeler for the War Department, and establishing in their place the U.S. Geological Survey (USGS). King rated these events above his three principal earlier accomplishments. He conceived and led the War Department's U.S. Geological Exploration of the Fortieth Parallel (1867–1879), which became the model and standard for fieldwork and publications by the three other western surveys. In 1872 King and his men also exposed a new but fraudulent Kimberley in Colorado, saving diamond investors far more than the cost of King's entire survey. His *Systematic Geology* (1878) innovatively synthesized the entire geologic and tectonic history of the fortieth-parallel country that flanked the transcontinental railroad's route between California's Sierra Nevada and Colorado's Front Range.

Founding the National Geological Survey. The yearlong process of founding a national geological survey for the United States began in March 1878 as the nation continued to recover from the economic depression that followed the financial panic in 1873. As part of wider efforts to reduce federal expenditures and to improve the civil service, the Forty-fifth Congress requested statements from Hayden, Powell, and Wheeler detailing their accomplishments, expenditures, and the nature and cost of any duplication. In June, Abram Hewitt, an iron manufacturer, a member of New Jersey State Geological Survey's

Board of Managers, and the representative from New York's Tenth Congressional District, followed King's suggestion for reform. Hewitt arranged in June for Congress to ask the National Academy of Sciences (NAS), its statutory adviser on subjects of science and art, for a plan to "secure the best possible results at the least possible cost" (20 *Stat. at Large of the USA* 230, 20 June 1878) of the scientific surveys by the War and Interior Departments and the land-parceling (cadastral) surveys of Interior's General Land Office (GLO).

The request from Congress, approved by President Rutherford Hayes, went to Yale's O. C. Marsh, acting president of the NAS since the death of Joseph Henry in May, and the author of *Odontornithes* (1880), the last volume of the King survey's final reports. Marsh decided the NAS must respond by the time Congress reconvened early in December. Between August and November, King, at Marsh's request, advised and aided the committee—Columbia's John Newberry and five other informed, not directly involved, and (seemingly) unbiased NAS members—appointed by Marsh to plan the reforms. Marsh served *ex officio*, as required by NAS bylaws.

The NAS committee proposed two new agencies to replace the surveys led by Hayden, Powell, and Wheeler. A new "Coast and Interior Survey," formed by transferring to the Interior Department the Treasury Department's U.S. Coast and Geodetic Survey, would conduct all federal cadastral, geodetic, and topographic surveys and prepare a national topographic map. A new geological survey, a bureau of practical geology, would classify, scientifically, the public domain—the lands mostly west of the Mississippi River to which the federal government still held title—and study its geological structure and economical resources to help the mineral industry aid the struggling economy by providing additional minerals needed for construction and currency. To protect the integrity and objectivity of the new geological survey's data and analyses, the founders barred employee speculation in the lands and minerals being studied and outside consulting. At the NAS meeting in New York during November 1878, the attending members approved their committee's proposals by a vote of thirty-one to one. Marsh hurried to Washington and obtained the approval of President Hayes, his reform-minded Interior Secretary Carl Schurz, Treasury Secretary John Sherman, Superintendent of the Coast and Geodetic Survey Carlile Patterson, and General of the Army William Sherman.

After Congress published the NAS report early in December 1878, Schurz asked Powell to rewrite it as legislation. When Powell added his land-reform clauses rejected by Congress in April, Schurz turned to Hewitt and King to prepare a new version. Hewitt introduced their bill, which substituted *national* for *public* domain to

Clarence King. SCIENCE, INDUSTRY & BUSINESS/NEW YORK LIBRARY/SCIENCE PHOTO LIBRARY.

allow operations nationwide, in the House during February 1879. After heated congressional debate and deft maneuverings by Hewitt to ensure that Hayden's supporters did not highjack the measure, the legislators and Hayes agreed to establish within the Interior Department the USGS, but not the proposed Coast and Interior Survey or to improve the GLO. On 3 March the new appropriations law for fiscal 1879-1880 (20 *Stat.* 394) established the office of director of the USGS, provided $106,000 (two-thirds the total given to the three surveys it replaced) for salaries and operations, specified the agency's publications, and required it to deposit its scientific collections in the U.S. National Museum.

King Becomes Director. The struggle for reform then passed to the selection of the initial director of the USGS. The framers, who specified that the director be appointed by the president and approved by the Senate, entrusted the success of their work to Hayes, who, while governor of Ohio, had reestablished that state's geological survey in 1869 and made Newberry its director. Powell, not yet a NAS member, decided that he alone could not defeat

Hayden, the senior candidate in years and experience. Powell joined the campaign for King, who had completed his survey and resigned his commission in January 1879 to promote the reforms and to seek the directorship. Hayes's choice then lay between the reactionary Hayden and the reformer King. Hayes initially favored Hayden, but requested and volunteered letters by and personal visits from Schurz, Newberry, many other NAS members, and several close friends convinced the president that King was the better scientist and manager. The Senate overwhelmingly confirmed King's nomination.

King, backed by Schurz, introduced to the USGS the modern methods and appliances used in, the high standards for appointments and work by, and the spirit of his fortieth-parallel survey. King's continued professional brilliance, personal magnetism, natural style of command, and personal sympathy with everyone who worked with him quickly exceeded his new staff's expectations, won their enthusiastic affection, and generated their best efforts. King urged American industries to utilize natural resources with technical skill and scientific economy. Responding to the USGS's twin mandates, King planned a scientific classification of the nation's lands and a comprehensive assessment of the extent, nature, and geological relations of their mineral resources, including large-scale mapping of mining districts to convince industry of geology's value in developing districts, locating new deposits, and determining their origins. The supporting basic studies in geochemistry, geophysics, and paleontology also would provide new science to apply and to increase knowledge of the Earth and its history. King directed the limited smaller-scale topographic mapping and general geologic investigations by the USGS primarily toward completing a more reliable national geologic map, a goal of federal agencies since the 1830s.

When the attorney general's decision confined USGS operations to the public domain, King extended the mineral program nationwide by co-funding and participating in a cooperative program with Interior's Tenth Decennial Census. King secured $156,000 for the USGS in fiscal 1880–1881. This was a fifty percent rise but still well short of the $350,000 that King and Schurz requested as an increase toward the $500,000 needed for the expanded work and parity with the sum long provided to the Coast and Geodetic Survey.

King's Later Activities. King intended to remain director, as Powell knew in 1879, "only long enough to appoint its staff, organize its work, and guide the force into full activity" (Benjamin Papers, SIA). King's failure to reach his full-funding and national-coverage goals, his family and personal financial requirements, and his struggles to avoid ethical compromises confirmed that decision, and he resigned in March 1881, a week after Hayes and Schurz left office. On King's recommendation, new president James Garfield nominated Powell to take over the USGS. On 7 August 1882, King's and Director Powell's friends in Congress and President Chester Arthur overcame the geographical restriction on USGS operations by authorizing the agency "to continue the preparation of a geological map of the United States" (22 *Stat.* 329). Under this rubric, and to King's dismay, Powell deemphasized USGS work in economic geology and confined it to the public lands. Powell quickly turned the USGS into an agency for topographic mapping (the needed national program) and staff-selected basic research in geology (less immediately required).

Between 1881 and 1893 King promoted, with mixed success, mines in Mexico and the United States, ranches in Wyoming, banks in Texas and California, and other financial ventures. These years also encompassed Powell's rise and fall in the USGS. After 1886, the failures of Powell's policies and programs to meet national needs led four Congresses and two presidents to enact a series of statutory restrictions and investigations intended to encourage Powell to return the USGS to its original practical work. The imposition of line-item budgets, the Irrigation Survey's demise, and then large reductions in USGS scientific statutory positions and specific appropriations for unwanted investigations, a USGS reduction-in-force, and requests for King to resume the directorship marked Powell's decline. King agreed to return if the change could be made without strife, but his physical ailments and responsibilities for his public and clandestine families led to a brief nervous breakdown in 1893 that prevented his restoration. The financial panic that year destroyed King's personal resources leaving him hopelessly in debt to his long-time friend John Hay.

King facilitated the called-for change in USGS management. On King's recommendation in 1893, Interior Secretary Hoke Smith appointed Charles Walcott (one of King's hires in 1879) as the agency's geologist-in-charge of geology and paleontology. Following King's suggestion in 1894, Smith, President Grover Cleveland, and the Senate advanced Walcott to director when Powell resigned before the appropriations law for fiscal 1894–1895 reduced his salary by one-sixth. Walcott reinstituted King's administrative control, mission orientation, and high standards for appointment. Walcott also restored congressional confidence and support by promoting and expanding USGS investigations to aid not only the mineral industry but also the wiser use of water supplies, land reclamation, and any other practical objective whose achievement depended on a greater knowledge of the Earth and its natural resources. Walcott, like King, thought applied and basic studies inseparable. Under Walcott's direction to 1907, and thereafter, the USGS undertook basic research not so much for its own sake, as done during Powell's years, but to meet specific needs for knowledge to solve specific problems.

SUPPLEMENTARY BIBLIOGRAPHY

WORK BY KING

Mountaineering in the Sierra Nevada. Boston: James R. Osgood, 1872. A facsimile reprint was edited by Thurman Wilkins, Philadelphia: Lippincott, 1963.

OTHER SOURCES

Bartlett, Richard. *Great Surveys of the American West.* Norman: University of Oklahoma Press, 1962.

———. "Scientific Exploration of the American West, 1865–1900." In *North American Exploration:* Volume 3, *A Continent Comprehended,* edited by John Allen. Lincoln: University of Nebraska, 1997. Chapter 22 summarizes the section on King's fortieth-parallel exploration found in Bartlett's 1962 volume.

Burich, Keith. "'Something Nobler Is Called into Being': Clarence King, Catastrophism, and California." *California History* 72 (1993), 234–249, 302. Burich concentrates on King's years with that state's survey led by Josiah Whitney.

Hausel, Dan, and Sandy Stahl. "The Great Diamond Hoax of 1872." In *Resources of Southwestern Wyoming. Wyoming Geological Association Guidebook 1995 Field Conference Guidebook, Rock Springs, Wyoming, 19–22 August 1995,* edited by Richard Jones. Guidebook 46, Pt. 1, pp. 13–27. Casper: Wyoming Geological Association, 1995. This is the forerunner of a book-length analysis of the swindle and also based on the results of Hausel's study of the discovery site and adjacent locales as a member of the Wyoming State Geological Survey.

Moore, James. *King of the 40th Parallel: Discovery in the American West.* Stanford, CA: Stanford General Books, 2006. USGS geologist Moore, who has worked extensively in the Sierra Nevada, reviews King's studies in that range, the Great Basin, and the Rocky Mountains. From a private collection, Moore prints several letters written by King's friend and colleague James Gardiner.

Nelson, Clifford. "Toward a Reliable Geologic Map of the United States, 1803–1893." In "Surveying the Record: North American Scientific Exploration to 1930," edited by Edward Carter II. *American Philosophical Society Memoir* 231 (1999), 51–74. Nelson evaluates King's contributions to this improving cartography.

———. "King, Clarence Rivers." In *The History of Science in the United States: An Encyclopedia,* edited by Marc Rothenberg. New York: Garland Publishing, 2001.

———."King, Clarence (Rivers)." In *The Development of the Industrial United States, 1870–1899,* edited by Ari Hoogenboom. New York: Facts on File, 2003.

———, and Mary Rabbitt. "The Role of Clarence King in the Advancement of Geology in the Public Service." In *Frontiers of Geological Exploration of North America,* edited by Alan Leviton. San Francisco: American Association for the Advancement of Science, Pacific Division, 1982. This is a briefer evaluation of King's role as a participant in and leader of mapping and science agencies.

———, and Mary Rabbitt. "King, Clarence Rivers." In *Biographical Dictionary of American and Canadian Naturalists and Environmentalists,* edited by Kier Sterling. Westport, CT: Greenwood Press, 1997.

O'Toole, Patricia. *The Five of Hearts: An Intimate Portrait of Henry Adams and His Friends, 1880–1918.* New York: C. Potter, 1990.

Rabbitt, Mary. *Minerals, Lands, and Geology for the Common Defense and General Welfare,* Vol. 1, *Before 1879.* Washington, DC: U.S. Geological Survey, 1979. Vol. 2, *1879–1904.* Washington, DC: U.S. Geological Survey, 1980. Rabbitt appraises King's role as a participant in and leader of mapping and science agencies.

Sandweiss, Martha. *Passing Strange: The Secret Life of Clarence King.* New York: Penguin Press, 2008. Sandweiss explores Wilkins's and O'Toole's views of King's last two decades.

Smith, Michael. *Pacific Visions: California Scientists and the Environment, 1850–1915.* New Haven, CT: Yale University Press, 1987.

Starr, Kevin. *Americans and the California Dream, 1859–1915.* New York: Oxford University Press, 1973.

White, Richard. *"It's Your Misfortune and None of My Own": A New History of the American West.* Norman: University of Oklahoma Press, 1991. Chapter 5 assesses King within a development context.

Wilkins, Thurman. *Clarence King, A Biography.* Albuquerque: University of New Mexico Press, 1988. Revised and expanded edition, completed with the assistance of Caroline Hinkley, significantly improves the groundbreaking version of 1958.

———. "King, Clarence Rivers." In *American National Biography,* edited by John Garraty and Marc Carnes. New York: Oxford University Press, 1999.

Wilson, Robert. *The Explorer King: Adventure, Science, and the Great Diamond Hoax—Clarence King in the Old West.* New York: Scribner, 2006. This journalistic retelling, like Moore's book, relies heavily on Wilkins's volume as a source.

Clifford M. Nelson

KINSEY, ALFRED CHARLES *(b.* Hoboken, New Jersey, 23 June 1894*; d.* Bloomington, Indiana, 25 August 1956*), biology, taxonomy, human sexuality.*

Kinsey was a biologist at Indiana University with a special interest in taxonomy and its application to the gall wasp. In 1938, at the age of forty-four, he was asked to teach a marriage course for students. This task confronted him with the extraordinary lack of scientific evidence relating to human sexual behavior, and led him to spend the rest of his life striving to fill this gap in knowledge. The extent to which human sexual behavior had been systematically studied previously was minimal, and in several respects Kinsey was a pioneer who broke through the social taboos to pursue his scientific goals, in the process carrying out a project that has not yet been equaled in size, breadth, or scope. His work also provoked considerable controversy, which has continued, at intervals, ever since. He started on this great undertaking relatively late in his career, and he died eighteen years later, when only sixty-two.

Early Years. Kinsey was born in Hoboken, New Jersey, on 23 June 1894, to Sarah Ann Charles Kinsey and Alfred Seguine Kinsey, a professor at the Stevens Institute of Technology. Both of his parents were devout Methodists with very conservative social values. Although Kinsey's childhood was marred by recurrent ill health that left him physically vulnerable and contributed to his early death, the young Kinsey loved being outdoors. He went camping with the local YMCA and joined the Boy Scouts. By the time he was in college, he followed an exhausting schedule of work, study and long nature expeditions.

A high school science teacher, Natalie Roeth, introduced Kinsey to the biological sciences, but his father insisted that he study engineering at Stevens. After two years, Kinsey rebelled and transferred to Bowdoin College, where he graduated magna cum laude and Phi Beta Kappa. Although Kinsey had to take an overload of courses to complete degrees in psychology and biology while working to earn his keep, he maintained his strong interest in the piano and continued to work with YMCA camps.

After graduating from Bowdoin, Kinsey entered the Bussey Institute at Harvard. There he studied with William Morton Wheeler, a prominent entomologist who studied ant societies. Kinsey chose gall wasps for the subject of his dissertation. After completing a PhD in zoology in 1919, he was awarded a Sheldon Traveling Fellowship, which subsidized a postdoctoral year in the field. He toured remote areas of the United States, adding to an already large collection of gall wasps. In 1920 he accepted the invitation of ichthyologist Carl Eigenmann to join the Department of Zoology at Indiana University.

Gall Wasps. For more than twenty years Kinsey devoted his academic life to an investigation of the taxonomy of gall wasps of the genus *Cynips*. By the time of his death he had studied more than five million specimens. When he began his work in graduate school, traditional taxonomy tended to focus on typical specimens. From this perspective, one only needed to look at a few "perfect" examples of any given species, and it would be a waste of time to collect thousands of wasps. But Kinsey took quite a different approach, one that reflected the growing influence of evolutionary biology.

From an evolutionary point of view, one needed to study populations of organisms, and the variations among them were of crucial interest. So Kinsey not only collected large numbers of specimens, he also measured them under the microscope, noting all sorts of differences between them. Because he was also interested in speciation and biogeography, Kinsey needed to collect specimens from diverse regions. Gall wasps do not travel far from their protective galls. Hence it is relatively easy for populations to become isolated, and this leads to an unusually large number of what Kinsey thought of as species (many in the early twenty-first century would be viewed as subspecies).

Starting in 1922 Kinsey published long papers on his findings, followed by two extensive books: *The Gall Wasp Genus* Cynips: *A Study in the Origin of Species* (1930) and *The Origin of Higher Categories in* Cynips (1936). Kinsey bluntly contrasted his approach with that of systematists who stick pins in a few so-called representative samples and place them in a box. His views on taxonomy won wide acceptance, and his 1926 textbook for undergraduates, *An Introduction to Biology*, was well received and went through three editions. In 1937 he was made a "starred scientist" in *American Men of Science*. And in 1938 Kinsey set out to offer a marriage course to Indiana undergraduates. Soon he was investigating the varieties of human sexual behavior with the same vigor as he had pursued gall wasps.

Sex Research. When Kinsey turned his research attention to human sexual behavior, his objective was to collect a large number of interviews covering extensive details of each interviewee's sexual life. His goal was 100,000 interviews—perhaps not surprising, considering the more than a million gall wasps he had amassed. By the time that his research team completed interviewing in 1963, the total amounted to more than 18,000, a long way from Kinsey's goal, but very large by any standards; Kinsey himself had conducted more than a third of these. He had planned a series of books based on this data, but only published two of them before his early death: *Sexual Behavior in the Human Male* (1948) and *Sexual Behavior in the Human Female* (1953). His research team, under the leadership of Paul Gebhard following Kinsey's death, published three further books based on this data: *Pregnancy, Birth and Abortion* (1958), *Sex Offenders: an Analysis of Types* (1965) and culminating in *The Kinsey Data: Marginal Tabulations of the 1938–1963 Interviews* (1979).

The two Kinsey volumes made a huge impact. The first, an extremely dry book filled with dense tables, sold more than 200,000 copies in its first six weeks. Within four months, it was at the top of the *New York Times* best-sellers list. Within the first year, there were translations in French, Spanish, Italian, and Swedish. By 1950 the attention of the public and media turned to anticipating the volume on females. Although at the outset this also received a massive response from the media (most of it in the three weeks before it was officially published), the reaction was brief by comparison with the first volume. Overall, reactions to both books ranged from outrage to admiration. The reactions from the scientific community were also mixed, mainly concerning issues of methodology, which will be considered further.

Kinsey's Sampling. There is general agreement that Kinsey's method of obtaining a sample of Americans did not meet modern standards of survey sampling. Probability sampling, in which random selection is used to improve representativeness, was in its infancy when Kinsey began his long-running study. He also believed probability sampling to be inappropriate because of the high refusal rate to be expected in a sex survey. Kinsey's alternative, reflecting his approach to the gall wasp, was to interview as large a number of individuals as possible, relying on the size of the sample to average out any sources of bias. However, he also deliberately over-sampled relatively rare varieties of sexual behavior in order to have enough of each variety from which to draw useful conclusions, a technique advocated today for the study of behavior in minority groups.

Kinsey was clearly sensitive to criticisms of his sampling approach and defended his methodology within the work itself. Once published, *Sexual Behavior in the Human Male* elicited a number of critical reviews from statisticians. In 1950 the National Research Council, which had been funding Kinsey's research (with money from The Rockefeller Foundation), requested that the American Statistical Association evaluate Kinsey's methodology. Following a lengthy period of assessment, a detailed report by the review group of three—William Cochran, Frederick Mosteller and John Tukey—was published. The group acknowledged the difficulties that Kinsey had faced, which were similar to but in many respects more formidable than those faced by many other large scale social surveys, and concluded that he had been justified in not using probability sampling in the earlier stages of his project. However, they did advise that he should do so, at least on a modest scale, in the future. (By then the data for *Sexual Behavior in the Human Female* had already been largely collected.) Given the potential for selection bias that his method involved, the review group was critical of his lack of caution in interpreting his findings, as well as his incorrect use of statistical procedures (e.g., the weighting procedure to produce "U.S. corrections"). On the other hand, they applauded his diligence, concluding that *Sexual Behavior in the Human Male* was "a monumental endeavor" and markedly superior to other studies in the field.

The sampling problem was most marked in *Sexual Behavior in the Human Male,* mainly because of the inclusion of large numbers of prisoners within the non-college sample. When confronted by his colleagues Paul Gebhard, Wardell Pomeroy, and Clyde Martin concerning the differences in the sexual behavior data between women with prison records and those without, Kinsey agreed to omit such special groups from the analysis, and they were excluded from consideration in *Sexual Behavior in the Human Female.* In addition, Kinsey conceded that the female sample was not appropriate for the "U.S. corrections" that he had employed in the first volume.

Subsequently, some years after Kinsey's death, the Institute staff re-analyzed the data, including additional interviews collected following the preparation of the second volume until 1963. They separated out and published in *The Kinsey Data* (1979) a "basic sample" of men and women who were never convicted of any offense other than traffic violations and who did not come from any source known to be biased in terms of sexual behavior (e.g., homosexual networks). This made clear the undersampling of non-college-educated men and women. In general, therefore, these samples were of most value in studying the college-educated part of the population. As a result of this "cleaning" of the data, Gebhard and Johnson concluded that "the major findings of the earlier works regarding age, gender, marital status and socioeconomic class remain intact. Adding to and cleaning our samples has markedly increased their value, but has not as yet caused us to recant any important assertion" (p. 9).

One issue that did look different as a result of this process was the incidence of male homosexual behavior. Whereas incidence figures for college-educated males did not change much, those for the non-college-educated, once those with criminal records were excluded, looked markedly lower. Gagnon and Simon (1973) reanalyzed the data from the college-educated group and found that, whereas 30 percent reported at least one homosexual experience, in more than half this experience occurred before the age of fifteen; an additional third had experienced all their homosexual acts during adolescence. This left about 3 percent with extensive and 3 percent with exclusive homosexual histories. Kinsey had not drawn attention to the fact that these reported same-sex experiences were predominantly occurring in early adolescence. Of some interest is the possibility that there may have been a substantial drop in this early adolescent male homosexual expression over the last fifty years.

The Kinsey Interview. This aspect of Kinsey's methodology has received little criticism, and in fact is widely regarded as of high quality. It focused on behaviors and responses and did not ask about feelings, attitudes, or

Kinsey

Alfred C. Kinsey. Kinsey interviewing a female subject. REPRODUCED BY PERMISSION OF THE KINSEY INSTITUTE FOR RESEARCH IN SEX, GENDER AND REPRODUCTION, INC.

values. In addition to obtaining a wealth of highly detailed information about each individual's sexual experiences, from childhood onward, Kinsey also sought fairly precise frequencies for a range of behaviors at specific time periods through life. Modern sex research would question the validity of such recalled frequencies, except for fairly recent time periods of recall. However, Kinsey's description of the principles of interviewing to obtain sexual information has probably never been bettered. Unfortunately, his method of interviewing, along with an elaborate method of coding answers, which needed to be memorized by the interviewer, required extensive training; not surprisingly, the method has not been used in more recent surveys.

There were two aspects of Kinsey's approach to the interview that remain of paramount importance: his ability to convey a nonjudgmental attitude, which enabled his subject to describe any sexual behavior, however stigmatized; and his ability to convince subjects that their records would remain completely confidential, a conviction that over the years has remained justified. Both of these issues, which are crucial methodologically as well as ethically, have contributed to the controversies around Kinsey's research, which will be considered below.

The Sexual Outlet. One of the principal criticisms of *Sexual Behavior in the Human Male* concerned its focus on orgasm and on the "total sexual outlet" (i.e., the number orgasms, from whatever source, in a particular time period). According to many critics, sex should not be reduced to the orgasm, and furthermore, orgasms from disparate sources should not be combined to derive a "total sexual outlet." This is one of the principal issues on which Kinsey has been misunderstood and misinterpreted. He regarded orgasm, at least in the male, as the most precise and specific indicator of a sexual experience. He acknowledged that there were many sexual situations or encounters that did not result in orgasm. "These emotional situations are, however, of such variable intensity that they are difficult to assess and compare." Furthermore, implicit in this approach was the assumption that the "total sexual outlet," defined in this way, provided some measure of "sexual drive" or "need for sexual release" for that individual, and as such would be an important measure of individual variability. Although somewhat of an oversimplification, this does have scientific heuristic value. His task, or anyone else's, in quantifying sexual activity, at least in the male, would be much more difficult if orgasm were not the defining characteristic.

By the time Kinsey wrote *Sexual Behavior in the Human Female,* his position had shifted somewhat on this issue. Along with this shift came a more sensitive attitude to female sexuality than had been apparent in the volume on men. Maybe this resulted from his confrontation with the mass of female data as it emerged, which challenged many of his male-oriented attitudes. Maybe he also responded to the criticisms of women whose opinion he clearly respected. Early in the volume on women, he comments that a considerable portion of female sexual activity does not result in orgasm, and he goes on to report incidences and frequencies of women's sexual experiences both with and without orgasm. Later in the volume, one finds the following: "It cannot be emphasized too often that orgasm cannot be taken as the sole criterion for determining the degree of satisfaction which a female may derive from sexual activity. … Whether or not she herself reaches orgasm, many a female finds satisfaction in knowing that her … partner has enjoyed the contact, and in realizing that she has contributed to the male's pleasure" (p. 371). But Kinsey retained his belief in the value of "total sexual outlet" as a useful measure of a woman's interest in or need for sex, and he presented this data, not as in the male volume as the first chapter, but rather as the last chapter of results.

Kinsey's Mission. Although he repeatedly asserted throughout both volumes that his task was to obtain and present the facts, leaving the sociopolitical and moral significance of such facts to others, it is difficult to escape the conclusion that he also had a mission. However, as he never explicitly stated a sociological or moral mission, it was left to others to suggest various agendas: for example, to change the pattern of sexual behavior in the United States, to bring about a "revolution" in sexual values, or even to undermine the social structure of the United States in such a way as to foster communism. (Kinsey was decidedly not a communist, but a Republican.) Scholars will continue to debate the nature and extent of Kinsey's mission beyond that of the socially aware scientist. Critics also debate the extent to which Kinsey's own sexuality and early negative sexual experiences influenced his research. It is noteworthy that in more recent times a scientist's emotional involvement with his or her field of research is more readily acknowledged, and is not necessarily regarded as a negative factor. To what extent Kinsey had insight into the impact of his personal sexual history on his research scholars will never know.

In *Sexual Behavior in the Human Male*, the central theme that may be said to relate to social change concerns the striking differences in patterns of male sexual behavior between what Kinsey summarized as the "upper and lower social levels." This was shown in a greater tendency for "upper level" males to engage in masturbation, premarital petting, and oral sex, and for "lower level" males to engage in premarital intercourse. Kinsey further described this social class difference as reflecting an awareness, at the upper level, of what is "right or wrong" (i.e., what is moral or immoral), and at the lower level of what is "natural or unnatural." In Kinsey's view there were two important consequences of this social class difference: First, there was a major lack of understanding by one class of the other, and resulting conflicts; secondly, many members of the upper social level "consider it a religious obligation to impose their code upon all other segments of the population" (p. 385). Thus Kinsey described how marriage counselors, most of whom came from the upper social level, imposed their concepts of sexual normality on lower-level couples, where they did not fit. More important, those who determined the laws came from the upper social level; thus, in Kinsey's analysis, most of the laws regulating sexual behavior, at the time of his writing, not only had a long background in religious doctrine, but were more consistent with the "sexual morality" of the upper social levels, and were inconsistent with accepted standards of "natural" sexual behavior in the lower social levels.

A recurring theme in both volumes was the extent to which such laws did not reflect actual sexual practice. At the time Kinsey was researching, virtually all forms of nonmarital sexuality were illegal, and some forms of sexual behavior within marriage (e.g., oral sex) were also illegal, at least in some states. "On a specific calculation of our data, it may be stated that at least 85 percent of the younger male population could be convicted as sex offenders if law enforcement officials were as efficient as most people expect them to be" (Male volume, p. 224). Yet "only a minute fraction of one percent of the persons who are involved in sexual behavior which is contrary to the law are ever apprehended, prosecuted or convicted ..." and "the current sex laws are unenforced and unenforceable because they are too completely out of accord with the realities of human behavior" (Female volume, pp. 18–20). Kinsey describes the consequences of this legal state of affairs, which consist not only of the impact of actual convictions, but much more frequently, the chronic effects of guilt about engaging in illegal activities that are, in Kinsey's view, part of the normal range of human sexual experience.

In the volume on women, the emphasis is different. Here the principal causes for concern are the differences in the sexuality of men and women, and the misunderstandings, conflicts, and interpersonal tensions that result from this apparent "mismatch," with Kinsey striving here for better understanding between men and women, and in the process, stabilizing marriage, which Kinsey saw as fundamental to any good social system.

It is inescapable that Kinsey chose to study human sexuality in an extremely behavioral fashion. Although he commented at length on the social processes that shaped sexual morality, and often referred to the anguish and guilt suffered by individuals whose sexual behavior contravened the sexual mores of their group, he confined himself to describing their behavior without attempting to assess its emotional concomitants. Kinsey's scientific training, and probably its interaction with his particular personality, led him to distance himself from the subject of his study. There was little or no consideration of love, intimacy, or tension. Much of his writing in these two volumes studiously avoids engaging with such concepts, leaving the text somewhat impersonal and incongruously cold considering its topic. This no doubt contributed to the discomfort that many felt when reading these books, and perhaps fueled the critics' attacks.

Kinsey and Controversy. It is not surprising that Kinsey's work has provoked controversy. At the time that his two volumes were published, there was little information about sexuality available to the general public, and he was criticized for making his findings available to ordinary people rather than restricting them to professionals and clergy. One of the assumed consequences of Kinsey's books, which has some validity, is that they made homosexual behavior not only more common but also less

Kinsey

pathological, an impact welcomed by the socially repressed homosexual communities, but not by many others. A continuing reason for political opposition to sex survey research is that it "normalizes" behaviors such as homosexuality. Another early reaction of outrage was to the revelation that women were much more sexual than they were "supposed" to be. This led to allegations that Kinsey's survey had excluded "normal, decent women." A concern that has continued ever since is that merely asking people about their sexual activity without passing moral judgment on it in some ways gives them license to continue the behavior. This has been a particular issue in relation to surveys of adolescents, where there is the added concern that research questions suggest ideas, leading to behavior that may not otherwise have happened.

A more pernicious campaign of criticism, which has continued unabated since the early 1980s, stems from concern within some sections of the "Religious Right" that there have been unacceptable changes in sexual mores and patterns of sexual behavior as a result of Kinsey's work. Although his data, together with an abundance of evidence from other sources, clearly indicate that such socio-sexual changes had started well before Kinsey published his findings, and have occurred extensively throughout the industrial world and are linked to a range of other important social changes, such as change in the status of women, critics nevertheless continue to blame Kinsey. A particular theme in this demonization has been to accuse him of sexual offenses against children. These allegations, which range from his being a pedophile, to his carrying out sexual experiments on children, or training other individuals to do so, are entirely without foundation, and are solely based on his reporting of the sexual, and in particular, orgasmic responses of children (Tables 31 to 34, *Sexual Behavior in the Human Male*). Kinsey to some extent was vulnerable to such allegations because he did not indicate clearly the source of this information. Subsequently, the Kinsey Institute made it clear that the information for these four tables came from one man, who had not only been involved throughout his adult life in numerous sexual activities with men, women, and children, but had also carefully documented his experiences. Kinsey interviewed this man close to the writing of the male volume and was clearly interested in his documented observations, which were made available to Kinsey. His preparedness to report this evidence, without passing judgment, is one of the more striking examples of his stance of "moral neutrality." He has also been criticized for not having exposed this man for his crimes, an example of the fundamental importance Kinsey gave to guaranteeing confidentiality to his research subjects. In retrospect, Kinsey's judgment in using this evidence can be questioned. However, the suggestion that some boys can experience "multiple orgasms" before they reach puberty raises crucial questions about normal sexual development, which, because of the difficulties in researching such themes, remain otherwise unanswered. In any case, whether ill judged or not, Kinsey cannot be held responsible for the sexual exploitation of these children, and there is no indication that he abused or promoted the abuse of any other children.

Alfred Kinsey. ARTHUR SIEGEL/TIME LIFE PICTURES/GETTY IMAGES.

Kinsey as a Scholar of Sexual Science. Many of Kinsey's ideas and interpretations in both volumes remain as thought-provoking and relevant as they were when first published. In *Sexual Behavior in the Human Female*, he reported important and striking differences in male and female patterns of behavior and attitudes. The most striking was the difference in accumulative incidence curves of various aspects of sexual behavior in the two sexes, with males showing a much more marked rise around puberty to an early peak and subsequent decline through adulthood, whereas females showed a much more gradual rise to a much later peak. The clear relationship in males between early age at puberty and higher subsequent levels of sexual activity was not found in females. Individual variability in sexual responsiveness and frequency of sexual activity was much greater among females. Males were generally more "promiscuous" than females.

In attempting to explain and interpret these sex differences, Kinsey clearly changed some of his opinions between the writing of the two volumes, moving to a state of greater uncertainty in the volume on women. Thus, the impact of socio-cultural influences was stressed in relation

Kinsey

to the male, with major differences between upper- and lower-level males in terms of premarital intercourse, petting, masturbation, and so on. Such differences were not found in women, which he interpreted in several places as indicating that women were less susceptible to socio-cultural influences. At the same time, his own data had shown that there had been important changes in certain aspects of female sexual behavior over time, with women born after 1900 being more likely to engage in premarital intercourse and petting than those born before 1900, a difference that apparently was consistent across the socioeconomic spectrum. At times Kinsey attempted to see biological determinants as more important in the female, yet this was inconsistent with his finding that age at puberty had more impact in males than females, suggesting more powerful biological determinants in males. Kinsey's writing suggests greater comfort with biological than with socio-cultural explanations, in spite of the considerable emphasis placed on socio-cultural differences in the volume on men. This presumably reflects his lack of training as a social scientist. As the female volume progressed, increasing attention was paid to the idea that there are basic psychological differences between the sexes. The picture of these undoubtedly complex differences became somewhat confused. Fifty years later, and with a fair amount of further evidence available, this confusion can only be reduced to a limited extent.

The last five chapters of the volume on women provide a masterly review of the evidence available at the time on the anatomy, physiology, psychology, neurophysiology, and endocrinology of sexual response. In Chapter 14, on the anatomy of sexual response, there are detailed descriptions of common patterns of muscle response and other responses during sexual activity and orgasm. The sources of this data were not given. It later became apparent, from Pomeroy's biography of Kinsey, that most of this observational data came from films of sexual activity involving volunteers, which Kinsey had made in the privacy of his own home. At the time of his writing it was clearly unwise to have revealed that such filming had been done. Kinsey was a scientist who was reluctant to rely solely on self-report; he wanted to be able to observe what happened during sexual activity. In the process he paved the way for William Masters and Virginia Johnson's important work on the physiology of human sexual response.

The Legacy. During the months before his death in 1956, Kinsey struggled to maintain his intense work schedule. He faced repeated disappointments in his attempts to secure funding for the Institute and, as everyone around him realized, his heart was failing. He died on 25 August, survived by his wife Clara (Mac) and three children, Anne, Joan, and Bruce. (Their first child, Donald, had died of juvenile diabetes in 1926 at the age of four.) The Institute also survived. The following year Paul Gebhard and Wardell Pomeroy were awarded the first grant from the National Institute of Mental Health.

In the field of sexual science, where intellectual heavyweights have been in short supply, Kinsey remains the preeminent sexual scientist. The fact that he was trained as a biologist, yet carried out a massive study of human sexual behavior using social science research methods, accounts for some of the mistakes and errors of judgment that he made. He was clearly a stubborn man with strongly held opinions, making it less likely that he would accept the advice of others. However, he showed himself responsive to criticism when it was backed up with good evidence, and perhaps his somewhat arrogant manner contributed to his success in what was an exceptionally courageous and monumental piece of pioneering research. While being sometimes hypercritical of his academic peers, he showed compassion for those who suffered as a result of their sexual lives. He doubted that the causes for such problems lay within the individual but rather the repressive social environment in which he or she had developed. The extent to which Kinsey is responsible for changes in sexual attitudes and behavior that have occurred since he published his two volumes is debatable. But what is beyond dispute is that he opened up the debate about sexual behavior and in several respects demystified it.

BIBLIOGRAPHY

A complete bibliography of Kinsey's published work is available in Gathorne-Hardy, Jonathan. Alfred C. Kinsey: Sex the Measure of All Things. *London: Chatto and Windus, 1998, pp. 494–495.*

WORKS BY KINSEY

An Introduction to Biology. Philadelphia: Lippincott, 1926.

The Gall Wasp Genus Cynips. *A Study in the Origin of Species.* Bloomington: Indiana University Publications, 1930.

A New Introduction to Biology. Chicago: Lippincott, 1933.

The Origin of Higher Categories in Cynips. Bloomington: Indiana University Publications, 1936.

With W. B. Pomeroy and C. E. Martin. *Sexual Behavior in the Human Male.* Philadelphia: W. B. Saunders, 1948.

With W. B. Pomeroy, C. E. Martin, and P. H. Gebhard. *Sexual Behavior in the Human Female.* Philadelphia: W. B. Saunders, 1953. Republished, Bloomington: Indiana University Press, 1998.

OTHER SOURCES

Bancroft, John. "Alfred C Kinsey and the Politics of Sex Research." *Annual Review of Sex Research* 15 (2004): 1–39. This review examines the political reaction to Kinsey's research and the periodic political reaction to sex research that has occurred in the United States since Kinsey.

Capshew, J. H.; M. H. Adamson; P. A. Buchanan; et al. "Kinsey's Biographers: A Historiographical Reconnaissance." *Journal of the History of Sexuality* 12 (2003): 465–486. This articles examines and compares the four biographies of Kinsey.

Christenson, C. V. *Kinsey: A Biography.* Bloomington: Indiana University Press, 1971.

Churchill, Frederick B. "The Evolutionary Ethics of Alfred C. Kinsey." In *Scientific Values and Civic Virtues,* edited by Noretta Koertge, pp. 135–153. New York: Oxford University Press, 2005. An analysis of how Kinsey avoided the naturalistic fallacy.

Gagnon, J. H., and W. Simon. *Sexual Conduct: The Social Sources of Human Sexuality.* Chicago, Aldine, 1973.

Gathorne-Hardy, Jonathan. *Alfred C. Kinsey: Sex the Measure of All Things.* London: Chatto and Windus, 1998. This biography, using much of the same evidence available to Jones (below), paints a very different picture of Kinsey as a man, and disputes Jones's assertion that Kinsey's research was flawed because of his own sexuality.

Gebhard P. H.; J. H. Gagnon; W. B. Pomeroy; and C. V. Christenson. *Sex Offenders: An Analysis of Types.* New York: Harper and Row, 1965.

Gebhard, P. H., and A. B. Johnson. *The Kinsey Data: Marginal Tabulations of the 1938–1963 Interviews Conducted by the Institute for Sex Research.* Philadelphia: W. B. Saunders, 1979.

Gebhard, P. H.; W. B. Pomeroy; C. E. Martin; and C. V. Christenson. *Pregnancy, Birth and Abortion.* New York: Harper and Brothers, 1958.

Gould, Stephen Jay. "Of Wasps and WASPS." In *The Flamingo's Smile: Reflections in Natural History,* pp. 155–166. New York: W. W. Norton, 1985, Provides a detailed comparison of Kinsey's approach and methodology in his studies of gall wasps and humans.

Jones, J. H. *Alfred C. Kinsey: A Public/Private Life.* New York: W. W. Norton, 1997. This thoroughly researched biography paints a negative picture of Kinsey's own sexuality with the assertion that it biased his research.

Morantz, R. M. "The Scientist as Sex Crusader: Alfred C. Kinsey and American Culture." In *Procreation or Pleasure?: Sexual Attitudes in American History,* edited by T. L. Altherr, pp. 145–166. Malabar, FL: R. E. Krieger, 1983.

Pomeroy, W. B. *Dr. Kinsey and the Institute for Sex Research.* New York: Harper & Row, 1972. These two biographies are by members of Kinsey's research team. As such they provide well informed accounts of Kinsey's work, but with no breach of the confidentiality that their working relationship with him incurred.

Robinson, P. *The Modernization of Sex: Havelock Ellis, Alfred Kinsey, William Masters, and Virginia Johnson.* New York: Harper & Row. 1976. An example of scholarly reactions to Kinsey's research and impact that provides a reasonably balanced perspective.

John Bancroft

KIRCH, MARIA MARGARETHA WINKELMANN

SEE **Winkelmann, Maria Margaretha.**

KIRCHER, ATHANASIUS (*b.* Geisa at the Ulster, Germany, 2 May 1602 [or 1601]; *d.* Rome, Italy, 27 November 1680). For the original article on Kircher see *DSB,* vol. 7.

Scholarly work since the 1980s on Kircher has considerably enriched people's understanding of the life and work of this seventeenth-century polymath. While an earlier literature primarily sifted through Kircher's encyclopedic publications to find individual instances in which he offered new information, provided the occasional insight regarding some unexplored aspect of the natural world, or introduced a novel machine, more recent scholarship has tried to assess the goals of Kircher's numerous projects to harmonize and synthesize knowledge. Rather than dismissing Kircher as an intellectual dilettante who wrote about the wonders of nature and the mysteries of science simply to amuse a baroque audience, historians in the early twenty-first century see Kircher as an exemplary figure in understanding the transition from ancient to modern ways of thinking about the world. He was a man who immersed himself in the currents of scholarship at the height of the seventeenth century while publicly proclaiming the value of traditional learning and faith. Most importantly, he was a fascinating by-product of the Society of Jesus: a Catholic natural philosopher in the age of Galileo, Descartes, and a young Newton, a Jesuit priest whose goal was to incorporate aspects of the new natural and experimental philosophy and a fuller understanding of ancient (especially Neoplatonic and Hermetic) philosophies of knowledge into the traditional Aristotelian-Ptolemaic worldview upheld by the Catholic Church following its condemnation of heliocentrism in 1616 and the trial of Galileo in 1633.

The results of Kircher's speculations were often far from orthodox. It is now known that Jesuit censors critiqued important works such as the *Oedipus Aegyptiacus, Mundus subterraneus,* and *Itinerarium exstaticum* prior to their publication on precisely these grounds. The *Itinerarium exstaticum,* in particular, continued to be criticized and revised after its initial appearance in 1656 for being too favorable toward Copernicanism and too willing to embrace other controversial ideas such as the infinity of space in the way in which it presented the truth of Tychonic astronomy, the official cosmology of the Society of Jesus since 1620. As has been recently argued by Carlos

Ziller Camenietski, Ingrid Rowland, and Harald Siebert, Kircher was a man trying to invent a new cosmology—one that incorporated the findings of early-seventeenth-century astronomy and physics into a philosophy deeply indebted to antiquity but also to philosophers such as Nicolas Cusanus and Giordano Bruno.

Research on Kircher since the 1980s has made greater use of unpublished manuscripts and his correspondence (with more than 760 patrons, natural philosophers, experimenters, Jesuit missionaries, and curious admirers) while also offering fresh readings of his published works. The foundational work of John Fletcher and other scholars helped to identify the importance of his letters in understanding his relationship to such important figures as Nicolas-Claude Fabri de Peiresc, Cassiano dal Pozzo, Marin Mersenne, Pierre Gassendi, Francesco Redi, Evangelista Torricelli, Johannes Marcus Marci, and Gottfried Wilhelm Leibniz. Subsequent work has tried to assess his place in the seventeenth-century republic of letters as a purveyor of interesting information, collector of curious artifacts, and disseminator of new ideas and discoveries in his numerous, highly illustrated encyclopedias. Modern facsimiles and translations of a number of Kircher's most important books have been published, and the nucleus of his manuscripts (fourteen folio volumes housed in the archive of the Pontificia Università Gregoriana in Rome, MSS 555–568) is available in a digital edition as *The Athanasius Kircher Correspondence Project.* Several museum exhibits (most notably, a temporary one in Rome at Palazzo Venezia in 2001 and a permanent one at the Museum of Jurassic Technology in Culver City, California) have attempted to reconstruct some of Kircher's machines and even to re-create his entire gallery in the Roman College. Rather than being perceived as a secondary figure of the Scientific Revolution—a man who amused many but enlightened few—Kircher might now be described as a scholar-impresario who sought to reveal the hidden connections among different kinds of learning in order to provide a grand, unified scheme of knowledge. Kircher's unbridled egoism, his failure to fulfill his vast ambitions, and his inability to see the shortcomings of his intellectual program infuriated his critics. At the same time, he also inspired many readers of his books and many visitors to his museum to consider the important scientific and antiquarian questions of his day. In short, Kircher is a fascinating reflection of why and how science mattered in the mid-seventeenth century.

Early Years. Admitted as a student at the Jesuit college in Paderborn on 2 October 1618, Kircher completed his novitiate in 1620. His early reputation as a scholar of ancient languages was soon coupled with a growing interest in mathematics and astronomy. A chance encounter in the Jesuit college library of Speyer with a book on Egyptian hieroglyphics, probably Johann Georg Herwath von Hohenburg's *Thesaurus hieroglyphicorum,* inspired Kircher to request a missionary posting in the Near East in 1628. He would repeat this request again in 1637 (earlier Kircher scholarship frequently stated that he had wanted to go to China, misunderstanding the Orient of his dreams because of his subsequent publication *China monumentis illustrata* in 1667). Both times the Society of Jesus refused to send him, considering Kircher to be better suited to a professorial career and more valuable as a publisher of missionary reports on the natural and human wonders of Asia, America, and the Near East than as an apostle in the field.

While teaching in Koblenz in 1623 Kircher built the first of many sundials. In 1630, during his tenure as a professor at the Jesuit college in Würzburg, he completed his first book: his unpublished mathematics textbook, *Institutiones mathematicae.* A year later the *Ars magnesia,* a slim pamphlet outlining the basic principles of universal magnetism as the underlying force in Kircher's natural philosophy, became his first publication. By the time Kircher arrived in Avignon, France, in 1632, where he was appointed professor of mathematics and Oriental languages at the Jesuit college and quickly befriended the Aix savant Nicolas-Claude Fabri de Peiresc, he had a strong sense of how to realize his ambitions. He built an even more elaborate sundial that demonstrated his skills in the science of optics as well as astronomy and mathematics and began demonstrating his fabled sunflower clock, a heliotropic plant whose ability to tell time allegedly demonstrated the inherent properties of magnetism in all parts of the natural world. Kircher also promised to show Peiresc a mysterious Arabic manuscript in his possession, attributed to the Babylonian rabbi Barachias Nephi and containing knowledge, according to Kircher, that would help him interpret the proper meaning of the Egyptian hieroglyphs and all the ancient secrets of Christianity.

Although Peiresc never did see the manuscript and began to express reservations about Kircher's abilities as a linguistic and experimental philosopher, he was nonetheless intrigued enough to persuade the Jesuit general Muzio Vitelleschi and Cardinal Francesco Barberini to have Kircher appointed as professor of mathematics and Oriental languages at the Roman College in November 1633, succeeding Christoph Scheiner in the chair made famous by the Jesuit astronomer Christopher Clavius. Peiresc certainly was fascinated with the new astronomy ushered in by Galileo's telescope, as was Kircher. However, he considered his Jesuit protégé's primary function in Rome to be antiquarian and Christian: Peiresc wanted Kircher to edit and translate Pietro della Valle's Coptic grammar and dictionary, considered the linguistic key to the interpretation of the hieroglyphs. Kircher eventually published it in his *Lingua Aegyptiaca restituta* (1643), one of many books advertising his ability to decode the hieroglyphs and, more generally, to penetrate the mysteries of Egyptian wisdom throughout his lengthy career. Once in Rome, however, he did not neglect to cultivate his reputation on all fronts, presenting himself as a natural philosopher, inventor, linguist, and antiquarian. News of the sunflower clock even reached the ears of Galileo, now under house imprisonment in Arcetri, whose advocacy of heliocentrism Kircher had discussed sympathetically with Peiresc and Pierre Gassendi in Aix, before Galileo's condemnation in June 1633.

Work in Earth Science. During a trip to Malta as the confessor of the recently converted Landgraf Friedrich of Hesse-Darmstadt, Kircher created his most spectacular instrument yet: the *Specula Melitensis* (1638), which contained a planisphere, kept track of the Julian and Gregorian calendars, told universal time, charted horoscopes, and condensed all important medical, botanical, alchemical, Hermetic, and magical knowledge into a single cube known as the "cabalistic mirror." Returning to Rome, he witnessed the eruptions of Etna and Stromboli and observed the crater of Vesuvius in 1638. His observations of nature's spectacle in southern Italy inspired a lifelong fascination with those aspects of nature that he would aptly call the "subterranean world." The *Mundus subterraneus* (1665) was the first encyclopedia to systematically explore the forces that shaped the world below the surface, including the nature and location of volcanoes, earthquakes, ocean currents, and the formation of fossils. Written from the reports of Jesuit missionaries throughout the world, it was a truly global natural history.

As Kircher developed a nascent interest in what is now called geology and paleontology, he was also bringing to fruition his work on universal magnetism and optics. The publication of his *Magnes sive de arte magnetica* (1641) cemented Kircher's reputation as the foremost "magnetic philosopher" since William Gilbert at the beginning of the century. Mersenne provided observations for the table of magnetic declination, forwarding data from his English correspondents. Gassendi, Jesuit mathematicians and natural philosophers such as Christoph Scheiner and Niccolò Cabeo, and Jesuit missionaries in Goa, Macao, Canton, and the West Indies also contributed to Kircher's project. The *Magnes* was the first work in which Kircher demonstrated his ability to create a global network of informants, using the combined resources of the Society of Jesus and the European republic of letters to gather information. It also contained his first report—an account he received from the Italian province of Puglia—of the use of music to cure a tarantula's bite. While highly criticized by some readers for its speculative leaps in attempting to create a full unified account of the forces of nature, it was sufficiently well read to warrant two further editions in 1643 and 1654.

Kircher

Athanasius Kircher. *Athanasius Kircher, circa 1666.* HULTON ARCHIVE/GETTY IMAGES.

On Light. Kircher's *Ars magna lucis et umbrae,* a fascinating encyclopedia on the subject of light and shadow, appeared in 1646, the year in which he ceased to have any regular teaching duties at the Roman College, which allowed him to pursue his publications and inventions without impediment. Filled with descriptions of optical and catoptrical devices such as Archimedes's burning mirror and the magic lantern that continues to be associated with Kircher more than almost any other invention, it enjoyed a similarly mixed reputation. While many readers lingered on the lavish illustrations, fantasizing about how they might re-create Kircher's machines, others, such as Torricelli and Mersenne, laughed themselves silly over the Jesuit's claim to have squared the circle through a demonstrably imprecise mathematical proof, which confirmed their suspicions that Kircher did not exactly represent the cutting edge of mathematics.

Understanding the nature of light was one of the central questions of natural philosophy in the mid-seventeenth century, as was the phenomenon of magnetism, which would ultimately interest Isaac Newton in his student days at Cambridge before he rejected its universality in favor of something less tangible and more universal: the idea of universal gravitation. Kircher's virtue was to recognize the significance of these questions. He attempted to collect all the relevant information and began to analyze it. While his conclusions often seemed flawed to the most critical and knowledgeable readers at the time, they nonetheless mined his books for useful data and specific insights that they might incorporate into their own natural philosophies. A similar tendency can be seen in Kircher's *Musurgia universalis* (1650), which attempted to build on the work of such well-known scholars as Kepler and Mersenne in exploring the idea of universal harmony. Like many of Kircher's other works, it was a book filled with fascinating descriptions and illustrations, in this case of sound-making machines and automata, including the hydraulic organ he helped to restore at the Palazzo del Quirinale in Rome. The machines made a much greater impression on readers than his theories.

Museum Curator and Antiquarian. One year after the publication of this book Kircher was offered the opportunity to become the custodian of the newly created Roman College museum. In 1651 Alfonso Donnino donated his collection of art and antiquities to the Society of Jesus, requesting that it be displayed in the Roman College. Kircher already had a collection of manuscripts, curiosities, and inventions which he had been showing to visitors in his private quarters at the college since the mid-1630s. The creation of an official Jesuit museum allowed him the opportunity to put his own things on display in the new exhibit space. Visitors had virtually nothing to say about Donnino's collection, since it was far less spectacular than the experience of meeting Father Athanasius in a gallery filled with miniature replicas of the obelisks he had deciphered for the popes; a rubbing of the Nestorian monument and numerous other antiquities; globes, clocks, and magnetic, optical, musical, and perpetual motion machines; his famous speaking tube and magic lantern; and numerous artifacts either gathered by Kircher or donated by Jesuits, patrons, and admirers throughout the world. Posthumously the collection would be known as the Kircherian Museum. It helped to cement Kircher's image as a kind of baroque Leonardo da Vinci who did not simply sketch his fantasies but attempted to realize them in three-dimensional form.

Although the previous DSB entry did not discuss the *Oedipus Aegyptiacus* (printed in several parts in 1652–1654 and published in final form in 1655) as part of Kircher's scientific corpus, subsequent research has persuasively argued that Kircher's science cannot be separated from his antiquarian pursuits. The most ambitious of Kircher's numerous books on Egypt, the *Oedipus* was a vast study of hieroglyphic wisdom made manifest across time and space. Filled with Kircher's research on such subjects as Jewish Kabbalah, Persian magic, Islamic alchemy, Chaldean astrology, Zoroastrian mysteries, and many other ancient sciences, it reflected his ongoing desire to identify the commonalities between ancient wisdom and modern knowledge. In the *Oedipus*, for example, Kircher upheld the fundamental importance of the *Corpus Hermeticum* as a document that demonstrated the essential connection between pagan Egypt and early Christianity (despite having been discredited by Isaac Casaubon as a late antique forgery in 1614). He demonstrated the Egyptian origins of John Dee's enigmatic *Monas hieroglyphica* (Hieroglyphic monad) and, more generally, offered an extensive critique of the flourishing of occult philosophy since the early Renaissance, dismissing philosophers who by his standards had abused this science and praising scholars such as Pico della Mirandola who had helped to lay its foundations as a pious pursuit. Kircher's quest for universal wisdom was also an exploration of the universality of Roman Catholicism. Jesuit censors critiqued the *Oedipus* for being far too admiring of its pagan sources—which it surely was. But they nonetheless let it appear, allowing readers to enjoy the fruits of Kircher's lengthy reflection on the ways in which Egyptian wisdom had shaped the nature of human knowledge across the centuries and how Jesuit missionaries in Asia and America had rediscovered the message of Christianity in the "hieroglyphic" writings of the Chinese and the Aztecs.

Publications. In 1652 Kircher renewed his acquaintance with his most important disciple, Kaspar Schott. Called to Rome by the Society of Jesus to assist Kircher with his numerous projects, Schott temporarily became the curator of machines at the Roman College museum. He ultimately published a series of important works that advertised the virtues of Kircher's inventions. Schott not only helped Kircher with the publication of the *Oedipus* but played an especially important role in the appearance of the *Itinerarium exstaticum* (1656), editing the manuscript substantially in the second edition of 1660 to tone down its favorable allusions to aspects of heliocentrism. Schott returned to his position as professor of mathematics at the Jesuit college in Würzburg in 1655, where he continued to edit and publicize Kircher's work for the rest of his career. He was the most important of several loyal disciples committed to promulgating Kircherian experimental and natural philosophy in the mid-seventeenth century.

Despite the controversies over the *Itinerarium exstaticum*, by the 1660s Kircher was at the height of his career as one of the best-known authors of his time. In 1661 he signed a contract with Joannes Jansson van Waesberghe of Amsterdam, who received the exclusive rights to publish Kircher's work in the Holy Roman Empire, England, and the Low Countries in the winter of that year in return for paying Kircher the princely sum of 2,200 scudi. In addition to his many and diverse publications on natural philosophy, Kircher began to publicize another dimension of his research: his ability to create universal and artificial languages. Publications such as his *Polygraphia nova et universalis* (1663) demonstrated numerous ciphers and codes but also explored the uses of machines in the creation and translation of language. His work in this area not only complemented the research of John Wilkins but also inspired a young Leibniz, who was equally intrigued by Kircher's writings on Egypt and China.

By the 1670s Kircher's findings and his philosophy were increasingly under attack and his health was failing, forcing others to step in to defend him. Kircher's disciple, Gioseffo Petrucci, countered the Tuscan physician and naturalist Francesco Redi's trenchant criticisms of the miraculous curative powers of the snakestone (a missionary artifact from Asia described in Kircher's *China monumentis illustrata*) in his *Prodomo apologetico alli studi Chircheriani* (1677). One year later his assistant Giorgio de Sepi published the *Romani Collegii Societatis Jesu Musaeum celeberrimum* (1678), as much a monument to Kircher as to his museum. In the final year of Kircher's life Johann Koestler produced a well-illustrated digest of Kircher's best experiments in his *Physiologia Kircheriana experimentalis* (1680). Two years after Kircher's death the Prague Jesuit Caspar Knittel offered a final tribute to Father Athanasius, especially to his highly criticized *Ars magna sciendi* (1669), by publishing the *Via Regia ad omnes scientias et artes. Hoc est: Ars universalis, scientiarum omnium artiumque arcana facilius penetrandi* (1682). Described in the subtitle as a "Universal Lullian-Kircherian Art of Knowing and Examining," it was the last treatise to openly advocate Kircherian natural philosophy as the key to unlocking the secrets of the universe. In the age of Newton—who, like Leibniz, was fascinated with many of the questions that animated Kircher's work while ultimately arriving at completely different conclusions—the age of the universal baroque polymath had passed.

SUPPLEMENTARY BIBLIOGRAPHY

Since 2000, the number of modern facsimile editions of Kircher's works, published primarily in Italy and Germany, has increased considerably. See especially the Editiones Neolatinae series published by A. W. F. Sommer in Vienna, which has printed many of these volumes since 2004. The only modern English translation of Kircher's work is the China Illustrata, *translated by Charles D. Van Tuyl (Muskogee, OK: Indian University Press, Bacone College, 1987), which should be used with caution in relation to the Latin original. For an introduction to Kircher's manuscripts in the archive of the Gregorian University in Rome, see "A Brief Survey of the Unpublished Correspondence of Athanasius Kircher S.J. (1602–80)," by John E. Fletcher (*Manuscripta *13 (1969): 150–160).*

WORKS BY KIRCHER

"Juan Caramuel: Su epistolario con Atanasio Kircher, S.J." Transcribed by Ramón Ceñal. *Revista de Filosofía* 44 (1953): 101–148.

"Drei unbekannte Briefe Athanasius Kirchers an Fürstabt Joachim von Gravenegg." Transcribed by John E. Fletcher. *Fuldaer Geschichtsblätter* 58 (1982): 92–104.

Astronomia e tecniche di ricerca nelle lettere di G. B. Riccioli ad A. Kircher. Transcribed by Ivana Gambaro. Genoa: Quaderni Centro di Studio sulla Storia della Tecnica del Consiglio Nazionale delle Ricerche 15, 1989.

"Le Lettere di Athanasius Kircher della Biblioteca Nazionale di Firenze." Transcribed by Alfonso Mirto. *Atti e Memorie dell'Accademia Toscana di Scienze e Lettere La Colombaria* 54, n.s. 40 (1989): 129–165.

La luz imaginaria: Epistolario de Atanasio Kircher con los Novohispanos. Edited and translated by Ignacio Osorio Romero. Mexico: Universidad Nacional Autonóma de México, 1993.

"Lettere di Athanasius Kircher dell'Archivio di Stato di Firenze." Transcribed by Alfonso Mirto. *Atti e Memorie dell'Accademia Toscana di Scienze e Lettere La Colombaria* 65, n.s. 51 (2000): 217–240.

Athanasius Kircher an Herzog August der Jünger: Lateinische Briefe der Jahre 1650–1666 aus den Sammlungen der Herzog August Bibliothek Wolfenbüttel—Transkription und Übersetzung. Edited by Thomas Stäcker. Available from http://diglib.hab.de/edoc/ed000005/start.htm.

The Athanasius Kircher Correspondence Project. Edited by Michael John Gorman and Nick Wilding. Available from http://kircher.stanford.edu/ (earlier version at http://archimede.imss.fi.it/kircher/).

OTHER SOURCES

Arecco, Davide. *Sogno di Minerva: La scienza fantastica di Athanasius Kircher.* Padua: Cooperativa Libraria Editrece Università di Padova (CLEUP), 2002.

Athanasius Kircher y la ciencia del siglo XVII. Madrid: Universidad Complutense de Madrid, 2001.

Bach, José Alfredo. "Athanasius Kircher and His Method: A Study in the Relations of the Arts and Sciences in the Seventeenth Century." PhD diss., University of Oklahoma, 1985.

Baldwin, Martha. *Athanasius Kircher and the Magnetic Philosophy.* PhD diss., University of Chicago, 1987.

———. "The Snakestone Experiments: An Early Modern Medical Debate." *Isis* 86 (1995): 394–418.

Bartòla, Alberto. "Alessandro VII e Athanasius Kircher S.I. Ricerche e appunti sulla loro corrispondenza erudita e sulla storia di alcuni codici chigiani." *Miscellanea Bibliothecae Apostolicae Vaticanae* 3 (1989): 7–105.

Belloni, Luigi. "Athansius Kircher: Seine Mikroscopie, die Animalcula und die Pestwürmer." *Medizinhistorisches Journal* 20 (1985): 58–65.

Camenietzki, Carlos Ziller. "L'Extase interplanetaire d'Athanasius Kircher: Philosophie, cosmologie et discipline dans la Compagnie de Jésus au XVIIe siècle." *Nuncius* 10 (1995): 3–32.

———. *L'harmonie du monde au XVIIe siècle: Essai sur la pensée scientifique d'Athanasius Kircher.* PhD diss., Université de Paris IV–Sorbonne, 1995.

Casciato, Maristella, Maria Grazia Ianniello, and Maria Vitale, eds. *Enciclopedismo in Roma barocca: Athanasius Kircher e il Museo del Collegio Romano tra Wunderkammer e museo scientifico.* Venice: Marsilio, 1986.

Chevalley, Catherine. "L'*Ars magna lucis et umbrae* d'Athanase Kircher: Néoplatonisme, hermétisme et 'nouvelle philosophie.'" *Baroque* 12 (1987): 95–109.

Corradino, Saverio. "Athanasius Kircher: 'Damnatio memoriae' e revisione in atto." *Archivum Historicum Societatis Iesu* 59 (1990): 3–26.

———. "L'*Ars magna lucis et umbrae* di Athanasius Kircher." *Archivum Historicum Societatis Iesu* 62 (1993): 249–279.

———. "Athanasius Kircher matematico." *Studi secenteschi* 37 (1996): 159–180.

Englmann, Felicia. *Sphärenharmonie und Mikrokosmos: Das politische Denken des Athanasius Kircher (1602–1680).* Cologne, Germany: Böhlau, 2006.

Evans, R. J. W. *The Making of the Habsburg Monarchy, 1550–1700: An Interpretation.* Oxford: Clarendon Press, 1979.

Feingold, Mordechai, ed. *Jesuit Science and the Republic of Letters.* Cambridge, MA: MIT Press, 2002.

Findlen, Paula. *Possessing Nature: Museums, Collecting, and Scientific Culture in Early Modern Italy.* Berkeley: University of California, 1994.

———. "Scientific Spectacle in Baroque Rome: Athanasius Kircher and the Roman College Museum." *Roma Moderna e Contemporanea* 3 (1995): 625–665.

———. "The Janus Faces of Science in the Seventeenth Century: Athanasius Kircher and Isaac Newton." In *Rethinking the Scientific Revolution,* edited by Margaret J. Osler. Cambridge, U.K.: Cambridge University Press, 2002.

———, ed. *The Last Man Who Knew Everything: Athanasius Kircher.* New York: Routledge, 2004.

Fletcher, John E. "Claude Fabri de Peiresc and the Other French Correspondents of Athanasius Kircher (1602–1680)." *Australian Journal of French Studies* 9 (1972): 250–273.

———. "Johann Marcus Marci Writes to Athanasius Kircher." *Janus* 59 (1972): 95–118.

———, ed. *Athanasius Kircher und seine Beziehungen zum gelehrten Europa seiner Zeit.* Wiesbaden: Harrassowitz, 1988.

Godwin, Joscelyn. *Athanasius Kircher: A Renaissance Man and the Quest for Lost Knowledge.* London: Thames and Hudson, 1979.

Gómez de Liaño, José Ignacio. *Athanasius Kircher: Itinerario del éxtasis o las imagines de un saber universal.* 2 vols. Madrid: Ediciones Siruela, 1986.

Gorman, Michael John. *The Scientific Counter-Revolution: Mathematics, Natural Philosophy, and Experimentalism in Jesuit Culture 1580–ca. 1670.* PhD diss., European University Institute, 1998.

———, and Nick Wilding. *La technica curiosa di Kaspar Schott.* Rome: Edizioni dell'Elefante, 2000.

Hankins, Thomas L., and Robert J. Silverman. *Instruments and the Imagination.* Princeton: Princeton University Press, 1995.

Hein, Olaf. *Die Drucker und Verleger der Werke des Polyhistors Athanasius Kircher S.J.* Cologne, Germany: Böhlau, 1993.

Hellyer, Marcus. " 'Because the Authority of My Superiors Commands': Censorship, Physics, and the German Jesuits." *Early Science and Medicine* 1 (1996): 319–354.

Kramer, Roswitha. " ' ... ex ultimo angulo orbis': Atanasio Kircher y el Nuevo Mundo." In *Pensamiento europeo y cultura colonial,* edited by Karl Kohut and Sonia V. Rose. Frankfurt: Vervuert, 1997.

Leinkauf, Thomas. *Mundus combinatus: Studien zur Struktur der barocken Universalwissenschaft am Beispiel Athanasius Kirchers SJ (1602–1680).* Berlin: Akademie Verlag, 1993.

Lo Sardo, Eugenio, ed. *Iconismi e mirabilia da Athanasius Kircher.* Rome: Edizioni dell'Elefante, 1999.

———, ed. *Athanasius Kircher: Il Museo del Mondo.* Rome: Edizioni de Luca, 2001.

Lugli, Aldagisa. "Inquiry as Collection: The Athanasius Kircher Museum in Rome." *RES* 12 (1986): 109–124.

Magie des Wissens: Athanasius Kircher (1602–80) Jesuit und Universalgelehrter. Fulda, Germany: Michael Imhof, 2003.

Marrone, Caterina. *I geroglifici fantastici di Athanasius Kircher.* Viterbo, Italy: Stampa Alternativa & Graffitti, 2002.

McCracken, George E. "Athanasius Kircher's Universal Polygraphy." *Isis* 39: 215–227.

Merrill, Brian L., ed. *Athanasius Kircher (1602–1680): Jesuit Scholar.* Provo, UT: Friends of the Brigham Young University Library, 1989.

Nocenti, Luca. "Vedere mirabilia: Kircher, Redi, anatre settentrionali, rarità orientali e mosche nel miele." *Rivista di estetica* n.s. 19 (2002): 36–60.

Pastine, Dino. *La nascita dell'idolatria: L'oriente religioso di Athanasius Kircher.* Florence: La Nuova Italia, 1978.

Rivosecchi, Valerio. *Esotismo in Roma Barocca: Studi sul Padre Kircher.* Rome: Bulzoni, 1982.

Rowland, Ingrid. *The Ecstatic Journey: Athanasius Kircher in Baroque Rome.* Chicago: University of Chicago Press, 2000.

Scharlau, Ulf. *Athanasius Kircher (1601–1680) als Musikschriftsteller: Ein Beitrag zur Musikanschauung des Barock.* Kassel: Bärenreiter-Antiquariat, 1969.

Siebert, Harald. *Die grosse kosmologische Kontroverse: Rekonstruktionsversuche anhand des Itinerarium exstaticum von Athanasius Kircher SJ (1602–1680).* Stuttgart: Franz Steiner Verlag, 2006.

Stolzenberg, Daniel, ed. *The Great Art of Knowing: The Baroque Encyclopedia of Athanasius Kircher.* Stanford, CA: Stanford University Libraries, 2001.

———. "*Lectio Idealis*: Theory and Practice in Athanasius Kircher's Translations of the Hieroglyphs." In *Philosophers and Hieroglyphs,* edited by Lucia Morra and Carla Bazzanella. Turin: Rosenberg & Sellier, 2003.

———. *Egyptian Oedipus: Antiquarianism, Oriental Studies, and Occult Philosophy in the Work of Athanasius Kircher.* PhD diss., Stanford University, 2004.

———. "Oedipus Censored: *Censurae* of Athanasius Kircher's Works in the Archivum Romanum Societatis Iesu." *Archivum Historicum Societatis Iesu* 73 (2004): 3–52.

———. "Utility, Edification, and Superstition: Jesuit Censorship and Athanasius Kircher's *Oedipus Aegyptiacus.*" In *The Jesuits II: Cultures, Sciences, and the Arts, 1540–1773,* edited by John O'Malley et al. Toronto: University of Toronto Press, 2006.

Strasser, Gerhard F. " 'Spectaculum Vesuvii': Zu zwei neuentdeckten Handschriften von Athanasius Kircher mit seinen Illustrationsvorlagen." In *Theatrum Europaeum: Festschrift für Elida Maria Szarota,* edited by Richard Brinkmann et al. Munich: Wilhelm Fink Verlag, 1982.

———. "La contribution d'Athanase Kircher à la tradition humaniste hiéroglyphique." *XVIIe Siècle* 40 (1988): 79–92.

———. "Science and Pseudoscience: Athanasius Kircher's *Mundus Subterraneus* and His *Scrutinum … Pestis.*" In *Knowledge, Science, and Literature in Early Modern Germany,* edited by Gerhild Scholz Williams and Stephan K. Schindler. Chapel Hill: University of North Carolina Press, 1996.

———. "Athanasius Kircher: Genie und Wagnis eines barocken Universalgelehrten." In *Fuldaer Geschichtsblätter: Zeitschrift des Fuldaer Geschichtsvereins* 79 (2003): 85–108.

Szczesniak, Boleslaw. "Athanasius Kircher's China Illustrata." *Osiris* 10 (1952): 385–411.

Wicki, Josef, S.J. "Die *Miscellanea Epistolarum* des P. Athanasius Kircher, S.J., in Missionarischer Sicht." *Euntes Docete* 21 (1968): 221–254.

Paula Findlen

KLAU, CHRISTOPH

SEE **Clavius, Christopher.**

KLEIN, CHUCK

SEE **Klein, Harold P.**

KLEIN, HAROLD P.

(*b.* New York, New York, 1 April 1921; *d.* Palo Alto, California, 15 July 2001), *microbiology, origin of life, astrobiology.*

Klein was a microbiologist who worked on lipid and carbohydrate metabolism in a wide range of microbes. But he is best known for his work as an administrator at the National Aeronautics and Space Administration's (NASA) Ames Research Center for more than twenty years, where he supervised work on the origin of life and on instruments to search for life on other planets. He served as scientific leader of NASA's Viking Biology Investigation on the *Viking 1* and *2* Mars landers, coordinating the development, construction, and execution of all the biology experiments aboard the spacecraft.

Harold (or "Chuck" as he was known to colleagues) Klein was born to Hungarian immigrant parents just a month after their arrival through Ellis Island into New York City. He received a BS in chemistry from Brooklyn College in 1942 and a PhD in microbiology from the University of California, Berkeley, in 1950. Klein served in the U.S. Army from 1943 to 1946, researching the effect of molds on electrical equipment in the humid South Pacific. In 1955, he joined Brandeis University in Waltham, Massachusetts, becoming associate professor in 1956 and full professor 1960 to 1963. He was chair of the Biology Department at Brandeis from 1956 to 1963, with the exception of 1960 to 1961 while on leave from Brandeis, when Klein was visiting professor of bacteriology at the University of California, Berkeley. Klein saw NASA's advertisement for the initial position at Ames and felt it would be an extraordinary challenge to be in on the development of a historic scientific and technological enterprise such as looking for life on other planets. NASA had begun to fund such research with the creation of its Life Sciences Office in March 1960; by 1962 a whole new research lab was being set up at Ames, to conduct in-house research, in addition to the studies by university scientists at their home institutions. In the heady days of the early "space race," NASA had high levels of funding from Congress for all its research goals; cuts did not begin until the late 1960s. After retiring from Ames he became scientist-in-residence at Santa Clara University, Santa Clara, California, as well as senior research scientist at the SETI (Search for Extraterrestrial Intelligence) Institute from 1985 until his death.

Science Administrator. As a young microbiologist, Klein studied the biochemical metabolism of carbohydrates and lipids in various microorganisms, from bacteria to yeast. During 1950 to 1951 he was a research fellow at the Biochemical Research Institute of Massachusetts General Hospital in Boston, working in Fritz Lipmann's research group there, just prior to Lipmann's receipt of the 1953 Nobel Prize. From 1951 to 1954 Klein was an instructor in the Department of Microbiology at the University of Washington, Seattle; he became assistant professor there from 1954 to 1955. Beginning with his start as chair of the Brandeis University Biology Department, however, Klein soon discovered that his greatest talents were as a science administrator and organizer. Thus, when he was approached by NASA in January 1963 and invited to become the division chief of a new set of labs in an exciting new science—exobiology—Klein felt the potential to be in on the beginnings of a major new field of exploration sufficient to overcome his reservations about leaving the freedom of academia for a bureaucratic civil service job. While he continued many of his former lines of research at NASA's Ames Research Center, this turning point marked the beginning of his work to coordinate an entirely different kind of investigation: designing, building, and operating devices to search for life on other planets, most of all Mars.

Validating his sense of his abilities as a science administrator, in less than a year at NASA, Klein was promoted to assistant director of all Life Sciences at Ames, by 1968 rising to director. Klein's twenty-one-year tenure at Ames encompassed the "boom days" of NASA, when the Apollo program was in full swing and planetary missions began to multiply, including Mariners to Mars and Venus, Pioneers and Voyagers to the outer planets, and Vikings to Mars.

Klein participated in the National Academy of Sciences workshop at Woods Hole, Massachusetts, in the summer of 1965, on scientific prospects for finding life on Mars. He oversaw the construction of a new laboratory building (completed in December 1965) and the training of many National Research Council postdoctoral scientists; in addition Klein presided over a time when a great many staff scientists were hired as civil servants. In the exobiology (soon to be called Planetary Biology) division of Life Sciences alone, there were three bureaucratic branches: Chemical Evolution, Biological Adaptation, and Life Detection Systems. Some of the talented young scientists Klein spotted and hired included microbiologist Ruth Mariner Mack, chemist Fritz Woeller, chemist Katherine Pering, biochemist Donald DeVincenzi, and geochemist Keith Kvenvolden. Thus, Klein's administrative abilities led to and catalyzed a large amount of research on the origin of life (overseen until 1971 by Cyril Ponnamperuma in the Chemical Evolution section), adaptations of living systems to extreme environments, and the design of actual instruments to search for life in the cosmos.

Viking Biology Leader. Nineteen sixty-eight saw the beginning of plans for the Viking Mars lander mission, and NASA advertised a competition among all submitted life-detection schemes, to decide which four experiments would be chosen to actually get built and sent to the Martian surface on the two Viking spacecraft. In December 1969, from more than fifty submissions the four experiments chosen were Norman Horowitz's pyrolytic release (PR), Gilbert Levin's labeled release (LR), Vance Oyama's gas exchange, and Wolf Vishniac's light scattering experiment. A committee was appointed by NASA to oversee the development of a workable Viking Biology Experiment Package, including all four experiments. Committee members included the four experimenters, but also microbiologist Joshua Lederberg and microbiologist Alexander Rich, scientists who it was believed could be more objective because they did not have experiments of their own at stake.

Vishniac was the first chair, but it soon turned out that he was too relaxed, willing to let everybody have his say. So work on the project bogged down amid disagreements; each experimenter thought his own approach the most important, yet all the experiments had to function in a common environment inside the same instrument. Horowitz, for example, argued that the design of all the other experiments was based on Earth-like conditions far too warm and wet to be realistic for Mars. He argued that the temperature inside the experiment package should be kept as low as possible; because his experiment did not involve any liquid water, he had no qualms about advocating a temperature of 0 degrees Celsius or less, despite the fact that this would render almost useless the other three largely aqueous experiments.

The tension between egos and differing experimental ideas led to regular deadlocks of the committee until Klein was asked to take over as new chair. He brought the same capable administrative talents that he had brought to directing the Ames Exobiology Program and then all of Life Sciences at Ames. Klein's managerial style worked, and though the Viking Biology Committee was noted by many as one of the most contentious groups of people ever assigned to work jointly, he managed to keep the group together and the project moving forward, if notoriously behind schedule and over budget. Klein's key talent in this context was as a diplomatic moderator more than as a scientist. His levelheaded calm would turn out to be most important of all in the days and weeks *after* Viking landed on Mars, and after results from the experiments began to come in.

Indeed, problems with the biology instrument were not limited merely to the difficulties of getting the team to work together. Fearing the complexities of getting all four experiments to function problem-free in a single instrument, NASA's Viking project manager issued a directive on 1 July 1971 declaring a new project policy that no single malfunction should cause the loss of data return from more than one scientific investigation. In November and December 1971, the instrument contractor, TRW, and NASA Ames personnel under Klein worked to simplify the biology instrument. It simply had too much going on in the space allotted. In 0.027 cubic meters—a box about the size of a gallon milk carton—were forty thousand parts, half of them transistors. Several items were eliminated.

By January 1972 administrators from NASA headquarters met with people from the Viking Project Office and the contractors to discuss the problems and especially the cost, which had escalated to $33 million for the

Klein

biology package alone. Soon NASA headquarters concluded that one of the four biology experiments would have to be dropped. Klein, Lederberg, and Rich, the biology team members who did not have a stake in any one of the experiments, met to discuss priorities; shortly afterward, by mid-March 1972 NASA headquarters had decided that Vishniac's light scattering experiment was based on the least Mars-like conditions and therefore it should be the one to be sacrificed. The entire Viking Biology Team met immediately, and showed rare cohesiveness in criticizing the decision at headquarters to drop Vishniac's experiment. But in the event, Viking carried only the remaining three biology experiments.

By the time the *Viking 1* and *2* spacecraft launched from Cape Kennedy, on 20 August and 9 September 1975, the team had written a description of the experiments for *Nature.* A special issue of the journal *Origins of Life* was also in preparation, describing the experiments in much greater detail. These articles clearly convey the scientists' sense of the historic nature of their enterprise; but also their awareness of how complex the experiments were and how limited their ability from Earth to check up on ambiguous results or run additional controls. Klein wrote an overview of the biology package and its development (Klein, 1976a). Knowing the sensational nature of the mission, Klein seemed to feel more than most the responsibility to educate the press and the public about a cautious, scientific attitude toward the experiments.

Life on Mars Controversy. Every one of the biology experiments yielded evidence of activity from the very first run on the Martian surface. The PR experiment gave one reading consistent with production of organic matter (e.g., by photosynthesis), and the reading was high enough, compared to his earlier-stated requirements, that even Horowitz was briefly shaken about his doubts over the existence of life on Mars. But this result was not repeatable. When wetted in the gas exchange experiment, the Martian "soil" (regolith) released significant amounts of oxygen. Heating the sample to 145 degrees Celsius for 3.5 hours reduced the amount of oxygen released by about 50 percent. There was a slow evolution of carbon dioxide when nutrient was added to the soil. However, by three days into the first run, the gas production had decreased considerably, leading some to suspect that the reaction was chemical rather than biological. That is, it may have been produced by a potent reactant present in the sample that was used up via chemical combination with the water or nutrients.

Levin's LR experiment showed the most potent reaction of all three. A nutrient solution added to the soil sample contained a mixture of the following acids: formic, glycine, glycolic, D-lactic, L-lactic, D-alanine, and L-alanine, each uniformly labeled with carbon-14, which would be detected as radioactive carbon dioxide (CO_2) gas if metabolized by microbes. There was an immediate peak of labeled CO_2 release in the first minutes after the nutrient solution was added, followed by a slow, continued release over many days as measurements continued. The amount of CO_2 released amounted to what would be expected if a single carbon atom had been cleaved at the same spot from the entire pool of a single substrate (see Figure 1). The plot of data looked somewhat like a bacterial growth curve (though it lacked an initial lag phase); furthermore, if the soil was first heated to 160 degrees Celsius for three hours the activity was completely destroyed. The effect was partially destroyed by incubating the soil at 40 to 60 degrees Celsius, and the activity was relatively stable for short periods at 18 degrees Celsius, but lost after long-term storage at 18 degrees C. All of these data seemed to Levin to be almost completely consistent with what one would expect from a biological reaction. He was tentative at first, but the subsequent controls convinced him that the best explanation of the LR results could well be the existence of microbial life on Mars.

Because results were being released to the press on practically a daily basis, the nation, indeed the world, was getting the chance to observe science in process in a new way. Viking officials, especially Klein, worked hard to explain the slow, deliberate process by which the experiments had to be checked, different kinds of controls tried,

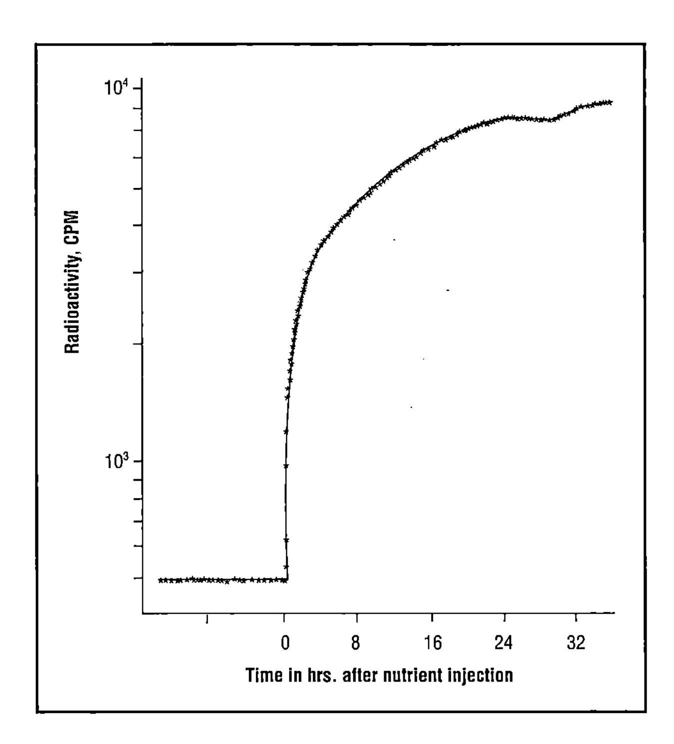

Labelled Release experiment results, Viking 1.

Klein

and so on. But the results were simply too unexpected; at each new trial that should have brought clarity in choosing between a chemical or biological explanation of the results, the ambiguity stubbornly persisted. Unused to doing science with an audience looking in at every step in the process, the scientists were exasperated at having to explain complex and ambiguous experimental results to a public and press corps that wanted to know simply: has Viking found life on Mars?

Shortly after, the gas chromatograph–mass spectrometer (GCMS) aboard each Viking spacecraft sent back the stunning results of its chemical analysis of the Martian soil: there were no detectable organic molecules of any kind. For most of the scientists, that immediately ruled out any possibility of life in those soil samples. In a press conference in which the GCMS results were announced, Klein also told the press about a new theory. This was that ultraviolet light getting from the Sun to the surface of Mars produces hydrogen peroxide, which oxidizes any organic compounds. Some laboratory experiments were carried out simulating Martian conditions; the result was that the half-life of any organic compound was at most two months. A team at the University of Maryland added peroxide to a sample of Levin's nutrient mixture that Klein sent them; they found a very similar response and amount of CO_2 evolution to what was seen in the Mars LR experiment. Oyama and his coworkers proposed, after some lab work, that γFe_2O_3 was the most likely oxidant, rather than hydrogen peroxide.

The data from further control experiments were as confused and ambiguous as ever, having some "chemical" and some "biological" features. But with so much at stake, not only life on Mars but the possibility of seeming impetuous, unscientific, or insufficiently cautious before a world audience, the double-sided nature of the public relations aspect of "Big Science" of the post–World War II period was visible in sharp relief. Particularly in the case of exobiology, to speak of the science artificially extracted from the public relations context that served as such nourishing soil for its development would be arbitrary indeed.

Levin and his coworker Patricia Straat continued to make the case that the interpretation of the biology results from Viking, at least the results from their LR experiment, were still open. By 1979, however, almost all other scientists concluded that the chemical explanation was more likely. In that context, Levin and Straat were viewed as being intransigent; they were rapidly marginalized. By the 1990s Straat was no longer writing on the subject, but Levin became still more convinced after results from the 1997 Mars *Pathfinder* spacecraft showed that water might exist in significant quantities not far below the surface of Mars; thus life was more likely. Similarly, he considered that the August 1996 announcement of the discovery of putative microfossils in a Martian meteorite gave broad support for the case for Martian biology, even if those possible organisms were from over three billion years in the past. Levin raised the possibility that Earth biota could have been seeded by Mars meteorites long ago when Mars was still habitable, or vice versa, now that it was recognized that meteorites were in fact moving at least in the Mars to Earth direction.

In 1997 a popular book appeared, championing Levin's cause and presenting him as a scientific genius suppressed by the Establishment. Levin's former Viking colleagues and the new generation of exobiology researchers had largely ignored Levin's writings for the previous fifteen years; however, the new book caused Harold Klein sufficient irritation that he felt compelled to respond (Klein, 1999), hoping to silence the argument once and for all.

Klein pointed out that Levin's argument consisted of two main propositions; only one of these had been properly and directly addressed, he said. The two main arguments were, as he saw them, first that the responses seen on Mars were practically indistinguishable from those shown by a variety of Earth microbes and second, that laboratory attempts to reproduce the LR results, based on nonbiological mechanisms, could not account for the results. Klein said all rebuttals had concentrated on the second argument, while little attention had been paid to the first. He went on to outline a number of characteristics the presumed Martian microbe or microbes must have, in order to fit with the data. First, they needed to live in an anaerobic environment devoid of liquid water at temperatures averaging (even at a sheltered depth of 5 centimeters below the surface) between –33 and –73 degrees Celsius.

Secondly, the organisms must survive after being brought from that ambient environment and placed in a storage container at an average of 15 to 18 degrees Celsius within the *Viking* lander. The samples were held at that temperature for eight days, at which time they were placed in an incubation chamber at 10 to 13 degrees Celsius. Two days later, ten days after being scooped up and dumped into the spacecraft, 0.115 milliliters of an aqueous solution of the organic carbon sources was added to the sample. After being put through those changes, the microbial species must *immediately* release gas (within the first four minutes, as the first measurement showed substantial gas already released by that time). Klein emphasized that the reaction took off immediately without the lag phase characteristic of most microbial growth curves. Then it leveled off after about twenty-four hours and ceased when an amount of carbon approximately equivalent to one of the added carbon atoms was released, while more than 90 percent of the added nutrients remained unaffected. Klein noted the further improbability for a living organism that had done all of the above: next, when the sample was treated with a second dose of nutrient solution, no further release of radioactive gas was seen. After still further criticisms, Klein insisted that to claim terrestrial organisms could reproduce all aspects of the LR data was not a plausible conclusion.

Not long after Klein's rebuttal, the case for life on Mars perked up with a prominent article in the *Proceedings of the National Academy of Sciences,* which argued that the Viking GCMS would have been unable to detect some of the most likely organic compounds delivered to the Martian surface by meteorites. In retrospect there is reason to believe that the GCMS was too insensitive to detect organic matter in amounts found in the number of cells suggested by Levin's interpretation of the LR data; it had merely been assumed in the instrument's design that if cells were able to grow, higher levels of organics must be present all around them. Further discoveries of subsurface water ice by Mars *Odyssey* in February and March 2002 continued to reveal, much like the observations of *Mariner 4* did in 1965, that Mars is a sufficiently complex place to repeatedly overturn past scientific certainties. The stunning discoveries at terrestrial hydrothermal vents, of the "third kingdom" of Archaea, and of the endosymbiotic behavior of bacteria that later turned into mitochondria, chloroplasts, and other cell organelles could suggest more caution than Klein displayed in predicting what microbes might and might not be capable of. At bottom, this turns upon a basic attitude toward the degree of adaptability of living organisms; what is more unlikely, life on a harsh planet such as Mars or Europa, or life (even complex multicellular animals) at many atmospheres of pressure and temperatures approaching 150 to 200 degrees Celsius near undersea hydrothermal vents? Though it may be a considerable time yet before any final answers about present or past life on Mars, Klein's role as scientist, and even more importantly as science administrator, was crucial for bringing the scientific investigation of the subject to its present state. Klein was the recipient of the NASA Exceptional Scientific Achievement Medal (1977), the NASA Medal for Outstanding Leadership (1981), and the Presidential Meritorious Service Award (1981).

BIBLIOGRAPHY

A box of Klein's personal papers donated to this author before Klein's death has been deposited at the NASA History Office.

WORKS BY KLEIN

With Michael Doudoroff. "The Mutation of *Pseudomonas putrefaciens* to Glucose Utilization and Its Enzymatic Basis." *Journal of Bacteriology* 59 (1950): 739–750.

"Fructose Utilization by *Pseudomonas putrefaciens.*" *Journal of Bacteriology* 61 (1951): 524–525.

With Fritz Lipmann. "The Relationship of Coenzyme A to Lipide Synthesis: I. Experiments with Yeast." *Journal of Biological Chemistry* 203 (1953): 95–99.

With Orr E. Reynolds. "The Utility of Automated Systems in the Search for Extraterrestrial Life." *Proceedings of the Second International Symposium on Basic Environmental Problems of Man in Space, Paris, 14–18 June 1965,* edited by Hilding Bjurstedt. New York: Springer-Verlag, 1967.

"Problems Involved in the Detection of Life on Extraterrestrial Bodies." *Proceedings of the Tenth International Congress for Microbiology* 241 (1970).

"Potential Targets in the Search for Extraterrestrial Life." In *Exobiology,* edited by Cyril Ponnamperuma. Amsterdam: North Holland, 1972.

With Joshua Lederberg and Alex Rich. "Biological Experiments: The Viking Mars Lander." *Icarus* 16 (1972): 139–146.

"Automated Life-Detection Experiments for the Viking Mission to Mars." *Origins of Life and Evolution of the Biosphere* 5 (1974): 431–441.

"General Constraints on the Viking Biology Investigation." *Origins of Life and Evolution of the Biosphere* 7 (July 1976a): 273–279. Klein's advance description of the Viking Biology Instrument package and its concept and design constraints.

"Life on Mars?" *Trends in Biochemical Sciences* 1, no. 8 (August 1976b): N 174–N 176.

With Norman Horowitz, et al. "The Viking Biological Investigation: Preliminary Results." *Science* 194 (17 December 1976): 1322–1329. A report on the earliest experimental data in the first weeks after the Vikings landed.

With Joshua Lederberg, et al. "The Viking Mission Search for Life on Mars." *Nature* 262 (July 1976): 24–27.

"The Viking Biological Investigation: General Aspects." *Journal of Geophysical Research* 82 (30 September 1977): 4677–4680. A more complete article on the Viking biology results a year after the experiments first began sending back data.

"The Viking Biological Experiments on Mars." *Icarus* 34 (1978a): 666–674. The definitive discussion of the Viking biology experiments.

"The Viking Biological Investigations: Review and Status." *Origins of Life and Evolution of the Biosphere* 9 (1978b): 157–160.

"Simulation of Viking Biology Experiments: An Overview." *Journal of Molecular Evolution* 14 (1979): 161–165. A review of laboratory attempts to simulate the Viking results, in order to choose between biological vs. chemical explanations as the best fit for those data.

A Personal History. Mountain View, CA: privately printed, 1998. Klein's personal and scientific autobiography.

"Did Viking Discover Life on Mars?" *Origins of Life and Evolution of the Biosphere* 29 (1999): 625–631.

OTHER SOURCES

Benner, Steven, Kevin Devine, Lidia Matveeva, et al. "The Missing Organic Molecules on Mars." *Proceedings of the National Academy of Sciences of the United States of America* 97: (2000) 2425–2430.

Dick, Steven J., and James E. Strick. *The Living Universe: NASA and the Development of Astrobiology.* New Brunswick, NJ: Rutgers University Press, 2004.

DiGregorio, Barry. *Mars, the Living Planet.* Berkeley, CA: Frog, 1997.

Ezell, Edward C., and Linda N. Ezell. *On Mars: Exploration of the Red Planet, 1958–1978.* NASA Special Publication 4212. Washington, DC: Scientific and Technical Information Branch, National Aeronautics and Space Administration, 1984.

Fry, Iris. *The Emergence of Life on Earth: A Historical and Scientific Overview.* New Brunswick, NJ: Rutgers University Press, 2000.

James E. Strick

KLIXBÜLL, CHRISTIAN

SEE **Jörgensen, Axel Christian Klixbüll.**

KLÜVER, HENRICH

(*b.* Schleswig-Holstein, Germany, 25 May 1897; *d.*, Chicago, Illinois, 8 February 1979), *cerebral lesions, eidetic vision, hallucinogenics, neurology, neuropsychology, occipital lobes, temporal lobes.*

Klüver is best known for describing what has come to be called the Klüver-Bucy syndrome, a constellation of behaviors that occurs when the temporal lobes are removed in monkeys and humans. Symptoms of this syndrome include failing to recognize objects by sight, hypersexuality, compulsive oral behaviors, attending to all visual stimuli, a marked change in dietary habits, and a great emotional passivity. Klüver's research demonstrated the fundamental importance of the limbic system for normal and abnormal behavior. He changed the way in which neurologists, psychiatrists, and psychologists thought about the neurological basis of emotion and behavior.

In addition, Klüver investigated eidetic imagery, mescaline and its hallucinatory effects, and the mechanisms of behavior in monkeys. He developed the method of equivalent and nonequivalent stimuli for studying behavior and determined some of what the brain does during vision, particularly the striate cortex. Finally, he discovered the presence of free porphyrins in the central nervous systems and invented the Klüver-Barrera method for staining neurons.

His Life. Born in Schleswig-Holstein, Henrich Klüver was seventeen years old when World War I started, and he joined the army as a private. He spoke little about his experiences in the military in later life, but it is clear that he thought the time spent in the army was quite dull. After the war, he became a student at the University of Hamburg and then at the University of Berlin, where he studied psychology with the Gestalt psychologist Max Wertheimer from 1920 to 1923.

In the summer of 1923, he boarded a German freighter as its sole passenger and traveled to the United States, where he enrolled in graduate school in psychology at Stanford University. He received his Ph.D. in 1924 for work in eidectic experiences—unusually vivid visual phenomena—in children. He subsequently was hired to teach psychology to undergraduates at the University of Minnesota. He remained in that position for two years. There he met the man he later referred to as the greatest of all neuropsychologists, Karl Spencer Lashley. In 1926 Klüver left Minnesota to become a fellow of the Social Science Research Council at Columbia University.

In 1928 Klüver and Lashley moved together to the Institute of Juvenile Research in Chicago to continue the collaboration they had started in Minnesota. In 1933 the University of Chicago hired them both, and Klüver remained at the University of Chicago for the rest of his career and his life, holding a variety of research positions. He was an associate professor in the Division of Psychiatry from 1936 though 1938, then a full professor in the Division of Biological Sciences from 1938 to 1957. He was named the Sewell L. Avery distinguished professor of biological psychology in 1957 and remained in that position until his mandatory retirement in 1962. Until his death in 1979, Klüver remained at the University of Chicago as the Sewell L. Avery Distinguished Service Professor Emeritus.

The University of Chicago supported Klüver and his research during his career, even though Klüver taught no students there—neither graduate nor undergraduate—nor did he participate in any of the functions of the Psychology Department (except for apparently serving as its chair for one twenty-four-hour period early in his tenure there). Toward his retirement and until his death, Klüver was left essentially alone in his Culver Hall laboratory, with not even a secretary to help him. A plumbing leak destroyed most of his papers, with only some material remaining, written in his German shorthand.

At Chicago, both Klüver and Lashley joined a group of neuroscientists coalescing around Percival Bailey, who moved to the University of Chicago from Boston in the fall of 1928. This group called themselves the "Neurology Club," and it was composed of all the people interested in the nervous system at Chicago: Anton J. Carlson and Arno Luckhardt, both physiologists; Ralph Gerard, an electroneurophysiologist; Stephen Polyak, a neuroanatomist; Fred Koch, a biochemist; Charles Judson Herrick, a comparative neuroanatomist; George W. Bertelmez, an endocrinologist; Robert Bensley, a cellular biologist; H. G. Well, a pathologist; Roy R. Grinker, a neuropathologist and psychiatrist; Paul Clancy Bucy, a neurosurgeon and future collaborator with Klüver; and, of course, Lashley, Bailey, and Klüver himself. Paul Bucy was a former student of Bailey and Grinker.

Even though he was a surgeon, Bucy spoke publicly about the importance of knowing as much as possible about the entire field of neurology. Klüver found in him an outstanding colleague and collaborator, and together they embarked on a series of experiments that led to their describing a set of pathological symptoms that subsequently bore their names, the Klüver-Bucy syndrome.

Discovery of the Klüver-Bucy Syndrome. The discovery of the Klüver-Bucy syndrome really occurred by accident. At the time, Klüver was experimenting with psychedelic drugs. He had heard of the hallucinogenic properties of the cactus *Lophephorus Williamsii* (commonly known as mescal) and extracted the drug mescaline from the "buttons," or peyote, on top of the plant. His interest in the drug came out of his earlier focus on eidetic visual experiences in which he concluded, contrary to others, that eidetic imagery does not resemble the visual hallucinations caused by ingesting mescal. He set out to prove this belief at first by taking the drug himself.

At the same time, Klüver became an authority on handling monkeys and monkey behavior. He developed the method of equivalent and nonequivalent stimuli, which he used to determine the range of stimuli that monkeys perceived as similar to the positively rewarded training stimulus. Eventually, he refined the equipment to measure monkeys' responses more precisely and to measure the number of conditions and stimuli that could be tested at the same time. These new tools let Klüver test monkeys under a wide range of conditions and using a variety of stimuli. It also gave Klüver an objective testing method for his later experiments on lesioned monkeys.

It was therefore natural that Klüver would extend his studies of mescaline to monkeys. He noticed that when he injected monkeys with mescaline, they continually moved their lips, tongue, and jaws, chewing, licking, and smacking. He and his colleague Paul C. Bucy tried to figure out what mechanisms were behind this rather odd behavior. If they could locate the neural underpinnings, then they would have located the places in the nervous system where mescaline acts.

They first lesioned the trigeminal nerves, then the facial nerves, then both of them together, but still the monkeys continued to manipulate their lips. They then thought that perhaps this behavior might be related to a similar behavior seen in patients with temporal lobe epilepsy. They decided to remove the temporal lobes in a monkey to see whether this would affect the lip manipulations of monkeys on mescaline. Fortunately for Klüver, who had no prior surgical experience, Bucy was an excellent surgeon.

Purely by coincidence, Klüver and Bucy removed the left temporal lobe on a particularly aggressive monkey named Aurora, previously donated to their lab. The next morning, Klüver noticed almost immediately that Aurora was unnaturally calm, a trait that did not diminish over time. They later learned through experimentation that most monkeys do not show such extreme behavioral changes with only one temporal lobe removed.

About six weeks later, the two removed Aurora's right temporal lobe. It was then that Klüver and Bucy first outlined the symptoms associated with the loss of the temporal lobes. Klüver then spent months observing and testing a whole series of monkeys with their temporal lobes removed, repeatedly confirming his earlier observations. His interest in this phenomenon displaced his earlier concern for locating the places where mescaline interacts with the nervous system. This was fortunate, indeed, since removing the temporal lobes did not stop the oral behaviors of mescalinized monkeys.

The six categories of symptom found in the Klüver-Bucy syndrome include:

1. "psychic blindness," or the inability to recognize objects for what they are, even though visual acuity remains unchanged;
2. "hypermetamorphesis," a condition in which subjects reach for an object as soon as it is presented visually, even if that object has been associated with negative rewards in the past;
3. extreme changes in feeding habits (in monkeys, this included ingesting large quantities of meat, which they almost never do when they are intact);
4. hypersexuality, including significant increases in masturbatory, homosexual, and heterosexual behaviors;
5. changes in emotional responses, including remaining calm under normally anxiety-producing circumstances;
6. oral manipulations, including both the manipulation of the lips, teeth, and gums mentioned above and examining objects by licking, biting, and chewing.

Klüver and Bucy first presented their results at the 1937 American Physiological Society Meeting and then published their findings in a series of articles over the next two years. It was only much later that Klüver formally described the actual damage to the temporal lobes. In

1940 he published a postmortem histology and then fifteen years later the full histological report. However, neither Klüver nor Bucy ever tried to localize their behavioral findings to any particular neuroanatomical structure. Even though Klüver was initially interested in finding the locus of activity for mescal, that concern morphed into wanting to understand the connection between the syndrome he identified and normal behavior and experiences.

Sanger Brown and Edward Albert Schäfer did publish similar findings about fifty years earlier, in 1888, in a report to the Royal Society of London; however, their article went largely unnoticed. It appears that Klüver was unaware of this publication until the late 1940s or early 1950s. Part of the difference in impact between reports is simply timing and the relative status of the scientists at the time; part of the difference concerns the precision of the techniques used by Klüver and Bucy. But most of the difference can likely be traced to the fact that while Brown and Schäfer largely dismissed the behavioral changes they saw, Klüver and Bucy gave an account of the phenomena that linked together theories from psychology and neuroscience, thus helping to unify what later became important subdisciplines in cognitive science.

In addition, Klüver and Bucy fortuitously capitalized on recent concerns in neurosciences and its cognate fields. Just four years before Klüver and Bucy publicized their findings, Charles Judson Herrick reported that removal of the rhinencephalon, which is near the temporal lobes, affects emotion and disposition. It turns out that Klüver and Bucy removed the rhinencephalon when they removed the temporal lobes, and their monkeys experienced similar effects to those that Herrick had described. In 1937 Karl Kleist hypothesized that the rhinencephalon was the center of affect, as well as the locus for sexual and food drives. Klüver relied on this hypothesis in his research. In that same year, just months after Klüver and Bucy published their first paper on the Klüver-Bucy syndrome, James Papez published his later well-known theory of the limbic system as neural mechanism underlying human emotions and emotive behavior. Papez considered Klüver and Bucy's work to be confirmation of his theory. Though Klüver and Bucy were more measured in their response, they did agree that Papez's work dovetailed with their own.

Klüver's Legacy. Though Klüver is best known for recounting what came to be called the Klüver-Bucy syndrome and for his work on mescaline and general monkey behavior, he also invented, with his colleague Elizabeth Barrera, a staining technique that simultaneously highlights neurons, glia cells, and myelin sheaths. In addition, due to his meticulous care of the monkeys in his laboratory, he was able to observe that, contrary to current beliefs, monkeys do indeed develop most of the same illnesses as man, including diabetes mellitus, brain tumors, carcinoma, and endometriosis. Until Klüver, monkeys had not lived long enough in a laboratory for anyone to observe these illnesses, often associated with aging.

Klüver worked as an associate editor for several professional journals in psychology, biology, and medicine during his career. He worked as a consultant for the National Institutes of Health and as a visiting professor to numerous universities. He was awarded the Lashley Award in neurobiology, the Hamilton Award in psychopathology, and the Gold Medal Award of the American Psychological Foundation. Although not a medical doctor, Klüver was awarded an honorary medical degree from the University of Basel in 1965 and an honorary doctorate from the University of Hamburg in 1969.

Klüver's work inspired both neuroanatomists and psychologists to localize the neurophysiological structures responsible for human thoughts and behaviors much more precisely, and soon the world knew which areas of the temporal lobe were responsible for which behavioral or emotional disturbances. Klüver's emphasis on phenomenology and the experiential side of human lives helped to change the way neuroscientists conceived of their research, such that they no longer exclusively focused on sensory-motor pathways.

BIBLIOGRAPHY

WORKS BY KLÜVER

"An Experimental Study of the Eidetic Type." *Genetic Psychological Monographs* 1 (1926): 71–230.

"Mescal Vision and Eidetic Vision." *American Journal of Psychology* 37 (1926): 502–515.

"Visual Disturbances after Cerebral Lesions." *Psychological Bulletin* 24 (1927): 316–358.

Mescal: The "Divine" Plant and Its Psychological Effects. London: Kegan Paul, 1928.

"Studies on the Eidetic Type and on Eidetic Imagery." *Psychological Bulletin* 26 (1928): 69–104.

"Fragmentary Eidetic Imagery." *Psychological Review* 37 (1930): 441–458.

"The Eidetic Child." In *A Handbook of Child Psychology*, edited by Carl Murchison. Worcester, MA: Clark University Press, 1931.

"The Equivalence of Stimuli in the Behavior of Monkeys." *Journal of Genetic Psychology* 39 (1931): 38–127.

"Eidetic Phenomena." *Psychological Bulletin* 29 (1932): 181–203.

Behavior Mechanisms in Monkeys. Chicago: University of Chicago Press, 1933.

"An Auto-multi-stimulation Reaction Board for Use with Sub-human P-Primates." *Journal of Psychology* 1 (1935): 123–127.

"A Tachistoscopic Device for Work with Sub-human Primates." *Journal of Psychology* 1 (1935): 1–4.

"Use of Vacuum Tube Amplification in Establishing Differential Motor Reactions." *Journal of Psychology* 1 (1935): 45–47.

"An Analysis of the Effects of the Removal of the Occipital Lobes in Monkeys." *Journal of Psychology* 2 (1936): 49–61.

"The Study of Personality and the Method of Equivalent and Non-equivalent Stimuli." *Character and Personality* 5 (1936): 91–112.

"Certain Effects of Lesions of the Occipital Lobes in Macaques." *Journal of Psychology* 4 (1937): 383–401.

With Paul Clancy Bucy. "'Psychic Blindness' and Other Symptoms Following Bilateral Temporal Lobectomy in Rhesus Monkeys." *American Journal of Physiology* 119 (1937): 352–353.

———. "An Analysis of Certain Effects of Bilateral Temporal Lobectomy in the Rhesus Monkey, with Special Reference to 'Psychic Blindness.'" *Journal of Psychology* 5 (1938): 33–54.

———. "Preliminary Analysis of Functions of the Temporal Lobes in Monkeys." *Archives of Neurology and Psychology* 42 (1939): 979–1000.

———. "Anatomic Changes Secondary to Temporal Lobectomy." *Archives of Neurological Psychiatry* 44 (1940): 1142–1146.

"Visual Functions after Removal of the Occipital Lobes." *Journal of Psychology* 11 (1941): 23–45.

"Functional Differences between the Occipital and Temporal Lobes with Special Reference to the Interrelations of Behavior and Extracerebral Mechanisms." In *Cerebral Mechanisms in Behavior*, edited by Lloyd A. Jeffress. New York: John Wiley and Sons, 1951.

"Brain Mechanisms and Behavior with Special Reference to the Rhinencephalon." *Lancet* 72 (1952): 567–574.

With Paul Clancy Bucy. "An Anatomical Investigation of the Temporal Lobe in the Monkey (*Macaca mulatta*)." *Journal of Comparative Neuroanatomy* 103 (1955): 151–251.

"'The Temporal Lobe Syndrome' Produced by Bilateral Ablations." In *Neurological Basis of Behavior*, edited by Gordon E. W. Wolstenholme and Cecilia M. O'Conner. Boston: Little, Brown, 1958.

"Neurobiology of Normal and Abnormal Perception." In *Psychopathology of Perception*, edited by Paul H. Hoch and Joseph Zubin. New York: Grune and Stratton, 1965.

Mescal and Mechanisms of Hallucinations. Chicago: University of Chicago Press, 1966.

OTHER SOURCES

Bucy, Paul Clancy. "Heinrich Klüver." In *Neurosurgical Giants: Feet of Clay and Iron*, edited by Paul Clancy Bucy. New York: Elsevier Science Publications, 1985.

———. "Heinrich Klüver." In *Psychopathology of Perception*, edited by Paul H. Hoch and Joseph Zubin. New York: Grune & Stratton, 1965.

———. "Henrich Klüver." *Surgical Neurology* 3 (1975): 229–231.

Nahm, Frederick K. D. "Heinrich Klüver and the Temporal Lobe Syndrome." *Journal of the History of the Neuroscience* 6 (1997): 193–208.

Valerie Gray Hardcastle

KODAIRA, KUNIHIKO (*b.* Tokyo, Japan, 16 March 1915; *d.* Kofu, Japan, 26 July 1997), *mathematics, complex manifolds, complex structures.*

Kodaira was one of the leading figures in complex algebraic geometry and function theory in the second half of the twentieth century. He was one of the first to apply modern topological methods to the classification of algebraic surfaces and then did pioneering work with Donald Spencer on the deformation of complex structures on a manifold. He was awarded a Fields Medal in 1954.

Early Career. Kodaira graduated from the Mathematics Department of Tokyo Imperial University in 1938 and proceeded to take a degree in the Department of Physics there in 1941. He became a professor of physics there in 1944, a position he retained until 1951, by which time he had become internationally recognized as a mathematician. He had by then obtained a PhD in mathematics, and a rewritten version of his thesis, titled "Harmonic Fields in Riemannian Manifolds (Generalized Potential Theory)" and published in the prestigious *Annals of Mathematics* (1949), came to the attention of Hermann Weyl. Weyl was then at the Institute for Advanced Study in Princeton, New Jersey, and he saw that Kodaira had made a significant new contribution to the study of harmonic integrals, one of the central topics in mathematics and one, indeed, that Weyl had himself worked on. Therefore, he invited Kodaira to the institute. Kodaira left for the United States in 1949, the start of his eighteen-year-long stay there. At Princeton he divided his time between the institute and Princeton University. In the 1960s he had positions at Harvard University in Cambridge, Massachusetts; Johns Hopkins University in Baltimore, Maryland; and Stanford University in Stanford, California, before returning to Tokyo in 1967. Early on in his American career he achieved high recognition with the award of a Fields Medal in 1954. In presenting the award to him, Weyl praised him for having the courage to attack the primary concrete problems in all their complexity and for having, as a result, found the right general concepts to resolve the difficulties and ease further progress. By the phrase "primary concrete problems," Weyl meant the fundamental questions concerning the existence of harmonic forms with prescribed singularities and periods.

Topology and the Riemann-Roch Theorem. In the early 1950s, Princeton was the setting for one of the most important mathematical developments of the twentieth century, and Kodaira played a leading role. These developments revolved around the introduction of structural methods to questions in the area of complex function theory (especially in several variables) and geometry. The new theories accomplished a number of things. They helped generalize results known in the one variable case to the much harder case of several variables, which in geometric terms amounts to passing from the case of curves to manifolds of higher dimension. They also gave quite precise answers to difficult existence questions. It had long been known, for example, that a Riemann surface generally admits a variety of different complex structures. The new methods permitted one to answer a similar question for higher dimensional complex manifolds, and much of Kodaira's later work was concerned with the theory of the deformation of complex structures on a complex manifold.

Kodaira continued to work with his American and European friends and colleagues on his return to Japan in 1967—he conducted a lifelong collaboration with Donald Spencer at Princeton—but back in Tokyo he also produced a considerable number of gifted students who did much to sustain Japan's reputation as a major mathematical center despite the isolation of the war years.

George Riemann's work in the 1850s on complex function theory, then in its infancy, can be thought of as exploiting the close ties that exist between a complex function of a single complex variable and a real-valued function of two real variables. Unfortunately, there is no such close identification of complex functions and harmonic functions in higher dimensions, and after a considerable amount of work by Émile Picard and Solomon Lefschetz, it was the great achievement of William V. D. Hodge in the 1930s to forge the first full general theory. One of the central results of his book *The Theory and Applications of Harmonic Integrals* (1941) was that harmonic forms exist in profusion on a complex manifold, indeed there is a unique one of any rank with preassigned periods. However, this proof of this result was found to contain a serious gap, and it was filled, independently in 1942–1943, by Weyl and Kodaira. It was Kodaira's generalization of this result to harmonic forms with singularities that caught Weyl's attention in 1949.

In order to develop the consequences of Hodge's theory, Kodaira turned to the study of the Riemann-Roch theorem. This result, in the one variable case discussed by Riemann in 1857 and his student, Gustav Roch, in 1862, gives real insight into the existence of meromorphic functions on a Riemann surface having prescribed singularities. It had long been known that there should be similar results in higher dimensions, and the Italian school of algebraic geometers around 1900 had found the right generalization for surfaces. Later, results were obtained for complex manifolds of dimension three, but at each stage the result was harder to understand. In the one variable case, the Riemann-Roch theorem expresses the dimension of the space of meromorphic functions in terms of various other numbers that are easy to compute in any given case. In particular, the answer is given in this form: the dimension is $d + 1 - g + h$, where d and h are numbers determined by the singularities and g is a number determined by the Riemann surface. In most cases, d is the number of singular points and h is necessarily positive, so one sees even without computing h that there will be nontrivial functions with these singularities whenever d is greater than g. However, when two variables are involved, the relevant formula is of the form $d + 1 - g + h - k$, so no simple argument is available and it is necessary to find h and k explicitly. This was often impossible, because no good geometric meaning attached to them. Furthermore, whereas the set of singularities of a function (like its zero set) is a set of points when the function is a function of a single variable, the set of singularities (its divisor) forms a curve on a surface (and a surface in a three-dimensional variety) so the geometry of the situation becomes more and more complicated with each increase in dimension.

Kodaira began by rederiving the Riemann-Roch theorem in dimension two, but in a much more perspicuous form. He then did the same for three-dimensional manifolds, each time applying and deepening the theory of harmonic forms. In the course of this work Kodaira, sometimes in joint papers with Spencer, showed that three arithmetic genera introduced by Francesco Severi in the 1920s and 1930s as fundamental quantities for describing a divisor on a complex surface were in fact the same. Kodaira's method was to solve the problem first for algebraic varieties, then for Kähler varieties, and then for arbitrary compact complex manifolds; he also had significant things to say about when a complex manifold is in fact algebraic. On the basis of the successful treatment of the low dimensional cases, the path to the general n-dimensional setting was finally open, as Weyl was to comment. What became visible was the extent to which purely topological considerations intervened in an analytic and geometric setting.

Algebraic Topological Methods. The methods of algebraic topology had been described by Samuel Eilenberg and Norman Steenrod in their influential book, *Foundations of Algebraic Topology* (1952). At about the same time, the Cartan seminar in Paris was developing the methods of sheaf theory and showing how they could be applied to the study of functions on complex manifolds. The geometry of such problems was also clarified by the introduction of the concept of fiber bundles, especially the simplest case of line and vector bundles. All these overlapping branches of mathematics share two common features: they had their origins in investigations begun in the 1930s, and further research on them was interrupted by World War II. As a result, it was often a new generation of young mathematicians that took up their study after the war. All of these concepts (cohomology, sheaves, and bundles) progressively appear in Kodaira's work in the 1950s, and they rapidly came to be associated with its success.

In 1954 Kodaira was able to characterize those complex manifolds that are algebraic: they are the ones that carry a Hodge metric. A topic that Kodaira studied in great detail, with the full rigor of modern methods, was the classification of complex surfaces. This had first been accomplished by Guido Castelnuovo and Federigo Enriques in the years before the World War I for algebraic surfaces. They correctly distinguished several major types, but their classification left the properties of some of these types of surface little understood. Kodaira reworked their accounts with a view to illuminating both the function theory of these surfaces and their geometry, and he extended it to complex two-dimensional manifolds that are not algebraic surfaces. A result of Wei-Liang Chow and Carl Ludwig Siegel showed that the dimension of the field of meromorphic functions on a complex surface is at most two. Kodaira, in work with Chow, showed that a complex surface is a nonsingular algebraic surface embedded in projective space if and only if this dimension is two.

This leaves the case where this dimension is either one or zero. Kodaira next showed that when the dimension is one, the complex manifold fibers as an elliptic bundle over a curve (the surface is then called an elliptic surface), and he then began the exploration of the dimension zero case. The final analysis of all complex surfaces depended on whether the surface carries an exceptional curve or not and upon its geometric genus. A curve on a surface is exceptional if there is a birational map of the surface that maps the curve to a point. The geometric genus measures the dimension of the space of holomorphic functions on the surface. If a surface does not have an exceptional curve, it is either a surface with geometric genus zero or a surface with geometric genus greater than zero and satisfying an inequality involving the first Chern class. Surfaces of this kind with geometric genus zero break into two types, according as the first Betti number is or is not one. If the surface carries an exceptional curve, it may be a K_3 surface, a complex torus, or an elliptic surface. Kodaira went on to make a detailed investigation of the difficult cases where the dimension of the field of meromorphic functions is less than two. Kodaira also showed that every K_3 surface is a deformation of a nonsingular quartic surface in projective three-space, and an elliptic surface is a deformation of an algebraic surface if and only if its first Betti number is even. All this work depended heavily on the Riemann-Roch theorem, which was the focus of much work by others, notably Michael Atiyah, Friedrich Hirzebruch, and I. M. Singer in the complex and differentiable settings and by Alexandre Grothendieck, Armand Borel, and Jean-Pierre Serre in the setting of the new algebraic geometry.

Deformations of Complex Structures. Starting in the late 1950s, Kodaira and Spencer switched their attention to the topic of varying or deforming the complex structure on a compact complex manifold, a vast generalization of the work done by Riemann a century before for Riemann surfaces, which are one-dimensional compact complex manifolds. In his book *Complex Manifolds and Deformation of Complex Structures* (1986, p. vii), Kodaira said: "The process of the development was the most interesting experience in my whole mathematical life. It was similar to an experimental science developed by the interaction between experiments (examination of examples) and theory." Their research, which was partly inspired by related work of Alfred Frölicher and Albert Nijenhuis, focused on an unexpected and initially inexplicable numerical coincidence. It was possible to argue that an infinitesimal deformation of the complex structure on a compact complex manifold M should be represented by an element of a certain cohomology group, while there seemed to be no reason to believe that every element of this group represented a deformation of the structure. However, as it happened, in many cases the dimension of this group turned out to equal the number of effective parameters involved in defining a structure on M, which suggested strongly that every element of the group did indeed represent a deformation. They then began working on proving the conjecture that this equality always holds.

Kodaira and Spencer asked themselves what the consequences would be of an element of the cohomology group not representing a deformation. Consequences of this kind are called obstructions in the language of cohomology theories, and they showed that the obstructions were elements of a second cohomology group. So if this group vanished there were no obstructions, and every element of the first cohomology group represented a deformation. They rederived Riemann's result in their setting. (Riemann had not presented his profound claims in a rigorous way.) They checked the correctness of the result they were seeking for complex analytic hypersurfaces and found a single counterexample to the claim. This led to the introduction of the concept of completeness for a family. All confirmations of the conjecture concerned complete families; the single counterexample was not complete. Finally, by 1960 Kodaira and Spencer were led to obtain sufficient conditions for the conjecture to be true. These were the vanishing of the second cohomology group mentioned above and also of another cohomology group. As Kodaira observed in his *Complex Manifolds,* these sufficient conditions were unduly restrictive, and the conjecture held in many cases when the conditions were not satisfied. In 1961 Masatake Kuranishi contributed an essential theorem on the existence of deformations that further strengthened belief in the conjecture, and Kodaira and Spencer now confirmed it for a wide class of complex, two-dimensional manifolds (complex surfaces). In 1962, David Mumford devised a new counterexample that involved a carefully constructed three-dimensional compact complex manifold. Then, in 1967, Arnold Kas found a counterexample among the family of elliptic surfaces, and writing in *Complex Manifolds*, Kodaira said: "In this way the conjecture … ultimately turned out to be false," and he and Spencer ultimately failed to find a "simple and useful criterion" for it to be true (1986, p. 319). A very modest way of concluding a program of work that had opened a whole new field in the study of complex manifolds.

BIBLIOGRAPHY

WORKS BY KODAIRA

"Harmonic Fields in Riemannian Manifolds (Generalized Potential Theory)." *Annals of Mathematics* 50 (1949): 587–665.

"The Theorem of Riemann-Roch on Compact Analytic Surfaces." *American Journal of Mathematics* 73 (1951): 813–875.

With D. C. Spencer. "On Deformations of Complex Analytic Structures, I-II." *Annals of Mathematics* 67 (1958): 328–466.

"On Compact Analytic Surfaces, II–III." *Annals of Mathematics* 77 (1963): 563–626.

Collected Works. 3 vols. Tokyo: Iwanami Shoten; Princeton, NJ: Princeton University Press, 1975. This contains a complete bibliography of Kodaira's works.

Complex Manifolds and Deformation of Complex Structures. Translated by Kazuo Akao. New York and Tokyo: Springer-Verlag, 1986.

OTHER SOURCES

Eilenberg, Samuel, and Norman Steenrod. *Foundations of Algebraic Topology.* Princeton, NJ: Princeton University Press, 1952.

Hirzebruch, Friedrich. "Kunihiko Kodaira: Mathematician, Friend, and Teacher." *Notices of the American Mathematical Society* 45, no. 11 (1998): 1456–1462.

Hodge, W. V. D. *The Theory and Applications of Harmonic Integrals.* Cambridge, U.K.: Cambridge University Press, 1941.

Jeremy Gray

KOEHLER, OTTO

(*b.* Insterburg [near Königsberg, now Kaliningrad], East Prussia, Germany, 20 December 1889; *d.* Freiburg, Breisgau, Germany, 7 January 1974), *animal physiology, ethology.*

Koehler is remembered as one of the pioneers in the study of animal behavior, or ethology. Together with Konrad Lorenz, he was a founding editor of the internationally recognized journal *Zeitschrift für Tierpsychologie* (Journal of animal psychology, now *Ethology*). He also contributed to general and methodological perspectives in the life sciences and did some seminal work in human ethology.

Life and Career. Koehler, son of the Lutheran minister Eduard Koehler and Karoline Koehler, was born in Insterburg, East Prussia. He attended secondary school in Schulpforta, a former monastery and (after 1543) a famous boarding school, Landesschule zur Pforta, in Sachsen-Anhalt. After he had passed the final examination, the *Abitur,* he studied mathematics, physics, zoology, and botany, first at the University of Freiburg, Breisgau, and then at the University of Munich. However, under the influence of August Weismann, Richard Goldschmidt, and Karl von Frisch, he paid attention mainly to biological sciences, especially zoology. He obtained his PhD in 1911 and took the position of an assistant at the University of Freiburg. In 1912 and 1913 he worked at the Stazione Zoologica di Napoli (Zoological Station in Naples), a prestigious institution that offered excellent conditions for students doing research work in different fields of biology. Koehler's objects of interest while in Naples were sea urchins.

After Koehler returned to Germany, he again held the position of an assistant, this time at the University of Munich. In 1916, however, he worked with the zoologist Alfred Kühn in a epidemiological laboratory in Strasbourg. In 1918 Koehler moved to the University of Wrocław in Poland, where—after spending two more years as assistant—he finished his *Habilitation* (in zoology and comparative anatomy and physiology) and worked as a privatdozent. In 1923 he was appointed associate professor at the University of Munich's Institute for Zoology and in 1925 full professor and head of the Zoology Department and Museum at the University of Königsberg (now Kaliningrad), the capital of East Prussia. In 1940 Konrad Lorenz, with the help of Koehler, was appointed full professor for comparative psychology in Königsberg. He and Lorenz were founding editors of the *Zeitschrift für Tierpsychologie,* which soon developed as the leading ethological journal in continental Europe.

From 1946 until his retirement in 1960, Koehler was full professor of zoology and director of the Zoological Institute at the University of Freiburg. During his first years there he spent much energy in rebuilding and rehabilitating the institute, in addition to considerable teaching obligations. Nevertheless, he continued his experimental research work. In Freiburg he died in 1974 at the age of eighty-four. Koehler was married twice. His first wife, Annemarie (née Deditius), died in 1944; his second wife, Amélie (née Hauchecorne), was twenty-six years old when he married her in 1955.

It should also be mentioned that Koehler, in his later years, was considered by many as the "doyen" of German zoologists. This "meant that in the last resort he had almost overwhelming power, through his advisory relations with the State, over all zoological posts and departments in the country" (Thorpe, 1979, p. 82).

Scientific Work. Koehler was deeply interested in various questions and problems in biology. Early in his career, soon after he had obtained his PhD, he worked with Franz Theodor Doflein on microorganisms. He wrote a considerable part of Doflein's chapter, "Überblick über den Stamm der Protozoen" (Survey of the phylum protozoa), which was published in a comprehensive, multivolume handbook of pathogenic microorganisms, *Handbuch der pathogenen Mikroorganismen* (2nd ed., 1912). As a medical orderly during World War I, he devoted his interest to questions of tropical medicine. Later, during his years at the University of Munich, he studied orientation in unicellular animals and color vision in lower crustaceans and was also concerned with certain general problems of ontogeny and genetics.

Koehler's further research work was strongly influenced by the young discipline of ethology, of which he became an influential promoter. Starting with investigations in sensory physiology and mainly concerned with phenomena like kinesis and taxis, he turned to the study of the number sense, or "counting ability," in animals and used pigeons, ravens, parrots, and squirrels for his experiments. He reached a most interesting and stimulating conclusion: some animal species have a number sense similar to that of humans.

> Previously, there had been attempts for the best part of a hundred years to show that animals were able to count, but there was always some deficiency in the experimental procedure, some loophole left unclosed which rendered the results suspect. It was the outstanding feat of Otto Koehler and his pupils to produce the final but absolutely unequivocal results which showed that animals, especially birds, can "think un-named numbers"—that is, they have a pre-linguistic number sense; to some extent, they think without words. This is an achievement with which Otto Koehler's name will always be linked. (Thorpe, 1979, p. 113)

His studies on "un-named numbers" and "un-named thinking" brought Koehler conveniently to some reflections that are anthropologically relevant.

Koehler was particularly interested in animal and human communication, and he tried to discover some basic patterns of "language" in various species. He maintained that there are prototypes of human communication in animals. Generally, his aim was to close the traditional gap between animals and humans and to show that any specific human ability has roots in some capacities displayed by animals. Thus, for example, he found parallels between music and birdsong. Some experimental studies led him to recognize the close connections and interrelations between hereditary and environmental factors in the ontogenetic development of animal and human behavior. In contrast to the empty-organism doctrine of behaviorism, he—like all other ethologists—was convinced that each living being is born with some specific innate capacities that are the results of the phylogenetic paths of the respective species. In his seminal paper on the smiling of the human infant as an innate expression, he gave some impetus to human ethology, which emerged as a subdiscipline of ethology, offering a new and broad perspective in the understanding of human nature.

Finally, it should be noted that Koehler also dealt with methodological questions of biology. He tried to separate sound ethological research from popular animal psychology and to establish clear and unambiguous concepts and notions in the behavior sciences. Also, he quite successfully combined the holistic perspective and the causal, analytical approach in the study of behavior and biology in general.

BIBLIOGRAPHY

WORKS BY KOEHLER

"'Zähl'-Versuche an einem Kolkraben und Vergleichsversuche an Menschen." *Zeitschrift für Tierpsychologie* 5 (1943): 575–712.

"Die Analyse der Taxisanteile instinktartigen Verhaltens." *Symposia of the Society of Experimental Biology* 4 (1950): 269–302.

"Der Vogelgesang als Vorstufe von Musik und Sprache." *Journal für Ornithologie* 93 (1951): 3–20.

"Vom unbenannten Denken." *Zoologischer Anzeiger* (Supplement) 16 (1952): 202–211.

"Vom Erbgut der Sprache." *Homo* 5 (1954): 97–104.

"Vom Spiel bei Tieren." *Freiburger Dies Universitatis* 13 (1966): 1–32.

"Prototypes of Human Communication Systems in Animals." In *Man and Animal,* edited by Heinz Friedrich. London: MacGibbon and Kee, 1972.

OTHER SOURCES

Hassenstein, Bernhard. "Otto Koehler—sein Leben und sein Werk." *Zeitschrift für Tierpsychologie* 35 (1974): 449–464.

Schmidt, Isolde. "Kurzbiographien." In *Geschichte der Biologie,* edited by Ilse Jahn. Heidelberg, Germany: Spektrum Akademischer Verlag, 2000.

Thorpe, William H. *The Origins and Rise of Ethology.* London: Heinemann, 1979.

Franz M. Wuketits

KOLBE, HERMANN (*b.* Elliehausen, near Göttingen, Kingdom of Hanover, 27 September 1818; *d.* Leipzig, Germany, 25 November 1884), *chemistry.* For the original article on Kolbe see *DSB,* vol. 7.

Kolbe was justly celebrated as one of the finest chemists of his generation, but he damaged his reputation by his obstinate retention of a theoretical viewpoint that both his contemporaries and his successors regarded as deficient and by his unfortunate polemics. Taking advantage of scholarship that has appeared subsequent to the original *DSB* article, the author of this postscript intends to provide a more nuanced assessment of Kolbe's genuine and substantial contributions to the science of his day and to suggest a contextualized understanding of his furious public diatribes.

Evolution as Theorist. Kolbe's life mission was to investigate the "rational constitutions" of organic molecules, and he was one of the greatest pioneers of this quest. According to Kolbe, experiment could be used as an inferential means to dissect schematically the formulas of organic substances, thus deducing the properties of the radicals that compose their molecules. His guide throughout was the dualistic electrochemical radical theory of Jöns Jacob Berzelius, which itself had been founded on the conviction that organic chemical theory must be based on analogies to inorganic chemistry. The evolution of Kolbe's theory is most clearly exhibited in his *Ausführliches Lehrbuch der organischen Chemie.* This remarkable textbook was published in installments that were issued periodically between 1854 and 1878, and in reading it one can follow the twists and turns of Kolbe's ideas, constantly being modified by new experiments and by the influence of the work of other scientists.

"Carbonic Acid" Theory. In the crucial decade of the 1850s Kolbe was in fact theoretically more flexible and more sensitive to the work of his self-declared "enemies" than has been appreciated. He accommodated increasingly to the ideas of the so-called type theorists, who emphasized the substitution of the hydrogen of organic molecules by chlorine and other atoms, and who deemphasized the relevance of inorganic chemistry for organic chemistry. Already in 1850 he had conceded that chlorine can substitute without dramatic change of properties of the substituted entity; in 1854 he agreed that oxygen could also substitute and began to develop a substitution-based theory much like that of the "typists." After initial strong resistance, in 1856 he adopted Edward Frankland's ideas of a "maximum saturation capacity" of radicals and atoms, which was the essence of what became known as valence theory. In 1857 he published a précis, and in 1860 a fully developed treatment, of a new "carbonic acid theory" that is well described in the original *DSB* article. What is not mentioned there is that Frankland later plausibly asserted that the theory had been worked out in 1856 in private correspondence between Kolbe and himself, and that the 1857 article should have been published as a joint-authored paper. Kolbe conceded parts of Frankland's claim but contested other parts, and the letters between them that would establish the truth of the matter have never been found.

Kolbe's theory—or the Kolbe-Frankland theory—was parallel in many ways to the emerging theory of chemical structure, which was being developed about the same time by August Kekulé, Archibald Couper, and others, and it was similarly powerful in its ability to gaze inferentially into the interiors of invisibly small molecules. What Kolbe could not countenance in the rival structure theory was the notion that individual atoms could form directed bonds to other atoms; he was utterly convinced that atoms could not do this, and that even if they could, knowledge of such detail at the atomic-molecular level was beyond human ken. Kolbe's molecules were formed of radicals or atomic groupings, combined (somehow) electrochemically and hierarchically, rather than "democratically," as in structure theory. As similar as Kolbe's and Kekulé's theories were, they generated predictions that differed in the two cases. Unfortunately for Kolbe, the experimental tests resulted in repeated victories of structural principles over Kolbean ideas. By about 1870 virtually no one in the collegial community, not even his own students, adhered any longer to Kolbe's "carbonic acid" radical theory or to his similar theory of the constitution of aromatic substances.

Contributions to Chemical Industry. During the 1860s, Kolbe's good friend August Wilhelm Hofmann benefited from lucrative dye patents derived from his scientific work, and Kolbe, too, was sensitive to possible technological applications. In the fall of 1873 he developed a new and much less expensive way to make salicylic acid from phenol, using sodium hydroxide rather than sodium metal. A few weeks later he entered into partnership with Friedrich von Heyden, a student of his former student Rudolf Schmitt. Heyden provided the capital and most of the development work to take the process from the laboratory to industrial scale; he built a factory for this purpose in Radebeul, near Dresden. In the meantime, Kolbe engaged in aggressive market-oriented research and

Hermann Kolbe. *Historical portrait of the German organic chemist Hermann Kolbe.* SCIENCE PHOTO LIBRARY.

development. He developed a tooth powder using salicylic acid and its methyl ester (synthetic oil of wintergreen), a mouthwash with a similar formulation, a foot powder, a bath salt product, and a tonic. He also championed two much larger potential applications of salicylic acid, as an antiseptic and a food preservative. Both had some success, though Joseph Lister's preference for the antiseptic carbolic acid (phenol) as both stronger and less expensive proved difficult for Kolbe's product to overcome, and there was significant market resistance to salicylic acid as a food preservative. But salicylic acid did become a widely used intermediate in a number of processes for perfumes, flavors, dyes, and pharmaceuticals—including, after Kolbe's death, the most widely used manufactured pharmaceutical in history, aspirin. Salicylic acid made Kolbe a great deal of money, and proved to be a significant element in the rise of the German fine chemical industry in the late nineteenth century.

Professional Isolation. The conundrum that has puzzled all later observers is how Kolbe could hold so tenaciously to a theory that seemed to virtually all others so clearly past its prime, and hold to it, moreover, with an utter confidence that lent a tone of the holy fire of truth to his ferocious voice and pen. One key was his scientific education and the models of behavior after which he patterned himself. Kolbe's earliest mentors were the superb chemists (and gentle personalities) Friedrich Wöhler and Robert Bunsen, but throughout his life his greatest heroes were Berzelius and Justus von Liebig. These two men were known for their occasionally caustic, sometimes even hyperbolic, critiques of the work of their contemporaries (in the case of the former in his annual *Jahresberichte*, and the latter in his journal *Annalen der Chemie und Pharmacie*). Moreover, in their middle age neither Berzelius nor Liebig stayed current with the latest research, and after about 1850 the outdated conservative views of Liebig in particular gave Kolbe the courage to continue the battle against what he considered the modish but utterly empty theory of chemical structure.

More broadly, this story must be placed in the larger context of the cultural history of the united German Empire after 1871, when Kolbe's critiques began in earnest. In his classic work *The Politics of Cultural Despair* (1961), Fritz Stern first explored the development in late-nineteenth-century Germany of illiberal currents of thought that preceded the rise of national socialism. Kolbe displayed many of the characteristics of Stern's protagonists: visceral political conservatism, fear and loathing of "modernity" in all its perceived manifestations, hyper-nationalism, anti-Semitism, and an abiding trust in authority and authoritarian political culture. For Kolbe, the theory of structure was false, meretricious, foreign (non-German) in its very essence, and it had been created and pursued especially by Frenchmen, liberals, Catholics, and Jews. Rightly or (more realistically) perversely, he associated structure theory with sensualism, materialism, liberalism, republicanism, and (possibly) irreligion. His beloved science of chemistry was being systematically destroyed by the structuralists (he thought), and he, at least, had the courage to speak up in its defense, even if he needed to do so alone. But his conduct destroyed much of his reputation, and in his last years his isolation from the peer community was nearly complete. The price of prejudice (or personality disorder?) is often high: in this case it was the self-inflicted demolition of a hitherto brilliant career.

SUPPLEMENTARY BIBLIOGRAPHY

Rocke, Alan J. *The Quiet Revolution: Hermann Kolbe and the Science of Organic Chemistry.* Berkeley: University of California Press, 1993. Cites additional articles by the author.

Alan J. Rocke

KOLMOGOROV, ANDREI NIKOLAEVICH

(*b.* Tambov, Russia, 25 April 1903; *d.* Moscow, Russia, 20 October 1987), *mathematics.*

Kolmogorov was one of the twentieth century's greatest mathematicians. He made fundamental contributions to probability theory, algorithmic information theory, the theory of turbulent flow, cohomology, dynamical systems theory, ergodic theory, Fourier series, and intuitionistic logic. Mathematical talent at this level of creativity and versatility is rarely encountered.

Early Development. Kolmogorov was born in western Russia. His mother having died as a result of his birth, he was brought up by his aunt. Kolmogorov's father was an agronomist who played little part in Kolmogorov's upbringing, and the name "Kolmogorov" was his maternal grandfather's, rather than his father's, name. Kolmogorov matriculated at Moscow University in 1920 to study mathematics, taking classes in set theory, projective geometry, and the theory of analytic functions in addition to Russian history. He studied real functions with Nikolai N. Luzin and early in his undergraduate career began to produce creative mathematics—most notably, in 1922, the construction of a summable function, the Fourier series of which diverged almost everywhere. This result brought him wide recognition at an early age. Following graduation in 1925 and a further four years as a research student after which he received his doctorate, Kolmogorov taught at Moscow University's Institute of Mathematics and Mechanics, being appointed professor there in 1931.

Probability Theory. Kolmogorov's most famous contributions are to the foundations of probability theory. From the mid-seventeenth century, probability had been explored in a somewhat unsystematic fashion. By bringing to bear on the topic the apparatus of measure theory, Kolmogorov's principal work in probability theory, *Grundbegriffe der Wahrscheinlichkeitsrechnung* (1933; *Foundations of the Theory of Probability,* 1956), established probability theory as a core area of rigorous mathematics. In so doing he transformed one-half of David Hilbert's sixth problem: "To treat in the same manner, by means of axioms, those physical sciences in which mathematics plays an important part; in the first rank are the theory of probability and mechanics." Besides its foundational importance, the monograph presented a framework for the theory of stochastic processes and, building on a result of Otton Nikodym, it gave a general treatment of conditional probabilities and expectations. The book was the culmination of an interest in probability that had begun as a collaboration with Aleksandr Y. Khinchin in 1924. This led in the ensuing four years to Kolmogorov's publishing his celebrated three-series theorem (Kolmogorov, 1928), which gives necessary and sufficient conditions for the convergence of sums of independent random variables, to his discovering necessary and sufficient conditions for the strong law of large numbers (1930) and to his proving the law of the iterated logarithm for sums of independent random variables (1929). His 1931 paper, on continuous time Markov processes with continuous states, is widely regarded as having laid the foundations of modern diffusion theory. His 1949 work, *Limit Distributions for Sums of Independent Random Variables,* co-authored with B. V. Gnedenko, was for many years the standard source on the central limit theorem and surrounding topics.

Other Work. Among Kolmogorov's other achievements is his introduction in 1935 of cohomology, the study of algebraic invariants on topological spaces (a field of which J. W. Alexander was an independent codeveloper). The year 1941 saw the publication of two papers on turbulent flow (Kolmogorov 1941a, 1941b). They contained the first clear quantitative predictions in the area of turbulence based on Kolmogorov's two-thirds law and described the equilibrium processes underlying the transfer of energy at different scales of the flow. The importance of this work has persisted as modern computational methods have allowed increasingly detailed investigations of this area of applied mathematics. As a result of this work Kolmogorov was appointed head of the Turbulence Laboratory of the USSR Academy of Sciences in 1946, having been elected to the academy in 1939. In dynamical systems theory, the widely used KAM theory (named after Kolmogorov, Vladimir Arnold, and Jürgen Moser) provides a foundation for the understanding of chaotic motions in Hamiltonian systems, another area in which the later development of computational resources was required for the full importance of Kolmogorov's work to be realized. In 1957 Kolmogorov made a major contribution to the solution of Hilbert's thirteenth problem—to find a proof of the hypothesis that there are continuous functions of three variables that are not representable by continuous functions of two variables—by giving a disproof of it.

Although Kolmogorov published only two papers in logic, the first in 1925 had considerable influence. In it he proved the consistency of classical logic relative to intuitionistic logic by translating formulae of classical logic into formulae of intuitionistic logic, showing that if intuitionistic logic was consistent, then so was classical logic. This is a restricted version of a result later proved by Kurt Gödel. A 1932 paper by Kolmogorov provides an objective reading of negation within intuitionistic mathematics.

In 1965 Kolmogorov unveiled a definition of a random sequence based on the idea that a sequence of integers is random just in case any algorithm that will generate that sequence has a length essentially equal to the sequence itself; that is, the information contained in the sequence cannot be compressed. This approach is often called Kolmogorov complexity, although it was independently arrived at by Gregory Chaitin and somewhat earlier by Ray Solomonoff. This work interestingly inverts Kolmogorov's earlier emphasis on probability in that it allows probabilistic concepts to be based on information theory rather than the reverse, which hitherto had been the standard approach.

Kolmogorov was actively involved for many years in teaching mathematically gifted children and served as the director for almost seventy advanced research students, many of whom became significant mathematicians in their own right. He had wide-ranging intellectual interests, including Russian history and Aleksandr Pushkin's poetry. Kolmogorov's fifty-three-year friendship with the topologist Pavel Sergeevich Alexandrov had an important influence on him. He maintained a deep commitment to the truth, clashing with Trofim Lysenko in 1940 over the interpretation of a geneticist's experimental data. In 1942 he married Anna Egorova. They had no children.

BIBLIOGRAPHY

For a full bibliography of Kolmogorov's published writings, see "Publications of A. N. Kolmogorov," Annals of Probability *17 (1989): 945–964.*

WORKS BY KOLMOGOROV

"On the 'Tertium non Datur' Principle." *Matemticheski Sbornik* 32 (1925): 646–667.

"Über die Summen durch den Zufall bestimmter unabhängiger Grössen *Mathematische Annalen* 99 (1928): 309–319.

"Über das Gesetz des iterierten Logarithmus." *Mathematische Annalen* 101 (1929): 126–135.

"Sur la loi fortes des grands nombres." *Comptes rendus de l'Acadmie des sciences* 191 (1930): 910–912.

"Über die analytischen Methoden in der Wahrscheinlichkeitsrechnung." *Mathematische Annalen* 104 (1931): 415–458.

"Zur Deutung der intuitionistischen Logik." *Mathematische Zeitschrift* 35 (1932): 58–65.

Grundbegriffe der Wahrscheinlichkeitsrechnung (*Foundations of the Theory of Probability*). Berlin: Springer, 1933.

"The Local Structure of Turbulence in an Incompressible Fluid with Very Large Reynolds Numbers." *Comptes rendus de l'Acadmie des sciences de l'URSS* 30 (1941a): 301–305.

"Dissipation of Energy under Locally Isotropic Turbulence." *Comptes rendus de l'Acadmie des sciences de l'URSS* 32 (1941b):16–18.

With Boris V. Gnedenko. *Limit Distributions for Sums of Independent Random Variables.* Cambridge, MA: Addison-Wesley, 1954.

Selected Works of A. N. Kolmogorov. Vol. 1: *Mathematics and Mechanics.* Edited by Vladimir M. Tikhomirov. Berlin: Springer, 1989. Vol. 2: *Probability Theory and Mathematical Statistics.* Edited by Albert N. Shirayayev. Berlin: Springer, 2001. Vol. 3: Information Theory and the Theory of Algorithms. Edited by Albert N. Shirayev. Berlin: Springer, 2006. Note variant spelling of Shirayev's name used by Springer.

OTHER SOURCES

Aleksandrov, Pavel S. "Pages from an Autobiography." *Uspekhi Matematicheskikh Nauk* 34 (1979): 219–249; 35 (1980): 241–278.

American Mathematical Society. *Kolmogorov in Perspective.* Translated by Harold H. McFaden. History of Mathematics Series, vol. 20. Providence, RI: American Mathematical Society; London: London Mathematical Society, 2000.

Kendall, D., G. K. Batchelor, N. H. Bingham, et al. "Andrei Nikolaevich Kolmogorov (1903–1987)." *Bulletin of the London Mathematical Society* 22, no. 1 (1990): 31–100.

Shiryaev, Albert. "A. N. Kolmogorov: Life and Creative Activities." *Annals of Probability* 17 (1989): 866–944.

Paul Humphreys

KÖPPEN, WLADIMIR PETER

(*b.* St. Petersburg, Russia, 25 September 1846, *d.* Graz, Austria, 22 June 1940), *climatology, meteorology, paleoclimatology.*

Köppen was a principal founder of modern meteorology and climatology. The word *founder* is used here in both its conceptual and organizational sense: Köppen produced a useful and durable scheme of descriptive climatology; a modified version of the system bearing his name remains in use more than a century after he proposed it. He was an early proponent of synoptic meteorology, and an active worker in developing the international network of observing stations that made this approach to meteorology possible. He was also the founding editor of the *Meteorologische Zeitschrift* and, during the half century of his contributions, maintained its position as the leading meteorology periodical in the world. In 1924 he produced a major survey of the climates of the Earth's past, and he is rightly considered one of the founders of modern paleoclimatology. The list of his publications includes more than five hundred journal articles and books, and his lifetime as a working scientist is one of the longest on record: His first publication appeared in 1868, and his last in 1940.

Childhood and Education. Köppen was born and raised in Russia. His grandfather had emigrated from Germany, in the time of Catherine II, to help set up a public health system. Köppen's father was a regional governor, economic officer, and epidemiologist in the service of the czar, and he was a member of the Imperial Academy of Sciences. Köppen's uncle, his father's brother, was later tutor to the young czar Alexander III. Köppen was thus a part of a hereditary caste of culturally and linguistically German scientists and civil officials permanently resident in Russia.

Until the age of fifteen Köppen lived in St. Petersburg, at which time his father, then sixty-seven years old, retired and took his family to his estate at Karabagh, in the Crimea. Köppen recalled, in a memoir written in 1931, being struck by the changes in the vegetation he encountered from the windows of the train as the family traveled south: from the boreal coniferous forests around St. Petersburg, to the Mediterranean, arid, and subtropical climate of the Crimea. Many subsequent trips between St. Petersburg and the Crimea confirmed his intuition that vegetation is controlled to a large degree by latitude. Completing his high school education in the Crimea, he returned to St. Petersburg for his university degree before continuing to the University of Heidelberg in Germany for his PhD—a path well trodden by Russian students with an interest in the sciences. Transferring to the University of Leipzig he produced (in 1870) a doctoral dissertation on "Heat and Plant Germination."

The German Marine Observatory. After a brief stint as a schoolteacher, Köppen began work in 1872 as an assistant at the Central Observatory in St. Petersburg, where he helped prepare the daily synoptic weather map, and he began to correspond with his counterparts elsewhere in Europe. Synoptic maps, which plot simultaneous values of meteorological elements at stations widely separated, are a staple of modern meteorology, but were just beginning to be produced at this time. In 1873 he attended the first international meteorological congress in Vienna and began his lifelong friendship with the Austrian meteorologist Julius Hann. Prospects for permanent employment in St. Petersburg pursuing meteorology were slim, and in 1875 Köppen accepted an offer from Georg Neumayer to head the meteorological section of the newly founded Deutschen Seewarte, the German Marine Observatory in Hamburg; he would remain there for the rest of his working life. From 1879 until 1919 he held the post of meteorologist of the Marine Observatory, which allowed him wide scope to pursue theoretical as well as practical investigations.

In addition to his practical work in maritime meteorology and his role in helping to establish an international network of meteorological stations, his publications in the 1870s and 1880s reveal a strong interest in the periodicity of meteorological phenomena, looking for patterns in rainfall amounts, wind, and barometric pressure on all time scales from daily through weekly and monthly up to annual and multiyear periods. He also worked (unsuccessfully) to establish the possible influence of the Moon and of the eleven-year sunspot cycle on weather phenomena.

Köppen was a night owl his whole life—a man who slept little and made a slow start in the morning. He would remain at the observatory late into the evening after his official duties were complete, compiling and plotting temperature data from the thousands of ships' logs for which the Marine Observatory was the official depository.

The Köppen Climate System. In 1884 he took the bold step of founding his own scientific journal, the *Meteorologische Zeitschrift,* and he edited it either alone or in combination with Julius Hann until 1891. He used this journal to publish his first climate scheme in 1884: "Die Wärmezonen der Erde, nach der Dauer der heissen, gemässigten und kalten Zeit, und nach der Wirkung der Wärme auf die organische Welt betrachtet" (The heat zones of the Earth, based on the duration of hot, temperate and cold periods and on the action of heat on the organic world). This article appeared almost simultaneously with two other significant contributions to climatology, Julius Hann's 1883 *Handbuch der Klimatologie* (Handbook of Climatology) and Alexsandr Ivanovich Voeikov's *Klimaty zemnago shara, v osobennosti Rossii* (Climates of Our Globe and Particularly Russia, 1884).

Köppen's initial climate system was the basis of every subsequent modification of his work until 1936. It was a latitude-zone scheme, which took as its basis the temperatures during the warmest month of the year (at a given latitude), deriving isothermal maps based on this data. The approach reflects his own PhD work in botany, as well as his experiences growing up in Russia: the intuition, acquired in childhood, that vegetation is largely determined by latitude. There are, of course, many ways to organize a climate scheme: One could use topographies and land forms, or soil types, or the nature and character of the air masses—that is, the dynamic weather—that passes over them. Köppen chose botany as his discriminating criterion, based on a particular discovery that he had made, that of the *Baumgrenze,* the "tree limit." The northernmost latitude for the growth of trees is well defined by the temperature of the warmest month. Where the mean monthly temperature is 10°C or less in the warmest month, trees cannot grow. Moreover, definable suites of vegetation show up when the mean monthly temperature remains above 10°C (50°F) for one, four, and twelve months. Using a 10°C (50°F) and a 20°C (68°F) isotherm denoting the warmest temperature for periods of one, four, and twelve months, he developed a seven-zone system of climate representing the whole globe. The initial version thus had seven zones per hemisphere: one tropical, one subtropical, three temperate, one cold, and one polar.

Throughout the 1880s and 1890s Köppen continued to work on practically every aspect of climate and weather, with increasing attention to cloud types and precipitation regimes around the world, constantly collating and bringing together work on the climate of all the continents. In 1887 he worked with Julius Hann to produce an *Atlas der Meteorologie* (Atlas of meteorology), and in 1890 published a *Wolkenatlas* (Atlas of cloud forms) with Neumayer and Hugo Hildebrandson.

By the later 1890s Köppen had become convinced that climate zones were differentiated not just by their temperature but by the entire weather picture, with a special attention to the amount and seasonal distribution of precipitation. This led to a new and more complex climate scheme, published in 1901 as *Versuch einer Klassifikation der Klimate. Vorzugsweise nach iheren Beziehungen zur Pflanzenwelt* (An attempt at a classification of climates chiefly according to their relation to the plant kingdom). Here Köppen discerned six basic climate types, five based on temperature and one based on aridity. When combined with different precipitation regimes, this led to twenty-four distinct climates, all with botanical and zoological names, and exact limits based on the temperature of the hottest and coldest months and the amount of precipitation.

Family Life and Personality. Simultaneously with this immense scientific labor, Köppen and his wife Marie raised a large family, with five children of their own, plus the two children of Marie's sister Sophie, who merged households with them in 1888, first in Hamburg proper and later in their large and rambling house in the suburb of Grossborstel close to the Marine Observatory's aerological station, from which Köppen launched both free and captive balloons and meteorological kites of his own design, carrying instrument packets up to altitudes of several kilometers. Köppen was quite gregarious and kept "open house" year-round, with a steady stream of friends and scientific visitors welcomed into the family circle; the Köppen home was also often filled with collateral relatives who had come for long visits.

Köppen was well known for his generosity to younger colleagues; he promoted their ideas and took an interest in their lives as well as their work. He seems to have been universally liked and admired for his good humor and his immense capacity for concentrated hard work, and his epithet "the Nestor of meteorology" refers to both his reputation for giving sage advice and his extremely prolific written output. He was a committed pacifist and internationalist, and he argued strenuously for the adoption of Esperanto as an international scientific language to help overcome national differences. His linguistic gifts were immense, and in addition to Greek, Latin, French, German, English, and Russian, he was also competent in Spanish and Italian. His promotion of the international language movement was a reflection not only of his belief that his ability to read in many scientific literatures had advanced his scientific understanding, but also of his desire for peace.

Köppen and Wegener. While Köppen remained committed to the basic plan of his descriptive climatology, he was not estranged from developments in dynamical meteorology, and indeed he was one of its great patrons; he had close ties to the Bergen School. Köppen was keenly aware of the growing pull of physics on studies of the Earth in all its aspects in the first decade of the twentieth century, and he vigorously promoted it. In fact, he worked extensively with a young atmospheric physicist for two years between 1908 and 1910, helping to write a textbook on thermodynamics of the atmosphere, and constantly insisting to the young man that it had to be about real gases in a real atmosphere. This work, *Thermodynamik der Atmosphäre* (Thermodynamics of the atmosphere), became the standard text in the subject for the next twenty years. Its young author married Köppen's daughter Else, and the two men forged a powerful scientific partnership. The younger man was Alfred Wegener.

The collaboration of Köppen and Wegener is of some scientific interest. It combined the sagacity, deep reading, and wide knowledge of the literature of the older man with the energy and strong physical intuition of the younger. While Köppen originally discouraged Wegener from pursuing his ideas about continental drift, he was gradually won over to his son-in-law's hypothesis, for which he became a strong advocate.

Paleoclimates. In the difficult period following World War I, the Wegeners and the Köppens merged households in Hamburg. Köppen used his influence to have Wegener named meteorologist at the Marine Observatory and procured for him an appointment at the new University of Hamburg. In these years, from 1920 to 1924, Köppen and Wegener collaborated in the production of a very significant work in paleoclimatology: *Die Klimate der geologischen Vorzeit* (The climates of the geological past). At some point Köppen and Wegener realized that a shift of latitude zones through geological time would be a necessary consequence of continental displacement, and would perhaps explain the anomalous appearance of tropical flora at very high latitudes; they therefore set about to reconstruct this history, which was not aimed as a proof of continental drift, but assumed its existence in order to map paleoclimates.

Die Klimate der geologischen Vorzeit relied on a simplified version of the Köppen zone system, with seven belts: a tropical equatorial belt, paralleled to the north and south by arid belts, these paralleled to the north and south by temperate moist zones, and each hemisphere capped by a cold polar climate zone. The climate reconstruction used reef limestones as proxies for equatorial climate; aeolian sandstones and evaporites for arid regions; peat, coal, and plant fossils for the temperate zones; and the glacial tills for the polar regions.

Using the stratigraphical data from the multivolume and internationally produced *Handbuch der regionalen Geologie* (Handbook of regional geology), Wegener and Köppen calculated paleopole positions and paleoequators for every period of the Paleozoic, Mesozoic, and Cenozoic eras. Notably, the book contained the first widely distributed version of the insolation theory of continental glaciations of Milutin Milankovitch.

In 1924 Wegener obtained a professorship of meteorology and geophysics at the University of Graz in Austria, and the Wegener and Köppen families moved to Austria that year. Köppen brought with him his quite considerable library, many thousand volumes in size, and continued to work on his climate system.

In the 1920s, with the appearance of a dynamic meteorology with real predictive power in the form of the theory of frontal weather, there was a great deal of interest in shifting the basis of climatology from descriptive to causal and dynamic, with climatology following meteorology in its proud mathematization. Much work in the 1920s shows the theoretical shift of climatology: Whereas previously the end products were climate maps, now the climate maps were the data and the end products were equations.

Ideal Climates. Köppen's response to this pressure to put climate into abstract form was interesting, and it had a lasting effect on climatology, which preserved his descriptive approach in the face of the challenge from dynamic meteorology. Köppen dealt with the question of theory not by putting climate in motion, but by inventing the idea of an ideal climate on an ideal continent on an ideal Earth and then showing how real climates departed from this ideal. This version of climatology and its idealization is an interesting version of the Germanic preference in these decades for theories of ideal type, whether climatological or sociological, as in the theories of Max Weber. It retained its influence in the depiction of climate and the aim of climatology until the appearance of electronic calculators and computers after World War II.

Köppen continued to expand and modify his climate system, with major revisions appearing in 1918, 1923, 1930, 1931, 1936, and 1939, the latter two versions appearing as part of a multivolume *Handbuch der Klimatologie* (Handbook of climatology) for which Köppen was the general editor, along with Rudolf Geiger. His final publication, a revised and enlarged version of his *Die Klimate der geologischen Vorzeit* appeared in 1940, shortly before his death. His daughter Else recalled in her biography of her father, published in 1955, that he sent an urgent telegram to the publishing house that was setting type for the enlarged edition, which read: "Please send corrections immediately, I am dying." He died a week later, in repose, at the age of ninety-three.

The Köppen system of descriptive climatology, based in temperature and rainfall and their effect upon vegetation, has survived for more than a hundred years and has been described as the least objectionable system ever devised. It has built into it Köppen's childhood, his employment history and his work habits, his feeling for plants, his background in Russian science, his commitment to theoretical idealization, and his relationship with his son-in-law. It was advanced and maintained by the length of his career, his placement in the world, his control of periodicals, and perhaps above all by its real and constructive utility in describing the climates of Earth.

BIBLIOGRAPHY

Köppen's correspondence to 1919 is housed in the Prussian State Library, Berlin; his later correspondence is lost. A preliminary and incomplete list of Köppen's scientific papers and his bound book publications may be found in Else Wegener-Köppen (1955), below. This list does not include Köppen's occasional and popular pieces or his publications in advocacy of Esperanto. Köppen's works remain untranslated into English.

WORKS BY KÖPPEN

"Die Wärmezonen der Erde, nach der Dauer der heissen, gemässigten und kalten Zeit, und nach der Wirkung der Wärme auf die organische Welt betrachtet." *Meteorologische Zeitschrift* 1 (1884): 215–226.

Versuch einer Klassifikation der Klimate. Vorzugsweise nach iheren Beziehungen zur Pflanzenwelt. Leipzig: B. G. Teubner, 1901.

Die Klimate der Erde. Berlin and Leipzig: Walter de Gruyter, 1923.

With Alfred Wegener. *Die Klimate der geologischen Vorzeit.* Berlin: Gebrüder Bornträger, 1924.

With Rudolf Geiger. *Handbuch der Klimatologie.* 5 vols. Berlin: Gebrüder Bornträger, 1940.

OTHER SOURCE

Wegener-Köppen, Else. "Wladimir Köppen. Ein Gelehrtenleben für die Meteorologie." In *Grosse Naturforscher*, edited by Hans Walter Frickhinger. Stuttgart: Wissenschaftliche Verlagsgesellschaft, 1955. The only biography of Köppen, written by his daughter, Alfred Wegener's widow.

Mott T. Greene

KORZHINSKII, DIMITRI SERGEYEVICH

(*b.* Saint Petersburg, Russia, 13 September 1899; *d.* Moscow, Russia, 17 November 1985), *theoretical petrology, chemical thermodynamics of rock recrystallization, metamorphism in open systems, Korzhinskii's mineralogical phase rule.*

Korzhinskii's major contribution was to apply the basic principles of physical chemistry to an understanding of the origin and evolution of igneous and metamorphic rocks. Beginning in the mid-1930s he introduced revolutionary new concepts of how the mineral suites that make up these rock formations may be recrystallized while maintaining conditions of thermodynamic equilibrium. His most radical departure from prevailing theory was to postulate that equilibration may take place in systems that are open to the gain or loss of components such as H_2O and CO_2. Korzhinskii's theories met with strong opposition from a majority of petrologists in the USSR and abroad, who believed that thermodynamic equilibrium is strictly limited to closed systems. By the end of his career, however, Korzhinskii's fundamental approach was being adopted worldwide.

Education, Fieldwork, and Professional Appointments. In 1900, the year after Dimitri's birth, his father, the academician Sergey I. Korzhinskii, chief botanist of the Saint Petersburg Botanical Garden, died while he was on a scientific expedition. His sudden loss left Dimitri, the youngest of four children, to be raised and educated by his mother with the help of private tutors. Dimitri's preparation was of such a high quality that at the age of twelve he was admitted directly into the fourth grade of the gymnasium. He completed courses there in a short time and entered a technical school from which he graduated in 1918, just as the Russian civil war began.

In 1919 Dimitri led a geological reconnaissance survey on the Kola Peninsula, where he was captured by the British and persuaded to join the White Army. He served as a telephonist until February 1920, when his unit rebelled and joined the Red Army. The Red Army transferred him to Saint Petersburg (Leningrad after 1924), where he continued working as a telephonist until 1921, when he took the examinations for the Leningrad Mining Institute and passed them with such distinction that he was accepted as a second-year student. That left him free to leave the army. At the institute, he encountered physical chemistry for the first time and was struck by its vital importance to petrology, a connection that had gone largely unappreciated in Europe and America.

Korzhinskii graduated from the institute in 1926 and spent the next ten years leading field parties sponsored by the Central Geological and Prospecting Institute in Leningrad, where he gave lectures as time permitted. He mapped several areas in Kazakhstan, and then moved to eastern Siberia, where he studied the Aldan Massif of Archaen metamorphosed marbles, the ancient quartz-bearing crystalline schists, and the phlogopite-lazulite deposits near Lake Baikal. Afterward, he recalled his experience in the phlogopite deposits as one of his happiest times in the field, when new ideas kept coming to him—one after another.

In 1935 and 1936 he published his first four papers linking petrology and physical chemistry. In 1937 he entered the Institute of Geological Sciences of the USSR Academy of Sciences in Moscow, where he was awarded the title of candidate of science. He then prepared a thesis for a doctoral degree, which he earned about six months later in 1938. Korzhinskii remained in Moscow for the rest of his life. During his years of field work in remote regions, Korzhinskii had begun formulating his original theories. All that time he was effectively isolated from contact with colleagues who were actively pursuing research and from publications in the Russian and international literature. The wonder is that, inasmuch as his ideas would prove to be of such fundamental importance, they were not anticipated by research teams in Europe or the United States. When he moved to Moscow in 1937, all the scientific support facilities he had lacked became available to him—laboratory equipment, libraries, colleagues, and students. Nevertheless, he continued to work independently.

Original Contributions. Metamorphism invokes the replacement of one mineral suite with another by recrystallization. The classical approach to metamorphism was governed by the theoretical work on equilibrium in heterogeneous systems published in 1878 by Josiah Willard Gibbs (1839–1903). Many petrologists and geologists were convinced that Gibbs's theorems were applicable only to rocks in closed systems that undergo no change in bulk chemical composition. They would not apply to rocks in open systems, in which the content of certain components vary during the process. However, Korzhinskii perceived that Gibbs's principles should apply equally well to rocks in both open and closed systems, and he took the bold step of publishing on heterogeneous equilibrium in metasomatic rocks, which are well known to behave as open systems. Equilibration in an open system is dependent on externally imposed values of temperature and pressure, and also on the externally imposed values of chemical potential or activities of fluid components, chiefly H_2O and CO_2.

In tracing rock-forming processes, Korzhinskii distinguished between two kinds of components that he designated as inert and mobile depending on whether they retained their initial content or were gained or lost. Korzhinskii saw under the microscope that changes in composition occur mainly along grain boundaries that form continuous networks throughout the rocks. These networks serve as pathways for the migration, by diffusion and percolation, of species that alter the chemical as well as the mineralogical composition of the rock. He observed that fully mobile behavior occurs in metamorphic rocks, magmatic rocks (including some granites), and ore deposits. By using the Gibbs phase rule, he showed that in these rocks the maximum number of minerals is unlikely to be more than the number of their thermodynamically inert components. This generalization, which eventually proved useful to petrologists, became known as Korzhinskii's mineralogical phase rule.

Korzhinskii published his first paper on metasomatism in 1935, and his first discussion of the activities of inert and mobile constituents in 1936. For the next twenty years, his ideas remained virtually unknown outside Russia, but they aroused great hostility inside Russia. Looking back on this situation, Korzhinskii wrote in 1980 that he had simply applied Gibbs's work in a logical way to open systems. So he had been greatly surprised by the fierce opposition that arose from many Russian geologists, and some physical chemists. These critics insisted that Gibbs's phase rule, by definition, could not be applied to open systems, and that thermodynamic calculations may be considered valid only for the final state of equilibrium and not, as Korzhinskii had done, for the intermediate stages of achieving it. In 1950 he was called before a special committee of scientists who subjected him to many sessions of hostile questioning for his heresies. Korzhinskii answered their charges, point by point. The committee continued its deliberations for six years and then, in 1956, it issued a final report saying that the physical chemists among them could find nothing wrong with Korzhinskii's scientific ideas. They added that geologists must make up their own minds about whether to accept his views.

Science does not always serve as an open-minded forum for debate and experiment, least of all when scientists feel that their canonical beliefs are being challenged. In 1955 James B. Thompson Jr. at Harvard University published the earliest paper in the United States to apply Gibbs's work on equilibrium to open systems. Thompson had written it independently and then learned of Korzhinskii's publications just in time to include references to them in his manuscript. For this paper Thompson was vilified by colleagues in the United States and Canada, some of whom even accused him of perpetrating a fraud. In time, however, his ideas became widely accepted, and he received apologies from all but a few of his most steadfast opponents.

Principal Publications. Between 1935 and 1985 Korzhinskii published nearly two hundred papers with virtually no coauthors. He was a stellar theorist but he earned a reputation as being a difficult teacher for all but the most advanced students because he could not express his ideas without using complex formulas.

Korzhinskii kept expanding his research interests and probing deeper into them. Through the 1940s he published at least one new paper every two years, and from 1950 to 1956 he published two or three papers every year. In 1957 he wrote a book on the physicochemical basis of mineral paragenesis. An English translation, issued in 1959, became the first book to circulate Korzhinskii's ideas among scientists in Europe and the United States. By the early 2000s it was ranked as a landmark work in the earth sciences, as was his 1936 paper on the mobility and inertness of components in metasomatism. In 1967 a portion of that paper, translated into English, was reprinted in the *Source Book in Geology: 1900–1950* compiled from the international literature by Professor Kirtley F. Mather (1888–1978) at Harvard University.

Honors. In 1953 Korshinskii advanced to the status of a full academician of the USSR, and became a member of the editorial boards of journals of science at home and abroad, including *Doklady Akademii Nauk S.S.S.R.*, *Izvestia Akademii Nauk S.S.S.R.*, *Proceedings of the Russian Mineralogical Society*, and the *Journal of Petrology* published in the United Kingdom. In 1969 he organized the Institute of Experimental Mineralogy and served as its director until 1979. From 1962 to 1970 he served as the vice president of the International Mineralogical Association, and from 1966 to 1974 he was the head of the USSR's National Committee of Geology. He also served as the vice president of the Mineralogical Society of the USSR and the president of the Metasomatic Section of the Committee on Ore Formation of the USSR. Over the years he received honors from several institutes and societies in Russia and elsewhere in Europe. In 1980, at a meeting of the Geological Society of America in Atlanta, James B. Thompson read the citation for presentation of the prestigious Roebling Medal of the Mineralogical Society of America to Korzhinskii. By that time Korzhinskii's fame had spread widely and his basic ideas on open systems had greatly influenced the thinking of petrologists around the world.

BIBLIOGRAPHY

WORKS BY KORZHINSKII

"Thermodynamics and Geology of Some Metamorphic Reactions with the Separation of a Gas Phase" [in Russian with German summary]. *Proceedings of the Russian Mineralogical Society*, ser. 2, 64, no. 1 (1935): 1–20.

"Archaen Marbles of the Aldan Platform and the Problem of Depth Facies" [in Russian with English summary]. *Transactions of the Central Geological and Prospecting Institute* 71 (1936).

"Mobility and Inertness of Components in Metasomatism" [in Russian]. *Bulletin of the Academy of Science, U.S.S.R.* 1 (1936): 35–60. Also in *Source Book in Geology: 1900–1950,* edited by Kirtley F. Mather. Translated by John B. Southard. Cambridge, MA: Harvard University Press, 1967.

"Paragenetic Analysis of Quartz-Bearing, Almost Calcium-Free Crystalline Schists of the Archaen Complex South of Lake Baikal" [in Russian]. *Memoirs of the Russian Mineralogical Society Series 2,* 65, no. 2 (1936): 247–280.

"Concept of Geochemical Mobility of the Elements" [in Russian]. *Memoirs of the Russian Mineralogical Society* 71, nos. 3–4 (1942): 160–168.

"Relationship between the Mineralogical Composition and the Chemical Potential Value of the Components" [in Russian]. *Proceedings of the Russian Mineralogical Society* 73, no. 1 (1942): 62–73.

"Metasomatic Zoning in Wall Rock Alteration and Veins" [in Russian]. *Proceedings of the Russian Mineralogical Society* 75, no. 4 (1946): 321–332.

"Phase Rule and Geochemical Mobility of Elements." *International Geological Congress, Report of 18th Session, Great Britain 1948, Proceedings of Section A, Part II* (1950a): 50–65 (in English); 66–73 (in Russian).

"Differential Mobility of Components and Metasomatic Zoning in Metamorphism." *International Geological Congress, Report of 18th Session, Great Britain 1948, Proceedings of Section A, Part III* (1950b): 65–72 (in English); 73–80 (in Russian).

"Granitization as Magmatic Replacement" [in Russian]. *Izvestia Akademii Nauk S.S.S.R. Seriya Geologicheskaya,* no. 2 (1952): 56–69.

Physicochemical Basis of the Analysis of the Paragenesis of Minerals. New York: Consultants Bureau, 1959. First published 1957 by the Russian Academy of Sciences.

Theory of Metasomatic Zoning. Translated by Jean Agrell. Oxford: Clarendon Press, 1970. First published 1969 in Moscow by Science Press.

"Response to Presentation of the Roebling Medal." *American Mineralogist* 66 (1980): 642.

With others. *A Petrologic Classic of the 20th Century: Centenary D. S. Korzhinskii; Reminiscences.* Moscow: Scientific World, 1999. This is a hardcover volume entirely in Russian except for one paragraph after the title page that gives the title in English and describes it as a collection of Korzhinskii's reminiscences on his times, colleagues, and of science, and those of his students and contemporaries.

OTHER SOURCES

Gibbs, J. Willard. "On the Equilibrium of Heterogeneous Substances." In *Scientific Papers of J. Willard Gibbs,* vol. 2, 55–349. New York: Longmans, Green, 1906. Reprint, London: Constable & Co., 1961.

Perchuk, Leonid L., ed. *Progress in Metamorphic and Magmatic Petrology: A Memorial Volume in Honor of D. S. Korzhinsky.* Cambridge, U.K.: Cambridge University Press, 1991.

Thompson, James B., Jr. "The Thermodynamic Basis for the Mineral Facies Concept." *American Journal of Science* 253 (1955): 65–103.

———. "Presentation of the Roebling Medal of the Mineralogical Society of America for 1980 to Dimitrii Sergeevich Korzhinskii." *American Mineralogist* 66 (1980): 640–641.

Ursula B. Marvin

KOVALEVSKAIA, SOFIA

SEE **Kovalevskaya, Sofya Vasilyevna.**

KOVALEVSKAYA, SOFYA VASILYEVNA (SONYA)

(*b.* Moscow, Russia, 15 January 1850; *d.* Stockholm, Sweden, 10 February 1891), *mathematics.* For the original article on Kovalevsky (the variant of her name used in that article) see *DSB,* vol. 7.

Interest in Kovalevskaya's life and work has increased considerably since the appearance of the original *DSB* article on her. As a result a number of new works about her have appeared in English and German, and a French work was completed in 2006, but has not yet appeared. Views of her position among late nineteenth-century mathematicians have altered slightly, and it seems that when she was mentioned among them, it was her talent rather than her gender that first came to mind. Her use of hyperelliptic functions to solve a complicated case of the equations of motion of a rigid body introduced techniques that found additional applications during the 1980s and thereby increased in importance, while the Cauchy-Kovalevskaya theorem in partial differential equations remains a standard result that everyone who works in this area must know.

Kovalevskaya's work in differential equations was so well respected that the name *Cauchy–Kovalevskaya theorem* was coined by Jacques Hadamard in his 1910 lectures. In the mid-1920s Felix Klein devoted a small amount of space to her work in his lectures on the development of nineteenth-century mathematics, unfortunately with a rather negative tone. Klein had a strong influence over Western European and American mathematical opinion, and the effect of his dismissal of Kovalevskaya was to put her in Karl Theodor Wilhelm Weierstrass's shadow for a generation. This low period in her posthumous reputation was reflected in the patronizing portrait of her painted by Eric Temple Bell in his 1937 book *Men of Mathematics,* who implied that she used her sex appeal to persuade Robert Bunsen to admit one of her friends to his laboratory, and went on to say that after returning to Russia and being unable to find work, "her sex had got the better of her ambitions."

Sofya Kovalevskaya. SPL / PHOTO RESEARCHERS, INC.

In the Soviet Union, however, the centenary of Kovalevskaya's birth in 1950 and the 125th anniversary of her birth in 1975 led to a host of new publications, providing both basic information and inspiration for further studies of her life and work. These publications brought her to the attention of mathematicians and historians around the world, and studies of her life and work began to appear outside the Soviet Union in the early 1980s.

Studies of Kovalevskaya's Life. Kovalevskaya's life is of interest from a number of points of view. She was part of the radical youth movement in Russia during the 1860s known as nihilism. A biography of Kovalevskaya as nihilist and scientist was published by Ann Hibner Koblitz in 1983. This work remains the definitive study of this aspect of Kovalevskaya's life. On a less political level, Kovalevskaya is deservedly a hero of the feminist movement. In her day there was determined opposition to careers for women, and if a woman defied expectation and discrimination and attempted to enter a career, she certainly found herself squeezed between the professional demand to work like the men in her field and the social demand to behave like other women in relation to her family and friends. For Kovalevskaya, such conflicts arose constantly.

Studies of Kovalevskaya's life include several scientific and personal biographies and a novel in which a decade of Kovalevskaya's life is depicted in fictionalized form. In addition, a highly inaccurate version of her years in Stockholm forms the subject of the 1983 Swedish film *Berget på månens baksida* (A mountain on the far side of the moon).

Studies of Kovalevskaya's Mathematical Work. Both of the major works that established Kovalevskaya's reputation in her lifetime were still regarded as outstanding achievements a century after her death and were still leading to new research. The fundamental place of the result known as the Cauchy-Kovalevskaya theorem in the theory of differential equations in the complex domain was established early, and this theorem remains a pillar of the subject.

Interest in her work on the equations of motion of a rotating rigid body continues for two reasons. First, it represents an exactly solvable case of the system of equations, that is, a set of parameters for which a complete set of "conservation laws" can be found to describe the motion. Second, to express the parameters of motion explicitly as functions of time, it is necessary to use theta functions of two variables.

The two exactly solvable cases of rigid-body motion that had been studied earlier by Leonhard Euler, Joseph-Louis Lagrange, and others were both important in that they covered the familiar cases of rotating heavenly bodies and gyroscopes. Sufficiently general to be of physical interest, they were also sufficiently symmetric that the motion could be described using, at worst, elliptic functions. For the Kovalevskaya case the degree of symmetry is less, so that physical cases where these equations apply are harder to find. At the same time, the computations are much more complicated. As a result, physicists paid little attention to this case for nearly a century. The late twentieth century, however, saw a revival of interest in this case as an exactly solvable model. Detailed studies of the phase portrait of the corresponding differential equations have been published, including one that was turned into a video.

Meanwhile new uses were found for theta functions in solving differential equations. The differential equations that describe comparatively simple mechanical systems, such as the motion of a pendulum, require elliptic functions for a closed-form solution. Mathematicians developed an intricate theory of more complicated algebraic functions as the natural extension of elliptic functions. Carl Jacobi introduced theta functions of one variable in the study of elliptic functions, and an analogous theory of theta functions of several variables was used to study more complicated algebraic functions. In developing this magnificent theoretical edifice, mathematicians had moved ahead of physical applications. For that reason, applications of theta functions of two variables were virtually nonexistent in mechanics during the nineteenth century. Some mathematicians went looking for such applications, notably Carl Neumann, who, in his 1856 dissertation, studied the motion of a point confined to a sphere and moving under a potential that is constant on ellipsoids. He found a system of two integrals of the type that occur in what is known as the Jacobi inversion problem, as stated by Weierstrass. This problem was solved by Weierstrass and Bernhard Riemann using theta functions of several variables. Although the problem is somewhat artificial, it does show that higher algebraic functions can be applied in mechanics.

Weierstrass had suggested that Kovalevskaya work along the same lines as Neumann in connection with the motion of a rigid body, and it was precisely by following this line of attack that she discovered and solved her case of the spinning top. Indeed, the system of equations that she finally developed was of exactly the same form as the system that Neumann had produced. But Neumann had not followed through with a thorough investigation of the solution. Kovalevskaya carried the solution through to completion using theta functions of two variables, thereby proving that theta functions could be useful in a "classical" problem of mechanics. This aspect of her work brought the highest praise from the Paris Academy.

In the 1970s and 1980s theta functions of two variables arose in connection with the Korteweg–de Vries equation describing the motion of a wave in a channel and in connection with George William Hill's equation for lunar motion. Due to modern computers, the complexity of an equation is not the barrier to study that it once was, and extensions of Kovalevskaya's work are once again being undertaken.

SUPPLEMENTARY BIBLIOGRAPHY

Bölling, Reinhard, ed. *Briefwechsel zwischen Karl Weierstrass und Sofja Kowalewskaja.* Berlin: Akademie-Verlag, 1993.

Cooke, Roger. *The Mathematics of Sonya Kovalevskaya.* New York: Springer-Verlag, 1984.

Dullin, Holger R., P. H. Richter, and Alexander P. Veselov. "Action Variables of the Kovalevskaya Top." *Regular and Chaotic Dynamics* 3, no. 3 (1998): 18–31.

Françoise, Jean-Pierre. "Sur les Action-angles de la toupée de Kowalevski." *Comptes Rendus de l'Académie des Sciences de Paris* 300 (1985): 427–430.

Hadamard, Jacques. *Lectures on Cauchy's Problem in Linear Partial Differential Equations.* 1923. Mineola, NY: Dover, 2003.

Hörmander, Lars. *The Analysis of Linear Partial Differential Operators.* Vol. 1, *Distribution Theory and Fourier Analysis.* Berlin: Springer-Verlag, 1983.

Klein, Felix. *Vorlesungen über die Entwicklung der Mathematik im 19. Jahrhundert.* Berlin: Springer-Verlag, 1926.

Koblitz, Ann Hibner. *A Convergence of Lives: Sofia Kovalevskaia, Scientist, Writer, Revolutionary.* Boston: Birkhäuser, 1983.

Kovalevskaya Top. Video C1961 of Institut für den Wissenschaftlichen Film, Göttingen.

Mittag-Leffler, Gustav. "Weierstrass et Sonja Kowalewsky." *Acta Mathematica* 39 (1923): 133–198.

Spicci, Joan. *Beyond the Limit: The Dream of Sofya Kovalevskaya.* New York: Forge, 2002.

Tuschmann, Wilderich, and Peter Hawig. *Sofia Kowalewskaja: Ein Leben für Mathematik und Emanzipation.* Basel, Switzerland: Birkhäuser, 1993.

Roger Cooke

KOVALEVSKY, SONYA

SEE **Kovalevskaya, Sofya Vasilyevna.**

KOWALEVSKY, SONJA

SEE **Kovalevskaya, Sofya Vasilyevna.**

KROPOTKIN, PETR ALEKSEYEVICH

(*b.* Moscow, Russia, 9 December 1842; *d.* Dimitrov, U.S.S.R., 8 February 1921), *geography, natural history, evolution.* For the original article on Kropotkin see *DSB,* vol. 7.

It is interesting to note that the original entry on Kropotkin concludes with a 1912 congratulatory address from the Royal Geographical Society of London in honor of Kropotkin's seventieth birthday. The address cited; "service in the field of natural sciences … contribution to geography and geology," and most importantly for this entry, "amendments to Darwin's theory." Kropotkin's contributions to geography and geology are well covered in the original entry by Oleg Naumov. This supplement focuses on Kropotkin's reassessment of Charles Darwin's theory, which he developed in response to what he saw as the excesses of some of evolution's most ardent supporters.

Kropotkin was attracted to natural history from a young age and developed that fascination throughout his young life as a member of a number of geographical expeditions and as a contributing member of various professional societies. After completing his studies in the corps of pages in 1862, he spent the next five years traveling through Siberia studying the geography and geomorphology of eastern Siberia. In the course of these travels,

Kropotkin

Kropotkin wrote several significant articles that established his reputation in the natural sciences. According to his memoir, during his travels he was intently studying a recent work by the British naturalist Charles Darwin, *On the Origin of Species.*

The year after his return from Siberia, Kropotkin won the Russian Geographical Society's gold medal for his account of the geography of the Olekmin-Vitim expedition. It was also during this time that Kropotkin's interest in socialist revolution and radical politics developed beyond private conversations with close associates to public statements about political reform. This activity led to his arrest and imprisonment in Saint Petersburg in 1874. A few months into his imprisonment Kropotkin received permission to continue his scientific work and completed *Investigation of the Ice Age* in 1876. Shortly thereafter, because of his failing health, he was moved to a military hospital, from which he escaped at the end of June in 1876. He settled initially in Edinburgh but within weeks moved to London. While in London he continued both his political and scientific pursuits, publishing a number of articles in *Nature* and regularly participating in Royal Geographical Society meetings. Kropotkin's involvement in the international anarchist movement led to his second arrest and imprisonment, in France in 1883. Though initially sentenced to five years, he was released in 1886 and returned to England.

A couple of years after his release, Kropotkin was invited by the editor of the popular journal *The Nineteenth Century,* to respond to an article written by Darwin's bulldog Thomas Henry Huxley. It was this series of articles that was later published as the collection *Mutual Aid: A Factor in Evolution.* Kropotkin remained in England until 1917, when he returned to Russia with the overthrow of the tsar. In his final years he worked on his book *Ethics* until his death in 1921.

The Nature of Nature. The influence of Kropotkin's early experience of nature in Siberia was to last a lifetime. Under the influence of his recent reading of Darwin, Kropotkin searched the steppes and Russian plains for the struggle for existence so vividly described in *The Origin.* In the first chapter of *Mutual Aid,* he writes:

> I recollect myself the impression produced upon me by the animal world of Siberia when I explored the Vitim regions in the company of so accomplished a zoologist as my friend Poliakov was. We were both under the fresh impression of the *Origin of Species,* but we vainly looked for the keen competition between animals of the same species which the reading of Darwin's work had prepared us to expect. ... We saw plenty of adaptation for struggling, very often in common, against the adverse circumstances of the environment, or against various enemies ... but even in the Amur and Usuri regions, where animal life swarms in abundance, facts of real competition and struggle between higher animals in the same species came very seldom under my notice, though I eagerly searched for them. (p. 9)

Petr Kropotkin. **HULTON ARCHIVE/GETTY IMAGES.**

This experience of nature was significantly different from the riot of life in the tropics that Darwin had described in *The Voyage of the Beagle* and later in his autobiography, recalling, "The glories of the vegetation of the tropics rise before my mind at the present time more vividly than anything else" (Darwin, 1958). The harsh environment of Siberia, where organisms struggled against seasonal extremes presented an entirely different tableau on which Kropotkin's ideas about evolution would be inscribed. In contrast to Darwin's focus on inter- and intraspecies competition, he emphasized the direct action of the environment. Kropotkin was also deeply influenced by the lack of Malthusian overpopulation in the Siberian context. Indeed, he was most impressed by the displays of organisms such as the musk oxen huddling together cooperatively against the physical hardship. Finally, Kropotkin wondered what the long-term effects of the harshest conditions meant for the long-term viability of the species.

He observed that under the most extreme environmental conditions (where the Darwinian would expect the most severe competition) the survival of the whole group was at risk and their overall fitness was often diminished.

Kropotkin's Amended Darwinism. As mentioned above, Kropotkin's development of his theory of mutual aid was a direct response to Huxley's article, "The Struggle for Existence: A Programme," originally published in the journal *The Nineteenth Century* in 1888 and later included in the collected volume *Evolution and Ethics* (1894). Drawing on materials and examples from anthropology, politics, philosophy, and economics, Huxley laid out the fundamental role that competition and natural selection had played in the development of human society from Stone Age cultures to the modern era.

Huxley's depiction of nature in "The Struggle for Existence" was "on about the same level as a gladiators show .... [In this case, however, the] spectator has no need to turn his thumbs down as no quarter is given." For Huxley, nature is a zero sum game and the metric of success is reproductive output. He continues:

> Let us be under no illusion then. So long as unlimited multiplication goes on, no social organization which has ever been devised, or is likely to be devised ... will deliver society from the tendency to be destroyed by the reproduction within itself ... of that struggle for existence, the limitation of which is the object of society. (pp. 211–212)

In the remainder of the article, Huxley appeals for both educational reform and worker's rights that while perhaps not sufficiently radical for Kropotkin might at least be consistent with his politics. Nevertheless, the emphasis on competition (to the exclusion of any form of cooperation) in nature struck Kropotkin as deeply misguided.

Kropotkin's first response to Huxley, on mutual aid in animals, appeared in *The Nineteenth Century* in 1890, followed in 1891 with a piece on early peoples, then on medieval city dwellers in 1892, and on contemporary societies in 1894. These essays and four others (along with Huxley's original article) were published in 1902 as a collected volume *Mutual Aid: A Factor of Evolution.* Kropotkin's interest in evolutionary theory did not, however, stop there. Between 1905 and 1919 he published seven additional articles in *The Nineteenth Century* and *The Nineteenth Century and After* that continued his attempt to broaden the scope of Darwinian theory against the constriction represented by Darwin's most strident supporters Huxley and later August Weismann. Indeed, Kropotkin would argue that his broader characterization of Darwin's theory that included cooperation and mutual aid was more consistent with Darwin's original intent.

In "The Theory of Evolution and Mutual Aid," Kropotkin (1910a) analyzed the historical development of Darwin's theory. Using excerpts from Darwin's correspondence and charting the editorial changes made over the six editions of the *Origin* published during Darwin's lifetime, Kropotkin presented his theory of mutual aid as the next step of the development of Darwin's theory. Kropotkin made the historical argument that Darwin was initially so focused on the power of natural selection largely as a response to the continued negative influence of the work of Jean-Baptiste de Lamarck and Robert Chambers.

Kropotkin further argued that the difference between his theory of mutual aid and that of Darwin's was exaggerated by the persistent influence of the Malthusian doctrine on Western Darwinists. His theory of mutual aid was supported by a shift in the focus from strict Malthusian or individual level competition to a point of view that emphasized the role of the cooperative struggle for existence against harsh environmental conditions. Kropotkin also argued in the 1910a article that Darwin was aware that the direct action of the environment played a more significant role in the process of evolution than he had originally allowed in the *Origin.* "He gradually came, in an indirect way, to attribute less and less value to the individual struggle inside the species, and to recognise more significance for the associated struggle against the environment" (p. 87).

In a second article published later that year, Kropotkin presented the next step in his theory of mutual aid. Again focusing on the collective struggle against the environment, Kropotkin argued that the more cooperative species would survive longer than their more individualistic, competitive rivals. On Kropotkin's expanded version of evolutionary theory, natural selection is no longer "a selection of haphazard variations, but becomes a physiological selection of those individuals, societies and groups which are best capable of meeting the new requirements by new adaptations of their tissues, organs and habits. It operates largely as a selection of groups of individuals, modified all at once, more or less, in a given direction" (1910b, p. 61, 75).

Kropotkin's insistence on the fundamental importance of the direct action of the environment precipitated another significant distinction between his theory and the neo-Darwinians. For Kropotkin, the direct action of the environment not only challenged the notion of individual level competition; it also modulated the claim of purely random variation. According to Kropotkin, the recent work of the biometricians had demonstrated that: "Whether we take the sizes of leaves of the same tree, or the stature of several thousand Englishmen at Cambridge

University ... everywhere we find that the laws of variation in organic beings are the same as those with which we are familiar in physical sciences under the name of laws of errors in the theory of probabilities" (1910a, pp. 105–106). Following this logic, he concluded that when we see consistent or directional deviation from the normal distribution this must be the result of some permanently acting cause, that is the direct action of the environment. Given such a cause, he continued, there is no need for an intense struggle between individuals to preserve the effects of variation, the influence of the environment will maintain and accumulate them in successive generations. Another development that clearly influenced Kropotkin's reassessment of Darwin's theory was the recognition of the importance of geographical isolation in the process of speciation. Kropotkin identified this development as further support for his diminished emphasis on intraspecies competition and individual level selection. Again, citing passages from Darwin's correspondence with Moritz Wagner and Karl Semper on the role of isolation, Kropotkin observed:

> Once we admit the successive migrations, in the course of ages, of certain species over several continents ... and once we realise the amount of segregation that ensued, we fully understand the necessary 'absence of intermediate forms.' And yet it was this absence which so much puzzled Darwin and for which he admitted 'extermination' during a severe struggle for life. With isolation, such an extermination is not necessary; and probably it did not take place at all. (1910a, p. 100)

Clearly, from Kropotkin's perspective the acknowledgement of the role of isolation decreased almost to the point of elimination, the significance of competition.

In his last article on his theory of mutual aid, published just two years before his death, Kropotkin summarized his findings. He concluded that Darwin's original theory had been misinterpreted and misapplied by his followers under the influence of Malthus. He had also become convinced that the neo-Darwinians had created a false dichotomy between Darwin's theory and Lamarck's. Finally, he was concerned that the neo-Darwinians had become so committed to the theory of Darwin that they had eschewed the importance of the naturalist tradition and thereby the connection with nature itself. Kropotkin, on the other hand, had developed his theory of mutual aid largely on the basis of his experience in Siberia, observing and recording animal behavior in nature. In a letter quoted by Daniel Todes, Kropotkin provided his characterization of evolutionary theory; one that moderated the competitive and selective focus of Weismann and Huxley and paid proper tribute to Darwin's original formulation:

> This is a theory of evolution which ... recognized the importance of Mutual Aid—that is, of the social instinct—for the preservation of the species, and which ... saw in it the primordial element of Ethics. ... This is above all a return to the Darwinism which saw in Evolution a spontaneous result of the forces of Nature, and not, as Weismann and his disciples wished, an Evolution predetermined (by the mechanisms of the Universe) by means of a substance possessed of an 'immortal soul'—this Hegelian creation of Weismann, his germ plasm. (quoted in Todes, 1989, p. 141)

Ultimately, Kropotkin's ideas had little lasting influence on Western biologists, except for some few who continued to champion some form of group selection theory. They did however, exert some significant influence in the Russian context, as demonstrated in the work of the historians Mark Adams and Daniel Todes.

SUPPLEMENTARY BIBLIOGRAPHY

WORKS BY KROPOTKIN

Memoirs of a Revolutionist. Boston and New York: Houghton Mifflin, 1899.

Mutual Aid: A Factor in Evolution. New York: McClure Philips, 1902.

"The Theory of Evolution and Mutual Aid." *The Nineteenth Century and After* 67 (1910a): 86–107.

"The Direct Action of Environment on Plants." *The Nineteenth Century and After* 68 (1910b): 58–77.

"Inheritance of Acquired Characters: Theoretical Difficulties." *The Nineteenth Century and After* 71 (1912): 511–531.

"Inherited Variation in Animals." *The Nineteenth Century and After* 78 (1915): 1124–1144.

"The Direct Action of the Environment and Evolution." *The Nineteenth Century and After* 85 (1919): 70–89.

OTHER SOURCES

Adams, Mark B. "The Founding of Population Genetics: Contributions of the Chetverikov School, 1924–1934." *Journal of the History of Biology* 1, no. 1 (1968): 23–39.

Adams, Mark B., ed. *The Evolution of Theodosius Dobzhansky: Essays on His Life and Thought in Russia and America.* Princeton, NJ: Princeton University Press, 1994.

Borrello, Mark E. "Mutual Aid and Animal Dispersion: An Historical Analysis of Alternatives to Darwin." *Perspectives in Biology and Medicine* 47, no. 1 (2004): 15–31.

Darwin, Charles. *The Autobiography of Charles Darwin 1809–1882.* With original omissions restored. Edited by Nora Barlow. New York: Norton, 1958.

Glick, Thomas F., ed. *The Comparative Reception of Darwinism.* Chicago: University of Chicago Press, 1988. First published 1974 by University of Texas Press.

Gould, Stephen J. "Kropotkin Was No Crackpot." *Bully for Brontosaurus: Reflections in Natural History,* pp. 325–339. New York: Norton, 1991.

Krumbein

Huxley, Thomas H. "The Struggle for Existence in Human Society." In *Evolution and Ethics and Other Essays*, pp. 195–236. London: MacMillan, 1894.

Todes, Daniel P. *Darwin without Malthus: The Struggle for Existence in Russian Evolutionary Thought.* New York: Oxford University Press, 1989.

Mark Borrello

KRUMBEIN, WILLIAM CHRISTIAN

(*b.* Beaver Falls, Pennsylvania, 28 January 1902; *d.* Los Angeles, California, 18 August 1979), *geology, quantitative methodology, sediment analysis, stratigraphic analysis.*

Krumbein is the pioneer in the application of quantitative methods in stratigraphy and sediment analysis. His many innovations include the logarithmic transform of particle size-distributions or phi scale, now almost universally used to describe the grain size of clastic sediments, the lithology tetrahedron to describe the complex interrelations between tectonic-sedimentary facies, the application of Markov chains to describe the stratigraphic memory of lithologic formations through geologic time, and the introduction of the general linear model in the statistical analysis of multidimensional geologic variables. Krumbein was a true trailblazer in the study of sediments and sedimentary rocks—years ahead of his time—and as a consequence, he was not recognized in his lifetime for his innovations by his more conventional colleagues.

Early Life. William Christian Krumbein, known as the father of mathematical geology, was born of German immigrants. An early marriage ended in failure and in 1946 he married Marjorie Kamm. Because Krumbein was a reserved person, little is known about his family, early childhood, and upbringing, other than that his parents, Carl and Hattie, moved to a largely German neighborhood in northwest Chicago when he and his older brother Henry J. were very young.

Academic Formation. Krumbein received his PhB in business administration from the University of Chicago in 1926 and worked for several years in the field of insurance adjustments with a finance company in downtown Chicago. Legend among his students had it that during the stock market crash of 1929, having witnessed a suicide fall from his skyscraper window, he returned to the University of Chicago and began graduate work in geology. His master's thesis was titled "A Key for the Determination of Minerals by Means of Structure, Form, and Texture," and he received an MSc in geology in 1930. His doctoral dissertation, written under J. Harlan Bretz and titled "The Mechanical Analysis of Related Samples of Glacial Till," earned him a PhD in geology from the University of Chicago in 1932.

Krumbein's interest in geology was stimulated when he was an undergraduate business student at the University of Chicago by Paul MacClintock, from whom he took an introductory geology course. Later, as a graduate student in geology taking Francis Pettijohn's first course in sedimentation, Krumbein recognized the potentialities of statistical analysis, already familiar to him from business school, to the description of sediments. While in graduate school, Krumbein shared an office with M. King Hubbert, who became famous for his petroleum resource prediction models; in hindsight this association proved to be very fruitful for petroleum geology.

Between 1933 and 1945, Krumbein rose from instructor to associate professor at the University of Chicago. During World War II he served with the Beach Erosion Board of the U.S. Army Corps of Engineers in Washington, D.C. (1942–1945), in the area of beach-landing intelligence, and then briefly worked for the Gulf Research and Development Corporation (1945–1946). In 1946 Krumbein became professor of geology at Northwestern University, and from 1960 until his mandatory retirement from Northwestern in 1970, he was William Deering Professor of Geological Sciences. In 1955 he was on leave for research at the National Bureau of Standards in Washington, D.C.

Scientific Production. Krumbein wrote more than 140 papers, including one published posthumously, and four books during his forty-seven years of productivity. His interests shifted appreciably from sediment analysis (39 publications) to stratigraphic analysis (27 publications) and finally to statistical methodology in geology (62 publications). His books followed the same trend: *Manual of Sedimentary Petrography* (1938), with Pettijohn, was followed by *Stratigraphy and Sedimentation* (1st ed., 1951; 2nd ed, 1963), with Laurence Sloss, and subsequently by *An Introduction to Statistical Models in Geology* (1965), with Franklin Graybill. Only six papers cannot be classified into these three categories; they are concerned mainly with educational matters and the use of computers in geology, as well as futuristic trends in geology.

He intended his first book, *Down to Earth,* coauthored with Carey Croneis in 1936, as an introductory text for a short course in geology. It attempted to develop an approach that was analytical instead of being solely a descriptive treatment of geological sciences. The purpose was to bring out the true relationship of geology to other sciences, following the "back to nature" trend observed by previous authors. In their 1936 preface Krumbein and

Croneis explain the "back to nature trend" as the growing public awareness to understand the earth sciences—which they can see and feel—as opposed to the "layman's relatively decent understanding of atoms electrons and genes which they cannot see." This book is a factual depiction and philosophical interpretation of the history of Earth and its inhabitants through geologic time. It ends with a prognosis of humankind's prospects on Earth that claims that "if he learns to control nature even reasonably completely, his mind and his studies will, indeed, have brought him immortality" (p. 484).

Krumbein's scientific productivity is legendary. His clear analytical ability, which helped him essentially to draft entire papers in his mind, and an iron discipline at the desk and typewriter, supported this productivity. He revised, edited, and corrected his own writing at least twice before giving it to a colleague for review, usually Laurence Sloss or Michael Dacey; he then produced the final version himself, seldom using the service of his part-time secretary. His papers are characterized by clear, careful prose, and he explained everything, but only once; there was no repetition in his papers.

As Krumbein often said in class, "it would be fun" to examine his scientific productivity with his own statistical methods. Arthur Howland published a complete bibliography of Krumbein in "William C. Krumbein: The Making of a Methodologist" (1975). This bibliography enables one to calculate the number of works and pages he published per year as well as the average length of his papers, and also to classify them according to the primary main topics: sediment analysis, stratigraphic analysis, and statistical methodology. Additionally, he published papers on Earth science teaching, reviews of the state of the art in computers used in geology, and analysis of geological bibliography. The bibliography data matrix shows that to 1975 he published 132 papers and four books on geology, for totals of 2,805 and 2,645 pages respectively. His papers average twenty-one pages in length, but this figure is somewhat misleading because it mixes soft-copy reports and printed papers.

As noted briefly above, Krumbein's research interests shifted through time. He concentrated from 1932 to 1944 on the analysis of sediments and on the statistical methodology it required. He published only two papers on the stratigraphy of clastic units in this period. During this time he worked at the University of Chicago and spent his sabbatical year at the hydraulics lab at the University of Iowa. In Iowa he worked with wave tanks and at the University of California–Berkeley on beach studies. This research focused on all aspects of grain-size parameters, including some interesting conclusions about permeability as a function of size parameters of unconsolidated sands, a pioneering work that remains relevant in the early twenty-first century. This period produced the previously mentioned books, *Down to Earth* and the classical *Manual of Sedimentary Petrography.* Few books have had so major an impact on a discipline as the *Manual,* which shaped the way sediments were analyzed from then on.

A major shift toward stratigraphic analysis occurred, apparently from mid-1945 to mid-1946, when Krumbein was working at the Gulf Research and Development Company labs. His interests probably changed because of the difference in expectations between the petroleum industry and academia and to the interchange of ideas and contacts with fellow researchers such as Charles Ryniker, Roy Hazzard, and Sigmund Hammer, who directed Krumbein's attention respectively to limestone, subsurface, and geophysical studies. He devoted the next eight years mainly to the production of stratigraphic and sedimentological papers, the most important being the development of the facies tetrahedron, *Lithofacies Maps: An Atlas of the United States and Southern Canada* (1960) with Laurence Sloss and Ed Dapples. He studied the concepts of recent sedimentation, diagenesis and oil exploration, and shales and their environmental significance, as well as the importance of following a single stratigraphic horizon over a considerable area showing the regional stratigraphic and diagenetic patterns. This activity led to the classic book, *Stratigraphy and Sedimentation* (1951), coauthored with Sloss.

From 1953 on, Krumbein published a markedly greater number of papers on statistical methodology. The topics of sampling, populations, Latin squares, regression, discriminant analysis, trend surfaces (both polynomial and Fourier), time series, and Markov processes, culminating in the unifying theory of the general linear model were explored and written about in this period. Another classic book, *An Introduction to Statistical Models in Geology* (1965), or the "yellow bible," as it was affectionately called by the students, was written with Graybill.

The Teacher. Krumbein was a passionate teacher and a relentless researcher. His twenty-five graduate students include many prominent scientists. He taught his students—of which this author was one—the value of questioning established theories and paradigms, and he also encouraged them to have fun doing so, because for him "statistics was fun." He showed them many ways to gather information, process it, and most important of all, to interpret the meaning of their findings. During one of my daily tutoring sessions with him for my master's thesis, I once commented, regarding some Venezuelan cyclothems, that "this does not mean anything." He corrected me with words that still, after almost forty years, ring in my ears: "There does not fall a leaf from a tree that does not have a meaning."

The Geology Department of Northwestern University at the end of this golden era had immensely influential professors: Krumbein (geostatistics), Robert Garrels (geochemistry), Laurence Sloss (stratigraphy), E. H. Timothy Whitten (tectonics), Arthur Howland (petrography), and others. In the department's Common Room, students and professors gathered informally over coffee and donuts for open discussion of everyone's scientific problems. Many ideas were tossed around, polished, and examined for consistency and truth. One left these informal gatherings fortified not only with coffee—or occasionally beer—but also with a clearer understanding of one's research problems. During a particularly severe winter storm over Lake Michigan, Krumbein showed us the importance of the seventh wave, a wave that is stronger and higher than the preceding ones, a theory that I later successfully applied to Caribbean beaches.

As Krumbein's research assistant while a master's candidate in the late 1960s, I was encouraged by him to work constantly on my thesis. The work was interrupted only for several daily trips to the computing center to hand in boxes of computer cards and retrieve sheets of printout, which he immediately inspected. There followed a half hour of tutoring, which did not feel like pedagogy at all, because he was thinking aloud, examining the meaning of the transition probability matrices and stabilization vectors of Markov simulations. After that he would invariably say, "wouldn't it be nice to ... " sending me away with new ideas and work instructions to keep me busy for the rest of the evening. Early the next morning he would knock at the door of my little office and inquire about the results of the night's computer runs. This work regime kept me busy and "off the streets," as he liked to say. In hindsight, it was also an invaluable training for my professional life.

One of the most remarkable qualities of this extraordinary teacher was that he followed the same discipline in his revisions of his students' master's theses and PhD dissertations. Chapters handed in during the afternoon (written over a couple of days) were returned early the next morning with corrections. He always had a word of encouragement, often along with new ideas to be developed during the week. Even when working on Saturday, he never closed his office door. Of the many ideas that he tossed out during these sessions was one that would shape my future research—the observation by M. L. Banks, a colleague during his short stay at Gulf Research and Development Labs in Pittsburgh, who had observed and published the intimate relationship of coal beds in the Oficina Formation of Eastern Venezuela to the amount of petroleum that occurred in adjacent sandstone reservoirs.

The annual Geology Department field trip in the fall quarter revealed other facets of Krumbein's personality. During long, tedious travel along highways of the North American Midwest, he kept us awake by counting makes and colors of passing vehicles. He elaborated transition probability matrices and discussed the value of the equilibrium vector as the end product of a customer survey.

We applied this methodology in the evenings to makes of beers and colors of ladies' skirts in local bars. He introduced many graduate students to time series and Markov processes this way. More serious applications to batholith crystallization patterns and thin-section analysis followed, producing lively discussions at outcrops. Krumbein liked to observe geologists on the outcrops, never getting too involved in the nitty-gritty of geology or paleontology, but always attentive to any theory that might arise and suggesting ways to prove or disprove the paradigm.

Krumbein considered that in geology, "most statistical methods are powerful enough to handle data distributed with a peak more or less in the middle." He was adamant about correctly identifying populations and sampling strategies: "[Sampling] is inherent to the model—you must have a conceptual picture in your mind—and the question you want to answer." He pointed out that "in many geological studies there is a large amount of biased observations—contacts, etc. and virtually nothing in between," that "statistical inferences are made from samples to sample populations," something frequently forgotten because "we take samples and literally beat them to death," without knowing "how much is my own variation—which I contribute to the study." In those days of large computer centers, when nobody envisioned the personal computer, he used to say that "the day will come when there is a computer at every drugstore," but if you put "garbage in—you get garbage out—no matter how much you beat it."

Krumbein philosophized in lectures that "a well trained geologist can get a pretty good mean value, but has larger variance than randomized samples," something that traditionally educated geologists absorbed with difficulty. He asked students to look "at your leisure" at the important papers on statistics in geology, and in the next class would begin with the statement that "the important attribute of the variance is that it be additive, that makes it much simpler to pay with." This latter statement stayed in my mind long after his death and was instrumental to produce Krumbein's posthumous paper "CORSURF—A Covariance-Matrix Trend-Analysis FORTRAN IV Computer Program" written with Daniel Merriam in 1995.

Awards and Honors. Krumbein was a charter member of the International Association for Mathematical Geology and was its first president. The association named its premier award the William Christian Krumbein Medal. He was a Fellow of Geological Society of America, the American Association for the Advancement of Science, and the American Statistical Association. He belonged to the Society for Sedimentary Geology (and was its president in 1950), the Society of Economic Geologists, the American Association of Petroleum Geologists, the American Geophysical Union, and the Illinois Academy of Science. He received a Guggenheim Fellowship, was a Fulbright lecturer, and was a President's Fellow at Northwestern University. In 1978 Krumbein received the William H. Twenhofel Medal from the Society of Sedimentary Geology.

In 1975 Krumbein's colleagues honored him with a Festschrift, *William C. Krumbein, The Making of a Methodologist,* in recognition of "his stimulating teaching and guidance ... and continuing leadership and research." It was published by the Geological Society of America. Krumbein died of a heart attack in 1979. He was survived by his second wife, Marjorie.

BIBLIOGRAPHY

For a complete bibliography of Krumbein's works to 1975, see A. L. Howland, "William C. Krumbein: The Making of a Methodologist," in Quantitative Studies in the Geological Sciences, a Memoir in Honor of William C. Krumbein, *edited by E. H. Timothy Whitten (Boulder, CO: Geological Society of America, 1975).*

WORKS BY KRUMBEIN

"A History of the Principles and Methods of Mechanical Analysis." *Journal of Sedimentary Petrology* 2 (1932): 89–124.

With Carey Croneis. *Down to Earth: An Introduction to Geology.* Chicago: University of Chicago Press, 1936.

With Francis J. Pettijohn. *Manual of Sedimentary Petrography.* NewYork: D. Appleton-Century, 1938.

With Laurence L. Sloss and Edward C. Dapples. *Integrated Facies Analysis.* Boulder, CO: Geological Society of America, 1949.

———. "Sedimentary Tectonics and Sedimentary Environments." *American Association of Petroleum Geologists Bulletin* 33 (1949): 1859–1891.

With Robert M. Garrels. "Origin and Classification of Chemical Sediments in Terms of pH and Oxidation-Reduction Potentials." *Journal of Geology* 60 (1952): 1–32.

With R. L. Miller. "Design of Experiments for Statistical Analysis of Geological Data." *Journal of Geology* 61 (1953): 510–532.

"Regional and Local Components in Facies Maps." *American Association of Petroleum Geologists Bulletin* 40 (1956): 2163–2194.

"Trend-Surface Analysis of Contour-Type Maps with Irregular Control-Point Spacing." *Journal of Geophysical Research* 64 (1959): 823–834.

"The 'Geological Population' as a Framework for Analysing Numerical Data in Geology." *Liverpool and Manchester Geological Journal* 2 (1960): 341–368.

With Edward C. Dapples and Laurence L. Sloss. *Lithofacies Maps: An Atlas of the United States and Southern Canada.* New York: John Wiley and Sons, 1960.

"Open and Closed Number Systems in Stratigraphic Mapping." *American Association of Petroleum Geologists Bulletin* 46 (1962): 2229–2245.

With Laurence L. Sloss. *Stratigraphy and Sedimentation.* 2nd ed. San Francisco, CA: W.H. Freeman, 1963.

With Franklin A. Graybill. *An Introduction to Statistical Models in Geology.* New York: McGraw-Hill, 1965.

"The Cyclothem as a Response to Sedimentary Environment and Tectonism." In *Symposium on Cyclic Sedimentation,* edited by Daniel F. Merriam. Lawrence: Kansas Geological Survey Bulletin, 1966.

With Wolfgang Scherer. *Structuring Observational Data for Markov and Semi-Markov Models in Geology.* Washington, DC: U.S. Office of Naval Research, Technical Report 15, Contract N00014.67-A0356-0018 (formerly Nonr-1228(36); NR 389-150), 1968.

"Markov Models in the Earth Sciences." In *Concepts in Geostatistics,* edited by Richard B. McCammon. New York: Springer-Verlag, 1975.

With Daniel F. Merriam and Wolfgang Scherer. "CORSURF: A Covariance-Matrix Trend-Analysis FORTRAN IV Computer Program." *Computers and Geosciences* 21 (1995): 1065–1089.

OTHER SOURCES

Howland, Arthur L. "William C. Krumbein: The Making of a Methodologist." In *Quantitative Studies in the Geological Sciences, a Memoir in Honor of William C. Krumbein,* edited by E. H. Timothy Whitten. Boulder, CO: Geological Society of America, 1975.

Merriam, Daniel F. "William C. Krumbein (1902–1979)." In *Encyclopedia of Sediments and Sedimentary Rocks,* edited by Gerard V. Middleton. Boston: Kluwer Academic, 2003.

Wolfgang Scherer

KŪHĪ, ABŪ SAHL WAYJAN IBN RUSTAM AL-

SEE **Qūhī, Abū Sahl Wayjan Ibn Rustam al-**.

KUHN, RICHARD

(*b.* Vienna-Dobling, Austria, 3 December 1900, *d.* Heidelberg, Germany, 31 July 1967), *chemistry.* For the original article on Kuhn see *DSB,* vol. 7.

More than thirty years after the death of Nobel Prize laureate Richard Kuhn, one of the most distinguished and successful biochemists in twentieth century, his role as a scientific administrator in the Nazi regime was critically debated for the first time. During his lifetime Kuhn's personal involvement in research on chemical Weapons of Mass Destruction and his collaboration in the persecution of Jewish scientists and his cooperation with the secret state police (Gestapo) was not discussed in public. It was not before the end of the Cold War that historical research was able to access declassified archival documents in the United States, Great Britain, and Germany, shedding light on Kuhn's role as an influential scientist in Nazi Germany.

At the time of the Nazi takeover in 1933, Kuhn headed the Institute for Chemistry, one of four institutes comprising the Kaiser Wilhelm Institute (KWI) for Medical Research in Heidelberg. Due to the acting director's illness, Kuhn became provisional head of all four institutes in March 1934. His appointment as overall director of the KWI was affirmed in January 1938. After the installation of the anti-Semitic "Law for the Restoration of Professional Civil Service" in April 1933, Kuhn did nothing to protect Jewish assistants at the KWI, unlike his colleague, the Nobel Prize laureate Otto Meyerhof, director of the Institute of Physiology. Even though Meyerhof himself faced persecution as a Jew, Kuhn, as provisional director, in a letter to the general administration of the Kaiser Wilhelm Society (KWS), demanded that Meyerhof should be instructed to dismiss three Jewish researchers working at the institute. Kuhn's demand was rejected by the general administration of the KWS, as it was not yet illegal to accept Jews as PhD candidates (Deichmann, 2000; Schmaltz, 2005).

Kuhn never joined the Nazi Party. However, party officials and the Gestapo described him as politically reliable and loyal to the regime. He rose to be one of the most influential scientific organizers of biochemistry in the Nazi era. Before World War II he held important positions in scientific organizations, expert councils, and governmental research organizations. He was a member of the executive boards of the Emil Fischer Society, which financed the KWI for Chemistry, and the Adolf Baeyer Society for the Promotion of Chemical Publications, and he served on the scientific committee of the Society of German Natural Scientists and Physicians. He was vice-chairman of the Justus Liebig Society, a foundation supporting German post-doctoral students in chemistry; chairman of the working group for organic chemistry of the Verein Deutscher Chemiker, and, in 1936, vice president of the Union Internationale du Chimie. After his 1937 appointment to the executive board of the German Chemical Society (Deutsche Chemische Gesellschaft, DChG), he participated in the society's nazification by actively adapting the society's statutes to the Nazi

Richard Kuhn. **THE LIBRARY OF CONGRESS.**

Führerprinzip (principle of leadership), abolishing the society members' right to elect their own executive board and president.

In March 1938 Kuhn was appointed president of the DChG, a post he held until 1945. In a speech as president of the DChG in 1942, Kuhn praised the "fighting front" of the nations that had joined the Anti-Komintern powers, hailing Mussolini, the Japanese emperor, and Hitler. One month after the German attack on Poland, Kuhn became head of the branch (*Fachspartenleiter*) for organic chemistry of the Reich Research Council (*Reichsforschungsrat*). Though sources are fragmentary, it is certain that Kuhn supported at least eight research projects on chemical warfare agents and gas defense as Fachspartenleiter in 1943–1944 alone. In 1942 Kuhn became senator of the Reichsfachgruppe Chemie of the Nationalsozialistischer Bund Deutscher Technik, a federation associated with the Nazi Party. In September 1944

Kuhn was appointed to the Wehrforschungs-Gemeinschaft, an agency coordinating war research efforts with the Armament Ministry led by Albert Speer and Four Year Plan agencies.

In 1938, during the intense preparation for the war, Kuhn's KWI for Medical Research cooperated with the Army Ordnance Office (*Heereswaffenamt*), doing chemical warfare research. His assistant, Christoph Grundmann, conducted an investigation on vitamin B_6 (adermin, pyridoxine). Grundmann was exploring the effectiveness of vitamin B_6 in the treatment of lesions caused by mustard gas. In March 1939 Grundmann tried to sell a patent on the synthesis of vitamin B_6 to the Swiss chemical company Hoffmann La Roche. In Kuhn's opinion, these efforts endangered his cooperation with the German military and chemical industry (I. G. Farbenindustrie and Merck). Kuhn accused his assistant of treason, denouncing him to the Gestapo. Grundmann was dismissed without notice, then was arrested and later tried in a secret trial, which ended in a dismissal. The charge of treason was dropped, as there had been a prior application for a similar patent in the United States, before Grundmann's imprisonment. Further studies showed that vitamin B_6 had no therapeutic effect on wounds caused by mustard gas. The vitamin research by Kuhn and collaborators, mentioned in the article on Kuhn by Dean Burk in the *Dictionary of Scientific Biography*, was therefore embedded in a military context. This is also true for the synthesis of "numerous analogues and reversibly competitive inhibitors ("antivitamins") mentioned by Burk.

Anti-vitamin research by Kuhn and some of his pupils in Heidelberg during World War II played an important role for investigations of a new group of nerve gases, discovered earlier in 1936 (Tabun) and in 1938 (Sarin) in the context of pesticide research by I.G. Farbenindustrie. After lengthy negotiations, collaboration between Kuhn and the Army Ordnance Office was finally transformed from assigned research contracts to the formal institutionalized establishment of a secret laboratory for chemical warfare research at the KWI for Medical Research. This laboratory, led by Kuhn, was set up in the period of preparations for the invasion of the Soviet Union in January 1941 as an outpost of the gas defense department of the Army Ordnance Office. One third of the academic staff at the institute for chemistry was taken over. Among those were Helmut Beinert, Otto Dann, Konrad Henkel, Dietrich Jerchel, Günter Quadbeck, and Friedrich Weygand. This new department was established in the rooms at the Institute of Physiology, which were available as a result of the displacement of Otto Meyerhof, who had fled Germany in September 1938 due to increased anti-Semitic harassment. Pharmacological studies on the treatment of mustard gas lesions were conducted there by Kuhn's assistant Günter Quadbeck, who detected a certain therapeutic effect of a cyclic carbonhydrogen on such wounds. But the main emphasis of the scientific research done at the chemical warfare department in Heidelberg consisted of investigating the effects of the nerve gases Tabun and Sarin. The experiments at the KWI also included the search for possible antidotes and other poisonous nerve gases.

Internationally, the KWI for Medical Research was one of the leading institutes in the field of vitamin research. Starting out with tested methods from vitamin research, the team led by Kuhn succeeded in determining the mode of action of nerve gases in 1943. They identified the strong inhibition of acetyl cholinesterase—which is vitally important for the stimulus-conducting function of the neurotransmitter acetylcholine in the brain—as the most significant effect. The hypothesis put forward by the Military Medical Academy scientists, stating that the effects of nerve gas were caused by additional enzyme systems, was therefore refuted. On the basis of the inhibition of acetyl cholinesterase, the team in Heidelberg developed a specific testing technique for the degree of toxicity of nerve gases. In the spring of 1944 Kuhn and his PhD candidate Konrad Henkel synthesized Soman, which proved to be stronger than Tabun and Sarin. Soman was not produced on an industrial scale until the end of the war. It is still one of the most effective chemical weapons to date.

In the spring of 1943 Kuhn asked the secretary-general of the KWS, Ernst Telschow, to support his search for the brains of "young and healthy men," presumably for nerve gas research. The sources indicate that these brains were most likely taken from execution victims. If and to what extent cerebrums were indeed delivered from morgues in Heidelberg and Stuttgart to the KWI for Medical Research has not yet been established with certainty. There is also no proof that KWI scientists participated directly in human experiments conducted in concentration camps. But Kuhn, as head of the special branch, together with Karl Brandt, who was the commissioner for public health and sanitation and largely responsible for the so-called Euthanasia program, authorized the use of funds from the Reich Research Council for the phosgene experiments on prisoners, carried out by the physician Otto Bickenbach, in the gas chamber of the concentration camp Natzweiler-Struhof in 1944. Bickenbach was tried as a war criminal at a French military tribunal in 1947. His experiments had caused at least four Roma or Sinti (Gypsies) to suffocate painfully. Kuhn justified the experiments as "scientifically outstandingly profound." Kuhn ignored the context of genocide and insisted that this research had pursued "a high aim for the whole of humanity" and were "beneficial for many" (Kuhn to Eber, 5 August 1947, Archives of the Max Planck Society, III. Abt., Rep. 25, no. 54).

After the liberation, Kuhn did not fully inform the experts of the Allied military intelligence about the chemical warfare research. His institute suffered only few importunities—but not, as Burk suggested, because of Kuhn's plant color experiments for a U.S. colonel's wife, but rather because of the lack of information about his role during the Nazi era, and because after 1945, his chemical warfare research was of great interest to the Chemical Warfare Service. In July 1945 Kuhn continued chemical warfare research under Allied authority, despite the strict prohibition of military research in Germany. When U.S. intelligence and military services set up a recruitment program for German and Austrian scientists and technicians under the codename Operation Paperclip, Kuhn denied two offers (1947 and 1949) to continue chemical warfare research in the United States.

In 2005, as a result of historical research, the Society of German Chemists (Gesellschaft Deutscher Chemiker, GDCh) declared their intention to no longer award the Richard Kuhn Medal: "The board of the GDCh intends to discontinue awarding the Medal named after the organic chemist, Nobel Prize laureate of the year 1938 and President of the GDCh in 1964–65, Richard Kuhn. The board thereby draws the consequences out of research on Richard Kuhn's behaviour during National Socialism. Even though the question of whether Kuhn was a convinced National Socialist or just a career-oriented camp follower is not fully answered, he undisputably supported the Nazi-regime in administrative and organizational ways, especially by his scientific work. Despite his scientific achievements, Kuhn is not suitable to serve as a role model, and eponym for an important award, mainly due to his unreflected research on poison gas, but also due to his conduct towards Jewish colleagues" (*Nachrichten aus der Chemie* 54, May 2006, p. 514).

BIBLIOGRAPHY

A bibliography of Kuhn's publications is provided by Selchow Christian, "Richard Kuhn (3 Dez. 1900 Wien–31 Juli 1967 Heidelberg)." Archiv zur Geschichte der Naturwissenschaften *10 (1984): 473–497.*

WORK BY KUHN

Kuhn, Richard. "Ansprache von Richard Kuhn auf der Besonderen Sitzung am 5. Dezember 1942." *Berichte der Deutschen Chemischen Gesellschaft* 75 (1942): 147–202.

OTHER SOURCES

Adamson, D. W., D. C. Evans, C. W. Scott, et al. *KWI für medizinische Forschung, Heidelberg—Target No. 8/58 & 24717 (2.5.1945).* CIOS Evaluation Report 10, 1945.

Baader, Gerhard, Susan E. Lederer, Morris Low, et al. "Pathways of Human Experimentation, 1933–45: Germany, Japan, and the United States." In *Politics and Science in Wartime: Comparative International Perspectives on the Kaiser Wilhelm Institute,* edited by Carola Sachse and Mark Walker. Chicago: University of Chicago Press, 2005.

Deichmann, Ute. "Kriegsbezogene biologische, biochemische und chemische Forschung an den Kaiser Wilhelm-Instituten für Züchtungsforschung, für Physikalische Chemie und Elektrochemie und für Medizinische Forschung." *Geschichte der Kaiser-Wilhelm-Gesellschaft im Nationalsozialismus: Bestandsaufnahme und Perspektiven der Forschung,* edited by Doris Kaufmann, vol. 2. Göttingen, Germany: Wallstein-Verlag, 2000.

———. *Flüchten, mitmachen und vergessen: Chemiker und Biochemiker in der NS-Zeit.* Weinheim, Germany, and New York: Willey-VCH, 2001.

Ebbinghaus, Angelika, and Karl Heinz Roth. "Vernichtungsforschung: Der Nobelpreisträger Richard Kuhn, die Kaiser Wilhelm-Gesellschaft und die Entwicklung von Nervenkampfstoffen während des 'Dritten Reichs.'" *1999. Zeitschrift für Sozialgeschichte des 20. und 21. Jahrhunderts* 17 (2002): 15–50.

Edson, E. F., D. C. Evans, R. E. F. Edelstein et al. *Interrogation of Certain German Personalities Connected with Chemical Warfare.* BIOS Final Report 542. London: H. M. Stationery Office, 1946.

Ruske, Walter. *100 Jahre Deutsche Chemische Gesellschaft.* Weinheim, Germany: Verlag Chemie, 1967.

Schmaltz, Florian. *Kampfstoff-Forschung im Nationalsozialismus: Zur Kooperation von Kaiser-Wilhelm-Instituten, Militär und Industrie.* Göttingen, Germany: Wallstein Verlag, 2005.

———. "Neurosciences and Research on Chemical Weapons of Mass Destruction in Nazi Germany." *Journal of the History of the Neurosciences* 15 (2006):186–209.

———. "Otto Bickenbach's Human Experiments with Chemical Warfare Agents at the Concentration Camp Natzweiler in the Context of the SS-Ahnenerbe and the Reichsforschungsrat." In *Man, Medicine and the State: The Human Body as an Object of Government Sponsored Research in the 20th Century,* edited by Wolfgang U. Eckart. Stuttgart, Germany: Steiner, 2006.

Florian Schmaltz

KUHN, THOMAS SAMUEL (*b.* Cincinnati, Ohio, 18 July 1922; *d.* Cambridge, Massachusetts, 17 June 1996), *philosophy of science, history of science, concept of paradigm.*

A physicist turned historian of science for philosophical purposes, Kuhn was one of the most influential philosophers of science in the twentieth century. In his famous book *The Structure of Scientific Revolutions,* first published in 1962, Kuhn helped destroy the popular image of science according to which science steadily and incrementally progresses toward a true and complete picture of reality. Relying on historical case studies, Kuhn argued that, ruptured by scientific revolutions, scientific

Thomas Kuhn. Multiple exposure portrait of Thomas Kuhn. BILL PIERCE/TIME LIFE PICTURES/GETTY IMAGES.

development was discontinuous and noncumulative and that scientific activity before and after a revolution was in some ways incommensurable, lacking a common measure. In this way Kuhn not only formed a startling picture of science, but also initiated a new way of doing philosophy of science informed by the history of science.

Life and Career. Thomas Kuhn was the son of Samuel L. Kuhn, who was trained as a hydraulic engineer at Harvard University and the Massachusetts Institute of Technology (MIT), and Annette Stroock Kuhn. Both parents were nonpracticing Jews. Kuhn attended several schools in New York, Pennsylvania, and Connecticut. Among them, Hessian Hills in Croton-on-Hudson, New York, a progressive school that encouraged independent thinking, made a particularly strong impression on him. He then attended Harvard University, graduating summa cum laude with a degree in physics in 1943. Despite the fact that his interest lay in theoretical physics, most of his coursework was in electronics, due to the orientation of his department. His professors included George Birkhoff, Percy W. Bridgman, Leon Chaffee, and Ronald W. P. King. He also took several elective courses in social sciences and humanities, including a philosophy course in which Immanuel Kant struck him as a revelation. He did not enjoy the history of science course that he attended, which was taught by the famous historian of science George Sarton.

After graduation, he worked on radar for the Radio Research Laboratory at Harvard and later for the U.S. Office of Scientific Research and Development in Europe. He returned to Harvard at the end of the war, obtained his master's degree in physics in 1946, and worked toward a PhD degree in the same department. He also took a few philosophy courses in order to explore other possibilities than physics. It was about this time that the legendary president of Harvard University, the chemist and founder of "Harvard Case Studies in Experimental Science" James Conant, asked Kuhn to assist his course on science, designed for undergraduates in humanities as part of the General Education in Science Curriculum. This event changed Kuhn's life. His encounter with classical texts, especially Aristotle's *Physics,* was a crucial experience for him. He realized that it was a great mistake to read and judge an ancient scientific text from the perspective of current science and that one could not really understand it unless one got inside the mind of its author and saw the world through his eyes, through the conceptual framework he employed to describe phenomena. This understanding shaped his later historical and philosophical studies.

In 1948 Kuhn became a junior member of the Harvard Society of Fellows upon Conant's recommendation.

A year later, he completed his PhD in physics under the supervision of John H. van Vleck, who won the Nobel Prize in 1977. Kuhn became an assistant professor of general education and the history of science in 1952 and taught at Harvard until 1956. During this period he trained himself as a historian of science, and Alexandre Koyré's works, especially his *Galilean Studies*, had a deep impact on him.

Between 1948 and 1956, Kuhn published three articles, one with van Vleck on computing cohesive energies of metals, derived from his PhD dissertation, and a number of historical works on Isaac Newton, Robert Boyle, and Sadi Carnot's cycle. He also wrote his first book, *The Copernican Revolution*, which was published in 1957. Nevertheless, Kuhn was denied tenure because the review committee thought that the book was too popular and not sufficiently scholarly.

Feeling disappointed, Kuhn accepted a joint position as an assistant professor in the history and philosophy departments at the University of California, Berkeley. Soon after, he published his masterpiece, *The Structure of Scientific Revolutions*. It was also here that he met Paul Feyerabend, who introduced a version of the thesis of incommensurability at the same time Kuhn did. But the interaction was not fruitful. The person who influenced him most at Berkeley was Stanley Cavell. Cavell introduced him to the philosophy of Ludwig Wittgenstein, whose view of meaning as use and idea of family resemblance had a lasting influence on Kuhn. He also heard Michael Polányi's lectures on tacit knowledge, a notion that also found its way into his influential book.

Between 1961 and 1964 he headed a project known as the "Sources for History of Quantum Physics," which contained interviews with, and manuscript materials of, all the major scientists who contributed to the development of quantum physics. These materials are now part of the Archive for History of Quantum Physics.

Kuhn was offered a full professorship at Berkeley in history, not in philosophy. Although disappointed, he accepted the offer. Not long after, however, he left Berkeley for the position of M. Taylor Pyne Professor of Philosophy and History of Science at Princeton University. He taught at Princeton from 1964 to 1979 and then, because of his divorce, he left Princeton and joined the philosophy department at MIT. In 1982 he was appointed to the Laurence S. Rockefeller Professorship in Philosophy, a position he held until 1991 when he retired. He became professor emeritus at MIT from then on until his death. He was survived by his second wife Jehane, his ex-wife Kathryn Muhs, and their three children.

Thomas Kuhn received the Howard T. Behrman Award for distinguished achievements in the humanities (1977), the History of Science Society's George Sarton Medal (1982), and the Society for Social Studies of Science's John Desmond Bernal Award (1983). He was a Guggenheim Fellow during 1954 to 1955, a member of the Institute for Advanced Study in Princeton (1972–1979), a member of the National Academy of Sciences, and a corresponding fellow of the British Academy. He also held honorary degrees from Columbia, Chicago, and Notre Dame universities in the United States, the University of Padua in Italy, and the University of Athens in Greece. He was the only person to have served as presidents of both the History of Science Society (1968–1970) and the Philosophy of Science Association (1988–1990).

The Structure of Scientific Revolutions. *The Structure of Scientific Revolutions* (*Structure* for short) opens with the sentence, "History, if viewed as a repository for more than anecdote or chronology, could produce a decisive transformation in the image of science by which we are now possessed" (1970, p. 1). According to that image, science progresses toward truth in a linear fashion, each new theory incorporating the old one as a special case. Scientific progress is due to the scientific method, whereby theories are tested against observations and experiments; those that fail are disconfirmed or get eliminated and those that pass the tests are considered to be confirmed, or at least not yet falsified.

This image was very popular among scientists, and in the philosophical world it was represented in various forms by logical positivists such as Rudolf Carnap, who emphasized confirmability and by Karl Popper, who emphasized falsifiability. Most logical positivists, though emphatically not Popper, also believed that observation provided neutral and secure grounds for the appraisal of scientific theories. It was generally agreed that scientific rationality and objectivity was a matter of compliance with the rules of scientific method, leaving little room for individual choices. Although *Structure* contained only one explicit reference to Popper and none to the logical positivists, clearly it targeted them, and together with the works of Norwood Hanson, Paul Feyerabend, and Stephen Toulmin, it destroyed the existing conception of science and scientific change.

The main thesis of Kuhn's book was that development in mature sciences typically goes through two consecutive phases: normal and revolutionary. Normal science is a paradigm-governed activity of puzzle solving. Based on settled consensus of the scientific community, normal scientific activity has little room for novelty that transcends the bounds of the paradigm. A paradigm provides a concrete model (called an "exemplar") for solving problems it has set out. Kuhn called these problems "puzzles" because the paradigm assures the members of the scientific community that with sufficient skill and ingenuity they can be solved within its resources. Thus, in case of failure to solve a puzzle it is the individual scientist, not the paradigm, that is to be blamed. When, however, puzzles resist persistent attempts at solution, they turn into anomalies; and anomalies lead to a crisis when they accumulate. Crisis is marked by a loss of confidence in the paradigm and a search for an alternative one. Rival accounts proliferate, the most fundamental commitments about nature get questioned, and in the end, the scientific community embraces the most promising alternative as the new paradigm. A scientific revolution has occurred. Consequently, a new period of normal science begins, and a similar cycle of normal science–crisis–revolution follows.

Whereas normal science is cumulative, revolutionary science is not. The new paradigm and the activity governed by it are in many ways incompatible with the old one. Kuhn expressed this point in terms of the thesis of incommensurability, which has several aspects. Both problems and the way they are solved change: there is a conceptual change, whereby certain terms acquire new meanings; because every observation is theory-laden, there is a perceptual change, a Gestalt switch, which causes the scientists to see the world differently; and, finally, there is even a sense in which the world itself changes after a revolution. For instance, according to Kuhn, the Aristotelian world contains swinging stones, but no pendulums. Accordingly, whereas the Aristotelian scientist sees constrained motion in a swinging stone, the Galilean-Newtonian scientist (who may as well be a transformed Aristotelian) literally sees a pendulum. In short, the new paradigm is incommensurable with the old one.

Scientists working under rival paradigms often talk past each other and experience a breakdown in communication. The switch from one paradigm to another is very much like a conversion experience rather than a rational choice dictated mechanically by scientific methodology. Furthermore, much that has been accepted as true is discarded, making it impossible to say that the new paradigm brings us closer to truth.

Not surprisingly, *Structure* sent shock waves through the philosophical community. Kuhn was accused of robbing science of its rationality and objectivity, turning it into a kind of mob psychology; he was charged with relativism, subjectivism, and outright idealism. Normal science was said to be dangerously dogmatic. The notion of "paradigm" was held to be too vague, lacking a definite meaning.

In the "Postscript" to *Structure,* which was added to the second edition in 1970, and in several subsequent articles, most notably "Objectivity, Value Judgment, and Theory Choice," collected in *The Essential Tension,* published in 1977, Kuhn defended himself against these charges, clarifying some of his earlier statements and retracting others. In this context the first thing he did was to clarify what he meant by "paradigm," for which he now preferred the term "disciplinary matrix." A disciplinary matrix consisted of four elements: metaphysical commitments; methodological commitments; criteria such as quantitative accuracy, broad scope, simplicity, consistency, and fruitfulness (which Kuhn called "values" since they are desired characteristics of scientific theories); and exemplars.

The most important of these is exemplars, that is, concrete problem solutions that serve as models. Exemplars are always given in use; they guide research even in the absence of rules; and the study of exemplars enables scientists to acquire an ability to see family resemblances among seemingly unrelated problems. Much knowledge that is acquired in this way is tacit, inexpressible in propositions. Normal science is dogmatic to some degree, since it does not allow the questioning of the paradigm itself, but this sort of dogmatism is functional: it allows the scientists to further articulate their paradigmatic theory and pay undivided attention to the existing puzzles and anomalies, the recognition of which is a precondition for the emergence of novel theories and subsequently a revolution. In this way Kuhn dispelled the charges of vagueness and dogmatism.

He also took pains to argue that incommensurability, the target of the greatest outrage, did not necessarily imply incomparability. Two paradigms, he said, often share enough common points to make it possible to compare them. For example, the astronomical data regarding the position of Mercury, Mars, and Venus were shared by both the Aristotelian-Ptolemaic and Copernican paradigms, and they both appealed to similar criteria ("values"). These commonalities provided sufficient grounds for paradigm comparison.

Kuhn pointed out, however, that two scientists working under rival paradigms may share the same criteria but apply them differently to concrete cases. When they are confronted with a new puzzle, they may disagree, for instance, about whether paradigm A or B provides a simpler solution, or they may attach different weights to the shared criteria. This is a perfectly rational disagreement, and the only way to resolve it is through the techniques of persuasion. It is for this reason that paradigm choice often involves subjective, though not arbitrary, decisions.

Rather than denying rationality, Kuhn developed a new conception of it. For him rationality is not just a matter of compliance with methodological rules. This is because the knowledge of how to apply a paradigm to a new puzzle is mostly learned not by being taught abstract rules but by being exposed to concrete exemplars. Yet this is a kind of tacit knowledge that is almost impossible to detach from the cases from which it was acquired. Thus, both paradigm choice and paradigm application often involve judgment and deliberation, a process akin to Aristotle's *phronesis;* each scientist must use her lifelong experience, her "practical wisdom," to make the best possible decision. In short, Kuhn urged a shift from a conception of rationality based on the mechanical application of determinate rules to a model of rationality that emphasizes the role of exemplars, deliberation, and judgment.

Kuhn also argued that science does progress, but not toward truth in the sense of correspondence to an objective reality, because later theories are incommensurable with the earlier ones. Scientific progress for Kuhn simply meant increasing puzzle-solving ability: later theories are better than earlier ones in discovering and solving more and more puzzles. Appealing to the existence of shared criteria for paradigm comparison and to an instrumental idea of scientific progress, Kuhn tried to defend himself against the charge of relativism.

The Linguistic Turn. In the 1980s and 1990s Kuhn wrote a number of articles, reformulating most of his philosophical views in terms of language, more specifically in terms of what he called taxonomic lexicons. These articles were published posthumously in the collection *The Road since Structure* (2000) and can be summarized as follows.

First of all, having abandoned the terms *disciplinary matrix* as well as the much-used and -abused term *paradigm* in favor of *theory,* Kuhn now underlined the point that every scientific theory has its own distinctive structured taxonomic lexicon: a taxonomically ordered network of kind-terms, some of which are antecedently available relative to the theory in question.

Second, lexicons are prerequisite to the formulation of scientific problems and their solutions, and descriptions of nature and its regularities. Hence, revolutions can be characterized as significant changes in the lexicons of scientific theories: both the criteria relevant to categorization and the way in which given objects and situations are distributed among preexisting categories are altered. Since different lexicons permit different descriptions and generalizations, revolutionary scientific development is necessarily discontinuous.

Third, the distinction between normal and revolutionary science now becomes the distinction between activities that require changes in the scientific lexicon and those that do not. Revolutions involve, among other things, novel discoveries that cannot be described within the existing lexical network, so scientists feel forced to adopt a new one. The earlier mentalistic description (i.e., Gestalt switches and conversions) disappears from Kuhn's writings.

Finally, incommensurability is reduced to a sort of untranslatability, localized to one or another area in which two lexical structures differ. What gives rise to incommensurability is the difference between lexical structures. Because rival lexical structures differ radically, there are sentences of one theory that cannot be translated into the lexicon of the other theory without loss of meaning. All other aspects of incommensurability that were present in *Structure* drop out.

Kuhn also gave a Kantian twist to these ideas. He argued that structured lexicons are constitutive of phenomenal worlds and possible experiences of them. In Kuhn's view a taxonomic lexicon functions very much like the Kantian categories of the mind. This in turn led him not only to embrace a distinction between noumena and phenomena, but also to claim that fundamental laws, such as Newton's second law, are synthetic a priori. The sense of a priori Kuhn had in mind is not "true for all times," but something like "constitutive of objects of experience." This is a historical or relativized a priori, like Hans Reichenbach's. Taxonomic lexicons do vary historically, unlike Kantian categories. Even the second law is revisable despite the fact that it is recalcitrant to refutation by isolated experiments. Accordingly, Kuhn's final position can be characterized as an evolutionary linguistic Kantianism.

Using first principles, as it were, regarding the structure of taxonomic lexicons of scientific theories, and having a developmental perspective not simply derivative from the historical case studies, Kuhn's linguistic turn enabled him to refine, add to, and unify his earlier views about scientific revolutions, incommensurability, and exemplars. He was also able to explain more clearly why incommensurability does not imply incomparability and why communication breakdown across a revolution is always partial. This is because incommensurability is a local, not global, phenomenon pertaining to a small subset of the scientific lexicon, and whatever communication breakdown exists can be overcome by becoming bilingual.

Furthermore, he was finally able to articulate the sense in which the scientist's world itself changes after a revolution. That sense is Kantian. Whereas the noumenal world is fixed, the phenomenal world constituted by a lexicon is not. Different lexicons "carve up," as it were, different phenomenal worlds from the unique noumenal world, so Kuhn could now respond to the charge of idealism by pointing out that the noumenal world does exist independently of human minds, though it remains unknowable.

History of Science. In the background of *The Structure of Scientific Revolutions* is *The Copernican Revolution,* Kuhn's first major contribution to the historiography of science. That book grew out of Kuhn's science course for the

Kuhn

humanities at Harvard in the 1950s and provided one of the key historical case studies that later enabled him to articulate his views about the development of science. *The Copernican Revolution* achieved several things at once. It showed above all that Nicolaus Copernicus was both a revolutionary and a conservative at the same time. Contrary to popular belief, the Copernican heliocentric system, with its rotating spheres, perfectly circular orbits, epicycles, and eccentricities, was in many ways a continuation of the Aristotelian-Ptolemaic tradition of astronomy. But this conservativeness also meant that the Aristotelian-Ptolemaic tradition was a respectable scientific enterprise, having its own conceptual framework, problems, and ways of solving them. When looked at retrospectively, however, the Copernican system did pave the way, albeit unintentionally, for a revolution in science through the works of Johannes Kepler, Galileo Galilei, and Newton.

Kuhn argued forcefully in his book that aesthetic considerations played an important role in Copernicus's placing the Sun at the center and thus turning Earth into an ordinary planet; the Ptolemaic system looked increasingly complicated, indeed "monstrous," in the eyes of Copernicus. Although his model did not automatically yield simpler calculations, it provided qualitatively more coherent interpretations of certain phenomena, notably, the retrograde motion of planets. In addition to these, Kuhn drew attention to social factors behind the Copernican Revolution as well, such as the need for calendar reform, improved maps, and navigational techniques. Kuhn also pointed out the larger ramifications of the heliocentric system—in particular, how it changed the conception human beings had of their unique place in the universe and what sense that conception had for them.

After *The Copernican Revolution,* Kuhn wrote a number of influential historical articles, including one on energy conservation as an example of simultaneous discovery, one on the difference between mathematical and experimental (dubbed as "Baconian") traditions in the development of physical sciences, and another, with John Heilbron, on the genesis of the Bohr atom. Most of these are conveniently collected in his book *The Essential Tension.*

Kuhn's final major contribution to the historiography of science was his controversial book *Black-Body Theory and the Quantum Discontinuity, 1894–1912,* published in 1978. It constituted a break with a longstanding historiographical tradition and undermined the consensus between physicists and historians that quantum physics originated in the works of Max Planck in 1900. According to the traditional interpretation, Planck was forced to introduce the idea of energy quanta, thus breaking with classical physics. More sophisticated versions of this interpretation, which recognized that Planck himself did not understand the exact meaning of the energy quanta, were also defended in various forms by historians of science. In his book Kuhn argued that Planck did not abandon the framework of classical physics until after Hendrik Lorentz, Paul Ehrenfest, and Albert Einstein in 1905 attempted to understand his theory of blackbody radiation.

Of the two historical books Kuhn wrote, the earlier one became a small classic of its own. Historians criticized the second one for exaggerating its case and ignoring certain developmental aspects of Planck's works, and philosophers were surprised that it did not contain any references to "paradigms," "normal science," "incommensurability," and the like. Kuhn defended himself in the second edition, arguing that many of the themes of *Structure* were there, though implicitly.

Kuhn wore two hats, but never simultaneously. He saw the history and the philosophy of science as interrelated but separate disciplines with different aims. He believed that no one could practice them at the same time. As a philosopher, he said, he was interested in generalizations and analytical distinctions, but as a historian he was trying to construct a narrative that was coherent, comprehensible, and plausible. For this latter task, the historian had to pay attention first to the factors internal to science, such as ideas, concepts, problems, and theories, and to external factors like social, economic, political, and religious realities. In his historical works Kuhn focused primarily (but not exclusively) on the internal factors, but believed that although the internal and the external approaches were autonomous, they were complementary. He saw the unification of them as one of the greatest challenges facing the historian of science.

Impact. Kuhn's immense impact on the philosophy of science was exclusively through his works, since he did not supervise any PhD theses in this field. He did have, however, a number of PhD students in the history of science, including John Heilbron, Norton Wise, and Paul Forman, though Forman, in the end, completed his PhD thesis officially under Hunter Dupree.

In historiography of science, Kuhn was a first-rate practitioner of the approach inaugurated by Alexandre Koyré, whom he admired deeply. Following Koyré, Kuhn believed that understanding a historical text necessarily involves a hermeneutical activity by which the historian interprets the text in its own terms and intellectual context. This means that the history of science should always be seen as part of the history of ideas, wherein the aim is to produce a maximally coherent interpretation. The historian is not someone who merely chronicles who discovered what and when. The projection of current conceptions onto past events is a cardinal sin often committed by the earlier positivistically inclined generations of historians of science, including Sarton. In the hands of Koyré, Kuhn, Rupert Hall, Bernard Cohen, Richard Westfall, and others, a new way of practicing historiography of science emerged. As a result, the Scientific Revolution of the sixteenth and seventeenth centuries became the topic that played a decisive role in historiographical developments.

Kuhn's influence was incomparably greater in the field of philosophy. *Structure* was translated into some twenty languages and sold over a million copies. It is still indispensable reading not only in philosophy of science, but also in philosophy generally. More than any other text, it was responsible for the overthrow of logical positivism both as a source of a certain image of science and as a philosophical practice. After *Structure,* the field of philosophy of science took a historical turn in the 1970s and 1980s, using historical case studies either to ground or to test "empirically" a given view of the development of science.

Kuhn's views also led to the Strong Programme in the Sociology of Scientific Knowledge founded by Barry Barnes and David Bloor, who argued that the very content and nature of scientific knowledge can be explained sociologically and a fortiori naturalistically. Kuhn, however, distanced himself from the Strong Programme, characterizing it as a "deconstruction that has gone mad." With its emphasis on the scientific community and its practices, Kuhn's philosophy eventually gave rise to what is called social studies of science, a subspecialty that attempts to unify philosophical, sociological, anthropological, and ethnographic approaches into a coherent whole. The feminist critique of science, too, that has emerged since the 1980s owes much to Kuhn's insights. Indeed, all of these studies are now routinely referred to as "post-Kuhnian."

Kuhn's views had virtually no impact on the practice of science itself, but they did catch the attention of both physicists and social scientists. While the former group was largely critical, the latter group was mostly sympathetic. The interest of social scientists was to a great extent methodological: they wondered whether sociology, political science, and economics were "mature sciences" like physics and chemistry, governed by a single paradigm at a given period, and whether they conformed to the pattern of normal science–crisis–revolution–normal science. One noticeable effect of such studies was that physical sciences came to be seen as being as interpretive as social sciences were, and in that respect not so different from them.

Were Kuhn's ideas as revolutionary as they were widely taken to be? Recent historical studies on the origins and development of logical positivism indicate that there are as many similarities and continuities as there are differences and discontinuities between that movement and Kuhn's views. Kuhn himself confessed later in life that he had fortunately very limited firsthand knowledge of logical positivist writings; otherwise, he said, he would have written a completely different book. But, as Alexander Bird put it, like Copernicus and Planck, Kuhn inaugurated a revolution that went far beyond what he himself imagined.

BIBLIOGRAPHY

WORKS BY KUHN

"Robert Boyle and Structural Chemistry in the Seventeenth Century." *Isis* 43 (1952): 12–36.

The Copernican Revolution: Planetary Astronomy in the Development of Western Thought. Cambridge, MA: Harvard University Press, 1957.

"The Function of Dogma in Scientific Research." *In Scientific Change: Historical Studies in the Intellectual, Social and Technical Conditions for Scientific Discovery and Technical Invention, from Antiquity to the Present,* edited by Alistair C. Crombie. London: Heinemann, 1963.

With John L. Heilbron, Paul Forman, and Lini Allen. *Sources for History of Quantum Physics: An Inventory and Report.* Memoirs of the American Philosophical Society, 68. Philadelphia: American Philosophical Society, 1967.

With John L. Heilbron. "The Genesis of the Bohr Atom." *Historical Studies in the Physical Sciences* 1 (1969): 211–290.

"Alexandre Koyré and the History of Science: On an Intellectual Revolution." *Encounter* 34 (1970): 67–69.

The Structure of Scientific Revolutions. 2nd enlarged ed. Chicago: University of Chicago Press, 1970. First published in 1962. The second edition contains the 1969 "Postscript."

"Notes on Lakatos." In *PSA 1970: In Memory of Rudolf Carnap; Proceedings of the 1970 Biennial Meeting, Philosophy of Science Association,* edited by Roger C. Buck and Robert S. Cohen. Boston Studies in the Philosophy of Science, vol. 8. Dordrecht, Netherlands: D. Reidel, 1971.

The Essential Tension: Selected Studies in Scientific Tradition and Change. Chicago: University of Chicago Press, 1977.

Black-Body Theory and the Quantum Discontinuity, 1894–1912. Oxford: Oxford University Press, 1978. 2nd ed. with a new "Afterword." Chicago: University of Chicago Press, 1987.

"History of Science." In *Current Research in Philosophy of Science,* edited by Peter D. Asquith and Henry E. Kyburg. East Lansing, MI: Philosophy of Science Association, 1979.

"The Halt and the Blind: Philosophy and History of Science." *British Journal for the Philosophy of Science* 31 (1980): 181–192.

The Road since Structure: Philosophical Essays, 1970–1993, with an Autobiographical Interview. Edited by James Conant and John Haugeland. Chicago: University of Chicago Press, 2000.

OTHER SOURCES

Barnes, Barry. *T. S. Kuhn and Social Science.* London: Macmillan, 1982.

Bird, Alexander. *Thomas Kuhn.* Princeton, NJ: Princeton University Press, 2000. A critical overview.

Darrigol, Olivier. "The Historians' Disagreement over the Meaning of Planck's Quantum." *Centaurus* 43 (2001): 219–239.

Friedman, Michael. "On the Sociology of Scientific Knowledge and Its Philosophical Agenda." *Studies in History and Philosophy of Science* 29 (1998): 239–271.

Fuller, Steve. *Thomas Kuhn: A Philosophical History for Our Times.* Chicago: University of Chicago Press, 2000.

Galison, Peter. "Kuhn and the Quantum Controversy." *British Journal for the Philosophy of Science* 32 (1981): 71–85.

Gutting, Gary, ed. *Paradigms and Revolutions.* Notre Dame, IN: University of Notre Dame Press, 1980. Written by eminent philosophers, social scientists, and historians of science, these essays assess Kuhn's pre-1980 writings and their impact in various fields.

Horwich, Paul, ed. *World Changes: Thomas Kuhn and the Nature of Science.* Cambridge, MA: MIT Press, 1993. An in-depth discussion of Kuhn's latest views; also contains Kuhn's long reply "Afterwords," which is his final statement.

Hoyningen-Huene, Paul. *Reconstructing Scientific Revolutions: Thomas S. Kuhn's Philosophy of Science.* Chicago: University of Chicago Press, 1993. Meticulous exposition, with a foreword by Kuhn.

Irzik, Gürol, and Teo Grünberg. "Carnap and Kuhn: Arch Enemies or Close Allies?" *British Journal for the Philosophy of Science* 46 (1995): 285–307.

Kindi, Vasso. "The Relation of History of Science to Philosophy of Science in *The Structure of Scientific Revolutions* and Kuhn's Later Philosophical Work." *Perspectives on Science* 13 (2006): 495–530.

Koyré, Alexandre. *Études galiléennes.* Paris: Hermann, 1939. Also 1966 and 1997. Translation by John Mepham as *Galilean Studies.* Atlantic Highlands, NJ: Humanities Press, 1978.

Lakatos, Imre, and Alan Musgrave, eds. *Criticism and the Growth of Knowledge.* London: Cambridge University Press, 1970. An early classic volume displaying the then-current state of debate among Kuhn, Popper, Lakatos, Feyerabend, and others.

Newton-Smith, W. H. *The Rationality of Science.* Boston: Routledge and Kegan Paul, 1981. A good overview of philosophy of science.

Nickles, Thomas, ed. *Thomas Kuhn.* Cambridge, U.K.: Cambridge University Press, 2003.

Sankey, Howard. *Rationality, Relativism and Incommensurability.* Aldershot, U.K.: Ashgate, 1997.

Sharrock, Wes, and Rupert Read. *Kuhn: Philosopher of Scientific Revolutions.* Cambridge, U.K.: Polity Press, 2002.

Westman, Robert S. "Two Cultures or One?: A Second Look at Kuhn's *The Copernican Revolution.*" *Isis* 85 (1994): 79–115.

Gürol Irzik

KUIPER, GERARD PETER (*b.* Harenkarspel, Netherlands, 7 December 1905; *d.* 24 December 1973, Mexico City, Mexico), *stellar astrophysics, binary stars, solar system astronomy, cosmogony, lunar and planetary studies, planetary probes.*

One of the most influential astronomers of the mid-twentieth century, Kuiper made significant contributions to the study of binary stars before he turned to solar system research in the mid-1940s. He discovered the atmosphere of Saturn's giant moon Titan, studied the characteristics of Mars and the outer planets, and worked out a cosmogonal model for the formation of the solar system that predicted the possibility of small bodies at the edge of the known solar system; although several of Kuiper's assumptions were later proven wrong, trans-Neptunian objects are commonly called Kuiper Belt objects in the early twenty-first century. The director of the Yerkes-McDonald Observatory, Kuiper founded the Lunar and Planetary Laboratory of the University of Arizona and served as a leading scientist on several National Aeronautics and Space Administration lunar projects in the 1960s.

Youth, Education, and Early Career. An ambitious, intellectually determined youth from a less-than-prosperous family, Kuiper (born Gerrit Pieter Kuiper) developed an early interest in astronomy. Successfully passing a particularly difficult entrance examination, Kuiper entered the University of Leiden, graduating with a BSc in science in 1927. He remained at Leiden for his graduate work, studying under the astrophysicists Ejnar Hertzsprung, Willem de Sitter, and Antonie Pannekoek as well as the theoretical physicist Paul Ehrenfest. In 1929 Kuiper spent eight months in Sumatra as part of the Dutch solar eclipse expedition.

On completing his doctoral thesis in 1933 on binary stars under Hertzsprung, and already fluent in English, he became a Kellogg Fellow at the Lick Observatory in California, then one of the largest observatories in the United States. A dedicated observational astronomer, Kuiper hoped to remain there but found a permanent appointment blocked by resentment against foreigners (exacerbated by his sharp, abrasive manners). He accepted a position at Harvard University in 1935 before becoming a permanent staff member the following year of the new McDonald Observatory in Texas, then operated by the Yerkes Observatory of the University of Chicago. Kuiper became a full professor in Chicago's Department of Astronomy in 1943.

Early Stellar Research. Although Kuiper apparently began thinking about planetary research while still at Leiden, when asked to review a theoretical work on the solar system's formation, his initial research involved stellar astrophysics. Beginning at Lick, Kuiper expanded his thesis research on physical double stars, discovering binary stars of extremely close periods (that is, in tight and rapid orbits). By 1942 Kuiper had discovered some twenty-one of the thirty-odd white dwarf stars then known. He also determined that some 50 percent of the stars closest to the Sun are binaries or members of multiple star systems. Writing up his work on the Beta Lyrae double-star system, Kuiper introduced the term *contact binaries,* and predicted that material drawn off from the larger star would create a ring about the smaller companion. Accretion disks caused by mass exchanges of closely orbiting stars would become a major focus of late-twentieth-century astrophysics.

Kuiper's appointment at Yerkes—brought about by Otto Struve, the Russian-born director of Yerkes-McDonald—put him into close contact with other new hires, including the Danish astrophysicist Bengt Strömgren and the theoretical astrophysicist Subrahmanyan Chandrasekhar; together they stimulated a renaissance of stellar astrophysics at Chicago. Working closely with them, Kuiper began studying stellar motions within several star clusters in neighboring regions of the Milky Way; his work provided an effective foundation for calibrating the stellar temperature scale, then a major research concern of stellar astrophysics. Kuiper also studied low-mass stars such as white dwarfs and faint blue stars, his interest in them kindled by Chandrasekhar's studies of degenerate matter. The aim of his research was to provide a basis for discriminating between theories of stellar energy production.

World War II and Solar System Research. Kuiper's switch to solar system research occurred in 1944 when, on research leave from wartime research at Harvard's Radio Research Laboratory, he used the 82-inch reflector telescope at the McDonald Observatory for scheduled observations. The McDonald Observatory was then the second-largest telescope in the United States, with greater sensitivity than the instruments at Yerkes, and Kuiper used the telescope both for stellar studies and to investigate the characteristics of planetary atmospheres. In the process he discovered that Titan, a large satellite of Saturn, possesses a methane-rich atmosphere. This was a surprising discovery, rich with implications for cosmogony. It suggested a fruitful avenue for research in solar system astronomy, but pursuing this would require him to abandon other promising avenues of stellar research in which he was deeply invested.

Kuiper considered these choices while he returned to wartime service. Because he was fluent in Dutch, German, and French, and had intimate knowledge of Western Europe, Kuiper was appointed a member of the secret Alsos mission in early 1945, which swept in behind advancing Allied troops to locate and interview Axis scientists about their wartime research, in particular seeking information about Germany's atomic project. During his Alsos service (which reinforced his already strong antipathy toward Germany, and led him after the war to publish articles identifying Nazi sympathizers among German astronomers), Kuiper came into contact with researchers who had pioneered studies of planetary atmospheres, including the Parisian Bernard Lyot (polarization studies) and Erich Regener, who had planned to place scientific instruments on board a German V-2 to study Earth's upper atmosphere.

On returning to the United States in late 1945, Kuiper still leaned toward returning to his stellar research. He soon decided against this for two reasons. His Alsos clearances led him to learn about wartime advances in lead sulfide cells that would enable astronomers to record infrared emissions far beyond the reach of infrared films, a great advantage for studying planetary bodies. He also became aware of new federal and military support for research involving planetary atmospheres. Kuiper decided to pursue solar system studies as his main research focus, a decision supported by Struve.

In the late 1940s, Kuiper made a number of significant planetary discoveries, including identifying carbon dioxide in Mars's atmosphere and discovering that Saturn's rings are composed of ice or covered with hoar frost, rather than the bare rock that he and other astronomers had expected. With visiting colleagues that Kuiper brought to Yerkes, he also studied the behavior of particles in planetary atmospheres and, with a graduate student, Daniel E. Harris III, made photometric measurements of planets, satellites, and asteroids. In 1948 Kuiper photographically discovered a fifth satellite of Uranus, later named Miranda, and one year later a second moon, Nereid, orbiting Neptune. He also initiated a survey of asteroids to obtain robust statistical data on all asteroids brighter than magnitude 16.5. By 1952, as a consequence of his emerging leadership in this field, Kuiper was elected president of the International Astronomical Union's (IAU) Commission 16, dedicated to the study of planets and satellites.

Kuiper's studies of planetary characteristics rekindled his interest in double stars, and he began to think of the solar system as an instance of an "unsuccessful" double-star system. Drawing on a recent nebular theory of cosmogony developed by German physicist Carl von Weizsäcker, Kuiper proposed, using insights from Struve and Chandrasekhar, that planets had formed in regions of gravitational instability within the solar nebula. Further drawing on his extensive studies of binary star systems, Kuiper introduced the then-startling notion that planetary systems were a fairly common consequence of star formation, not a rare event as many astronomers then believed. By the early 1950s Kuiper's cosmogonic

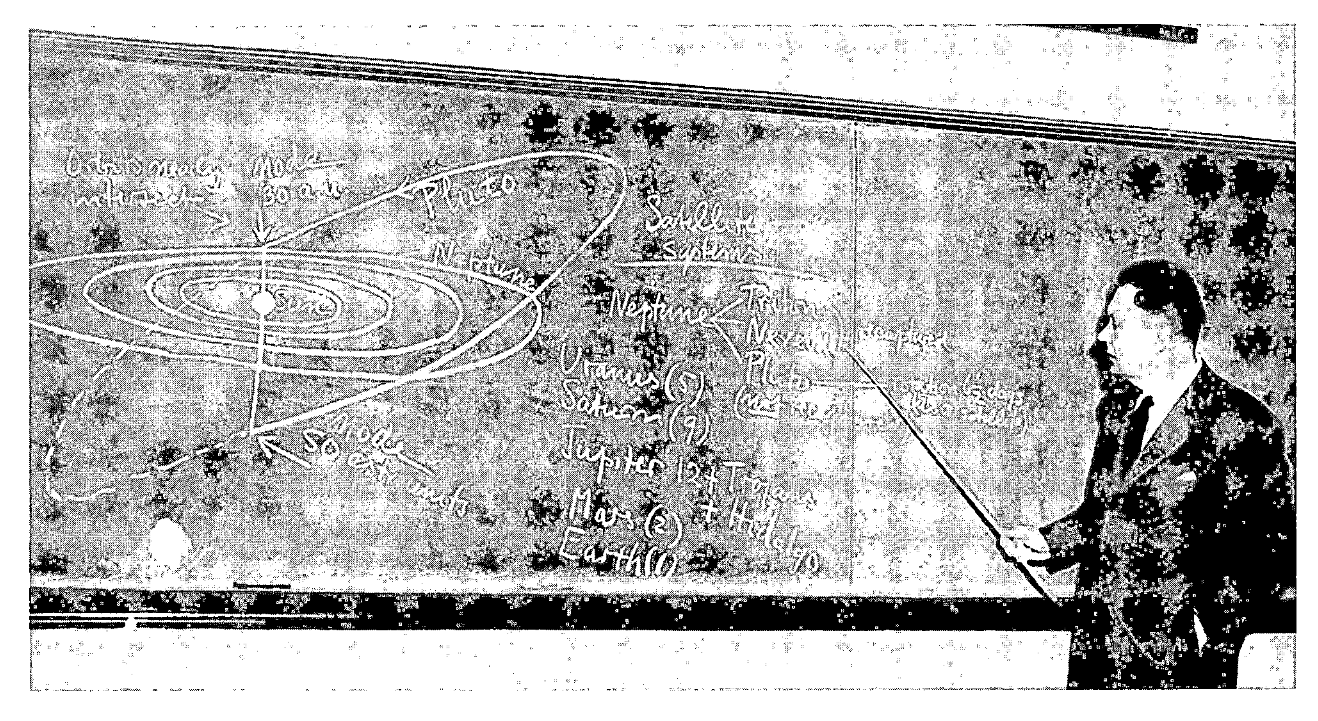

Gerard Peter Kuiper. © BETTMANN/CORBIS.

theory emerged as a leading model among American astronomers. Contained in one published version of his theory in 1951 was a terse assertion that billions of comet-like bodies would be found at the periphery of the solar nebula at distances of 35 to 60 astronomical units (one astronomical unit being the mean distance of Earth from the Sun), on the assumption that Pluto was sufficiently massive to have diverted comets into a still more remote distribution of comets known as the Oort Cloud. While contemporary astronomers now accept a much smaller mass for Pluto, in 1992 the first object orbiting beyond Neptune was discovered (within a decade, more than a thousand additional bodies were identified). Many astronomers use the term *trans-Neptunian objects* to describe these remote bodies, but *Kuiper Belt object* has come into common use as well.

Already by the late 1940s Kuiper realized that the most significant advances in solar system astronomy would come through interdisciplinary cooperation with neighboring scientific disciplines, including chemistry, physics, and geology. By the early 1950s Kuiper began a close collaboration with a fellow Chicago scientist, the Nobel Prize–winning chemist Harold C. Urey, who had become interested in the geochemical evolution of Earth and the planets. Astronomers and geochemists in the 1950s both sought to determine the absolute abundances of the elements, since many scientists believed that concentrations of radioactive potassium, uranium, and thorium would indicate whether planetary interiors had become sufficiently hot to cause core formation and global melting. Initially Urey accepted Kuiper's nebular cosmogony as the starting point for his own geochemical studies and advanced models of planetary geochemistry within the broad outlines of Kuiper's failed-binary-star model. For his part, Kuiper accepted Urey's geochemical arguments, which held that Earth and the Moon had formed and remained at relatively cool temperatures. But by 1954 Kuiper's studies of the lunar surface convinced him that the Moon had been molten early in its history. One year later, in the pages of the *Proceedings of the National Academy of Sciences,* Urey strongly criticized Kuiper's solar system research and privately sought to have him removed as chair of IAU's Commission 16. Urey's anger was not primarily over their scientific disagreement but instead over what he perceived (with justification) as Kuiper's unwillingness to credit Urey's priority in establishing a number of geochemical ideas.

The dispute left Kuiper convinced that astronomical rather than geochemical evidence was paramount in solving the puzzle of the solar system's origin. But it also illuminated the frailty of interdisciplinary research programs that stretched across distinct disciplines and professional societies, where no means existed for resolving priority disputes, as well as the challenge of finding common frameworks for evaluating evidence from astrophysical and geochemical sources. Briefly Kuiper considered turning to a different field of research, and his then graduate student Carl Sagan recalled feeling during the controversy's aftermath like "the child of divorced parents." But by the mid-1950s Kuiper's studies had helped to forge a new consensus among American astronomers about the formation and general characteristics of the solar system, and he remained firmly engaged in the field.

Planetary Research after *Sputnik*. The launch of *Sputnik* in October 1957 caused government officials and the general public to become much more interested in the Moon and the planets. Kuiper recognized the sea change and quickly began to take advantage of the situation to obtain more federal patronage for solar system studies.

Throughout the late 1950s Kuiper built up a large grant-supported group at Chicago's Yerkes Observatory, where he had again become director. Using U.S. Air Force and National Science Foundation funds, he launched new programs to chart the lunar surface and to expand studies of the physical characteristics of planets and asteroids. While he also sought funding to aid stellar and galactic research (he was instrumental in securing a promise from the U.S. Air Force to build what became the first telescope placed at Cerro Tololo, Chile, later home to the Cerro Tololo Inter-American Observatory), Kuiper focused increasingly on developing knowledge about the Moon and planets, and sharing this information with the newly formed National Aeronautics and Space Administration (NASA) as well as an increasingly interested U.S. government.

Indeed, as the space program took shape amid the Cold War—and particularly as the space race became a means for the Soviet Union and the United States to demonstrate the technological superiority of their political systems—Kuiper found himself having to fulfill an additional role beyond active scientist, leader of solar system astronomy, and director of a major astronomical observatory: he was increasingly called on to interpret Soviet scientific advances to his government patrons. In 1959 Kuiper successfully secured a contract from the Central Intelligence Agency (CIA) to interpret Soviet research in astronomy, hiring a visiting Yugoslavian astronomer to do this work. He also struggled to interpret a controversial finding by the Soviet astronomer Nikolai Kozyrev suggesting that the Moon remained active volcanically, a critical issue for spacecraft engineers if accurate (Western scientists accepted that the Moon was seismically quiescent). Cold War limitations initially kept Western astronomers from evaluating Kozyrev's evidence directly, and it was not until late 1960, when Kuiper could finally travel to Leningrad, that he resolved the issue (genuine observation, but misinterpreted). Convinced that science was more easily corrupted in authoritarian regimes, Kuiper embraced a public role as interpreter of the proper ethical boundaries of science.

The rapid growth of solar system astronomy at Chicago caused stellar and galactic astrophysicists at Yerkes-McDonald to grow anxious that their specialties would become overwhelmed by this increasingly popular field. Escalating tension between Kuiper and his colleagues climaxed in what contemporaries later called a "civil war," a simmering professional dispute heightened by Kuiper's haughty demeanor and intransigence in handling professional disputes. In 1960 Lawrence Kimpton, president of the University of Chicago, fired Kuiper as director of the Yerkes-McDonald observatories. As a tenured faculty member, Kuiper was entitled to remain at Chicago as a full professor. But several weeks later, Kuiper relocated his research associates, graduate students, and associated staff, ten people altogether, to the University of Arizona, with a new appointment split between Arizona's Department of Astronomy and its Laboratory of Atmospheric Physics. His invitation had come from university president Richard Harvill, who perceived that federal funds for space exploration (and the new, nearby Kitt Peak National Observatory) would permit rapid expansion of lunar and planetary studies. Kuiper indeed created a major research center at the Laboratory of Atmospheric Physics (soon renamed the Lunar and Planetary Laboratory, and eventually the Institute of Atmospheric Physics), whose success became apparent during the 1960s. This was a significant moment for American astronomy, for it marked a bifurcation in the once-unified discipline of U.S. astronomy into increasingly specialized fields of planetary astronomy and stellar/galactic astrophysics. Thereafter the fields were supported and nurtured by distinct institutions, patrons, instruments, professional organizations, and graduate training programs.

At Arizona, Kuiper moved swiftly to develop new research programs and new instruments adapted for planetary research. Kuiper served as chief scientist for NASA's lunar Ranger series, which radioed back photographs of the lunar surface until they crashed into the Moon. He and his colleagues used these photographs to determine potential landing sites for NASA's unmanned lunar Surveyor program (he served as an experimenter) and for the later manned Apollo program. In the process Kuiper identified numerous ancient multiringed basins on the lunar surface, aiding interpretation of the Moon's early history. While he remained committed to ground-based astronomy, promoting the development of an infrared telescope (funded by NASA) atop Mauna Kea in Hawaii, Kuiper also began using a telescope-equipped Convair 900 aircraft to make infrared observations of planets and stars from 40,000 feet (12,192 meters), above most of Earth's atmosphere; in 1975 this facility was posthumously named the Kuiper Airborne Observatory.

Kuiper made numerous professional contributions as well. He edited two major multivolume text series (*The*

Solar System and *Stars and Stellar Systems*), seeing them as vehicles for stimulating interdisciplinary research, and trained a handful of graduate students in planetary science between the 1950s and 1970s. Several of them (the most famous was Sagan) became leaders in the field in the first decades of the space age.

In 1950 Kuiper was elected a member of the National Academy of Sciences, three years after receiving the Janssen Medal of the Astronomical Society of France. He also received the Kepler Gold Medal at a joint meeting of the American Association for the Advancement of Science and the Franklin Institute in 1971. For his participation in Alsos, Kuiper received a high award from the Order of Orange and Nassau by the queen of the Netherlands.

Throughout his career Kuiper was viewed by colleagues and his graduate students as formal and distant, rarely away from work, prone to sudden coolness and acerbic comments when challenged. As a mentor, Kuiper let graduate students find their way to individual thesis questions, influencing them primarily through his memorable physical stamina and scientific style. On two-week observing runs, Kuiper slept just three to four hours per night and refreshed himself through brief catnaps; when confronting problems in planetary physics, Kuiper demonstrated a highly intuitive approach, making first-order computations on paper from physical principles.

Kuiper met the former Sarah Parker Fuller while in Cambridge (her family donated the land in Harvard, Massachusetts, where Harvard University placed its Oak Ridge Observatory). They married in 1936 and had two children, Paul Hayes and Sylvia Lucy. Kuiper died in Mexico City on Christmas Eve, 1973, just after his sixty-eighth birthday, while on a trip with his wife and Fred Whipple, a fellow astronomer and longtime friend.

BIBLIOGRAPHY

Kuiper's papers are housed in the University of Arizona Library Special Collections.

WORKS BY KUIPER

"The Empirical Mass-Luminosity Relation." *Astrophysical Journal* 88 (1938): 472–507.

"On the Interpretation of βLyrae and Other Close Binaries." *Astrophysical Journal* 93 (1941): 133–177.

"Titan, a Satellite with an Atmosphere." *Astrophysical Journal* 100 (1944): 378–383.

Editor. *Atmospheres of the Earth and Planets: Papers Presented at the Fiftieth Anniversary Symposium of the Yerkes Observatory, September, 1947.* Chicago: University of Chicago Press, 1949.

"On the Origin of the Solar System." In *Astrophysics: A Topical Symposium Commemorating the Fiftieth Anniversary of the Yerkes Observatory and a Half Century of Progress in Astrophysics*, edited by J. Allen Hynek, 357–424. New York: McGraw-Hill, 1951.

Editor. Nine volumes in the series *Stars and Stellar Systems.* Chicago: University of Chicago Press, 1960.

Editor. Four volumes in the series *The Solar System.* Vol. 1, *The Sun;* vol. 2, *The Earth as a Planet;* vol. 3 (with Barbara M. Middlehurst), *Planets and Satellites;* and vol. 4 (with Barbara M. Middlehurst), *The Moon, Meteorites, and Comets.* Chicago: University of Chicago Press, 1953–1963.

OTHER SOURCES

Cruikshank, Dale P. "20th-Century Astronomer." *Sky and Telescope* 47 (March 1974): 159.

———. "Gerard Peter Kuiper, December 7, 1905–December 24, 1973." *Biographical Memoirs of the National Academy of Sciences* 62 (1993): 258–295. This is the only comprehensive biography of Kuiper available.

Davidson, Keay. *Carl Sagan: A Life.* New York: Wiley, 1999.

Doel, Ronald E. "Evaluating Soviet Lunar Science in Cold War America." *Osiris*, 2nd ser., 7 (1992): 238–264.

———. *Solar System Astronomy in America: Communities, Patronage, and Interdisciplinary Science, 1920–1960.* New York: Cambridge University Press, 1996.

Sagan, Carl. "Gerard Peter Kuiper (1905–1973)," *Icarus* 22 (1974): 117–118.

Tatarewicz, Joseph N. *Space Technology and Planetary Astronomy.* Bloomington: Indiana University Press, 1990.

Ronald E. Doel

L

LA METTRIE, JULIEN OFFRAY DE

(*b.* Saint-Malo, France, 19 December 1709; *d.* Berlin, Germany, 11 November 1751), *medicine, physiology, psychology, philosophy of science.* For the original article on La Mettrie see *DSB,* vol. 7.

Since the publication of the *Dictionary of Scientific Biography,* the scholarly understanding of La Mettrie has undergone slight shifts of emphasis rather than dramatic reconsiderations or revisions. La Mettrie's philosophy is no longer appreciated simply or even primarily as a thoroughgoing application of Cartesian mechanics to human beings, despite the title of his best-known work, *L'homme machine* (1747). Appraisals of his philosophy have been amplified through critical studies of a number of his works, which have not only deepened appreciation of the content and impact of his philosophical writings but also placed these works in a richer scientific and philosophical context. The importance of medicine to La Mettrie's entire corpus and to his contribution to the Enlightenment is also widely acknowledged, although his relationship to the broader movement is usually considered problematic or peripheral.

Historians of science have increasingly explored the social context of the development of science and the careers of scientists. La Mettrie, who left no correspondence or memoirs and whose career took place too early and on the periphery of Parisian culture and the emerging Enlightenment movement, has largely proved difficult to place culturally or to integrate into the social history of science. However, several additional features of his biography warrant mention. His initial education took place at the provincial colleges of Coutances and Caen, where he was influenced by Jansenism and is reputed to have written a Jansenist text, although it has never been found. In 1725, La Mettrie went to Paris to study philosophy and natural science at the Collège d'Harcourt, which was the first to make Cartesianism central to its curriculum. These early exposures to Jansenism and Cartesianism offer tantalizing insights into La Mettrie's intellectual development.

It is also worth noting that La Mettrie, like most young men from the French provinces seeking to make their way in professional society, benefited from hometown contacts. His fellow citizen of Saint-Malo, Pierre-Louis Moreau de Maupertuis, was president of the Berlin Academy of Sciences and crucial in gaining a position for La Mettrie as a court physician to Frederick II of Prussia when his radical ideas led to his expulsion from Holland after the publication of *L'homme machine.* In La Mettrie's case, his provincial roots may have heightened his outsider status, although his radical ideas were the most significant cause of his alienation from most of his contemporaries.

Recent scholarship not only has emphasized the medical roots of La Mettrie's science and philosophy but also continues to refine that appreciation. After completing his studies at the Collège d'Harcourt, La Mettrie studied medicine at the University of Paris for the next five years but (to avoid high graduation fees) took his degree from the University of Reims. It is important to note, in light of his subsequent vociferous critique of the medical profession, that it was his dissatisfaction with his medical education, specifically as a foundation for medical practice, that took him to the University of Leiden to study with Hermann Boerhaave, a renowned teacher of physiology and chemistry and an innovative clinical practitioner.

Julien Offray de La Mettrie. THE LIBRARY OF CONGRESS.

La Mettrie was the French translator of a number of specific texts written by Boerhaave and translator of and commentator on his fundamental eight-volume *Institutions de médecine.* Historians of science, who have reappraised Boerhaave's role in the history of medicine, have acknowledged that his understanding of human physiology was less clearly iatromechanical and more nuanced than previously thought.

Like Boerhaave's, La Mettrie's view of human physiology was much less strictly mechanistic than previous scholars have assumed. As La Mettrie worked through Boerhaave's medical and physiological works, he hardened what might be seen as materialist tendencies in Boerhaave's work into a more consistent materialist physiology. Where Boerhaave suggested that Lockean accounts of mental states might parallel brain activity, La Mettrie was willing to assert a causal connection. His exposure to Boerhaave was crucial to La Mettrie's development both as a philosophical and medical writer; it gave him a philosophical position and a critical perspective from which to evaluate contemporary medicine. He returned from Leiden a staunch proponent of a cautious, empiricist, and utilitarian approach to knowledge and an opponent of rationalist metaphysics.

Recent scholarship has been more appreciative of the positive contributions of La Mettrie's satires to his overall medical agenda. The awareness of professional issues that La Mettrie gained as a medical student led him to lampoon the ignorance and venality of Parisian medical practitioners. In his satires, he not only supported the surgeons in their dispute with the Faculty of Medicine but also honed his attack on the metaphysical foundations of medicine and argued instead for a medical practice rooted in empirical observation and clinical teaching and dedicated to public health.

La Mettrie also wrote five medical treatises on specific diseases, such as smallpox and venereal disease, which were (as the *DSB* article noted) not especially original treatments. These texts were nonetheless important for their reliance on case studies as the most legitimate to approach the study of diseases and for their intent to raise public awareness of heath issues by conveying information to the public.

The original *DSB* article emphasized the significance of La Mettrie's appreciation of Boerhaave in bringing medicine into the philosophical and intellectual context of the eighteenth century. However, as scholars of eighteenth-century culture have moved beyond the great luminaries of the Enlightenment to less prominent figures and to scientific and literary institutions, they have recognized that medical writings and ideas had a greater impact on the Enlightenment than earlier scholarship acknowledged. Thus La Mettrie was important but not unique in appreciating the empirical foundation and method medicine could offer to philosophical investigations. However, as scholars increasingly appreciate, medicine was the decisive foundation for La Mettrie's philosophy. For him, medicine demonstrated the dependence of mind on body, the variety of human constitutions, and the different responses of individual constitutions to external stimuli. Medicine also gave La Mettrie his distinctive philosophical style. He wrote with polemical zest, his points corroborated by physiological evidence and medical case studies, and addressed to issues of public health.

The most significant reappraisal of La Mettrie has been the emergence of a scholarly consensus about his best-known work, *L'homme machine.* Previous scholars often considered the text an attempt to apply Cartesian mechanism to human beings. Scholars now recognize that the text asserts that the smallest particles of organic matter are characterized by life and mobility; thus, in essence, La Mettrie's materialism is vitalistic. The text vehemently rejects René Descartes's *bête machine* as an absurd characterization of animals and also refutes human dualism, documenting (as extensively as possible) the effects of physical states on human behavior and the comparability between human and animal anatomy and behavior. He presents physiological and clinical examples as privileged evidence and defers to the authority of the physician over the metaphysician and the theologians.

Scholars have also recognized the significance of La Mettrie's *Discours préliminaire.* Written in 1751 to introduce a collection of his philosophical works, La Mettrie explicitly identified his work in medicine and philosophy with the reformist agenda of the nascent intellectual movement we now call the Enlightenment. He defined the *médecin-philosophe* as the ideal intellectual, embodying the astute empirical observation of surgeons, the thorough training in physiology of an idealistic physician, and the zeal of the reform-minded *philosophe.* The *médecin-philosophe* was thus the most effective practitioner of critical analysis and reform.

Despite La Mettrie's self-identification with the Enlightenment, scholars have difficulty defining his relationship to that movement. In many ways, La Mettrie defies easy classification. He wrote from a medical perspective, very early in the movement, and without some of the stylistic verve and innovation of the later Enlightenment. He deliberately positioned his more radical ideas in the context of the established philosophical tradition. In *Histoire naturelle de l'âme* (1747), La Mettrie placed his arguments against a distinctly human, immortal soul in the context of Aristotle's three souls. He set his ideas about humanity's possible evolution from an elemental mud in a discussion of Epicurus, and presented his dangerous moral ideas as a response to Stoicism.

However, more recent studies acknowledge the originality and prescience of some of La Mettrie's ideas. For example, he offered a systematic materialism twenty years before Paul-Henri d'Holbach's *Système de la nature* and asserted the unity of organic life before Georges-Louis de Buffon. La Mettrie was one of the earliest to speculate about human evolution, although he postulated no explanatory mechanism.

The misapprehension of La Mettrie as a mechanist made him a figure often cited as a forerunner to modern interests in artificial intelligence and computer simulations of mental functions, and, to some degree, he is still invoked as a progenitor in these fields. In more recent years, La Mettrie has been more accurately acknowledged as a proponent of comparative anatomy and organic models of human life. As a significant advocate for the physiological bases of human behavior, La Mettrie's work resonates in modern discussions of neurobiology and the genetic sources of human behavior. Although the moral implications of his philosophy were considered extremely dangerous in the eighteenth century, moderns have come to appreciate his endorsement of human pleasure, his concern with public health, and his humanitarianism, especially as reflected in his calls for tolerance.

SUPPLEMENTARY BIBLIOGRAPHY

Several modern editions of some of La Mettrie's works have become available.

WORKS BY LA METTRIE

Œuvres philosophiques. Hildesheim, Germany: G. Olms Verlag, 1974.

Œuvres philosophiques. Paris: Fayard, 1987.

Man the Machine and Other Writings. Edited by Ann Thomson. Cambridge, U.K.: Cambridge University Press, 1996. Several English translations of La Mettrie's work have appeared.

De la volupté: Anti-sénèque ou le souverain bien; L'École de la volupté; Système d'Epicure. Edited by Ann Thomson. Paris: Desjonquières, 2000.

Ouvrage de Pénélope, ou, Machiavel en médecine. Paris: Fayard, 2002.

OTHER SOURCES

Corpus: Revue de philosophie. Paris, 1987. Vol. 5/6 is dedicated to La Mettrie.

Jauch, Ursula. *Jenseits de Maschine: Philosophie, Ironie und Ästhetik bei Julien Offray de La Mettrie.* Munich, Germany: Carl Hanser Verlag, 1998.

Roggerone, Giuseppe Agostino. *Controilluminismo: Saggio su La Mettrie ed Helvétius.* 2 vols. Lecce, Italy: Milella, 1975.

Stoddard, Roger. *Julien Offray de La Mettrie: A Bibliographical Inventory.* Cologne, Germany: Verlag Jurgen Dinter, 2000.

Thomson, Ann. *Materialism and Society in the Mid-Eighteenth Century: La Mettrie's "Discours préliminaire."* Geneva: Droz, 1981. Excellent critical edition. The text is in French and the monograph-length introduction in English.

Verbeek, Theodorus. *Traité de l'âme de La Mettrie.* 2 vols. Utrecht, Netherlands: OMI-Grafisch Bedrijf, 1988. Excellent critical edition, with monograph-length introduction.

Wellman, Kathleen. *La Mettrie: Medicine, Philosophy, and Enlightenment.* Durham, NC: Duke University Press, 1992.

Kathleen Wellman

LA RAMEE, PIERRE DE

SEE **Ramus, Petrus.**

LADD, HARRY STEPHEN

(*b.* St. Louis, Missouri, 1 January 1899; *d.* Bethesda, Maryland, 30 November 1982), *paleoecology, coral reefs, Pacific Islands geology and paleontology.*

An expert on the geology of the islands of the Pacific, Harry Ladd spent many years studying that ocean's atolls and coral reefs. The *Treatise on Marine Ecology and Paleoecology* (1957), the second volume of which he edited, created the foundation for decades of paleoecology work. He also promoted a project that produced *Bikini and nearby Atolls,* which became the authoritative work on the geology, biology, and paleontology for much of the Pacific. His drilling at Enewetak Atoll essential confirmed Charles Darwin's postulate that the atoll shape was the result of subsidence.

Early Life and Career. Harry Ladd was interested in nature as a youth, and the summers that his family spent at Bar Harbor, Maine, whetted his interest in marine invertebrates. This led in turn to an interest in fossils, encouraged during his years at Washington University in St. Louis. Ladd received his master's degree in 1924 and doctorate from the University of Iowa in 1925, his thesis work based on study of an Ordovician rock formation in Iowa and its contained fossils, published as "Stratigraphy and Paleontology of the Maquoketa Shale of Iowa" (1929). During this interval he was also an assistant geologist on the Iowa Geological Survey. Intrigued by the Devonian fossil coral reefs of Iowa, he obtained a fellowship from Yale University and the Bernice P. Bishop Museum in Honolulu to examine Cenozoic and Recent reefs. He traveled to Fiji and marked this trip as the year he had no birthday, as a result of the ship crossing the International Date Line just at the close of 31 December. Following his first efforts in the South Seas, he taught from 1926 to 1929 at the University of Virginia. Disliking teaching, he then worked for two years in Venezuela as a paleontologist for the Gulf Oil Company of Venezuela. Living on money he had saved, Ladd spent a year studying his collections of fossils from the Fiji Islands at the National Museum of Natural History in Washington, D.C. During a second year, he worked as an assistant for the U.S. Geological Survey (USGS) and continued his paleontological investigations in the evenings.

In the Pacific. During his first ten years of activity in the Pacific, nearly three were spent in fieldwork. In 1934 Ladd and J. Edward Hoffmeister of the University of Rochester returned to the Fiji Islands and laid the basis for all future detailed mapping of the island group. More importantly, they studied the reefs and concluded that subsidence was not necessary for coral growth and that water level fluctuation due to glaciation was not a significant factor in reef formation. Ladd persuaded his fiancée, Jane Mahler, to travel to Fiji, where they were married. He was presented with a sperm whale tooth that thereafter graced their double bed. They had two sons and, at the time of his death, four grandchildren.

U.S. Geological Survey. Upon the couple's return to the United States in 1936, Ladd joined the National Park Service, first as a district geologist for Atlanta and Richmond and then rising to regional geologist in 1938. The work included examining water supplies in various regions, drilling water wells, and locating sites of construction material, these investigations engendered in part by the Tennessee Valley Authority. During 1940 he transferred to the USGS, retiring from that organization in 1969.

Ladd's first USGS assignment was to study the Late Cenozoic and Recent marine invertebrates of the Texas Gulf Coast. This study was probably arranged by T. Wayland Vaughan, a pioneer in the study of corals, who had left the USGS to head the Scripps Institution of Oceanography in San Diego, California, and later returned to the National Museum, where he advised Ladd in his study of the Fiji fossils. In 1940 Vaughan organized the Subcommittee on the Ecology of Marine Organisms under the Committee on Geologic Research of the National Research Council. Impressed by Ladd's organizational skills and enthusiasm, Vaughan arranged for his appointment as chairman of the subcommittee.

The subcommittee was moribund during the crisis of World War II, which also ended Ladd's investigations of Gulf Coast marine organisms. Instead, in 1941 and 1942 he investigated manganese deposits in the Appalachian Mountains. During 1942 he was transferred to Rolla, Missouri, as a regional geologist to coordinate investigations for strategic minerals in midcontinent. After the war Ladd returned to Washington, D.C., and served as an assistant chief geologist of the USGS from 1946 to 1949.

Despite his official administrative activities, Ladd made the time to reconstitute his earlier panel into a Committee on Marine Ecology and Paleoecology. In scarcely more than a decade, the *Treatise on Marine Ecology and Paleoecology* (1957) was published. It consisted of two volumes, the second of which, titled *Paleoecology,* was edited by Ladd. At 2,372 pages, the two volumes were the largest single publication of the Geological Society of America. It laid the basis for half a century of studies in paleoecology.

Studying Pacific Atolls. As part of his official duties, Harry Ladd recommended and outlined a project for the USGS to map a number of island groups in the Pacific. The project began in 1946 and lasted for fifteen years, producing invaluable geologic data. More or less simultaneously, he was engaged in fieldwork, surveying Bikini Atoll in advance of atomic bomb testing. Ladd returned in 1947, following the tests. This project led to the publication of USGS Professional Paper 260, *Bikini and nearby Atolls.* Its thirty-five chapters of more than one thousand pages and more than three hundred plates make it by far the largest USGS professional paper. It is *the* standard reference for many aspects of the geology, biology, and paleontology of a vast portion of the Pacific region.

In connection with investigations at Enewetak Atoll, Ladd made his third major contribution to geology. To better understand the formation of atolls, considerable core drilling was undertaken, most of it under the direct supervision of Ladd, who logged wells and collected cuttings with the rest of the crew. Drilling was done on a barge towed from Hawaii carrying a Failing 1500 Holemaster rig. The actual drilling at Bikini and Enewetak Atolls was done by Virgil Mickel, an Oklahoma driller Ladd had employed during his time with the National Park Service.

At Enewetak, Mickel drilled through more than 4,000 feet of reef rock before reaching the basalt from a sunken volcano. Ladd hung up a sign "Darwin was Right." Charles Darwin had postulated that the atoll shape was the result of subsidence. Careful study of the cores indicated fluctuations in subsidence and changes in sea level, however, so those who argued against Darwin's hypothesis were correct in a few details.

Project Mohole. Because of this experience of drilling from a barge in a lagoon, Ladd was the logical person to try test drilling in open water for scientific purposes. This explains Ladd's involvement in the proposed Project Mohole, an attempt to drill down to Earth's mantle layer. The effort was never completed, but continuous drilling from 1958 to 1966 resulted in vastly expanded geologic knowledge of the world's ocean basins.

Honors and Awards. After his official retirement in 1969, Harry Ladd continued his investigations at the National Museum for another full decade until his activities were curtailed by Parkinson's disease. Ladd was president of the Paleontological Society in 1954, a vice president of the Geological Society of America in 1955, and a vice president of the American Association for the Advancement of Science in 1965. He received the Department of Interior's Distinguished Service Award in 1965 and was awarded the Paleontological Society Medal in 1981.

BIBLIOGRAPHY

WORKS BY LADD

"Stratigraphy and Paleontology of the Maquoketa Shale of Iowa." *Annual Report of the State Geologist* 34 (1929): 305–448.

Editor. *Treatise on Marine Ecology and Paleoecology.* Vol. 2, *Paleoecology.* Geological Society of America, memoir 67. New York: Geological Society of America, 1957.

OTHER SOURCES

"Paleontologist Harry S. Ladd Dies." *Washington Post,* 2 December 1982.

U.S. Geological Survey. *Bikini and nearby Atolls,* Marshall Islands. Professional Paper 260 (parts A-II). Washington, DC: Government Printing Office, 1954–1969.

Whitmore, F. C., and J. I. Tracey Jr. "Memorial to Harry Stephen Ladd, 1899–1982." *Geological Society of America Memorial* 14 (1984): 1–7.

Yochelson, E. L. "Presentation of the Paleontological Society Medal to Harry Stephen Ladd." *Journal of Paleontology* 56 (1982): 826–827.

Ellis Yochelson

LADYZHENSKAYA, OLGA ALEXANDROVNA (*b.* Kologriv, U.S.S.R., 7 March 1922; *d.* St. Petersburg, Russia, 12 January 2004), *mathematics, partial differential equations.*

Ladyzhenskaya was one of the very few outstanding female mathematicians of the twentieth century. The general theory of partial differential equations, governing fluids, gases, elasticity, electromagnetism, and quantum physics was developed during the twentieth century. Ladyzhenskaya was a major figure in the treatment of parabolic (typified by $u_t - u_{xx} = 0$) and elliptic ($u_{xx} + u_{yy} = 0$) equations. She obtained pioneering results in the spectral theory of general elliptic operators and in diffraction. With her student, Nina Ural'tseva, she analyzed in depth the regularity of quasilinear elliptic equations and with Ural'tseva and Vsevolod A. Solonnikov, the regularity of parabolic equations.

Life and Career. Ladyzhenskaya grew up in Kologriv in European Russia, where her father was the high school principal and a teacher of mathematics and art. Ladyzhenskaya's mathematical education began in the Kologriv high school. After school, her father gave her lessons at home. Her father was denounced to the Soviet authorities in 1937, declared an enemy of the people, imprisoned, and then executed. Ladyzhenskaya, unlike her older sisters, was permitted to finish high school, from which she graduated in 1939, but she was not admitted to Leningrad University because of her father. (He was completely exonerated after Khrushchev's secret speech in 1956, three years after Stalin's death.) Her mother, Anna Mikhailovna, had to struggle to keep the family housed and fed.

Olga attended a pedagogical institute, taught school in Kologriv after the German invasion of June 1941, and was finally admitted to Moscow University in 1943. She attended one of Israel Gelfand's first seminars with Mark

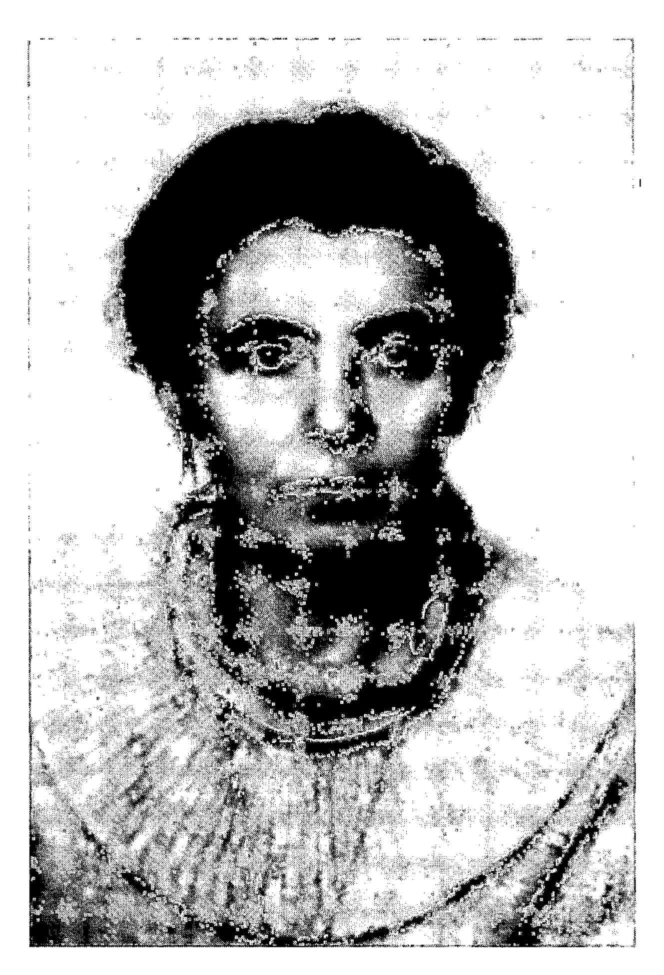

Olga Ladyzhenskaya. COURTESY OF CATHLEEN SYNGE MORAWETZ.

Visik and Olga Oleinik, who later became very well known mathematicians and professors at Moscow University. Gelfand's seminar later became the most renowned of Russian mathematical seminars. All three of them did their student theses with Ivan Petrovsky. Ladyzhenskaya remained close to Visik but had a long rivalry with Oleinik. She graduated in 1947, moved to Leningrad, and was married briefly to her fellow student, Andrei Kiselev. She completed her PhD with Sergei Sobolev and Vladimir Smirnov in 1951, and defended her DSc at Moscow University in 1953. She began teaching in the Physics Department of the University of Leningrad in 1950; she became a member of the Mathematical Physics Laboratory of the Leningrad branch of the Steklov Institute of Mathematics starting in 1954 and was made head of the laboratory in 1961.

Mathematical Work. Ladyzhenskaya's university thesis, under the supervision of Ivan Petrovsky, was a result on the approximation of hyperbolic ($u_{tt} - u_{xx} = 0$) equations, but her earliest important success was her elegant estimates for elliptic operators involving spaces of generalized functions. Her best-known "sharp" estimate is $\mathrm{P}u\ \mathrm{P}_{W_2^2(\mathrm{W})} £\ C(\mathrm{W})\left(\mathrm{PL}u\,\mathrm{P}_{L^2(\mathrm{W})} + \mathrm{P}u\,\mathrm{P}_{L^2(\mathrm{W})}\right)$. Here u is a function defined in a domain on whose boundary $u = 0$ and Λ is an elliptic operator acting on u. For example, u is an electric potential and $\Lambda\ u$ is the charge density, the operator Λ is the Laplacian, $\mathrm{PF\ P}_{L^2(\mathrm{W})}$ measures the function by the mean square of its value on the domain W, and $\mathrm{P}u\mathrm{P}_{W_2^2(\mathrm{W})}$ measures the mean square of all the derivatives up to order 2. From such estimates one can prove uniqueness and existence for solutions of equations involving Λ. "Sharp" means that the constant C can be chosen so that the inequality is an equality for some function u for a given domain W.

For Navier-Stokes equations (incompressible viscous flow), Ladyzhenskaya proved unique global solvability for the two-dimensional case, refining the earlier work of Jean Leray in "Sur le movement d'un liquide visqueux emplissant l'espace" (1934; On the motion of a viscous fluid filling space) and Eberhard Hopf in "Uber die Anfangswertaufgabe fur die hydrodynamischen Grundgleichungen" (1951; On the initial value problem for the fundamental equations of fluid dynamics). For the three-dimensional case, Ladyzhenskaya introduced innovative modifications to Navier-Stokes equations to treat large velocity fluctuations. She also established the first "attractor" results. This means that a solution of an initial value problem, under certain restrictions, "eventually" is very close to a particular solution, "the attractor," where it stays. Here she was one of the important innovators.

Legacy and Honors. Ladyzhenskaya authored or coauthored seven books and 250 papers. Her most influential works were her books, which contain many original theorems.

Her honors include a corresponding membership in the Academy of Sciences U.S.S.R. in 1981; and full membership in the Russian Academy of Sciences in 1991. She was made a foreign member of the Lincei National Academy in Rome in 1989 and a member of the American Academy of Arts of Sciences in 2001. She received an honorary doctorate from the University of Bonn in 2002. In the latter year Ladyzhenskaya also received the Lomonosov Medal of the Russian Academy of Sciences. She gave the Noether lecture at the 1994 congress of the International Mathematical Union.

Noted everywhere for her great charm, beauty, culture, and depth of feeling, Olga Ladyzhenskaya several times risked herself and her career to help individuals oppressed by the Soviet regime.

A number of her students, in particular Ludwig Faddeev, Nina Ural'tseva, and Vsevolod A. Solonnikov have made major contributions to physics, to partial differential equations, and to the Navier-Stokes equations.

BIBLIOGRAPHY

For a complete bibliography of Ladyzhenskaya's writings, see G. A. Seregin and N. N. Ural'tseva, "Ol'ga Aleksandrovna Ladyzhenskaya (on her 80th birthday)." Russian Mathematical Surveys *58 (2003): 395–425.*

WORKS BY LADYZENSKAYA

With N. N. Ural'tseva. *Lineinye i kvazilineinye uravneniia ellipticheskogo tipa.* Moscow, USSR: Nauka, 1964. Translated by Scripta Technica as *Linear and Quasilinear Equations of Elliptic Type* (New York: Academic Press, 1968).

With V. A. Solonnikov and N. N. Ural'tseva. *Lineinye i kvazilineinye uravneniia parabolicheskogo tipa.* Moscow, USSR: Nauka, 1967. Translated by S. Smith as *Linear and Quasilinear Equations of Parabolic Type* (Providence, RI: American Mathematical Society, 1968).

OTHER WORKS

Friedlander, Susan, Peter Lax, Cathleen Morawetz, et al. "Olga Alexandrovna Ladyzhenskaya." *Notices of the American Mathematical Society* 51, no. 11 (December 2004): 1320–1331.

Hopf, Eberhard. "Uber die Anfangswertaufgabe fur die hydrodynamischen Grundgleichungen." *Mathematische Nachrichten* 4 (1951): 213–231.

Leray, Jean. "Sur le movement d'un liquide visqueux emplissant l'espace." *Acta Mathematica* 63 (1934): 193–248.

Struwe, Michael. "Olga Ladyzhenskaya—a Lifelong Devotion to Mathematics." In *Geometric Analysis and Nonlinear Partial Differential Equations,* edited by Stefan Hildebrandt and Hermann Karcher, pp. 1–10. Berlin: Springer, 2003.

Cathleen Synge Morawetz

LAMARCK, JEAN-BAPTISTE

(*b.* Bazentin-le-Petit, Picardie, 1 August 1744; *d.* Paris, 28 December 1829), *botany, invertebrate zoology and paleontology, evolution.* For the original article on Lamarck see *DSB,* vol. 7.

Leslie Burlingame's article on Lamarck, published in the first edition of the *Dictionary of Scientific Biography,* provides a lucid and reliable account of Lamarck's life and work. The present notice is intended to supplement that article by surveying the trends in Lamarck scholarship since the original article appeared, and by updating the bibliography.

Lamarck studies since 1973 have addressed a number of tasks. These have included: analyzing Lamarck's diverse theoretical ventures and the interrelations among them; reconstructing how Lamarck's practice as a naturalist connected to his transformist thinking; developing a more detailed picture of Lamarck's context with respect to the influences on his work and his influence on others; and correcting persistent misconceptions about Lamarck's evolutionary theory.

System Building. With respect to the multiple dimensions of Lamarck's scientific work, Lamarck's evolutionary thinking and his work as a zoologist and paleontologist have attracted the largest share of historical attention. Nonetheless, his botany, chemistry, meteorology, and geology have also been subject to continued scrutiny. Scholars have sought to deepen the understanding of these diverse intellectual enterprises and also to illuminate the relations and differences between them. Thus in the growing literature on Lamarck's physicochemical system (for example, Burlingame, 1981; Conry, 1981; Goux, 1997), one finds both an effort to understand that system on its own terms and an assessment of how Lamarck's ideas about chemistry related to his later transformist biology. Each of these enterprises represent Lamarck's intellectual predilection to system-building (*"l'esprit de système"*), and similar patterns of thought show up in each. As L. J. Burlingame has pointed out, the action of what Lamarck called "the matter of fire," which was so central to Lamarck's physicochemical system, played a key role again when Lamarck came to explain the action of subtle fluids in the formation of life.

However, the logic of Lamarck's broad physicochemical system as set forth in 1794 did not lend itself to an analysis of life. Instead, Lamarck at this point in his career represented life as an incomprehensible principle. Furthermore, in postulating that all minerals were produced by the successive decomposition of the remains of once-living things, his system offered no way to account for life itself. The same was true for his system of geology. Though the geological system he set forth in 1802 afforded all the time necessary for small-scale changes to become large ones, for ocean basins to be displaced, and for climatic change to occur at all points of the Earth, Lamarck continued to promote the notion that all minerals were created by the decomposition of organic bodies, leaving him no way to address how life originated on the Earth in the first place (a point stressed by Corsi, 1983).

In short, the way he structured his physicochemical and geological systems offered no vantage point from which to confront the topic of life's initial formation. When Lamarck came to the idea of spontaneous generation, he came to it from a different angle, the vantage point of his new duties as professor of the "insects, worms, and microscopic animals" at the Muséum d'Histoire Naturelle in Paris, explaining to his students the nature of life in the simplest forms endowed with it. Adopting the idea of spontaneous generation as of 1800 was a critical new step in his thinking, a precondition for the broader explanation of the successive production of life he began presenting in 1802 (Burkhardt, 1977; Corsi, 1983, 1988; Tirard, 2006).

Work as a Naturalist. Scholars have also pursued the relations between Lamarck's broad theorizing and his practice as a naturalist. Burlingame, in her 1973 article, describes Richard Burkhardt's argument that Lamarck's expertise as a conchologist was intimately related to the inspiration of his belief in species change. Briefly stated, Lamarck found himself faced with the problem of explaining the differences between fossil and living shells. Unwilling to endorse the idea that fossils represented species that had all become extinct as the result of a global catastrophe, he concluded that the forms had changed over time. Having once concluded that species are mutable, he then called upon the familiar idea of the inheritance of acquired characters to explain the close relation between habits and forms that, especially in the case of birds, had been a staple of naturalists' commentaries for decades. But it was not the case that Lamarck's broader theory of organic change was simply an extrapolation from the idea of change at the species level: His broader theory reflected his abiding concern with animal classification, his efforts as a teacher to explain to his students the importance of studying invertebrate zoology, and his conclusion that the basic differences among the various classes of invertebrate animals could be explained as the long-term results of moving fluids acting on organic forms and structures. He invoked what he called the "power of life" or "the cause which tends to make organization increasingly complex" to account for the general, linear series that in his view best represented how the different animal classes were related to each other.

Looking at Lamarck's later work in invertebrate zoology and paleontology has likewise proved instructive. Goulven Laurent's (1987) comprehensive study of French paleontology from 1800 to 1860 has looked at Lamarck's broader theorizing in conjunction with the details of his "Memoirs on the Fossils of the Paris Region" (1802–1806) and his great, seven-volume treatise, *Histoire Naturelle des Animaux sans Vertèbres* (Natural history of the invertebrates, 1815–1822), paying attention to the later volumes of the work instead of just the first volume, where Lamarck gave his last major presentation of his whole zoological theory. Laurent highlights the tenacity with which Lamarck pursued the task of describing and naming hundreds of previously unidentified species of fossil, and he emphasizes the importance of the way Lamarck described fossils and living forms together instead of representing them as wholly separate creations. A similar emphasis on Lamarck's practice as a naturalist can be found in Burkhardt's overview of Lamarck's work with species over the course of his entire career (1985) and his discussion of the relations between Lamarck's work as a cabinet naturalist and the information collected by the field naturalist François Péron (Burkhardt, 1997).

The relations between Lamarck's broad theorizing and his ongoing experience as a naturalist have also been explored by Stephen Jay Gould. Gould describes how Lamarck's decision to make separate classes of the annelid and parasitic worms led him over time to give up his original commitment to a linear arrangement of the different animal classes and to adopt a truly branching scheme instead (2000). Gould indicates that in the very last of his publications Lamarck reversed his decades-long belief that the "power of life" was greater than the influence of environmental circumstances, admitting to the contrary that the force of circumstances was ultimately stronger than nature herself. Gould represents this as a fundamental change in Lamarck's philosophy of nature that must have greatly distressed him.

In fact, however, Lamarck's late writings on classification and on nature reveal no philosophical discomfort on his part. His later writings about the linear progression (or lack of it) in the animal scale do not represent this as fundamental to his philosophy of nature but instead as a pedagogical or taxonomic principle, to be followed as long as it facilitated one's studies, but not beyond that. Thus in 1818 in Volume 5 of his *Histoire Naturelle des Animaux sans Vertèbres,* in discussing classifying the annelids, he states, "Who does not feel here the inconvenience of being obliged to form a simple series, when nature could not make a similar one in the order of her productions!" (p. 276). That said, Gould's essay is a fine contribution to the ongoing study of how Lamarck's theorizing and his practice as a naturalist interrelated. Giving up his linear arrangement of living things for a strongly branching view of life's development, and acknowledging that the influence of the environment had had a greater role in evolution than the "cause which makes organization increasingly complex," Lamarck showed his willingness to let the evidence of natural history be the final arbiter in determining the complex path nature had traced, over time, in bringing all the different forms of life into existence.

Influences. Another significant aspect of Lamarck scholarship over the three decades since Burlingame's original article has been the enlarging of the cast of characters who expressed ideas of a transformist or quasi-transformist character in the late-eighteenth and early-nineteenth centuries—or whose ideas in other ways help illuminate Lamarck's thinking. Pietro Corsi's book *The Age of*

Lamarck (1988) stands out in this regard. Corsi identifies in particular a "Buffonian" camp of naturalists and writers who felt shut out by the newer, more technical, positivistic orientation promoted by Georges Cuvier and who, unlike Cuvier, did not scorn the kind of broad, systematic views of nature that Lamarck sought to construct. The picture that emerges is of a complex and diversified world of professional and popular natural history where Cuvier's power was not so dominant as earlier scholars were disposed to suggest—a point also made by Dorinda Outram (1984)—and one which also shows Lamarck's aspirations as a *naturaliste philosophe* to have been less unique than Lamarck himself was inclined to suggest.

Unfortunately, Lamarck's habit of setting forth his theorizing in a deductive fashion, without identifying his intellectual debts or otherwise specifying other writers to whom he may have been responding, continues to make it difficult to reconstruct precisely which potential influences were in fact significant for him. Be that as it may, a 2006 article by Corsi makes an excellent case that Lamarck abandoned his 1802 project of developing a new science of "biology" not because of ill health, as Lamarck maintained, but because he recognized that the new political climate would not look with favor on his materialistic explanation of all vital phenomena. Corsi argues that Lamarck revived his hopes when the Restoration seemed to promise a greater liberty of expression, but that these hopes were soon dashed, and Lamarck's last writings reflected his efforts to avoid being accused of materialism or atheism.

Corsi (1997) has also pursued the question of Lamarck's influence on others, including organizing a project to identify and track the intellectual careers of as many as possible of the students who registered over the years in Lamarck's course on invertebrate zoology at the museum. Attendance in Lamarck's classes was no guarantee of conversion to Lamarck's ideas, but it is an index of some exposure to these ideas, and Corsi has identified a number of individual scientists and students of Lamarck, such as the Italian Giosué Sangiovanni, who can be seen as Lamarck disciples (1984). More generally, scholars have identified increasing numbers of writers who, prior to Darwin's publication of *The Origin of Species*, knew of Lamarck's ideas and were sympathetic to the notion of species transformation (e.g., Laurent, 1987; Desmond, 1989; Secord, 1991; Corsi, 2005).

Correcting Misconceptions. With respect to common misconceptions about Lamarck's thinking, two are sufficiently prevalent to require recurrent attention. The first is that the inheritance of acquired characters was Lamarck's primary explanation of organic change. The second is that Lamarck believed that animals could gain new organs as

Jean Baptiste Lamarck. © BETTMANN/CORBIS.

the result of wishing for them. Burlingame's *DSB* article of 1973 properly observes that neither caricature of Lamarck's thinking is correct. For readers who may be coming to Lamarck for the first time, however, this observation is worth repeating. Although the idea of the inheritance of acquired characters did play a role in Lamarck's theorizing, it was neither an idea for which he claimed credit nor the keystone of his evolutionary theory. As for the assertion that "wishing" played a role in his explanation of organic change, that is simply not true.

Beyond these misconceptions about Lamarck's theorizing, the scholarly literature has seen another characterization of Lamarck's views acquire more credence than the evidence warrants. This is the idea that Lamarck, in coming to believe in evolution, converted the scale of nature into what Charles Gillispie engagingly called an "escalator of being" (1959, p. 271). Peter Bowler has carried this notion further, saying Lamarck believed that "each point of the scale of being we observe today has been derived by progression from a separate act of spontaneous generation" (1984, p. 80). Although this might appear to be a logical conclusion from some of Lamarck's statements about the effects of "the power of life," it is not a conclusion that Lamarck ever advanced himself, nor does it correspond to his conclusion that the force of circumstances was ultimately superior to the tendency to increased complexity.

Lamarck liked to think of himself as a naturalist-philosopher. Throughout his career, there was always a tension between his ambitions as a theorist and his experience as a naturalist. For historians of science, Lamarck will continue to pose the challenge of understanding how his career and thinking were shaped by the specific scientific, institutional, cultural, and political circumstances in which he operated.

SUPPLEMENTARY BIBLIOGRAPHY

Bange, Christian, and Pietro Corsi. "Œuvres et rayonnement de Jean-Baptiste Lamarck." Available from http://www.lamarck.cnrs.fr. Includes an elaborate chronology of Lamarck's career and provides ready access to complete transcriptions of the greater part of Lamarck's books and to many of his papers and other writings. It also provides a selected bibliography of historical studies of Lamarck.

Barsanti, Giulio. *Dalla storia naturale alla storia della natura: Saggio su Lamarck.* Milan: Feltrinelli, 1979.

———. "Lamarck and the Birth of Biology." In *Romanticism in Science: Science in Europe, 1790–1840,* edited by Stefano Poggi and Maurizio Bossi. Dordrecht and Boston: Kluwer, 1994.

Bowler, Peter J. *Evolution: The History of an Idea.* Berkeley: University of California Press, 1984.

Burkhardt, Richard W., Jr. *The Spirit of System: Lamarck and Evolutionary Biology.* Cambridge, MA: Harvard University Press, 1977. Published with a new preface, 1995.

———. "Lamarck and Species." In *Histoire du concept d'espèce dans les sciences de la vie,* edited by Scott Atran, et al. Paris: Fondation Singer-Polignac, 1987.

———. "Unpacking Baudin: Models of Scientific Practice in the Age of Lamarck." In *Jean-Baptiste Lamarck,* edited by Goulven Laurent. Paris: Éditions du CTHS, 1997.

Burlingame, Leslie J. "Lamarck's Chemistry: The Chemical Revolution Rejected." In *The Analytic Spirit,* edited by Harry Woolf. Ithaca, NY: Cornell University Press, 1981.

Conry, Yvette. "Une lecture newtonienne de Lamarck. Est-elle possible?" In *Lamarck et son temps; Lamarck et notre temps: colloque international dans le cadre du Centre d'Études et de Recherches interdisciplinaires de Chantilly.* Paris: Vrin, 1981.

Corsi, Pietro. "The Importance of French Transformist Ideas for the Second Volume of Lyell's *Principles of Geology.*" *The British Journal for the History of Science* 11 (1978): 221–244.

———. *Oltre il mito: Lamarck e le scienze naturali del suo tempo.* Bologna: Il Mulino. 1983. Published as *The Age of Lamarck: Evolutionary Theories in France, 1790–1830,* translated by Jonathan Mandelbaum. Berkeley: University of California Press, 1988.

———. "Lamarck en Italie." *Revue d'Histoire des Sciences* 37 (1984): 47–64.

———. "Les élèves de Lamarck: un projet de recherché." In *Jean-Baptiste Lamarck,* edited by Goulven Laurent. Paris: Éditions du CTHS, 1997.

———. "Before Darwin: Transformist Concepts in European Natural History." *Journal of the History of Biology* 38 (2005): 67–83.

———. "Biologie." In *Lamarck, philosophe de la nature,* edited by Pietro Corsi, Jean Gayon, Gabriel Gohau, and Stéphane Tirard, Paris: Presses Universitaires de France, 2006.

Desmond, Adrian. *The Politics of Evolution: Morphology, Medicine, and Reform in Radical London.* Chicago: University of Chicago Press, 1989.

Duris, Pascal. "Lamarck et la botanique linnéenne." In *Jean-Baptiste Lamarck,* edited by Goulven Laurent. Paris: Éditions du CTHS, 1997.

Gayon, Jean. "Hérédité des caractères acquis." In *Lamarck, philosophe de la nature,* edited by Pietro Corsi, Jean Gayon, Gabriel Gohau, et al. Paris: Presses Universitaires de France, 2006. A penetrating analysis of the phrasing and formulation of the idea of the inheritance of acquired characters, relating to Lamarck's own work and time and the development of the concept of heredity in the nineteenth century.

Gillispie, Charles C. "Lamarck and Darwin in the History of Science." In *Forerunners of Darwin: 1745–1859,* edited Bentley Glass, Owsei Temkin, and W. L. Straus. Baltimore: Johns Hopkins University Press, 1959.

Gohau, Gabriel. "L'Hydrogéologie et l'histoire de la géologie." In *Jean-Baptiste Lamarck,* edited by Goulven Laurent. Paris: Éditions du CTHS, 1997.

Gould, Stephen Jay. "A Tree Grows in Paris: Lamarck's Division of Worms and Revision of Nature." In *The Lying Stones of Marrakech: Penultimate Reflections in Natural History.* New York: Harmony Books, 2000.

Goux, Jean-Michel. "Lamarck et la chimie pneumatique à la fin du XVIIIe siècle." In *Jean-Baptiste Lamarck,* edited by Goulven Laurent. Paris: Éditions du CTHS, 1997.

Jordanova, Ludmilla J. *Lamarck.* Oxford: Oxford University Press, 1984.

Laurent, Goulven, *Paléontologie et évolution en France 1800–1860: une histoire des idées de Cuvier et Lamarck à Darwin.* Paris: Editions du Comité des travaux historiques et scientifiques, 1987.

———. "Idées sur l'origine de l'homme en France de 1800 à 1871 entre Lamarck et Darwin." *Bulletins et Mémoires de la Société d'Anthropologie de Paris* 1 (3–4, 1989): 105–130.

———. "Lamarck, Jean Baptiste Pierre Antoine de Monet, chevalier de (1744–1829)." In *Dictionnaire du Darwinisme et de l'Évolution,* edited by Patrick Tort. Paris: Presses Universitaires de France, 1996.

Laurent, Goulven, ed. *Jean-Baptiste Lamarck (1744–1829).* Paris: CTHS, 1997. An important collection of scholarly papers on all aspects of Lamarck's life and work.

Outram, Dorinda. *Georges Cuvier: Vocation, Science, and Authority in Post-Revolutionary France.* Manchester, U.K.: Manchester University Press, 1984.

Richards, Robert J. *Darwin and the Emergence of Evolutionary Theories of Mind and Behavior.* Chicago: University of Chicago Press, 1987.

Secord, James A. "Edinburgh Lamarckians: Robert Jameson and Robert E. Grant." *Journal of the History of Biology* 24 (1991): 1–18.

Tirard, Stéphane. "Générations spontanés." In *Lamarck, philosophe de la nature,* edited by Pietro Corsi, Jean Gayon, Gabriel Gohau, et al. Paris: Presses Universitaires de France, 2006.

Richard W. Burkhardt Jr.

LAMB, HUBERT HORACE

(*b.* Bedford, United Kingdom, 22 September 1913; *d.* Holt, Norfolk, United Kingdom, 27 June 1997), *synoptic meteorology, historical climatology, paleoclimatology, climatic change.*

Through his wide knowledge and deep understanding of climatology Lamb was universally regarded as one of the leading scientists working in the field during the late twentieth century. His research principally brought to light the constantly changing nature of climate and its effect on both the planet and human beings. In 1971, after a long and distinguished career in the U.K. Meteorological Office, Lamb founded and became the first director of the Climatic Research Unit in the School of Environmental Sciences at the University of East Anglia, Norwich, England.

Family Background. Lamb came from a strict Anglican and highly talented Stockport (Manchester) family. His grandfather was the mathematical physicist, Sir Horace Lamb, whose book, *Hydrodynamics,* became a classic text for students of meteorology; his father, Ernest, was a professor of engineering and his mother was Lilian (née Brierley). An aunt was a tutor in charge of one of the halls at Newnham College, Cambridge; another was a leading archaeologist; while Lamb's uncle, Henry Lamb, was an accomplished painter.

When, in 1914, Lamb's father took up a professorship in the East London College (now Queen Mary College), the family moved from Bedford to Hampstead Garden Suburb. Only 5 miles from central London, this was a pleasantly situated suburb with then open country nearby; it was also an interesting political community with leading members of all the three main parties, Conservative, Labour, and Liberal, living close by the Lamb household.

School and University. In his early school years, Lamb's best friend was Trevor Huddleston, who later, as an Anglican priest, became a staunch supporter of civil rights; as adults they always kept in touch. Lamb was also at school with Olaf, the eldest son of Lewis Fry Richardson, pioneer of numerical weather forecasting. Lamb and Olaf also became good friends and between the ages of eight and twelve Lamb often visited the Richardson household. Although Lamb did not come into contact with Olaf's father in the field of meteorology, the elder Richardson's Quakerism and his interpretation of Christianity influenced Lamb in later life.

Lamb's secondary schooling was at Oundle, a school in Rutland, which had a strong influence upon him. Lamb later expressed disapproval of boarding schools, except when they were absolutely necessary, and thought children should live at home, as he had done, up to the age of thirteen. Because the discipline of Oundle was more congenial than that imposed by his father, Lamb had two conflicting authorities in his life at that time. For instance, on returning to school after a summer holiday, the fifteen-year-old found that he had been transferred, without warning or explanation, to the science side. Although pleased to study science, Lamb resented having to abandon history and languages. Earlier, through an adopted aunt who was Danish, together with the Norwegian folktales that his father's generation had been brought up on, Lamb had become interested in Denmark and Norway and was keen to learn Scandinavian languages. A bonus from his school transfer was a German lesson once a week reading *Grimm's Fairy Tales.*

As a teenager Lamb was concerned about living up to family expectations. He went up to Cambridge to study natural sciences. Having won mathematics prizes at Oundle he was expected to read the subject at the university. However, after two years a further crisis arose because he wanted, again against his father's wishes, to abandon natural sciences and enter the geography school. Nevertheless, Lamb completed the two-year geography course in one year and graduated with an MA in the first parts of both the natural science and geography courses. The broad interdisciplinary platform for his career began to take shape.

Early Career. After leaving Cambridge, Lamb applied to join the U.K. Meteorological Office. He was interviewed by the director, Sir George Simpson, and sent for training, under Sidney P. Peters, to Croydon Airport, the main London airport of the 1930s. After Croydon, Lamb was posted to Montrose, which suited him admirably, since, when off duty, he was able to enjoy exploring the Scottish mountains and hills. Lamb had developed an early love for hill walking and claimed that observing the variations of weather in upland areas had triggered his interest in meteorology. He also spent a holiday sailing around Iceland and during this period published his first paper, "Climate and Legend in Norway."

Hubert Horace Lamb. *Hubert Horace Lamb drawing.* COURTESY OF B. MOIRA LAMB.

In 1939, he was elected Fellow of the Royal Meteorological Society and became a regular contributor to its publications, the *Quarterly Journal* and *Weather*. Later Lamb served on its council (1956–1959) and as vice president (1975–1977); he was elected honorary member in 1985 and two years later awarded the Symons Gold Medal.

During the late 1930s Lamb, with his views as a conscientious objector, became concerned about the increasing threat of war. This situation came to a head in July 1939 when he received notice to take part in a poison-gas spraying exercise. As he refused, the Meteorological Office demanded that he should immediately resign. Nevertheless, he was later asked to reconsider his position and was offered a posting, under secondment (a temporary transfer), to the Meteorological Service of neutral Ireland to work at Foynes, a flying-boat base on the Shannon estuary. Here he was appointed instructor of a forecasting course for Irish graduates who had been expecting to complete their training at the Massachusetts Institute of Technology; he also undertook the training of the first intake of Irish meteorological assistants in weather observing duties.

From 1941 to 1945 Lamb was in charge of the Meteorological Office at Foynes preparing forecasts for transatlantic flights; he also developed analysis and prediction techniques for these very long-range missions. This transatlantic service provided a vital wartime link for the Allies; many VIPs, including Anthony Eden, traveled to and from America via Foynes. In 1942 Lamb himself took a flight on this route when he attended meetings in Toronto and New York.

In 1946, following his return to the U.K. Meteorological Office, Lamb showed his enterprising spirit by volunteering to be the meteorologist on an expedition to the Antarctic. This was undertaken in the whaling ship *Balaena,* accompanied by two amphibious aircraft for making sorties to spot bad weather and ice. Lamb obtained valuable firsthand experience of weather types and circulation patterns in the Southern Hemisphere from this voyage.

Climatic Research. On his return to the United Kingdom Lamb joined the Long-Range Forecasting Research Division of the Meteorological Office, where he carried out research on subjects such as natural seasons, persistent weather spells, and recurrent episodes or singularities. After overseas forecasting duties in Germany and the Mediterranean, Lamb was posted to the Climatological Division of the Meteorological Office in the United Kingdom, where he discovered an immense archive of virtually untapped historical weather data. By processing and analyzing this material into a systematic framework during the 1950s and 1960s he found that it was possible to reconstruct meaningful circulation patterns for past climatic periods. This work stimulated further studies that opened up new avenues and was the beginning of the period when Lamb became seriously involved in climatic research.

By defining timescales of past climatic periods and reconstructing circulation patterns, Lamb discovered that change, sometimes occurring quite rapidly, was the norm. This challenged received opinion, which up to the mid-twentieth century had generally postulated that the present climate was more or less stable and that events such as ice ages and the melting of polar ice caps were matters of the distant past or future.

Once he had established firm evidence of variations in climate, Lamb's central concern lay in developing a rigorous methodology that would provide a better understanding of climatic history, for this, he argued, would also illuminate trends visible in recent records. Through his all-embracing study of climatic change, Lamb came to realize that before future climate could be predicted, there was a need to understand the behavior of past and present climate.

Lamb recognized that large-scale spatial and long-term temporal atmospheric variations that characterize climatic change must be as deeply rooted in physics as the short-term transient and dynamic weather systems studied by meteorologists. Though his awareness of the huge energy budgets of climatic systems made him cautious about assigning human agency a major role in the precipitation of change, his research ultimately focused on the consequences of both natural and human-driven changes.

This work revealed his remarkable gift for identifying and extracting trends and new significance from what otherwise had been regarded as routine climatic data. It was mainly in recognition of this pioneering research that in 1963 Lamb was awarded a special merit promotion within the Meteorological Office to senior principal scientific officer. This prestigious position in the scientific civil service is equivalent to that of a university research professorship and, in a similar way to academic life, provides release from routine duties to allow full-time concentration on a specialized area of research.

Such a promotion embodied an implicit guarantee of research assistance and adequate administrative support. Had this been forthcoming after the retirement of Lamb's supportive leader, Sir Graham Sutton (a highly respected and perceptive postwar director-general), the U.K. Meteorological Office would have established itself a decade or so earlier as one of the world leaders in the study of climatic change.

In the 1970s Lamb calculated that by 2000, atmospheric changes resulting from human activities would begin to affect the course of natural climatic changes and later stated that, whatever its cause, climatic change has an enormous impact on human affairs. In this sense Lamb was a visionary, although he was far too modest to see himself in this light.

Sadly, Lamb's appeals for assistance with his research were repeatedly ignored. Although many contacts with scientists outside the Meteorological Office confirmed the value of his research, the situation within the organization was very different. While financial resources were generally available, it was decided that they should be mostly channelled into installing powerful new computers for short-term forecasting.

Academic Career. Stymied by the organization he had served for over thirty years, Lamb, despite being at an age when most people are thinking of retirement, sought outside help during the late 1960s to continue his research. Although this exercise took one or two years, with the timely founding of the School of Environmental Sciences under Professor Keith Clayton at the University of East Anglia, and financial backing from the Nuffield Foundation and North Sea oil companies, Lamb resurfaced in 1971 as professor and founding director of the Climatic Research Unit in Norwich.

This was a courageous and risky academic venture. The unit had no secure core funding, salaries were uncertain, and cynics expected the whole venture to collapse within a decade. However, Lamb's enterprising personality was a source of inspiration both to colleagues and students and after only six years, when Lamb retired, the Climatic Research Unit was enjoying a prestigious reputation as one of the world's leading centers in the field. Lamb undoubtedly achieved a great and lasting success with this venture.

As a man, Lamb was respected for his kindness and loyalty, and professionally for his vision and ability to synthesize a mass of diverse material into a meaningful whole, as best exemplified by his monumental two-volume work, *Climate: Present, Past and Future.* With undiminished energies in retirement, Lamb still found time to keep in contact with the Climatic Research Unit and produce a steady stream of notable publications as well as becoming president of the North Norfolk Liberal Democrats. After his death, Lamb left a widow, Moira, and three grown-up children, Kirsten, Catherine, and Norman; the last following in his father's footsteps politically, is the Liberal Democrat Member of Parliament for North Norfolk.

BIBLIOGRAPHY

The Lamb papers are held in subject-listed boxes and files at the Climatic Research Unit Library. A complete list of Lamb's publications is given on the Climatic Research Unit Web site: http://www.cru.uea.ac.uk/.

WORKS BY LAMB

"Climate and Legend in Norway." *Quarterly Journal of the Royal Meteorological Society* 65 (1939): 510.

"Types and Spells of Weather around the Year in the British Isles: Annual Trends, Seasonal Structure of the Year, Singularities." *Quarterly Journal of the Royal Meteorological Society* 76 (1950): 393–438.

With A. I. Johnson. *Secular Variations of the Circulation since 1750.* Geophysical Memoirs (Great Britain. Meteorological Office), 14, no. 110. London: Her Majesty's Stationery Office for Meteorological Office, 1966.

"Volcanic Dust in the Atmosphere; with a Chronology and Assessment of Its Meteorological Significance." *Philosophical Transactions of the Royal Society,* series A, 266 (1970): 425–533.

British Isles Weather Types and a Register of the Daily Sequence of Circulation Patterns 1861–1971. Geophysical Memoirs (Great Britain. Meteorological Office), 16, no. 116. London: Her Majesty's Stationery Office for Meteorological Office, 1972.

Climate: Present, Past and Future. Vol. 1, *Fundamentals and Climate Now.* London: Methuen, 1972.

Climate: Present, Past and Future. Vol. 2, *Climatic History and the Future.* London: Methuen, 1977.

Climate, History and the Modern World. London: Methuen, 1982. 2nd ed. London: Routledge, 1995.

Weather, Climate and Human Affairs. London: Routledge, 1988.

With Knud Frydendahl. *Historic Storms of the North Sea, British Isles and Northwest Europe.* Cambridge, U.K.: Cambridge University Press, 1991.

Through All the Changing Scenes of Life: A Meteorologist's Tale. East Harling, U.K.: Taverner Publications, 1997. Lamb's autobiography.

OTHER SOURCE

Hulme, Mike, and Elaine Barrow, eds. *Climates of the British Isles: Present, Past and Future.* London: Routledge, 1997. Dedicated to Lamb.

John A. Kington

LANDSBERG, HELMUT ERICH (*b.* Frankfurt am Main, Germany, 9 February 1906; *d.* Geneva, Switzerland, 6 December 1985), *geophysics, climatology.*

Landsberg had a significant impact on the development of the geophysical sciences in the mid- to late twentieth century. He raised the status of climatology in the United States from an exercise in geographic description to a well-developed applied physical science. His broad-ranging interests produced pioneering studies linking atmospheric and social phenomena. In his work both in academia and government, he used his organizational talents to advance the disciplines of meteorology and climatology.

Origins, Education, and Emigration to the United States. Helmut Erich Landsberg was born in Frankfurt am Main. His father, Georg Landsberg, was a physician who died of tuberculosis when Landsberg was about three years old. An only child, Landsberg was raised by his mother, Clare Zedner Landsberg, whom he described as a housewife, but with a good education. He attended the city's Woehler Realgymnasium High School, and then he went on to study at the University of Frankfurt, where he received a solid grounding in physics, mathematics, and geosciences. After further studies at Frankfurt's Institute of Meteorology and Geophysics, he received his PhD in 1930. Landsberg wrote his doctoral dissertation under Professor Beno Gutenberg on the subject of seismographic instrumentation for measuring earthquakes. He held two positions at the University of Frankfurt, in both cases working under Professor Franz Linke, professor of meteorology. The first position was as a postdoctoral assistant to Linke in seismology and climatology; his main task was to set up a weather station in the Rhineland to study frosts in vineyards and ways to mitigate their effects with heaters. He then served, from 1931 to 1934, as chief of forecasting at the Taunus Observatory near Frankfurt, run by the university's Institute of Meteorology and Geophysics; among other duties, he produced weather forecasts for a local airline between Cologne and Frankfurt.

Landsberg emigrated to the United States in 1934, at age twenty-eight, when he accepted a position at Pennsylvania State College (Penn State; since 1953, Pennsylvania State University) as assistant professor of geophysics. He had inquired with his dissertation advisor about the possibility of a position in the United States, and Gutenberg, who had moved to the United States in 1930, recommended him for the position at Penn State. Landsberg continued to teach there until the end of the 1930s. His establishment of meteorological and geophysical laboratories, as well as some interdisciplinary studies, at Penn State expanded the curriculum. In 1941 Landsberg moved to the University of Chicago to become an associate professor in the Department of Meteorology.

Pre–World War II Researches. The scope of Landsberg's research interests were evident in his first writings published in Germany, and they continued to develop steadily in the United States in the late 1930s. At the Taunus Observatory, he had the opportunity to make upper air observations and (in cold and foggy weather) to read the scientific literature in its excellent library. There Landsberg began the prolific and varied output that continued throughout his long career, publishing an article approximately every two months over the course of five decades. His early papers already exhibited his characteristic interest in exploring the interactions between meteorological and atmospheric phenomena and a range of human activities.

One strand of Landsberg's work was a series of papers on earthquakes and related issues of mining safety, including the hazard of dust. This led him into the broader topic of atmospheric condensation nuclei, on which he published a monograph in 1938. Going beyond their meteorological role in cloud formation to their biological effects, the study established the importance of microscopic particles in air pollution and noted the retention of submicron particles in human lungs. Another strand of Landsberg's work was his interest in observing phenomena such as air masses as part of the local climatology. The papers reflect his focus on observational science, its instrumentation and methodology, and statistical analysis of data. In 1941 Landsberg published a landmark textbook titled *Physical Climatology,* which introduced an American student audience to the concept of climatology as an applied physical science utilizing statistical data. It was reprinted several times and was issued in a revised edition in 1958.

Wartime Career and Government Service. Beginning in 1941 Landsberg worked in some very different environments, first in the strenuous climate of World War II and then in postwar government agencies, both military and civilian. The American academic meteorological community was called on to support the United States during the war through expanding forecasting capabilities for military operations. Starting in 1941, five major American departments of meteorology (at the University of Chicago, California Institute of Technology, Massachusetts Institute of Technology, University of California at Los Angeles, and New York University) began offering postgraduate courses in weather forecasting to Army Air Corps cadets.

Besides participating in the teaching program, Landsberg supervised Chicago's large Military Climatology Project, run by the Institute of Meteorology. Under contract to the Weather Directorate at Headquarters, Army Air Forces, the project prepared climatological atlases for bombing missions. Landsberg was editor of a number of these reports for areas in Europe. In 1944 he also authored a climatic study of cloudiness over Japan. In conjunction with preparing these climatological reports, Landsberg traveled widely in order to evaluate weather statistics for air routes and various types of operations. Later on he worked full time as an operations analyst for the Eighth and the Twentieth Air Forces, making extended visits to both the European and Pacific theaters. Landsberg met his future wife, A. Frances Simpson, a registered nurse, during World War II. They married in 1946 and had a son, Bruce S. Landsberg.

Helmut Landsberg. COURTESY OF BRUCE LANDSBERG AND RUTH LIEBOWITZ.

After the war, Landsberg joined the U.S. Weather Bureau, where he became chief of the Section of Industrial Climatology. After only a few months there, in late 1946 he moved over to the government's military side, to join what historians of science such as Stuart Leslie have termed the "permanent mobilization of science" in support of future defense that followed the end of World War II. Over the next eight years, he held two administrative positions relating to this effort. The first was with the new Joint Research and Development Board (JRDB) that the secretaries of war and the navy had created in 1946 to fill the void left by the rapid phaseout of the wartime Office of Scientific Research and Development. The JRDB proceeded to create a set of panels and committees to cover various technical areas, one of them being a Committee on Geophysical Sciences. Landsberg became the deputy executive director for this committee, which had its first meeting on 18 December 1946. One of his tasks was to prepare the 1948 *Survey of Scientists Engaged in Geophysical Researches* to assess the need to train more scientists in the field. He also prepared the classified report *Geophysics and Warfare* (1948), which was published in 1954 in an updated, unclassified version. By mid-1951, when Landsberg once again moved on, the Department of Defense had been created, the JRDB had been replaced by the more powerful Research and Development Board (RDB), his committee had been expanded to include geography, and he had become its executive director. He also became associate editor of the *Journal of Meteorology* in 1951 and continued in this capacity until 1960.

The second position that Landsberg held in the Department of Defense was as director of the Geophysical Research Directorate (GRD). This directorate, together with an Electronic Research Directorate, made up the new Air Force Cambridge Research Laboratories (AFCRL). In his time there, between mid-1951 and mid-1954, Landsberg had the responsibility of implementing the geophysical research and development program for the air force that he had worked to plan and fund while at the RDB. From all accounts, GRD flourished under Landsberg's management. By 1954 he had expanded GRD's contract program with universities, hired high-quality staff to build up its in-house capabilities (including its library), and developed new program areas. Landsberg was an able publicist for his organization, shown most notably in his *Geophysics and Warfare.* He also became editor for the new series, *Advances in Geophysics,* published by Academic Press. Its first annual volume came out in 1952. Landsberg continued as the primary editor through the first nineteen volumes of the series.

In 1954 Landsberg decided to return to the U.S. Weather Bureau, where he had been offered the position of director of the Climatological Services Division (after 1956 the Office of Climatology), and he remained there until 1965. Landsberg made it his business to upgrade the collection of climatological data and to better organize the available records. For example, after sponsoring a study to determine the required density of a network to more accurately measure rainfall, he expanded the number of stations set up for volunteer observers to collect climate data. He also brought in computers to assist in organizing, centralizing, and maintaining climatological records. One of his major projects at the Weather Bureau was to work toward establishing state and regional climatologists across the country. Landsberg also prepared new *World Maps of Climatology* that were published in two editions in 1963 and 1965.

During these middle decades of Landsberg's career, his research continued to explore a range of atmospheric phenomena and their interactions with human society. Even with his full-time administrative work, he continued to publish, if at a somewhat slower rate. The appearance of his textbook *Physical Climatology,* followed by definitive articles on climatology in the 1945 *Handbook of Meterology* and the 1951 *Compendium of Meteorology,* made him a recognized authority in the field. Landsberg's 1946 paper, "Climate as a Natural Resource," evinced his growing interest in the relevance of climatology for many human activities and the need to consider it in plans for industry, construction, and agriculture. At the same time, Landsberg delved into related weather phenomena and air pollution. He began to study these interactions on various scales ranging from microclimates to local and regional climates and global phenomena. His researches made a significant contribution to the emerging new sciences of bioclimatology and biometeorology. One end product of these studies was his widely read introductory book *Weather and Health,* published in 1969.

The University of Maryland and "Retirement" Years. In the mid-1960s Landsberg made a transition from government work back into academia. When his Office of Climatology at the U.S. Weather Bureau was transformed into the Environmental Data Services component of the newly founded Environmental Science Services Administration, he became director of Environmental Data Services for about a year. Then, in 1967, he accepted an appointment as research professor at the University of Maryland at College Park in the Institute for Fluid Dynamics and Applied Mathematics. During 1974–1976, the last two years before he retired from the university, he served as the director of the institute. At Maryland, Landsberg was instrumental in founding a separate Department of Meteorology and nurturing a graduate program in the discipline. He also took on another major editorial task in the mid-1960s as editor-in-chief of the *World Survey of Climatology.* After his official retirement, he became a professor emeritus at the university, a position he held until his death in 1985. His "retirement" years were ones of intense professional activity. Landsberg continued his affiliation with the World Meteorological Organization (WMO), which had begun in 1969 when he had become president of the WMO's Commission for Special Applications of Meteorology and Climatology. In December 1985 Landsberg was attending a congress of the WMO in Geneva, Switzerland, when he suddenly collapsed; he died on 6 December, just two months short of his eightieth birthday.

Landsberg's research in the last two decades of his life yielded many fruitful results. One was his series of studies starting in the 1950s on the climatology of cities. He published a seminal paper in 1979 titled "Atmospheric Changes in a Growing Community (The Columbia, Maryland, Experience)," showing how the creation of a new city had affected the local climate of the region. In his book *The Urban Climate* (1981), published when he was seventy-five, Landsberg presented a major research monograph discussing the last forty years of studies on the dynamics of urban heat islands. Another area in which Landsberg made an important contribution was his investigations of larger regional and global trends in climate. To try to ascertain whether there had been long-term climatic fluctuations or alterations in the past, scientists had recourse to historical records. Landsberg compiled a number of extended series of regional climate data, including a series on the precipitation record in the Boston area that went back to 1751, which he published in 1967.

The subject of climate change was coming to public attention by the 1970s, and one question raised was whether human activities were altering the climate. In an important article published in *Science* in 1970, Landsberg reviewed the scientific literature regarding human influences on climate. His overall conclusion was that he regarded the available historical records and studies to date as insufficient evidence to demonstrate any significant human effect on global climate. He was, by contrast, quite definite that the evidence showed that human activities were making major alterations to the climate on regional and local scales and, as one example, he cited the climatological studies on cities. Some dramatic global meteorological events over the course of the 1970s (droughts, abnormal cold and heat waves, famine in Africa), together with the oil crisis, increased public concerns in this area. As J. Murray Mitchell, one of Landsberg's memorialists, pointed out, climatologists were now in the position of being pressed for answers on global climate change (Baer, Canfield, and Mitchell, 1991). He noted that, despite this pressure, Landsberg continued to deliver balanced and judicious comments on the issue and the state of current research. His invited scientific lecture at the 1983 World Meteorological Congress, "The Value and Challenge of Climatic Predictions," represented his final magisterial review of the subject.

Honors and Awards. During his lifetime, Helmut Landsberg held leadership positions in a number of scientific and professional organizations including the American Geophysical Union (AGU), serving as the AGU's vice president from 1966 to 1968 and its president from 1968 to 1970. His accomplishments in many scientific fields are reflected in his election to the National Academy of Engineering and his selection as a Fellow by the Royal Meteorological Society, the American Academy of Arts and Sciences, the American Association for the Advancement of Science, the AGU, the American Meteorological Society (AMS), and the Washington Academy of Sciences. Among the most important awards Landsberg received were the William Bowie Medal from the AGU in 1978, and the International Meteorological Organization prize from the WMO in 1979. The following year, the German Meteorological Society presented Landsberg with the Alfred Wegener Medal. In 1983 he received the Cleveland Abbe Award from the AMS. Landsberg was honored with the William F. Peterson Foundation Award for outstanding accomplishments in the field of biometeorology in 1983. The Presidential Medal of Science was presented to him by President Ronald Reagan at a White House ceremony on 27 February 1985.

Landsberg's colleagues had been planning a symposium in his honor for his eightieth birthday. When his death intervened just before the event, it was changed to a memorial tribute; the proceedings of the February 1986 symposium were published five years later. After Landsberg's death, several new awards were named in his honor. As part of the National Weather Service's Cooperative Weather Observer Awards Program, the Helmut E. Landsberg Award was established to recognize individuals who had completed sixty years of service as cooperative observers.

BIBLIOGRAPHY

The papers of Helmut E. Landsberg, an extensive collection comprising 20.25 linear feet and covering his entire life from 1906–1985, are deposited in the Archives and Manuscripts Department, University of Maryland Libraries, Hornbake Library, College Park, MD. A chronological listing of Landsberg's scientific writings is included in Baer, Ferdinand; Norman L. Canfield; and J. Murray Mitchell, eds., 1991, cited below.

WORKS BY LANDSBERG

"Origin and Occurrence of Earthquakes." *Proceedings of the Pennsylvania Academy of Sciences* 11 (1937): 88–92.

"Atmospheric Condensation Nuclei." *Ergebnisse der Kosmischen Physik III* (Gerlands Beiträge zur Geophysik, Dritter Supplementband), (1938): 155–252.

Physical Climatology. DuBois, PA: Gray Printing, 1941 (1st ed.); reprintings 1942 (slightly revised), 1943, 1947 (revised), and 1950; 2nd ed., 1958.

With E. R. Biel, eds. *Preliminary Climatic Atlas of the World.* Prepared by the Military Climatology Project, Institute of Meteorology, University of Chicago, under the direction of the Climatological Section, Directorate of Weather,

Headquarters Army Air Forces, November 1942. Washington, DC: U.S. Government Printing Office, 1942.

"Climate as a Natural Resource." *Scientific Monthly* (October 1946): 293–298. Reprinted in *Managing Climatic Resources and Risks,* pp. 17–22. Washington, DC: National Academy Press, 1981.

Editor. *Survey of Scientists Engaged in Geophysical Researches.* National Military Establishment, Research and Development Board: Digest Series no. 11, GS 62/1, 25 June 1948. There is a copy of the survey in the National Archives.

Editor. *Advances in Geophysics.* New York: Academic Press, 1952–1971. Landsberg edited the first nineteen volumes of this annual series, which presented the latest results and reviews of geophysical research.

Geophysics and Warfare. Office of Assistant Secretary of Defense: Research and Development Coordinating Committee on General Sciences, CGS 202/1, March 1954. The original classified report from 1948 does not seem to have survived.

"The Climate of Towns." In *Man's Role in Changing the Face of the Earth,* edited by William L. Thomas, 548–606. Chicago: University of Chicago Press, 1956.

With H. Lippman, K. Paffen, and Carl Troll. *World Maps of Climatology.* Berlin: Springer-Verlag, 1963; 2nd ed., 1965.

"Two Centuries of New England Climate." *Weatherwise* 20, no. 2 (April 1967): 52–57. Presents the precipitation record from 1751 for the vicinity of Boston.

Weather and Health. Garden City, NY: Doubleday, 1969.

Editor. *World Survey of Climatology.* Amsterdam, NY: Elsevier Scientific, 1969. Landsberg was editor-in-chief for the first fifteen volumes of the survey.

"Man-Made Climatic Changes." *Science* 170, no. 3964 (18 December 1970): 1265–1274.

"Weather, Climate, and Human Settlements." Special Environmental Report no. 7 (WMO, no. 448). Geneva: World Meteorological Organization, 1976.

"Atmospheric Changes in a Growing Community (The Columbia, Maryland, Experience)." *Urban Ecology* 4 (1979): 53–81.

The Urban Climate. Vol. 28, International Geophysics Series. New York: Academic Press, 1981.

"The Value and Challenge of Climatic Predictions." *Invited Scientific Lectures Presented at the Ninth World Meteorological Congress.* WMO, no. 614 (1985): 20–32.

With Stephen G. Brush and Martin Collins, eds. *The History of Geophysics and Meterology: An Annotated Bibliography.* New York: Garland, 1985.

OTHER SOURCES

Baer, Ferdinand. "Helmut E. Landsberg, 1906–1985." *Bulletin of the American Meteorological Society* 67, no. 12, Necrology (December 1986): 1522–1523.

———. "Symposium on Climate in Memory of Helmut E. Landsberg, 10 February 1986, College Park, Maryland." *Bulletin of the American Meteorological Society* 67, no. 12 (December 1986): 1493–1500.

———. "Helmut E. Landsberg: Leadership through Vision, Breadth and Depth." In *Advances in Geosciences: Selected Papers from the Symposia of the Interdivisional Commission on the History of the International Association of Geomagnetism and Aeronomy (IAGA) during the IAGA General Assembly, held in Exeter, UK, 1989,* edited by Wilfried Schröder, 241–260. Bremen-Roennebeck, Germany: 1990.

———. "Helmut E. Landsberg, 1906–1985." *Memorial Tributes: National Academy of Engineering* 5 (1992): 152–157.

Baer, Ferdinand, Norman L. Canfield, and J. Murray Mitchell, eds. *Climate in Human Perspective: A Tribute to Helmut E. Landsberg.* Dordrecht, Netherlands: Kluwer Academic Publishers, 1991. The volume is the published proceedings of the February 1986 symposium cited above. In addition to the chronological listing of Landsberg's scientific writings, the volume includes a 1981 oral history interview with Landsberg, plus extensive notes on his activities, honors, and awards.

Liebowitz, Ruth P. "Post-War Military Sponsorship of Geophysical Research: The Role of Helmut E. Landsberg, 1946–1954." In *Advances in Geosciences: Selected Papers from the Symposia of the Interdivisional Commission on the History of the International Association of Geomagnetism and Aeronomy (IAGA) during the IAGA General Assembly, held in Exeter, UK, 1989,* edited by Wilfried Schröder, 261–277. Bremen-Roennebeck, Germany: 1990.

Robock, Alan. Videotaped interview with Helmut E. Landsberg, University of Maryland, 1983.

Taba, H. Interview with Professor H. E. Landsberg, 29 July 1981, for the *Bulletin of the World Meteorological Organization,* reprinted in *Climate in Human Perspective: A Tribute to Helmut E. Landsberg,* edited by Ferdinand Baer, Norman L. Canfield, and J. Murray Mitchell, 97–109. Dordrecht, Netherlands: Kluwer Academic Publishers, 1991.

Ruth Prelowski Liebowitz

LANZ, JOSÉ MARÍA DE

LANZ, JOSÉ MARÍA DE (*b.* Campeche, Mexico, 26 March 1764; *d.* Paris, France, c. 1839), *abstract theory of machines, classification of machines, mechanisms design, naval sciences, scientific exploration.*

Lanz belongs to an exceptional generation of Spanish exact scientists, all born in the 1760s. Most of them joined the Spanish Royal Navy, and several were later sent to Paris to acquire further experience in science or engineering. In Paris, Lanz worked at national scientific projects and became a professor at the École des Géographes. In 1808 Lanz wrote a seminal book on an abstract classification of mechanical machines. Later he moved to his native Spanish America, playing a leading role in the transmission of the exact sciences to Argentina and Colombia. In Paris, where he returned, he was associated with the firm founded by his friend, the clockmaker Abraham-Louis Breguet (1747–1823).

Origins and Studies. José María de Lanz was born in Campeche, Mexico, where his Spanish father had moved on crown business; on both sides his family had a long Basque ancestry, a connection that would have some relevance in his life.

At the age of fourteen he was sent to Spain to be educated at the Real Seminario in Bergara, in the Basque country. This was a newly established institution for the training of the young nobility, with a definite emphasis on areas of science and technology, perhaps with a view to a more efficient exploitation of natural resources in America. The chemical element Wolfram (or tungsten) was first isolated in Bergara in 1873, and large-scale techniques for the purification of platinum were also developed there; the chemist Joseph-Louis Proust (1754–1826) taught in Bergara by the end of the 1770s.

Work in Spain and France. Around 1780 the Spanish navy supported an observatory and an outstanding group of scientifically oriented officers at its naval base near Cadiz; they were attempting to develop technical facilities relevant to navigation and naval warfare. Graduates from Bergara's Real Seminario had the privilege of joining the navy immediately after graduation; this is the route Lanz followed. He became a Guardia Marina in 1781 and in the following two years received further naval training, seeing action in naval combat against Great Britain at a time when Spain had the support of France. In 1783, already an Alférez de Fragata, he was sent to Cuba and Mexico to report on vegetal fibers used to make ropes for naval use.

In 1784 Lanz started a new stage in his navy life: He was moved to the scientific navy unit near Cadiz, joining an elite cartographic unit. At the time the eminent astronomer and mathematician Joseph de Mendoza Ríos, later a member of the Académie des Sciences, Paris, and of the Royal Society, London, was planning the creation of a more advanced scientific institution within the navy, and in 1788 chose Lanz, then his protégé, as his assistant for a two-year journey through Europe, in 1789–1791. The purpose of this journey was to evaluate new scientific techniques of interest to the navy that were being developed outside Spain, as well as to buy new books, tables, maps, and instruments. During this commission, which was later extended, he worked on problems of mathematics related to the calculation of mathematical tables. He also wrote a treatise on infinitesimal calculus.

Back in Madrid in July 1792 he requested permission to return to Paris, which was denied because of the possibility of a war with France. Disobeying his orders, and expecting permission would be granted later, he returned to France, where he married, probably in October that year. In 1794, after long deliberations, as his talent was widely recognized, his name was struck from the navy list. There were, however, other reasons recommending his removal from the navy: Lanz had become a supporter of the ideals of the French revolutionaries; officially, his defection was attributed to his marriage to a French woman.

In a rapidly changing world, the political situation in Spain changed, and Lanz's sins were forgotten; however, he was never reinstated to the navy. He was invited to return to Madrid, which he did for a short period in 1796. In that visit he acquired new powerful political friends. Back in Paris he worked on calculations for Gaspard Riche de Prony's mathematical tables project, of considerable interest to the navy; he also was a professor at the École des Géographes from 1796 until 1802, when the school was closed.

In 1802 Lanz returned to Spain as a professor of the newly founded Escuela de la Inspección General de Caminos, later Escuela de Caminos y Canales, modeled on the Paris' École des Ponts et Chaussées. There he worked under the Spanish engineer Agustín de Betancourt (1758–1825); in 1806 he returned to Paris on a one-year leave of absence.

The *Essai sur la composition des machines*. His celebrated *Essai sur la composition des machines* was published by the École Impériale Polytechnique in 1808 under the authorship of Lanz and his Madrid chief, Betancourt. The reversal of the alphabetic name of authors suggests that Lanz had the larger responsibility in the making of this work; later editions of the *Essai* were updated by Lanz alone. In this book, his main work, he attempted to present an abstract classification of machines based on the composition of a limited number of "atomic" or elementary machines. He followed the linguistic ideas of Étienne Bonnot de Condillac (1715–1780), already adapted by Antoine-Laurent Lavoisier, Carl von Linné, and others to different scientific contexts. The idea of finding the "atomic" components of machinery was first advanced by Gaspard Monge (1746–1818) and sketched by his student Jean N. P. Hachette (1769–1834). For a good part of the nineteenth century Lanz's book remained an important source; in the early twenty-first century it is considered a classic. Lanz was short-listed for nomination as a corresponding member of the Académie des Sciences, Paris, in 1811 and again in 1813, but was not elected; on both occasions, the scientists considered were internationally renowned.

Back in Spain. In 1809 Lanz returned to Spain, where he worked in different technical departments under Joseph Bonaparte, imposed as king of Spain after the French invasion. From late 1809 to early 1810 Lanz was in

Madrid as director of the Conservatorio de Artes y Oficios and, briefly, also of the Servicio Hidrográfico; in 1811 he served as town governor of Córdoba. By the end of 1812, as French troops retreated from Spain, he moved back to France.

Lanz was an *afrancesado* (Francophile), a name reserved in Spain for supporters of the Enlightenment and, later, of the French Revolution and Napoleon. They included a good part of educated Spaniards of the time. Among them was the Colombian-born botanist Francisco Antonio Zea (1770–1822), director of Madrid's Botanic Garden and a friend of Lanz.

Scientific Work in Latin America. In 1816, either in London or in Paris, Lanz accepted an invitation of Bernardino Rivadavia, future first president of Argentina, to move to Buenos Aires as a professor of mathematics and mechanics. There he founded a mathematical institute with a modern organization, which should be regarded as the true origin of advanced mathematics teaching in Argentina. A year later, back in France, he worked for the reconciliation of Spain and independent Argentina, but with little success. He reestablished contact with Zea, who was then in touch with Simón Bolivar (1783–1830) and other Colombians and Venezuelans working for the independence of the countries at the north of South America, so-called Grand Colombia. Diverting funds from a loan obtained in London for the repayment of war debts, Zea recruited Lanz as head of a scientific and cartographic expedition. Possibly with the help of Alexander von Humboldt, Lanz assembled a formidable group of young scientists eager to help the new republics and study their natural world; they moved in strict secrecy to Colombia in 1822. Among others in that group was Jean Boussingault (1802–1887), who later became a member of the French Académie des Sciences in Paris and a world leader on agricultural chemistry; in his memoirs he referred to Lanz with affection and respect. As a result of their work, important new maps, some compiled by Lanz himself, and more accurate descriptions of fauna, flora, and minerals in the area became available. One of Zea's goals in financing this work was to attract the attention of European investors to the wealth of his country. The expedition also stimulated interesting institutional developments in science and engineering in Colombia, and the creation of an academy of science in Bogotá.

Lanz remained in Colombia for some two years, but his health was affected by Bogotá's altitude and he was forced to return to France in 1824. From Paris he worked as an agent of the Grand Colombian government, trying to negotiate France's recognition of its independence. From intense discussions at the highest levels of government, which involved the president of the council, or premier, Joseph de Villèle (1773–1854), and the foreign minister, Ange-Hyacinthe-Maxence Damas (1785–1862), it became clear to Lanz that, at the time, the French government, even if sympathetic to open commerce with the new republics, could not possibly recognize the independence of Colombia before Spain, or at least Great Britain, did so. This view was probably misinterpreted in Colombia, and Lanz was relieved from his duties in Paris in 1826. He did return to Colombia, where he had been made a colonel in the army, a frigate captain in the Grand Colombian navy, and a member of the new academy of science. As with so many other personalities of the time, Lanz had Masonic connections; possibly, this was mainly a line of personal communications and contacts in Europe and America.

Last Years in Paris. After breaking with Colombia, Lanz lived modestly in Paris, possibly designing mechanisms for Bréguet's works. In the early 1830s, when changes began to take place in Europe, he attempted to return to Spain. In 1832 he applied for a teaching position at the Conservatorio he had helped to create in Madrid; those charged with judging applications dismissed his, alleging he was unknown.

He died, possibly soon after his wife, his life companion, toward the end of the 1830s, possibly around 1839. No picture of him seems to have survived. However, Lanz has been described at various times in his life, and in different countries, as a deeply intelligent man, sensible, with a sense of humility and dignity, and as a trustworthy friend. Even in Paris police reports he is described as respectful man with deep and elevated convictions.

BIBLIOGRAPHY

Primary sources for the works of Lanz are: Archivo General de Indias, Seville, Spain; Archivo Histórico Nacional, Buenos Aires, Argentina; Archivo Histórico Nacional, Bogotá, Colombia; Archivo Histórico Nacional, Madrid, Spain; Archives Nationales, Paris, France; Archivo General de la Marina Álvaro de Bazán, Viso del Marqués (Ciudad Real), Spain; Archivo Naval, Madrid, Spain; Archivo de Simancas, Simancas, Valladolid, Spain; Archivos de la Universidad de Buenos Aires, Buenos Aires, Argentina; Bibliothèque de l'Institut, Paris, France; Bibliothèque Nationale, Paris, France; Bureau des Longitudes, Paris, France; Quai d'Orsay, Paris, France.

WORK BY LANZ

Essai sur la composition des machines. Paris: L'Imprimerie Impériale, 1808. English translation: *Analytical Essay on the Construction of Machines* (London: Ackermann, 1820).

OTHER SOURCES

Bret, Patrice, and Eduardo L. Ortiz. "On Lanz's Numerical Work in M. de Prony's Project." *Revista de Obras Publicas* 3305 (1991): 63–66.

García Diego, José-Antonio. *En busca de Betancourt y Lanz.* Madrid: Castalia, 1985.

García Diego, José-Antonio, and Eduardo L. Ortiz. "On a Mechanical Problem of Lanz." *History of Technology* 5 (1988): 301–313. A discussion on a mathematical manuscript by Lanz on the reversible pendulum.

Ortiz, Eduardo L. "Mathematics in Spain, Portugal, and Latin America." In *Companion Encyclopedia of the History and Philosophy of Mathematical Sciences,* edited by I. Grattan-Guinness, Vol. 2. London: Routledge, 1994. A brief overview of the exact sciences in the Spanish- and Portuguese-speaking world in Lanz's period.

———. "Geometría, lógica y teoría de las máquinas: El ensayo de Lanz y Betancourt, de 1808, sobre la teoría de máquinas." *Fórmula, Société d'Études Basques* 5 (1999): 261–272. A technical discussion on the *Essai.*

———. "Joseph de Mendoza y Ríos: Teoría, observación y tablas." *Gaceta de la Real Sociedad Matemática Española* 4, no. 1 (2001): 155–183. A discussion of the problems considered by Mendoza y Ríos's and Lanz's Cadiz group.

———. "Joseph de Mendoza Ríos." In *Oxford Dictionary of National Biography: In Association with the British Academy; From the Earliest Times to the Year 2000,* edited by H. C. G. Matthew and Brian Harrison, Vol. 37. Oxford: Oxford University, 2004.

———. "Babbage and the French Idéologie: Functional Equations, Language, and the Analytical Method." *Episodes in the History of Modern Algebra (1800–1950),* edited by Jeremy Gray and Karen Hunger Parshall, chapter 2. Providence, RI: American Mathematical Society, 2007. A discussion on the philosophical background supporting the approach used in Lanz-Betancourt's *Essai,* and in similar classification attempts.

———, and Patrice Bret. "José María de Lanz and the Paris-Cadiz Axis." In *Naissance d'une communauté internationale d'ingenieurs: Actes des Journées d'étude, 15–16 décembre 1994,* edited by Irena Gouzévitch and Patrice Bret. Paris: Centre de Recherche en Histoire des Sciences et des Techniques, Cité des Sciences et de l'Industrie, 1997. On Lanz in France.

Eduardo Ortiz

LASHLEY, KARL SPENCER (*b.* Davis, West Virginia, 7 June 1890; *d.* Poitiers, France, 7 August 1958), *psychology, neurophysiology.* For the original article on Lashley see *DSB,* vol. 8.

A major biographical treatment of Karl Lashley since the original *DSB* article is Nadine Weidman's *Constructing Scientific Psychology: Karl Lashley's Mind-Brain Debates* (1999). Weidman's interpretation builds on earlier biographical work on Lashley by Darryl Bruce and Donald Dewsbury but also differs from it in noteworthy ways. Bruce has written two biographical articles on Lashley. One of them, "Lashley's Shift from Bacteriology to Neuropsychology, 1910–1917, and the Influence of Jennings, Watson, and Franz" (1986), is focused on Lashley's early career and traces the path that took him from bacteriology and genetics to his life's work in neuropsychology. The other, "Integrations of Lashley" (1991), is a portrayal of Lashley as a pioneering psychologist. Dewsbury has examined a correspondence between Lashley and Lashley's mentor and friend from the 1950s, John B. Watson, late in both men's lives, in "Contributions to the History of Psychology XCIV: The Boys of Summer at the End of Summer" (1993).

Karl Lashley.

While Bruce and Dewsbury portray Lashley as a pure, disinterested scientist whose work was devoid of social or political meaning, Weidman argues that Lashley's neutral stance and avowed opposition to psychological theorizing were themselves political statements and that Lashley's concept of brain function and his lifelong hereditarianism were correlated with his views on race and on the social order. Weidman's biography emphasizes Lashley's hereditarianism—his belief that intelligence and brain function were innate properties rather than influenced strongly by environment—and uses that emphasis to explain Lashley's scientific stance, his opposition to certain traditions in psychology and biology, and his sociopolitical leanings.

According to this interpretation, Lashley's doctoral work in genetics, done under the supervision of the biologist Herbert Spencer Jennings, encouraged Lashley's interest in uncovering the biological and genetic bases of behavior. Lashley allied himself with the behaviorist Watson up to the mid-1920s, but he broke with behaviorism just as Watson was asserting the total shaping power of the environment on behavior. Lashley's mature work involved the ablation, or destruction, of different areas of the rat's cerebral cortex and the assessment of its effect on the rat's behavior. He devised ingenious methods for testing the rats' abilities to thread a maze, solve a puzzle box, and discriminate between different patterns, before and after ablation. The principles of equipotentiality and mass action that Lashley derived from his experimental work emphasized the dynamic functioning of the cortex and were opposed to the theory that memories, thoughts, or abilities were localized in discrete cells. This experimental work received thorough treatment in *Brain Mechanisms and Intelligence* (1929), the only book Lashley ever wrote. Lashley's emphasis on dynamic functioning was also, according to Weidman, correlated with his innatist stance.

Lashley's hereditarianism and interest in elucidating the biological bases of behavior brought him into conflict with two important traditions in twentieth-century American biology and psychology. His career was shaped by his debates with representatives of these two traditions, psychobiology and behaviorism. The first conflict pitted Lashley against his colleague at the University of Chicago, the psychobiologist and neurologist Charles Judson Herrick. For Herrick, consciousness and free will were emergent properties of the nervous system that could never be fully explained by physics and chemistry. Lashley dismissed Herrick's emergentism as mystical, arguing for the reduction of consciousness to the physico-chemical workings of the nervous system and against the notion of progress in evolution that underlay Herrick's theory. Sharon Kingsland explores the assumptions of Herrick's science in "A Humanistic Science: Charles Judson Herrick and the Struggle for Psychobiology at the University of Chicago" (1993). In *Constructing Scientific Psychology* (1999), Weidman develops Lashley's side of the debate and examines his interactions with Herrick.

The second major debate of Lashley's career was with Clark Hull, the neo-behaviorist psychologist at Yale University's Institute of Human Relations in New Haven, Connecticut. In Hull, Lashley faced an opponent who was quite as opposed to Herrick's emergentism as was Lashley himself. Both Hull and Lashley claimed the mantle of mechanistic, deterministic psychology but disagreed on what that meant. For Hull, behavior was the product of reflex connections between stimulus and response; how these connections were actually achieved in the brain mattered to him less than that behavioral outputs could be manipulated by changing the stimulating sensory inputs. Hull also believed that intelligent behavior could be simulated by a machine and spent some effort designing and building such "thinking machines." According to Weidman, Lashley objected to such mechanical analogies for the mind and to Hull's neo-behaviorist theory as well as to the emphasis on the environmental shaping of behavior that underlay Hull's science.

Weidman's interpretation of the Lashley-Hull debate in her book expands on her earlier article, "Mental Testing and Machine Intelligence: The Lashley-Hull Debate" (1994). Bruce criticizes Weidman's interpretation of the debate in this article in "The Lashley-Hull Debate Revisited" (1998). Weidman's response to Bruce ("A Response to Bruce") and Bruce's reply ("Lashley's Rejection of Connectionism")—in which he argues that a change in evidence from Lashley's experiments underlay his break from behaviorism—can both be found in the May 1998 issue of *History of Psychology.*

Dewsbury compares Bruce's and Weidman's views of Lashley in his article "Constructing Representations of Karl Spencer Lashley." There Dewsbury argues, in a critique of Weidman, that Lashley's theories of dynamic cortical function and his emphasis on innate ability could be interpreted as correlated with his idiosyncratic and iconoclastic personality rather than with his sociopolitical views. Dewsbury's article, Weidman's response to it ("The Depoliticization of Karl Lashley"), and Dewsbury's reply to her ("The Role of Evidence") can all be found in the summer 2002 issue of the *Journal of the History of the Behavioral Sciences.*

Lashley ended his career as the director of the Yerkes Laboratories of Primate Biology in Orange Park, Florida. In keeping with his career-long interests, Lashley guided the work of the laboratories toward an investigation of the biological and ultimately genetic basis of intelligence and of sexual behavior in anthropoid apes.

SUPPLEMENTARY BIBLIOGRAPHY

Bruce, Darryl. "Lashley's Shift from Bacteriology to Neuropsychology, 1910–1917, and the Influence of Jennings, Watson, and Franz." *Journal of the History of the Behavioral Sciences* 22 (1986): 27–44.

———. "Integrations of Lashley." In *Portraits of Pioneers in Psychology*, Vol. 1, edited by Gregory A. Kimble, Michael Wertheimer, and Charlotte White. Hillsdale, NJ: Erlbaum, 1991.

———. "The Lashley-Hull Debate Revisited." *History of Psychology* 1, no. 1 (February 1998): 69–84.

———. "Lashley's Rejection of Connectionism." *History of Psychology* 1, no. 2 (May 1998): 160–164.

Dewsbury, Donald A. "Contributions to the History of Psychology XCIV: The Boys of Summer at the End of

Summer: The Watson-Lashley Correspondence of the 1950s." *Psychological Reports* 72 (1993): 263–269.

———. "Constructing Representations of Karl Spencer Lashley." *Journal of the History of the Behavioral Sciences* 38, no. 3 (Summer 2002): 225–245.

———. "The Role of Evidence in Interpretations of the Scientific Work of Karl Lashley." *Journal of the History of the Behavioral Sciences* 38, no. 3 (Summer 2002): 255–257.

———. *Monkey Farm: A History of the Yerkes Laboratories of Primate Biology, Orange Park, Florida, 1930–1965.* Lewisburg, PA: Bucknell University Press, 2006.

Kingsland, Sharon. "A Humanistic Science: Charles Judson Herrick and the Struggle for Psychobiology at the University of Chicago." *Perspectives on Science* 1 (1993): 445–477.

Weidman, Nadine M. "Mental Testing and Machine Intelligence: The Lashley-Hull Debate." *Journal of the History of the Behavioral Sciences* 30 (April 1994): 162–180.

———. "Psychobiology, Progressivism, and the Anti-Progressive Tradition." *Journal of the History of Biology* 29 (1996): 267–308.

———. "A Response to Bruce (1998) on the Lashley-Hull Debate." *History of Psychology* 1, no. 2 (May 1998): 156–159.

———. *Constructing Scientific Psychology: Karl Lashley's Mind-Brain Debates.* New York; Cambridge, UK: Cambridge University Press, 1999.

———. "The Depoliticization of Karl Lashley: A Response to Dewsbury." *Journal of the History of the Behavioral Sciences* 38, no. 3 (Summer 2002): 247–253.

Nadine M. Weidman

LAURENT, AUGUSTE (originally Augustin)

(*b.* La Folie, near Langres, France, 14 November 1807; *d.* Paris, 15 April 1853), *chemistry.*

A founder of modern organic chemistry, Laurent was one of the most important chemists of the nineteenth century. He considered the behavior of matter to be a manifestation of its intimate internal structure, which one cannot determine with certainty but which one has to investigate if one wants to understand. Laurent's preoccupation was to construct a method that could guide the chemist forward along this path, from facts to their causes. He was the first chemist to intimately associate crystallographic data and chemical studies. Louis Pasteur and Charles Friedel later followed the way.

Laurent is almost forgotten in his fatherland but is slightly better known outside its borders. His memoirs first appeared in the *Poggendorffs Annalen,* the *Journal für Praktische Chemie,* and the *Annalen der Chemie und Pharmacie.* When the French *Annales de Chimie et de Physique* refused them, Justus von Liebig offered to publish the entirety of his ideas and work in a supplementary volume of his *Annalen.* Laurent's treatise *Méthode de chimie,* published posthumously in France in 1854 thanks to the tenacity of Jérôme Nicklès and Jean-Baptiste Biot, was immediately translated into English by William Odling.

Laurent's early work introduces his lifelong interests: minerals and products of earth. Laurent was an alumnus of the École des Mines (1830), then employed at the École Centrale des Arts et Manufactures and at the royal porcelain factory at Sèvres. His early publications in geology, mineralogy, and crystallography exhibit his observational and measuring talents, as well as his experimental skills and imaginative genius. His earliest work appeared in Jean-Baptiste Dumas's *Traité de chimie appliquée aux Arts* (vol. 3, 1833) and later in Alexandre Brongniart's *Traité des arts céramiques* (1844), sometimes without being acknowledged.

In 1838 Laurent was appointed professor at the Faculté des Sciences de Bordeaux. In 1845 he obtained a leave from his position and moved to Paris, the center of action in the scientific world. Two years later his leave (and half-salary) was canceled, and Laurent faced penury. After the revolution of 1848, he was appointed assayer at the Paris Mint. In 1850, supported by Jean-Baptiste Biot, he was a candidate to be Théophile-Jules Pelouze's successor at the Collège de France. Although he won the vote of professors in the Collège, he lost the second vote in the Académie des Sciences, and Antoine-Jérôme Balard won the position. Laurent died three years later, from tuberculosis.

Although he never had a decent laboratory at his disposal, Laurent was an exceptional experimenter. He corrected a great number of current results, including the exact composition and formula of many alkaloids and mellon, in which Liebig had found no hydrogen. Leading authorities in the sciences, including Jöns Jakob Berzelius, greatly respected Laurent's experimental accuracy and reliability. With his pupils, in his private school at the mint, or in Pelouze's or Balard's laboratory, in every place he could work and discuss, with fellow chemists such as Gerhardt, August Wilhelm von Hofmann, Alexander Williamson, and Gustave Chancel, or students such as Pasteur, Laurent debated and discussed, passing on his enthusiasm.

Partly because of the lack of laboratory facilities, Laurent also became a leading theoretician. His first theoretical efforts were largely taxonomic in character. He then began to develop a pictorial model based on atomistic representations as considered by earlier French crystallographers. Laurent's "nucleus theory" or "theory of derived radicals" located every substance at the intersection of two kinds of transformations: substitutions, which operate on the matter inside the fundamental radical and do not affect its general chemical behavior, and external modifications, which influence various chemical functions.

Laurent

Hence, a single fundamental radical provided the theoretical basis for the chemical parentage that binds in one unit the empirical diversity of all its members.

Laurent was a theoretician in the modern sense: He wanted to understand, to relate, and to predict. His speculations about the role and the place of the atoms within the molecule were not suppositions of reality, not something that could be proven true or false. They were a means of constructing arrangements of atoms according to definite operations, a rational arrangement of perfectly abstract entities that were intended ultimately to guide and corroborate experience.

To connect theory that deals with atoms and experiment that concerns properties, Laurent invented a method: his system of formulation, classification, and nomenclature. We can know nothing for certain about atomic arrangements, he thought, but these must be responsible for the properties observed. Let us assume the minimum hypothesis: to similar properties correspond similar arrangements. Formulas that represent experience will demonstrate atomic arrangements.

Laurent concentrated his efforts on certain organic substances, especially the products of distillation of coal tar (such as naphthalene and anthracene) and their derivatives, the products of oxidation of indigo, and organic nitrogen compounds. He had two kinds of preferred reagents. First, he used halogens systematically, and these led him to distinguish two types of reactions, (equivalent) substitutions and additions, and to develop his nucleus theory. He also used ammonia, which complicated compounds: Laurent prepared a great number of nitrogen compounds of benzoyl hydride (benzaldehyde, oil of bitter almonds), the constitutions of which were revealed half a century later. In Laurent's hands, ammonia became a tool to analyze constitutions and distinguish certain groups of atoms. He demonstrated that acids did not contain water and that ammonia compounds could contain NH_4, NH_2, or NH, but not NH_3.

He gave special attention to crystallizations, not only as a model to construct his theory of organic compounds, but as a means to physically identify and separate new substances. That was of an eminent benefit considering the delicate mixtures obtained by this kind of manipulation. Pasteur took advantage of this lesson.

Laurent's was a lonely path, and his work was highly original. Atoms became a necessary hypothesis for the organic chemist. But more than the atom, Laurent emphasized the molecule, groups of atoms, as the central entity in organic chemistry: the smallest quantity of a simple body that we need to operate on a compound, this quantity being divisible during the act of combination. This idea of divisibility, connected with that of double decomposition, emerged as a new conception of radicals, different from that of Joseph-Louis Gay-Lussac or Robert Bunsen or Justus von Liebig, as a group of atoms that moves from one to another molecule but cannot exist in a free state.

Contrary to Liebig, Laurent did not seem to care about his popularity. Contrary to Dumas, he profited from no personal or institutional protection. His novel ideas on chemistry finally passed to posterity through the work of such followers as Gerhardt, Williamson, Charles-Adolphe Wurtz, and Friedrich August Kekule von Stradonitz (original surname Kekulé), but too often misleadingly confounded with Gerhardt's work.

BIBLIOGRAPHY

About 215 papers are listed by Jean Jacques in "Essai bibliographique sur l'oeuvre et la correspondance d'Auguste Laurent," in Archives, Institut Grand-Ducal de Luxembourg; Section des Sciences naturelles, physiques et mathématiques *(1955): 11–35.*

WORKS BY LAURENT

"Chalumeau," "Chimie," "Cobalt," "Combinaison," "Combustion." In *Encyclopédie nouvelle*, vol. 3, edited by Pierre Leroux and Jean Reynaud. Paris: Gosselin, 1837.

Précis de cristallographie suivi d'une méthode simple d'analyse au chalumeau (d'après les leçons particulières de M. Laurent). Paris: Masson, 1847.

Méthode de chimie. Paris: Mallet-Bachelier, 1854.

Chemical Method, Notation, Classification, and Nomenclature. Translated by William Odling. London: Cavendish Society, 1855.

OTHER SOURCES

Blondel-Mégrelis, Marika. *Dire les choses: Auguste Laurent et la méthode chimique.* Paris: Vrin; Lyon: Institut Interdisciplinaire d'Études Épistémologiques, 1996.

———. "Auguste Laurent et les alcaloïdes." *Revue d'Histoire de la Pharmacie* 49 (2001): 303–314.

Brooke, John Hedley. *Thinking about Matter: Studies in the History of Chemical Philosophy.* Aldershot, U.K.: Variorum, 1995.

Byers, Twig. "The Radical, Dualism, and Auguste Laurent." *Synthesis* 3, no. 1 (1975): 22–37.

de Milt, Clara. "Auguste Laurent: Guide and Inspiration of Gerhardt." *Journal of Chemical Education* (1951): 198–204.

———. "Auguste Laurent, Founder of Modern Organic Chemistry." *Chymia* 4 (1953): 85–114.

Grimaux, Edouard, and Charles Gerhardt Jr. *Charles Gerhardt: Sa vie, son œuvre, sa correspondance, 1816–1856; document d'histoire de la chimie.* Paris: Masson, 1900.

Jacques, Jean. "Auguste Laurent et J.-B. Dumas d'après une correspondance inédite." *Revue d'Histoire des Sciences* 6 (1953): 329–349.

Lauritsen

———. "La thèse de doctorat d'Auguste Laurent et la théorie des combinaisons organiques." *Bulletin de la Société Chimique de France* (1954): D31–39.

Kapoor, Satish C. "The Origins of Laurent's Organic Classification." *Isis* 60 (Winter 1969): 476–527.

Mauskopf, Seymour. *Crystals and Compounds: Molecular Structure and Composition in Nineteenth-Century French Science.* Philadelphia: American Philosophical Society, 1976.

Nicklès, Jérôme. "Auguste Laurent." *American Journal of Science* 2, no. 16 (1853): 103.

Novitski, Marya. *Auguste Laurent and the Prehistory of Valence.* Philadelphia: Harwood Academic, 1992.

Tiffeneau, Marc, ed. *Correspondance de Charles Gerhardt.* 2 vols. Paris: Masson, 1918–1925.

Williamson, Alexander. "Laurent's Biography." *Journal of the Chemical Society* 7 (1855): 149.

Marika Blondel-Mégrelis

LAURITSEN, CHARLES CHRISTIAN

(*b.* Holstebro, Denmark, 4 April 1892; *d.* Los Angeles, California, 13 April 1968), *physics, physical and therapeutic properties of high-energy x-rays, nuclear physics, astrophysics, weapons systems.*

Lauritsen was an experimental nuclear physicist of remarkable ability and vision. From 1926 until his death he was at the California Institute of Technology (Caltech), first as a graduate student and then as a faculty member. His groundbreaking PhD research on cold field-emission enabled him to design and build x-ray tubes that achieved record voltages. He used these tubes to study properties of x-rays and their physiological effects. Then, in response to British developments in 1932, he converted one of his x-ray tubes into an accelerator of ions that he then used to probe the interiors of atomic nuclei. He thus became one of America's earliest pioneers of accelerator-based nuclear research and established at Caltech what became a world famous laboratory for research in nuclear physics. During and after World War II, Lauritsen made significant contributions to the development of important weapons systems and advised the U.S. government on military and defense matters.

Charles Lauritsen. *Charles Lauritsen sitting in his office.* LEO ROSENTHAL/PIXINC./TIME LIFE PICTURES/GETTY IMAGES.

Danish Origins. Charles Christian Lauritsen was born in Holstebro, Denmark, in 1892, the son of Thomas Lauritsen and Marie Lauritsen (née Nielsen). The father, a sawmill owner, committed suicide in 1903; the mother, with young Charlie and his older brother Laurits to care for, soon remarried. In 1911 Charlie graduated from Odense Tekniske Skole with a degree in structural engineering and certification to supervise construction work. After completing his required military service in the Danish infantry (1911–1912), he studied architecture and sculpture at the Royal Danish Academy of Fine Arts. In 1914 and again in 1915 he was called back into service when Denmark mobilized to protect its neutrality during World War I. On 21 May 1915 he married Sigrid Henriksen, a medical student and the daughter of Niels Henriksen, a farmer. In July 1916 Lauritsen emigrated from Denmark to the United States, and Sigrid, with their infant son Thomas, joined him in Florida three months later.

Over the next ten years Lauritsen worked in Florida, Massachusetts, Ohio, California, and Missouri as a draftsman, radio designer, fluids engineer, and chief engineer for the Kennedy Corporation, a manufacturer of radio sets in St. Louis, Missouri. During this time he patented several inventions and taught himself the new technology of radio.

Earliest Work in Physics. In 1926 Lauritsen quit his job in St. Louis and drove his wife and child to Pasadena, California. There at age thirty-four, with little background in formal mathematics or physics, he persuaded doubtful faculty to admit him to Caltech's graduate physics program. Lauritsen made rapid progress based on his exceptional physical insight coupled with his remarkable talents for designing and building apparatus. These qualities first became apparent after Robert A. Millikan, the Nobel Prize–winning head of Caltech, suggested that for a PhD thesis, Lauritsen examine how electrons are pulled off the surface of a metal by strong electric fields—an effect known as cold field-emission. Millikan had been unable to make progress on this problem, but within a few months Lauritsen found experimental arrangements and created apparatus that yielded the first reliable, reproducible measurements of the electron currents drawn from metals by different strengths of electric field. He showed from his data that cold field-emission currents rise exponentially with the field strength. This behavior was explained by J. Robert Oppenheimer, who visited Caltech in 1928 just after returning from Europe where, working with Max Born, he had mastered the new theory of quantum mechanics. Oppenheimer recognized that Lauritsen's data were a consequence of quantum tunneling. This occasion began a lifelong friendship between Oppenheimer and Lauritsen.

Lauritsen quickly showed remarkable ability to do physics both in the classroom and in the laboratory. In his first year at Caltech, he completed the research he would use for his PhD. In his second year, working with Ralph Bennett, he used the experience and insights from his cold field-emission work to invent and build a 750,000-volt x-ray tube, which for a short time was the highest voltage tube in the world. He also became an American citizen. In 1929 he received his PhD and was appointed to the faculty; the next year he became an assistant professor. He built a 1-million-volt x-ray tube that became the principal instrument for a cancer treatment center, the Kellogg Radiation Laboratory, built on the Caltech campus in 1931. That year he was named director of the Kellogg lab and was promoted to associate professor.

Lauritsen had a gift for using the apparatus that was at hand. For his x-ray research, he took advantage of four 250,000-volt transformers left over from a discontinued testing program of an electrical utility company, and his tubes used glass cylinders from the gasoline pumps typical of that era. His first tube cost less than one hundred dollars.

He studied the physical properties of the very short wavelength x-rays produced with his tubes, which he soon improved to sustain one million volts. Recognizing that energetic x-rays can have useful therapeutic properties, he studied their physiological effects in collaboration with physicians. In 1931, in recognition of his work, the American College of Radiology made him an honorary fellow and awarded him its gold medal.

Lauritsen's work led Caltech to establish a cancer treatment clinic on its campus. Lauritsen served as its technical director, and he and his students did radiological research, developed improved x-ray tubes, and maintained the equipment of the clinic. Around 1930 he invented an ingenious pocket-sized electroscope for measuring doses of radiation. This Lauritsen dosimeter, or Lauritsen electroscope, which he patented in 1935, became widely used for monitoring levels of exposure to radiation. Lauritsen and others also used it in their nuclear research to detect and identify particles.

Pioneering Nuclear Physics. In 1932 the British physicists John D. Cockcroft and Ernest T. S. Walton reported that they had built an apparatus which accelerated hydrogen ions, that is, protons, to energies sufficient to penetrate lithium nuclei and produce nuclear reactions. On learning of this achievement, Lauritsen converted one of his x-ray tubes into a crude alternating-voltage, positive-ion accelerator and began doing nuclear physics. One of the first three Americans to study atomic nuclei by bombarding them with accelerated particles, Lauritsen broke new ground and helped to create the modern field of nuclear physics. For the rest of his career, his principal research was the study of the properties of light nuclei.

For nuclear physicists, the mid-1930s was an exciting time. As Lauritsen's son Tommy remembered in a 1967 interview by Barry Richman and Charles Weiner, "every day you went to the laboratory you found something new." From 1933 to 1935, Lauritsen and his students published twenty-six notes and papers. The first of these reported how they produced neutrons by accelerating helium ions into beryllium. It was the first use of an accelerator to produce neutrons (just discovered in 1932 by James Chadwick). They detected the neutrons using an ingenious adaptation of Lauritsen's electroscope. A few weeks later they became the first to produce neutrons using accelerated ions of deuterium, the heavy isotope of hydrogen just discovered in 1931.

Lauritsen responded quickly when, in January 1934, French physicists Irène Curie and Frédéric Joliot reported the first artificial production of radioactivity. They had bombarded boron with alpha particles from a polonium source and produced nitrogen-13 nuclei that decayed with a half-life of ten minutes into stable carbon-13 nuclei. They noted that nitrogen-13 nuclei could also be produced by bombarding carbon-12 with deuterons. Lauritsen and Horace R. Crane put a carbon target into the beam of deuterons that they were accelerating for their neutron studies, and in late February they reported the first observations of the production of radioactivity with an accelerator.

Lauritsen and Crane were also the first to see evidence of what came to be known as nuclear resonance. They observed that when accelerated to a certain energy,

Electroscope Radiation Dosimeter. *Lauritsen's 1933 electroscope radiation dosimeter.* COURTESY OF THE ARCHIVES, CALIFORNIA INSTITUTE OF TECHNOLOGY.

protons would suddenly be absorbed by lithium-7 nuclei, which then emitted a gamma ray rather than a particle. They saw a similar effect for carbon-12. Ultimately, this behavior was understood to show that a nucleus had well-defined internal energy states, but because of the limits of Lauritsen and Crane's crude accelerator, their observations were greeted with a skepticism that they themselves shared. The existence of such nuclear resonances was widely recognized only after they had been observed by other physicists working either with very low energy neutrons or with accelerator beams possessing very well-defined energy—in one case relying on a Lauritsen electroscope provided by Lauritsen himself.

Lauritsen and his students built a cloud chamber, a device in which ions, electrons, and gamma rays (very short wave-length electromagnetic radiation) revealed their presence by visible tracks in a supersaturated vapor. By measuring the energies of gamma rays emitted from nuclei, they sought to identify and understand the internal states of light nuclei, often called energy levels. Their first measurements resulted in a mélange of falsely identified levels. Only when they changed to inferring gamma-ray energies from measurements of the energies of electron-positron pairs produced by gamma rays did they get reliable values.

Their cloud chamber measurements of the energies of positrons emitted from various radioactive nuclei revealed some of the first evidence that the nuclear force between two protons is the same as the nuclear force between two neutrons—that is, nuclear forces exhibit charge symmetry. They were also among the first to determine nuclear masses from measurements of reaction energies; that is, Q-values, a technique of great importance for establishing the relative energies of nuclear ground states and for determining what reactions might be possible in different circumstances—such as deep inside stars, for example.

Lauritsen's work was known and admired for its reliability and ingenuity. Interpretation of his results benefited from the advice and insights he received from his friends, the brilliant theorist Oppenheimer and fellow Dane and world-famous theoretical physicist and Nobel laureate Niels Bohr. Physicists came from all over the world to work in Lauritsen's lab. Within a few years he had made Caltech an internationally recognized center of nuclear physics research.

Nuclear Physics Matures. In 1936 Hans Bethe published the first of his three review articles that became the "Bethe's Bible" of nuclear physicists. This comprehensive account of what had been learned in the preceding four

Lauritsen and Bennett in the High Voltage Laboratory. *Lauritsen and Bennett in 1928 in the High Voltage Laboratory by the 250kV transformers that powered the x-ray tube mounted in the wooden framework on the right.* **PHOTO BY EYRE POWELL PRESS SERVICE/COURTESY OF CALTECH ARCHIVES.**

years marked the end of the pioneering era of nuclear physics. As well as summarizing the basic knowledge of the field, the articles showed experimentalists many promising lines of research. These would require improved precision and accuracy of measurements and higher energy accelerators with better quality beams.

Lauritsen understood this, and aided by his students—one of which was his son, Tommy—and by his former student, William A. Fowler, a Caltech faculty member and future Nobel laureate, he built two Van de Graaff accelerators. The first was a low energy accelerator built in 1937, and the second was a larger machine built in 1938–1939 that could accelerate ions through a potential difference of up to 1.7 million volts. Construction of a planned 5-million-volt machine had to wait until after World War II.

The cancer treatment clinic closed in 1939, due to a study that showed the radiation treatments were either ineffective or did more damage than good. Lauritsen, still director of the Kellogg Radiation Laboratory, took over the building for nuclear physics. He moved his accelerators and research program into Kellogg that summer. However, World War II began in Europe in September, and during the spring of 1940 he moved to Washington, D.C., to work with other scientists on preparations for America's involvement in the war. From 1940 until 1945 no nuclear physics research was done at Kellogg.

In Washington, Lauritsen helped develop the proximity fuse, but in May 1941 on a visit to Britain, he observed British rocket weapons in use and became convinced that American military forces should also have rocket weapons. He persuaded the federal government's Office of Scientific Research and Development (OSRD) to set up at Caltech a large program to develop rocket weapons. He returned to Pasadena and, using the Kellogg Laboratory, much of the rest of Caltech, and large parts of Pasadena, he directed the invention, development, production, and testing of a variety of rocket munitions, mostly for the U.S. Navy. Toward the end of the war, Oppenheimer, who was directing the effort to make an atomic bomb, asked Lauritsen to go to Los Alamos, New Mexico, to assist with the project. Lauritsen complied with the request.

Lauritsen's work on rocketry led to close friendships with key naval officers, and he was influential in the navy's decision to create the Office of Naval Research (ONR) in 1946. Immediately after the war, ONR became the leading agency through which the federal government began supporting basic research at a level that vastly increased the pace and scale of scientific work in the United States.

Lauritsen used ONR support to reestablish Kellogg's research program, and he decided to continue its prewar focus on understanding the properties of light nuclei. This was not an obvious decision, because technological advances and generous funding allowed nuclear physicists in the postwar years to work with larger accelerators at higher energies and probe the deep structure of nuclei and their constituent particles.

Lauritsen, however, saw that there was still much to learn by studying light nuclei at modest accelerator energies. In 1939 Bethe had proposed that two particular series of nuclear reactions—one called the p-p cycle, and the other called the CNO cycle—could explain energy generation in stars, a dramatic example of how nuclear physics could be used to understand stellar processes. In 1945 Lauritsen, Fowler, and Caltech astronomers held a series of seminars that showed that studies of light nuclei could answer important questions of astrophysics. Lauritsen included this line of research in Kellogg's postwar plans; he and his colleagues would continue to study light nuclei to discover and understand their properties, but they would also open a new line of research by doing nuclear physics experiments that answered astrophysical questions. This set Kellogg Laboratory researchers on a path along which they created a new field of physics: nuclear astrophysics. It was a field in which Fowler would work with imagination and insight; for his work on how stars synthesize elements, he shared the Nobel Prize in Physics for 1983.

As the Cold War with the Soviet Union intensified in the postwar years, Lauritsen contributed extensively to American military efforts. His low-key, nonconfrontational emphasis on reasoned discourse based on reliable facts made him an influential advisor. After 1945 and for the rest of his life, he typically spent forty-five days a year or more consulting, advising, and evaluating defense efforts and programs for the army, navy, air force, and Department of Defense.

Lauritsen and Millikan in the High Voltage Laboratory. *Lauritsen and Millikan in the High Voltage Laboratory with a tube that could sustain 1 million volts of potential.* PHOTO BY WORLD WIDE PHOTOS/COURTESY OF CALTECH ARCHIVES.

In 1950 Lauritsen was deeply upset by President Harry Truman's decision to develop thermonuclear weapons. Lauritsen believed that a weapon a thousand times more powerful than the atomic bombs dropped on Japan had no justifiable military use. He felt that American defense strategy had become dangerously dependent on large bombs. In summer long special study groups such as Project Charles, Project Vista, and the ad hoc Lincoln summer study group, he worked with fellow scientists to identify weapons technologies that would support flexible military strategies offering alternatives to all-out nuclear war. Thus, he strongly advocated the development of tactical nuclear weapons that would permit American forces to fight small wars against numerically superior forces, and he worked strenuously with others to develop a credible system for air defense of the continental United States, believing that adequate warning would reduce the chance of preemptive nuclear war. He was part of a group of scientists that persuaded President Truman and, after him, President Dwight D. Eisenhower, to create a vast air defense system that included a long chain of manned radar stations above the Arctic Circle. The two presidents supported this program over the objections of air force leaders who feared that a diversion of resources from the Strategic Air Command would weaken America's ability to mount a massive nuclear retaliation in response to any Soviet attack. Partly because of these fears, air force leaders supported efforts to take away Oppenheimer's security clearance, and they reacted with suspicion and hostility to recommendations made by Vista scientists. Lauritsen testified for Oppenheimer at the hearings in 1954. These ended with the cancellation of Oppenheimer's clearance and his exclusion from government counsels. Although deeply unhappy with this outcome, Lauritsen continued to be a valuable advisor to government agencies for the rest of his career.

Honors. Lauritsen was honored both in America and in his native Denmark. In 1939 he was elected to the Royal Danish Academy of Sciences and Letters, and in 1953 he was awarded the Commander's Cross of the Order of Dannebrog by the king of Denmark. In America, in addition to his 1931 gold medal from the American College of Radiology, he was elected to the National Academy of Sciences in 1941; received the President's Medal for Merit in 1948; served as president of the American Physical Society in 1951; and received the society's Tom W. Bonner Prize in 1967 for his work in nuclear physics. In 1954 he was elected a member of the American Philosophical Society; in 1958 he was the first recipient of the navy's Captain Robert Dexter Conrad Award for Scientific Excellence; and in 1965 the University of California at Los Angeles awarded him an honorary degree of doctor of laws. Two libraries, a street, and two buildings are named for him, as well as a crater on the far side of the moon.

BIBLIOGRAPHY

The Charles Christian Lauritsen Papers are in the archives of the California Institute of Technology. The collection contains both personal and professional correspondence, research notes and data, manuscripts, reprints, patents, and photographs.

WORKS BY LAURITSEN

With R. D. Bennett. "A New High Potential X-Ray Tube." *Physical Review* 32 (1928): 850–857. In 1935 Lauritsen was issued U.S. patent 1995478 for his "High Potential X-Ray Tube."

With R. A. Millikan. "Relations of Field-Currents to Thermionic-Currents." *Proceedings of the National Academy of Sciences of the United States of America* 14 (1928): 45–49.

"Energy Considerations in High Voltage Therapy." *American Journal of Roentgenology and Radium Therapy* 30 (1933): 380–387.

"Energy Considerations in Medium and High Voltage Therapy." *American Journal of Roentgenology and Radium Therapy* 30 (1933): 529–532. The work reported in this article and the preceding one constituted an important part of the basis for the standard of dose adopted internationally in the late 1940s.

With H. R. Crane and A. Soltan. "Production of Neutrons by High Speed Deutons." *Physical Review* 44 (1933): 692–693. The word *deuton* was one of several proposed names for the nucleus of the mass-2 isotope of hydrogen. By 1935 the name *deuteron* was coming into general acceptance.

———. "Artificial Production of Neutrons." *Physical Review* 44 (1933): 514; 45 (1934): 507–512.

With H. R. Crane and W. W. Harper. "Artificial Production of Radioactive Substances." *Science* 79 (9 March 1934): 234–235. The authors published in *Science* to get into print before their competition at Berkeley and at the Carnegie Institution of Washington.

With W. A. Fowler and L. A. Delsasso. "Radioactive Elements of Low Atomic Number." *Physical Review* 49: (1936): 561–574. Fowler says that Oppenheimer and his student, Robert Serber, saw in these results the first evidence for charge symmetry of nuclear forces.

With Thomas Lauritsen. "Simple Quartz Fiber Electrometer." *Review of Scientific Instruments* 8 (1937): 438–439. This article describes the device for which Lauritsen had received U.S. patent 2022117 in 1935. It is a more compact version of the earlier device that he developed around 1930.

With Tom Lauritsen and W. A. Fowler. "Application of a Pressure Electrostatic Generator to the Transmutation of Light Elements by Protons." *Physical Review* 59 (1941): 241–252.

In the Matter of J. Robert Oppenheimer: Transcript of Hearing before Personnel Security Board, Washington, D.C., April 12, 1954 through May 6, 1954. Washington, DC: U.S. Government Printing Office, 1954. Contains Lauritsen's verbatim testimony.

OTHER SOURCES

Christman, Albert B. *History of the Naval Weapons Center, China Lake, California.* Vol. 1, *Sailors, Scientists and Rockets.* Washington, DC: U.S. Government Printing Office, 1971. Several chapters describe the importance to the navy of Lauritsen's wartime work. His face is prominent on the cover of the 1992 paperback edition of this book.

Fowler, William A. "Charles Christian Lauritsen: April 4, 1892 to April 13, 1968." *Biographical Memoirs of the National Academy of Sciences* 46 (1975): 221–233. This lively, affectionate, and informed account of Lauritsen's life was written by his former student and colleague of more than thirty years. The article contains a complete bibliography of Lauritsen's scientific publications.

Goodstein, Judith R. *Millikan's School: A History of the California Institute of Technology.* New York: W. W. Norton, 1991. The chapter titled "The Rockets' Red Glare" gives an account of the wartime rocket project at Caltech.

Holbrow, Charles H. "The Giant Cancer Tube and the Kellogg Radiation Laboratory." *Physics Today* 34 (July 1981): 42–49. This article describes the founding of the Kellogg Radiation

Laboratory and shows how Millikan promoted Lauritsen's achievements to obtain funding during the Great Depression.

———. "Charles C. Lauritsen: A Reasonable Man in an Unreasonable World." *Physics in Perspective* 5 (2003): 419–472. This is the most carefully researched biography available. It describes Lauritsen's research, his style of leadership, his active and extensive advising on weapons and defense technologies, and his efforts to foster flexible military strategies offering alternatives to all-out nuclear war.

———. "Scientists, Security, and Lessons from the Cold War." *Physics Today* 59 (July 2006): 39–44. Describes how Lauritsen and others made the case for constructing the chain of radar stations above the Arctic Circle.

Lauritsen, Thomas. Interview by Barry Richman and Charles Weiner, 16 February 1967. American Institute of Physics Center for History of Physics, College Park, Maryland.

Charles H. Holbrow

LAVOISIER, ANTOINE-LAURENT

(*b.* Paris, France, 26 August 1743; *d.* Paris, 8 May 1794), *chemistry, physiology, geology, economics, social reform.* For the original article on Lavoisier see *DSB*, vol. 8.

While Henry Guerlac's article in the original *DSB* offers a reliable and useful guide to the life and works of the French scientist, since 1973 new and important documentary evidence on Lavoisier has come to light that has made a reassessment of his contributions to science necessary. This contribution offers a brief survey of the new evidence in chronological order.

Education. In October 1754 Lavoisier entered the Collège des Quatre Nations, popularly known as the Collège Mazarin, in Paris. While there he was awarded two prizes for Latin and Greek translations in 1755 and 1759. As early as the autumn of 1760, Lavoisier was taking the course in mathematics and physics taught by the astronomer Nicolas Louis de Lacaille. In April 1761, in a report for a prize to be awarded by the Académie Besançon, Lavoisier exalted the contributions of such scientists as Archimedes, Bacon, Descartes, and Newton as positive examples of establishing a good reputation through beneficial and useful works. In 1761 he began to attend the chemical lectures of Guillaume François Rouelle and of the Parisian apothecary Charles Louis La Planche. At about the same time he followed a course in experimental physics taught by Jean Nollet. In an autobiographical note written around 1792, Lavoisier recalled this intense period of study thus:

> When I began for the first time to attend a course in chemistry, I was surprised to see how much obscurity surrounded the first approaches to the science, even though the professor I had chosen [Rouelle] was regarded as the clearest and most accessible to beginners, and even though he took infinite pains to make himself understood. I had taken a useful course in physics, I had followed the experiments of the Abbé Nollet, I had also studied elementary mathematics with some success in the works of the Abbé La Caille had attended his lectures for a year. (Beretta, 1994, pp. 15–16)

Surprising as it may seem, Lavoisier regarded La Planche as "the clearest" chemical teacher in Paris. La Planche's preference for beginning his course with analysis of the mineral kingdom instead of the vegetable one, as was customary, was regarded by the young Lavoisier as an innovation that would eventually prove important in his classification of chemical operations. While following Rouelle's lectures on the vegetable kingdom in 1761, Lavoisier managed to get a copy of Denis Diderot's notes, and he apparently made a copy of the course for himself. Lavoisier followed Rouelle's course for three years until 1763, when he wrote a note (a brief paper) on chemistry that revealed his preference for a quantitative and instrumental approach to the science and that showed little deference to his teachers.

After Lacaille's and Nollet's courses, Lavoisier became interested in the precision achieved with various instruments and in experimental physics and chemistry. Between 1761 and 1766 he regularly made barometric observations at his Parisian residence and during his natural-historical excursions outside Paris. In 1765 and 1766, following a meticulous and assiduous series of experiments, he perfected a light-reflecting lamp to improve the lighting of the streets of Paris, and his first attempts to improve chemical apparatuses were made in 1767.

Interest in Minerology. In 1763 the distinguished naturalist Jean Étienne Guettard, an old friend of Lavoisier's father, was advising Lavoisier and may have taken the latter under his wing as early as 1761. Guettard criticized the traditional approach to natural history and advocated a science of mineralogy supported by chemistry, topography, and physics. Since 1746 Guettard had been collecting material for a mineralogical map of France, but the task proved to be too great for a single naturalist. After Lavoisier's apprenticeship, Guettard, in 1763, decided to take the young scientist along as his assistant during his geological and mineralogical excursions outside Paris. Guettard's interdisciplinary approach to mineralogical research became evident in the writings of the young Lavoisier at a very early stage.

Lavoisier's budding interest in chemical mineralogy is revealed in a note dated 16 August 1763, in which he discussed a stone collected at Saint-Germain-en-Laye, outside of Paris. About a year later, guided by Guettard, he intensified his mineralogical survey of the regions around Paris, Mézières, and Champagne. He began to carry a barometer around with him, which he used to measure the levels of rock layers. It is not clear where he got the idea of using the barometer in geology, but this probably led to the idea of studying mineral ores in relation to their stratigraphic positions.

In July 1764 Lavoisier began to record his experiments with gypsum in a journal. This research was important, not only because chemists and mineralogists were interested in determining this mineral's composition, but also because Guettard considered the distribution of gypsum to be a good indicator of the mineralogical composition of the areas surrounding Paris that he had studied in the early 1750s. During his field surveys Lavoisier collected enough gypsum specimens to subject them to a comprehensive chemical study. On his travels of 1763–1765 he collected about one hundred gypsum specimens, most of which are still preserved at the Muséum d'histoire naturelle Henri-Lecoq in Clermont-Ferrand. Unlike Johann Friedrich Pott, who had subjected his gypsum specimens to fire, Lavoisier preferred to dissolve his specimens in water, because this was a simpler and more natural method of analysis. This choice of method is explained by Lavoisier's desire not only to break gypsum down into its constituent parts, but also to produce an artificial sample by combining vitriolic acid and calcareous earth. Toward this end, in August 1764 he began to use a hydrometer to achieve exact measurements of specific gravity. From then on he used this instrument to measure the specific gravities of components of chemical solutions. In 1768, in his first report devoted to this instrument, Lavoisier defended the originality of his approach in the following words:

> It is to the art of combination that the knowledge of the specific gravities of fluids can bring most light. *This aspect of chemistry is much less advanced than we thought*, we possess barely the rudiments of it. …
>
> If it is possible for the human spirit to penetrate these mysteries [connected to chemical combination], it is by means of research into the specific gravity of fluids that one may hope to achieve this. The quantity of real saline matter contained in the two fluids to be combined, their mean specific gravity with that resulting from their mixture, in other words the result of the same experiment, repeated on the same mixture combined with all the others, may produce a considerable quantity of data leading to the solution of the problem. (Lavoisier, 1862–1893, 3, pp. 448, 450)

Therefore, it was no coincidence that in the gypsum experiments at the beginning of 1765, Lavoisier began to use balance sheets to compare the weights of reactants before and after distillation.

In 1766 Lavoisier became acquainted with Johann Friedrich Meyer's theory that the causticity of lime is due to *acidum pingue*, a theory that was published in French translation under the title *Essais de chimie sur la chaux vive* (1766; Chemical essays on quicklime). In a memorandum dated May 1766, Lavoisier writes of deciding to try to verify Meyer's theory by embarking on a new series of experiments on the calcination of calcareous earths. This was several years earlier than the autumn of 1772, when, according to Guerlac in *The Crucial Year* (1961), "Lavoisier acquired most of his knowledge of the work done abroad on the chemistry of air" (p. 71). In the same year Lavoisier purchased numerous chemical and mineralogical books from the library of the mineralogist Jean Hellot (who had died in 1766), among which was a Latin manuscript version, with notes and comments by Hellot, of Georg Ernst Stahl's treatise on sulfur.

After his geological travels in Alsace with Guettard in 1767, Lavoisier began to focus on chemical experiments in a more systematic way, and between 1768 and 1774 he was entirely absorbed in his research on pneumatic chemistry. In November 1774 the Italian natural philosopher Giambattista Beccaria reported to Lavoisier concerning his experiments on calcinations. In *Elettricismo artificiale* (1772; Artificial electricity) Beccaria outlined an original application of the concept of electricity for understanding chemical combinations and operations. According to Beccaria, experiments performed by submitting metals and calxes to the action of an electrical machine showed that calcined metals could be revivified (reduced) by a discharge of electrical sparks. Within the Stahlian interpretation of matter, the revivification of metals was due to the addition of phlogiston and its consequent combination with the calx. This interpretation contradicted the gravimetric data, which indicated a loss of weight during reduction. Furthermore, phlogiston, which was regarded by Georg Stahl as an earthy and heavy substance, had not yet been isolated. Beccaria's experiments thus hinted at the identification of phlogiston with electricity. Electricity was also supposed to cause the calcination of metals and the release of their phlogiston. Beccaria's identification of phlogiston and electricity had a large impact on the European chemical community, and many naturalists welcomed his experiments as an authoritative demonstration of Stahl's principle of inflammability. Lavoisier was influenced and inspired by Beccaria's experiments on the calcination of metals and attentively followed the research on electricity. Sometime after his encounter with Beccaria, he began to make observations and experiments on his own.

The Synthesis and Analysis of Water. The origins of Lavoisier's experiments from 1783 to 1785 on the synthesis and analysis of water are rather obscure, and since the nineteenth century, many historians and chemists have raised doubts about the originality of Lavoisier's contributions to this crucial breakthrough. While the research by Henry Cavendish and James Watt in this field is well documented, their impact on Lavoisier remained as of 2007 far from clear. Even less well-known was the pervasive influence exerted by the pneumatic experiments performed by two Italian natural philosophers: Felice Fontana and Alessandro Volta. In 1777 Fontana invented an instrument composed of a sort of upside-down test tube immersed in a tray containing mercury, and it was included by Jacques-Louis David in the famous 1788 portrait of Lavoisier and his wife. Using this instrument, Fontana found that extinguishing red-hot charcoal in mercury contained in a glass tube containing mercury and immersed in a bath caused the absorption of a great quantity of air. Subsequently, other European naturalists used this method to experiment with red-hot charcoal and various types of air, and in 1782 Fontana himself discovered that if one extinguished red-hot charcoal in a glass bell full of water, inflammable air (hydrogen) was liberated.

This was a significant discovery because, if taken to its theoretical conclusion, it would have shown Fontana the compound nature of water. Lavoisier certainly knew of this experiment, because in 1783, in his famous *Mémoire dans lequel on a pour objet de prouver que l'eau n'est point une substance simple* (Report which seeks to prove that water is a simple substance), he declared, "The Abbé Fontana, having extinguished the red-hot charcoal in water, under a bell filled with water, drew therefrom a significant quantity of inflammable air.... As Abbé Fontana had shown with charcoal, [this method] proved that red-hot iron extinguished with water, under a bell, also produced inflammable gas" (1862–1893, 2, p. 341). Fontana's device was thus one of the fundamental instruments that led Lavoisier to the threshold of achieving his chemical revolution.

In the spring of 1782 Lavoisier met with Alessandro Volta in Paris and collaborated in a series of experiments on the absorption of electricity and the vaporization of fluids. The experiments on the vaporization of water and the use of several electrical instruments suggest that during his stay in Paris, Volta had demonstrated for Lavoisier and other French scientists his electrical eudiometer, or as he also called it, electrical gun. In the spring of 1777 Volta used this instrument for the first time in experiments in which he employed electrical discharges in a closed receiver to measure the inflammability of various types of air. During one of these experiments Volta observed that in the combustion of inflammable air (hydrogen) in presence of dephlogisticated air (oxygen), these airs ceased to be gases and left dew in the receiver. Because this result was completely unexpected, he did not identify the dew with water. Interestingly, it was not until 1784, after the publication of key reports by Cavendish and Lavoisier, that Volta began to understand the meaning of his experiments; only in the early 1790s, however, did he accept that water is composed of oxygen and hydrogen.

Sometime between 1782 and 1784 Lavoisier used Volta's electrical eudiometer, and in August 1784 the chemist Jean Darcet, at Lavoisier's request, sent Volta a report on the latest experiments by Lavoisier and Jean-Baptiste Meusnier de la Place on the decomposition of water. In the letter accompanying the report, Darcet showed that he was aware not only of Volta's eudiometer but also of its possible consequences for Lavoisier's work on the synthesis of water. From Darcet's testimony it is clear that Volta, while in Paris, had demonstrated his electrical gun and performed the experiments that allowed him to synthesize water. It is very unlikely that Volta showed these experiments to Darcet but not to Lavoisier; even if he did, however, Darcet probably would have told his colleague at the Académie des sciences about them. Moreover, three of Volta's electrical guns are in an inventory of Lavoisier's laboratory, compiled in November 1794. Thus, sometime after 1782, Lavoisier used Volta's eudiometer to replicate the experiments for synthesizing water with a different and much cheaper apparatus. With Fontana's and Volta's devices it was simple and easy to analyze and synthesize water, but these successes were not enough to persuade a skeptical European chemical community that water was a compound of two gases. After all, neither Fontana nor Volta understood the consequences of their experiments before Lavoisier's experiments on a grand scale.

When Lavoisier sought to design an experiment demonstrating both the synthesis and analysis of water in one process, it became necessary to construct a large gasometer. Lavoisier went to great expense to construct this apparatus, built by the Parisian instrument maker Pierre Bernard Mégnié between 1783 and 1787, not only to persuade a skeptical public but also to bring a high degree of accuracy to chemistry.

Historians have debated the role and efficiency of physical instruments such as the hydrometer (1768), the calorimeter (1782), the gasometer (1785), and the precision balances made by Nicolas Fortin (1788). While such contributions by historians have thrown light on important aspects of Lavoisier's approach to experimental procedures, little has been done to assess the effective quality of these devices, all of which are still preserved at the Musée des arts et métiers in Paris. One notable exception is the late twentieth-century reconstruction of the ice calorimeter of Lavoisier and Pierre-Simon Laplace and reenactment of

Antoine-Laurent Lavoisier. *Antoine-Laurent Lavoisier at work.* HULTON ARCHIVE/GETTY IMAGES.

their experiments on specific heat of 1783. The results, obtained after a laborious reconstruction of the experimental settings, were remarkably accurate and not far from historic levels of precision.

Chemical Nomenclature. The unfavorable reception of Lavoisier's new chemical nomenclature in the French press in 1787 inspired one member of his laboratory, Pierre-Auguste Adet, to propose the creation of a new chemical journal. To obtain government authorization to publish, the new journal was initially proposed as a French translation of Lorenz Crell's *Chemische Annales,* founded in 1784. While Lavoisier initially was not directly involved in the project, at the end of 1788 the *Société des Annales de chimie* was founded; it included Lavoisier as treasurer as well as many of his collaborators, among whom Claude Louis Berthollet and Louis-Bernard Guyton de Morveau became particularly active. In 1789 the first issue of the *Annales de chimie* was published by Lavoisier's printer, Gaspard-Joseph Cuchet, and it soon became a formidable means of propagating the new chemistry. After the publication in 1789 of his *Traité élémentaire de chimie* (which presented the theory, nomenclature, and apparatuses of the new chemistry), Lavoisier undertook an ambitious campaign of persuasion.

Human Respiration. In 1790 Lavoisier began an intensive series of experiments on human respiration, the results of which were only partly published in the *Annales de chimie.* In this enterprise Lavoisier was assisted by Armand Séguin, a promising young scientist who had been introduced to Lavoisier by Antoine François Fourcroy in 1785 during the large-scale experiments on the synthesis and analysis of water. Lavoisier became very fond of Séguin and relied heavily on his assistance and ingenuity during the experiments on respiration and transpiration. The background of this project went back almost two decades. Between 1773 and 1774 Lavoisier ascertained that animals absorbed a part of air through the lungs and that it was fixed there. In 1775 he understood that during animal respiration, oxygen was converted into fixed air, and that the oxygen then combined with blood. And in 1777 he was able to conclude that respiration was a slow combustion of carbon and hydrogen similar in every way to what took place with a lit candle, so that a breathing animal could be compared to a combustible body that was burning. During the experiments on heat carried out with Laplace in 1783, Lavoisier undertook quantitative calorimetric observations of the respiration of guinea pigs, comparing the ratios of the heat produced during respiration with that released during the combustion of charcoal.

In 1790 Lavoisier and Séguin finally decided to explore the physiology of human respiration and to test further Lavoisier's idea that respiration and combustion are analogous. Séguin was the first to observe that the increase in pulse rate was proportional to the bodily effort expended and that the ratio between pulse rate and effort could easily be quantified. Séguin became a human guinea pig, measuring with remarkable accuracy his consumption of oxygen in different situations, such as effort, rest, and digestion.

Because Lavoisier was increasingly involved in public affairs, the publication of these results was partial and largely unexploited. Apart from an unnoticed Italian translation of two reports published by the Venetian apothecary Vicenzo Dandolo in 1792, the revolutionary work of Lavoisier and Séguin on the physiology of human respiration appeared only in fragments published between 1793 and 1814.

Publishing Projects. In April 1793 Lavoisier renewed a previous contract with the Parisian publisher Charles-Joseph Panckoucke to include the "Régie des poudres" (The control of powders) in the *Dictionnaire d'artillerie* of the *Encyclopédie méthodique.* Lavoisier prepared entries for "Coal" and "Detonation" but these, together with entries prepared by other contributors, were not published until 1997.

In 1791 Lavoisier undertook the preparation of a new, more comprehensive work to be titled *Mémoires de physique et chimie* (1793; Memoir on physics and chemistry) in order to accomplish his long-cherished project of making chemistry as exact as physics was. In a manuscript note of 1792 (Archives de l'Académie des sciences, Paris) intended to serve as an introduction, he wrote:

> It is easy to see that these two sciences overlap at a good many points and that they have a lot in common; it is impossible to present a good physics course without introducing certain aspects of chemistry and, vice versa to create a good chemistry course without beginning with a few elementary notions of physics. These points of juncture between the two sciences increase day by day, since physicists and chemists have adopted a common approach, taken from that of the mathematicians, because they have rejected supposition and they no longer accept as truth that which is not proven through experimentation. (Beretta, "Lavoisier and His Last Printed Work," 2001, p. 334)

The publication of this work was suddenly interrupted in the summer of 1793 after the Académie des sciences closed down. The first page proofs of the planned five volumes had arrived at Pierre Samuel Dupont's printing house on 10 March 1793, and it appears that the printing went ahead under the direct supervision of Lavoisier, and probably also Séguin, until July 1793. In the summer of 1805, Madame Lavoisier started to distribute the first copies of a collection of proofs of the *Mémoires de physique et chimie* to selected friends and acquaintances; of the five volumes envisaged, 416 pages of the first volume (a nearly complete set), the whole of the second volume (413 pages), and 64 pages of the fourth volume were produced. The late distribution of this work by Madame Lavoisier affected both its diffusion and impact, so much so that it has often been neglected by historians of the chemical revolution. Yet the *Mémoires* are in fact crucial to understanding the development of Lavoisier's latest chemical investigations.

The work was intended to be divided into three parts. In the first, Lavoisier and Séguin selected those reports that, owing to their theoretical and experimental value, presented the most important chemical facts analytically. In the second part, Lavoisier and Séguin presented a dictionary of the main topics treated, thus facilitating the task of a reader who wished to study a particular subject. Finally, in the third part they presented a summary of the elementary truths presented in the first two parts, thus providing a synthetic idea of what had been set out analytically in the first part. This was evidently a highly ambitious project, the precise outline of which cannot be determined without further evidence.

The *Mémoires* contain forty reports overall, of which only twenty-eight were written by Lavoisier, eleven appearing for the first time. Even to those reports previously published, Lavoisier sometimes made significant changes, and it is unfortunate that Jean Baptiste Dumas and Edouard Grimaux preferred to include in the national edition of Lavoisier's *Oeuvres* (published between 1862 and 1893) the original reports without taking any notice of Lavoisier's revised and corrected versions. Of the remaining reports, ten were authored by Séguin; one by Séguin, Fourcroy, and Louis-Nicolas Vauquelin; and one by Louis Charles Henri Macquart and Fourcroy. Lavoisier and Séguin were thus the main authors. Lavoisier's most original contribution in the *Mémoires* was his reassessment of the role of chemical affinities and caloric. To explain why some bodies, when subjected to the action of caloric, remained solid and did not decompose, he hypothesized that a force in some way maintained them in this form and compensated for the dilating force of heat:

> We must therefore admit the existence of a force whose effects are opposed to the preceding one, which restrains the molecules of bodies and binds them one to another, and this force, whatever the cause of it may be, is *universal gravitation,* the force by virtue of which a molecule of matter tends to combine with another molecule, in a word: *attraction.*
>
> We must thus consider the molecules making up a body as obeying two forces; caloric, which continuously tends to separate them, and attraction, which counterbalances this force. As long as the latter, counterbalancing force, attraction, is victorious, the body remains solid. And if these two forces are in a state of equilibrium? The body becomes liquid. Finally, when the separating force of caloric is stronger, the body passes into a gaseous state. (*Mémoires,* 1, pp. 5–6.)

Even if his point of view was closer to that of a physicist rather than to that of a chemist, Lavoisier had decided to approach the problem of chemical affinities, a theme that he had previously dismissed because it relied on a qualitative approach to chemical reactions. To explain the different degrees of affinity between different substances within a quantitative framework, Lavoisier resorted to the theory of their atomic configurations. Early in his scientific career Lavoisier supported an atomistic philosophy of matter. In a manuscript fragment dated 1768, he outlined a molecular hypothesis on the structure of matter. Probably aware of the experimental difficulties of proving such a hypothesis for explaining chemical reactions, Lavoisier waited until 1793, when, after developing his theory on chemical heat, he felt more confident of the validity of his early idea on matter. Accepting theories that René Just Haüy had successfully applied to mineralogy, Lavoisier by 1793 believed that differences in atomic configurations effectively explained why bodies have different capacities to withstand heat and also explained the greater or lesser affinities between chemical substances.

Lavoisier closed his essay with an important distinction between aggregative and integrant molecules. He felt that he had sufficiently demonstrated the principle that the molecules in bodies never touched each other and that the distance between them was maintained by a given quantity of caloric. This principle he now regarded as valid also for aggregative molecules—in other words, those that made up a mixed body. Examples of such molecules are the molecules of a salt or an acid, because in these kinds of substances, two or three different elements can be isolated, and hence also two or three different kinds of molecules (these "molecules" are actually atoms, but such a distinction was unknown at the time). Lavoisier thought that caloric also combined with integrant and elementary molecules:

> It is more than possible, perhaps even probable, that there exist types of combinations where the elementary molecules touch, and it is undoubtedly in these kinds of combinations that the caloric enters as an integrant part in the form of combined caloric.
>
> It is high time today, now that the main phenomena that accompany releases and absorptions of caloric are well known, that geometricians try to test by calculation the various hypotheses that could be made to explain these phenomena. (*Mémoires*, 2004, 1, p. 27)

Lavoisier's tentative distinction between aggregative, integrant, and elementary molecules was new and preceded the debate on the atomic structure of matter by almost a decade. This corpuscular view was complemented by Lavoisier's assumption that atoms are indivisible and that, therefore, their nature is radically different from what he formerly attributed to chemical elements. Admittedly, when Lavoisier set forth this hypothesis, he was well aware that experimental practice available in chemical laboratories was still too primitive to allow an appreciation of the nature of constituent molecules. Despite these limitations, Lavoisier was the first scientist to attempt a chemical definition of the molecule. An in-depth study of the numerous unpublished manuscripts of the period 1792–1793 preserved in the Archives de l'Académie des sciences in Paris is anticipated to cast new light on the theoretical significance of this attempt.

Lavoisier's Public Career. From his youth onward, Lavoisier was eager to have a good reputation in the academic world. He participated in the 1761 prize competitions of the Académie d'Amiens and the Académie de Besançon. In 1766 he collaborated on a draft for reforming the Académie royale des sciences in Paris. After successful efforts in his scientific endeavors, he joined the Académie des sciences, also in Paris, in 1768, and he thereafter became a member of more than twenty scientific academies. His keen interest in the institutional organization of science was often rewarded with prominent directive positions at the Académie des sciences, as well as at other French academies. Lavoisier's career within these academic institutions was rooted in his ideas on the practice of science, which emphasized the importance of teamwork and of the organization of experimental endeavors.

In his youth Lavoisier worked for years with Guettard. When he finished his apprenticeship, Lavoisier often collaborated with other chemists. In 1771 he prepared his first experiments on the combustion of diamond in the laboratory of Guillaume François Rouelle, which were continued with a large lens one year later with Pierre Joseph Macquer, Jacques Mathurin Brisson, and Louis Claude Cadet de Gassicourt. In 1776 Lavoisier organized a prize competition on improving the quality of saltpeter together with Patrick d'Arcy, Cadet de Gassicourt, Macquer, Herni François de Paul Lefèvre d'Ormesson, and Balthazar Georges Sage. In 1773 Lavoisier began an ambitious experimental program on the nature of various airs together with Jean Baptiste Michel Bucuqet, and, occasionally, Jean Charles Philibert Trudaine de Montigny, whose laboratory was often used as the site of their experiments. In 1782 Lavoisier and Laplace undertook research on the nature of heat, and while in Paris in the spring of the same year, Volta regularly joined Lavoisier in his laboratory for more than three months. On 24 June 1783, Lavoisier carried out some experiments on the production of water by detonating oxygen and hydrogen under a bell jar, this in the presence of Laplace, Fourcroy, Charles Blagden, Alexandre Théophile Vadermonde, Meusnier, Adrien Marie Legendre, and Jean Baptiste Le Roy. The later experiments of 1785 involved Séguin and Vauquelin. In 1787 Lavoisier presented his new nomenclature together with Fourcroy, Berthollet, Guyton de Morveau, Adet, and Jean-Henri Hassenfratz. In 1788 he published a French translation, with commentary, of Richard Kirwan's *Essay on Phlogiston* (1787), on which he collaborated with his wife, Marie Anne Pierrette Lavoisier, née Paulze, Gaspard Monge, Berthollet, Guyton de Morveau, and Fourcroy. For Lavoisier, such collaborations were part of his pervasive strategy to change the structure of chemistry.

SUPPLEMENTARY BIBLIOGRAPHY

Most of Lavoisier's papers (more than 4,000 documents) are preserved at the Archives de l'Académie des science in Paris. A significant collection of books and manuscripts coming from Lavoisier's collection is kept at the Kroch Library at Cornell University (Ithaca, New York). The instruments are preserved at the Musée des Arts et métiers in Paris. Four thousand minerals and a few manuscripts are preserved at the Musée Lecoq at Clermont Ferrand. The chemical part of Lavoisier's library is preserved at the Bibliothèque de l'Institut in Paris. See Panopticon Lavoisier *(below) for a complete bibliography.*

WORKS BY LAVOISIER

Oeuvres de Lavoisier. 6 vols. Paris: Imprimerie impériale, 1862–1893. This national edition of Lavoisier's collected works is both incorrect and largely incomplete: The original spelling published in Lavoisier's first and original editions of his work was modernized, and most of his papers preserved in the archives, including the laboratory notebooks, are not published.

Oeuvres de Lavoisier: Correspondance. Vols. 1–3, edited by René Fric. Paris: Albin Michel, 1955–1964. Vol. 4, edited by Michelle Goupil. Paris: Belin, 1986. Vols. 5–6, edited by Patrice Bret. Paris: Académie des Sciences, 1993 and 1997. Volume 7 is forthcoming in 2007. An additional volume, including Lavoisier's correspondence with Jean Etienne Guettard and the numerous unpublished letters of the period 1763–1780, will follow.

With Pierre Simon Laplace. *Memoir on Heat.* Translated by Henry Guerlac. New York: Neale Watson Academic Publications, 1982.

De la richiesse territoriale de la France. Edited by Jean-Claude Perrot. Paris: Editions du C.T.H.S., 1988.

Mémoires de physique et de chimie. 2 vols. Bristol, U.K.: Thoemmes Continuum, 2004.

Panopticon Lavoisier. Edited by Marco Beretta. Available from http://moro.imss.fi.it/Lavoisier. The most comprehensive edition of Lavoisier's material. Includes a catalog of all the manuscripts, Lavoisier's bibliography, the bibliography on Lavoisier, the catalogue of the minerals, the instruments, the library and the iconography. Also contains the publication in text format of Lavoisier's national edition of his works as well thousands of digital reproductions of manuscripts, minerals, instruments, books, and other materials.

OTHER SOURCES

Bandinelli, Angela. "1783, Lavoisier and Laplace: Another Crucial Year—Antiphlogistic Chemistry and the Investigation on Living Beings between the Eighteenth and the Nineteenth Centuries," *Nuncius: Journal of the History of Science,* 18 (2003): 127–139.

Bensaude-Vincent, Bernadette. "A View of the Chemical Revolution through Contemporary Textbooks: Lavoisier, Fourcroy, and Chaptal." *British Journal for the History of Science* 23 (1990): 435–460.

———. *Lavoisier: Mémoires d'une revolution.* Paris: Flammarion, 1993. A critical and comprehensive survey of Lavoisier's historiography.

———, and Ferdinando Abbri, eds. *Lavoisier in European Context: Negotiating a New Language for Chemistry.* Canton, MA: Science History Publications-USA, 1995.

Beretta, Marco. "Chemists in the Storm: Lavoisier, Priestley, and the French Revolution." *Nuncius: Journal of the History of Science* 8 (1993): 75–104.

———. *The Enlightenment of Matter: The Definition of Chemistry from Agricola to Lavoisier.* Canton, MA: Science History Publications-USA, 1993.

———. *A New Course in Chemistry: Lavoisier's First Chemical Paper.* Florence, Italy: Leo S. Olschki, 1994.

———. *Bibliotheca Lavoisieriana: The Catalogue of the Library of Antoine Laurent Lavoisier.* Florence, Italy: Leo S. Olschki, 1995.

———. "Pneumatics vs. 'Aerial Medicine': Salubrity and Respirability of Air at the End of the Eighteenth Century." In *Nuova Voltiana: Studies on Volta and His Times.* Vol. 2, edited by Fabio Bevilacqua and Lucio Fregonese. Milan, Italy: Editore Ulrico Hoepli, 2000.

———. "From Nollet to Volta: Lavoisier and Electricity." *Revue d'histoire des sciences* 54 (2001): 29–52.

———. *Imaging a Career in Science: The Iconography of Antoine Laurent Lavoisier.* Canton, MA: Science History Publications-USA, 2001.

———. "Lavoisier and His Last Printed Work: The *Mémoires de physiques et de chemie* (1802)." *Annals of Science* 58 (2001): 327–356.

———. "Lavoisier's Collection of Instruments: A Checkered History." In *Musa Musaei: Studies on Scientific Instruments and Collections in Honour of Mara Miniati,* edited by Marco Beretta, Paolo Galluzzi, and Carlo Triarico. Florence, Italy: Leo S. Olschki, 2003.

———. "Collected, Analyzed, Displayed: Lavoisier and Minerals." In *From Private to Public: Natural Collections and Museums,* edited by Marco Beretta. Sagamore Beach, MA: Science History Publications-USA, 2005.

———, ed. *Lavoisier in Perspective.* Munich, Germany: Deutsches Museum, 2005.

Bret, Patrice, ed. "Débats et chantiers actuels autour de Lavoisier et de la révolution chimique." *Revue d'histoire des sciences* 48 (1995): 1–2. Double issue devoted specifically to Lavoisier.

———. *Lavoisier et l'Encyclopédie Méthodique: Le manuscrit des régisseurs des poudres et salpêtres.* Florence, Italy: Leo S. Olschki, 1997.

———. "Les origines et l'organisation éditoriale des Annales de chimie (1787–1791)." In *Lavoisier: Correspondance.* Vol. 6. Edited by Maurice Bret. Paris: Académie des sciences, 1997.

———. *L'état, l'armée, la science: L'invention de la recherche publique en France (1763–1830).* Rennes, France: Presses universitaires de Rennes, 2002.

———. "Dévenir des héros : Le mémoire inédit de Lavoisier au concours d'éloquence de l'Académie de Besançon en 1761." *Dix-huitième siècle* 37 (2005): 329–346.

Cotty, Gaspard Hermann. *Dictionnaire d'artillerie, Encyclopédie méthodique.* Paris: Agasse, 1822.

Demeulenaere-Douyère, Christiane. "A propos d'une entreprise intellectuelle: La publication des oeuvres et de la correspondance de Lavoisier." *La vie des sciences, comptes rendus,* série générale, 11 (1994): 319–332.

———. *Il y a 200 ans Lavoisier: Actes du Colloque organisé à l'occasion du bicentenaire de la mort d'Antoine Laurent Lavoisier le 8 mai 1794.* Paris: Lavoisier TecDoc, 1995

Crosland, Maurice. "Science and Secret Weapons Development in Revolutionary France, 1792–1804: A Documentary History." *Historical Studies in Physical and Biological Sciences* 23 (1992): 35–152

———. *In the Shadow of Lavoisier: The Annales de Chimie and the Establishment of a New Science.* London: British Society for the History of Science, 1994.

Dhombres, Nicole, and Jean Dhombres. *Naissance d'un nouveau pouvoir: Sciences et savantes en France, 1793–1824.* Paris: Payot, 1989.

Donovan, Arthur, ed. "The Chemical Revolution: Essays in Reinterpretation." Osiris, second series 4 (1988), special issue.

———. *Antoine Lavoisier: Science, Administration, and Revolution.* Oxford: Blackwell, 1993. An insightful biography.

Durand, Yves. *Les fermiers généraux aux XVIIIe siècle.* Paris: Presses Universitaires de France, 1971.

Duveen, Denis I. *Supplement to A Bibliography of the Works of Antoine Laurent Lavoisier (1743-1794).* London: Dawsons of Pall Mall, 1965.

———, and Herbert Klickstein, A Bibliography of the Works of Antoine Laurent Lavoisier. London: Dawsons, 1954.

Gillispie, Charles Coulston. *Science and Polity in France at the End of the Old Regime.* Princeton, NJ: Princeton University Press, 1980.

———. *Science and Polity in France: The Revolutionary and Napoleonic Years.* Princeton, NJ: Princeton University Press, 2004.

Gough, Jerrry B. "The Origin of Lavoisier's Theory of the Gaseous State." In *The Analytic Spirit: Essays in the History of Science in Honor of Henry Guerlac,* edited by Harry Woolf. Ithaca, NY, and London: Cornell University Press, 1981.

———. "Lavoisier's Memoirs on the Nature of Water and Their Place in the Chemical Revolution." *Ambix* 30 (1983): 89–106.

Guerlac, Henri. *Lavoisier—The Crucial Year: The Background and Origin of His First Experiments on Combustion in 1772.* Ithaca, NY: Cornell University Press, 1961.

———. *Antoine-Laurent Lavoisier: Chemist and Revolutionary.* New York: Scribner, 1975.

———. "Laplace's Collaboration with Lavoisier." In *Essays and Papers in the History of Modern Science.* Baltimore: Johns Hopkins University Press, 1977.

———. "The Lavoisier Papers—A Checkered History." *Archives internationales d'histoire des sciences* 29 (1979): 95–100.

Goupil, Michelle, ed. *Lavoisier et la révolution chimique.* Paris: Sabix, 1992.

Heering, Peter. "Weighing the Heat: The Replication of the Experiments with the Ice-calorimeter of Lavoisier and Laplace." In *Lavoisier in Perspective,* edited by Marco Beretta. Munich, Germany: Deutsches Museum, 2005.

Holmes, Frederic Lawrence. *Lavoisier and the Chemistry of Life: An Exploration of Scientific Creativity.* Madison: University of Wisconsin Press, 1985.

———. *Antoine Lavoisier, the Next Crucial Year, or the Sources of His Quantitative Method in Chemistry.* Princeton, NJ: Princeton University Press, 1998.

Holmes, Frederic L., and Trevor H. Levere, eds. *Instruments and Experimentation in the History of Chemistry.* Cambridge, MA: MIT Press, 2000.

Kim, Mi Gyung. "Public Science: Hydrogen, Balloons, and Lavoisier's Decomposition of Water." *Annals of Science* 63 (2006): 291–318.

Palmer, Louise Yvonne. "The Early Scientific Work of Antoine Laurent Lavoisier: In the Field and in the Laboratory, 1763–1767." PhD diss., Yale University, 1998.

Pelucchi, Stéphane. "Histoires parallèles: Histoire de la collection de minéralogie d'Antoine Laurent Lavoisier." *Nuncius: Journal of the History of Science* 18 (2003): 705–732.

Perrin, Carlton E. "Lavoisier's Thoughts on Calcination and Combustion, 1772–73." *Isis* 77 (1986): 647–666.

———. "Document, Text, and Myth: Lavoisier's Crucial Year Revisited." *British Journal for the History of Science* 22 (1989): 3–25.

Poirier, Jean-Pierre. *Lavoisier: Chemist, Biologist, Economist.* Translated by Rebecca Balinski. Philadelphia: University of Pennsylvania Press, 1996.

———. "De la Ferme générale et la Caisse d'escompte à la Tresorerie nationale: Lavoisier, financier de l'Ancien Régime et grand commis de l'Etat moderne." In *Lavoisier: Correspondance.* Vol. 6. Edited by Maurice Bret. Paris: Académie des sciences, 1997.

———, ed. *De la situation du Trésor public au 1er juin 1791.* Paris: Editions du C.T.H.S., 1997.

Prinz, Johann Peter. *Die experimentelle Methode der ersten Gasstoffwechseluntersuchungen am ruhenden und quantifiziert belasteten Menschen (A. L. Lavoisier und A. Séguin 1790): Versuch einer kritischen Deutung.* Sankt Augustin, Germany: Academia Verlag, 1992.

Rappaport, Rhoda. "Lavoisier's Theory of the Earth." *British Journal for the History of Science* 6 (1972–1973): 247–260.

Roberts, Lissa. "A Word and the World: The Significance of Naming the Calorimeter." *Isis* 82 (1991): 198–222.

Smeaton, William A. "Madame Lavoisier, P. S. and E. I. Du Pont de Nemours and the Publication of Lavoisier's 'Mémoires de chimie.'" *Ambix* 36 (1989): 22–30.

Marco Beretta

LE CLERC, GEORGE-LOUIS

SEE **Buffon, George-Louis Le Clerc, Comte de**.

LEAKEY, MARY DOUGLAS NICOL

(*b.* London, United Kingdom, 6 February 1913; *d.* Nairobi, Kenya, 9 December 1996), *paleoanthropology, Paleolithic archaeology, early hominid stone toolmaking.*

Kenyan archaeologist and paleoanthropologist, particularly noted for her work on the earliest stone toolmaking traditions at Tanzania's Olduvai Gorge, Mary Leakey's name is invariably linked with that of her charismatic husband Louis Leakey; but although the careers of the two were almost inextricably intertwined over more than forty years, Mary's contribution was a clearly distinctive one. Indeed it is fair to observe that Mary Leakey was the mainstay of the pathbreaking work in East African ancient prehistory that the two carried out together from the 1930s onward.

Early Interest in Archaeology. Mary Douglas Nicol was born in London on 6 February 1913, to Erskine Nicol and Cecilia Marion Frere. On her maternal side she traced her ancestry to the eighteenth-century British antiquarian John Frere (who had presciently recognized the great antiquity of the stone tools found near his home at Hoxne, in Suffolk), while her father was a well-known and widely respected painter. Mary thus moved in privileged social circles from an early age. Traveling widely with her peripatetic parents to exotic locales in Europe and Africa, Mary rapidly developed a devotion to both drawing and prehistory. Early on she became fascinated with the Ice Age decorated caves of France's Vézère Valley, and later fondly recalled crawling through the cave of Pech Merle with the renowned French prehistorian Abbé Lemozi, in search of some of the greatest art of the Pleistocene epoch.

At school, Mary was an unenthusiastic student whose imagination readily strayed from the classroom to far-flung archaeological sites where her imagination and gift for drawing could bring alive the past. The beginnings of her ambition to become an archaeologist can be traced back to a visit with her mother to Stonehenge in the 1920s. To the end of her life Mary vividly recalled the nearby Neolithic site of Windmill Hill that was nominally under the control of the gentleman archaeologist Alexander Keiller, but where the excavations themselves were run by his sister-in-law Dorothy Liddell. It was meeting Liddell that first convinced the young Mary Nicol that a career as a professional archaeologist might be attainable.

As the child of an affluent English family Mary was expected to attend a university, her mother's choice being Oxford (her father having died in 1926 when Mary was thirteen). However, in light of her mediocre academic record the eminent Oxford geologist William Johnson Sollas strongly discouraged her mother from pursuing this idea. In her autobiography Mary herself noted the irony when, in 1981, Oxford joined many other universities in presenting her with an honorary doctorate in recognition of her lifetime of contributions to the archaeological sciences. Undeterred by her failure to gain formal university admission, Mary continued doing what she loved most, digging and drawing, while also attending lectures at the University of London and at the London Museum. She worked at numerous archaeological sites, her first major experience of excavation being at the Roman site of Verulamium, under the direction of Mortimer Wheeler. Mary then moved in the early 1930s to the Neolithic site of Hembury in Devon, where the excavations were run by the same Dorothy Liddell who had originally inspired her ambition to be an archaeologist. In early 1933 the drawings of flint tools Mary had done for Liddell caught the attention of British archaeologist Gertrude Caton-Thompson, who set her to work illustrating stone tools from the Egyptian region of the Faiyûm for her book *The Desert Fayum* (1934). And it was Caton-Thompson who shortly thereafter introduced Mary Nicol to Louis Leakey, as a potential illustrator for the latter's book *Adam's Ancestors* (1934).

Mary's final "training" season as a young archaeologist came in 1934 at Jaywick, near Clacton in Essex. Mary had already become intrigued by the primitive Clactonian stone tools years earlier, and Jaywick was a site run by Kenneth Oakley, who was principally a geologist, although he later became best known for his seminal work on Paleolithic archaeology, *Man the Toolmaker* (1949). The two worked well together, and Jaywick became the first archaeological site for which Mary assumed full responsibility, although in 1934 she also began work with Louis Leakey at Swanscombe. During the three seasons that Mary Nicol worked with Liddell, she took a short break to work at Meon Hill, an Iron Age and Saxon site; and in early 1935, on her way to join Leakey on an expedition to East Africa, she even dug at Oakhurst Shelter in South Africa with Astley John Hilary Goodwin. Thus by the middle of the 1930s Mary had already gained experience in the archaeology of a gamut of periods of prehistory ranging from the early Paleolithic to post-Roman times, and had been exposed to a variety of techniques of excavation. And despite her lack of formal qualifications she had already interacted extensively with many of the leading English-speaking archaeologists of the day.

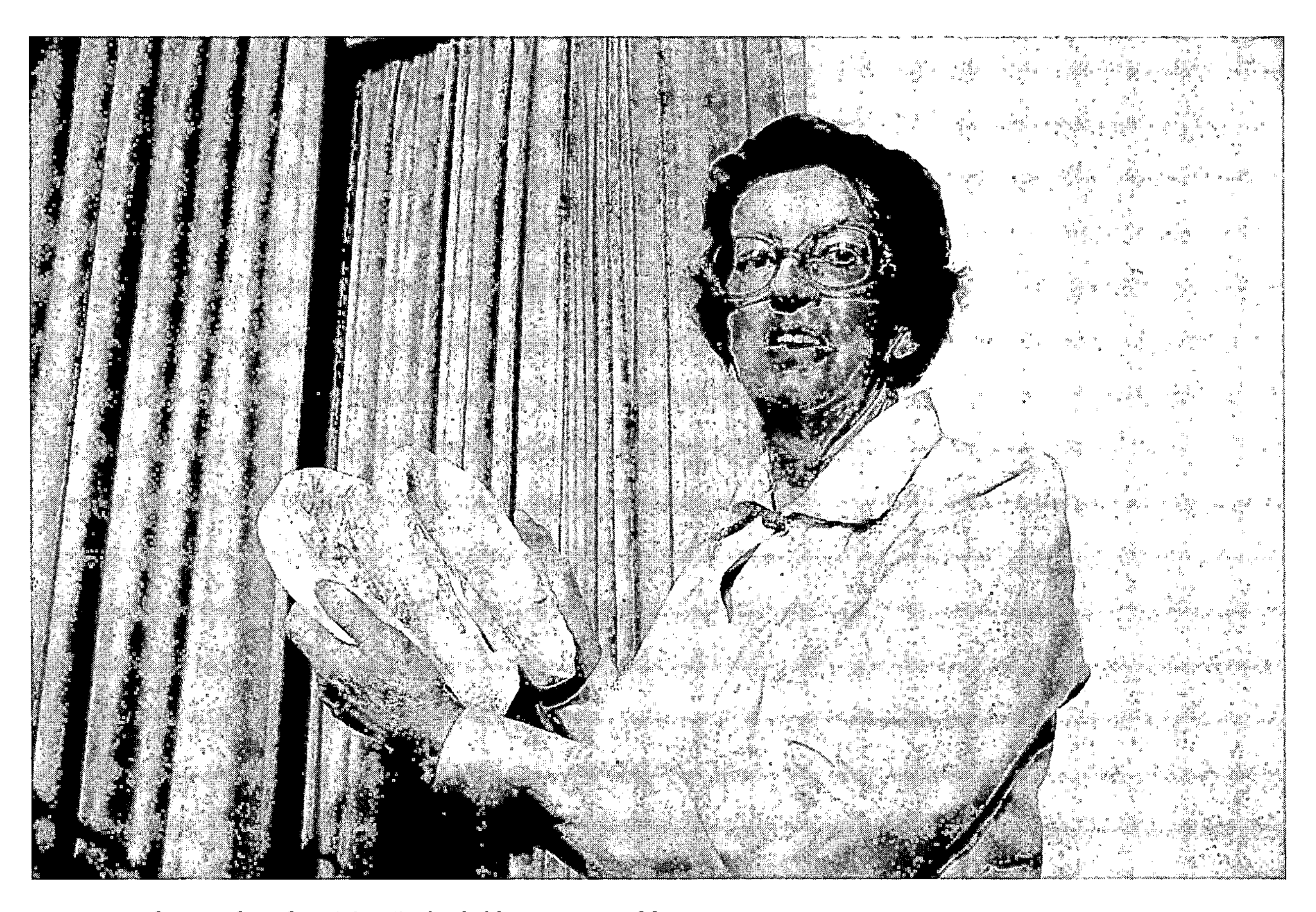

Mary Douglas Nicol Leakey. Mary Leakey holding two casts of footprints. © BETTMANN/CORBIS.

Louis Leakey and Kenya. When Mary first met Louis Seymour Bazett Leakey in 1933, Louis was married to his first wife Frida and was working at St. John's College, Cambridge. Rapidly Mary and Louis became close, and in 1936 Louis was divorced by Frida in an atmosphere of scandal that brought his career at Cambridge to an end. At the end of 1936 Mary and Louis were married, and shortly thereafter they moved to Kenya, where Louis had been born, where his family still lived, and where he had already carried out several paleontological expeditions. Thus was born a scientific partnership that was to last until Louis's death in 1972 and that was symbolized by the joint work the couple conducted over decades at Olduvai Gorge, in northern Tanzania, where the German geologist Hans Reck had initially discovered a human skeleton (Olduvai Hominid 1, now known to be only 17,000 years old) in 1913. They had first visited Olduvai together in 1935, when Mary discovered two hominid skull fragments (Olduvai Hominid 2 [OH2]); on the same expedition they also visited the nearby site of Laetoli, to which Mary was ultimately to devote the final years of her career.

Louis Leakey's principal employment when the couple returned to Kenya in 1937 was on a study of the Kikuyu tribe financed by the Rhodes Trust, but Mary soon set to work excavating the Neolithic site of Hyrax Hill, though by the end of the year she had shifted her attention to the Late Stone Age site of Njoro River Cave. Shortly thereafter World War II intervened, although it did not alter the Leakeys' path. For her part Mary contrived to continue her archaeological activities, including the discovery of the famous implement-rich surface at the Acheulean site of Olorgesailie; and in 1940 Louis assumed the curatorship of the Coryndon Museum in Nairobi, which was thereafter their base. During the war years Mary Leakey also made her first visit to Rusinga Island in Lake Victoria, where in 1948 she discovered the only known cranium of the Miocene hominoid *Proconsul africanus.* By 1948 the Leakeys' numerous hominoid finds at Rusinga had already garnered considerable press attention, and had led to financing from the entrepreneur Charles Boise, whose generosity was also to make possible an important survey and documentation of East African rock art, and a serious return to Olduvai.

During the period from 1951 to 1958 the Leakeys, accompanied by their three sons—Jonathan, Richard, and Philip—and a pack of Dalmatians, focused their efforts on the gorge's Bed II, in which Mary had found the OH2 fossils. Numerous artifacts and mammalian fossils were found, but only a couple of hominid teeth. However, in 1959, after spending most of the season at Laetoli, the group began to survey Bed I of the gorge. On July 17, near the bottom of Bed I, Mary discovered Olduvai Hominid 5, the famous *Zinjanthropus boisei* (now *Australopithecus boisei*) cranium. Two years later Louis Leakey and colleagues shocked the paleoanthropological community by announcing that dating by the new potassium/argon method had shown this specimen to be a previously unimaginable 1.7 million years old.

The publicity attendant on this and subsequent spectacular finds (including the type materials of *Homo habilis* in 1964) vastly increased the financial support available to the Leakeys for their fieldwork; and while Louis described most of the Olduvai hominid fossils, starred in numerous documentaries, and went on extensive speaking tours, Mary stayed for increasing amounts of time at the gorge and quietly made progress on understanding its archaeology. As early as 1931 Louis had reported the existence in the lowest levels of the gorge of crude stone artifacts that had acquired the designation "Oldowan" by virtue of their provenance. Higher in the section were found more complex bifacially flaked artifacts comparable to "Acheulean" implements from Europe, but usually made from fine-grained volcanic materials rather than the flints and cherts more familiar to European archaeologists. In 1966 Mary published a characterization of the various implement types she recognized within the Oldowan culture, noting the preponderance among the tools of "choppers," fist-sized cobbles altered by the knocking-off of a few flakes—although subsequent researchers have concluded that the small sharp flakes that Mary regarded mostly as "waste" may in most cases have been the primary products the toolmakers sought. Significantly, Mary observed the frequent presence at hominid activity sites of "manuports"—stones deliberately carried in by hominids from remote source areas.

This preliminary review was followed in 1971 by the magisterial *Olduvai Gorge*, vol. 3, *Excavations in Beds I and II, 1960–1963*, a massive volume in which Mary Leakey analyzed nearly forty thousand artifacts and twenty thousand fossil animal bones from the bottom two layers of the gorge, and documented the twenty or so hominid fossils by then known from Olduvai. She showed how Oldowan tools in the most ancient rock layers exposed in the sides of the gorge yielded higher up to a "Developed Oldowan A" industry and then to a "Developed Oldowan B," and how toward the upper part of Bed 2 the Acheulean intruded. She analyzed the associations in the deposits of stone tools with animal and hominid fossils, and suggested that some occurrences represented "living floors," where hominids had lived and butchered animal carcasses. At one site she suggested that a ring of stones represented the remains of a windbreak deliberately constructed by early hominids. And although later work has cast doubt on some of her detailed interpretations, Mary Leakey's meticulous documentation in this work of the archaeology of Olduvai Gorge set a new standard by which all future research of the kind would be judged.

As soon as she completed work on this volume Mary set out for Olduvai again, this time to work on the uppermost Beds III and IV. And she left without her by then ailing husband. As the Leakey family's biographer Virginia Morell has observed, "Olduvai was very much her project now" (1995, p. 318). In 1972 Louis Leakey died, and Mary's career once again took a new turn. In 1974 she returned to Laetoli, where she and her team found several upper and lower jaw bones and other fragmentary hominid fossils now known to date between about 3.7 million and 3.5 million years ago. Her coworkers made these specimens the type materials of the species *Australopithecus afarensis* (to which the Ethiopian "Lucy" skeleton is also assigned), but Leakey herself refused to make the assignment to *Australopithecus* and withdrew from authorship of the new species.

The Laetoli Beds, within which these fossils were found, consist of a succession of volcanic ashfalls on a relatively treeless open plain, and the most unusual finding in the area consists of diverse animal trackways, first identified in 1976. Some of these tracks were made by hominids, and the most spectacular discovery of this kind was made in 1978. In that year were found the 80-feet-long trails left by at least two individuals who strode across the open landscape after a rainfall that had moistened fresh ash to the consistency of wet cement. Although their exact interpretation is still disputed, the footprints were unquestionably made by upright bipeds: a unique confirmation that on the ground at least hominids were moving around on their hind feet by 3.5 million years ago. Unsurprisingly, there are no stone tools in the ancient deposits at Laetoli, since stone tools do not occur elsewhere until about 2.6 million years ago at the earliest. However, in younger deposits in the Laetoli region a cranium some 120,000 years old is associated with Middle Stone Age tools.

Over a career of more than sixty years Mary Nicol Leakey compiled an extraordinary record of discovery and achievement not only as an archaeologist but as a paleontologist. She was, in other words, a true paleoanthropologist, whose legacy lives on not only in her many publications and in the fossils and artifacts she discovered, but in the dynasty she and her husband Louis founded.

For not only did her son Richard follow her into a distinguished career in paleoanthropology, but Richard's daughter Louise has now picked up the mantle too.

BIBLIOGRAPHY

WORKS BY LEAKEY

With Louis S. B. Leakey. *Excavations at the Njoro River Cave: Stone Age Cremated Burials in Kenya Colony.* Oxford: Clarendon Press, 1950.

Olduvai Gorge. Vol. 3, *Excavations in Beds I and II, 1960–1963.* Cambridge, U.K.: Cambridge University Press, 1971.

Olduvai Gorge: My Search for Early Man. London: Collins, 1979.

Africa's Vanishing Art: The Rock Paintings of Tanzania. Garden City, NY: Doubleday, 1983.

Disclosing the Past: An Autobiography. Garden City, NY: Doubleday, 1984.

With John M. Harris, eds. *Laetoli: A Pliocene Site in Northern Tanzania.* New York: Oxford University Press, 1987.

OTHER SOURCES

Leakey, Louis S. B. *By the Evidence: Memoirs, 1932–1951.* New York: Harcourt, 1974.

Leakey, Richard. *One Life: An Autobiography.* Salem, NH: Salem House, 1984.

Morell, Virginia. *Ancestral Passions: The Leakey Family and the Quest for Humankind's Beginnings.* New York: Simon & Schuster, 1995.

Roe, Derek Arthur. *The Year of the Ghost: An Olduvai Diary.* Bristol, U.K.: Beagle, 2002.

Ken Mowbray
Ian Tattersall

LECONTE, JOSEPH

(*b.* Liberty County, Georgia, 26 February 1823; *d.* Yosemite Valley, California, 6 July 1901), *natural history, physiology, geology.* For the original article on LeConte see *DSB,* vol. 8.

LeConte's major contributions to science include works on the physiology of vision, geology, and the theory of evolution in relation to religious beliefs. Although much of his geological work was theoretical, it also included several original studies. LeConte wielded widespread influence through his college and high school textbooks in geology, his success as a teacher, and his publications on the harmony of evolution and religion. Renewed attention to the scientific contributions and evolutionary views of LeConte commenced during the 1970s, and culminated in 1982 with the publication of *Joseph LeConte: Gentle Prophet of Evolution* by Lester D. Stephens. Corrections of earlier biographical details also appear in the same work.

Biography. Joseph LeConte was one of seven children born to Louis and Ann Quarterman LeConte. A descendant of Huguenots, Louis LeConte moved to Liberty County, Georgia, around 1809, and settled on a 3,350-acre rice and cotton plantation purchased by his father many years earlier. Although Louis LeConte briefly studied medicine, he devoted his career to the plantation, on which he cultivated a flower garden that attracted many botanists and other visitors.

Joseph LeConte entered the University of Georgia in Athens in January 1838, and received an AB degree in August 1841. He was enrolled in New York's College of Physicians and Surgeons from January to May 1844, and took an extensive trip to the Midwest during the summer of 1844. In order to improve his training, he voluntarily took the medical course again from 1844 to 1845. After practicing medicine in Macon, Georgia, from the winter of 1848 to August 1850, he moved to Cambridge, Massachusetts, to study with Louis Agassiz at the Lawrence Scientific School of Harvard College. He and his cousin William Louis Jones completed the program in 1851, and became the first two students to receive degrees from the new school.

In 1852 LeConte was a member of the faculty of Oglethorpe College, then located in Milledgeville, Georgia. He joined his brother John, a physicist, at the University of Georgia in January 1853, and taught there through December 1856. Along with his brother, he was associated with South Carolina College in Columbia from 1857 to 1869. Joseph served the Confederate government from 1863 to 1865, first in the manufacture of medicines and later as a chemist in the Nitre and Mining Bureau. Dissatisfied with Reconstruction policies, he accepted a faculty position at the University of California in Berkeley late in 1869, joining his brother John, who had gone there a few months earlier as the first faculty member hired by the new institution. Joseph was the third person appointed to the University of California faculty, and continued his association with that institution until his death in 1901. John LeConte served as the president of the University of California on two occasions, but Joseph never indicated any interest in academic administration.

Married to Caroline Elizabeth Nisbet on 14 January 1847, he was the father of four daughters, Emma, Sarah, Josephine, and Caroline, and one son, Joseph Nisbet. Josephine died shortly before her second birthday. The death of Joseph LeConte occurred on 6 July 1901, while he was on a pleasure trip to Yosemite.

LeConte's Contributions. By the time LeConte published his *Outlines of the Comparative Physiology and Morphology of Animals* in 1900, his knowledge of the subject was dated in many respects. More influential was his *Sight: An*

Exposition of the Principles of Monocular and Binocular Vision, published originally in 1881 and revised in 1897. Although the revised version indicated that LeConte had been unable to keep current on the topic, he was a pioneer in explicating the physiology of vision.

More enduring and more widely distributed were LeConte's *Elements of Geology, A Compend of Geology,* and *Evolution and Its Relation to Religious Thought* and the revised version of the same under a modified title, *Evolution: Its Nature, Its Evidences, and Its Relation to Religious Thought.*

First published in 1877 and revised three times by LeConte and once, in 1903, by Herman L. Fairchild, *Elements of Geology* was a highly popular college textbook that included an extensive treatment of the theory of evolution. *A Compend of Geology,* originally published in 1884 and revised in 1898, was used in many high schools. It also devoted considerable attention to evolution, and thus introduced a large number of students to the theory before most school systems dropped geology from the curriculum and before the theory became so controversial that many schools excised it from their programs. In geology, LeConte was especially interested in the ancient river beds of the High Sierras, geological deformation, and the origin of continents and mountains. He wrote extensively on the contractional theory of mountain building, and became, along with the noted geologist James Dwight Dana, one of America's major advocates of the idea.

A highly influential teacher and popular lecturer, LeConte played a significant role in the effort to reconcile the theory of evolution with biblical accounts. Calling himself a theistic evolutionist, he was among the most ardent champions of evolution, and referred to the theory as "the grandest of modern ideas" (1887, p. 123). His book on evolution and religion, first published in 1888 and revised in 1891, was perhaps the most popular American work on the subject for many years. Although he followed the ideas of Charles Darwin in part, LeConte was, like many of his American contemporaries, more attuned to the theory of evolution espoused by the French naturalist Jean-Baptiste Lamarck.

In any case, LeConte delivered numerous public lectures and also published many articles on the topic of evolution. Although he was opposed to the notion of survival of the fittest, LeConte was nonetheless a committed evolutionist who believed he could harmonize the theory with Christian teachings. Identifying himself as a "theistic evolutionist" ("evolution in relation to materialism," *Princeton Review,* 4th series [March 1881], pp. 149–174]), LeConte argued that God created the first animate beings and then allowed speciation to occur by the process of evolution. He rejected the deistic philosophy, and maintained that the creator was constantly involved in an "eternal unfolding of the original conception" (ibid.). In LeConte's view, God designed evolution as a progressive process that operated through natural law. His efforts compelled him to reject the literal occurrence of biblical revelations and miracles, but he attracted many followers, especially because of his pleasant personality and persuasive arguments. In 1895, a San Francisco newspaperman aptly dubbed him "the gentle prophet of evolution" (*San Francisco Examiner,* 26 May 1895, p. 12).

Elected to membership in the National Academy of Sciences in 1875, LeConte served as president of the American Association for the Advancement of Science in 1891 and as president of the Geological Society of America in 1896. A friend of John Muir, he was a founding member of the Sierra Club, which erected the Joseph LeConte Lodge in Yosemite in 1904 in tribute to his influence as a teacher and his devotion to the region.

SUPPLEMENTARY BIBLIOGRAPHY

WORKS BY LECONTE

Elements of Geology. New York: D. Appleton, 1877.

Sight: An Exposition of the Principles of Monocular and Binocular Vision. New York: D. Appleton, 1881. Revised edition, 1897.

A Compend of Geology. New York: D. Appleton, 1884. Revised New York: American Book Company, 1898.

"Relation of Biology to Society." *Berkeleyan* 23 (1887): 123.

Evolution and Its Relation to Religious Thought. New York: D. Appleton, 1888.

Evolution: Its Nature, Its Evidences, and Its Relation to Religious Thought. New York: D. Appleton, 1891.

Outlines of the Comparative Physiology and Morphology of Animals. New York: D. Appleton, 1900.

The Autobiography of Joseph LeConte. Edited by William Dallam Armes. New York: D. Appleton, 1903. A useful source, but omits some information from the original manuscript, which is in the Southern Historical Collection, University of North Carolina, Chapel Hill.

'Ware Sherman: A Journal of Three Months' Personal Experience in the Last Days of the Confederacy. Edited and introduced by Caroline LeConte. Berkeley: University of California Press, 1938. Reissued with a new introduction by William Blair, Baton Rouge: Louisiana State University Press, 1999.

OTHER SOURCES

Anderson, Richard LeConte. *LeConte Family History and Genealogy.* 2 vols. Macon, GA: privately printed, 1981. A thorough work.

LeConte, Emma. *When the World Ended: The Diary of Emma LeConte.* Edited by Earl Schenck Miers. New York: Oxford University Press, 1957. Reissued with a foreword by Anne Firor Scott. Lincoln: University of Nebraska Press, 1987.

Stephens, Lester D. "Joseph LeConte on Evolution, Education, and the Structure of Knowledge." *Journal of the History of the Behavioral Sciences* 12 (April 1976): 103–119.

———. "Joseph LeConte's Evolutional Idealism: A Lamarckian View of Cultural History." *Journal of the History of Ideas* 36 (July–September 1978): 465–480.

———. "Joseph LeConte and the Development of the Physiology and Psychology of Vision in the United States." *Annals of Science* 37 (1980): 303–321.

———. *Joseph LeConte: Gentle Prophet of Evolution.* Baton Rouge: Louisiana State University Press, 1982. A comprehensive biography, with a complete list of manuscript sources and works published by LeConte.

———. "Joseph LeConte." In *American National Biography,* edited by John A. Garraty and Mark C. Carnes. New York: Oxford University Press, 1999.

Lester D. Stephens

LEDERER, EDGAR (*b.* Vienna, Austria, 5 June 1908; *d.* Sceaux, near Paris, 19 October 1988), *chemistry of natural products, biochemistry, immunology.*

Lederer was one of France's leading mid- and late-twentieth-century chemists and biochemists. He left his mark on the chemistry of natural products (plant and animal pigments and odorants) and saw to a rebirth of chromatography as a tool for microanalysis. His elucidation of the structure and role of lipids and peptidolipids from mycobacteria led to the discovery of novel immunostimulants.

Early Life and Turmoil. The turmoil of the twentieth century did not spare Edgar Lederer and his family. Fleeing anti-Semitism in his native Vienna, he went in 1930 to Heidelberg, Germany. He had to leave Germany in a rush when the Nazis came to power in 1933. A refugee in Paris, he accepted a call to work in Leningrad. His stay there, too, was curtailed by the Stalinist purges. He was unaware to some extent of the magnitude of the danger, but his wife was politically more critical, and her good advice prevailed. Lederer returned to France—only to be drafted as a soldier in the so-called Phony War, and to be subsequently hounded as a Jew by the Vichy police and by the fascist thugs of the Milice. Only at age thirty-six could he resume his scientific work in relative peace.

Regardless of the turbulence of his life, Lederer was always dapper, with the appearance and demeanor of a British gentleman. Tall and bony, he seemed devoid of muscle and fat. His blue-eyed face, somewhat equine and crowned with blond hair, had a kind look. Furthering the impression of Britishness was the fact that English was his first language; he learned it from his English nurse. A polyglot, he also spoke and wrote German, French, Russian, and Italian.

Lederer came from a Jewish family in Vienna. His father was a business attorney. His mother, highly cultured, taught him French. The influence of his maternal uncles, Hans Przibram, a biologist, and Karl Przibram, a physicist, together with that of his natural history teacher in the gymnasium, caused him to opt for the sciences. At the University of Vienna he went on to receive his PhD in 1930 after two years in Ernst Späth's laboratory, working on the synthesis of indole alkaloids. Meanwhile, he began lifelong friendships with fellow students Erwin Chargaff and Percy Lavon Julian, who went on to distinguished careers in the United States. After he had obtained his PhD degree, his advisor told him bluntly that he had to leave the university. His being Jewish precluded any university appointment, even as a research assistant. Anti-Semitism in Austria was indeed virulent.

The Rediscovery of Chromatography. Compared to Austria, Germany seemed more lenient. In September 1930 Lederer joined Richard Kuhn's group at the Kaiser Wilhelm Institut für Medizinische Forschung (Kaiser Wilhelm Institute for Medical Research) in Heidelberg. As a postdoctoral fellow in Kuhn's laboratory, Lederer applied himself to clarifying the structure of vitamin A. His first results contradicted the structure proposed by Paul Karrer. In early 1931 he isolated two isomers, alpha- and beta-carotene. Two years later, he discovered a third isomer, gamma-carotene.

In doing this work, Kuhn and Lederer rediscovered partition chromatography; Michael Tswett, a botanist, described it in a book published in 1906 in Warsaw. The method of chemical separation fell into disrepute among chemists, however, especially after Richard Willstätter was unable to purify chlorophyll with it. Ironically, when in 1930 Kuhn was struggling with impure polyene pigments, Willstätter sent him a German translation of Tswett's book. Kuhn assigned Lederer the task of purifying carotene using Tswett's methodology. Lederer made significant technical changes, especially in regard to adsorbent materials. Kuhn and Lederer then were also able to isolate and purify a large number of carotenoids, such as astaxanthin, which gives lobsters and shrimps their orange-red color when cooked.

While in Heidelberg, Lederer met two people who would become important in his life and career. Hélène Fréchet was the daughter of a French mathematician and a professor at the Sorbonne. She married Lederer in June 1932. André Lwoff, who would win the Nobel Prize in Physiology or Medicine in 1965, was working in the laboratory of Otto Meyerhof, who had won that prize in 1922. Lwoff would remain a lifelong friend.

Two Dangerous Decades. In March 1933, two months after the Nazis came to power, the Lederers precipitously left for Paris. Lederer—already having two children—had to support his family. He was helped by Harry Plotz, a Boston banker, medical doctor, and philanthropist and was able to find a small laboratory on rue Pierre-Curie. In 1935, Caisse de la recherche scientifique (the then-existing Fund for Scientific Research) gave him a small grant to investigate the polio virus. He then moved into a nearby laboratory in the Institut de biologie physico-chimique, otherwise known as Fondation Edmond de Rothschild.

Because of his need for a greater income and his strong Communist convictions, Lederer approached the Soviet scientific attaché in Paris in 1935. He was offered and accepted a fellowship in the Vitamins Institute in Leningrad. He left for the Soviet Union with his family in October 1935 with a three-year contract. After only two years, with the Moscow trials and the Stalinist purges in full gear, the Lederers hurried back to France in December 1937.

The seed for the Centre national de la recherche scientifique (National Center for Scientific Research), known as CNRS, had been sown in the 1920s with the Caisse nationale des sciences (National Fund for the Sciences), started by Jean Perrin with funding by Baron Edmond de Rothschild. The French Popular Front government of 1936–1937, comprising the Nobel Prize–winners Irène Joliot-Curie and Jean Perrin, with its socialist belief in science for the common good, greatly increased funding for Caisse nationale des sciences, which would become CNRS in 1939, with an initial budget of one million francs. Its aim, not unlike that of the Kaiser Wilhelm Society in Germany (later known as the Max Planck Society), was to launch institutes in promising disciplines neglected by French universities. As a side effect, a significant one in the wake of World War II, CNRS offered a shelter to foreign scientists, often persons displaced from their countries by the war who lacked the credentials for a university career in France. In later years, Lederer would welcome numerous such refugees in the two institutes he headed.

In January 1938, Lederer returned to the Institut de biologie physico-chimique. In April, having started to write his dissertation in Russian, then translating it into French and completing it, he was awarded his French doctorate in the sciences. The dissertation presented his work on a dozen carotenoid pigments of invertebrates and plants which he had isolated and characterized while in Leningrad. He became a French citizen in December 1938. Soon afterward, he was drafted as a private into an artillery regiment. At the start of World War II, French cannons were still drawn by horses. Lederer was in charge of taking care of the animals and cleaning their stables. After the French capitulation in 1940, he resumed his work at CNRS as a hired *attaché de recherches* (research assistant). Then, Claude Fromageot, a professor of biochemistry in Lyon, helped by getting him promoted from *attaché* to *chargé de recherches* (group leader) and transferred to his laboratory. Lederer was fired from CNRS in 1941 because of the Vichy laws against Jews. He would be reintegrated only in the spring of 1944, because the Vichy authorities realized that the war tide was turning and on the basis of his having a French wife, and hence an Aryan family.

Meanwhile, the war years were filled with threats for the Lederers and their four children. They lived near Lyon until 1947. Toward the end of the war, Lederer protected himself with a forged identification. He narrowly escaped an arrest by the Milice. On 25 May 1944, when Lyon was bombed, five persons were killed in Fromageot's institute. Lederer was away in central France, where he had gone to find a shelter for his children.

Moves to Orsay, Gif. After France was liberated, Lederer continued work in Fromageot's Lyon laboratory. Fromageot received a professorial appointment at the Sorbonne, and Lederer moved back to Paris and to the Rothschild Foundation in March 1947.

After Fromageot died in January 1958, Lederer succeeded him in his chair at the Sorbonne, as well as heading a large biochemical laboratory on Boulevard Raspail in Paris. Moreover, his high reputation granted him the building of a brand-new CNRS institute, to be devoted to the chemistry of natural products, in Gif-Sur-Yvette, a distant southern suburb of Paris. In December 1960, Lederer moved into this new building.

At that time he was contacted by a physicist, André Guinier, the dean of the new Faculté des sciences (Faculty of Sciences) that had opened in Orsay, also in the southern suburbs of Paris, rather close to Gif. Many of the professors in Orsay leaned toward the Left in politics, which was not the norm at the Sorbonne. Lederer thus all the more readily accepted the offer. In 1963 he was able to move his biochemistry staff from Paris to a new institute in Orsay.

Thus wearing several hats signals membership in the power elite. Lederer's industrial contacts, on the one hand, gave him precious support during difficult times in the 1930s and early 1940s but, on the other hand, were frowned upon by the administrative officers at CNRS. Before the onset of World War II, he had been contacted by Max Roger, who directed the perfume factory Roure-Bertrand in Argenteuil, near Paris, and the Justin Dupont enterprise in Grasse, near the Riviera. Roger gave him funding for two coworkers together with a small amount

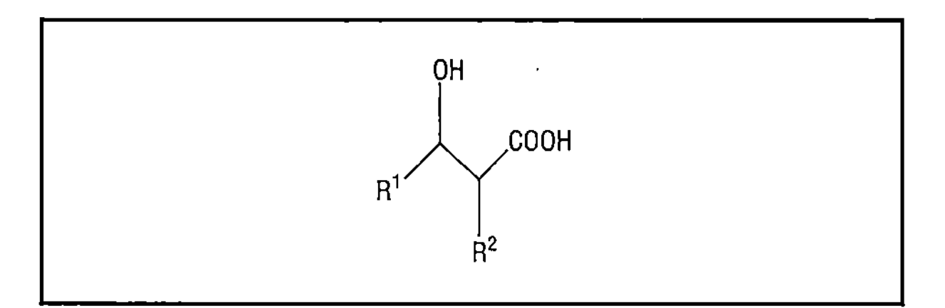

Mycolic acids.

for Lederer himself. They were to work on animal ingredients for perfumes, such as whale ambergris and castoreum, derived from the scent gland of the Canadian beaver. Lederer showed that ambrein, a constituent of ambergris, is a triterpene with a squalene-based biosynthesis. Lederer also showed that most other ingredients of ambergris were oxidation products of ambrein.

Another source of funding that allowed Lederer to survive while cut off from CNRS during the war years was from Henri Pénau, the director of research of the Roussel Laboratories, near Paris. He contacted Lederer in 1942, offering a contract for the isolation of cholesterol from sheep wool, needed to replace imports from Argentina that the war had dried up.

Lederer's work on ambergris came to the notice of Leopold Ruzicka at the Eidgenössiche Technische Hochschule (ETH) in Zürich. Ruzicka was a consultant to the Firmenich perfumery company in Geneva. In 1949 a collaboration began between Lederer and Firmenich, whose laboratories were headed by talented scientists, first Max Stoll and then Günther Ohloff. This collaboration included, besides natural products with an amber fragrance, the constituents of the essential oil of jasmine, the sesquiterpenes from that of bourbon geranium, and in subsequent years, the aroma of cocoa.

Research on Mycobacteria. Soon after his return to Paris in 1947, Lederer started work on mycobacteria. These are aerobic, rod-shaped germs (bacilli). The genus *Mycobacterium* includes the bacteria causing both tuberculosis and leprosy. Lederer perchance had met Dr. Nine Choucroun in Paris, who before the war had isolated from *Mycobacterium tuberculosis,* a complex lipidic fraction with interesting properties for the immunization of guinea pigs against tuberculosis, induction of delayed hypersensitivity, and so on. Lederer and his coworker, Jean Asselineau, began the study of the lipids from the cell wall of mycobacteria.

Tuberculosis was at that time still an endemic disease in western Europe. Half a century later, it still kills more than two million people per year worldwide. Antibiotics have to some extent muzzled *Mycobacterium tuberculosis.*

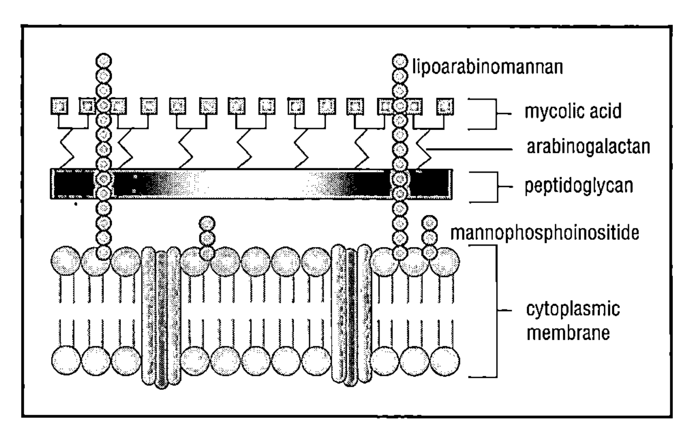

Cs4-Mycolic acid.

However, strains have become increasingly resistant to antibiotics.

Two French bacteriologists, Albert Calmette and Camille Guérin, after joining the Institut Pasteur in Paris in 1919, found a growth medium for the bacillus that made it less virulent. In so doing, they produced the BCG vaccine against tuberculosis, with which infants and children are inoculated in many countries.

Asselineau and Lederer established that mycolic acids are major and specific components of the cell envelope of mycobacteria, including *M. tuberculosis.* The cell wall skeleton of *M. bovis* bacillus Calmette-Guérin (BCG-CWS), a purified noninfectious material, consists of peptidoglycan, arabinogalactan, and mycolic acids.

Having benefited in 1950 from a contract awarded by the Ciba pharmaceutical company in Basel for this line of research, which would be renewed over the next twenty years, Lederer subsequently studied the enzymes that are involved in the production of mycolic acids. Among these are the S-adenosylmethionine-dependent methyltransferases that catalyze the introduction of key chemical modifications in defined positions of mycolic acids. Some of these subtle structural variations are crucial for both the virulence of the tubercle bacillus and the permeability of the mycobacterial cell envelope.

Colonial morphology of pathogenic bacteria is often associated with virulence. For *M. tuberculosis,* the causative agent of tuberculosis, virulence is correlated with the formation of serpentine cords, a morphology that was first noted by Robert Koch.

Cord factor (trehalose 6,6'-dimycolate, TDM) is a unique glycolipid with a trehalose and two molecules of mycolic acids in the mycobacterial cell envelope. Since TDM consists of two molecules of very long branched-chain 3-hydroxy fatty acids, the molecular mass ranges widely and in a complex manner. Trehalose (alpha-D-glucopyranosyl-alpha'-D-glucopyranoside) is essential for the growth of the human pathogen *M. tuberculosis* but not for the viability of the phylogenetically related corynebacteria. To isolate and determine the structure of peptidolipids and glycopeptidolipids in mycobacteria and their relatives, nocardiae and corynebacteria would become Lederer's major research target, not only until he retired but afterward, until death claimed him.

Opening up French Chemistry. At the time Lederer was elected at the Sorbonne, French chemistry was still reeling from the consequences of World War I, when promising young scientists died in the trenches. (The British and the Germans were smarter, bringing theirs back to laboratories.) With few exceptions, therefore, members of the ensuing generation received their university appointments without competition. Distant from the Anglo-Saxon mainstream—few knew any English—and ruled over by a few Parisian professors, a mandarinate which saw to it that its students were appointed to the few vacant positions, French chemistry closed in upon itself and thus became insular, provincial, and mediocre.

Lederer's contribution was to bring fresh air and ideas, drawing upon his international network of friends and colleagues. During the 1960s, the weekly seminars at his Gif institute, together with Alain Horeau's lectures at the Collège de France in Paris, were a source of renewal. The Institut de chimie des substances naturelles (Institute for Chemistry of Natural Products), under the joint directorship of Maurice-Marie Janot, a pharmacist, and Lederer, became one of the prize holdings of CNRS, at the leading edge of French chemistry.

In such positions of leadership he enjoyed in French chemistry and biochemistry, Lederer had several interrelated qualities: he was forward-looking and optimistic, he had an imaginative intuition, he had foresight. Just as he had pioneered chromatography in the early 1930s, he knew how to bank on both nuclear magnetic resonance and mass spectrometry in the early 1960s. He realized that these would become analytical tools important to chemists, and he invested in them early on.

Political and Social Activism. During his whole career, Lederer was a political activist of the Left. He lent his name and prestige to many petitions and international campaigns. Being a Communist sympathizer, albeit a critical one, did not endear him to the U.S. government. In 1951, when Lederer was denied a visa by the United States to receive a prize awarded by the American Chemical Society, it became a cause célèbre. In the late 1960s he became a member of the Russell Tribunal investigating war crimes by the United States in the Vietnam War. In 1967 he authored a report castigating the Pentagon for its use of chemical weapons in that war. Together with the

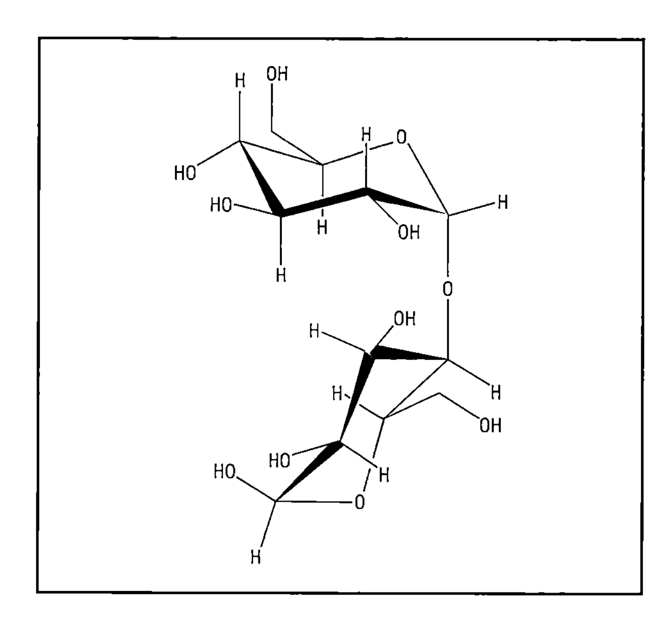

Trehalose.

French Fields medalist Laurent Schwartz, who became a friend, Lederer also helped dissident scientists to emigrate from the Soviet Union.

Lederer's political conscience went with an equally acute social conscience. He saw to it that quite a few of his coworkers, recruited as technicians from low-status families, progressed and received their degrees, whether as engineers or as doctorate-holding scientists, with the attendant salary and status. He prided himself on facilitating this type of social advancement within the rather rigid French class system.

In the late 1960s, while on a visit to America after the State Department granted him a visa (following many years of denial because of his Communist sympathies), he visited this author's home in Princeton, New Jersey. After dinner, we sat in the garden to chat. It was summer. There were fireflies about, which he had never seen before. Out of the corner of his eyes, Lederer watched them. He then remarked that they flashed only in upward flight. This was typical of his keen sense of observation. In this respect, he belonged to the grand tradition of natural historians, while practicing biochemistry and molecular biology in the modern manner.

Later Work. His retirement in 1978 from the directorship of the institute in Gif did not end his scientific career, to which he continued to apply himself with his usual zest and dynamism. About forty-five post-1978 publications bear his name. After his "retirement," this organic chemist and biochemist taught himself immunology.

The study of mycobacteria rewarded his attention with a bounty of results with much pharmacological promise. Lederer found that the already-mentioned cord factors, consisting of esters of mycolic acids and sugars such as trehalose, were immunostimulants. He also found that muramyl dipeptides modulate or stimulate the body's immune defense against hostile bacterial or viral invaders and against parasitic protozoans as well. In tumor immunology they can serve as immunoadjuvants when administered simultaneously with tumor antigens.

Lederer made yet another contribution to science: Edgar and Hélène Lederer had seven children, all of whom attended universities, one in the humanities and six in the sciences. Five of them espoused academic careers.

Among his numerous awards, he was proudest of the gold medal from CNRS in 1974 and of his belated (because of anti-Semitism) election to the French Academy of Sciences in 1982.

The pivotal character in Jean Renoir's film masterpiece, *La règle du jeu* (1939; *The Rules of the Game*) is a French Jew. His wealth opened high society for him. He outshone the aristocrats around him with his humanity. Lederer, with his moral elegance, with his international outlook, with his concern for political refugees of every ilk, reminds one of this character.

BIBLIOGRAPHY

WORKS BY LEDERER

"Biochemistry of the Natural Pigments." *Annual Review of Biochemistry* 17 (1948): 495–520.

With J. Asselineau. "Structure of the Mycolic Acids of Mycobacteria." *Nature* 166 (1950): 782–783.

With Michael Lederer. *Chromatography: A Review of Principles and Applications.* Amsterdam; Houston, TX: Elsevier Publishing, 1953.

"Chemistry and Biochemistry of Some Biologically Active Bacterial Lipids." *Pure and Applied Chemistry* 2 (1961): 587–605.

"The Origin and Function of Some Methyl Groups in Branched-Chain Fatty Acids, Plant Sterols, and Quinones." *Biochemical Journal* 93 (1964): 449–468.

Report on Chemical Warfare in Vietnam. (Second Session, 20 November–1 December 1967). In *Reports from the Sessions of the International War Crimes Tribunal Founded by Bertrand Russell.* London: Russell Tribunal, 1967–1971.

"The Mycobacterial Cell Wall." *Pure and Applied Chemistry* 25 (1971): 135–165.

"Cord Factor and Related Trehalose Esters." *Chemistry and Physics of Lipids* 16 (1976): 91–106.

With L. Chedid, F. Audibert, P. Lefrancier, et al. "Modulation of the Immune Response by a Synthetic Adjuvant and Analogs." *Proceedings of the National Academy of Sciences of the United States of America* 73 (1976): 2472–2475.

With L. Chedid. "Past, Present, and Future of the Synthetic Immunoadjuvant MDP and Its Analogs." *Biochemical Pharmacology* 27 (1978): 2183–2186.

"Synthetic Immunostimulants Derived from the Bacterial Cell Wall." *Journal of Medicinal Chemistry* 23 (1980): 819–825.

"Adventures and Research." In *Selected Topics in the History of Biochemistry: Personal Recollections,* edited by Giorgio Semenza. Amsterdam: Elsevier, 1985.

"New Developments in the Field of Synthetic Muramyl Peptides, Especially as Adjuvants for Synthetic Vaccines." *Drugs and Experimental Clinical Research* 12 (1986): 429–440.

"Edgar Lederer, la chimie des substances naturelles." In *Cahiers pour l'histoire du CNRS 1939–1989, 1989–1992,* edited by J. F. Picard and E. Pradoura. Paris: Editions du CNRS, 1989.

OTHER SOURCES

Schmidt, M., ed. *Hommes de Science. 28 Portraits.* Paris: Hermann, 1990.

Tswett, M. S. "Physikalische-chemische Studien über das Chlorophyll. Die Adsorptionen." *Berichte der Deutschen Botanischen Gesellschaft* 24 (1906): 316–328.

Witkop, B. "Paul Ehrlich and His Magic Bullets—Revisited." *Proceedings of the American Philosophical Society* 143 (1999): 540–557.

Pierre Laszlo

LEDYARD, GEORGE

SEE **Stebbins, George Ledyard, Jr.**.

LEE, SARAH EGLONTON WALLIS BOWDICH

(*b.* Colchester, England, 10 September 1791; *d.* Kent, United Kingdom, 23 September 1856), *natural history.*

The first European woman systematically to collect plants in tropical West Africa, Lee can be credited with the discovery and description of six new genera and two new species of plants, in addition to six new species of fish. She also established a productive career as author of a dozen articles and five books on natural history, illustrator of another three, plus sixteen books for children or young adults, mostly fiction. Sarah Bowdich Lee was acknowledged for her achievements in natural history by her contemporaries with a Civil List Pension and inclusion in the *Dictionary of National Biography,* one of very few women to be so honored.

Background and Early Career. Sarah Eglonton Wallis was the daughter of an heiress and a well-to-do Nonconformist merchant; she enjoyed a largely privileged childhood until her father went bankrupt in 1802. The family moved to London, and nothing is known of her adolescence, except that she was almost certainly tutored. In

Lee

1813, Sarah married Thomas Edward Bowdich, who soon thereafter became a junior officer in the Africa Company, and was sent to Cape Coast Castle in tropical West Africa in 1815. In 1816 she independently sailed to Cape Coast, only to find that her husband had temporarily gone back to England. She decided to stay, and wait for his return.

Dazzled by the light and color of Africa, and enchanted by the exotic plants and animals, especially parrots and monkeys, Sarah became fascinated with her surroundings. Although she did not then realize it, her stay in Africa had catalyzed a passion for natural history that would become the central theme of her life.

The Bowdichs returned to England in 1817, but by April 1819 had moved to Paris, so that Thomas could study the natural sciences he needed in order to become an African explorer. Sarah worked and studied along with Thomas; both became familiar with Baron Georges Cuvier, who was their virtual mentor, giving them the use of his personal library. To support themselves, the Bowdichs translated French natural history into English. Sarah illustrated a number of texts abridged from Cuvier's work: *Mammalia, Conchology, Ornithology.*

In 1821, she produced and illustrated *Taxidermy,* a successful volume whose sixth and last edition was issued in 1843. During their Paris stay she was elected a member of the Wetterauische Gesellschaft für die Gesamte Naturkunde, in Hanau (now in Germany). In July 1822, the Bowdichs set out on their second African expedition, stopping at Madeira and doing natural history for a year before continuing to the Gambia, where Thomas caught fever, and on 10 January 1824 died in his wife's arms.

Natural History Writing and Illustrating. On returning to England, Sarah had to find a way to support her three children, which she finally did both by writing and by a second marriage in 1826, to Mr. Robert Lee. But first she edited and published Thomas's last manuscript, *Excursions in Madeira and Porto Santo,* to which she added a botanical and zoological appendix. Soon after, she was persuaded to write stories about Africa for R. Ackermann's *Forget-Me-Not,* to which she contributed nearly every year until 1844.

Meanwhile, again probably through Ackermann, she began the project that eventuated in *Fresh-Water Fishes of Great Britain* (1828–1838), a book produced for fifty subscribers, with hand-colored watercolor paintings of fish, drawn from life in the most precise and careful detail. As late as the 1950s, a curator of cichlid fishes at the British Museum of Natural History would note her surprise at the scientifically valuable contents of *Fresh-Water Fishes,* a book she had at first taken to be little more than a collection of pretty paintings. It would take Lee eleven years and more than three thousand paintings to complete the book

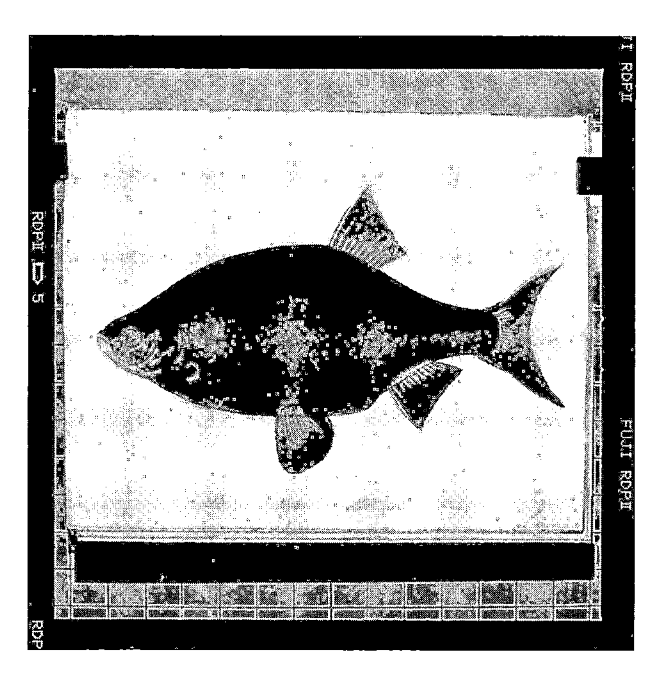

***Rudd* Scardinius erythropthalmus.** *Lee's illustration of a rudd fish.* THE ACADEMY OF NATURAL SCIENCES, EWELL SALE STEWART LIBRARY.

for her fifty subscribers, among whom were Roderick Impey Murchison and John Herschel.

In addition, between 1829 and 1831 she wrote twelve articles for John Claudius Loudon's newly created *Magazine of Natural History,* sending reviews from Paris, and offering original articles and comments, several of them signed only as "B." Several of her original articles were included in the Royal Society's *Catalogue of Scientific Papers 1800–1863.*

Lee traveled frequently between London and Paris, often conveying messages or running errands for some of the eminent men of science, amongst them Cuvier, Robert Brown, and Charles Babbage. In addition to these and her subscribers, she was well acquainted with a number of prominent scientists, including Alexander von Humboldt, Richard Owen, and Cuvier's collaborator, Achille Valenciennes.

For Cuvier, she traced a number of sketches of fish, particularly those of Johann Forster and Sydney Parkinson, deposited in Sir Joseph Banks's library. Cuvier would later acknowledge her help in more than twenty different places in his and Valenciennes's classic and encyclopedic *Histoire naturelle des poissons.* She also gave him some of her original sketches of fish from St. Jago, made on her second voyage to Africa.

An accomplished artist, Lee painted many pictures of plants and flowers, in addition to fish and mollusks. In general, she painted painstakingly accurately, with little embellishment. Her paintings of plants from Gabon are so faithful that one can easily determine their genus, if not immediately the species.

As the 1830s progressed, Lee turned from original and creative natural history to more didactic and entertaining forms. After 1840 she published what were essentially two introductory textbooks in natural history and botany, for adolescents or (young) adults: *Elements of Natural History* and *Trees, Plants, and Flowers.* Two other books turned out to be very popular: *Anecdotes of the Habits and Instinct of Animals,* and *Anecdotes of the Habits and Instincts of Birds, Reptiles, and Fishes.* In all, she wrote twenty books in her lifetime. Largely for her later didactic and popular works on natural history, but also in part for her earlier achievements, Lee finally received a Civil List Pension in 1854, an award relatively rarely given to scientists, still more rarely to women, and almost never to women in recognition of their own achievements.

Lee is among the ranks of the earliest female travelers to Africa. She did not just travel there, but made novel observations in natural history. Her discoveries in natural history, like her paintings, attest to a keenly observant eye, and a style devoted more to scientific realism than interpretive license. So, too, with her literature, even including her fiction, whose natural history content is often annotated with explanatory footnotes.

Although recognized by her contemporaries for her excellence, Lee gradually disappeared from the history of science. Perhaps her career was too fragmented, or the demands of rearing a family prevented her from establishing and maintaining a salient identity. Nonetheless, her story is illuminative of what was possible for an intelligent, capable, and determined woman in early-nineteenth-century Britain.

BIBLIOGRAPHY

WORKS BY LEE

Taxidermy. Initially attributed to T. E. Bowdich. Paris, 1821.

As edited. *Excursions in Madeira and Porto Santo,* by Thomas Bowdich. London: G. B. Whittaker, 1825.

The Fresh-Water Fishes of Great Britain. London: R. Ackermann, 1828–1838.

Elements of Natural History. London: Longman, 1844.

Anecdotes of the Habits and Instinct of Animals. London: Grant and Griffith, 1852.

Anecdotes of the Habits and Instincts of Birds, Reptiles, and Fishes. London: Grant and Griffith, 1853.

Trees, Plants, and Flowers. London: Grant and Griffith, 1854.

OTHER SOURCES

Beaver, Donald deB. "Writing Natural History for Survival, 1820–1856: The Case of Sarah Bowdich, later Sarah Lee." *Archives of Natural History* 26, no. 1 (1999): 19–31.

Bowdich, T. E. *An Analysis of the Natural Classifications of Mammalia.* Illustrated by Sarah Bowdich. Paris, 1821.

———. Introduction to *The Ornithology of Cuvier.* Illustrated by Sarah Bowdich. Paris, 1821.

———. *Elements of Conchology.* Illustrated by Sarah Bowdich. Paris, 1822.

Cuvier, G., and A. Valenciennes. *Histoire Naturelle des poissons.* 22 vols. Paris, 1828–1831.

Mabberley, D. J. "Edward and Sarah Bowdich's Names of Macaronesian and African plants, with Notes on Those of Robert Brown." *Botánica Macaronésica* 6 (1978): 53–66.

Strickrodt, Silke. *Those Wild Scenes: Africa in the Travel Writings of Sarah Lee (1791–1856).* Cambridge, MA: Galda + Wilch Verlag, 1998.

Donald deB. Beaver

LEHMANN, INGE (*b.* Copenhagen, Denmark, 13 May 1888; *d.* Copenhagen, 21 February 1993), *geophysics, seismology.*

Lehmann is first and foremost known as the discoverer of Earth's inner core in 1936, but she is also highly regarded for her studies of Earth's mantle, carried out during many visits to the United States in the 1950s and the 1960s. In these studies she identified a low velocity layer between 130 and 220 kilometers (80 to 140 mi) below Earth's surface. The bottom of this layer, at which the velocity of seismic waves rises abruptly, is now called the Lehmann discontinuity.

Childhood and Education. Inge Lehmann grew up and lived almost all her life in Copenhagen. She came from an influential family of academic traditions. Her paternal grandfather laid down the first Danish telegraph cable in 1854, and her father, Alfred Lehmann, became the first professor of experimental psychology at Copenhagen University in 1919. The family also included several prominent women. Lehmann's mother, Ida Sophie Tørsleff, had a sister who was an active proponent of women's rights, and her daughter, Lis Groes, became Danish minister of commerce in the 1950s. Lehmann had a younger sister, Harriet, who became an actress and who had family and children in contrast to Lehmann, who lived by herself all her life.

A strong influence on the young Inge Lehmann was her schooling at the coeducational *Fællesskolen,* a school run by Hanna Adler, an aunt of Niels Bohr. At Adler's school, no differentiation based on sex or social status was accepted, and both girls and boys were taught needlework and played soccer. In an obituary of Adler, Lehmann wrote in 1947 that "there was no unnecessary discipline, and we were not burdened by the prejudice [regarding gender, race, or social status], which makes life difficult for so many people." In 1906 Lehmann left Adler's school after passing the university entrance examination.

In 1907, Lehmann studied mathematics at Copenhagen University, and she continued her studies at Newnham College in Cambridge in 1910–1911. Here the relationship between men and women was very different from what she had experienced at Adler's school. Even though the many restrictions on young women's behavior dissatisfied Lehmann, she enjoyed her stay in Cambridge. However, in 1911 Lehmann returned from Cambridge overworked, and so she put her university studies on standby. For a few years she worked at an actuary's office, where she acquired good computational skills, before resuming her studies at Copenhagen University in 1918. Two years later she completed the candidates magisterii degree in mathematics and physical science. In the autumn of 1922 she studied mathematics with Professor Wilhelm Blaschke in Hamburg, Germany, and when she returned to Denmark in 1923, she accepted a position as assistant to the professor in actuarial science at Copenhagen University, J. F. Steffensen.

Introduction to Seismology. A turning point in Lehmann's career came in 1925, when she was appointed assistant to Niels Erik Nørlund, the newly appointed director of Gradmaalingen, a Danish geodetic institution. Nørlund had plans to establish seismological stations in Denmark and Greenland, and Lehmann's job was to run the stations and interpret and publish the observations. Research work was not a part of her job description, but she was free to take it on. Lehmann supervised the establishment of one seismological station in Copenhagen, and she helped prepare instruments for two stations in Greenland. The Copenhagen station was in the old fortress that encircled the old part of the city, while the Greenland stations were placed in the mining city Ivittuut on the west coast and at Ittoqqortoormiit on the east coast. The station on the east coast turned out to be a particular challenge to run, because the only contact with the station was by boat once per year.

Why Lehmann was named to the position at Gradmaalingen is unknown, but one might speculate that Hanna Adler played a role through Niels Bohr, who was married to Nørlund's sister. In any case Lehmann quickly engaged in the new field, and in the summer of 1927 she visited seismological stations around Europe and had an extended stay in Darmstadt with Professor Beno Gutenberg, who in 1914 had determined the depth of Earth's core. Based on her studies of seismology, Lehmann in 1928 acquired the magister scientiarum degree (equivalent to an MA) in geodesy. The same year she was appointed state geodesist at the Danish Geodetic Institute, which had been established with the merging of Gradmaalingen and the general staff's Topographic Department in early 1928.

In 1927 Lehmann participated in the meeting of the International Union of Geodesy and Geophysics in Prague, even though "it was not customary for a person in my position," as Lehmann wrote in "Seismology in the Days of Old" (1987, p. 33). Both Gutenberg and the renowned British seismologist Harold Jeffreys participated, and an important topic at the conference was the determination of travel times of seismic waves through the interior of Earth. Many attempts had been made to construct travel-time curves that described the travel time of seismic waves as a function of epicentral distance. At the heart of these attempts were problems with the accuracy of seismographic measurements. The network of seismological stations was uneven, and the instrumentation and methods of reading seismograms were very heterogeneous. Lehmann realized that an assessment of the accuracy of different stations and a consistent interpretation of seismograms was decisive in the construction of travel-time curves. She took on the task of evaluating the European stations, concluding that Copenhagen's was among five particularly accurate stations on the continent. She also made the thorough and consistent analysis of seismograms from several stations the trademark of her own research, and within a few years her hard work would result in an important discovery.

The Inner Core. Following the conference in Prague, Lehmann had a lively correspondence with Harold Jeffreys on the problem of travel-time curves. Lehmann was interested in how the observations of seismic waves varied with epicentral distance, a relationship that is also reflected in the travel-time curves. After an earthquake, two main types of seismic waves that have traversed Earth's interior are observed: compressional *P*-waves or pressure waves, and transverse *S*-waves or shear waves. Because *P*-waves have higher velocity than *S*-waves in Earth's mantle, *P*-waves arrive ahead of *S*-waves. Both *P*- and *S*-waves are observed up to an epicentral distance around 103°. Above this epicentral distance no waves are observed until about 143°, and after this point only *P*-waves are seen. (These waves are called *PKP* or *P'* indicating that they have passed through Earth's core). The reason for this is that seismic waves have a lower velocity inside the core, and therefore the core acts as a converging lens, thus making a shadow zone between 103° and 143°, where no direct *P*-waves are observed. The absence of *S*-waves also above an epicentral distance of 143° was taken as evidence of a fluid core, since transverse waves cannot penetrate a fluid medium.

All this had been known since 1914, when Gutenberg had used these observations to estimate the depth of Earth's core at 2,900 kilometers (1,800 mi). But in the 1920s many observations of weaker phases of *P'* between 103° and 143° were made. Lehmann discussed these observations in a letter to Jeffreys in May 1932: "But it remains to explain *P'* at smaller distances [than 143°]. I suppose they could be explained by the assumption of a discontinuity surface within the core at which the velocity increases. There is hardly anything to disprove the existence of such a surface in present observational data." (Hjortenberg & Larsen, 2004). Despite this early hint of a possible inner core, Jeffreys indicated in the Jeffreys-Bullen travel-time tables that were published in 1935 that the *P'*-waves between 103° and 143° were caused by a diffraction phenomenon on the surface of the core. Gutenberg had in his travel-time tables made a similar interpretation of the unexpected *P'*. He designated them *gebeugte Wellen* (bent waves) with no further explanation of their origin.

The problem of the unexplained *P'*-waves had not been considered very seriously, because the amplitudes of the observed phases had been rather small. Lehmann, though, in her meticulous examination of seismograms from all over Europe of a New Zealand earthquake in 1929, realized that the reason for the weak *P'*-phase was that at many stations, only the horizontal component had been measured. The vertical component, by contrast, turned out to be significant, and Lehmann rejected the possibility that it could be caused by diffraction. She instead hypothesized that inside the core there is an inner core, in which the velocity of seismic waves is higher than in the outer core. The *P'*-waves observed in the shadow zone would then be caused by *P'*-waves being refracted at the inner core. Lehmann supported her hypothesis by thorough analysis of seismograms from four different stations, and in her 1936 article with the conspicuously short title "P'," she concluded: "It cannot be maintained that the interpretation here given is correct since the data are quite insufficient.... However, the interpretation seems possible, and the assumption of the existence of an inner core is, at least, not contradicted by the observations; these are, perhaps, more easily explained on this assumption" (p. 115).

Gutenberg quickly recognized Lehmann's discovery of an inner core, while Jeffreys was more reluctant. Within a few years, though, Jeffreys showed that the diffraction theory could not explain the observations and accepted the inner core. In 1938 Beno Gutenberg and Charles F. Richter determined the radius of the inner core at 1,200 kilometers (750 mi) as well as the velocity of *P*-waves in the inner core at 11.2 kilometers per second (6.9 miles per second). Lehmann's hypothesis was thus accepted broadly in the seismological community within a few years.

Earth's Upper Mantle. Inge Lehmann frequently participated in international meetings, and she engaged actively in several societies. She regularly attended meetings of the International Union of Geodesy and Geophysics, beginning with the Prague meeting in 1927. In 1936 she was one of the founders of the Danish Geophysical Society, and she chaired the organization in 1941 and 1944. She also participated in the establishment of the European Seismological Federation (ESF) in 1950 and was elected its first president. The European Seismological Commission, of which she was a member, succeeded ESF in 1951.

In 1951 Maurice Ewing of Lamont Geological Observatory in Palisades, New York, visited the seismological station in Copenhagen. Ewing was a close friend of Lehmann and valued her special skills in reading seismograms. He invited her to come to Lamont to do research on a newly discovered surface wave, *Lg* (a transverse surface wave in Earth's crust. Lehmann went to Lamont for several months in 1952, bringing European seismograms to compare with the American observations of *Lg*. She succeeded in estimating travel times for *Lg* and demonstrated significant differences in the European and American records, which reflected different structures of the upper mantle under each continent. The same year Lehmann was considered for a professorship in geophysics at Copenhagen University. She was judged fully qualified for the position by the evaluation committee, but Niels Bohr, who was on the committee, had his own candidate for the position. Other members of the committee did not find Bohr's candidate nearly as qualified as Lehmann, and the result was that neither got the position, which was not filled until a decade later. This must have been a disappointment to Lehmann, and might have been part of the reason she retired from her position at the Geodetic Institute in 1953, five years before the mandatory retirement age of seventy.

Retirement for Lehmann was not retirement from research. Instead, it opened new opportunities for research and international collaboration. As Francis Birch said when Lehmann was awarded the American Geophysical Union's Bowie Medal in 1971: "Since her retirement from the Geodetic Institute, Dr. Lehmann has increased her rate of publication, which is understandable, since she no longer has to worry about keeping someone on the job at Scoresbysund [Ittoqqortoormiit]!" (Bolt and Hjortenberg, 1994, p. 231). In the 1950s and 1960s Lehmann had many extended stays in the United States at the Lamont Observatory in New York State, at the Seismographic Stations at the University of California at Berkeley, and with Gutenberg at the Seismological Laboratory in Pasadena, California. She also spent time at the Dominion Observatory in Ottawa, Canada.

Lehmann's stays in North America coincided with a period when seismology began to receive much attention after decades of neglect. As a response to the shift from atmospheric to underground testing of nuclear bombs, a research program named Vela Uniform was established, to develop improved and standardized seismographs to detect underground explosions. These standardized instruments were installed at two hundred seismological stations around the world, constituting the Worldwide Standardized Seismographic Network. The combination of more accurate measurements and Lehmann's unique analytical skills made possible more detailed analysis of Earth's upper mantle. Also, Lehmann benefited from the measurements of the seismic waves caused by underground nuclear explosions. Because their time and place of origin were well defined, travel times could be determined very accurately. Lehmann paid particular attention to measurements of *S* and *P*-waves at small epicentral distances and found evidence of a low velocity layer beginning a little below 100 kilometers's (62 mi's) depth and extending down to a depth of around 220 kilometers. At the bottom of this low velocity layer Lehmann found indications of an abrupt velocity increase. Jeffreys had already found indications of a change in the velocity gradient at a depth of 220 kilometers (136 mi), but Lehmann showed that there was actually a discontinuity in the velocity profile. This discontinuity, as well as the one at the surface of the inner core, have become known as Lehmann discontinuities.

Skills and Personal Qualities. As a seismologist Lehmann stood out with her exceptional analytical skills and her ability to identify and compare phases in seismograms from stations all over the world. Perhaps her aunt's grandson, Niels Groes, has described these qualities the best:

> I remember Inge one Sunday in her beloved garden on Søbakkevej; it was in the summer, and she sat on the lawn at a big table, filled with cardboard oatmeal boxes. In the boxes were cardboard cards with information on earthquakes and the times for their registration all over the world.... With her cardboard cards and her oatmeal boxes, Inge registered the velocity of propagation of the earthquakes to all parts of the globe. By means of this information, she deduced new theories of the inner parts of the Earth. (Bolt, 1997)

Lehmann was also a very shy person and disliked being the center of attention. When in Denmark she enjoyed spending time at her quiet summer cottage, where many colleagues visited her over the years. Despite her shyness she maintained an extensive network of colleagues, and many considered her a close friend. She was a very active person, loved to ski in the Alps or Norway, and to climb mountains in the summertime.

As a woman in a male-dominated science, Lehmann's career was often an uphill struggle. Groes reports that she said, "You should know how many incompetent men I had to compete with—in vain" (Bolt, 1997).

Lehmann wrote her last article, "Seismology in the Days of Old," in 1987 at the impressive age of ninety-nine. The following year she celebrated her 100th birthday at a reception at the Geodetic Institute, attended by several internationally renowned geophysicists. In February 1993, Lehmann died at the age of 104.

Honors and Awards. During her career Lehmann received great international recognition. From 1936 to 1948 she was a member of the executive committee of the International Seismological Association and in two periods, 1951–1954 and 1957–1960, was a member of the executive committee of the International Association of Seismology and Physics of Earth's Interior (IASPEI). From 1963 to 1967 she was vice president of the executive committee of IASPEI. She was elected associate of the Royal Astronomical Society, London, in 1957 and Honorary Fellow of the Royal Society, Edinburgh, in 1959.

Lehmann received several travel awards, and in an unusual honor, she received the Danish Tagea Brandt travel award twice, in 1938 and 1967. In 1964 she received the Deutsche Geophysikalische Gesellschaft's (German Geophysical Society's) Emil-Wiechert Medal and in 1965 was awarded the Gold Medal from the Royal Danish Academy of Sciences and Letters. In 1971 she received the American Geophysical Union's Bowie Medal for "outstanding contributions to fundamental geophysics and unselfish cooperation in research" in 1971. Finally, in 1977 she received the Medal of the Seismological Society of America. Though her recognition in Denmark came late, she was especially pleased when she received an honorary doctor of philosophy degree at Copenhagen University in 1968. In 1964 she received the honorary degree of doctor of science at Columbia University.

Lehmann's name lives on in two awards. One is the Inge Lehmann Medal, which was established by the American Geophysical Union in 1997 and is awarded every other year for "outstanding contributions toward the understanding of the structure, composition, and/or dynamics of Earth's mantle and core." The other is a travel award, which was instituted by Lehmann herself and given in alternate years to a psychologist and a geophysicist.

BIBLIOGRAPHY

Inge Lehmann bequeathed her personal papers to her colleague Erik Hjortenberg, who has systematized and scanned the documents. The material is available through the Danish state archives and through Storia Geofisica Ambiente in Bologna, Italy. A complete list of Inge Lehmann's publications is given in Bolt (1997).

WORKS BY LEHMANN

"P.'" *Publications du Bureau Central Seismologique International,* series A 14 (1936): 87–115.

"Rektor Hanna Adler in Memoriam." *Kvinden og Samfundet* 63 (1947): 29.

"*S* and the Structure of the Upper Mantle." *Geophysical Journal of the Royal Astronomical Society* 4 (1961): 124–138.

"Recent Studies of Body Waves in the Mantle of Earth." *Quarterly Journal of the Royal Astronomical Society* 3 (1962): 288–298.

"Seismology in the Days of Old." *Eos* 68, no. 3 (1987): 33–35.

OTHER SOURCES

Bolt, Bruce A. "Inge Lehmann. 13 May 1888–21 February 1993." *Biographical Memoirs of Fellows of the Royal Society* 43 (1997): 286–301.

Bolt, Bruce A., and Erik Hjortenberg. "Memorial Essay, Inge Lehmann (1888–1993)." *Bulletin of the Seismological Society of America* 84, no. 1 (1994): 229–233.

Brush, Stephen G. "Discovery of the Earth's Core." *American Journal of Physics* 48, no. 9 (1980): 705–724.

Hjortenberg, Erik, and Tine B. Larsen. "The Scientific Correspondence between Inge Lehmann and Harold Jeffreys." Poster at the European Seismological Commission, XXIX General Assembly, Potsdam, Germany. September 2004.

Kölbl-Ebert, Martina. "Inge Lehmann's Paper: 'P'" (1936)." *Episodes* 24, no. 4 (2001): 262–267.

Maiken Lolck

LEHMANN-NITSCHE, ROBERT

(*b.* Radonitz, Posen, Prussia, 9 November 1872, *d.* Berlin, Germany, 9 April 1938), *physical anthropology, paleoanthropology, ethnology, folklore.*

Lehmann-Nitsche lived in La Plata, Argentina, from 1897 to 1930. He worked as head of the anthropological department of the museum of the city, which in 1906 became part of the new Universidad de La Plata. There, as well as at the Universidad de Buenos Aires, he held the first South American university professorships in anthropology. He undertook numerous journeys through Argentina and worked on anthropology, mythology, and ethnology, as well as folklore and Creole ethnic studies in the River Plate (Río de la Plata) region.

Early Years. Paul Adolf Robert Lehmann-Nitsche was born in Radonitz, Posen, where he attended primary school. Only his mother's name, Ida, is known. He obtained a PhD in philosophy and another one in medicine. He completed his dissertation, "Über die langen Knochen der südbayerischen Reihengräberbevölkerung," in July 1893 at the Ludwig Maximilian Universität of Munich, Germany. This "study on the long bones found in aligned burials in Southern Bavaria" was awarded half of the Ernest Godard Prize established at the Societé d'Anthropologie of Paris (Paupillault, 1897). For many years this work stood as one of the standard references on the significance of indices and the measurements of femora in the field of comparative anthropological studies (Hrdlička, 1912).

Argentinean Studies (Museo de La Plata). In 1897, Lehmann-Nitsche arrived in Argentina, where he had been appointed head of the anthropological section of the Museo de La Plata, established in 1884. He proceeded to arrange the collections gathered in the expeditions of Francisco P. Moreno (1852–1919) and his collaborators. With that purpose he classified the collections following the geographical and political regions of Argentina as established by the head of cartographic section of the museum. This system was later adopted by the anthropological and archaeological collections of all Argentinean museums (Podgorny, 1999).

As a member of the staff of the Museo de La Plata (the Museum of La Plata), Lehmann-Nitsche undertook several anthropological journeys in Argentina. He traveled to Tierra del Fuego in 1902 and to the Argentinean Northwest in 1906. There he proceeded to record measurements and qualitative characteristics of 160 individuals.

University Teaching. Lehmann-Nitsche taught the first Argentinean university courses in physical anthropology (University of Buenos Aires, 1903). His syllabus presented anthropology as the comparative study of humankind on the basis of physical characteristics such as pigmentation and skull and head shape in various human groups. He also analyzed "fossil and contemporary aborigines." In 1904, he taught another course on paleoanthropology, devoted to the evidence of Argentinean fossil man, the Neanderthal hominid, and the issue of *Pithecantropus erectus.* Lehmann-Nitsche defined "paleoanthropology" as the "physical and psychological anthropology of fossil man, i.e. man from past geological ages" (1907, p. 193-194, fn.). In 1905, he was appointed professor of anthropology in Buenos Aires, where he organized his course of "physical anthropology of the human races." He taught the same syllabus in his courses given at the new University of La Plata, established in 1906. In this capacity, he directed several doctoral dissertations on topics such as fossil man and the analysis of collections of crania.

Contribution to South American Fossil Man. In 1907, Lehmann-Nitsche compiled "Nouvelles recherches sur la Formation Pampéenne et l'Homme Fossile de la République Argentine," an important summary of the evidences relating to man of the so-called Pampean Formation in Argentina. This work was a reaction to Florentino Ameghino (1854?–1911), who postulated the Tertiary age of both that formation and the fossil humans supposedly found in those deposits (Podgorny, 2005). The "Nouvelles recherches" was the first work in Argentina that gathered together professionals from several disciplines (geologists, physical anthropologists, and mineralogists) to analyze such a contested issue. In this work Lehmann-Nitsche accepted the conditions of the modern practices of paleoanthropology: that it was necessary to agree upon both the association of the human bones with geologic deposits of well-determined age and the significance of the morphologic characteristics defining a clear stage of evolution. Contrary to Ameghino, Lehmann-Nitsche concluded that part of the Pampean Formation was of Quaternary age. He analyzed a human-like atlas (the first vertebra of the neck) of small size, found in the 1880s, which for many years was left unattended in the collections of the La Plata Museum. By comparing this bone with sixteen atlases of native people from South America and with those of orangutans and gorillas, he defined *Homo neogaeus*, a Tertiary South American species of humankind that "must have approached very closely the *Pithecanthropus*" (Lehmann-Nische, 1907, p. 399). He concluded that the Pampean fossil man was of Pliocene age (Middle Pampean).

Contribution to Folklore and Ethnology. Lehmann-Nitsche turned gradually to South American folklore and mythology. Attributed to him are the oldest sound ethnographic and folkloric documents recorded in Argentina on Edison cylinders between 1905 and 1909: 107 recorded in the Museum of La Plata (1907); 98 in the province of Jujuy (1906); 8 in Buenos Aires (1909), and 126 folk songs in La Plata (1905). Lehmann-Nitsche also recorded texts in the Araucanian language. These texts include narratives on historically important Indian leaders, myths, dialogues, letters, fairy tales, fables, and songs. In 1910 he published a collection of Creole riddles, later complemented with a volume devoted to scatological texts from the La Plata region. The latter was published in Leipzig in 1923, in a series of researches on the history of development of moral habits. Lehmann-Nitsche analyzed different objects and characters of the folklore of the La Plata region. His later years in Argentina were dedicated to the study of South American mythology and ethno-astronomy. In these works he classified the different topics of the legends, proposing connections throughout South America.

Lehmann-Nitsche's collection of folklore was accomplished through the work of German immigrants who inhabited the Argentinean countryside and articulated as a network of observers. By the same means, Lehmann-Nitsche gathered archaeological and anthropological data from all over the continent. He maintained an epistolary exchange with archaeologist Max Uhle and ethnologist Eduard Seler, among others, which is of interest for research work concerning the history of the German Latin America investigations. During World War I, he was very engaged in defending the name of Germany as the main and real center of culture, science, and civilization (García and Podgorny, 2000). In the 1920s, he became an active opponent to the German Social Democratic government and became affiliated with the Argentine section of the Conservative German National Popular Party.

He returned to Germany in May 1930. In October 1934, he was appointed to lecture on Ibero-American cultures at the philosophy department of the University of Berlin. He died of cancer in Berlin-Schöneberg in 1938. In Argentina, he had married his former student, anthropologist Juliane Dillenius. Lehmann-Nitsche's bibliography includes more than 370 titles (Torre Revello, 1947).

BIBLIOGRAPHY

The most complete bibliographical index of Lehmann-Nitsche's is Torre Revello, José. "Contribución a la bio-bibliografía de Robert Lehmann-Nitsche." Boletín del Instituto de Investigaciones Históricas *29, no. 101–104 (1947): 724–805. His correspondence, scientific, and political papers are kept in the archives of the Ibero-Amerikanisches Institut, Preussischer Kulturbesitz, Berlin. Part of this entry is based on work done at Ibero-Amerikanisches Institut in 1994.*

WORKS BY LEHMANN-NITSCHE

"Über die langen Knochen der südbayerischen Reihengräberbevölkerung." *Beiträge zur Anthropologie und Urgeschichte Bayerns* 11 (1894).

"Antropología y craneometría. Conferencia dada en la sección antropológica del primer congreso científico Latino-Americano." *Revista del Museo de La Plata 9* (1899): 121–140.

"Quelques observations nouvelles sur les indiens Guayaquis du Paraguay." *Revista del Museo de La Plata* 9 (1899): 399–408.

"Catálogo de las antigüedades de la provincia de Jujuy conservadas en el Museo de La Plata." *Revista del Museo de La Plata* 11 (1904): 75–120.

"Études anthropologiques sur les indiens Takshik (groupe Guaicuru) du Chaco Argentin." *Revista del Museo de La Plata* 11 (1904): 261–314.

"Estudios antropológicos sobre los Chiriguanos, Chorotes, Matacos y Tobas (Chaco occidental)." *Anales del Museo de La Plata* 1 (1907): 53–149.

"Nouvelles recherches sur la Formation Pampéenne et l'Homme Fossile de la République Argentine." Recueil de Contributions Scientifiques de MM. Burckhardt, A. Doering, J. Fruh, H. von Ihering, H. Leboucq, R. Lehmann-Nitsche, R. Martin, S. Roth, W. B. Scott, G. Steimann et F. Zirkel. *Revista del Museo de La Plata* 14 (1907): 143–488.

"Patagonische Gesänge und Musikbogen." *Anthropos* 3, 5–6 (1908): 916–940.

Catálogo de la sección antropológica del museo de La Plata. Buenos Aires: Coni, 1910.

Folklore argentino. Adivinanzas Rioplatenses. Biblioteca Centenaria 6. Buenos Aires: Universidad Nacional de La Plata, 1911.

"El Grupo Lingüístico Tshon de los Territorios Magallánicos." *Revista del Museo de La Plata* 22 (1913): 217–276.

"La Bota de Potro." *Boletín de la academia nacional de ciencias de Córdoba* 21 (1916): 183–300.

"Santos Vega." *Boletín de la academia nacional de ciencias de Córdoba* 22 (1917).

"Texte aus den La Plata-Gebieten in volkstümlichem Spanisch und Rotwelsch." *Beiwerke zum Studium der Anthropophyteia* 8 (1923).

"Mitología Sudamericana. La Astronomía de los Mocoví. La Astronomía de los Chiriguanos. La constelación de la Osa Mayor y su concepto como huracán o dios de la tormenta en la espera del Mar Caribe. La astronomía de los tobas. La astronomía de los Vilelas." *Revista del Museo de La Plata* 28 (1924–1925): 66–79; 80–102; 103–145; 181–209; and 210–233.

"Arqueología peruana: Coricancha. el templo del sol en el Cuzco y las imágenes de su altar mayor." *Revista del Museo de La Plata* 31 (1928): 1–260.

Studien zur südamerikanischen Mythologie, die ätiologischen Motive. Hamburg: Friederichsen, De Gruyter, 1939.

OTHER SOURCES

Bilbao, Santiago A. *Rememorando a Roberto Lehmann-Nitsche.* Buenos Aires: La Colmena, 2004.

Cáceres Freyre, Julián. "Homenaje al doctor Roberto Lehmann-Nitsche (1872–1972)." *Cuadernos del Instituto Nacional de Antropología* 8 (1972–8): 7–19.

Hrdlička, Alex. *Early Man in South America.* In Collaboration with W. H. Holmes, B. Willis, E. Wright, and C. Fenner. Smithsonian Institution Bureau of American Ethnology Bulletin 52. Washington, DC: Government Printing Office, 1912. Contains a good summary of Lehmann-Nitsche's contributions to the debate on "early man."

García, Susana, and I. Podgorny. "El sabio tiene una patria: La primera guerra mundial y la comunidad científica en la Argentina." *Ciencia hoy* 10, no. 55 (2000): 32–34.

Paupillault. "Rapport sur le Prix Godard." *Bulletin de la Societé d'Anthropologie* (1897): 482–483.

Podgorny, Irina. "De la antigüedad del hombre en el Plata a la distribución de las antigüedades en el mapa: Los criterios de organización de las colecciones antropológicas del Museo de La Plata entre 1890 y 1930." *História, Ciências, Saúde-Manguinhos* 6, no. 1 (1999): 81–100.

———. "Bones and Devices in the Constitution of Paleontology in Argentina at the End of the Nineteenth Century." *Science in Context* 18, no. 2 (2005): 249–283. The most accessible discussion in English of the South American fossil man.

———. "La derrota del genio: Cráneos y cerebros en la filogenia argentina." *Saber y Tiempo* 5 (20) (2006): 63–106.

Irina Podgorny

LEHNINGER, ALBERT LESTER

(*b.* Bridgeport, Connecticut, 17 February 1917, *d.* Baltimore, Maryland, 4 March 1986), *biochemistry, energy metabolism* (*bioenergetics*).

Lehninger is perhaps most widely known for his synoptic and lucid textbook, *Biochemistry* (1970), which inspired many students in the field. Through his research, as well as through substantive contributions in leadership roles and education (including two other books: *The Mitochondrion,* 1964; and *Bioenergetics,* 1965), he helped pioneer a new interdisciplinary field, bioenergetics, in the mid-twentieth century. His most significant research achievements were to identify the mitochondrion as the site of the most important energy reactions in the cell and to characterize and quantify many features of that system (including the burning of fats, calcium transport, and proton stoichiometries). His focus on the mitochondrion also laid a foundation for studies on the movement, storing, and regulation of calcium in the cell.

From Writing to Science. Albert grew up in relative economic security in Bridgeport and Hartford, Connecticut. He attended nearby Wesleyan University, a small liberal arts college for men, from 1935 to 1939. Originally he intended to write stories and poetry, and he was a member of the Scrawlers' Club. However, one of his teachers, Ross Fortner Jr., introduced him to the emerging field of biochemistry and to the recent discoveries of Otto Warburg and Hans Krebs on cellular metabolism. Lehninger's interests and major soon shifted to chemistry, and he targeted a new career in medicine and biochemistry.

Lehninger earned his PhD from the University of Wisconsin in 1942. His dissertation research focused on the metabolism of fats. When World War II started, he joined the Plasma Fractionation Program. His task was to develop methods to extend blood plasma by modifying its globulin proteins (a project later abandoned as ill conceived). After the war, Lehninger settled into a position at the University of Chicago, where he enjoyed the mentorship of Charles Huggins (who later won a Nobel Prize for work on cancer treatment).

From Fatty Acids to the Mitochondrion. At Chicago, Lehninger continued his investigations of fatty acids, the long chain molecules that make up fats. Fats supply roughly one-third of the body's energy, yet in the mid-1940s no one yet knew quite how. Lehninger methodically characterized all the conditions (reactants, temperature, pH, cofactors, etc.) that affect the reactions. In the process, he found in 1945 that the breakdown of fats was linked to another set of well-known energy reactions: Krebs' tricarboxylic acid cycle (the discovery of which had earlier inspired Lehninger). Lehninger had discovered how fat and carbohydrate metabolisms importantly converge. Each breaks food molecules into two-carbon fragments, which then share a common energy pathway.

Lehninger had been working with whole cell extracts. A next major aim was to isolate the specific set of enzymes that catalyzed fat metabolism. In 1948 George Hogeboom, Walter Schneider, and George Palade at the Rockefeller Institute reported a new method for suspending broken cells in a dense sugar solution. They had successfully separated one organelle, the mitochondrion, undamaged, using differential centrifugation. Lehninger saw an opportunity. He and one of his first graduate students, Eugene Kennedy, set up their own chilled centrifuge—in a refrigerator normally used for storing urine samples. Within weeks they found that isolated mitochondria could break down the fatty acid chains, while the remaining parts of the cell did not. The mitochondria also contained the enzymes for the tricarboxylic acid cycle, as well as for the subsequent production of adenosine triphosphate, or ATP, the unit molecule of energy in the cell. The mitochondria could not, however, begin the breakdown of glucose. In a remarkably short period, Lehninger had identified the location of nearly all the major energy reactions in the cell. It was a major discovery, foundational for further studies.

Lehninger's characterization of mitochondrial reactions was also significant conceptually for cell biology. For the first time, a specialized function of a cellular organelle had been experimentally demonstrated. The discovery supported the widespread view that particular functions of the cell were sorted in different parts of the cell, and that one could investigate them separately. At the same time, as Lehninger would note in subsequent publications, compartmentalization was important to understand on its own. How did the structure of reactions organized in a membraned unit relate to their function and integration in the cell?

The Mitochondrion: Energy. Lehninger would ultimately devote over three decades to the study of the mitochondrion, and his expertise on it earned him wide acclaim. First, he continued to further localize and identify the vital energy reactions. The primary focus became the final stage of the energy pathways, oxidative phosphorylation (or ox phos). Here, a highly energized phosphate is added to adenosine diphosphate (ADP) to generate the final ATP while, at the same time, cells consume oxygen. Severo Ochoa had already determined in 1943 that for each oxygen used, three ATP were produced. Yet the nature of the reactions was still unclear. In 1948 Lehninger showed definitively, with his student Morris Friedkin, the suspected role of electron transport through the cytochrome system. They thereby clarified just how the Krebs cycle yielded ATP: via the high-energy electrons of nicotine adenine dinucleotide (NADH). In 1949 Lehninger further showed how one could generate NADH without the Krebs cycle by using β-hydroxybutyrate; the method would be applied widely in subsequent investigations. In both cases, to measure oxygen consumption, Lehninger had relied on a classic apparatus designed by Warburg, who had earlier inspired his turn to biochemistry. In his typically methodical style, Lehninger had also noted inhibition by calcium and the effect of ion concentration on reaction rates, apparently modest findings that would later become central to his studies.

Lehninger's renown was growing, and in 1952, after a year as a Guggenheim fellow and Fulbright scholar in England, he moved to the Johns Hopkins School of Medicine. There, at the mere age of thirty-five, while continuing his research, he assumed leadership of a department and the development of educational programs.

Lehninger added another important method to the biochemists' toolkit in 1956. He and postdoctoral fellow Cecil Cooper and student Thomas Devlin were searching for ways to further disassemble the mitochondrion into functional parts. They tried extracts obtained by "vibration, exposure to butanol-water mixtures, drying with acetone, exposure to hypotonic media, grinding, and treatment with cholate or deoxycholate" (Cooper and Lehninger, 1956, pp. 502-503), but all were found to be totally inactive. However, digitonin effectively disrupted the membrane. The resulting fragments, or submitochondrial particles (SMPs), exhibited oxidative phosphorylation, but not the Krebs cycle. They concluded that the electron transport chain and ATP-synthesizing enzyme were located in the mitochondrial membrane, which they had isolated, and that the Krebs cycle was in the fluid interior, or matrix—another landmark localization. At the same time, they began to identify functional segments of the electron transport chain. Later that year Briton Chance and his lab, using spectrographic data, would complete this task, identifying three sites that each yielded one ATP.

By the late 1950s, ox phos had become the premier research topic among biochemists. Lehninger's lab was a leading contributor. In 1958 Lehninger and Charles Wadkins developed evidence for the terminal reaction in ox phos. Their claim of a phosphorylated high-energy intermediate (based on the ADP-ATP and ATP-P exchange reactions) was widely accepted, a welcome benchmark in an increasingly perplexing field. (Much later, the data would be reinterpreted in an alternative theoretical context.) Beginning in 1959, Lehninger also drew attention to the swelling of the mitochondrion and its relation to energy changes, indicating that ion movements and osmotic changes were also significant in interpreting mitochondrial function. In 1963 such concerns led to precise measurements of calcium transport across the mitochondrial membrane fueled by the electron transport system (see below). Based on these findings and earlier conclusions about compartmentalization, Lehninger prominently advocated a role for mitochondrial structure in understanding the energy reactions.

As a respected leader in the field and an effective communicator, Lehninger became an important interpreter of research on how cells transform energy. For example, he coauthored review articles for *Science* in 1958 and for the *Annual Review of Biochemistry* in 1962. He also wrote for a more general audience in *Scientific American* in 1960 and 1961. In 1964 he published *The Mitochondrion,* the first monograph on the organelle, widely noted for its completeness and clarity. Lehninger expanded his focus the following year in another book, *Bioenergetics.* The trim 258-page volume would serve both as a reference and advanced text. Lehninger valuably consolidated and organized information, effectively profiling an emerging scientific field, which ultimately adopted the name he had given it.

As the study of oxidative phosphorylation unfolded, intense controversy emerged. While Lehninger typically addressed theoretical issues thoughtfully, he distanced himself from the theoretical fray and focused instead on what experiments could concretely demonstrate. Nevertheless, Lehninger's theoretical impact was significant, especially in contributing to the development of the chemiosmotic hypothesis. Lehninger's influence began even before the concept was published by Peter Mitchell in 1961. Mitchell's ideas were still incomplete when he attended a conference in Stockholm in 1960. There he met Lehninger and quizzed him about several aspects of oxidative phosphorylation, which at the time was outside his own expertise. Not long after, Mitchell wrote a colleague about the value of the exchange. Lehninger, steeped in research on mitochondrial swelling and ion movements, was well primed to appreciate Mitchell's unconventional idea that electron transport might be coupled to proton movements across the mitochondrial membrane. In 1962, when most investigators were still unaware of Mitchell's ideas, Lehninger endorsed their significance in his review. He discussed them again in his 1964 book, lucidly explaining even Mitchell's generalized concept of vectorial metabolism.

Lehninger was not without criticism, however. Here, his important role was to stimulate further research. Lehninger—ever one to get the numbers straight—noted that Mitchell's original proposal did not properly account for the energy levels as actually measured. That prompted Mitchell and his colleague Jennifer Moyle to redo their measurements with more care (and ultimately to articulate the theory further). Lehninger also observed that according to Mitchell's model, based on membrane gradients, ox phos should occur only if the membranes remained intact and unbroken. Yet Lehninger's own submitochondrial particles, presumably membrane fragments, seemed obvious counterexamples. In response, Mitchell and Moyle would soon argue, ultimately correctly, that the SMPs were indeed sealed vesicles, albeit turned inside out from the mitochondrion's natural orientation. Closed compartments were indeed critical, as Lehninger had originally claimed.

Lehninger's lab also developed evidence for key chemiosmotic claims. In 1966—when Mitchell's ideas were still largely considered peripheral—Lehninger, with Carlo Rossi and Jozef Bielawski, showed that electron transport (induced by calcium) led not only to an increase in external protons (already documented), but also to a decrease in internal protons: the net shift supported Mitchell's notion of a proton "pump," or the separation of H^+ and OH^- across the membrane. That same year Lehninger, Bielawski, and Thomas Thompson addressed the functional role of membrane integrity. They showed that dinitrophenol, an insecticide regularly used to disrupt ox phos, increased the electrical conduction of simple membranes. That would, as Mitchell had claimed, allow protons to "leak" across the membrane, dissipating the postulated energy gradient and thereby interfering with ox phos. Lehninger's work thus provided important support for elements of chemiosmotic theory in its early development. Yet Lehninger himself guarded against broad theoretical conclusions.

Lehninger's caution was perhaps warranted, as illustrated in his later and perhaps most challenging work on characterizing the energy reactions of ox phos. In his earlier work Lehninger had studied how electron transport could fuel both calcium movements and the production of ATP. In the early 1970s discrepancies on the magnitude of the intermediate proton gradient led him to reevaluate their relationship. In particular, Lehninger wondered whether undocumented ion movements invalidated earlier measurements of proton extrusion. Assisted by Baltazar Reynafarje and Martin Brand, he remeasured the proton movement with additional controls. Whereas earlier measurements had indicated two protons for each electron transport site, Lehninger now found an $H^+/2e^-$ ratio of three, possibly four. That would require radical revision in Mitchell's theoretical models. The team cross-checked the measurements with two other methods. They thoroughly analyzed the flaws in earlier methods. When they presented their findings at a conference in Bari, Italy, in 1975, Mitchell disputed them, remarking on the need to distinguish fact from opinion. Lehninger felt his experimental competence had been challenged, and his once cordial relationship with Mitchell soon became formal and strictly professional. The two tried to resolve their differences over the next decade. The mitochondrial reactions were quite complex, and it was not easy finding methods and controls for calcium flux that satisfied everyone. Details were pursued on many fronts in many labs. By 1986 Mitchell had finally conceded. By then, more than ten years later, the ox phos community had generated a much deeper and more robust characterization of proton movements. For example, the $H^+/2e^-$ ratio was found to differ for each site of electron transport, reflecting different mechanisms for translocating protons. The ultimate results certainly reflected Lehninger's lifelong standards for precision and experimental rigor.

The Mitochondrion: Calcium. As reflected in the controversy over proton measurements, biochemists recognized the importance of calcium in the energy economy of the mitochondrion. The capacity of the mitochondrion to accumulate calcium was discovered in the early 1950s, and its link to electron transport in the early 1960s. Lehninger had encountered a role for calcium in ox phos as early as 1949, and then noted in 1956 that it did not affect his digitonin particles. Calcium uptake would affect osmotic balance, possibly accounting for the mitochondrial swelling that Lehninger observed. So, beginning in 1963, with students Carlo Rossi and John Greenawalt, Lehninger set about precisely quantifying the energy used for calcium uptake. They dramatically confirmed the connection to electron transport. They also found that the energy from electron transport was more readily used for calcium transport than for ATP production—a puzzling discovery, given the prevailing view of the mitochondrion as the "power plant of the cell." Calcium was not a mere peripheral reaction, Lehninger concluded, but likely integral to the mitochondrion's function. Interpreting the role of mitochondrial calcium became another major theme in his research.

In 1963 Lehninger's lab showed further that phosphate is taken up along with the calcium. Over the next several years they found that inside the mitochondria the two substances formed amorphous granules of tricalcium phosphate, a precursor to hydroxyapatite (the stuff of bones and teeth). That led to speculation about the role of the mitochondrion in regulation of cell calcium and perhaps in biological calcification.

For further clues, Lehninger turned to other species. One colleague suggested land crabs, as they salvage calcium from their exoskeletons before molting. Lehninger ultimately found the common blue crab more practical, since it was readily obtainable from a fish market on the Baltimore waterfront. In 1974 he, Chung-ho Chen, and Gerald Becker found that the crab's liver mitochondria indeed concentrated a great deal of calcium. Moreover, as in rats, the cells stored calcium phosphate in an amorphous form and contained a substantial amount of tightly bound ATP. (ATP, biochemists later learned, helped bind calcium phosphate.) Lehninger and another student also showed that the mitochondrion could concentrate calcium carbonate, indicating its prospective contribution to calcification in mollusks and corals. The role of the mitochondrion in regulating cellular calcium levels seemed phylogenetically widespread.

Another thread of investigation considered the fate of the mitochondrion's calcium phosphate granules: Why did they not spontaneously crystallize, forming bone? A clue came unexpectedly from a colleague at the Johns Hopkins School of Medicine. His research indicated that patients with urinary stones seemed deficient in a calcification inhibitor normally present in urine. Lehninger saw the analogy with mitochondria: Was a similar inhibitor at work? Several colleagues were indeed able to extract such a substance from both urine and the mitochondrion—and the blue crab as well—and show their similarities. In 1981, after the urine inhibitor was identified as phosphocitrate, Lehninger helped show that it could inhibit calcification in mitochondria *in vivo*. He consolidated the available information into a prospective account of biological mineralization, and he stressed the unique ability of mitochondria to concentrate calcium, along with phosphate, yet to not crystallize them irreversibly. However, Lehninger never saw evidence for his scheme fully realized. His work left important ideas (and measurements) for others to pursue.

Teacher, Author, Leader. Lehninger's research was paralleled by equally significant contributions in education and professional service. He led the Department of Physiological Chemistry at the Johns Hopkins School of Medicine for twenty-five years. During his early years there, he helped modernize their teaching of medicine and strengthened the school's graduate program across all the sciences. In 1958 he earned the distinction of receiving the very first graduate teaching grant from the National Institute of Health. Lehninger's skill for well-organized and vivid lectures was renowned. He also understood unique teachable moments. On the first day of class one year, he set aside his prepared lecture to discuss how the newly announced discovery of reverse transcriptase would revolutionize biochemistry. In 1977 Lehninger was named University Professor of Medical Science, a position created to honor his outstanding service.

Lehninger's influence extended well beyond his own students and home institution. His 1970 textbook, *Biochemistry*, became the standard in the field for many years. More than a half million copies of the first two editions were sold, and it was translated into several languages. Lehninger had become a writer after all, and reviewers acknowledged his extraordinary skills. They found the book "eminently readable" and "vibrant with scientific enthusiasm." "With a felicity of diction," one wrote, "he has woven a meaningful pattern into the fabric of the science which is characterized by beauty and accuracy" Even the opening lines evoked a sense of Lehninger's appreciation for the topic: "Living things are composed of lifeless molecules. … Yet living organisms possess extraordinary attributes not shown by collections of inanimate matter." By the time a second edition appeared, the first edition had achieved "an almost unprecedented world-wide popularity and acclaim" (all comments are quoted and cited in Talaley and Land, 1986). His editor, too, was deeply impressed on another level with his "skill and attention to details of content, level of discussion, writing style, illustrations, and all the other elements that shape a book and determine its success." (Neil Patterson, as stated at http://www.cambridge.org/, confirmed in personal interview by Author with Mr. Patterson) Lehninger was writing the third edition when he died. His book remained in print (with supplemental revisions, but still exhibiting Lehninger's character) at least two decades after his death.

Lehninger's skills in organization, his discipline, and his sense of thoroughness led to numerous leadership roles. At different times, he served on the editorial boards of eight journals, most notably the *Journal of Biological Chemistry* (1954–1966) and the *Journal of Membrane Biology* (1968–1986). He served on various advisory groups, including panels at the National Academy of Sciences, a presidential panel on biomedical research, and the board of trustees of his alma mater, Wesleyan University, as well as assuming leadership positions in several professional organizations. He was elected to the National Academy of Sciences (1956), the American Academy of Arts and Sciences (1959) and the American Philosophical Society (1970), and he received seven honorary degrees, among numerous other awards.

Although congenial with students, Lehninger was a private person. When he died of complications from asthma, few colleagues even knew that he had managed the disease for many years. While he worked in downtown Baltimore, he resided with his wife and two children in a rural area north of the city. He enjoyed sailing on the Chesapeake Bay.

BIBLIOGRAPHY

WORKS BY LEHNINGER

"Fatty Acid Oxidation and the Krebs Tricarboxylic Acid Cycle." *Journal of Biological Chemistry* 161 (1945): 413–414.

With Morris Friedkin. "Esterification of Inorganic Phosphate Coupled to Electron Transport Between Dihydrodiphosphopyridine Nucleotide and Oxygen. I." *Journal of Biological Chemistry* 178 (1949): 611–623. Links ATP production from oxidative phosphorylation definitively to electron transport (of NADH). Follows letter announcing results, *JBC* 174 (1948): 757–758.

"Esterification of Inorganic Phosphate Coupled to Electron Transport between Dihydrodiphosphopyridine Nucleotide and Oxygen. II." *Journal of Biological Chemistry* 178 (1949): 625–644. Introduces hydroxybutyrate as substrate to generate NADH.

With Eugene P. Kennedy. "Oxidation of Fatty Acids and Tricarboxylic Acid Cycle Intermediates by Isolated Rat Liver Mitochondria." *Journal of Biological Chemistry* 179 (1949): 957–972. Follows letter announcing results *JBC* 172 (1948): 847–848.

With Cecil Cooper. "Oxidative Phosphorylation by an Enzyme Complex from Extracts of Mitochondria. I. The span β-hydroxybutyrate to Oxygen." *Journal of Biological Chemistry* 219 (1956): 489–506. Introduces phosphorylation-competent digitonin treated submitochondrial particles.

With Charles L. Wadkins. "The Adenosine Triphosphate-Adenosine Diphosphate Exchange Reaction of Oxidative Phosphorylation." *Journal of Biological Chemistry* 233 (1958): 1589–1597.

"Water Uptake and Extrusion in Mitochondria and Its Relation to Oxidative Phosphorylation." *Physiological Reviews* 42 (1962): 467–517.

With Charles L. Wadkins. "Oxidative Phosphorylation." *Annual Review of Biochemistry* 31 (1962): 47–78.

With Carlo S. Rossi and John W. Greenawalt. "Respiration-dependent Accumulation of Inorganic Phosphate and Ca^{++} by Rat Liver Mitochondria." *Biochemical and Biophysical Research Communications* 10 (1963): 444–448.

With Carlo S. Rossi. "Stoichiometry of Respiratory Stimulation, Accumulation of Ca^{++} and Phosphate, and Oxidative Phosphorylation in Rat Liver Mitochondria." *Journal of Biological Chemistry* 239 (1964): 3971–3980.

The Mitochondrion. New York: W. A. Benjamin, 1964.

Bioenergetics. New York: W. A. Benjamin, 1965. A second edition appeared in 1972.

With Jozef Bielawski and Thomas E. Thompson. "The Effect of 2,4-dinitrophenol on the Electrical Resistance of

Phospholipid Bilayer Membrane." *Biochemical and Biophysical Research Communications* 24 (1966): 948–954.

With Carlo Stefano Rossi and Jozef Bielawski. "Separation of H^+ and OH^- in the Extramitochondrial and Mitochondrial Phases during Ca^{++}-activated Electron Transport." *Journal of Biological Chemistry* 241 (1966): 1919–1921. Early evidence supporting chemiosmotic hypothesis.

With Ernesto Carafoli and Carlo Stefano Rossi. "Energy-linked Ion Accumulation in Mitochondrial Systems." *Advances in Enzymology* 29 (1967): 259–320.

"Mitochondria and Calcium Ion Transport." *Biochemical Journal* 119 (1970): 129–138.

Biochemistry. New York: Worth, 1970. A second edition appeared in 1975. An abridged adaptation appeared in 1973, an expanded version (retitled *Principles of Biochemistry*) in 1982. Revised versions published since his death still bear his name as primary author.

With Gerald Becker, Chung-ho Chen, and John Greenawalt. "Calcium Phosphate Granules in the Hepatopancreas of the Blue Crab (*Callinecte sapidus*)." *Journal of Cell Biology* 61 (1974): 316–326.

With Martin D. Brand and Baltazar Reynafarje. "Stoichiometric Relationship Between Energy-dependent Proton Ejection and Electron Transport in Mitochondria." *Proceedings of the National Academy of Sciences USA* 73 (1976): 437–441.

With Baltazar Reynafarje, A. Vercesi, and William P. Tew. "Transport and Accumulation of Ca^{2+} in Mitochondria." *Annals of the New York Academy of Sciences* 307 (1977): 160–176.

With Baltzar Reynafarje, P. Davies, A. Alexandre, et al. "The Stoichiometry of H^+ Ejection Coupled to Mitochondrial Electron Flow, Measured with a Fast-responding Oxygen Electrode." In *Mitochondria and Microsomes,* edited by Chaun Pin Lee, Gottfried Schatz, and Gustav Dallner. Reading, MA: Addison-Wesley, 1981.

OTHER SOURCES

Harvey, A. McGehee. "The Department of Physiological Chemistry: Its Historical Evolution." *Johns Hopkins Medical Journal* 139 (1976): 257–273.

Talalay, Paul, and M. Daniel Land. "Albert Lester Lehninger (1917–1986): A Perspective." *Trends in Biochemical Sciences* 11 (1986): 356–358.

Douglas Allchin

LEHRMAN, DANIEL SANFORD

(*b.* New York, New York, 1 June 1919; *d.* Santa Fe, New Mexico, 30 August 1972), *comparative psychology, ethology, behavioral endocrinology, instinct.*

Lehrman was a comparative psychologist who made important contributions to the study of animal behavior both as a research scientist and as a critic of theory. He pioneered the field of behavioral endocrinology. His critique of ethological instinct theory led to an exchange of views between European and American scientists that did much to shape the subsequent evolution of the study of animal behavior.

Career. Lehrman grew up in New York City, where he attended public schools, including the elite Townsend Harris High School in the Bronx. He enrolled in the City College of New York, but his undergraduate studies were interrupted when he enlisted in the U.S. Army in 1942. During World War II his linguistic skills enabled him to serve as a translator of German and a cryptographer. Returning to City College he completed his BS degree, majoring in psychology and biology, in 1946. He then took up graduate study at New York University. There he worked under the direction of Theodore C. Schneirla, a noted comparative psychologist connected with the American Museum of Natural History. During this time he also worked as an assistant psychologist at the Haskins Laboratories (1945–1947), and held a summer fellowship at the Bronx Zoo. He completed his PhD in psychology in 1954. Meanwhile, he was appointed lecturer in psychology at City College (1947–1950), and then, in 1950, assistant professor in the Psychology Department of the Newark campus of Rutgers, the State University of New Jersey. He continued to give evening classes at City College until the early 1960s, partly because this served as a recruiting ground for the research program he was developing at Rutgers. In 1957–1958 he was a visiting professor at Yale.

In 1958 Rutgers promoted him to associate professor. In the same year he applied for and was awarded a sizable U.S. government grant to establish an animal behavior research facility in Newark. Thus he founded the Institute of Animal Behavior (IAB) at Rutgers, which grew to be internationally recognized for the quality of its science. This had Lehrman's work at its center, but comparable contributions came from the several colleagues who joined him, and the numerous graduate students who did PhD research in the institute.

Despite proposals to move the institute closer to the main Rutgers campus in New Brunswick, and similar invitations from other universities, including Harvard, Lehrman remained loyal to his first employer and elected to stay in Newark. He was rewarded by enlarged quarters in a new building when the Newark campus was relocated as part of an urban renewal development in 1969. The new institute facility, which Lehrman took a considerable part in designing, was described by one visitor as "the Taj Mahal of animal behavior" (Richard Michael, personal communication to Colin Beer).

In 1963 Lehrman was appointed associate editor of the *Journal of Comparative and Physiological Psychology,* which enabled him to take a major part in promoting the publication of comparative work in animal behavior. He also collaborated (with Robert Hinde and Evelyn Shaw) in founding and editing a series of volumes, *Advances in the Study of Behavior,* for Academic Press. Beginning in 1955, he became a major figure in the biennial International Ethological Conferences, and was involved in numerous other meetings between representatives of European ethology and American comparative psychology. In 1970 he was invited to join the Salk Institute in La Jolla, California, as a resident fellow, but declined, accepting instead the status of a visiting fellow.

Lehrman's contributions to science were recognized by the granting of a lifetime Research Career award from the U.S. Public Health Service, election to membership of the National Academy of Sciences, and a fellowship of the American Academy of Arts and Sciences. He was also honored by the Animal Behavior Society, the American Ornithologists' Union, and other institutions.

Biological Background. Although "Danny" Lehrman's academic training was mainly in psychology, his first love was natural history, especially that of birds. As a schoolboy he responded to the encouragement and example of his scoutmaster by joining bird walks with local ornithologists in such places as Van Cortlandt Park and the Brooklyn Botanic Garden. He developed a passionate interest in bird life, in addition to the birder's affliction of needing to include as many species as possible on a life list of birds seen (he was still preparing to add to the tally on the day he died). Indeed this absorption with birds sometimes interfered with his college career, as he cut classes to indulge it so often, especially during migration times, that he was temporarily suspended.

During his first year at City College Lehrman began working part time at the American Museum of Natural History as a volunteer assistant to Kingsley Noble, curator in the Department of Experimental Biology (later the Department of Animal Behavior), whose work included bird studies. For instance, he was involved in a series of field experiments on what a laughing gull might treat as an egg, for which Lehrman did most of the tests. This earned him coauthorship of the resulting publication, although he was chagrined at having his interpretations of the observations ignored or overruled by the senior man. Indeed there are records suggesting that Noble consistently took advantage of the enthusiasm and enterprise of his young assistant. Nevertheless, Lehrman gained much from this involvement at the museum. He later recalled how the influences of other people there, including Frank Beach, William Etkin, Libbie Hyman, Ernst Mayr, and Ted Schneirla, turned his attention toward animal behavior and encouraged his pursuit of it. In their different ways Hyman and Schneirla introduced him to the conception of levels of organization in the animal kingdom, a viewpoint that influenced his thinking throughout the rest of his life.

It was also at the museum that Lehrman had his first close encounter with ethology, the European-based approach to the biology of behavior. In 1938 Niko Tinbergen, one of the "fathers" of ethology, made his first trip to the United States, which included an extended visit to the museum. He and the young ornithologist met and got to know one another during bird walks in New York parks. In 1947 Tinbergen returned to New York at the invitation of the museum and Columbia University to give a series of lectures titled "The Study of Innate Behavior in Animals." When the text of these lectures was accepted by Oxford University Press for publication as *The Study of Instinct* (1951), Danny Lehrman was given the task of polishing the Dutchman's English into what remains one of the most readable texts in the animal behavior literature.

The other major figure in ethology's founding also came to Lehrman's notice during his "apprenticeship" at the museum. He began reading and translating Konrad Lorenz's work, becoming so involved with it that Margaret Morse Nice, one of the luminaries of American ornithology at the time, invited him to review one of Lorenz's major articles ("Vergleichende Verhaltensforschung," 1939) for the journal *Bird Banding.* The review was published in 1941. Though critical of some points, such as Lorenz's proneness to sweeping generalizations that ignored "the great differences in organization of the nervous system at different evolutionary levels," Lehrman enthusiastically promoted the Lorenzian approach to the causes of behavior as "superior and more fruitful than any that can be obtained otherwise" (1941, p. 87). Lehrman even offered to lend his translation of Lorenz's paper to anyone interested in the new orientation. There was little indication of the attack to be launched twelve years later.

Research: Interacting Causes of Bird Behavior. When Lehrman began work for his PhD under the direction of Schneirla in 1948, he naturally took up a project on birds. He settled on the breeding behavior of ring doves (*Streptopelia risoria*), partly because these birds were available at the museum and can be easily bred and observed in captivity. To begin with he took an amusingly wrong turn. He drew up a plan to study how incubation behavior might be meshed with changes in the brood patches. These are areas of ventral skin that become denuded of feathers and extensively vascularized in the service of conveying heat to the eggs when a bird is sitting on them. They are almost ubiquitous in birds that incubate their eggs this way, as does the ring dove. To his surprise Lehrman found his

Lehrman

species to be an exception: there were no detectable changes in dermal state or sensitivity of the kinds he expected to be connected with a dove's tendency to sit on eggs.

Consequently he turned to another feature that also sets pigeons and doves apart from other birds: the way they feed their young. At around the time the eggs hatch, the epithelial lining of the crop of the sitting birds begins to change texture, exuding a substance having the consistency and nourishing properties of mammalian milk. It is referred to as crop milk, and the hungry nestlings (squabs) ingest it by inserting their bills into the parent's mouth and having the parent regurgitate. Lehrman set about investigating how the production of crop milk is induced, and the joint contributions of parent and young to the feeding process. This led to his getting involved with behavioral endocrinology, since crop milk production, like mammalian lactation, is due to secretion of the hormone prolactin from the anterior pituitary gland. Lehrman looked at how the hormone might otherwise be involved in the regurgitation feeding behavior. He found that when he injected prolactin into adult doves with previous breeding experience they responded to hungry squabs by feeding them, in contrast to control birds injected with the hormone vehicle, which ignored the squabs.

However, hormone-treated birds lacking previous breeding experience also failed to respond to the squabs. And so did experienced birds whose crops had been anesthetized. These results supported the conjecture that prolactin promotes feeding behavior in the doves by causing engorgement of the crop tissue, a condition from which the bird is relieved by regurgitation. This is initially invoked by tactile stimulation from a newly hatched squab's thrusting its head upward against the parent's breast. As a consequence the parent dove comes to associate the other features of a squab—its appearance and vocalization—with the relief from crop engorgement afforded by feeding it (Lehrman, 1955). This kind of story, in which hormones, stimulation, experience, and behavior are linked in dynamic and reciprocal relation, became a Lehrman leitmotif as he went on to experiment with other aspects of the ring dove breeding cycle, and apply its lessons to comparable systems in other kinds of animals.

For example, further study of the role of prolactin in incubation showed that its secretion is a consequence of the initiation of sitting on eggs. This is induced by another hormone, progesterone, produced by the gonads, together with the visual stimuli from the nest and eggs. The consequent tactile stimulation leads to changes in the brain affecting the release of prolactin. This part of the process involves a vascular link, discovered by Geoffrey Harris (1955), which conveys a humoral agent from the hypothalamus to the anterior pituitary. The prolactin has the effect of sustaining incubation, as well as preparing for the next phase of the cycle by inducing crop milk production. Tracing the chain of causal command still further back, Lehrman was able to show how the gonadal development responsible for progesterone secretion depends upon the previous events of courtship, copulation, and nest building. These involve visual, auditory, and tactile stimulation, which act via the hypothalamic-pituitary link to mobilize gonad-stimulating hormone; this, in addition to gametogenesis, induces secretion of sex steroids, which feed back to the brain to produce changes in behavior and modulate pituitary output. Thus internal and external factors—stimuli, brain states, endocrine secretion, behavior—take turns as cause and effect in the complex progressive succession that is the ring dove breeding cycle (e.g., Lehrman, 1965).

Lehrman's work on the doves bore comparison to that of his friend Robert Hinde at Cambridge University. Hinde's studies of the breeding behavior of canaries revealed a similar pattern of reciprocal relations along with differences of detail. Likewise Lehrman's close colleague at the IAB, Jay Rosenblatt, demonstrated in numerous ways the intertwined and detailed connections between behavioral development of the young and associated maternal behavior in rats, cats, and other mammals.

When William C. Young invited Lehrman to contribute a chapter on "Hormonal Regulation of Parental Behavior in Birds and Infrahuman Mammals" for a new edition of *Sex and Internal Secretions* (1961), he was very well placed to do so and readily undertook the task. In this now-classic survey Lehrman reviewed in depth and detail the extensive relevant literature in behavioral endocrinology, bringing to bear his unique combination of synthetic and analytic judgment, making connections and drawing distinctions; for the chapter continues to be consulted, in contrast to the ephemeral life of most such scientific writing, and in spite of the newer work it has in part inspired.

At the International Ethological Conference held at Cambridge University in 1959 Lehrman gave a talk indicating that he was off on a new tack. He described a method by which behavioral data recorded by an observer using a keyboard might be coded and fed into a computer for immediate data reduction and statistical analysis. Despite considerable expense of time and effort this ambitious project proved too cumbersome to be practical and never got off the ground.

This gravitation to computer technology was consistent with one of Lehrman's predilections, but the grandiose scale of his scheme was at odds with the kind of stance he generally took toward programs of comparable scope. He was critical of learning theorists, such as Clark

Hull and Edward Chace Tolman, for their assumption that learning conforms to universal laws, and hence their failure to take into account the profound differences consequent on the levels of organization that separate different kinds of animals. He also found fault with European ethology for failure to draw necessary distinctions. But this was only part of an extended critique that shook up "classical" ethology to such an extent that it underwent profound regrouping. It also made Daniel Lehrman a voice to be reckoned with in the larger world of behavioral science.

Critique of Ethology: Instinct. Lehrman was still working on his PhD when he published "A Critique of Konrad Lorenz's Theory of Instinctive Behavior" (1953). Many people regarded it as an audacious move on the part of a young man who had yet to make his mark as a figure in the field of behavioral studies. It would have been thought even more so had he not been dissuaded from including passages of invective directed at Lorenz's pro-Nazi, anti-Semitic wartime writings, which Lehrman had translated.

Lehrman saw ethological instinct theory, Tinbergen's along with Lorenz's, as a product of the mischief of ambiguity of the term *instinct.* The word can have several distinct meanings, which do not logically entail one another: genetically transmitted behavior, behavior that does not depend upon experience for its development (i.e., is not acquired by learning), and motivational impulsion or blindingly compelling urge (e.g., maternal instinct), action based on impulse rather than deliberation. Lorenz (1950) had claimed that ethology owed its existence to discovery of a type of behavior, the *Instinkthandlung* or fixed action pattern, to which all the salient senses of "instinct" apply: it is innate both in the sense of being genetically inherited and in the sense of being independent of learning; its performance is both endogenously impelled by centrally generated energy for which it serves as outlet and goal, and endogenously patterned or controlled (in contrast to peripheral reflexes). Tinbergen's (1951) version expanded "instinct" to the status of a whole hierarchically organized motivational system serving one of the major functional divisions, such as foraging and reproduction, which incorporated fixed action patterns as terminal components. It too was supposed to be inborn, developmentally preprogrammed, and endogenously driven.

Lehrman took issue with the central assumption of these systems, according to which innate in the sense of heredity implies innate in the sense of not learned: evidence of genetic transmission is not, ipso facto, evidence bearing on the role of experience in development. The fact that Rudolf Serkin was a superb concert pianist did not absolve his comparably gifted son Peter from having to practice. Lehrman argued that the ethological position on questions of behavioral development was usually based either on invalid extrapolation from genetic considerations, or flimsy observations or experiments, such as the Kasper Hauser procedure of raising animals in social isolation (named for a youth treated this way who was found wandering the streets of Nürnberg in 1882), which attempts to exclude from experience whatever might be thought to be relevant. Lehrman contended that insufficient consideration was given to what deprivation might have failed to exclude, pointing to cases of possible prenatal learning or an animal's using part of its own body in lieu of what it had been kept from. Admittedly some of his examples were far fetched, but the general point that questions of development require their own study, and cannot be answered by breeding experiments, was compellingly made. Similarly he maintained that the fixity of fixed action patterns was usually more a matter of assertion than experimentally established fact.

Furthermore, Lehrman challenged the physiological plausibility of the motivational mechanisms proposed by Lorenz and Tinbergen. On the one hand the main evidence for them appeared to be the alleged features of observed behavior they were supposed to explain. Apart from some appeal to work by Charles Sherrington, Erich von Holst, and a few others with physiological credentials, the mechanisms were unsupported by independent physiological investigation, and hence could be regarded as "reifications." On the other hand the relevant physiology was at odds with the hydraulic analogies informing the motivational models: neural excitation is not the kind of thing that can accumulate, overflow, or be "dammed up."

Finally, as with monolithic learning theory, Lehrman deplored the undiscriminating inclusiveness of instinct theory: "reification of the concept of 'instinct' leads to a 'comparative' psychology which consists of comparing levels in terms of *resemblances* between them, without that careful consideration of *differences* in organization which is essential to an understanding of evolutionary change, and of the historical emergence of new capacities" (1953, p. 351). He was particularly bothered by the "patently shallow" (p. 353) attempts to comprehend the human case within instinct's purview.

Although Lehrman's critique landed like a bombshell in the precincts of ethology, the ground had been prepared for it to some extent. In 1950 there had been a meeting on ethological terminology, which had recommended against continued talk of "action specific energy" with its controversial material connotations. Especially in Britain, a younger generation of ethologists, led by Robert Hinde, had begun questioning the utility of the hydraulic models, taking a tough-minded, often statistical stance toward the grounding of concepts, in contrast to the tender-minded speculations of the founding fathers. Lehrman's arrival on the scene thus encountered a mixture of hostility and welcome.

It did not take long for confrontation to reach the conference table. In 1954 two important meetings on the issues dividing European ethology and American comparative psychology took place, the first in Paris under the auspices of the Singer-Polignac Foundation, the second in Ithaca, New York, sponsored by the Macy Foundation. Lehrman was a key figure in both of these. The following year an American contingent, again including Lehrman, was invited to the Third International Ethological Conference in Groningen, Netherlands.

These and other such opportunities to try to come to terms did much to reduce transatlantic tension and misunderstanding. "Hard core" classical ethologists, such as Gerard Baerends and Jan van Iersel, were surprised to have their "rat runner" stereotype of the American animal psychologist confounded by a Lehrman who knew as much about wild birds as they did. By 1967, when the Ethology Conference convened in Stockholm, much of the dissension had dissipated, thanks in part to Lehrman's consistent attendance at these meetings. Apropos the Stockholm conference, Lorenz wrote in a letter to Tinbergen: "I believe Danny Lehrman has now finally understood me, and we have both in many discussions astonishingly come to exactly the same thing" (quoted in Burkhardt, 2005, p. 405).

Nevertheless, some of the old divisions had merely been pasted over, and it did not take much to make the cracks show. In 1965 Lorenz published *Evolution and Modification of Behavior,* which returned to the nature/nurture issue, albeit with a new twist, which relocated the dichotomy to genetic and environmental sources of "information" contributing to behavioral development. He argued that whenever behavior shows adaptive design it must be a product of naturally selected specification encoded in the genome; hence if learning is developmentally involved it must be due to an "innate schoolmarm" (p. 80). He also accused "N. Tinbergen and many other ethologists writing in English" (p. 1) of apostasy, and lumped Lehrman indiscriminately with the behaviorists.

Lehrman responded to this book in the chapter he wrote for a memorial to his mentor T. C. Schneirla. He objected to the way he and Schneirla had been represented as ignorant of biology, and to Lorenz's tactic of defending his position on the innate/learned dichotomy by shifting the goal posts. However, his main point was that the issues were less about matters of fact than due to deep-seated ideological differences, which were abetted by conceptual and semantic confusions. He maintained that Lorenz's obsession with questions of adaptive function had blinded him to the true nature of questions about proximate causation and individual development:

> It is not necessary that all problems fit into the same conceptual framework. It is not required of any theory based on watching intact lower vertebrates that it explain the causes of war, the physiology of the nervous system, and *also* the mode of action of the genes; and it is not an affront to any theory to point out that there are some questions that it cannot answer because it has not asked them. (Lehrman, 1970b, pp. 47–48)

Beyond Behavior. A charismatic teacher in the classroom and forceful speaker on the podium, Lehrman sometimes took advantage of the stage to express opinions on topics of broader concern than the issues of behavioral science. Two examples will have to suffice.

Lehrman's political sympathies tended to favor left-wing causes. Among these was feminism, to which his wife Dorothy Dinnerstein was to make a now classic contribution with her book *The Mermaid and the Minotaur* (1976). In September 1970, Lehrman gave a talk at a symposium in Canada on "The Application of Ethology to Human Growth and Development." The audience included a substantial number of psychiatrists. Using birds as his main examples, but also drawing on monkey comparisons presented by previous speakers, he made the point that even closely related species can differ profoundly in their ecological and social relationships in ways affecting behavioral development. Hence it can be erroneous to generalize from one species to another without taking into account how natural selection might have differentially designed their domestic arrangements, especially when appealing to animal comparisons to support views about what is natural for the human case: from the fact that a rhesus monkey infant suffers lasting psychological impairment from a period of separation from its mother, it does not necessarily follow that woman's place is in the home (Lehrman, 1974).

As part of the centenary celebrations of the American Museum of Natural History, Lehrman found himself sharing the limelight with Burrhus Frederic Skinner. He contrasted his own "natural history" approach to animal behavior with Skinner's behavioral engineering, which he took to be essentially concerned with using animals to study how contingencies of reinforcement can serve as means of controlling, shaping, and predicting behavior in general, including the behavior of people. For Lehrman the study of animal behavior was something to be undertaken for its own sake, more akin to the appreciation of poetry than the pursuit of power. Of science in general he said:

> In addition to (or instead of) serving a function like that of an engineer, the scientist can also serve

> a function like that of an artist, of a painter or poet—that is, he sees things in a way that no one has seen them before and finds a way to describe what he has seen so that other people can see it in the same way. This function is that of widening and enriching the content of human consciousness, and of increasing the depth of the contact that human beings, scientists and nonscientists as well, can have with the world around them. (Lehrman, 1971, p. 471)

On 27 August 1972, in Santa Fe, New Mexico, Danny Lehrman suffered a heart attack and died three days later. He was only fifty-three. The following year the International Ethological Conference was held in Washington, D.C. The program included a full-day plenary session in tribute to Lehrman's work and character, the only such memorial in the history of these meetings. The man who entered the ethological scene as an antagonist left it as a hero.

BIBLIOGRAPHY

WORKS BY LEHRMAN

With C. K. Noble. "Egg Recognition by the Laughing Gull." *Auk* 57 (1940): 22–43.

"Comparative Behavior Studies." *Bird Banding* 12 (1941): 86–87. This was a review of K. Z. Lorenz, "Vergleichende Verhaltensforschung." *Verhandlungen der deutschen zoologischen Gesellschaft,* edited by C. Apstein. Zoologische Anzeiger, supp. 12. Leipzig, Germany: Akademische Verlagsgesellschaft m.b.H., 1939.

"A Critique of Konrad Lorenz's Theory of Instinctive Behavior." *Quarterly Review of Biology* 28 (1953): 337–363.

"The Physiological Basis of Parental Feeding in the Ring Dove (*Streptopelia risoria*)." *Behaviour* 4 (1955): 241–286.

"Induction of Broodiness by Participation in Courtship and Nest-Building in the Ring Dove (*Streptopelia risoria*)." *Journal of Comparative and Physiological Psychology* 51 (1958): 32–36.

"Hormonal Responses to External Stimuli in Birds." *Ibis* 101 (1959): 478–496.

With P. N. Brody and R. P. Wortis. "The Presence of Mate and of Nesting Material as Stimuli for the Development of Incubation Behavior and for Gonadotropin Secretion in the Ring Dove (*Streptopelia risoria*)." *Endocrinology* 68 (1961): 507–516.

"Hormonal Regulation of Parental Behavior in Birds and Infrahuman Mammals." In *Sex and Internal Secretions,* 3rd ed., edited by W. C. Young. Baltimore: Williams and Wilkins, 1961.

"Interaction of Hormonal and Experiential Influences on Development of Behavior." In *Roots of Behavior,* edited by E. S. Bliss. New York: Harper and Row, 1962.

With J. S. Rosenblatt. "Maternal Behavior of the Laboratory Rat." In *Maternal Behavior of Mammals,* edited by H. L. Rheingold. New York: Wiley, 1963.

"On the Initiation of Incubation Behavior in Doves." *Animal Behaviour* 11 (1963): 433–438.

"Control of Behavior Cycles in Reproduction." In *Social Behavior and Organization among Vertebrates,* edited by W. E. Etkin. Chicago: University of Chicago Press, 1964.

"Interaction between Internal and External Environments in the Regulation of the Reproductive Cycle of the Ring Dove." In *Sex and Behavior,* edited by F. A. Beach. New York: Wiley, 1965.

"Experiential Background for the Induction of Reproductive Behavior Patterns by Hormones." In *Biopsychology of Development,* edited by E. Tobach, L. R. Aronson, and E. Shaw. New York: Academic Press, 1970a.

"Semantic and Conceptual Issues in the Nature-Nurture Problem." In *Development and Evolution of Behavior,* edited by L. R. Aronson, E. Tobach, D. S. Lehrman, and J. S. Rosenblatt. San Francisco: Freeman, 1970b. Lehrman's response to Lorenz's *Evolution and Modification of Behavior.*

"Behavioral Science, Engineering, and Poetry." In *The Biopsychology of Development,* edited by E. Tobach, L. R. Aronson, and E. Shaw. New York: Academic Press, 1971.

"Can Psychiatrists Use Ethology?" In *Ethology and Psychiatry,* edited by Norman F. White. Toronto: McMaster University Press, 1974.

OTHER SOURCES.

Beer, C. G. "Was Professor Lehrman an Ethologist?" *Animal Behaviour* 23 (1975): 957–964. Text of a talk given in the Lehrman Memorial Plenary Session at the 13th International Ethological Conference in Washington, DC, August 1973.

Burkhardt, Richard W. *Patterns of Behavior: Konrad Lorenz, Niko Tinbergen, and the Founding of Ethology.* Chicago: University of Chicago Press, 2005.

Dinnerstein, D. *The Mermaid and the Minotaur: Sexual Arrangements and Human Malaise.* New York: Harper and Row, 1976.

Harris, G. W. *Neural Control of the Pituitary Gland.* London: Edward Arnold, 1955.

Lorenz, Konrad Z. "The Comparative Method in Studying Innate Behaviour Patterns." *Symposia of the Society for Experimental Biology* 4 (1950): 221–268. Lorenz's best-known version of his instinct theory in English.

———. *Evolution and Modification of Behavior.* Chicago: University of Chicago Press, 1965.

Rosenblatt, J. S. "Daniel Sanford Lehrman 1919–1972." *Biographical Memoirs,* vol. 66. Washington, DC: National Academy of Sciences, 1995.

Tinbergen, N. *The Study of Instinct.* Oxford: Clarendon Press, 1951. The classic thesis of classical ethology.

———. "On Aims and Methods of Ethology." *Zeitschrift für Tierpsychologie* 20 (1963): 410–433. Famous for its statement of "the four questions of ethology" with which Lehrman was in harmony.

Colin Beer

LEIBNIZ, GOTTFRIED WILHELM

(*b.* Leipzig, Germany, 23 June 1646, d. Hanover, Germany, 14 November 1716), mathematics, philosophy, metaphysics. For the original article on Leibniz see *DSB,* vol. 8.

Since the original *DSB* article, a more nuanced and complex picture of G. W. Leibniz's scientific work has emerged. This is in part due to the hard work of Leibniz's editors at the German Academy of Sciences, who continue to uncover rich new material (as of 2007, they have succeeded in bringing before the public well under half of Leibniz's total writings, including the massive edition of Series 6, Volume 4 of the philosophical writings, released in 1999). In part, however, the new picture that has emerged in recent decades is the result of changing historiographical concerns among scholars of early modern science and philosophy. Whereas earlier scholarship had been largely content with a triumphalist account of the history of science, placing Leibniz at the beginning of a few lines of scientific inquiry that have been, as it happens, successful ones, the new historiography has been intent on bringing to light all of the interests of the heroes of the scientific revolution, even those that have turned out in the interim to be dead ends.

One significant part of this change in recent decades has been a growing sensitivity to the different ways in which the various scientific disciplines have been divided up in different times and places. Often—as is the case with Leibniz—proposals for new ways to divide the sciences themselves reflect philosophical convictions about the structure of the world. It is illuminating to consider Leibniz's division of the sciences presented in his 1700 memorandum to the Elector of Brandenburg for the creation of a scientific society in Berlin. There Leibniz identifies two fundamental branches of the "real sciences," mathematics and physics, and in turn divides these two as follows: mathematics consists of

1. geometry, including analysis;
2. astronomy and its related fields, including geography, chronology, and optics;
3. civil, military, and naval architecture (along with painting and sculpture); and
4. mechanics.

Physics, in turn, includes

1. chemistry;
2. study of the mineral kingdom, including mining and smelting;
3. study of the vegetable kingdom, including agriculture and forestry; and
4. study of the animal kingdom, including anatomy, the science of hunting, and animal husbandry. (Aiton, 1985, p. 251)

We need not follow Leibniz's schema here, but it can at least serve as a reminder that the traditional view of Leibniz's scientific concerns, which places a mathematized physics at the foundation of all things, and leaves those things largely untouched, surely does not capture the full range of his concerns. When we say that for Leibniz mathematics underlies physics, what we mean to say is not, as the schema suggests, inter alia that painting underlies forestry, but that Leibniz included much of what we would call "physics" under the heading of mathematics because he, in keeping with the revolution begun by Galileo and others a century earlier, believed that the natural world could best be understood in quantitative terms. But the natural world is carved up into more than just homogeneous bodies in motion: it also contains chemical compounds, crystals, organic bodies, embryos, and so on, none of which traditional mechanical physics had ever been up to the task of describing, and all of which, by the second half of the seventeenth century, had risen to the top of the list of phenomena in need of explanation in scientific terms. Leibniz managed to contribute more to some of these fields than others. Here we shall consider his contributions to chemistry, biology, geology, and ethnography (as these terms are understood in the twenty-first century).

Chemistry. In part because it was not easily subsumable into the new mechanical science—which sought to explain everything in terms of mass, figure, and motion alone—chemistry continued to be inflected by the mystical ideas associated with alchemy well into the seventeenth century. Many mystical thinkers made important discoveries in chemistry, such as Jean-Baptiste van Helmont, the first scientist to correctly describe the motion of gas.

Leibniz was careful to maintain at least a public distance from the alchemists, but it has been proven that he was involved in an alchemical society in 1666–1667, and statements he made at least through 1698 indicate an enduring belief in the possibility of manufacturing gold. He believed in this possibility for sound reasons: as he writes in the *Protogaea,* "material, which is everywhere identical with itself, can take on any form, since there are no ultimate, non-interchangeable elements" (§ 3]). He himself had seen evidence of radical transformations of physical substances, of the transformation of urine into phosphor, for example.

Leibniz inherited from van Helmont the alchemical view that bodies possess a lightest distillation fraction, which could be obtained through a process of chemical sublimations and which constitutes the core of that entity's corporeal being. He transforms this to serve his own doctrine that no corporeal substance can ever be taken out of existence, but always remains in some subtle or reduced form. Thus he writes in explicitly alchemical terms: "We shall put off the body, it is true, but not entirely; and we shall retain the most subtle part of its substance (quintessence), in the same way as chemists are able to sublimate a body or mass" (*Otium hannoveranum,* 411 R 164fn.)

Leibniz is intent to reject the Kabbalistic theory that the spirit of a being may be located in some very hard bone in the body, the *luz* in Hebrew. He prefers the alchemical doctrine, according to which there always remains some core or *flos* of a corporeal being, which may be resurrected at any time (hence Leibniz's interest in experiments on insect palingenesis), but that this core cannot be identified with some particular, indestructibly hard part. Such a view would be too close to traditional atomism, and Leibniz imagines instead that "the clothing or covering" of the body "is in constant fluctuation, and at one time is evaporated, at another is again enlarged by the air or by food" (*Sämtliche Schriften und Briefe* II 1: 118f).

Biology. Palingenesis, or the regeneration of supposedly dead animals, is an example of a problem of largely theological interest that was long studied empirically by the alchemists. It was also, clearly, a biological problem, but biology did not exist as an independent science, and indeed would not for quite some time.

If biology was not an independent discipline, this does not indicate an absence of scientific interest in the phenomena of living nature. Leibniz, in particular, was intensely interested in the problems of organic structure and the origins, development, and motion of living bodies. Even if he habitually denied that he borrowed any ideas directly from the research of the microscopists Anton van Leeuwenhoek, Jan Swammerdam, Marcello Malpighi, and others, insisting instead that his metaphysical views were derived from higher principles, nonetheless he often credits these researchers for having empirically corroborated what he knew to be true on a priori grounds. Swammerdam's discovery of insect metamorphosis, and Leeuwenhoek's of the spermatozoon, seemed to Leibniz to confirm the view that no substance ever exists in a fully non-corporeal state, and that all apparent generations and destructions of corporeal substances are in fact just radical transformations.

Leibniz's theory of corporeal substances is a theory of nested individuality, according to which there are individual substances constituting the organic bodies of other individual substances. "Every animated thing," Leibniz writes to Antoine Arnauld in a letter of 30 April 1687, "contains a world of diversity in a true unity." This view of organic structure—and Leibniz is the first thinker in history to distinguish between organism on the one hand, which he sees as parts within parts to infinity, and mechanism on the other, understood as any structure decomposable in a finite series of steps—also appears to be inspired by the microscopic discovery that what look to be individuals often are but colonies of smaller individuals. In Aristotle's metaphysical biology, there had been a basic conviction that where there is one organic body, there is only one substance, such as a horse or a man. For Leibniz, in contrast, in the organic body dominated by the soul of the horse there are infinitely many other souls conspiring, each with its own organic body, and so on without end.

For Leibniz, there could be no lower level at which we arrive at rock-bottom, basic living entities. Cells would not be discovered for some time after Leibniz's death, and in some sense Leibniz's vision excludes the possibility of these biological building blocks. This would not stop some, such as Charles Bonnet in the eighteenth century, from interpreting Leibniz's theory of monads materialistically as an anticipation of the view that every part of an organic body contains the code responsible for the generation of the entire body. Yet Leibniz's theory of nested individuality may be seen as anticipating some trends in biology, to the extent that it calls into question the common-sense view of spatiotemporally separable organisms as the true individuals in living nature, and instead suggests that individuality may be a relative matter, just as today evolutionists identify variously the gene, the organism, and the population each as the unit of selection upon which adaptive forces might work.

Geology. The family of Leibniz's employer, Duke Johann Friedrich von Braunschweig-Lüneberg, gained much of its revenue from the mining of valuable metals. Leibniz thus was able to try his skills as a mining engineer, attempting to develop a system for the extraction of silver ore from the Harz Mountains. This practical activity, combined with another responsibility his employer placed on him—the writing of the history of the House of Brunswick—yielded a major work on geology. Leibniz, wishing to begin his history of the royal family from the very beginning, and seizing on the fact that the House of Brunswick had its own financial interest in understanding how mountains are formed, wrote his speculative *Protogaea,* intended to be the first part of his uncompleted royal history, on, among other things, the evolution of the earth, the formation of continents, oceans, and mountains, the origins of fossils, and so on.

By the early eighteenth century, significant evidence had been accumulated to call into doubt the accuracy of the biblical account of cosmogony. Among the most important evidence were the remains of unknown animal species, and seashells discovered at high altitudes. Some, intent on defending the traditional account, argued that these were tricks of the devil, while a naturalistic but still creationist account had it that these were "forms" imposed by astral influx and seared into receptive matter. Leibniz was cautious not to overtly deny the biblical account, but nonetheless sought a consistently naturalistic way of accounting for puzzling natural phenomena such as these. He argues that "fishes expressed in slate are from true fishes, and this proves that they are not tricks of nature" (§ 20]). Leibniz believes that under certain circumstances enclosed soil can be "cooked" within the Earth as in an oven, and rapidly turned into stone. If animal remains happen to be trapped within, these will turn into fossils. Thus Leibniz correctly discerns the source of the fossil remains, but greatly underestimates the length of time required for them to be produced.

Some of his reflections in the *Protogaea* also indicate a grasp of the epistemological problem of scientifically accounting for processes completed in the distant past, a problem that affects paleontology, cosmology, geology, and archaeology equally. Leibniz believed that fossil evidence could be used together with what was currently known of mechanics and chemistry in order to arrive at a plausible account of the earth's history. While in some respects speculative, Leibniz's contribution to the earth sciences is nonetheless noteworthy for his consistent effort to stay within the bounds of the demonstrable, even when the subject in question makes this difficult. While Leibniz continues to presume that a deluge early on changed the face of the Earth, in many passages the *Protogaea* is also an early example of the uniformitarian approach to geological processes that would gradually come to be favored over the cataclysmic approach.

Ethnography. As with geology, in the early modern period significant new ethnographic evidence was rapidly being accumulated that seemed to dispute the biblical account of origins. On the one hand, new discoveries, particularly in the Americas, made it increasingly difficult to believe that enough time had elapsed from the biblical creation for human beings to wander so far from the presumably Near Eastern Garden of Eden, and to change so much with respect to physical appearance. On the other hand, increasing awareness of the technological achievements of other civilizations, particularly the Chinese, made it increasingly difficult to believe that the revealed truth of Christianity gave Europeans any greater access to scientific truth than their pagan neighbors enjoyed. Interest was also piqued by the fact that some cultures, such as the Chinese, the ancient Persians, and the Mexica (Aztecs) also had alternative chronologies of world history that placed the origin much further back than the Old Testament had it. In view of these problems, certain libertines advocated a doctrine of multiple creations, holding that revealed scripture was only of relevance for those created from Adam and Eve.

Gottfried Leibniz. GEORGE BERNARD/SCIENCE PHOTO LIBRARY.

Leibniz was very interested in the Jesuit reports from China, and in the speculations of the missionaries as to the nature and origins of Chinese science and technology. Many Jesuits believed that the Chinese had strayed from the Near East long ago, and they were seen to have a developed legal system and sophisticated machines, but no understanding of the principles underlying either of these. Confucianism was thus portrayed as a system of laudable rules, the reasons for which had been forgotten in the flow of centuries. Early on, Leibniz too entertained the strong monogenetic hypothesis that all human beings have a Middle Eastern pedigree. He writes in the mid-1670s, for example: "Whether the Chinese are from the Egyptians or the latter from the former, I dare not say; certainly the similarity of their institutions and hieroglyphics along with their shared kind of writing and

Leibniz

philosophizing suggests that they are consanguineous peoples (*Sämtliche Schriften und Briefe,* IV i 270–1).

Later, Leibniz grows increasingly agnostic as to the origins of Chinese civilization, but also grows thoroughly convinced of the innate capacity of the Chinese to arrive at the same basic truths that revealed theology would have us believe could only come from genealogical connection to Christ. In the *Discourse on the Natural Theology of the Chinese,* still unfinished at his death in 1716, he writes that the Chinese, unlike the English, who are inadvertently lapsing into paganism by reintroducing a meddling God or a world soul into nature, are right to "reduce the governance of Heaven and other things to natural causes and distance themselves from the ignorance of the masses, who seek out supernatural miracles" ([§ 2]). Here, it seems almost that Leibniz believes that precisely the isolation of the Chinese from the scriptural tradition, and their consequent need to rely on nature alone for their understanding of the divine, is itself the fortunate cause of their theological superiority to the English.

Ultimately, Leibniz sides with the Jesuits in the so-called "rites controversy," in which the pope insisted against the missionary sect that the Chinese could not continue their traditional ancestor worship while identifying themselves as Christians. Leibniz and the Jesuits believed that Chinese ritual did not imply any particular theological convictions or other, and thus that it was compatible with Christian dogma. The position Leibniz takes up on this issue reveals a sharp understanding of the nature of cultural distinctiveness and of the still problematic question of the boundary between culture and religion. As with biology, anthropology was not a scientific discipline in Leibniz's time, and he had no systematic approach to it. But many of his ideas in this area were subtle and prescient.

SUPPLEMENTARY BIBLIOGRAPHY

WORKS BY LEIBNIZ

Otium hannoveranum, sive Miscellanea. 2nd edition. Edited by J. F. Feller. Leipzig: G. G. Leibnitii, 1737.

Sämtliche Schriften und Briefe. Edited by the Prussian Academy of Sciences (later German Academy of Sciences). Darmstadt: O. Reichl, 1923–1999. Among important recent editions of Leibniz's writings, the most recent volume in the Akademie edition, begun by the Prussian Academy in 1923 and still underway, of Leibniz's writings deserves first mention. The fourth volume of series six (dedicated to the philosophical writings) appeared in 1999. This volume itself consists in four separate volumes, and is well over a thousand pages long.

Oeuvres de Leibniz. 7 vols. Edited by A. Foucher de Careil. Paris: Firmin Didot, 1861–1865. Reprint, Hildesheim: Olms, 1969.

Philosophical Essays. Edited and translated by Roger Ariew and Daniel Garber. Indianapolis, IN: Hackett, 1989. An edition of Leibniz's English writings, very useful for instructional purposes.

De Summa Rerum: Metaphysical Papers, 1675–1676. Edited and translated by G. H. R. Parkinson. New Haven, CT: Yale University Press, 1992. A series of bilingual editions (English along with Latin, French or German), organized thematically.

Protogaea: de l'aspect primitif de la terre et des traces d'une histoire très ancienne que renferment les monuments mêmes de la nature. Edited and translated by Bertrand de Saint-Germain. Toulouse: Presses Universitaires du Mirail, 1993. French-Latin bilingual edition of Leibniz's geological treatise.

The Labyrinth of the Continuum: Writings on the Continuum Problem, 1672–1686. Edited and translated by Richard T. W. Arthur. New Haven, CT: Yale University Press, 2001.

Discours sur la théologie naturelle des Chinois. Edited and translated by Wenchao Li and Hans Poser. Frankfurt: Vittorio Klostermann, 2002. Published as *Discourse on the Natural Theology of the Chinese.* Translated with an introduction by Henry Rosemont Jr. and Daniel J. Cook. Honolulu: University Press of Hawaii, 1977. Leibniz's treatise on China.

Confessio Philosophi: Papers concerning the Problem of Evil, 1671–1678. Edited and translated by Robert C. Sleigh Jr. New Haven, CT: Yale University Press, 2004.

The Leibniz-Des Bosses Correspondence. Edited and translated by Brandon Look and Donald Rutherford. New Haven, CT: Yale University Press, 2007.

OTHER SOURCES

Aiton, Eric J. *Leibniz: A Biography.* Boston: A. Hilger, 1985. A good biography chronicling many of the details of Leibniz's life and work.

Antognazza, Maria Rosa. *Leibniz on the Trinity and the Incarnation: Reason and Revelation in the Seventeenth Century.* Translated by Gerald Parks. New Haven, CT: Yale University Press, 2008. Intellectual biography.

Ariew, Roger. "Leibniz on the Unicorn and Various Other Curiosities." *Early Science and Medicine* 3 (4, 1998): 267–288. Leibniz's activity in geology and paleontology.

Brown, Stuart. "Some Occult Influences on Leibniz's Philosophy." In *Leibniz, Mysticism, and Religion,* edited by Allison P. Coudert, Richard H. Popkin, and Gordon M. Weiner. Boston: Kluwer, 1998.

Coudert, Allison P. *Leibniz and the Kabbalah.* Dordrecht: Kluwer, 1995. For Leibniz's connection to the Kabbalists.

Duchesneau, François. *Les modèles du vivant de Descartes à Leibniz.* Paris: Vrin, 1998. Leibniz's place in the history of biology.

Nachtomy, Ohad, Ayelet Shavit, and Justin Smith. "Leibnizian Organisms, Nested Individuals, and Units of Selection." *Theory in Biosciences* 121 (2002): 205–230.

Perkins, Franklin. *Leibniz and China: A Commerce of Light.* Cambridge, U.K.: Cambridge University Press, 2004. Excellent study of Leibniz's Sinological activity, as well as some insight into his ethnographic ideas in general.

Phemister, Pauline. *Leibniz and the Natural World: Activity, Passivity, and Corporeal Substances in Leibniz's Philosophy.* Dordrecht: Springer, 2005.

Ross, George MacDonald. "Leibniz and Alchemy." *Studia Leibnitiana Sonderheft* 7 (1978): 166–77. Explores relation of Leibniz to the alchemists.

Smith, Justin E. H., ed. *The Problem of Animal Generation in Early Modern Philosophy.* Cambridge, U.K.: Cambridge University Press, 2006.

Justin E. H. Smith

LELOIR, LUIS FEDERICO

(*b.* Paris, France, 6 September 1906; *d.* Buenos Aires, Argentina, 2 December 1987), *biochemistry, nucleotide sugars, oligo- and polysaccharide synthesis.*

Leloir received the 1970 Nobel Prize in Chemistry for his discovery of nucleotide sugars and the role of those compounds in the interconversion of monosaccharides and as precursors in the synthesis of oligo- and polysaccharides. His many contributions to biochemistry also include the first description of fatty acid oxidation in a cell-free system, of angiotensinogen and its conversion to angiotensin upon incubation with rennin, and the role of lipid-bound saccharides as intermediates in protein glycosylation.

Medical and Biochemical Training. Luis F. Leloir was born in Paris, France. His Argentine parents had traveled to the French capital seeking medical treatment for the father, a lawyer who never practiced his trade. Two years later his mother (his father died before Leloir's birth) took him to Buenos Aires. He studied medicine at the University of Buenos Aires, becoming a medical doctor in 1932. From 1932–1934 he was an internist at the University Hospital while doing the experimental work for his doctoral thesis under the supervision of Bernardo A. Houssay. Houssay was director of the Institute of Physiology of the University of Buenos Aires and had performed remarkable studies on the role of the hypophysis on carbohydrate metabolism for which he received the Nobel Prize in Physiology or Medicine in 1947. Leloir's dissertation, "The Adrenal Gland and Carbohydrates," received the 1934 School of Medicine award for the best thesis. In 1935 Leloir went to Sir Frederick Gowland Hopkins's Biochemical Laboratory at the University of Cambridge for postdoctoral training in biochemistry. In the United Kingdom, Leloir became familiar with current biochemical techniques.

Fatty Acid Oxidation and Hypertension. Back in Buenos Aires, Leloir resumed his work at the Institute of Physiology and tackled two projects (fatty acid oxidation and hypertension) in the years from 1937 to 1942. With Juan M. Muñoz he studied the metabolism of ethanol and fatty acids. In a series of papers, they reported that a particulate fraction obtained from liver homogenates, if supplemented with a tetracarbon dicarboxylic acid, cytochrome C, and adenosine monophosphate, was able to sustain oxidation of fatty acids. This was a remarkable contribution, as until then it had been believed that the process required cell integrity. The lack of a refrigerated centrifuge nearly frustrated the successful preparation of the particulate fraction, but Leloir's craftsmanship, one of his strengths, provided an ingenious and money-saving solution to the obstacle as he wrapped several car inner tubes filled with freezing mixture around an old pulley-driven centrifuge. His practical ingenuity was extremely useful in a country such as Argentina, where research funds were often scarce.

The second problem tackled by Leloir in those years was that of malignant renal hypertension. His study of it was carried on in collaboration with Muñoz, Eduardo Braun Menendez, and Juan C. Fasciolo. It was then known that constriction of the renal artery of dogs resulted in permanent hypertension. Fasciolo then showed that an increase in blood pressure also resulted from grafting a constricted kidney into a normal dog, thus indicating that the effect was due to a substance that the treated kidney secreted to the blood. One early finding of the group was that an aqueous acetone extract of a constricted kidney was able to produce a transient increase in blood pressure. The aqueous acetone-soluble substance was different from renin, an already-known pressor substance that could also be extracted from kidneys. Leloir and coworkers found then that incubation of renin (a protease) with blood plasma (that contained what is now called angiotensinogen) produced the pressor substance soluble in aqueous acetone (now called angiotensin). Leloir's knowledge of biochemistry was fundamental in these findings.

Nucleotide Sugars. Argentina experienced a military coup d'état in Argentina in 1943. Soon after, Houssay, together with many other important personalities, sent a public letter to the authorities demanding a return to "constitutional normality, effective democracy, and American solidarity." "American solidarity" was a euphemism for siding Argentina with the Allies in World War II. The new government, which had a certain sympathy with the Axis, reacted by dismissing all signers that were public employees. Houssay, as a university professor, fell into this category. His dismissal was followed by the resignation of most of the scientific staff of the Institute of Physiology,

The initial members of the Instituto de Investigaciones Bioquímicas Fundación Campomar. *From left are Caputto and Paladini (using a home made Warburg apparatus for measuring oxygen consumption), Cardini, Trucco and at the extreme right is Leloir opening a refrigerator (ca. 1948).* COURTESY INSTITUTO LELOIR FUNDACION.

Leloir among them, in solidarity with Houssay. Leloir then traveled to the United States, where he first worked with Ed Hunter on the formation of citric acid at Carl and Gerty Coris's laboratory at Washington University in St. Louis and then with David Green (with whom he had already worked in Cambridge) at Columbia University in New York City on the separation of aminotransferases. In 1943 Leloir married Amelia Zuberbuhler. The outcomes of this happy marriage were a daughter and nine grandchildren.

Returning to Argentina, Leloir in 1945 affiliated with the Institute of Experimental Biology and Medicine, which had been recently inaugurated in Buenos Aires to host Houssay and most of the former members of the Institute of Physiology. Leloir then started recruiting collaborators, the first of which was Ranwel Caputto, a medical doctor who, like Leloir, had received postdoctoral training in biochemistry with Malcolm Dixon at the University of Cambridge. Raúl Trucco, a microbiologist, was invited next, as the intention was to study fatty acid oxidation in bacteria. Carlos Cardini and Alejandro Paladini soon joined the group. At that time Jaime Campomar, a textile industrialist, had decided to fund a new institution dedicated to biochemical research and approached Houssay for advice on choosing its director. Houssay quickly suggested Leloir for the job, and the latter was to be the director of the Instituto de Investigaciones Bioquímicas Fundación Campomar (Institute for Biochemical Research Campomar Foundation; now called the Fundación Instituto Leloir [Leloir Institute Foundation]) for the rest of his life. The institute was inaugurated in November 1947 in a small, one-story, and rather old house that was adjacent to the institute directed by Houssay. Figure 1 depicts the initial research group at the central patio of the institute.

Because experiments on fatty acid oxidation produced dubious results, the group decided to switch its efforts to studying the synthesis of lactose (a disaccharide composed of galactose plus glucose, Gal and Glc, respectively). Caputto claimed that during his thesis work he had been able to synthesize the disaccharide on incubating glycogen (see below) with a mammary gland extract. As those results could not be repeated, Leloir suggested studying instead lactose degradation, as this process might

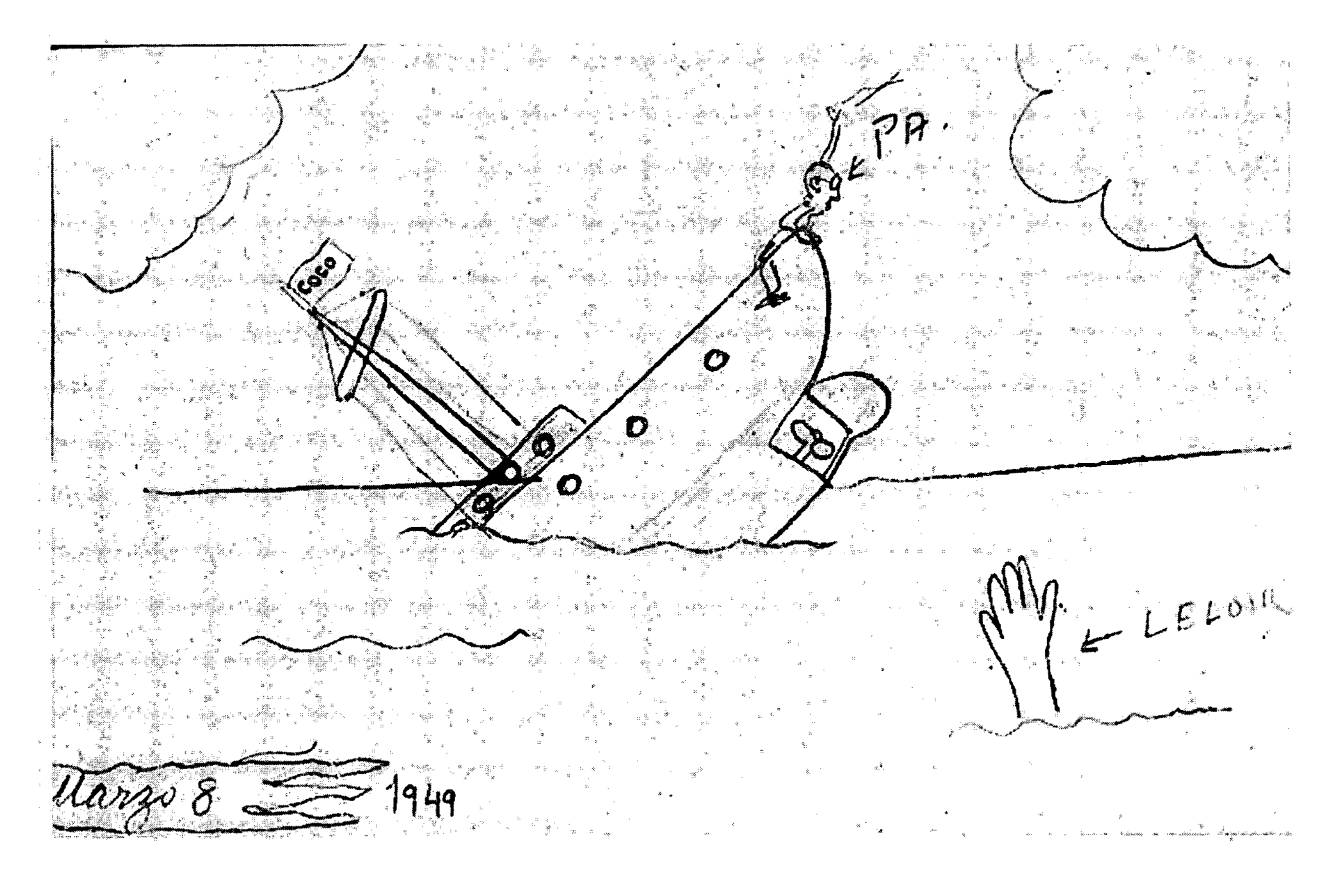

Cartoon drawn by Leloir. *Cartoon drawn by Leloir on 8 March 1949, showing the profound despair of the group at their up to then inability to characterize the structure of the second cofactor (CoCo, now UDP-Glc). PA stands for Paladini, the youngest member of the team.* COURTESY INSTITUTO LELOIR FUNDACION.

provide information about the synthetic pathway. This prediction proved to be correct. In hindsight, it may be speculated that what Caputto had observed was probably the formation of maltose (another disaccharide but containing two glucoses) by amylolytic degradation of the polysaccharide. Identification of disaccharides was at that time based on unreliable methods, as the morphology of the crystals of derivatives formed upon reaction with phenylhydrazine (osazones).

For the planned research, the group decided to use a cell-free extract derived from a yeast (*Saccharomyces fragilis*) adapted to grow on lactose as a carbon source. The scientists first detected a lactase that degraded the disaccharide to its monosaccharide constituents and then a galactokinase that phosphorylated galactose to galactose 1-phosphate (Gal 1-P). Conversion of the last compound to glucose 6-P (Glc 6-P) required two unknown thermostable factors: glucose 1,6 diphosphate (Glc 1,6 diP) for the conversion of glucose 1-P (Glc 1-P) to Glc 6-P, and uridine diphosphate glucose (UDP-Glc) for the conversion of Gal 1-P to Glc 1-P. Determining the structure of UDP-Glc was a real tour de force, given the minimum amounts of reagents and equipment available. They determined that the compound had one Glc residue per two phosphates and that it absorbed in the ultraviolet light, but the absorption spectrum was that of an unknown substance (only spectra of adenosine-containing compounds were known by then). One day Caputto came to the institute in an excited state, carrying the last issue of the *Journal of Biological Chemistry* depicting in one of the articles the spectrum of uridine, which coincided with that of the new substance. Figure 2 is a cartoon drawn by Leloir showing the mood of the group before solving the structure of UDP-Glc.

The so-called Leloir's Pathway may be then represented as:

$$\text{Gal} + \text{ATP} \rightarrow \text{Gal 1-P} + \text{ADP}$$

$$\text{Gal 1-P} + \text{UDP-Glc} \rightarrow \text{UDP-Gal} + \text{Glc 1-P}$$

$$\text{UDP-Gal} \rightleftarrows \text{UDP-Glc}$$

$$\text{Glc 1-P} \rightarrow \text{Glc 6-P}$$

where UDP-Gal stands for uridine diphosphate galactose.

UDP-Glc was the first nucleotide sugar to be described, and the pathway shows the first role of these novel compounds, that of being involved in monosaccharide interconversion. In the case described above, conversion of UDP-Gal to UDP-Glc was shown to require NAD (nicotinamide adenine dinucleotide) as inversion of the OH group in C_4 proceeds by an oxidoreduction reaction. Furthermore, description of the pathway provided a full explanation of galactosemia, a human congenital disease that is characterized by the inability to metabolize galactose. Most patients are deficient in the enzyme involved in the second reaction depicted above, whereas in a milder form of the disease, galactokinase is the deficient enzyme.

As reported also by Leloir and collaborators, the general pathway for nucleotide sugar synthesis can be represented as:

$$\text{NTP} + \text{monosaccharide 1-P} \rightarrow \text{NDP-monosaccharide} + \text{PP}$$

where NTP stands for nucleoside triphosphates (ATP, UTP, GTP, and CTP) and PP for pyrophosphate. Of the nearly one hundred nucleotide sugars now known, several of them (GDP-Man, UDP-GlcNAc, UDP-GalNAc, and ADP-Glc) were first described by Leloir and co-workers.

The question as to whether UDP-Glc had an additional role beside that of being involved in monosaccharide interconversion was raised because UDP-Glc was detected also in yeast strains unable to use galactose for growth. The method then used for UDP-Glc quantification (acceleration of Gal 1-P to Glc 1-P transformation) provided evidence for a second role of nucleotide sugars, that of being intermediates in monosaccharide transfer reactions. The disappearance of UDP-Glc incubated with a yeast extract proceeded at a higher rate in the presence of added Glc 6-P. This effect was quickly traced by Leloir and a new collaborator, Enrico Cabib, to the formation of treahalose 6-P. (Trehalose is a disaccharide composed of two glucoses, different from maltose.) The reaction can be described then as:

$$\text{UDP-Glc} + \text{glucose 6-P} \rightarrow \text{trehalose 6-P} + \text{UDP}$$

Similarly, soon after Leloir described the synthesis of sucrose 6-P (sucrose is a sophisticated name for our daily sugar, composed of two monosaccharides, Glc and fructose), using wheat germ extracts:

$$\text{UDP-Glc} + \text{fructose 6-P} \rightarrow \text{sucrose 6-P} + \text{UDP.}$$

Curiously, synthesis of lactose, which was the first aim of the project, was described not by Leloir and his coworkers but in 1961 by Winifred M. Watkins and W. Z. Hassid, working with mammary gland extracts ("The Synthesis of Lactose by Particulate Enzyme Preparations from Guinea Pig and Bovine Mammary Glands"). It proceeds as follows:

$$\text{UDP-Gal} + \text{Glc} \rightarrow \text{lactose} + \text{UDP}$$

Glycogen is a reserve polysaccharide constituted by many glucose units present in a variety of organisms, from bacteria to mammals. Plants have, additionally, another polysaccharide (starch) structurally closely related to glycogen. In the late 1930s and early 1940s, Carl and Gerty Cori, in a series of articles, described the incubation of Glc 1-P with a mammalian cell extract resulting in glycogen synthesis according to the following (reversible) reaction:

$$\text{Glc 1-P} + (\text{Glc})_n \rightleftarrows (\text{Glc})_{n+1} + \text{P}$$

where $(\text{Glc})_n$ stands for glycogen containing n Glc molecules. The participating enzyme (glycogen phosphorylase) was crystallized. For many years this was taken as the pathway of glycogen synthesis, although some conflicting reports soon appeared. For instance, the application of adrenaline to animals, which resulted in phosphorylase activation, led to glycogen degradation, not to glycogen synthesis. Moreover, the P levels in live tissues suggested that the equilibrium was displaced from right to left in the above equation. In 1957 Leloir and Cardini reported in "Biosynthesis of Glycogen from Uridine Diphosphate Glucose" that incubation of UDP-Glc with a mammalian cell extract resulted in the synthesis of glycogen. The synthetic reaction was then:

$$\text{UDP-Glc} + (\text{Glc})_n \rightarrow (\text{Glc})_{n+1} + \text{UDP.}$$

The activity of the enzyme involved (glycogen synthetase) was found to be highly regulated to allow accumulation of the polymer in times of plenty and its degradation in times of need.

Changing Times: 1958–1970. Jaime Campomar died in 1956, and as his heirs decided to discontinue funding the institute, Leloir almost decided to close it for good. Fortunately, however, a series of events helped to dissuade Leloir from taking such a drastic decision. First, Leloir applied for and received a generous grant from the National Institutes of Health in the United States. Second, the government of General Juan Peron, which was not friendly to higher education and research, was deposed in 1955. The new government reinstated the autonomy of public universities and created the National Research Council, an institution that provided funds for research and full-time positions. Houssay became its first president and held that post until his death in 1971. The building that housed Leloir's institute was almost completely dilapidated. (He had constructed with his own hands a series of interior channels to prevent leaking rainwater from damaging journals and books in the library.) The post-Peronist government, however, offered Leloir a much larger building (a former nun's school), and both he and Houssay moved their institutes into it. In addition, a fruitful association between Leloir's Instituto de Investigaciones Bioquímicas Fundación Campomar and the School of Sciences of the University of Buenos Aires was initiated in 1958. Leloir was designated research professor of the university's School of Sciences.

Leloir devoted the first years in the new building to the study of the regulation of glycogen synthetase by metabolites (mainly by Glc 6-P) and by interconversion between active and inactive forms triggered by phosphorylation and dephosphorylation. He also determined that the precursor in the synthesis of starch was not UDP-Glc but a new nucleotide sugar, adenosine diphosphate glucose (ADP-Glc). As UDP-Glc showed a poor incorporation of glucose into the polysaccharide in cell-free assays, he tried several synthetic nucleotide sugars and found that ADP-Glc was by far the best precursor. He then isolated the compound from natural sources (sweet corn). Another problem that caught Leloir's attention was the synthesis of high molecular weight (particulate) glycogen. He found that the polysaccharide synthesized in the test tube either by glycogen phosphorylase from Glc 1-P or by glycogen synthetase from UDP-Glc was differentially decomposed by a series of physical and chemical treatments. The fact that the polysaccharide formed by the synthetase from UDP-Glc in the test tube showed decomposition features identical with those of the native molecules was a definitive demonstration that UDP-Glc was the true glycogen precursor *in vivo.*

The Nobel Prize and Afterward. On 20 October 1970 the Swedish Academy of Sciences announced the awarding of the Nobel Prize in Chemistry to Luis F. Leloir for his discovery of sugar nucleotides and their role in the biosynthesis of carbohydrates. By then he had already begun what was going to be his last great contribution to biochemistry.

Phillips Robbins at the Massachusetts Institute of Technology and Jack Strominger at Harvard University had, in the mid-1960s, determined that lipid bound mono- and oligosaccharides behaved as biosynthetic intermediates between nucleotide sugars and several polysaccharide components of the bacterial cell wall. The lipid moiety was identified as a polyprenol (undecaprenol) phosphate. Leloir then described how the incubation of rat liver membranes with UDP-Glc resulted in the formation of dolichol-P-Glc, dolichol being a polyprenol containing 20-21 isoprene units in mammalian cells. Further incubation of dolichol-P-Glc led to the transfer of the monosaccharide to a compound tentatively identified as dolichol-P-P-oligosaccharide (later shown to be $GlcNAc_2Man_9Glc_3$). Further incubation of this last lipid derivative with the membranes resulted in the transfer of the whole oligosaccharide to proteins. Leloir thus established the basis of the pathway leading to the synthesis of glycoproteins in eukaryotic cells and also provided the first evidence indicating that the protein-linked oligosaccharide was processed (i.e., that monosaccharides were both removed from it and added to it).

In 1978 the mayor of the city of Buenos Aires donated land and formed and presided over a committee to gather funds for constructing a new building. At the end, equal amounts of public and private funds were received. The move to the new premises took place in December 1983. Leloir was to continue doing research there for four years.

Leloir the Man. Leloir belonged to an old and wealthy Argentinean family, a circumstance that allowed him fully to devote his time to basic research when funding for such endeavors was almost nonexistent in the country. He personally covered funds for most subscriptions of scientific journals received at the Institute's library and fully donated his salary as research professor of the University of Buenos Aires to the Institute. Leloir had a low profile personality; he shunned public exposure and displayed an extremely subtle and exquisite sense of humor. Leloir was a hard worker who, until a few years before passing away, had continued working at the bench. Research was for him the best of hobbies. (He never had a private office, instead receiving visitors and doing institute paperwork at the laboratory.) He was extremely courteous and treated everybody, independently of their social position, in the same unassuming way. He had a calm personality and showed a degree of discomfort only when somebody behaved rudely or showed bad manners. Leloir's personality greatly contributed to the substantial pleasure of working in his group or his institute.

BIBLIOGRAPHY

WORKS BY LELOIR

With Carlos E. Cardini. "Biosynthesis of Glycogen from Uridine Diphosphate Glucose." *Journal of the American Chemical Society* 79 (1957): 6340-6341.

"Two Decades of Research on the Biosynthesis of Saccharides." *Science* 172 (25 June 1971): 1299–1303. The text of Leloir's Nobel lecture.

"Biosynthesis of Polysaccharides Seen from Buenos Aires." In *Biochemistry of the Glycosidic Linkage*, edited by Romano Piras and Horacio G. Pontis. New York: Academic Press, 1972.

"Far Away and Long Ago." *Annual Review of Biochemistry* 52 (1983): 1–15.

OTHER SOURCES

Cabib, Enrico. "Research on Sugar Nucleotides Brings Honor to Argentinian Biochemist." *Science* 170 (6 November 1970): 608–609.

Cori, Carl F.; Gerty T. Cori; and Albert H. Hegnauer. "Resynthesis of Muscle Glycogen from Hexosemonophosphate." *Journal of Biological Chemistry* 120 (1937): 193-202.

Cori, Gerty T.; Carl F. Cori; and Gerhard Schmidt. "The Role of Glucose-1-Phosphate in the Formation of Blood Sugar and Synthesis of Glycogen in the Liver." *The Journal of Biological Chemistry* 129 (1939): 629-639.

Cori, Gerty T., and Carl F. Cori. "Crystalline Muscle Phosphorylase: IV. Formation of Glycogen." *Journal of Biological Chemistry* 151 (1943): 57-63.

Kresge, Nicole; Robert D. Simoni; and Robert L. Hill. "Luis F. Leloir and the Biosynthesis of Saccharides." *Journal of Biological Chemistry* 280 (13 May 2005): 158-160.

Myrbäck, Karl. "Presentation Speech of the Nobel Prize in Chemistry 1970." In *Les Prix Nobel en 1970*, edited by Wilhelm Odelberg. Stockholm, Sweden: Nobel Foundation, 1971.

Parodi, Armando J. "Leloir, su Vida y su Ciencia." *Ciencia Hoy* 16 (August-September 2006): 23-30.

Watkins, Winifred M., and William Z. Hassid. "The Synthesis of Lactose by Particulate Enzyme Preparations from Guinea Pig and Bovine Mammary Glands." *Journal of Biological Chemistry* 237 (1962): 1432-1440.

Armando J. Parodi

LEMIEUX, RAYMOND URGEL

(*b.* Lac La Biche, Alberta, Canada, 16 June, 1920; *d.* Edmonton, Alberta, Canada, 22 July 2000), *carbohydrate chemistry, synthesis, conformational analysis, NMR coupling constants, stereoelectronic effects, molecular recognition, glycobiology.*

Lemieux was one of the outstanding chemists of the second half of the twentieth century. His contributions spanned a wide area that included stereochemistry, nuclear magnetic resonance (NMR)-based methods for structural and configurational assignments, synthetic methods and molecular recognition. Lemieux's insight into chemical problems resulted in numerous seminal discoveries such as the first synthesis of sucrose, the correlation of vicinal coupling constants with configuration, the anomeric effect, the first chemical synthesis of the human blood group substances, and numerous contributions to molecular modeling and molecular recognition of carbohydrates by proteins. His work influenced organic chemistry extensively, and was a determining factor in converting carbohydrate chemistry from an academic specialization to one of great practical importance in chemistry, biology and medicine. Lemieux's influential role was recognized when, with twenty-one world-renowned chemists, he was invited by the American Chemical Society to contribute to the autobiographical series, "Profiles, Pathways and Dreams," which documents the development of modern organic chemistry through the research careers of chemists who made seminal contributions to the field. Lemieux's contribution, *Explorations with Sugars, How Sweet It Was,* is an excellent account of his research from 1946–1990.

Lemieux was widely recognized for his scientific achievements. Amongst his most notable awards were the Gairdner Foundation International Award (1985), the Rhône-Poulenc Award of the Royal Society of Chemistry (1989), the King Faisal International Prize in Science (1990), and the Wolf Prize for Chemistry (1999).

Family Background and Formative Development. Raymond Urgel Lemieux was born on 16 June 1920, in the small prairie community of Lac La Biche (population 75–300 during the 1921–1931 period), two hundred kilometers northeast of Edmonton, Alberta. Both his parents traced their roots to France. His mother died when he was seven and sometime later the family moved from Lac La Biche to Edmonton (the provincial capital of Alberta, with a population in 1935 of about 80,000). Lemieux was a good student and began his education in the fall of 1939 at the University of Alberta, Edmonton. In the spring of 1943 he graduated and later the same year he left Edmonton for Montreal, Quebec (a three-day train trip) and McGill University, where he registered for graduate studies with Clifford Purves at the Pulp and Paper Research Institute of Canada.

By 1946 Lemieux had completed studies for his doctoral dissertation and the possibility of postdoctoral studies attracted him. He had become completely entranced by stereochemistry and the discovery that the antibiotic streptomycin was a carbohydrate was of great interest. When he discovered that research on streptomycin was going on in the laboratory of Melville Wolfrom, Lemieux applied for postdoctoral studies in the famous carbohydrate group at Columbus, Ohio. The move to Ohio held far greater significance, as it was at Ohio State University that Raymond met Virginia McConaghie, who was studying for her PhD degree in high-resolution infrared spectroscopy. They were married in New York City in 1948, and over the ensuing years in locations from Saskatoon, Saskatchewan to Ottawa, Ontario and back to Edmonton, they raised five daughters and one son.

Early Research. At Ohio State University Lemieux was involved in the structural elucidation of streptomycin. He also became fascinated by the configurational correlation of sugars and amino acids. Techniques he was using to elucidate the structure of streptomycin appeared to be well suited to the controlled degradation of D-glucosamine to give L-alanine thereby providing a correlation with the relative configuration of D-glyceraldehyde (Figure 1).

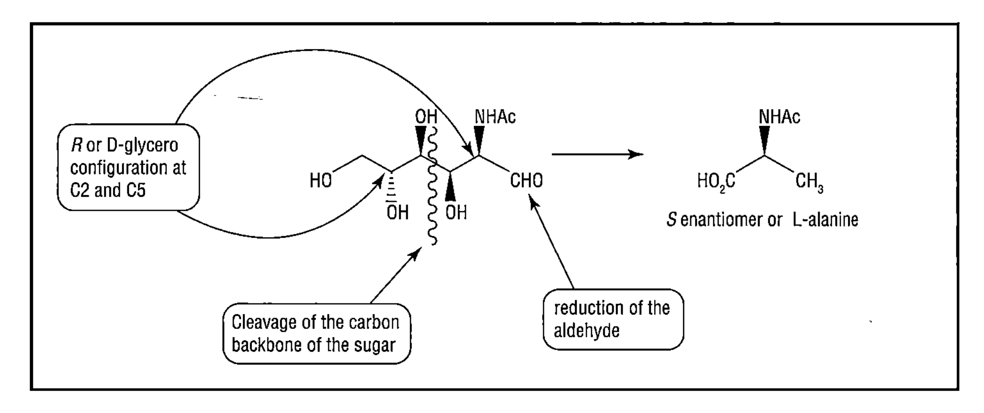

***Figure 1.** Degradation of N-acetyl-**D**-glucosamine with retention of configuration at carbon 2 to give **L**-alanine.*

This work served as a milestone in stereochemistry by linking the stereochemical notation for amino acids with that for sugars.

In 1947 Lemieux became an assistant professor at the University of Saskatchewan, and two years later he joined the National Research Council's (NRC) Prairie Regional Laboratory, also in Saskatchewan. During this period he attracted considerable public and scientific attention with the first rational synthesis of sucrose. During the 1947–1954 period he began his studies and lifelong interest in the chemistry of the anomeric center (the aldehyde or ketone carbon atom of a sugar that is the most reactive and the atom through which one sugar is joined to another to create oligosaccharides and polysaccharides). During this period he attracted considerable public and scientific attention with the first rational synthesis of sucrose.

It was obvious that the methods available at the time for determination of the stereochemistry at the anomeric center were certainly laborious and left a great deal of uncertainty. It was also clear to him that there were special effects in play when it came to the conformational preference of certain pyranose derivatives. These caused large electronegative substituents at C-1 of the pyranose ring to occupy the sterically crowded axial orientation rather than the expected equatorial position (Figure 2). In 1953, however, there was no way to obtain direct evidence for the preferred conformations of such molecules in solution. The resolution of this problem coincided with his move to the University of Ottawa, Ontario in 1954.

The president of the National Research Council of Canada, Edgar W. R. Steacie had strongly urged the young Lemieux to consider a move to Ottawa to help build the faculty and an "atmosphere of research." Lemieux became professor and chairman of the Department of Chemistry at the University of Ottawa in 1954, and served as the vice-dean of the Faculty of Pure and Applied Science. During his tenure he established a flourishing research environment.

Conformational Analysis by NMR Spectroscopy. When Lemieux first heard a presentation on NMR at the National Research Council (NRC) he immediately began to speculate on the steric effects that might influence the chemical shifts of protons attached to the carbon atoms of the pyranose ring. With Rudolf Kullnig (a graduate student in his group) and under the guidance of Harold Bernstein (NRC), the first NMR spectra of sugar acetates were obtained at 40 MHz. The work provided the long-sought definitive assessment of the preferred conformations of the sugar acetates in solution. Expansion of the approach led to the first application of proton NMR (^{1}H NMR) spectroscopy for the establishment of the relative configurations of chiral centers in organic compounds and thus the foundation of the Karplus relationship. In 1958, Lemieux presented this work, prior to publication, in the Karl Folkers lectures at the University of Illinois, where Martin Karplus was in the audience. It was his equations that provided a theoretical basis for the quantitative correlation of three bond coupling constants with torsional angle, one of organic chemistry's most potent stereochemical probes. Karplus later wrote, "Just as I finished the work on vicinal coupling constants, I heard a lecture by R. U. Lemieux on the conformations of acetylated sugars. I do not remember why I went to the talk because it was on organic chemistry. Lemieux reported results for vicinal coupling constants and noted that there appeared to be dihedral angle dependence, although the details of the behavior were not clear. However, it was evident that these experimental results confirmed the theory even before it was published" (Karplus, 1996).

In the same year Lemieux discovered the anomeric effect, the preference of large electronegative constituents at C-1 of the pyranose ring to preferentially adopt the

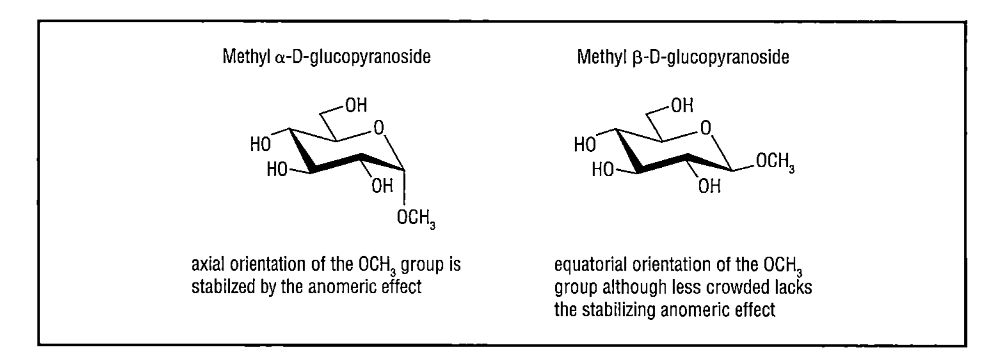

Figure 2. *The anomeric effect favors the axial orientation of electronegative substituents at carbon-1 of the pyranose ring.*

axial orientation, now recognized as a fundamental stereoelectronic phenomenon. The phenomenon extends to acetals in general and to electronegative substituents at the C-2 position of saturated heterocycles. However, it was not until 1971 that Lemieux made a formal publication on this topic in a paper entitled, "Effects of Unshared Pairs of Electrons and Their Solvation on Conformational Equilibria."

In 1961 Lemieux received an offer of a professorship from the University of Alberta in Edmonton. His group in Alberta in the early 1960s undoubtedly represented one of the high points of his career. Several outstanding PhD students and postdoctoral fellows during the 1961–1973 period helped him consolidate an undisputed reputation as a world leader in his field. Key advances in the chemistry of orthoesters, glycals and their nitrosyl chloride adducts represented major contributions during this period. (For details and references to the primary publications, see Lemieux, 1990, and the biographical essay by Bundle, 2003). Against this backdrop of new synthetic chemistry and an increasing understanding of the anomeric effects, the exploitation of ^{1}H NMR spectroscopy to solve conformational and configurational questions was now routinely applied in his group. During this period further studies of conformational equilibria in solution, using both NMR and chiroptical approaches, added to the appreciation of the importance of the anomeric effect in dictating not only the anomeric preference of electronegative substituents, but also the conformation of glycosides (exo-anomeric effect).

Rationale Synthesis of Complex Oligosaccharides. With the increasingly sophisticated understanding of reactions at the anomeric center and the capability to contemplate synthetic targets that few others could consider in the late 1960s, Lemieux turned his attention to the selection of challenging targets. He felt the oligosaccharide chains of glycoproteins and glycolipids could no longer be ignored as these structures carry messages essential for the control of many crucial cellular functions. The study of these new phenomena was critically hampered by the enormous difficulties encountered in trying to obtain even milligram quantities of structurally well-characterized carbohydrates.

The most direct solution was to synthesize the required complex oligosaccharides, but this had not been attempted because of the difficulties involved. In the late 1960s, the synthesis of a disaccharide was considered a major undertaking, and the preparation of more elaborate oligosaccharides must have appeared as an unrealistic project. The successful completion of such a program required at a minimum the development of new glycosylation methods. The controlled synthesis of a glycosidic bond between two sugars was especially challenging at this time, particularly for the stereochemical arrangement where, for glucose and galactose, the two most frequently encountered hexoses, the hydroxyl group at C-2 and the newly formed bond at C-1 are in a 1,2- *cis* relationship. For all practical purposes there was no reliable method to achieve this outcome in reliable yield and high stereoselctivity. In order to be able to follow the outcome of such reactions new methods for the structural analysis of both protected oligosaccharide intermediates and of the final synthetic products were required.

Intrigued by the biological activities of natural occurring glycoconjugates such as the glycolipids and glycoproteins, Lemieux set himself the formidable goal of achieving the synthesis of the human blood groups. This required mastery of reactions at the anomeric center of hexoses. Because the challenges of stereocontrolled glycoside synthesis had been a continuing interest since the synthesis of sucrose in 1953 it was not surprising that Lemieux's laboratory would develop novel methodologies for the stereospecific formation of the glycosidic linkage. These discoveries and advances came to a climax in the early 1970s. For the first time, the synthesis of oligosaccharides with the complexity of the naturally occurring

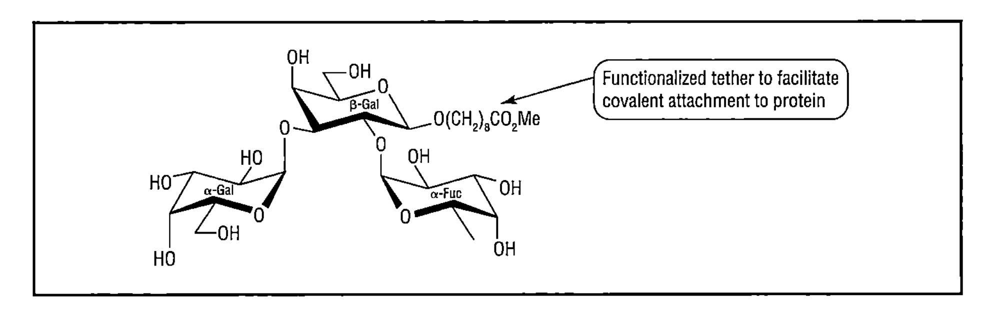

Figure 3. *The structure of human blood group B trisaccharide as derivatized by Lemieux for chemical synthesis of oligosaccharide-protein conjugates.*

structures could be accomplished and their structures confidently confirmed by NMR spectroscopy. These new synthetic reactions included a method for the preparation of α-linked 2-amino-2-deoxy glycosides and later a procedure for the preparation of β-linked 2-amino-2-deoxy glycosides (Lemieux, 1990). Most importantly, the development of the halide-ion glycosylation reaction permitted the synthesis of the hitherto elusive α-glycosidic linkage (Lemieux, Henricks, et al., 1975a). These achievements were reported in four publications (Lemieux et al., 1975a–1975d) dealing with the syntheses of the trisaccharide antigenic determinants of the B and Lewis-a human blood groups (Figure 3).

Synthetic Antigens and Immunoadsorbents. In itself, the laboratory preparation of these antigenic determinants would have been a remarkable achievement. However, Lemieux's insight into the potential utility of these compounds was of such clarity that he foresaw their use as artificial antigens. His syntheses were conducted in such a way that the completed oligosaccharides incorporated a tether to allow covalent attachment to appropriate carrier molecules. He also recognized that attachment to solid supports would provide biospecific adsorbents that would be of exceptional value in medical research. Ideas began to take root for the formation of what would later be called a biomedical start-up or spin-off company. However, Lemieux's conception of a company that combined leading edge chemistry with immunological applications was years ahead of its time, although subsequent and comparable ventures would later emerge in the financially more adventurous climate of the United States. In 1979 the new company, ChemBiomed, was formed but after approximately ten years in search of viable markets it ceased business.

Determination of Oligosaccharide Conformation. The exploitation of NMR to answer stereochemical problems had become a hallmark of Lemieux's publications and with the advent of the pulsed, Fourier-transform technique, fast digital computers and cryomagnets in the mid 1970s, the capabilities of the technique again offered unique opportunities. It was evident that measurement of inter-proton distances by either T_1 measurements or nuclear Overhauser (nOe) experiments was now relatively straightforward and in combination with the estimation of torsional angles, conformational analysis of oligosaccharides was a tangible objective. Lemieux began a collaboration with Klaus Bock (Technical University, Lyngby, Denmark) that laid the foundations for a large body of pioneering studies on the determination of oligosaccharide solution conformation using NMR methods and semi-empirical calculations, based on the HSEA (Hard-Sphere-Exo-Anomeric) algorithm. Lemieux had anticipated that the relative orientation of contiguous sugar residues in oligomeric structures was governed by the exo-anomeric effect. Confirmation of this idea was achieved through the observation of near-invariant *vicinal* coupling constants between the anomeric hydrogen and aglyconic carbon atoms in the nuclear magnetic resonance spectra of appropriately C-13 enriched synthetic model glycosides (Lemieux et al., 1979). The development of a molecular modeling program, the HSEA forcefield, was basically an extension of this work. These calculations were subsequently refined into a computer program that was made widely available during the 1980s. The initial work was largely completed by 1980 and published in a comprehensive paper describing the three dimensional properties of the human blood group oligosaccharides.

Molecular Recognition in Sugar-Protein Complexes. Knowledge of the three-dimensional shapes of oligosaccharides was a prerequisite for appreciation of their biological activities and the availability of these oligosaccharides in quantities sufficient for systematic study was an essential component in the successful research program that evolved during the 1980–1990 decade. These developments allowed Lemieux to apply his discoveries to the human blood-group specific oligosaccharide determinants, including those with specificities designated

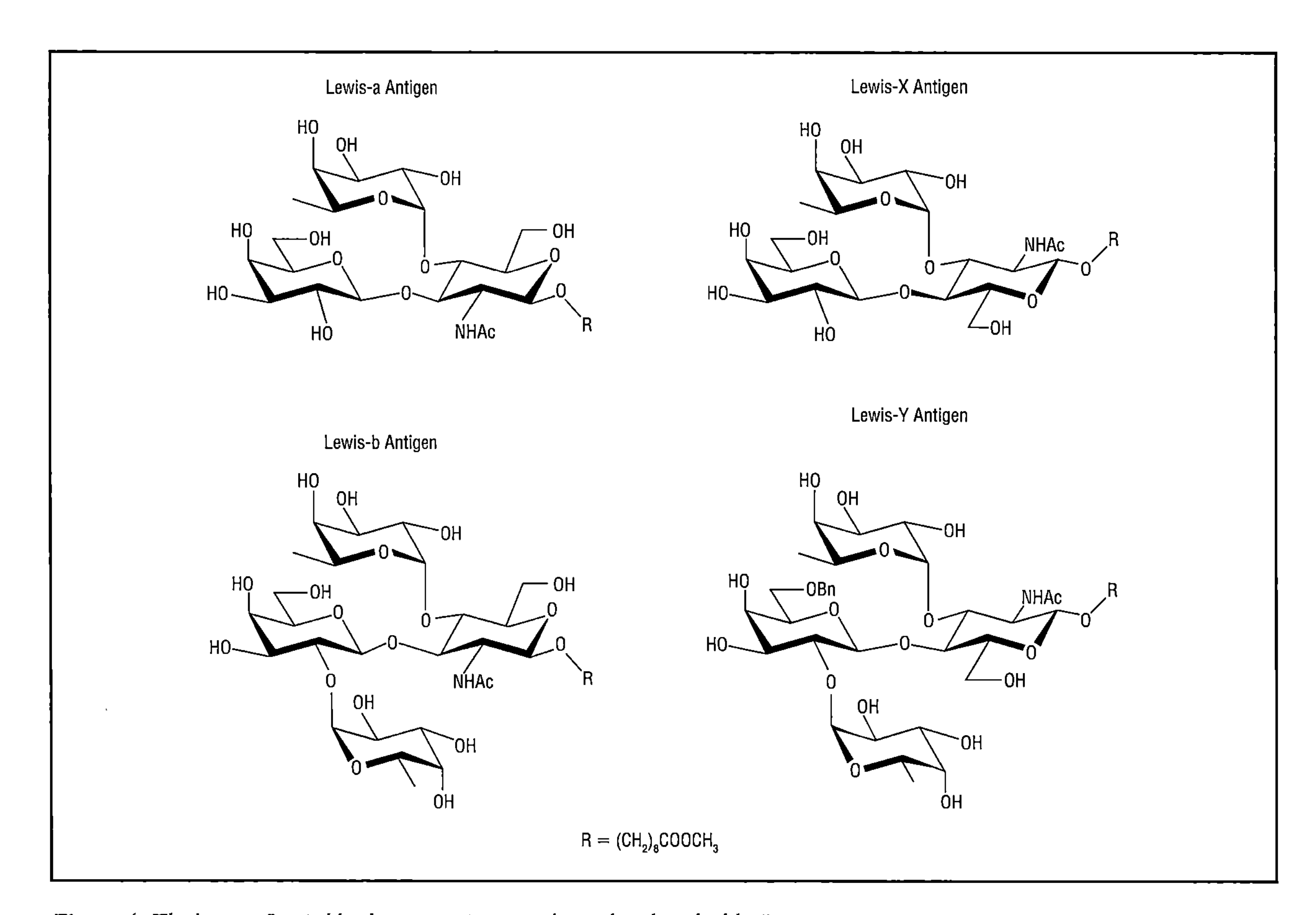

Figure 4. *The human Lewis blood group antigens synthesized and studied by Lemieux.*

serologically as A, B, O(H), Lewis-a, Lewis-b, Lewis-X, Lewis-Y and related antigens (Figure 4). With knowledge of the three-dimensional shapes and flexibility of these important biologically active oligosaccharides, their binding to antibodies, lectins and enzymes could, for the first time, be examined in solution at the molecular level.

A strategy based on functional group replacement was developed to dissect the contribution of individual hydroxyl groups to the free energy of binding. Lemieux initiated a vigorous synthetic program that produced more than one hundred tri- and tetrasaccharide structures. These synthetic oligosaccharides were all analogs of the natural blood group determinants, which had been modified, either through removal of hydroxyl groups or by their replacement with other substituents. The specific alterations permitted the highly specific recognition of oligosaccharides by protein receptors to be dissected in molecular detail. During this period two research associates played leading roles in the molecular recognition studies. Beginning late in 1981, Ole Hindsgaul helped lead Lemieux's group until 1986 in studies on the lectins, *Ulex europaeus* and *Griffonia simplicifolia* IV, as well as Lewis-b and blood group B monoclonal antibodies. Ulrike Spohr, who had joined Lemieux's group as a postdoctoral fellow in 1982, became his research associate from 1985 until he closed his research laboratory in 1995 and led in depth studies of *Griffonia simplicifolia* IV and other lectins that recognized the Lewis or H-type antigens.

Through a systematic study of the binding of these analogs Lemieux was able to define the precise molecular features required for the specific recognition of complex carbohydrate determinants by these proteins. An account of this work summarizes the development of his thinking during the 1975–1996 period, beginning with the "hydrated polar-group gate effect" as the key to the specificity in the recognition of complex carbohydrates through to the idea of water reorganization as a major driving force for complexation. The work culminated in the high resolution crystal structure of the lectin *Griffonia simplicifolia* complexed with the human Lewis b tetrasaccharide, data that substantiated the crucial inferences that had been drawn from congener mapping of the binding site. The most important was the confirmation, for this and several other systems, that only a very limited number of hydroxyl groups—often 2–3 out of some 10–12 present in an oligosaccharide epitope (the part of an antigen recognized by an antibody)—are essential for acceptor recognition and biological activity.

Water in Sugar-Protein Complexes. The results of Monte Carlo calculations on the hydration of oligosaccharide surfaces led Lemieux to conclude that the principle source of binding energy between protein receptor and oligosaccharide epitope derived not from polar interactions between solutes but from the collapse of perturbed water about the interacting, and polyamphiphilic surfaces. The return of these energetically disadvantaged water molecules from the closest hydration layers to bulk water would then provide a large source of free energy change. These controversial ideas were refined over several years, and about the time Lemieux was closing his laboratory supporting experimental evidence for this interpretation appeared. Calorimetric studies showed that 25–100 percent of the observed binding enthalpy could arise from solvent reorganization.

Industrial Activities. Lemieux founded three companies: R&L Molecular Research, Raylo Chemicals, and Chembiomed Ltd., which sought to apply the creativity of his university-based research on antibiotics and complex carbohydrates. Several practical applications arose from this work. One intriguing idea at Chembiomed was the development of immunoabsorbents to remove ABO iso-antibodies from a patients blood, thereby permitting organ transplantation in spite of the ABO histocompatibility barrier. From the mid 1980s into the early 2000s, the shortage of organs for transplant and the increasing attention being given to xenotransplants suggests that the Lemieux technology or its subsequent embodiments may find expanded application. Much of the subsequent activity in the search for carbohydrate based therapeutics, in laboratories around the world, had its origin in his pioneering work.

Lemieux's profound influence on organic chemistry derived in large part from his enduring interest in the basic physical characteristics of molecules. Theoretical support for the concept of the anomeric effect followed many years after their recognition and acceptance as general, stereoelectronic effects in organic chemistry. This fundamental understanding of the chemistry of the anomeric centre paved the way for new methods of 1,2-*cis*-glycoside synthesis, a career-long interest. These major creative steps first generated methods for assignment of structural details, then provided tools for the assembly of complex structures, and finally placed him in a position to explore the subtleties of carbohydrate recognition phenomena.

Awards. Lemieux's academic and research accomplishments have been recognized in Canada, the United States, and Europe through honorary degrees and numerous awards: the American Chemical Society's Claude S. Hudson Award in 1966; fellowship in the Royal Society of London in 1967; and the Haworth Memorial Medal of the Chemical Society of London in 1978. In Canada, he was appointed an Officer of the Order of Canada in 1968, and in 1996 was elevated to the highest level of recognition, Companion of the Order of Canada. He was the first recipient of the Izaak Walton Killam Memorial Prize and of the Science and Engineering Research Council's Canada Gold Medal for Science and Engineering. Among his many awards, Lemieux was especially proud of his highest Canadian honors, the Companion of the Order of Canada and the Canada Gold Medal for Science and Engineering. In order to perpetuate and nurture the research Lemieux began, the University of Alberta established an endowed chair in his name of which he and Mrs. Lemieux where major benefactors.

In his early years Lemieux had a reputation as a demanding, tough supervisor and lecturer. He had intense drive, did not tolerate nonsense, and always came to the point quickly, even at times brusquely. He was committed to excellence in himself and from those around him and he did not hesitate to let his coworkers know if they were wrong about something. However, his criticism was given without rancor or harshness. Lemieux was an unassuming individual who was devoid of pretensions and was regarded by colleagues from around the world as good-humored and fun to be around, especially when he was relaxing at the pub or bar.

Although formally retired, Lemieux maintained his research group well into the 1990s, and despite an ongoing battle with failing eyesight caused by macular degeneration, he maintained his interest in carbohydrate-protein recognition until his death in July 2000. He was without doubt one of the most outstanding Canadian chemists of the twentieth century, whose contributions to organic chemistry profoundly influenced the development of the discipline in the second half of the century.

BIBLIOGRAPHY

WORKS BY LEMIEUX

With Melville L. Wolfrom and S. M. Olin. "Configurational Correlation of L-(*levo*)-Glyceraldehyde with Natural (*dextro*)-Alanine by a Direct Chemical Method." *Journal of the American Chemical Society* 71 (1949): 2870–2873.

With Georg Huber. "A Chemical Synthesis of Sucrose." *Journal of the American Chemical Society* 75 (1953): 4118.

With Rudolf K. Kullnig, Harold J. Bernstein, and William G. Schneider. "Configurational Effects in the Proton Magnetic Resonance Spectra of Acetylated Carbohydrates." *Journal of the American Chemical Society* 79 (1957): 1005–1006.

With Rudolf K. Kullnig, Harold J. Bernstein, and William G. Schneider. "Configurational Effects on the Proton Magnetic

Resonance Spectra of Six-Membered Ring Compounds." *Journal of the American Chemical Society* 80 (1958): 6098–6105.

"Rearrangements and Isomerizations in Carbohydrate Chemistry." In *Molecular Rearrangements,* edited by Paul de Mayo, Interscience, 1964.

"Effects of Unshared Pairs of Electrons and Their Solvation on Conformational Equilibria." *Pure and Applied Chemistry* 25 (1971): 527–548.

With Shinkiti Koto. "The Conformational Properties of Glycosidic Linkages." *Tetrahedron* 30 (1974): 1933–1944.

With K. B. Hendriks, Robert V. Stick, and Kenneth James. "Halide-Ion Catalyzed Glycosidation Reactions. Syntheses of α-Linked Disaccharides." *Journal of the American Chemical Society* 97 (1975a): 4056–4062.

With Hughes Driguez. "The Chemical Synthesis of 2-Acetamido-2-deoxy-4-*O*-(α-L-fucopyranosyl)-3-*O*-(β-D-galactopyranosyl)-D-glucose. The Lewis a Blood-Group Antigenic Determinant." *Journal of the American Chemical Society* 97 (1975b): 4063–4069.

With Hughes Driguez. "The Chemical Synthesis of 2-*O*-(α-L-fucopyranosyl)-3-*O*-(α-D-galactopyranosyl)-D-galactose. The Terminal Structure of the Blood-Group B Antigenic Determinant." *Journal of the American Chemical Society* 97 (1975c): 4069–4075.

With David R. Bundle and Donald A. Baker. "The Properties of a 'Synthetic' Antigen Related to the Human Blood-group Lewis a." *Journal of the American Chemical Society* 97 (1975d): 4076–4083.

"Human Blood Groups and Carbohydrate Chemistry." *Chemical Society Review* 7 (1978): 423–452.

With Shinkiti Koto and D. Voisin. "The *Exo*-Anomeric Effect." In *Origin and Consequence of the Anomeric Effect, American Chemical Society Symposium Series, No. 87.* Washington, DC: American Chemical Society, 1979.

With Klaus Bock, Louis T. J. Delbaere, Shinkiti Koto and V. S. Rao. "The Conformations of Oligosaccharides Related to the ABH and Lewis Human Blood Group Determinants." *Canadian Journal of Chemistry* 58, (1980): 631–653.

Explorations with Sugars. How Sweet It Was. Profiles, Pathways and Dreams. Autobiographies of Eminent Chemists, Jeffrey I. Seeman, series editor. Washington, DC: American Chemical Society, 1990.

With Louis T. J. Delbaere, Margaret Vandonselaar, Lata Prasad, et al. "Molecular Recognition of a Human Blood Group Determinant by a Plant Lectin." *Canadian Journal of Chemistry,* 68 (1990): 1116–1121.

With Pandurang V. Nikrad and Herman Beierbeck. "Molecular Recognition X. A Novel Procedure for the Detection of the Intermolecular Hydrogen Bonds Present in a Protein•Oligosaccharide Complex." *Canadian Journal of Chemistry* 7.0 (1992): 241–253.

"How Water Provides the Impetus for Molecular Recognition in Aqueous Solution." *Accounts of Chemical Research* 29 (1996): 373–380.

OTHER SOURCES

Bannett, A. D., William I. Bensinger, R. Raja, et al. "Immunoadsorption and Renal Transplant in Two Patients with a Major ABO Incompatibility." *Transplantation* 43 (1987): 909–911.

Bundle, David R. "Raymond Urgel Lemieux 16 June 1920–22 July 2000." *Biographical Memoirs of Fellows of the Royal Society* 48 (2002) 251–273. Contains a complete list of Lemieux's publications.

———. *Advances in Carbohydrate Chemistry and Biochemistry* 58 (2003): 1–33. Contains a complete list of Lemieux's publications.

Karplus, Martin. "Theory of Vicinal Coupling Constants." In *Encyclopedia of Nuclear Magnetic Resonance,* edited by David M. Grant and Robin K. Harris. New York: John Wiley & Sons, 1996.

David R. Bundle

LEONICENO, NICOLÒ (*b.* Vicenza, Italy, 1428; *d.* Ferrara, Italy, 9 June 1524), *medicine, philology.* For the original article on Leoniceno see *DSB,* vol. 8.

In 1991, the catalogue of Nicolò Leoniceno's library was brought to light, something that renewed interest in Leoniceno's activity. Compared with earlier work, this new research emphasized epistemology, which opened new avenues for fresh research.

Leoniceno received in his youth a classical education that deeply influenced his activity, because of the emphasis put on philological rigor and a practical orientation: reading of classical texts aimed at knowing the things (*res*) referred to by the words, rather than the words themselves (*verba*).

In his collection of more than 300 volumes (manuscripts and printed), he had dictionaries, lexica, and philological works for a right understanding of texts, as well as ancient and medieval philosophical, medical (particularly Galen) and mathematical texts (above all, in Greek). The driving force behind the library was not collectionism or antiquarianism, but the attempt to renew science and knowledge, mainly by bringing to light previously unknown texts. To this end, Leoniceno collaborated actively with the printing activity of Aldo Manuzio in Venice by lending his manuscripts as models for printed editions and participating in the *emendatio* of texts. This he did with Aristotle (published in five volumes, 1495–1498), Dioscorides (1499), and Galen (published after his death, in 1525).

The epistemological and scientific principles underpinning Leoniceno's teaching, scientific production and editorial activity were explicitly stated in his 1492 booklet *De Plinii aliorumque in medicina erroribus.* As the introductory letter by Angelo Poliziano suggests, Leoniceno's principles were inspired by the method used by humanist-philologists to restore ancient texts. Best represented by Poliziano's *Centuriae*, this method consisted of identifying the most ancient version of literary texts, correcting the mistakes resulting from hand-copying by means of a careful philological scrutiny and analysis, and eliminating all other possible transformations and additions introduced into the texts in successive accretions. Leoniceno transferred and adapted the method to scientific texts: On the basis of a comparison of classical Latin texts (mainly the books of Pliny's *Natural History* devoted to botany and pharmacology), Salernitan and post-salernitan translations of Arabic medical works (especially pharmacological treatises), and classical Greek medico-pharmaceutical texts (particularly Dioscorides' *De materia medica*), he concluded that the former two descended from the latter and thus could be omitted, all the more because they often contained mistakes, mainly incorrect translations and interpretations of plant and disease names. Such mistakes were of particular concern at Leoniceno's time, as they exposed patients to medical mistakes (for example, confusions of plants to be administered as medicines or wrong diagnosis of medical conditions). Consequently, Leoniceno proposed abandoning the Latin and Arabic pharmacological and scientific literature as later epiphenomena of an earlier textual body, thus returning to Greek science. Since even Greek texts were corrupted because of manual reproduction, he suggested submitting them to a rigorous philological examination aimed at restoring them in their purity.

Leoniceno's attack against Pliny provoked a harsh polemic, as it seemed to undermine the scientific authority and the role played by the *Natural History* in teaching and science during the Middle Ages. To overcome the opposition and to show the validity of his methodological proposals and scientific enterprise, Leoniceno collaborated in the publication of Dioscorides' *De materia medica* in Greek by Aldo Manuzio in 1499 by lending one of his manuscript copies of the text and probably also by establishing the text. Such editing made it possible to systematically compare Dioscorides' and Pliny's works and to recognize the best value of the Greek work.

Over time, Leoniceno's method became more of an epistemological nature. After the opposition to his *De Plinii* ceased, he published four other works in which he gradually exposed his theory of knowledge. Again, he insisted very much on the exigence of using exact terms of an unambiguous nature, that is, words (*verba*) allowing an unequivocal identification of the thing referred to (*res*). But, at the end of this group of works and of his personal evolution, he arrived at the conclusion that the words used (*verba*) had no importance provided that the things referred to (*res*) were properly known. Though apparently

Nicolò Leoniceno. SCIENCE PHOTO LIBRARY.

logical, such a conclusion was paradoxical, for it was anti-lexical and maybe also anti-philological, and put more emphasis on the knowledge itself than on the discourse aimed at communicating it.

Leoniceno's scientific and epistemological method—including the shift in the focus—had a deep impact on contemporary science, not only because it led to the production of several printed editions of important scientific texts (many of which were previously unknown, not even in Latin translations), but also—if not more—because it stressed the primacy of Greek scientific literature and, as a consequence, required understanding it correctly and translating it in an unambiguous way. Such a method required time before producing its effects, as is shown by the fact that no botanical work depending on classical sources was published until 1532 with Brunfels's *Herbarum vivae eicones.* Furthermore, it could not be equally applied to all fields. Although a pupil of Leoniceno, Antonius Musa Brasavola, made a systematic inventory of all medicines then used in pharmacies in order to submit them to examination, the natural substances described and prescribed in ancient texts were not properly identified and known until late sixteenth century, thanks to the *Republic of Botanists* and the travelers to the Eastern Mediterranean. Whereas Leoniceno was right in considering from a philological viewpoint that Latin and Arabic translations of Greek texts did not reproduce *ad litteram*

Leoniceno

the original and, as a consequence, he did not take into consideration the scientific nature of the differences in the texts. Apart from containing obvious mistakes resulting from textual tradition, they were in many cases adaptations of the texts to the context of the societies in which they were received, translated and used, specifically different natural resources and epidemiological conditions. Returning to Greek texts for philological reasons would have required eliminating all the new data and updates introduced into ancient texts up to Leoniceno's time. This was probably impossible to implement and, in any case, might have put at risk the population's health. Traditional (that is, medieval) handbooks of therapeutics and the traditional practice of pharmaceutical therapy thus continued to be used until late in the sixteenth century.

SUPPLEMENTARY BIBLIOGRAPHY

WORK BY LEONICENO

De Plinii in medicina erroribus. A cura di Loris Premuda. Milano, Italy: Edizione de Il Giardino di Esculapio, 1958. On the controversy on Pliny.

OTHER SOURCES

Bylebyl, Jerome J. "The School of Padua: Humanistic Medicine in the Sixteenth Century. " In *Health, Medicine and Mortality in the Sixteenth Century,* edited by Charles Webster. Cambridge, U.K.: Cambridge University Press, 1979. See especially pp. 339–342.

Edwards, William F. "Niccolo Leoniceno and the Origins of Humanist Discussion of Method." In *Philosophy and Humanism. Renaissance Essays in Honor of Paul Oskar Kristeller,* edited by Edward P. Mahoney. Leiden, Netherlands: E. P. Brill, 1976.

Ferrari, Giovanna. *L'esperienza del passato: Alesandro Benedetti filologo e medico umanista.* Biblioteca di Nuncius, XXII. Florence, Italy: L .S. Olschki, 1996. See especially pp. 256–296.

Franceschini, Adriano, ed. *Nuovi documenti relativi ai docenti dello studio di Ferrara nel sec. XVI.* Deputazione Provinciale Ferrarese di Storia Patria, Serie Monumenti, Vol. 6. Ferrara, Italy: Stab artistico tip. Editoriale, 1970. On Leoniceno's teaching.

French, Roger K. "*Pliny and Renaissance Medicine.*" In *Science in the Early Roman Empire: Pliny the Elder, His Sources and Influence,* edited by Roger French and Frank Greenaway. London: Croom Helm, 1986.

Hoffmann, Phillipe. "Un mystérieux collaborateur d'Alde Manuce : l'Anonymus Harvardianus." *Mélanges de l'École française de Rome, Moyen age, Temps modernes,* 97 (1985): 45–143. See pp. 133–138 on the manuscripts Leoniceno asked to be copied.

———. "Autres données relatives à un mystérieux collaborateur d'Alde Manuce: l'Anonymus Harvardianus." *Mélanges de l'École française de Rome, Moyen age, Temps modernes* 98 (1986): 673–708. See pp. 704–708 on the manuscripts Leoniceno asked to be copied.

Mani, Nikolaus. "Die griechische Editio princeps des Galenos (1525), ihre Entstehung und ihre Wirkung." *Gesnerus* 13 (1956): 29–52. See p. 38 and notes 34–36 on Leoniceno's role in the editing of ancient texts.

Mugnai Carrara, Daniela. "Fra causalità astrologica e causalità naturale. Gli interventi di Nicolo Leonico e della sua scuola sul morbo gallico." *Physis* 21 (1979): 37–54.

———. "Profilo di Nicolò Leoniceno." *Interpres 2* (1979): 169–212.

———. "Una polemica umanistico-scolastica circa l'interpretazione delle tre dottrine ordinate di Galeno." In *Annali dell'Istituto e Museo di storia della scienza di Firenze,* VIII (1983), pp. 31–57.

———. "La polemica 'de cane rabido' di Nicolò Leoniceno, Nicolò Zocca e Scipione Carteromaco: un episodio di filologia medico-umanistica." *Interpres* 9 (1989): 196–236.

———. *La biblioteca di Nicolò Leoniceno. Tra Aristotele e Galeno, cultura e libri di un medico umanista. Accademia Toscana di Scienze e Lettere "La Colombaria," "Studi,"* 118. Florence, Italy: L.S. Olschki, 1991. On Leoniceno's library and his philosophico-medical interests.

———. "Nicolò Leoniceno e Giovanni Mainardi: aspetti epistemologici dell'Umanesimo medico." In *Alla Corte degli Estensi. Filosofia, arte e cultura a Ferrara nei secoli XV e XVI.* Atti del Convegno Internazionale di studi, Ferrara, 5–7 marzo 1992. A cura di M. Bertozzi. Ferrara, Italy: Università degli studi, 1994.

Nutton, Vivian. "Hellenism Postponed: Some Aspects of Renaissance Medicine, 1490–1530." *Sudhoffs Archiv* 81 (1997): 158–170.

———. "The Rise of Medical Humanism: Ferrara, 1464–1555." *Renaissance Studies* 11 (1997): 2–19.

Premuda, Loris. *Un discepolo di Leoniceno tra filologia ed empirismo: G. Mnardo e il "libero esame" dei classici della medicina in funzione di più spregiudicati orientamenti metodologici.* In *Atti del Convegno Internazionale per la Celebrazione del V Centenario della nascita di Giovanni Manardo, 1462–1536,* Ferrara, 8–9 dicembre 1962. Ferrara, Italy: Università degli studi di Ferrara, 1963. On the influence of Leoniceno's teaching.

Samoggia, Luigi. *Manardo e la Scuola umanistica-filologica tedesca con particolare riguardo a Loenardo Fuchs.* In *Atti del Convegno Internazionale per la Celebrazione del V Centenario della nascita di Giovanni Manardo, 1462–1536,* Ferrara, 8–9 dicembre 1962. Ferrara, Italy: Università degli studi di Ferrara, 1963. On the influence of Leoniceno's teaching.

———. *Le ripercussioni in Germania dell'indirizzo filologico-medico Leoniceniano della scuola ferrarese per opera di Leonardo Fuchs* (*Quaderni di Storia della Scienza e della Medicina,* IV), Ferrara, Italy: Università degli studi di Ferrara, 1964. On the influence of Leoniceno's teaching.

Santoro, M. "*La polemica pliniana fra il Leoniceno e il Collenuccio.*" *Filologia romanza* 3 (1956): 162–205. On the controversy on Pliny.

Sicherl, Martin. *Handschriftliche Vorlagen der Editio princeps des Aristoteles.* Mainz, Germany: Akademie der Wissenschaften und der Literatur Mainz, 1976, p. 14, note 33. (Reproduced in Martin Sicherl, *Griechische Erstausgaben des Aldus Manutius. Druckvorlagen, Stellenwert, kultureller Hintergrund.*

Studien zur Geschichte und Kultur des Altertums, Neue Folge, 1. Reihe: Monographien, 10. Band. Paderborn, Germany: F. Schöningh, 1997.) On Leoniceno's role in the editing of ancient texts.

Streeter, Edward C. "Leoniceno and the School of Ferrara." *Bulletin of the Society of Medical History of Chicago* 1 (1916): 18–22. On the controversy on Pliny.

Alain Touwaide

LEOPOLD, ALDO (*b.* Burlington, Iowa, 11 January 1887; *d.* Baraboo, Wisconsin, 19 April 1948), *wildlife management, ecology, game surveys, forestry.*

During his lifetime, Leopold was recognized as an innovator in the field of wildlife management and an important contributor to ecology. After his death his writings became widely known and influential to a growing community of environmental thinkers. He undertook the first large-scale game survey in the United States, was the author of the first textbook on game management, and became a founding member of the first academic department in wildlife management. Leopold incorporated concepts from plant and animal ecology as he worked to reform state and federal wildlife policy during a period of intense growth in knowledge of wild populations and concern over human interactions with those populations.

Aldo Leopold. AP IMAGES.

The Development of a Sport Hunter. Leopold grew up in the town of Burlington, Iowa, in a house overlooking the Mississippi River. He was the oldest of four children. His sister, Marie, was a year younger, and his two brothers, Carl Jr. and Frederic, were five and eight years younger, respectively. The bluffs along the river provided a training ground for natural history, and the Leopold children came to know the flora and fauna well. The boys were especially well versed in the knowledge of nature that made for success in hunting. Carl Leopold, Aldo's father, taught them to recognize the grasses, shrubs, trees, birds, and mammals of the hills and wetlands surrounding the town. Under this tutelage, Aldo developed an understanding of the interconnections among species and an awareness of how human communities relied upon the natural world.

Even with all the time he spent outdoors with his father and siblings, Aldo's mother, Clara Leopold, made sure that he was immersed in subjects including poetry, philosophy, and German literature. She took pride in his achievements at school and encouraged his writing. She aspired to have all of her children attend finishing school, and Aldo would be the first, in 1904, to venture eastward.

The Training of a Forester. When Leopold recognized forestry as a legitimate career option in the first decade of the twentieth century, the educational opportunities in that field were only just emerging. Yale University offered the only program in the country that would provide a young forester with the credentials to manage public forest lands. President Theodore Roosevelt appointed Gifford Pinchot to administer the nation's forests as chief of the newly formed Forest Service in 1905. Together, Roosevelt and Pinchot formulated a conservation plan that focused on a utilitarian philosophy for protecting resources from excessive harvesting that might deplete the public value of lands. Pinchot's inspiration came from Europe, after receiving his own forestry training in Germany. Programs such as the one at Yale adopted this conservation philosophy for forestry. Forests would be managed by people trained to recognize their short-term productivity as well as the need to sustain them for the public good over generations. Leopold therefore entered forestry at a pivotal time in the history of the profession.

In order to apply to the Yale program with more than his Burlington public school credentials, and to meet his mother's ambitions, Leopold entered the Lawrenceville Preparatory School in New Jersey. He spent a year there, which proved to be a valuable experience in his personal as well as academic development. Upon leaving his family in Iowa, he was instructed to write home regularly and to include in his letters every detail of his new experiences. Those letters became a vital part of his development as a writer; his mother regularly corrected grammar and made suggestions for improving the narrative flow of his writing. For his part, Leopold sought to relate his experiences in an engaging literary fashion. He also continued and expanded his habit of daily writing in a journal that included personal reflections and natural history observations.

After a year at Lawrenceville, he applied and was admitted to the Yale forestry program. There he received the training that would prepare him for a career in Pinchot's Forest Service. He completed his coursework and received a bachelor's degree in the spring of 1908. He returned to Yale in the fall for a final year of training as a master's student. Students at Yale were immersed in the utilitarian approach that became a motto for the conservation movement in those years: "for the greatest good for the greatest number over the long run." The field experience provided in the last year also gave Leopold and his peers a practical sense of the work that lay ahead.

Raising Questions about Conservation. Through his training at Yale Leopold had adopted the views of the Forest Service and its administrators. However, once he began to experience the work of forest management he articulated some innovative notions, incorporating his perspective as a sport hunter.

His first months as a forester included a move to New Mexico, where he was stationed on the Apache National Forest. There he attempted to familiarize himself with a diverse landscape of mountains, rivers, forests, and wildlife. Amid surroundings very different from the countryside of his youth, he relished the challenge, but his inexperience also took its toll. After just a few weeks, Leopold was placed in charge of a reconnaissance crew, a team of men who rode through the forest making estimates of the timber. Their daily tasks included arduous horseback rides and careful measurement of forest and geographical parameters. In the evening, Leopold worked out calculations of the day's measurements. His crew, lower ranking but more experienced, found him rather incompetent. Leopold had struggled with mathematics during his education, and the pressure of his new responsibilities amplified his shortcomings. He responded by denying his errors and further alienating his crew. The reconnaissance survey lasted three months, and in the end, Leopold was as enthusiastic as ever about fieldwork, but his supervisors began to recognize he lacked the skills to manage a Forest Service crew in the field.

Over the years, Leopold moved into positions that involved less direct fieldwork. The challenges of developing new policies for management of the public forests suited the young forester, especially as continued reconnaissance provided a more detailed picture of the existing forest conditions. He became particularly interested in linking emerging concerns about game and soil conditions in the forests. He observed firsthand how natural communities had changed in recent decades. Since the arrival of settlers in the foothills and valleys of the mountain west, farmers and ranchers had taken a toll on native game populations. Extending Pinchot's resource management philosophy to include game, Leopold began a reconsideration of the relationship between public lands and hunting practice that would involve private landowners and government policy at every level. He recognized that the best management practices for government forests did not necessarily provide adequate conservation of game habitat. Although Roosevelt had hoped to create hunters' paradises in specific locations throughout the West, as a hunter, naturalist, and president, he lacked a sound understanding of what it would take to maintain viable game populations in areas where timber harvests and other resource extraction enterprises took their toll on wildlife. It would take Leopold, and a generation of range managers and ecologists, decades to work out the details of effective game management.

In 1912, Leopold married Estella Bergere. Together they had five children, two girls and three boys. Leopold's wife's family included several generations of prominent sheep ranchers. Her Spanish heritage and family connections to practical range management broadened Leopold's personal and professional perspective. As much as her family cemented his bond to the Southwest, he also longed to be closer to his own family in Iowa. Another personal experience that shaped his view of natural resource management came while he was traveling alone in a remote area of the Carson National Forest, where he was superintendent, in 1913. A rainstorm created floodwaters that soaked his provisions as he rode, then the weather turned cold, and he was caught in a blizzard. Wet and freezing, he contracted an inflammatory disease. He nearly died and spent months recovering at home. During his slow recovery, he began to think more systematically about game conditions and resource management.

A key motivation in Leopold's rethinking of game policy came from the well-known and bombastic writings of William T. Hornaday, a naturalist who established his reputation as the superintendent of the New York Zoological Park. Hornaday decried the wasteful practices of eastern hunters who had depleted the forests and marshes of game. He railed against poachers, especially recent immigrants who, he claimed, lacked an appreciation of American hunting ideals. In 1915, Leopold met Hornaday in Albuquerque and became an immediate disciple. Declining game populations in the West seemed to be following the pattern established in the eastern states since colonial times, and Leopold believed that it was time to reverse that trend. Leopold read Hornaday's books and the two men corresponded regularly. More than any scientific or rational understanding of the conditions, Leopold gained from Hornaday a sense of urgency that sparked him to action. He worked actively with the Albuquerque Game Protective Association for nearly a decade, and from his position with the Forest Service, helped to establish the Gila Wilderness Area, the first designated wilderness in the national forest system.

Contributions to an Emerging Field. Health problems and a persistent yearning to get back to the Midwest provided the impetus for Leopold to take a position in Madison, Wisconsin, at the Forest Service's Forest Products Laboratory. The move, in 1924, kept Leopold in the employ of the federal government, but meant a significant shift from the responsibilities that had enabled him to be involved in game issues in New Mexico. His new tasks of managing and directing research gave him more time than ever to devote to his growing interests in conservation. Game conservation related to forest management, but not to the work he was doing in Madison. Wisconsin residents, however, had a much more vital awareness of conservation as a political and social concern. The state was populated by hunters who had a particularly keen interest in deer.

Leopold's professional interests continued to shift from research at the forest lab to game management. He began planning a survey of game, which soon expanded to include collaborations with game experts in several midwestern states. In the midst of his planning, an offer from the Sporting Arms and Ammunition Manufacturers' Institute gave Leopold an alternative to the Forest Service job. Work for the sport-hunting industry, which had a vested interest in the conservation of game that would perpetuate sales of their products, gave Leopold freedom to explore his interests.

Leopold spent much of 1928 learning game protection, preservation, and conservation at the national level. His connection to the Sporting Arms and Ammunition Manufacturers' Institute placed him in a prominent position. He met with experts on the East Coast, including Hornaday. Leopold's plans to conduct research into the conditions for game in various states initially drew criticism from Hornaday, who believed that the time for action was long overdue, and that research would only delay policy makers. Yet Leopold persisted with the plan and soon began meeting systematically with state officials and university scientists across the Midwest. This approach yielded a groundbreaking publication, a *Report on a Game Survey of the North Central States,* in 1931.

So detailed that some reviewers found it overwhelming, the publication's gaps also highlighted the lack of knowledge in certain areas. The overriding message was that game populations were either in decline or undergoing significant shifts that needed local as well as broader attention in the form of policy and regulation. Leopold noted that local conditions needed ongoing study to determine how game populations might respond to increased hunting pressure, decreased quality habitat, ongoing predator control, or short-term weather effects. Game management assumed that certain factors could be controlled, but Leopold pointed out that in most cases, the factors were not yet defined, and the control was not at all understood.

The report also appeared against the backdrop of national awareness of the failures of past game protection practices. In some places those failures had been spectacular. Leopold was particularly aware of the deer management efforts on the Kaibab Plateau, an isolated forest region north of the Grand Canyon. Deer protection, initiated by Roosevelt in 1906, led to a population explosion that many observers blamed on the cessation of hunting coupled with ongoing predator control. These practices resulted from conflicting state and federal goals for the area. Leopold's experience in the Southwest, where similar instances of game mismanagement had scarred the landscape, suggested that game experts needed to provide a more sophisticated solution than protection. He coined the term *irruption* to describe what happened when game populations were allowed to expand without control, and the Kaibab case became his key example. Although he had not visited the Kaibab and seen the conditions there firsthand, he corresponded regularly with foresters there, and invited one, S. B. Locke, to write a chapter on the events for a textbook Leopold was preparing. Locke never wrote the chapter, but their correspondence revealed the persistence of the Kaibab case in Leopold's thinking about game management.

In 1933, Leopold published *Game Management.* The book was widely hailed as the first textbook in the field, and it remains a classic reference for wildlife management. In it, Leopold acknowledged his debt to the Sporting Arms and Ammunitions Manufacturers' Institute, stating that the support he had received to conduct the game surveys had made possible his articulation of empirical principles of game management. He referenced European principles and practices, as well as the expertise of North American ecologists.

The book, which contained eighteen chapters in three sections, united game theory, practice, and administration into a comprehensive, empirically based profession. Leopold introduced principles from animal and plant ecology, including population growth and fluctuations.

He suggested new terminology for certain population cycles, most notably *irruption* for the rapid increases that led to dramatic declines in game. His admiration for rational management for sustainability, as found in European forests, inspired suggestions that game could be similarly maintained. In the early 1930s, he did not yet advocate the end of predator control, but his enthusiasm for the practice had waned. He proposed more research in order to determine the proper role of predators in rational game management practices.

Another outcome of Leopold's success with this book was establishment of a faculty position at the University of Wisconsin in Madison. The Alumni Research Foundation provided funding for a program in game management, including a salary and other expenses for Leopold. He became the first professor in such a program in the country. In that capacity, he interacted more frequently with his academic colleagues and developed closer ties to research ecology. By the late 1930s, he had integrated ecological views more explicitly into his articulation of game management as a profession. In 1939, he presented a paper to the Society of American Foresters and the Ecological Society of America at their joint meeting in Milwaukee. He titled the paper "A Biotic View of Land" and incorporated recent ecological concepts, which had appeared only marginally in *Game Management.* For example, Leopold used the "pyramid of numbers" to illuminate ecological relationships and demonstrate how management tended to oversimplify those relationships. Leopold also dismissed principles that had previously guided game managers. Ecology began to replace "economic biology," and the role of species in an ecological discussion could no longer be assigned to purely utilitarian categories such as "useful, harmless, or injurious to man." Instead, species served as links in complex food chains, the prevailing metaphor in ecology by that time.

Legacy. As a testament to Leopold's stature in fields affiliated with game management, he was named an honorary vice president by the American Forestry Association. The Ecological Society of America, much to Leopold's surprise, elected him as their president in 1947. Because he was not an active member of either group and rarely attended those meetings, these elections discomfited him to a degree, but he acknowledged the privilege, stating that he felt the responsibility of serving in spite of his previous lack of direct involvement.

Leopold died suddenly the following year. On 19 April 1948 Leopold, along with his wife and daughter, joined neighbors to fight a grass fire that had spread from a trash pile in a farmyard adjacent to the Leopold property near Baraboo, Wisconsin. This weekend property had become a family retreat for the Leopolds. They referred to its sole building as "the shack," and studied its surroundings as a kind of experiment in restoring prairie to a landscape depleted by farming. Leopold was anxious to protect his land that day, but carrying a water tank on his back, he suffered a heart attack.

In the last decade of his life, Leopold wrote a series of essays, mostly at the shack, considering the proper role of humans in managing and appreciating the natural world. These essays blended natural history and philosophy with a call to action. For example, having realized the importance of predators in ecological systems, he wrote with regret about his own role in killing wolves in the Southwest. He expanded his definition of wilderness, and advocated for preservation of more such areas. After his death foresters, game managers, and ecologists recognized Leopold's contributions as central to their profession. His philosophical contributions to ecology and conservation inspired the conservation minded, but were somewhat marginal to researchers in those fields. The influence of Leopold's general works, however, became established nearly twenty years later, after the publication of Rachel Carson's *Silent Spring,* in the midst of growing concern about the degradation of the natural environment. Leopold's *A Sand County Almanac* became a major source of that message. In essays that included "Thinking Like a Mountain" and "The Conservation Ethic," Leopold had described the complex and essential role of humans in terms that resonated with an expanding public awareness of environmental degradation. His ability to connect ecological concepts with the sense that humans should bear responsibility for their activities and impacts on future generations has prompted some to refer to him as a "prophet" for conservation.

BIBLIOGRAPHY

WORKS BY LEOPOLD

"Wanted—National Forest Game Refuges." *Bulletin of the American Game Protective Association* 9 (January 1920): 8–10, 22. An early example of Leopold's appeals for the Forest Service to adopt game protective policies.

"Determining the Kill Factor for Black-Tail Deer in the Southwest." *Journal of Forestry* 18 (February 1920): 131–134. Leopold's avocation as a hunter demonstrated in the context of his professional role as a forester.

"Game Methods: The American Way." *American Game* 20 (March–April 1931): 20, 29–31. An example of Leopold's emerging professional views of game management.

Game Management. New York: Charles Scribner's Sons, 1933. The first "textbook" of game management, in which Leopold referred to some basic ecological principles along with considerable speculation and experiential explanations of population phenomena.

"Deer and Dauerwald in Germany: I. History." *Journal of Forestry* 34 (April 1936): 366–375. An overview of Leopold's experience touring forests in Germany and considering the history of game and forest management practices.

"Deer and Dauerwald in Germany: II. Ecology and Policy." *Journal of Forestry* 34 (May 1936): 460–466. A further demonstration of the influence of European forestry had on Leopold, including discussion of recent ecological developments and the implementation of policy.

"A Biotic View of Land." *Journal of Forestry* 37 (September 1939): 727–730.

"Deer Irruptions." *Wisconsin Conservation Bulletin* 8 (August 1943): 1–11. A discussion of how game management efforts had gone awry in various locations, pointing to the need for more scientific management.

A Sand County Almanac and Sketches Here and There. New York: Oxford University Press, 1949.

OTHER SOURCES

Callicott, J. Baird, ed. *Companion to A Sand County Almanac: Interpretive and Critical Essays.* Madison: University of Wisconsin Press, 1987. This standard volume examines a wide range of the philosophical and naturalistic views of Leopold in his best-known work.

Flader, Susan L. *Thinking Like a Mountain: Aldo Leopold and the Evolution of an Ecological Attitude toward Deer, Wolves, and Forests.* Madison: University of Wisconsin Press, 1974. This book remains the best analysis of Leopold's shifting views on game, ecology, and resource management across his career.

Lorbiecki, Marybeth. *Aldo Leopold: A Fierce Green Fire.* Helena, MT: Falcon Publishing, 1996. This biography is the most readable, detailed account of Leopold's life. Special emphasis is placed on the philosophical aspects of his writings.

Meine, Curt. *Aldo Leopold: His Life and Work.* Madison: University of Wisconsin Press, 1988. Includes the most complete bibliography currently available in published form.

Meine, Curt, and Richard L. Knight, eds. *The Essential Aldo Leopold: Quotations and Commentaries.* Madison: University of Wisconsin Press, 1999.

Ripple, William J., and Robert L. Beschta. "Linking Wolves and Plants: Aldo Leopold on Trophic Cascades." *BioScience* 55 (July 2005): 613–621. A discussion of the persistence and significance of Leopold's influence on ecological thinking about connections between vegetation, herbivore populations, and predators.

Christian C. Young

LEPTINES (*fl. c.* 200 BCE?), *astronomy.*

Among the comparatively small number of extant astronomical manuscripts from Ptolemaic Egypt, the largest and most substantial is a two-meters-long papyrus roll (*P. Par.* 1, as of 2007 P. Louvre N 2388 Ro + Paris, Louvre N 2329 Ro) commonly known as the "Eudoxus Papyrus" or *Ars Eudoxi* because its back bears fourteen acrostic verses, the first letters of which spell out in Greek "art of Eudoxus." The true author or compiler of this didactic treatise appears to be named in the final column of prose text on the front side as Leptines. Calendrical information in the text dates it to about 190 BCE, making Leptines's elementary survey of astronomy the earliest surviving example of its genre in Greek.

An uncertain number of columns of text have been lost from the beginning of the manuscript, and what remains is rather chaotic in the arrangement of topics. Leptines adapted at least as much from earlier sources as he wrote himself. Some sections were evidently composed originally in iambic verse, and there are verbal parallels with the introduction to an astronomical calendar found in another papyrus manuscript (*P. Hibeh* 27) that was written about 300 BCE. Among the topics reviewed is the division of the solar year into astronomical seasons, with a simple arithmetical scheme for the varying length of daylight, the cycle of phases of the moon, and the cycle of risings and settings of constellations. The relative sizes of the Earth, Moon, and Sun, the Moon's phases, and both kinds of eclipse are explained in terms of a simple geocentric cosmology.

Discussion of the planets is limited to crude and obvious periodicities. The text mentions the eight-year calendrical cycle sometimes (but not here) associated with Eudoxus, but not the more accurate nineteen-year *Metonic* cycle, which was certainly preferred by astronomers of this period. The revolutions of the Sun and Moon are described in terms of passage through zodiacal signs, though there is no mention of division of the signs into degrees, or for that matter of solar or lunar anomaly. The absence of any exposition of spherical astronomy or geometrical modeling of celestial motions is to be attributed to the elementary level to which Leptines's work aspires and should not be taken as a fair representation of contemporary astronomical theory.

A distinctive and puzzling feature of the papyrus is the interspersing of rather crude diagrams amidst the text. Though they are clearly attempts to represent spherical heavenly bodies, these figures are for the most part neither self-explanatory nor related except in the most oblique way to the passages that they accompany. Nevertheless they entitle the papyrus to the claim of being the oldest surviving illustrated Greek manuscript.

BIBLIOGRAPHY

Blass, Friedrich. "Eudoxi ars astronomica qualis in charta aegyptiaca superest denuo edita." *Diei natalis nonagesimi serenissimi et potentissimi principis Guilelmii germanorum imperatoris regis borussiae faustissima sollemnia...*Kiel, Germany: Program of University of Kiel, 1887. 3–25. Reprinted in *Zeitschrift für Papyrologie und Epigraphik* 115 (1997) 79–101.

Letronne, Jean Antoine; Wladimir Brunet de Presle; and Émile Egger. *Notices et texts des papyrus du Musée du Louvre et de la*

Bibliothèque. Paris: Impériale, 1865. Reprinted as *Notices et Extraits des Manuscrits de la Bibliothèque impériale et autres bibliothèques* 18 part 2. Paris, 1865. 25–76, with plates 1–6 and 9 [in accompanying volume of plates]. The first edition of the papyrus, still useful for its respect for the manuscript's layout. and for the facsimile.

Neugebauer, Otto. *A History of Ancient Mathematical Astronomy*, 3 vols. Berlin: Springer, 1975. vol. 2, pp. 686–689.

Alexander Jones

LERAY, JEAN (*b.* Chantenay [now in Nantes], France, 7 November 1906; *d.* La Baule, France, 10 November 1998), *mathematics, particularly algebraic topology and elasticity.*

Leray made fundamental contributions to fluid mechanics, in particular the existence of classical and weak solutions of Navier-Stokes equations, to partial differential equations, especially linear hyperbolic or analytic equations, and nonlinear elliptic Dirichlet problems, nonlinear functional analysis, with Leray-Schauder degree and monotone-like operators; algebraic topology, introducing sheaves and spectral sequences; elasticity; functions of several complex variables, with the Cauchy-Fantappié-Leray formula; and Lagrangian analysis.

Early Life and Career. Jean Leray was the son of two teachers, Francis Leray and Baptistine Pineau. After graduating at the École Normale Supérieure in Paris in 1929, Leray defended his PhD thesis, "Étude de diverses équations intégrales non linéaires et de quelques problèmes que pose l'hydrodynamique" (A study of various nonlinear integral equations and of some hydrodynamical problems), at the Faculté des sciences de Paris in 1933. His advisor was Henri Villat. One year before, Leray had married Marguerite Trumier, with whom he had three children, Jean-Claude, Françoise, and Denis. When World War II started, Jean Leray was a professor at the Faculté des sciences of Nancy. He served as a reserve officer in the French army but was captured by the Germans in June 1940 and sent to the prison camp Oflag XVIIA at Edelbach, near Austerlitz (in what was later Czechoslovakia). There, he organized and led a university until the liberation of the camp in May 1945. After two years at the Faculté des sciences de Paris, Leray became a professor at the Collège de France in Paris (holding the chair of differential and functional equations) until his retirement in 1978.

A few years before his death, Leray left Sceaux for la Baule, near Nantes. He became a Fellow of the French Académie des sciences (mechanics section) in 1953, and later of most of the prestigious science academies in Europe and the United States. Besides receiving several prizes from the French Académie des sciences, Leray was awarded the prestigious Malaxa (1938), Feltrinelli (1971), and Wolf (1979) prizes.

Leray's Personality. Leray's mathematical work extended over an exceptionally long period; more than sixty years separated his first and his last papers. His contributions are highly original and cover an unusually wide spectrum of mathematics, from algebraic topology to elasticity. Leray's independence of mind prevented him from allying himself with a political party, school, group, or ideology. He was among the few important mathematicians to react publicly against some excesses in the use of "modern mathematics" in French high school programs. His incisive sarcasm is reminiscent of Voltaire and Henri Poincaré, and the severe elegance of his language follows from a constant care for concision. Leray's lectures, austere and deprived of embellishments, required a great deal from his listeners.

Fluid Mechanics. Leray's work can for the most part be described in a chronological way, because after moving on to a new subject, he almost never came back to an earlier interest. In his PhD thesis, Leray successfully combined the Lyapunov-Schmidt reduction method with the Arzelà-Ascoli theorem and the existence of a priori bounds to obtain global existence results for nonlinear integral equations and stationary solutions of the equations of hydrodynamics through analytical continuation. The corresponding Navier-Stokes evolution equations are masterly treated in two memoirs published in 1934, introducing for the first time fundamental concepts such as weak solutions (called turbulent solutions by Leray) and the Sobolev space $H^1(\mathbf{R}^3)$. For the Cauchy problem, he proved the existence of at least one global weak solution, which is regular and unique near the initial time. The global uniqueness and regularity is still open.

In 1933 Leray discovered, at a meeting in Paris with the Polish mathematician Julius Schauder, the topological techniques required by his thesis. The consequence was a joint paper worked out in two weeks in the city's Jardin du Luxembourg (Luxembourg Garden). Topological degree theory in infinite-dimensional Banach spaces was born, as well as the global theory of nonlinear elliptic partial differential equations. The Leray-Schauder degree and the continuation theorem have inspired the entire development of nonlinear functional analysis. Leray-Schauder's fixed point theorem, in its simplest version, ensures the existence of at least one fixed point for a completely continuous mapping T on a Banach space, when the set of possible fixed points of ΛT is a priori bounded independently of Λ. In the remaining years before World War II, Leray initiated algebraic topology in infinite dimensional Banach spaces through his product formula for degree, and he also successfully applied the new continuation method to the theory of wakes and bows and to fully nonlinear Dirichlet problems.

Afraid that his expertise in fluid mechanics could lead the Germans to force him into collaboration with their war effort, Leray concentrated his teaching in the Oflag on algebraic topology, reconstructed through an original approach. In order to avoid finite-dimensional approximations and the restriction to linear spaces, Leray tried to determine the relation between the cohomology of the source, the target, and the fiber of a continuous mapping. To achieve this aim, he introduced the seminal concept of "sheaf" on a topological space (a general tool to go from local to global results) and the powerful method of spectral sequences. In his hands and those of other mathematicians, such as Henri Cartan, Jean-Pierre Serre, Armand Borel, Jean-Louis Koszul, and Alexandre Grothendieck, those tools not only revolutionized algebraic topology but also the theory of functions of several complex variables; homological algebra; algebraic geometry; and more recently, algebraic analysis. Leray applied his new ideas to the cohomology of closed continuous maps, fiber spaces, and Lie groups in what came to be called the Leray-Hirsch theorem.

Hyperbolic Partial Differential Equations. In the early 1950s, Leray shifted his interest to linear hyperbolic partial differential equations of arbitrary order, treated by Jacques Hadamard and Julius Schauder in the second order case. When the coefficients were constants, he applied Schwartz's distributions and algebraic geometry to construct the elementary solutions in what became known as the Herglotz-Petrovsky-Leray formula. In the case of variable coefficients, Leray corrected and extensively extended the results of Petrovsky based upon the energy method. The notes of lectures that he delivered at Princeton University in New Jersey in 1953 and Rome in 1956 have inspired much of the subsequent work.

Analytic Partial Differential Equations. Between 1955 and 1965, Leray initiated the study of Cauchy's problem for analytic partial differential equations that are singular on the manifold carrying the initial data. The goal was to prove that the singularities of the solution belong to the characteristics issued from the singularities of the data or tangent to the manifold carrying them. This program, not fully realized yet, led Leray to new continuations of the Laplace transform, to original insights on asymptotic wave theory (with Lars Gaarding and Takeshi Kotake), and to an important generalization of Cauchy's formula and of the residues theorem to analytic functions of several complex variables now reffered to as Cauchy-Fantappié-Leray formulas.

Elasticity and Monotone Operators. The decade from the mid-1950s to the mid-1960s was also enriched by important work in elasticity theory—motivated by new techniques in the construction of bridges—and in fixed-point theory, where Leray simplified and extended his earlier work by introducing what is now called the Leray trace. With Jacques-Louis Lions, the theory of monotone-like operators in Banach spaces was developed, leading to the important class of Leray-Lions operators, and paving the way for Haim Brezis's fruitful concept of pseudo-monotone operator.

Nonstrict Hyperbolic Systems. The next five years were mainly devoted to the study of nonstrict hyperbolic systems, which are important, for example, in relativistic magnetohydrodynamics. In collaboration with Yujiro Ohya and Lucien Waelbroeck, Leray solved them in some Gevrey spaces, which are intermediate between the spaces of holomorphic and of smooth functions.

At the beginning of the 1970s, Vladimir I. Arnold called Leray's attention to Victor P. Maslov's work on asymptotic solutions of partial differential equations, connected to the WKB method in quantum mechanics. Leray brought in the techniques of pseudo-differential operators and a new structure based upon symplectic geometry and called Lagrangian analysis. His approach leads mathematically to a constant, which, in the special cases of Schrödinger, Klein-Gordon, and Dirac equations, can be identified to Max Planck's one.

Schrödinger's equation for one electron was solved using Fuchs's theorem. In the early 1980s, Leray extended it to cover the case of several electrons, and describes the behavior of the solutions near the atomic nucleus. In the meantime, Leray had proved, together with Yusaku Hamada, Claude Wagschal, and Akira Takeuchi, several extensions of the Cauchy-Kovalevskaya theorem, including ramified data, and the analytic continuation of the solutions. Motivated by soil mechanics, Leray also contributed, around the beginning of the 1990s, to the propagation of waves in an elastic half-plane, through a new technique called the Laplace-d'Alembert transform.

BIBLIOGRAPHY

A comprehensive bibliography of Leray's work can be found in each volume of Jean Leray, Oeuvres scientifiques. *3 vols. Paris: Société Mathématique, 1998.*

WORKS BY LERAY

Hyperbolic Differential Equations. Princeton, NJ: Institute for Advanced Studies, 1953.

Notice sur les travaux scientifiques de M. Jean Leray. Paris: Gauthier-Villars, 1953.

La théorie de Garding des équations hyperboliques linéaires. Rome: Istituto di Alta Matematica, University of Rome; Varenna, Italy: CIME, 1956.

Analyse lagrangienne et mécanique quantique. Séminaire du Collège de France (1976–1977), I, Exposé No. 1, 303 p., Paris: Collège de France. Translated by Carolyn Schroeder as *Lagrangian Analysis and Quantum Mechanics: A Mathematical Structure Related to Asymptotic Expansions and the Maslov Index.* Cambridge, MA: MIT Press, 1982.

Oeuvres scientifiques. 3 vols. Paris: Société Mathématique, 1998.

OTHER SOURCES

Andler, Martin. "Jean Leray." *Proceedings of the American Philosophical Society* 144 (2000): 469–478.

Guillopé, Laurent. *Actes des journées mathématiques à la mémoire de Jean Leray.* (Nantes 2002), Séminaires et congrès no. 9, Société mathématique de France, 2004.

"Jean Leray (1906–1998)." *Gazette des Mathématiciens* 84, suppl., (2000). This is a special issue of the *Gazette des Mathématiciens,* issued by the Société mathématique de France as a supplement to no. 84. It contains articles about Leray by Kantor, Choquet-Bruhat, Siegmund-Schultze, Miller, Houzel, Yger, Serrin, Chemin, and Malliavin.

Lions, Jacques-Louis. "Les travaux de Jean Leray en mecanique des fluids." *Gazette des mathematiciens* 75 (1988): 7–8.

Mawhin, Jean. "In Memoriam Jean Leray (1906–1998)." *Topological Methods in Nonlinear Analysis* 12 (1998): 199–206.

———. "Jean Leray (1906–1998)." *Académie. Royale de Belgique. Bulletin de la Classe des Sciences,* series 6, 10 (1999): 89–98.

———. "Leray-Schauder Degree: A Half Century of Extensions and Applications." *Topological Methods in Nonlinear Analysis* 14 (1999): 195–228.

Schmidt, Marian, ed. *Hommes de Science: 28 Portraits.* Paris: Herrmann, 1990.

Jean Mawhin

LERNER, I(SADORE) MICHAEL

(*b.* Harbin, Manchuria, 14 May, 1910; *d.* Berkeley, California, 12 June 1977), *genetics, evolutionary theory, poultry husbandry.*

I. Michael Lerner is best known for formulating the influential concept of *genetic homeostasis,* usually defined as the tendency of a Mendelian population to maintain a constant genetic composition in the face of external pressure. The concept was formally articulated in 1954 in a widely read book titled *Genetic Homeostasis.* Lerner's interests in evolutionary and genetical theory largely grew out of his interests in the practical breeding of poultry. He later turned to theoretical genetics and behavior genetics and explored the social implications of biology. Over the course of his career, he made notable contributions to both practical and theoretical aspects of breeding, genetics, evolution, and understanding the relationship between biology and society.

Early Life and Education. Lerner was born in Harbin, Manchuria in 1910, of Russian Jewish parents. Manchuria at that time was a Chinese territory under Russian government lease. Lerner's father was a businessman who ran a successful import export firm. The family was fairly prosperous and supported artistic and cultural activities popular at the time. Lerner, therefore, grew up in an environment that encouraged the love of theater, music, and the performing arts in general. He was the second of two children (his sister was two years older), who were tended to first by Russian nurses and then by German governesses. His facility with the English language owed much to this early Manchurian influence, which stressed English rather than French as a second language to Russian.

Largely tutored at home, the Lerner children began to attend formal school only after the Russian Revolution in 1917. This event had a dramatic effect on life in Harbin and on the Lerner family especially, as they suffered enormous financial reversals. Lerner and his sister were sent to private schools because the family could no longer afford tutors at home. In the fall of 1922, Lerner was sent to the Harbin Public Commercial School, which was a cross between a classically oriented gymnasium and a technically oriented *Realschule.* For Lerner, one of the few positive outcomes of the revolution was that many of his teachers were university-level instructors who were Russian émigrés. This was especially true of his teachers in the humanities and social sciences, who taught secondary school courses with the sophistication usually seen at the university level: Lerner took courses in philosophy, literary criticism, and history at a much earlier age and taught by individuals with advanced specialties in the area.

The other positive outcome of the revolution was that Harbin itself became a kind of refuge for Russian émigré artists as well as intellectuals who formed a lively and bustling community rich in the arts, including opera, symphony, ballet, and theater companies. Lerner's lifelong interest in the arts and humanities owes much to this early family influence and to his educational experience in Harbin. He showed no marked interest in the sciences initially, and even well after he had earned himself a formidable reputation in genetics he continued to identify with the humanities, even going to far as to call himself a kind of "historian manqué" in one interview (Hall, 2005).

At the time of his graduation in 1927, Lerner was unsure of the direction he would take. His family had tried to emigrate to Switzerland in 1923, but had failed; Lerner leaned toward using his education as the opportunity to leave Harbin. His sister had earlier left for Russia to become a physician, but Lerner dreaded the required military service in Russia. He also feared what the new regime would make of his "bourgeois" family background, the expense of a Russian education, and the political uncertainty that clouded any future direction he might take there. Because of his comfort with the English language, he considered immigrating to the United States, but because immigration restriction laws were tightened by 1927, he decided on Canada because it was accepting immigrants, especially if they showed an interest in agricultural sciences.

Lerner's entry into what would develop into a brilliant scientific career thus began as a result of an opportunistic decision to leave Harbin for Canada. He later described it as "the path of least resistance." By September of 1927, Lerner landed in Vancouver, British Columbia, a preferred port of entry for people on the Pacific rim, without funds or a passport, but with the job of digging ditches and caring for the chickens on the poultry farm of the University of British Columbia. He was paid the sum of two dollars per day for this work. He eventually served as assistant to Vigfus F. Asmundson, an assistant professor in the poultry genetics department who encouraged Lerner to continue in the area by obtaining advanced degrees. Asmundson even went so far as to help pay for his education by lending him funds when needed. Lerner thus obtained his B.S. and M.S. degrees at the University of British Columbia with a specialty in poultry genetics. He also met and married a fellow classmate, Ruth Stuart, with whom he shared a number of interests.

Theodosius Dobzhansky, also a Russian émigré, would also play a critical role in Lerner's life. They met in 1931 when Dobzhansky, who had been then working as a fellow in the laboratory of American geneticist Thomas Hunt Morgan, visited the University of British Columbia and found himself stranded there awaiting a return visa to the United States. Because both Dobzhansky and Lerner spoke Russian and because they had similar interests in genetics, the two bonded in what would become a fruitful, lifelong scientific and personal relationship. Dobzhansky's famous charismatic influence played itself out on Lerner as his interest in genetics and its applications in evolution began to grow in the 1930s.

In 1933, Lerner received an offer of an assistantship to continue graduate work at the University of California, Berkeley, in the poultry husbandry department working under the direction of L. W. Taylor. The timing was propitious, because the University of British Columbia was shutting down its own poultry program. He received his PhD in 1936, and was appointed instructor in poultry husbandry in Berkeley. He received a series of accelerated promotions to professor. In 1958, he moved to the Department of Genetics as chair. Lerner became emeritus professor in 1973 and remained associated with the Berkeley campus until the end of his life.

Scientific Work. Lerner's scientific work is generally divided into three periods. The first grew out his involvement with practical breeding, primarily with poultry. His primary interests were in improving poultry stocks and egg production: He first examined poultry growth and disease resistance, then moved on to egg production and the effects of artificial selection and inbreeding in chickens. Lerner was especially interested in the effects of selection combined with inbreeding, and devised a number of tests for the theoretical projections of gains from the selection of several different characteristics simultaneously. He formulated what is known as the "optimum selection index," which was a way of optimizing the effects of selection that could lead to rapid increases in egg production. This latter concept was applied widely and led to the doubling of egg production in commercial flocks of domestic chicken.

For much of this research, Lerner collaborated with Berkeley colleague Everett R. Dempster, a mathematically inclined population geneticist, and Dorothy C. Lowry, his technical assistant. It was an especially productive phase of Lerner's career, culminating with the publication of his first book, *Population Genetics and Animal Improvement,* in 1950. This book established Lerner as one of the leaders and original thinkers in the field of animal breeding as a whole. His reputation and influence in animal breeding were enhanced further in 1958 when he published *Genetic Basis of Selection.* Both books helped to take what was long thought to be folk knowledge into the realm of a quantifiable and testable science. Some of his work examining poultry growth also took him to collaborative ventures with pioneers of allometry (the science of differential growth) such as Joseph Needham and Julian Huxley.

Lerner's second research period grew out of his direct knowledge of animal breeding and the interplay of selection, inbreeding, and their effects on the genetic composition of populations. It was also aided and abetted by a growing circle of acquaintances in the San Francisco Bay Area with an interest in exploring genetics and evolution and the integration of the two as it was emerging during the period of the evolutionary synthesis. Beginning as a graduate student, Lerner had organized a group of active younger researchers on the Berkeley campus with a keen interest in integrating knowledge of the new genetics in both practical and theoretical terms within a dynamic evolutionary framework. Berkeley genetics itself was booming at the time, with workers such as Ernest Brown

Babcock, the plant geneticist who focused on the genus *Crepis,* and Roy Elwood Clausen and Thomas Harper Goodspeed, who were working on the genetics of *Nicotiana tabacum.* With Dempster, Lerner organized a fortnightly journal club, Genetics Associated. By the late 1930s the group included Donald Cameron (then a research assistant to Clausen) and James Jenkins and G. Ledyard Stebbins (junior geneticists on the *Crepis* project with Babcock), along with plant breeders such as Francis Smith and Alvin Clark. Lerner also interacted with German émigré physiological geneticist Richard Goldschmidt when he later joined the faculty at Berkeley, though he disagreed with Goldschmidt's notion of sudden or abrupt evolution through macromutations. Located nearby at the California Institute of Technology, Lerner's old friend Dobzhansky was also a frequent visitor to the Bay Area and infused colleagues with his enthusiasm for evolutionary genetics as it was emerging from his researches on varied species of *Drosophila.* The Berkeley campus was thus a hub for research into evolutionary genetics in the late 1930s and 1940s; Lerner's interactions with his colleagues outside his primary department of poultry husbandry and applied genetics inspired him to think in broader and more theoretical terms.

Lerner's increasing confidence with theoretical genetics combined with his empirical knowledge in practical animal breeding came to fruition during his second period of research into theoretical genetics. In 1954 he published the book setting forth the conceptual theory of genetic homeostasis for which he is best known. Written in 1953 over a period of six months, while on a Guggenheim Fellowship at the Institute of Genetics in Pavia and at the Istituto Italiano di Idrobiologia in Pallanza, Italy, the book was dedicated to his long-time collaborator Dempster, who helped guide him with mathematical modeling. A small book of only 134 pages, *Genetic Homeostasis* packed a punch and was widely discussed in the biological sciences, and not without some controversy.

The novel concept set forth in the book and captured in the title was inspired by the concept of physiological homeostasis as it had been set forth by Walter Cannon in his celebrated *The Wisdom of the Body* in 1932. It referred to organisms regulating their internal physiological mechanisms as a way of buffering themselves or resisting external pressures. By *genetic homeostasis,* Lerner had in mind the same kind of regulatory or buffering capacity in populations of organisms that enabled them to preserve optimal fitness despite external pressures. Lerner defined it in his book as the property or ability of a population of organisms "to equilibrate its genetic composition and to resist sudden changes" (p. 2). The concept bore some resemblance to the concept of "genetic inertia" formulated by C. D. Darlington and Kenneth Mather in 1949, but Lerner strongly opposed the notion that they were identical or that he should cede priority to them.

As he pointed out repeatedly, homeostasis referred to self-regulatory capacity or the ability of organisms to buffer themselves against external pressures in a way that inertia, borrowed from physics, did not. It was a component he viewed as critical to his theory. As he extended the genetic conceptualization further, Lerner stressed that the most adapted types within a population were not those that conformed to phenotypic norms, but those that would have the highest fitness by their ability to adapt to changing environments. For Lerner, heterozygosity conferred just such an advantage: Heterozygotes generated two kinds of gene products and were therefore better buffered than homozygotes against environmental changes. The argument for genetic homeostasis thus hinged on Lerner's argument for heterozygote superiority, the mechanics of which he took great pains to explore in his book. In espousing heterozygote superiority, Lerner was echoing colleagues such as Dobzhansky, but he also drew notable critics, who pointed out that attempts to find the mechanisms for gene action had proven unsuccessful. The data simply could not support it, while other genetic phenomena such as pleiotropy (interactive gene effects) and linkage disequilibrium offered more plausible explanations.

Other elements of Lerner's conceptual theory were well received, especially by biologists stressing integrative approaches to evolution, genetics, and development. Fundamentally, Lerner's conceptualization of genetic homeostasis depended on developmental homeostasis, because he began with the assumption that all sexual organisms had genotypes that produced self-regulating developmental patterns and phenotypes. The relationship was explicitly stated. "Error," he wrote in his concluding paragraph, "is minimized in successful populations by developmental homeostasis, genetic homeostasis arises as an after-effect." The relationship between the two was so close that to some the terms homeostasis, assimilation, buffering, inertia, and canalization (terms usually associated with development) are used interchangeably. But what made Lerner's conceptualization new to developmental biologists at the time was its linked dependency on the heterozygous condition; as Lerner himself stated, heterozygotes were better able to "stay within the norms of canalized development" (1954).

Lerner's integration of population genetics with developmental biology was instantly appreciated by Conrad Hal Waddington, the noted developmental biologist, who had been famously frustrated by the absence of embryology or developmental biology in the synthetic theory of evolution. Population geneticists such as Dobzhansky, whose own preference was for heterozygote superiority, regarded the thesis as brilliant. Indeed, at a time when workers such as Ernst Mayr charged mathematical population genetics with being simplistic, reductionist, and "bean-bag," Lerner offered an ambitious and satisfying alternative that integrated genetics, evolution, and developmental biology in an entirely novel way. Though critics have charged that empirical data have not supported the argument for heterozygote superiority, Lerner is still regarded as a pioneer who made a "valiant" effort in seeking to integrate "genes, organisms, and development" (Hall, 2005). In this respect, Lerner may join workers such as Waddington in being one of the early pioneers of the new field known as "evo-devo," seeking to integrate evolutionary biology with developmental biology. Lerner is also recognized for being one of the first to appreciate the importance of the co-adapted gene pool or co-adapted gene complexes (West-Eberhard, 2003). Whatever the durability of all or parts of the concept of genetic homeostasis, the 1954 book was in its day important and influential, read by a wide circle of biologists. It was deemed an "acknowledged masterpiece" by many (Glass, "I. Michael Lerner Papers"), and deemed so important that even his critics described it as "speculative, imaginative, controversial, and influential" (Allard, 1996, p. 171).

Lerner's final period of research began with a shift to behavior genetics and led to his entry into the field of behavioral psychology. He adopted as his model organism the common flour beetle, *Tribolium spp.*, and designed a series of elegant and elaborate experiments to understand the interplay of genetics and behavior in the competitive behavior of this organism. This work eventually led to his appointment to the Institute of Personality Assessment and Research on the Berkeley campus, a part-time appointment that enabled him to work with graduate students in psychology. His general turn toward the interplay of genetics, evolution, and social behavior became increasingly the centerpiece of his work in the 1960s. It was also the opportunity for Lerner to integrate his lifelong passions for the humanities and social sciences. A committed teacher, Lerner designed an enormously popular course on the Berkeley campus to explore the interplay of genetics, evolution, and society, for which he developed the textbook *Heredity, Evolution, and Society* in 1968. The book explored contentious topics that included the extension of evolutionary and genetic theory to human affairs such as intelligence, race, and the role of selection in human social behavior broadly construed. Lerner skirted strict genetic determinism, but nonetheless believed that genetics played a powerful role in governing human affairs. The course also explored issues of world hunger, conservation, and population control, all major topics of concern to the social implications of biology. With a liberal, and indeed a markedly leftist political orientation, Lerner managed to teach this course all through the period of campus unrest and the radical politics at Berkeley through the 1960s, drawing huge crowds of students to it and becoming one of the most celebrated teachers at Berkeley. His political commitments were also apparent in his fight with Lysenkoism (an anti-genetics political campaign) in the Soviet Union; he was an outspoken critic of this regime and its effect on population genetics, writing works with provocative titles such as *Genetics in the U.S.S.R.: An Obituary*, and "Marxist Biology Viewed Dimly." One of his major contributions to disseminating knowledge of the oppressive effects of Stalinist control of science was his translation and editing of Zhores A. Medvedev's *The Rise and Fall of T. D. Lysenko* in 1969.

Even though Lerner had as technical and exacting a scientific career as one could imagine, he managed to interweave his extensive love of the humanities, including history, literature, and philosophy with his science throughout much of his career. He was renowned as a skilled and indeed an erudite writer, and a master of the synthetic and original thought-piece. He also brought some of his insights into the human condition not only to his teaching, but also to his administrative work, which began to occupy his time beginning in the late 1960s and 1970s. Lerner had an amiable personality and was much liked and respected at Berkeley. He held a number of prominent positions there, along with roles on committees and boards in the University of California system. He was also an active organizer of the International Congresses of Genetics and was elected to prominent offices, including membership in the National Academy of Sciences in 1959. In the last decade of his life, he was plagued with poor health that included cataracts, a detached retina, and emphysema, and he required a number of abdominal operations. He worked through many of these illnesses but finally succumbed in 1977.

BIBLIOGRAPHY

WORKS BY LERNER

Genetics in the U.S.S.R.: An Obituary. Vancouver: University of British Columbia, 1950.

Population Genetics and Animal Improvement as Illustrated by the Inheritance of Egg Production. Cambridge, U.K.: Cambridge University Press, 1950.

Genetic Homeostasis. New York: Wiley, 1954.

The Genetic Basis of Selection. New York: Wiley, 1958.

"Marxist Biology Viewed Dimly." *American Naturalist* 93 (1960): 45–55.

Heredity, Evolution, and Society. San Francisco: Freeman, 1968.

As editor and translator. Zhores A. Medvedev. *The Rise and Fall of T. D. Lysenko.* New York: Columbia University Press, 1969.

OTHER SOURCES

Allard, Robert W. "Israel Michael Lerner." *Biographical Memoirs National Academy of Sciences* 69 (1996): 167–173. Name is erroneously presented. Includes excerpts from an oral history memoir with a description of Lerner's early life.

Cannon, Walter B. *The Wisdom of the Body.* New York: Norton, 1932.

Darlington, Cyril D., and Kenneth Mather. *The Elements of Genetics.* London: Allen and Unwin, 1949.

Dempster, Everett R., M. M. Green, and S. Washburn. "I Michael Lerner, 1910–1977." *Genetics* 88 (1978): 139–140.

Futuyma, Douglas J. *Evolutionary Biology,* 3rd ed. Sunderland, MA: Sinauer Associates, 1998.

Glass, Bentley. "Israel Michael Lerner." *Yearbook of the American Philosophical Society* (1984–1985): 130–135. Name is erroneously presented.

———. "I. Michael Lerner Papers." *Guide to the Genetics Collections at the American Philosophical Society Library.*

Hall, Brian K. "Fifty Years Later: I. Michael Lerner's Genetic Homeostasis (1954): A Valiant Attempt to Integrate Genes, Organisms and Environment." *Journal of Experimental Zoology* 304B (2005): 187–197. Includes a detailed explanation of genetic homeostasis and offers a historical assessment of its applicability as well as its scientific rigor.

Mayr, Ernst, and William B. Provine, eds. *The Evolutionary Synthesis: Perspectives on the Unification of Biology.* Cambridge, MA: Harvard University Press, 1980.

Patrikeeff, Felix. *Politics in Exile: Manchuria and the Balance of Power in Northeastern Asia, 1924–1931.* New York: Palgrave Press, 2002.

Siegel, Paul B. "I. Michael Lerner (1910–1977): Specialist and Generalist." *Behavior Genetics* 8, no. 3 (1978): 225–226.

Smocovitis, Vassiliki Betty. "Keeping Up with Dobzhansky: G. Ledyard Stebbins, Plant Evolution, and the Evolutionary Synthesis." *History and Philosophy of the Life Sciences* 8, no. 8 (2006): 9–48.

West-Eberhard, Mary Jane. *Developmental Plasticity and Evolution.* Oxford, U.K.: Oxford University Press, 2003.

Wolff, David. *To the Harbin Station: The Liberal Alternative in Russian Manchuria, 1898–1914.* Stanford, CA: Stanford University Press, 1999.

Woolf, Charles M., and Therese A. Markow. "Genetic Models for Developmental Homeostasis: Historical Perspective." Tucson: University of Arizona, 1994. Available from http://eebweb.arizona.edu.

Vassiliki Betty Smocovitis

LERNER, ISRAEL MICHAEL

SEE **Lerner, Isadore Michael.**

LEWIN, KURT

(*b.* Mogilno [present-day Poland], 9 September 1890; *d.* Newtonville, Massachusetts, 11 February 1947), *psychology, social science, philosophy of science, psychological field theory, topological psychology, group dynamics.*

Lewin was the creator of psychological field theory, a pioneer of action research in psychological social science, and a founder of group dynamics. He combined thinking from psychology and philosophy of science throughout his career. His aims were to link theoretical insight with empirical research in the study of motivation, child development, and social behavior, as well as to humanize the workplace and the school with the help of social science. To him these tasks were not opposed; as he often said, nothing is as practical as a good theory.

Lewin was born in Mogilno (now Poland), which was then in the Prussian province of Posen. He was the second child and eldest son of a Jewish family, and knew Yiddish and Hebrew as well as German. His father owned a small general store and a farm outside the town. He was sent to Gymnasium in Breslau and studied medicine briefly in Freiburg im Breisgau before transferring to Berlin, where he studied with the philosopher and psychologist Carl Stumpf and the neo-Kantian philosophers of science Ernst Cassirer and Alois Riehl. He received his doctorate in 1916, while on leave from military service during World War I, and earned the right to teach (*Habilitation*) in Berlin in 1921. From that time until he resigned for political reasons in 1933, he was *Dozent* and senior assistant (*Oberassistent*) in the department of applied psychology at the Psychological Institute of the University of Berlin; he received the title of associate professor in 1927. He married Maria Landsberg in 1917 and had two children with her; after they divorced he married Gertrude Weiss in 1929 and had two children with her. His career after 1933 will be detailed below.

Philosophy of Science. In the 1920s Lewin elaborated a comparative theory of science. Instead of establishing ideal norms for the sciences and humanities according to their subject matter, he based his approach on the concepts scientists actually construct. He was inspired to do this by Cassirer's comparative and historical treatment of scientific concepts in *Substance and Function* (1910). His first and most extensive attempt to realize this program was a book on the concept of time series in physics and biology published in 1922. For Lewin, when physicists refer to a particle of matter persisting in a series of instants, this constituted as much a "genetic" (meaning a temporal) series as a reference to the path of energy from a lump of coal to a power plant and thence to a light bulb. In both cases what he called "genidentity" (*Genidenität*) is attributed to the object in question. Correspondingly, when an egg develops into a chicken, biologists speak of the life history of a single organism, even though egg and chicken might have no molecules in common, except perhaps in the germ cell; moreover, in evolutionary theory biologists construct historical series of species linked by descent, even though there may be no proven material linkages between them. Thus, in physics and biology entities defined as existing continuously over time differ according to the point of view required by the scientific task at hand. In the 1920s Lewin extended this pluralistic, pragmatic analysis to other disciplines, but chose not to publish the resulting texts.

Lewin was in contact with the founders of the logical empiricist movement, particularly Hans Reichenbach, with whom he had been involved in the Socialist youth movement before World War I. When Reichenbach organized the Society for Scientific Philosophy in Berlin in the late 1920s, Lewin participated actively. He shared the logical empiricists' interest in illuminating the conceptual foundations of science by examining actual scientific concepts, but rejected Rudolf Carnap's and Otto Neurath's call for a unified science based on physical language.

Lewin presented his view of psychology's place in his philosophy of science in two essays: "Law and Experiment in Psychology" (1927), and a paper on the transition from "Aristotelian" to "Galilean" thinking in psychology (1931), which first appeared in the journal *Erkenntnis*, the organ of the logical empiricist movement. In the 1927 paper he noted that what appears to be a unitary behavioral event may be the result of multiple psychological processes. Though this statement appears to deny the possibility of causal explanation in psychology, it could also be true of physical events and processes. A rolling ball, for example, appears to be a single series of events, yet a complete physical analysis shows it to be the product of multiple forces in interaction. A child's behavior in a given situation can also be seen as a product of interacting forces. Thus, in Lewin's view, it is possible to derive causal laws for psychology without reducing psychical phenomena to physical events. Instead, he posited "event types" (*Geschehenstypen*) as the appropriate explanatory objects for psychology, and hoped that laws for such "event types" would eliminate factors such as previous experience or heredity from psychological explanations.

In the 1931 paper, Lewin opposed the idea that psychology is or ought to be limited to statistical laws. He called such claims "Aristotelean," because they referred to typological categories such as "the obstreperous three-year-old," or to specific populations, such as one-year-old children in Vienna and New York in 1928. Modern physics, in contrast, derives universal mathematical laws from concrete, albeit ideal cases. Lewin proposed to create a "Galilean" or dynamic psychology, in which, instead of computing statistical averages from as many given cases as possible, researchers would recreate and analyze ideal-typical person-environment interactions in the laboratory. For him analysis of such interactions was a necessary basis for deriving formal, ultimately mathematical, descriptions of their dynamics. He understood this procedure to be analogous to the way in which Galileo had deduced the laws of free fall and projectile motion from mathematically derived "pure cases."

Research and Theory on Volition and Motivation. Lewin's early research challenged associationist theories of volition. Narziss Ach had suggested in 1910 that "determining tendencies" stimulated by an experimenter's instruction inhibit subjects' ability to recall associative connections they had already learned; the resulting delay in carrying out the instruction would thus be a measure of will. In his dissertation (published 1917) Lewin set out to improve this measure by asking observers to learn lengthy series of meaningless syllables, then instructing them either to reverse or rhyme the syllables. The prediction was that they would either take longer to complete the second task or give wrong answers. To his surprise, there was generally no inhibitive delay, and only a few errors. In further studies, using different instructions (for example, to rhyme syllables in a specific way), he found that subjects made few errors even with only a few repetitions during the training period. Lewin concluded that Ach's associationist view of will was untenable, because the predicted effects failed to occur even under optimal conditions, and decided that more work was needed on the relation of motivation and volition.

Lewin's subsequent studies of motivation and action during the 1920s combined affiliation with and independence from the Berlin school of Gestalt theory, whose leading thinkers Wolfgang Köhler and Max Wertheimer were Lewin's colleagues at the time. In papers titled "Preliminary Remarks on the Structure of the Mind" (author's translation) and "Intention, Will, and Need" (both 1926), Lewin accepted the Gestalt theorists' claim that actions, like acts of perception, are structured wholes (his term was "action wholes"), but he enriched their conception of behavior in two respects. First, he focused on the way situations appear to the actor at a given time, which he called their "psychological reality"; he suggested, further, that the psychical person is itself a complex, "layered" whole. The needs that influence a person's interaction with the (perceived) environment can come from more superficial or deeper layers of the self. Thus, the totality of forces present in the psychical field at a given time controls the direction of action; and this totality is not limited to perceived objects and their relations to an actor, but can include objects and needs of which the actor is not conscious.

Kurt Lewin. THE LIBRARY OF CONGRESS.

Lewin did not call this approach "field theory" at this time, but the expression "field forces" is ubiquitous in his German work. A frequently cited example of the impact of such forces is what he called the "demand character" of objects. The roots of this concept are already visible in an essay called "Kriegslandschaft" (1917; Warscape), written while he was at the front. That a house might appear to someone as a source of firewood, for example, would be barbarous in peacetime, but quite normal and maybe even necessary in war. The example he used in 1926 was a "peace thing": a mailbox has a different relation to me when I have a letter in my hand than when I do not. In the former case, the mailbox seems almost to jump out of the environment and announce its presence.

Lewin attributed such phenomena to what he called "quasi-needs," contending that objects related to them exert greater psychological "force" at particular times than at others. To account for these he suggested that "tension systems" emerge in specific "regions" of the self; these function in the same way as the tensions caused by real needs, transforming the psychical environment in accordance with a person's current intentions. The satisfaction of such needs reestablishes personal equilibrium at a lower level of tension. Of course, the "tension" in such systems is not directly measurable, as is the tension in a coiled spring.

Lewin's students in Berlin elaborated these ideas in empirical studies published in the journal *Psychologische Forschung*, in a series he edited entitled "Studies on the Psychology of Action and Emotion" (author's translation). Among the studies were Bluma Zeigarnik's investigation of memory for completed and uncompleted tasks (1927), the work of Anitra Karsten on "psychical satiation" in repetitive tasks (1928), Tamara Dembo's study of the dynamics of anger (1931), and Ferdinand Hoppe's work on the role of "level of aspiration" in task completion. These studies contained richly detailed descriptions of motivated actions, achievement, and task interruption, derived with the help of an interactive methodology of Lewin's invention. Dembo's experimental design, for example, involved an actual struggle between subject and experimenter, who deliberately frustrated subjects' efforts to complete the assigned task, then prevented them from leaving the room.

Psychology in Practice. Lewin's choice of topics clearly indicated his desire to connect scientific psychology with practical issues. He expressed that wish as early as 1920, in an essay entitled "The Socialization of the Taylor System" (author's translation), published just after the abortive German revolution in a series entitled "Practical Socialism" (author's translation), edited by his friend and independent Marxist thinker Karl Korsch. Lewin did not object in principle to Taylorism's attempt to discover quantitative laws of performance that could rationalize production and thus increase output. Instead he criticized capitalism's use of that effort to maximize profit rather than workers' well-being. Under socialism, he argued, workers could be assigned to jobs according to their abilities in a cooperative effort involving management, workers, and psychologists; thus both productivity and job satisfaction would be enhanced.

Humanizing the workplace and the school remained Lewin's aim throughout the 1920s and beyond, and his choice of basic research topics was clearly related to this purpose. In a 1928 essay, for example, Lewin suggested on the basis of Karsten's study of "satiation" that the psychological meaningfulness of a task to a worker can vary significantly even if productivity in output remains the same. This can have significant impact on the quality of performance, and even on physical fatigue. Thus, monotonous factory or school work alone does not cause psychological satiation; the decisive difference is the involvement of the person's self or ego. In the same year, in a paper on the textile industry published together with applied psychologist Hans Rupp, Lewin elaborated an analysis of work as a process, an "action whole" (*Geschehensganze*) that constitutes man and machine as a dynamic unity. In his view, it was important to consider the work process as a whole and not only to measure results or the times of individual motions of workers, because the purpose is to reshape that process itself.

At this stage, Lewin confined himself to the behavior of individuals in simply structured environments. This was true also of the film of a small child's problem-solving behavior, with which he introduced himself to American colleagues at the International Congress of Psychology in New Haven, Connecticut, in 1929. However, he always made it clear that other people are important parts of such environments. In the late 1920s and the early 1930s, he extended his thinking to pedagogy, speaking, among other things, of the importance of the "social atmosphere" of a school for educational success. He also described the behavior of children in conflict situations in ways that included relations with significant other people such as parents and teachers within the (subjective) field of children's action. However, he did not investigate social psychological questions or have the idea of working with groups as units before leaving Germany.

After 1933. Because Lewin had served at the front during World War I, he was nominally exempt from the provisions of the Nazi civil service law of 7 April 1933, which mandated the dismissal of persons of Jewish descent from state employment. His institute head, the Gestalt psychologist Wolfgang Köhler, wished to retain him in Berlin, but Lewin recognized the danger for Jews who remained in Germany. In a moving letter to Köhler dated 20 May 1933, which he never sent but which was discovered in his papers after his death, he wrote, "Everything within me rebels against the idea of leaving Germany despite all logical arguments," and yet,

> The actual loss of civil rights of the Jews has not abated, (but) is increasing daily and will no doubt be carried out completely in the peculiarly systematic German way, whether slowly and methodically, or in periodic waves ... I cannot imagine how a Jew is supposed to live a life in Germany at the present time that does justice to even the most primitive demands of truthfulness.

Shortly after he left his position in Berlin, Lewin received a stipend at Cornell University, where he worked on children's eating habits with support from the Emergency Committee in Aid of Displaced Foreign Scholars and the Rockefeller Foundation. Lawrence K. Frank, a foundation official who had met Lewin in Berlin and had been impressed by his experiments with children, then obtained a new grant in 1935 that sent Lewin to the Child Welfare Research Station at the University of Iowa. There he soon received a tenured appointment, rose to the rank of full professor in 1939, and remained until 1944. Both the Cornell center and the Iowa station were participants in a large-scale research program in child development that had been maintained with Rockefeller funding since the mid-1920s.

Because of his rapid integration into this network, Lewin refused the offer of a professorship from the Hebrew University in Jerusalem, which, as a Zionist, he would have preferred to accept. For this position he devised an ambitious research program, including, for example, studies of the relations between Jews and Arabs in Palestine. But the Jerusalem offer included no laboratory facilities, and Lewin's efforts to raise money for these from private donors failed.

Nonetheless, Lewin took up the problems of minorities soon after his emigration. In a 1935 paper on social-psychological problems of a minority group, the topic was clearly the Jews. Here, Lewin extended the concept of "life space," which he had already employed before 1933 to describe the subjective location of human-environment interactions, to human-human relationships under the heading "social space." He argued that precisely Jews who wished to assimilate to a predominant culture had difficulty in forming clear identities, because their location on the boundary between groups did not allow them to develop a feeling of belonging to either group. Roots of the concept of "marginal man," later articulated by Seymour Martin Lipset and others, can be seen here.

After moving to Iowa, Lewin took up the topic of cultural differences in education. This issue had obvious relevance in a research center for child welfare, but biographical factors plainly contributed to the choice: His younger children were reaching school age and the international situation literally demanded such comparisons. In this context, what he called the "range of free movement" became the fundamental feature of educational systems. The presence of hierarchical structures even in the "democratic" educational style of the United States, and the reliance of American teachers on externally mandated teaching plans and techniques gave American children, in Lewin's opinion, the support they needed to act independently in a heterogeneous social system, while rigidity and strict obedience were the educational norms in the comparatively homogeneous German social system.

From such considerations, and also on the basis of conversations that Lewin had with American collaborators in Iowa, came the studies of "democratic" and "authoritarian" leadership styles in children's play groups that made Lewin famous in America. In the "authoritarian" group both the task—making theater masks—and the way it was to be accomplished were defined step by step by the group leader, who intervened only to criticize the children's work. In the "democratic" group, the leader participated as a fellow group member, for example in decisions about how and with what materials to make the masks; he was allowed to give technical advice, but only when asked and then only in the form of presenting alternatives from which the group then chose.

In this work Lewin transferred the approach he had called "Galilean" in the early 1930s—the construction of ideal-typical person-environment interactions—to the behavior of groups. The "Lewin, Lippitt, and White" study, as it came to be known, acquired an almost mystical aura as the first group experiment in the history of social psychology. To visualize their approach, Lewin and his collaborators made a demonstration film that presented the behavior of the children's groups in often amusing scenes and was soon much in demand. The ideological resonances of this research were obvious in the late 1930s; one reason for the rapid success of the Iowa group's work was the support it seemed to provide for the hope that "democratic" leadership is indeed possible.

The politically progressive psychologists who founded the Society for the Psychological Study of Social Issues in 1936 shared this hope. Moved in part by impatience with their discipline's slow response to the problems of the Depression, the organization's members advocated social research for social change, if necessary by abandoning professional objectivity and distancing methodologies. Lewin was among the founders, and was elected president of the society in 1942–1943.

Building on this foundation, Lewin developed an ambitious program in the late 1930s and 1940s that he called "action research," to be conducted not in laboratories but in factories and communities. Early work along these lines at the Harwood Manufacturing Corporation reflected the roots of this approach in Lewin's Taylorism study of 1920, but he soon applied it to minority group issues as well. That program, organized within the framework of a Commission on Community Interrelations (CCI) and funded largely by the politically liberal American Jewish Congress, aimed both to study the social psychology of racism and anti-Semitism and at the same time to work toward changing racist and ethnically prejudiced social relations by deriving concrete practical guidelines from observations of group behavior and reflections on that behavior by the group members themselves. In a 1946 paper Lewin himself described all this as "research for social engineering."

By this time, Lewin had already moved from Iowa to the East Coast. Beginning with a visiting professorship at Harvard in 1939–1940, he expanded his contacts through work on morale research during World War II, including a programmatic essay on transforming Germany after Allied victory. In 1944 he accepted a professorship established for him at the Massachusetts Institute of Technology (MIT), where he founded an interdisciplinary Research Center for Group Dynamics.

Lewin never gave up hope of unifying theoretical and applied psychology. The means was to be topology, with the aid of which he hoped to achieve a mathematically rigorous representation of psychological dynamics. By the late 1920s, he had begun to transform this abstruse branch of mathematics into a device for the formal representation of psychological field forces and concrete psychological situations as well as the structure and internal dynamics of personality. In *Principles of Topological Psychology* (1936) he elaborated this approach in detail, with the hope of moving eventually to a process—rather than a performance-oriented concept of psychological measurement.

Impact. After Lewin died from a heart attack in 1947, at the age of fifty-seven, his prestige reached its high point. Edward Tolman went so far as to call him the most important thinker in the history of psychology after Sigmund Freud. Many Lewinian terms, including "level of aspiration," "life space," and "marginal affiliation," and slogans such as "nothing is as practical as a good theory" entered the vocabulary of American psychology, and later returned to Europe via translations into German and other languages. The Research Center for Group Dynamics, which moved from MIT to the University of Michigan at Ann Arbor shortly after his death, still existed in the early twenty-first century; Lewin's reputation as a founder of experimental social psychology seems secure. Nonetheless, his experiments with "authoritarian" and "democratic" groups, though greatly admired, did not become exemplars for research design. Rather than study groups as wholes, mainstream social psychologists generally examine the influence of groups on the behavior of individuals. However, some of Lewin's collaborators were instrumental in establishing the T-group and group dynamics movements in the 1950s, and others were among the founders of the approach called "ecological psychology" in the 1960s.

Lewin's idea of a "topological and vector psychology" has come to be regarded as a blind alley. In the 1930s few psychologists outside Lewin's immediate circle understood what he was talking about, and rivals willingly seized on disparaging remarks by mathematicians about his unsophisticated use of topology. Seen in historical context, his "Galilean" research program paralleled and competed with Yale psychologist Clark Hull's equally ambitious, and disappointing, effort to derive general laws of behavior deductively in a manner allegedly analogous to Isaac Newton's system of the natural world.

In the 1970s and 1980s Lewin's program for making psychology an agent of social change was sharply criticized from the left as a reformist project that would not change fundamental power relations. Since the 1990s action research has experienced a comeback as a results-oriented approach to understanding political conflict.

Lewin's fecund metaphors and brilliant individual insights, as well as his ability to inspire talented researchers, made him a success in Berlin, in the United States, and then internationally. He established an independent research base in America, but it was his cogent criticism of predominant styles of thought and practice in American psychology and his effort to develop concrete alternatives that gained him a hearing. At the same time, his support for U.S. democracy and his optimism about the practical potential of social science impressed the progressive segment of his discipline. His early work in the philosophy of science has never been translated into English and thus remains largely unknown outside Germany. For historians of science, his career exemplifies the deep connection of modern social science with social practice and also shows how a Jewish scientist created new science after reflecting on his own persecution under Nazism.

BIBLIOGRAPHY

*A collection called the Kurt Lewin Papers is located in Ohio at the Archives of the History of American Psychology, University of Akron. This consists primarily of materials used by Alfred Marrow to prepare his biography of Lewin (see below). Other collections of Lewin manuscripts and correspondence are located at the Institute for History of Modern Psychology in Passau, Germany, and the Distance University (*Fernuniversität*) in Hagen, Germany.*

WORKS BY LEWIN

Die Sozialisierung des Taylorsystems: Eine grundsätzliche Untersuchung zur Arbeits- und Berufspsychologie. Praktischer Sozialismus 4. Berlin: Weltkreisverlag, 1920.

Der Begriff der Genese in Physik, Biologie und Entwicklungsgeschichte: Eine Untersuchung zur vergleichenden Wissenschaftslehre. Berlin: Springer-Verlag, 1922. Reprinted in *Kurt-Lewin-Werkausgabe,* vol. 2, pp. 47–318.

Gesetz und Experiment in der Psychologie. Berlin: Weltkreis-verlag, 1927. Reprinted in *Kurt-Lewin-Werkausgabe,* vol. 1, pp. 279–320.

Die Entwicklung der experimentellen Willenspsychologie und die Psychotherapie. Leipzig: Hirzel, 1928.

Die psychologische Situation bei Lohn und Strafe. Leipzig: Hirzel, 1931. Reprinted in *Kurt-Lewin-Werkausgabe,* vol. 6, pp. 113–168.

A Dynamic Theory of Personality: Selected Papers. Translated by Donald K. Adams and Karl E. Zener. New York: McGraw-Hill, 1935.

Principles of Topological Psychology. New York: McGraw-Hill, 1936.

Experimental Studies in the Social Climates of Groups, Parts I and II. 1938. This film is available in several locations. The original is located at the Herbert Hoover Presidential Library and Museum, West Branch, Iowa. Restored copies are available at the University Archives, University of Iowa, Iowa City. An unrestored copy is located at the Archives of the History of American Psychology, University of Akron, Ohio.

With Ronald Lippitt and Robert K. White. "Patterns of Aggressive Behavior in Experimentally Created 'Social Climates.'" *Journal of Social Psychology* 10 (1939): 271–299.

Resolving Social Conflicts, Selected Papers on Group Dynamics [1935–1946]. Edited by Gertrude Weiss Lewin. New York: Harper, 1948.

Field Theory and Social Science: Selected Theoretical Papers. Edited by Dorwin Cartwright. New York: Harper, 1951.

Kurt-Lewin-Werkausgabe. Edited by Carl-Friedrich Graumann. Vols. 1, 2, 4, and 6. Stuttgart: Klett-Cotta; Bern: Huber, 1981–1982. Four of eight projected volumes published.

"Everything within Me Rebels: A Letter from Kurt Lewin to Wolfgang Köhler." Translated by Gabriele Wickert and Miriam Lewin. *Journal of Social Issues* 42 (1986): 40–47.

The Complete Social Scientist: A Kurt Lewin Reader. Edited by Martin Gold. Washington, DC: American Psychological Association, 1999.

OTHER SOURCES

Ash, Mitchell G. *Gestalt Psychology in German Culture, 1890–1967: Holism and the Quest for Objectivity.* Cambridge, U.K.: Cambridge University Press, 1995. See especially chapter 16.

———. "Cultural Contexts and Scientific Change in Psychology: Kurt Lewin in Iowa." *American Psychologist* 47 (1992): 198–207. Reprinted in *Evolving Perspectives on the History of Psychology,* edited by Wade E. Pickren and Donald A. Dewsbury. Washington, DC: American Psychological Association, 2001.

———. "Learning from Persecution: Émigré Jewish Social Scientists' Studies of Authoritarianism and Anti-Semitism after 1933." In *Jüdische Welten,* edited by Beate Meyer and Marion Kaplan. *Juden in Deutschland vom 18. Jahrhundert bis in die Gegenwart.* Göttingen: Wallstein, 2005.

Back, Kurt. *Beyond Words: The Story of Sensitivity Training and the Encounter Movement.* Baltimore, MD: Penguin, 1973.

Bargal, David. "Kurt Lewin and the First Attempts to Establish a Department of Psychology at the Hebrew University." *Minerva* 36 (1998): 49–68.

———. "Personal and Intellectual Influences Leading to Lewin's Paradigm of Action Research." *Action Research* 4, no. 4 (2006): 367–388.

Cravens, Hamilton. *Before Head Start: The Iowa Station and America's Children.* Chapel Hill: University of North Carolina Press, 1993.

Danziger, Kurt. *Constructing the Subject: Historical Origins of Psychological Research.* Cambridge, U.K.: Cambridge University Press, 1990.

———. "The Project of an Experimental Social Psychology: Historical Perspectives." *Science in Context* 5 (1992): 309–328.

———. "Making Social Psychology Experimental: A Conceptual History, 1920–1970." *Journal of the History of the Behavioral Sciences* 36 (2000): 329–347.

De Rivera, Joseph, ed. *Field Theory as Human-Science: Contributions of Lewin's Berlin Group.* New York: Gardner, 1976.

Habermas, Tilmann. "Eine nicht ganz zufällige Begegnung: Kurt Lewins Feldtheorie und Siegfried Bernfelds Psychoanalyse im Berlin der späten 20er Jahre." *Zeitschrift für Psychologie* 209 (2001): 416–431.

Journal of Social Issues 42, nos. 1–2 (1986). Special issues on Kurt Lewin.

Lück, Helmut E. *Kurt Lewin: Eine Einführung in sein Werk.* Weinheim: Beltz, 2001.

Marrow, Alfred. *The Practical Theorist: The Life and Work of Kurt Lewin.* New York: Basic, 1969. Appendices include an incomplete biography of Lewin's works; summaries of the Berlin experiments; members of the "Topology Group" (1935); studies done under Lewin's direction in Iowa; CCI publications; and publications of the Research Center for Group Dynamics, 1945–1950.

Métraux, Alexandre. "Kurt Lewin: Philosopher-Psychologist." *Science in Context* 5 (1992): 373–384.

Patnoe, Shelley. *A Narrative History of Social Psychology: The Lewin Tradition.* Berlin: Springer-Verlag, 1988.

Schönpflug, Wolfgang, ed. *Kurt Lewin: Person, Werk, Umfeld: Historische Rekonstruktionen und aktuelle Wertungen aus Anlass seines hundertsten Geburtstags.* Vol. 5 of *Beiträge zur Geschichte der Psychologie.* Frankfurt am Main: Lang, 1992.

Schwermer, Josef. *Die experimentelle Willenspsychologie Kurt Lewins.* Meisenhein am Glan: Hain, 1966.

Stivers, Eugene H., and Susan A. Wheelan, eds. *The Lewin Legacy: Field Theory in Current Practice.* New York: Springer-Verlag, 1986.

Tolman, Edward C. "Kurt Lewin (1890–1947)." *Psychological Review* 55 (1948): 1–4.

Van Elteren, Mel. "Karl Korsch and Lewinian Social Psychology: Failure of a Project." *History of the Human Sciences* 5 (1992): 33–61.

Wittmann, Simone. *Das Frühwerk Kurt Lewins: Zu den Quellen sozialpsychologischer Ansätze in Feldkonzept und Wissenschaftstheorie.* Frankfurt am Main: Lang, 1998.

Mitchell G. Ash

LEWIS, DAVID

(*b.* Oberlin, Ohio, 28 September 1941; *d.* Princeton, New Jersey, 14 October 2001), *metaphysics, philosophy of language, philosophical logic, philosophy of mind, epistemology.*

Lewis was one of the most important philosophers of the twentieth century, working in the Anglo-American analytic tradition. His corpus is extraordinary for its breadth of subject matter and for its systematicness. For both these reasons, it is difficult to do justice to his work in a short space—there are rich interconnections among his myriad writings and numerous possible entry points. This article approaches Lewis and his work in three passes: first, a biographical tracing of his intellectual influences; second, a summary of his metaphilosophy; and third, a survey of his more specific philosophical views, mostly following their order of conceptual dependence.

Intellectual Biography. While Lewis was a strikingly original thinker, a number of others helped to shape his eventual intellectual outlook to varying degrees, as he acknowledged. He was born in Oberlin, Ohio, in 1941, the eldest child of John Donald Lewis, a professor of government at Oberlin College, and Ewart K. Lewis, a medieval historian. Shortly before his sixteenth birthday he entered Swarthmore College, where his teachers, Jerome Schaffer and Michael Scriven, and such fellow students as Gilbert Harman, Allan Gibbard, Barbara Hall (later Barbara Hall Partee), and Peter Unger, were early philosophical interlocutors and influences. Lewis initially planned to major in chemistry. That changed during a year in Oxford in 1959–1960, where he was tutored by Iris Murdoch and went to lectures by Gilbert Ryle, Peter Frederick Strawson, Paul Grice, and John Langshaw Austin. (Lewis returned to Oxford to deliver the John Locke Lectures in 1984.)

After graduating from Swarthmore in 1962, he went to Harvard University for his doctoral studies under Willard van Orman Quine. Lewis's dissertation, revised to become his first book, *Convention* (1969), was partly a rehabilitation of the analytic-synthetic distinction in the face of Quine's famous rejection of it. Lewis's application of game theory to analyzing conventions, understood as coordination problems, was to some extent inspired by Thomas Schelling (a Harvard economics professor at the time), who had deployed game theory in his study of the strategy of conflict. Lewis was attracted to Nelson Goodman's egalitarianism about properties and to Donald C. Williams's four-dimensionalist approach to time, with its analogizing of time to space (and temporal parts to spatial parts). These figures were also fine philosophical writers, and as such they arguably served as models for Lewis, who went on to earn a reputation himself as a master of philosophical prose.

In 1963 Lewis befriended John J. C. Smart, who visited Harvard from Australia, consolidating Lewis as an "Australian materialist" regarding the mind, which meant that he believed that mental states are physical states—specifically, neurochemical states—that play certain causal roles. An argument for this position is expounded in his first publication, "An Argument for the Identity Theory" (1966). In Smart's graduate seminar, Lewis met Stephanie Robinson, whom he married in 1965. Lewis's friendship with Smart led to an ongoing connection with Australia and annual trips there for thirty years. David M.

Armstrong became a longtime friend and correspondent of Lewis's, playing an important role in what Lewis later described as "a big turning point in my philosophical position": recognizing the distinction between natural properties (such as "green") and gerrymandered properties (such as "grue").

Lewis was hired by the University of California at Los Angeles, working there from 1966 to 1970. The philosophy department was a hothouse for formal semantics, with colleagues such as Rudolf Carnap, Donald Kalish, Hans Kamp, Richard Montague, David Kaplan, and Barbara Hall Partee, whose collective influence can be discerned in some of Lewis's early publications in the philosophy of language and philosophical logic. He began teaching at Princeton in 1971, where he was associate professor, then full professor from 1973, subsequently holding a series of endowed chairs. Richard Jeffrey's Bayesian decision theory proved influential on Lewis, underpinning both his early work on interpretivism in the philosophy of mind and his later work on rational credence and decision.

Metaphilosophy. Lewis's metaphilosophy is intertwined with his first-order philosophy. He takes seriously Quine's maxim regarding ontological commitment—we should regard as existing everything over which we quantify in our best theories—and the pragmatism that underlies his related "indispensability" argument for mathematical objects. Shades of the indispensability argument are recognizable in the Lewisian doctrine that philosophical positions should be judged by how well their costs trade off against their benefits. Their costs are measured by the extent to which their consequences are unintuitive or in tension with our best science (including mathematics); their benefits are measured by their ability to systematize folk theory and the findings of science. But despite these fundamental regulative roles played by common sense and science, they are not to be deferred to unquestioningly. Like Carnapian explications, philosophical theories may not perfectly respect our pretheoretical judgments, and science itself (particularly quantum mechanics) may need further refinement. According to Lewis, then, philosophical analysis looks to folk theory, identifies the theoretical roles of target terms in this theory, and looks for the best "deservers" of those words—things in the world that best play the specified roles. To be sure, nothing may play these roles perfectly; but imperfect candidates may suffice for many philosophical purposes.

The best-systematization approach that Lewis exemplifies in his philosophical methodology resonates with some of his more specific philosophical positions. He analyzes laws of nature, for example, as the regularities that appear in the true theories of the universe that best balance simplicity and strength. Likewise, he contends that mental states should be attributed to others by balancing various standards for best interpreting their behavior.

The ghost of Carnapian positivism is found in Lewis's thesis of Humean supervenience: "all there is to the world is a vast mosaic of local matters of particular fact." The natural properties discovered by science (many familiar to common sense) are crucial here: everything is determined by the distribution of perfectly natural intrinsic properties of space-time points. Here again, Lewis's metaphilosophy shades seamlessly into his philosophy proper.

Philosophical Positions. Lewis's Quinean methodology yields some positions quite at odds with Quine's. Applied to modality, it results in Lewis's most notorious doctrine: his modal realism, as defended in *On the Plurality of Worlds* (1986). On this view, for each way that a world (universe) *could* be, there is a world that *is* that way—we may call these worlds *concrete* to emphasize that they are of a kind with our world, not merely abstract objects or linguistic entities. Our world is privileged only insofar as it is actual (for us); but worlds in which donkeys talk, or in which there are alien properties, are just as real as ours. Indeed, Lewis believes that gods exist in infinitely many worlds, even though our world happens to be godless! He argues that modal realism strikes the best overall balance of costs and benefits among philosophical accounts of modality. Its primary cost is its offense to common sense, enshrined in what he calls the "incredulous stare" objection. But this cost is more than offset by the manifold ways in which the doctrine is serviceable to philosophy, and that, Lewis insists, is a reason to believe that it is true. It provides elegant analyses of such philosophical bugbears as possibility and necessity, supervenience, counterfactuals, verisimilitude, mental content, and properties. Moreover, he maintains that these benefits cannot be realized as well by any rival philosophical theory—in particular, by any theory that offers some "ersatz" alternative to concrete possible worlds.

Many of Lewis's specific philosophical positions can be fitted into a bravura chain of reductions or dependences. Several of them appeal to the notion of causation. For example, he regards things (objects, persons) that persist through time as consisting of temporal parts, typically united by causal continuity: later parts depend causally on earlier parts for their existence and nature. In the case of persisting persons, such causal dependence will be among psychological states. Psychological states underpin his analyses of conventions, as we find in languages—for example, semantic facts exist in virtue of the mutual expectations of members of a linguistic community. And psychological states, in turn, are definable as the occupants of certain causal roles. Beliefs and desires—or more generally, degrees of belief and degrees of desire, understood decision-theoretically—are analyzed in terms of their functional role. Perception is likewise defined in terms of appropriate causal relations between external scenes and an agent's representations of them.

Causation, for its part, is analyzed by Lewis in terms of patterns of counterfactual dependence among events. Indeed, it was his concern to secure proper foundations for that analysis that prompted Lewis to write his book *Counterfactuals* (1973), a seminal work on the truth conditions and logic of conditionals that are typically expressed in the subjunctive mood—for example, "if kangaroos didn't have tails, then they would topple over." Lewis's analysis of counterfactuals invokes relations of comparative similarity among possible worlds. Roughly, "if it were the case that X, it would be the case that Y" is true (at a given world "w") if and only if there is a world in which X and Y are true that is more similar (to "w") than any world in which X is true and Y false. Similarity of worlds is determined by closeness of match of matters of particular fact; the sharing of laws of nature is an important respect of such match, because they codify much information about what is true at the relevant worlds.

This brings us back to the worlds themselves. They are individuated by spatiotemporal connectedness: you and I are world-mates because we are spatiotemporally related to each other. But distinct possible worlds are isolated from each other, bearing no spatiotemporal relation to each other; likewise, parts of worlds, such as you and I, are isolated from parts of other worlds. Lewis infers from this that modal claims about individuals are made true by corresponding facts about counterparts of these individuals in other possible worlds, rather than by facts about the individuals themselves in these other worlds. "You could have been a movie star" is thus analyzed roughly as: There is a possible world in which a counterpart of yours—someone who plays a very similar role in that world to the one you play in the actual world—*is* a movie star.

Lewis's views on other topics respect his broader philosophical commitments, especially to modal realism and to Humean supervenience. While his early work is concerned more with issues in the philosophy of language, philosophical logic, and the philosophy of mind, he moves in later work more towards metaphysics and (to a lesser extent) ethics-social philosophy, philosophy of mathematics, and epistemology. This essay closes with a quick overview of some further distinctive themes.

An integrated set of papers concerns probability and decision theory. Lewis regards opinion as coming in degrees—"credences"—and he follows Bayesians in modeling rational credences as subjective probabilities. These are constrained by the usual probability axioms, but Lewis adds a further principle that links one's credences to one's beliefs about corresponding objective chances. He provides a novel defense of a particular rule (conditionalization) for updating rational credences in the face of new evidence. He famously proves various "triviality results" against the equating of probabilities of conditionals with conditional probabilities. On the side of decision, Lewis offers a version of causal decision theory according to which rational decisions maximize expected utilities of actions, with probabilities and utilities assigned to various "dependency hypotheses" about how outcomes depend causally on one's actions. He proves further "triviality results" against an anti-Humean thesis that decision-theoretically reduces "desires" to "beliefs" of a certain kind.

In metaethics, Lewis argues for a subjectivist position that portrays our values as those properties that we are disposed to desire to desire, when suitably apprehending them. He has a number of papers on more specific topics in social philosophy, including deterrence, punishment, and tolerance. *Parts of Classes* (1990) is an important contribution to the philosophy of mathematics, in which Lewis reduces set theory to mereology (the theory of the "part-whole" relation). In epistemology, his "Elusive Knowledge" (1996) is a groundbreaking work that gives a new analysis of "S knows that p," accompanied by a detailed contextualist analysis of the pragmatics of knowledge ascriptions. And then there are many papers on sundry other topics: holes, properties, dispositions, truth, vagueness, fiction, quantum mechanics, and more.

Lewis's work continues to be enormously influential, and often agenda-setting. He was renowned as a great teacher and supervisor. His sudden death in 2001, due to complications arising from many years of diabetes, ended a remarkable career.

BIBLIOGRAPHY

The Lewis archive is in the care of his wife, and literary executor, Stephanie R. Lewis. There is a hope that the archive will be deposited someplace, but that has yet to be arranged. There is not much of Lewis's work that is unpublished; efforts are in progress to see the remaining unpublished works into print. The major source of unpublished work by Lewis is in his correspondence—there are many hundreds of letters of philosophical import in his files. Ms. Lewis is working on publishing a series of volumes of Lewis's correspondence.

WORKS BY LEWIS

"An Argument for the Identity Theory." *Journal of Philosophy* 63 (1966): 17–25.

Convention: A Philosophical Study. Cambridge, MA: Harvard University Press, 1969.

Counterfactuals. Oxford: Blackwell and Cambridge, MA: Harvard University Press, 1973.

Philosophical Papers. 2 vols. New York: Oxford University Press, 1983–1986.

On the Plurality of Worlds. Oxford and New York: Blackwell, 1986.

Parts of Classes. Oxford and New York: Blackwell, 1990.

"Elusive Knowledge." *Australasian Journal of Philosophy* 74 (1996): 549–567.

Papers in Philosophical Logic. Cambridge, U.K.: Cambridge University Press, 1998.

Papers on Metaphysics and Epistemology. New York, and Cambridge, U.K.: Cambridge University Press, 1999.

Papers in Ethics and Social Philosophy. New York, and Cambridge, U.K.: Cambridge University Press, 2000.

OTHER SOURCES

Jackson, Frank, and Graham Priest, eds. *Lewisian Themes: The Philosophy of David K. Lewis.* Oxford: Oxford University Press, 2004.

Nolan, Daniel. *David Lewis.* Montreal: Acumen Publishing; and Kingston, Ontario: Chesham and McGill-Queen's University Press, 2005.

O'Grady, Jane. "David Lewis: Princeton Philosopher Who Formulated Ground-Breaking Theories on Everything from Language to Identity to Alternative Worlds." *Guardian,* 23 October 2001. Obituary. Available from http://books.guardian.co.uk/news/articles/0,6109,579258,00.html.

Alan Hájek

LEWIS, WARREN KENDALL

(*b.* Laurel, Delaware, 21 August 1882; *d.* Plymouth, Massachusetts, 9 March 1975), *chemical engineering, distillation, fluidized-bed catalytic cracking, continuous automatic chemical processing.*

According to Ralph Landau, a distinguished chemical engineer who began his career at the M.W. Kellogg Company, one of the first American engineering firms to specialize in the design and development of plants for the modern chemical and oil industries, Lewis "virtually single-handedly created modern chemical engineering and its teaching methodology" (1991, p. 49). "Doc" Lewis converted nineteenth-century industrial chemistry into twentieth-century chemical engineering. He did so by using the new physical chemistry initiated in Europe by Wilhelm Ostwald (and taught in the United States by his student, Arthur A. Noyes), and crucially, by establishing the new concept of unit operations, developed by William H. Walker.

In 1920, Lewis was appointed chairman of the new Chemical Engineering Department of the Massachusetts Institute of Technology (MIT). A good university administrator, a successful and respected teacher, a sought-after technical consultant with scores of patents to his name, and an enthusiast for the scientific method, he promoted the role of the engineer in society while acknowledging the importance and difficulties of human relations in industry. Lewis's seminal work on the distillation and the cracking of petroleum provided the foundations of the modern oil processing industry. A religious man, Lewis never stopped trying to reconcile his faith with his profound belief in science. His numerous honors include membership in the National Academy of Sciences (1938), the Priestley Medal of the American Chemical Society (1947), the President's Medal for Merit (1948), the Gold Medal of the American Institute of Chemists (1949), and honorary doctor of science degrees from the University of Delaware, Harvard University, and Princeton University.

Childhood and Education. Lewis was born into a middle-class farming family, the only child of Martha Ellen Kinder Lewis, who ran her own millinery business, and Henry Clay Lewis. As he grew up in the 1880s and 1890s, Warren and his parents expected that he would inherit the farm that had been in the family since the eighteenth century. To prepare young Warren for a college education in the agricultural arts, he was sent at age fifteen to live with a cousin in Newton, Massachusetts, so that he could attend the local high school, which had a good reputation. Lewis was not immediately a star pupil, but he thrived on competition and eventually chose to enroll at MIT in Cambridge. Because no agronomy course was offered, Lewis opted for a mechanical engineering major, changing later to chemical engineering.

William H. Walker, a graduate of Pennsylvania State College with a PhD in organic chemistry from the University of Göttingen in Germany, taught analytical chemistry at MIT until 1900, when he left to enter into a partnership with the consultant Arthur D. Little. Walker and Little were convinced that the chemistry departments of MIT and other institutions should change their curricula to fit better with the needs of various industries, based on the chemical processes that they used. At the time, MIT offered several chemical engineering options, each one tailored to only one type of chemical manufacturing.

Walker believed that the teaching method of his predecessor at MIT, Frank H. Thorp, was wrong. "What an industry needed," Walker wrote, "was not a man who had been taught what that industry already knew, but rather a man who was trained to do what the industry had not been able to do" (p. 2). Walker's vision of a chemical engineer was a man who had been trained not only in physics and chemistry, but also in applying his knowledge to solve whatever industrial problems arose. Walker met opposition in academia and industry, but he returned to teach at

Warren Kendall Lewis. Warren Kendall Lewis in laboratory talking to colleague. ALFRED EISENSTAEDT/TIME LIFE PICTURES/GETTY IMAGES.

MIT in 1903 "to prove the soundness of [his] idea" (p. 2). He reconstructed the institute's Course X as a "general education course without options" (p. 2)

Writing in 1934 about his changes to the chemical engineering course, Walker recalled that he "cut such courses as mechanical drawing, analytical chemistry, shop and foundry practice, and introduced all the physical chemistry that was then available and greatly strengthened the courses in organic and advanced inorganic chemistry. I then organized a laboratory course in industrial chemistry which was designed to teach method of attack in the solution of industrial problems through the application of chemical engineering already acquired. This was the beginning of the [modern] course in chemical engineering" (p. 2). Walker used George F. Davis's *A Handbook of Chemical Engineering* (2 vols., 1901–1902) as a model for organizing the material for his revised Course X.

Walker hired Lewis as a research assistant before Lewis had graduated in 1905. With a Swett Fellowship and an Austin Traveling Fellowship from MIT, Lewis moved to Breslau University in Germany for his PhD work, supervised by Richard Abegg (a pupil of Ostwald, Svante Arrhenius, and Walther Nernst). He successfully defended his fifty-five-page dissertation on physical chemistry, "Die Komplexbildung zwischen Bleinitrat und Kaliumnitrat" (The complexes of lead and potassium nitrates) in July 1908.

Early Career. In 1910, Lewis was appointed assistant professor of industrial chemistry at MIT. As Walker said later, the new Course X became synonymous with modern chemical engineering and "was copied more or less by many other institutions" (p. 2). Asked in 1905 to give an industrial chemistry course at Harvard, Walker decided to introduce the concept known (from about 1915) as unit operations, standard processes such as crushing and grinding, filtration, distillation, crystallization, and drying, which had applications in chemical industries with many different products. Walker's lecture notes were a foundation for the groundbreaking textbook, *Principles of Chemical Engineering,* coauthored with Lewis and William H. McAdams, a student of Lewis, and eventually published in 1923.

In 1911 Walker and Lewis published an article in the *Journal of the American Chemical Society* titled "A Laboratory Course of Chemical Engineering," a description of a four-year undergraduate course intended to demonstrate general chemical engineering principles (such as the economic importance of minimizing heat losses). Walker was in charge of chemical engineering at MIT from 1912 to 1920; he believed that the institute should concentrate on applied sciences and the training of builders and leaders of industry. With those objectives, Walker and Lewis, supported by Little, founded the MIT School of Chemical Engineering Practice in 1916 to give undergraduates "the engineering equivalent of the hospital internship" (Lewis, 1953, p. 700). Lewis realized that simply learning the scientific principles of physics, chemistry, and engineering would not easily enable students to apply their knowledge to complex problems in industrial chemical engineering, but in the practice school, especially by focusing on unit operations, they would understand how commercial processes could be analyzed and quantified, and thereby improved and made more profitable. The practice school required significant commitments from MIT, the students, and industry. George Eastman agreed to provide three hundred thousand dollars for equipment, and a number of MIT stations were established at chemical plants in various industries, staffed by MIT teachers.

Well after his own retirement, Lewis said that the payoff from Walker's reorganization of the MIT chemical engineering curriculum in the early years of the twentieth century came during World War I. German shipping was subject to an Allied blockade, so that many imports from Europe had to be produced in America; furthermore, the demand for explosives multiplied as Europe's Allied powers turned to the United States for supplies. At the same time, the growth of the automobile industry (total car production in the United States in 1918 exceeded 800,000) was pushing up the demand for gasoline.

The provision of the Hague Convention of 1899 that outlawed projectiles containing poison gas held only until World War I; by then, most European governments had a gas capability. The *Boston Evening Transcript* of 11 January 1919 proclaimed that "poison gas will remain indefinitely one of the weapons of civilized warfare." Two months before the United States entered the war, Van H. Manning, director of the U.S. Bureau of Mines, advised the War Department to prepare for gas warfare; the National Research Council (NRC) formed a committee on gases, chaired by Manning. MIT academic staff were co-opted. The physical chemist Arthur A. Noyes chaired the NRC in Washington, D.C.; Walker, a commissioned colonel in charge of the new Chemical Warfare Service, built and ran the Edgewood Arsenal near Baltimore, the U.S. Army's first chemical warfare facility; and Lewis became the civilian head of the Service's Gas Defense Production Division.

In 1920 a separate Chemical Engineering Department was created at MIT under Lewis's chairmanship. Three years later the pioneering chemical engineering text, *Principles of Chemical Engineering*, was published, and for years afterward, chemical engineering was particularly associated with MIT. In the preface, the authors emphasized that "the treatment is mathematically quantitative as well as qualitatively descriptive"; chemical engineers, they said, should themselves design the industrial apparatus they need, rather than building it first and relying on trial and error to make it work (Lewis, et al., 1923, p. v).

Writing in 1953, Lewis acknowledged the considerable achievement of Davis's *Handbook*, but noted that "his quantitative treatment of operations was limited by lack of data" (Lewis, 1953, p. 699). This was not a criticism; simply to calculate the heat lost by a fluid flowing through a pipe required a considerable amount of experimental data. Aware of the importance of chemical engineering data, Lewis was an advocate of the undergraduate thesis, which trained MIT students to produce valuable empirical data that could be rapidly diffused to professional engineers and consultants. Much of the utility of the *Principles* lay in the useful data that it contained. E.I. du Pont de Nemours & Company (DuPont), recognizing the need for chemical engineering data, began a program of fundamental research in 1927, and in 1934 John H. Perry issued the first edition of the *Chemical Engineer's Handbook*, the work of sixty contributors.

Warren K. Lewis's teaching style was idiosyncratic, but among his papers are numerous affectionate testimonies to its effectiveness from grateful former students. Furthermore, a remarkable number of his students became eminent chemical engineers. Three of them were eventual heads of MIT's department of chemical engineering. In the classroom, Lewis could be merciless when faced with a student who came badly prepared; nevertheless, a collection of anecdotes about himself and stories that he told his students was published by them under the title *A Dollar to a Doughnut* (1953)—a bet that he often made (and sometimes lost) in scientific disputes with students.

Continuous Distillation. In 1911 the two largest refiners of petroleum were the Standard Oil Company of Indiana, later Amoco and part of British Petroleum from 1998, and the Standard Oil Company of New Jersey, or Esso, later part of ExxonMobil. (Hereafter, these firms are referred to as Amoco and Esso, respectively.) Natural petroleum was separated into various fractions by distillation in the later decades of the nineteenth century, when the greatest demand was for kerosene for lighting. Batches were heated and the lightest hydrocarbons were vaporized first and then were condensed by passing them through cooled tubes. Eventually, gasoline became the most important product. Higher boiling fractions could be transformed into lower boiling fractions, such as gasoline, by a process called cracking. Amoco had the larger research group and developed a thermal cracking process, using heat to break large molecules (heavier oils or tars) into smaller ones, thereby increasing the yield of gasoline, for which a post-1918 boom was accurately predicted.

Distillation and cracking dominate oil refining. Early in the twentieth century, Lewis worked at MIT on the theory of continuous distillation (extending the work of Ernest Sorel in France in the 1890s) in order to make precision distillation a continuous and automatic process. Lewis first realized that industrial practice was well behind his own and his university colleagues' understanding of the distillation of multicomponent mixtures such as petroleum when an MIT colleague visited a refinery. In the late nineteenth century and the early decades of the twentieth century, chemical engineers developed the theory of fractional distillation in terms of multiple stages, and it was found that multistage distillation could be achieved with a fractionating column in which there was a series of perforated plates: vapor bubbled up through the column from plate to plate and liquid flowed down. With experience and more theoretical study, material to be distilled was fed in part way up the column and part of the

Lewis

vapor leaving the top was condensed and returned to the column as reflux, improving the separation of the components of the feed. Different fractions of product could be withdrawn from different points in the column. Lewis and his colleague (and former doctoral student) Edwin R. Gilliland acquired international reputations in this field.

Esso lacked a development facility but soon hired an engineer from its competitor, Amoco; realizing that patent protection of any new methods would be vital, Esso also hired Amoco's patent attorney, Frank A. Howard. In 1919 Esso created a new Development Department, headed by Howard, who immediately engaged Lewis, the best consultant he could find—by then, Lewis had a strong record with the Goodyear Tire and Rubber Company and the Humble Oil and Refining Company. From 1914 to 1927, Esso's gasoline yield doubled to 36 percent of the crude oil input, partly through Lewis's efforts. Howard now made contact with the German chemicals giant I. G. Farben, initiating agreements that gave Esso access to Farben's work on coal hydrogenation. He hoped that the technical information would improve gasoline yields and increase Esso's activities in chemicals. Howard realized that Esso would need a new research group and asked Lewis for advice; Lewis recommended hiring Robert Haslam, head of MIT's School of Engineering Practice.

Haslam set up a team of fifteen researchers (all MIT faculty members and graduates) in Baton Rouge, Louisiana. By applying German know-how to oil refining, American petrochemical firms closed the gap between industrial practice and university and industrial research. From 1935 Gilliland (who was the fourth author of the third edition of the *Principles of Chemical Engineering* [1937]) took on Lewis's role of consultant to the Baton Rouge group. Lewis, however, had been pivotal: it was he who focused chemical engineering on the design of continuous automated processing, leading to enormous growth in the world petrochemical industry. Lewis showed that university professors with hands-on industrial experience were good teachers of chemical engineering at MIT, and their Chemical Engineering Department enjoyed high worldwide prestige during the 1920s and 1930s.

Fluidized-Bed Catalytic Cracking. Petroleum contains a complex mixture of molecules, ranging from the light gases (one to four carbon atoms), through gasoline (five to ten carbon atoms) and lubricating oils (twenty to fifty carbon atoms), to heavy bituminous residues whose molecules contain more than seventy carbon atoms. Cracking transformed the heavy fractions into more valuable lighter ones. A short two-and-a-half page memorandum in the MIT archives, written by Lewis probably in the 1930s, is typical of his scientific approach to new problems. He listed eleven points that he had deduced from his experience of cracking. For example, he learned that the stability of a hydrocarbon decreases in any series as the molecular weight increases and that the C-H bond is the most stable; therefore, it is the C-C bonds that tend to break. Thermal cracking produced sixty to seventy octane gasoline, which required a tetraethyl lead additive to boost the octane number and so prevent gasoline engine pre-ignition. By 1930, however, the French engineer Eugène Houdry had invented a method of producing high-octane fuel by cracking heavy tars using a silica-alumina catalyst. Houdry succeeded in selling the process in the United States to the Sun Oil Company, which worked on its development with Socony-Vacuum; half a dozen plants were in operation in the early 1940s. Lewis was consulted about catalytic cracking by Esso, and serious work began at the Baton Rouge research center in 1936.

The catalyst was quickly poisoned by carbon deposits. This was at first overcome by using two separate beds of catalyst: the vaporized petroleum was passed (at high temperature and pressure) through the first catalyst bed and after a predetermined time, the petroleum stream was automatically switched to a second bed, while the first bed was regenerated with an air blast; the process, however, was complex and costly. Small-scale experiments at MIT showed that gasoline could be successfully produced if powdered catalyst were mixed with petroleum vapor and passed under pressure through a hot pipe. Lewis and Gilliland's innovation was the fluidized catalyst bed: Particles of catalyst were suspended in the flowing petroleum vapor, giving time for the chemical reactions to take place; with the fluidized catalyst in suspension, it was easier to automate the process, passing the catalyst rapidly between a reaction zone and a regeneration zone. Lewis's design was in operation in 1940; the first plant cost $4.5 million and processed 13,300 barrels of oil per day. Soon, the production of American high-octane aviation fuel was increased a hundredfold and quantities were shipped to Europe, helping the Allied effort in World War II. By 1962 there were 222 fluidized-bed catalytic crackers worldwide.

Polymers: The Manhattan Project. In 1929, after nine years in the post, Lewis resigned his chairmanship of the MIT Chemical Engineering Department to concentrate on research and teaching; he was nearly forty-eight years old. However, his interests continued to expand with scientific developments, and in 1942 he published *Industrial Chemistry of Colloidal and Amorphous Materials,* with coauthors Lombard Squires of DuPont and Geoffrey Broughton of Eastman Kodak. The book, reprinted twice that year, included discussions of the newly synthesized polymers (such as polyester and nylon fibers, polystyrene, and the rubberlike polyisobutylene) as well as the commercially important natural polymers: wool, cotton, and silk. The illustrations included the kind of x-ray diffraction photographs of fibers that were important in establishing the structure of large organic molecules such as DNA a decade later. Lewis and his coauthors used basic science to explain large-scale phenomena in terms of the forces between atoms and molecules. In the 1930s and 1940s, this was a new approach to engineering, and the term *engineering science* was soon current. Lewis noted that polymer chemistry had developed to the point that "industry [could] produce materials of almost any required physical characteristics" (Lewis, et al., 1942, p. 469).

In 1941 Lewis was asked in his capacity as a leading consultant in the process industries to join a National Academy of Sciences committee to review existing atom bomb research; the Manhattan project received presidential approval later that year. From September 1942, the project was managed by General Leslie R. Groves, a senior army engineer. Groves appointed a review committee under Lewis's chairmanship to monitor progress and priorities; the other committee members were the ordnance design engineer E. L. Rose, the Harvard physics professor John H. Van Vleck, and the physicist Richard C. Tolman (a former high-school friend of Lewis). Lewis and his committee traveled constantly during World War II among the various Manhattan research laboratories and development sites, including Berkeley, Clinton, Hanford, Los Alamos, and the naval research laboratory in Washington, D.C. The Lewis committee approved the nuclear physics research program proposed by J. Robert Oppenheimer and his staff, and they recommended that work on a thermonuclear bomb should have lower priority than the atomic bomb. Lewis was among the select group invited to witness the first self-sustaining nuclear chain reaction on 2 December 1942 in Chicago; then sixty years old, Lewis at the last minute generously gave up his place to a younger man (probably Edwin Gilliland). The successful production of the atomic bomb irrevocably changed the world: all-out war would in the future be either unthinkable or suicidal. Later, Lewis hardly ever talked about the Manhattan Project, although he was haunted by it; he had believed that it was necessary to build the bomb, but in his very old age he "worried that he could not wash the radiation off his hands" (Williams, 2002, p. 6).

Social Responsibility of Engineers. A year after the Hiroshima bomb, MIT vice president James R. Killian Jr., in a mood of postwar reevaluation, suggested to the MIT president, Karl Compton, that a committee of the faculty be convened to study educational objectives, organization, and operations at the institute. Some fundamental questions needed answers. For example, were courses too theoretical? Was there enough time for humanistic studies? Killian suggested that Lewis was the logical person to chair the committee, in view of "his great prestige and his strong interest in teaching" (Williams, 2002, p. 67). The *Report of the Committee on Educational Survey,* known as the Lewis Report, was published in December 1949; Lewis was sixty-seven and had retired the previous year.

The nub of the *Report* lay in chapter 3, "A Broader Educational Mission." Acknowledging the increasing complexity of society, it stated that science and technology could not be separated from their human and social consequences, that the postwar generation's most difficult and complicated problems lay in the humanities and the social sciences, and that these problems reflected the impact of science and technology on society. MIT now had the opportunity to make a larger contribution to solving social problems and to giving scientists and engineers a better understanding of the forces at work in society. Conversely, the school could also give social science and humanities students a deeper understanding of the implications of science and technology. After the publication of the Lewis Report, the School of Humanities and Social Sciences was added to MIT's three existing schools (Engineering, Science, and Architecture and Planning). The Lewis Committee believed that MIT was in a position to make contributions "to education and the advancement of knowledge" in all four spheres (Lewis, et al., 1949b, pp. 42–43).

Lewis wrote and spoke about the role of the engineer in society. Not infrequently, he made the point that found its way into the Lewis Report, namely, that the most difficult problems were not those of engineering and the exact sciences, but those of sociology, economics, and political science. In articles by Lewis published in the 1940s and 1950s, he identified the essential product of engineering: the huge increase in human efficiency. Lewis recalled his job in a tannery in New Hampshire about fifty years earlier, when the working week was seventy-eight hours; by the 1950s it was down to forty hours and the standard of living of the worker had more than doubled. In the 1850s, fewer than 5 percent of children attended high school, whereas in the 1950s the figure was 85 percent. And yet, Lewis noted, the operators of the machinery that produced unprecedented wealth and leisure were often less happy than the subsistence workers of the past. Lewis's point was that engineers who failed to attend to the social consequences of technological developments were not members of a profession, but mere technicians who were uninformed about the social, economic, political, and international environment.

Lewis was fond of defining the engineer as "someone who can do for a dollar what any damn fool can do for two" (Williams, 2002, p. 30). He had faith in the utility of science and engineering and believed them to be the agents for social progress. Engineering students, virtually all men, many of them (even as late as the 1960s) from the poorer and less-well-educated segments of the middle class, were able to obtain a professional degree in four years. Such students had a strong belief in meritocracy and accepted capitalist society. Lewis too was a staunch believer in the profit system as an objective measure in a competitive capitalist economy of the success of an enterprise and its economic contribution to the community, but he despised the profit motive—undertaking an enterprise with the sole purpose of making a profit. The irony for Lewis was that his own profession, chemical engineering, helped fuel the automobile age and later supplied the antibiotics required to support factory farming, so that it was instrumental in destroying his childhood way of life on a farm near a small town (Williams, 2002, p. 12).

BIBLIOGRAPHY

There is a great deal of unpublished, uncataloged archival material pertaining to Warren K. Lewis at MIT. Most of it resides in the MIT archives (building 14). The author of this article was informed that the majority of the papers were classified, presumably because of Lewis's wartime association with the Manhattan Project; no one had asked for them to be declassified. At the opposite side of the campus is located the MIT Museum (building 52), which holds fewer papers—again uncataloged. Professor Rosalind H. Williams (Lewis's granddaughter) is a senior faculty member at MIT and has a quantity of papers, though many have been handed over to MIT. Although this author knows of no single exhaustive bibliographic list, a combination of the one for this article and Hottel's would be fairly comprehensive.

WORKS BY LEWIS

"The Theory of Fractional Distillation." *Industrial and Engineering Chemistry* 1 (1909): 522–533.

With William H. Walker. "A Laboratory Course of Chemical Engineering." *Journal of the American Chemical Society* 33 (January–June 1911): 618–624.

With Frank Hall Thorp. *Outlines of Industrial Chemistry: A Text-Book for Students.* 3rd ed. New York: Macmillan, 1916.

With William H. Walker and William H. McAdams. *Principles of Chemical Engineering.* New York: McGraw-Hill, 1923.

With Lombard Squires and Geoffrey Broughton. *Industrial Chemistry of Colloidal and Amorphous Materials.* New York: Macmillan, 1942.

"The Professional Responsibilities of the Technical Man." *Chemist* (June 1949a): 205–211. This was Lewis's acceptance address on the occasion of his receipt of the American Institute of Chemists Gold Medal, 7 May 1949, in Chicago.

With John F. Loofborouw, Ronald H. Robnett, C. Richard Soderberg, et al. *Report of the Committee on Educational Survey to the Faculty of Massachusetts Institute of Technology,* Cambridge, MA: Technology Press, 1949b. Copies are in the MIT Museum.

"Chemical Engineering—A New Science." In *Centennial of Engineering, 1852–1952,* edited by Lenox R. Lohr. Chicago: Museum of Science and Industry, 1953.

"The Future of Engineering as a Profession." *Technology Review* 59, no. 7 (May 1957): 351–354.

"Evolution of the Unit Operations." *Chemical Engineering Progress Symposium Series* 55, no. 26 (1959): 1–8.

OTHER SOURCES

Brigham, W. E. "We Were Ready...." *Boston Evening Transcript,* 11 January 1919.

Cohen, Clive. "The Early History of Chemical Engineering: A Reassessment." *British Journal for the History of Science* 29 (1996): 171–194.

A Dollar to a Doughnut, or Doc Lewis, as Remembered by His Former Students. New York: American Institute of Chemical Engineers, 1953.

Furter, William F., ed. *History of Chemical Engineering.* Washington, DC: American Chemical Society, 1980.

Hottel, Hoyt C. "Warren Kendall Lewis, August 21, 1882–March 9, 1975." *Biographical Memoirs of the National Academy of Science (U.S.A.)* 70 (1996): 204–218. Also available from http://stills.nap.edu/html/biomems/wlewis.html.

Hounshell, David A., and John Kenly Smith. *Science and Corporate Strategy: DuPont R&D, 1902–1980.* Cambridge, U.K.; New York: Cambridge University Press, 1988.

Landau, Ralph. "Academic–Industrial Interaction in the Early Development of Chemical Engineering at MIT." *Advances in Chemical Engineering* 16 (1991): 41–49.

Landau, Ralph, and Nathan Rosenberg. "Successful Commercialization in the Chemical Process Industries." In *Technology and the Wealth of Nations,* edited by Nathan Rosenberg, Ralph Landau, and David C. Mowery. Stanford, CA: Stanford University Press, 1992.

Mattill, John. *The Flagship: MIT School of Chemical Engineering Practice 1916–91.* Cambridge, MA: MIT Press, 1991.

Peppas, Nicholas A. *One Hundred Years of Chemical Engineering.* Boston, and Dordrecht, Netherlands: Kluwer Academic Publishers, 1989.

Perry, John H., ed. *Chemical Engineer's Handbook.* New York and London: McGraw-Hill, 1934.

Rhodes, Richard, *The Making of the Atom Bomb.* New York: Simon & Schuster, 1986.

Servos, John W. "The Industrial Relations of Science: Chemical Engineering at MIT, 1900–1939." *ISIS* 71 (1980): 531–549.

Spitz, Peter H. *Petrochemicals: The Rise of an Industry.* New York: Wiley, 1988.

Weber, Herman C. *The Improbable Achievement: Chemical Engineering at MIT.* Washington, DC: American Chemical Society, 1980.

Williams, Rosalind H. *Retooling: A Historian Confronts Technological Change.* Cambridge, MA: MIT Press, 2002. Williams is Lewis's granddaughter.

Clive Cohen

LIAIS, EMMANUEL-BERNARDIN

(*b.* Cherbourg, France, 15 February 1826; *d.* Cherbourg, 5 March 1900), *astronomy, meteorology, instrumentation, scientific institutions and expeditions.*

Liais is known for his initiative in the creation and organization of scientific institutions such as Cherbourg's National Institute of Natural Sciences and Mathematics, the French meteorological telegraphic network, and the National Observatory of Brazil. He also contributed to the improvement of scientific instruments such as the recording barometer, the electric clock, and the alt-azimuth. Finally, under the influence of Alexander von Humboldt, he devoted himself to the exploration of Brazil's nature and to the popularization of natural sciences in Europe.

Early Works. The only son of Anténor Liais and Mathilde-Françoise Dorey, a bourgeois couple, Emmanuel-Bernardin Liais was born in Cherbourg, an important seaport situated in Normandy. In the local secondary school he received awards for his achievements in mathematics and natural sciences, but he had no formal scientific training. His scientific interests during those early years focused on a variety of subjects, particularly in the domains of instrumentation and meteorology. Many experiments and observations were described in papers sent to scientific societies. Among them, a series of regular meteorological observations made in his homeland attracted the attention of François Arago, the secretary of the Paris Academy of Sciences and director of the Paris Observatory.

Liais entered the Paris Observatory in the beginning of 1854 on Arago's recommendation, even though the observatory was already under the leadership of Arago's successor Urbain Le Verrier. There is no empirical evidence that Liais entered the Paris Observatory before 1854, when Le Verrier was nominated its director, and it was under Le Verrier's patronage that he achieved successive promotions and the Légion d'Honneur. Liais helped the latter implement a telegraphic meteorological network centered at the observatory and spread throughout France, adapting the meteorological instruments for the use of telegraph operators. During this short stay at the observatory Liais developed a recording barometer and an electric chronograph. Most significantly, under the influence of Adolphe Quetelet, he applied the concept of atmospheric waves in a famous study on the path of the Balaklava storm, presented at the Paris Academy of Sciences on 31 December 1855. Based on this work, Le Verrier justified the wide institutional reform then underway, demonstrating the feasibility of providing weather forecasts to French stations and seaports.

Voyages to Brazil. Liais left the Paris Observatory in the beginning of 1858, after a disagreement with Le Verrier. With the excuse of observing the total solar eclipse of 7 September 1858, he traveled to Brazil, initially at his own expense, and worked there for almost two decades. The eclipse observation gave birth to different scientific studies, including pioneer experimentation on the use of photography for determining longitudes during such events. In 1860 he became involved in another controversy with Le Verrier. Liais was one of the first scientists to deny the existence of a new planet, named "Vulcan," mathematically predicted by Le Verrier and supposedly observed by Edmond Modeste Lescarbault.

Once in Brazil, Liais took the opportunity to explore its territory in expeditions, funded by the Brazilian government, toward the northeastern coast and the inland of Minas Gerais. In the town of Olinda, where a temporary observatory was erected, he discovered a new comet on 26 February 1860. During the exploration of the tropical forests around Rio de Janeiro, he identified and described a new botanical genus, which he named *Pradosia,* rendering homage to his closest Brazilian friend and collaborator, Camilo Maria Ferreira Armond, the viscount of Prados.

The Brazilian government sponsored the publication of three books based on the results of his expeditions. During short visits to France, Liais also published two books addressed to the general public: *L'Espace Céleste et la Nature Tropicale,* a voluminous text that was both a scientific treatise on physical astronomy and a picturesque narrative of his travels in Brazil, and *Suprématie intellectuelle de la France,* a pamphlet against the myth of German racial superiority, written under the impact of the French defeat to Prussia. On the basis of his previous works and especially the hydrographical and cartographical surveys accomplished in Minas Gerais, he meanwhile applied to the Paris Academy of Sciences in January 1866, without success.

The last years spent in Brazil were dedicated to the organization of the National Observatory, located in Rio de Janeiro. Under the patronage of the Brazilian emperor Dom Pedro II, Liais was director of this institution between 26 August 1870 and 1 May 1881. He improved its facilities, acquired new instruments, and succeeded in training a small number of employees, including the Belgian Louis Cruls, who would become his successor. These achievements were acknowledged on many occasions, such as the International Exhibition of Vienna in 1873, when an alt-azimuth designed by him was granted an award. However, international recognition did not prevent the eruption of bitter controversy with a group of Brazilian engineers, which finally led to his decision to leave Brazil.

Political Activities. Liais returned to France in 1881, but after preparing a new edition of *L'Espace Céleste* and securing the participation of the Brazilian Observatory in the International Transit of Venus Conference, he abandoned his scientific career. Back in Cherbourg he turned to politics. He was mayor between 1884 and 1886, when he resigned, and again between 1892 and 1900, having been reelected for the post in 1896.

In this late period of his life, he was nominated president of Cherbourg's Institute of Natural Sciences and Mathematics, a local scientific society that he cofounded in 1852 with Théodose du Moncel and Auguste Le Jolis. By that time he was already a member of many other scientific societies, such as the French Meteorological Society and the Brazilian Historical and Geographical Institute. More significantly, the patronage of Le Verrier in his youth and of Dom Pedro II during his stay in Brazil rendered him the nomination to the Légion d'honneur and to the Imperial Order of the Rose.

When Liais died in 1900, he was a widower. His wife Margaritha Trouwen had died in 1874 from a tropical fever, and they never had children. His properties in Cherbourg were donated to the city, under the condition of being used for the benefit of science. In fact, the Emmanuel Liais Park hosts the scientific society that Liais helped to create, and still has tropical greenhouses and an unfinished astronomical tower.

BIBLIOGRAPHY

The Emmanuel Liais Archives are in the Cherbourg National Institute of Natural Sciences and Mathematics, Cherbourg, France. Official documents can be found in the archives of the observatories of Paris and Rio de Janeiro. There is also a voluminous correspondence between Liais and Dom Pedro II deposited in the Archives of the Imperial Museum of Brazil, Petrópolis, Brazil.

WORKS BY LIAIS

"Sur la tempête de la mer Noire, en novembre 1854." *Comptes rendus hebdomadaires des séances de l'Académie des Sciences* 41 (1855): 1197–1204.

"Relation des travaux exécutés par la Commission astronomique chargée par le Gouvernement brésilien d'observer dans la ville de Paranagua l'éclipse totale du soleil qui a eu lieu le 7 septembre 1858." *Comptes rendus hebdomadaires des séances de l'Académie des Sciences* 47 (1858): 786–792.

Influence de la mer sur les climats; ou Résultats des observations météorologiques faites à Cherbourg en 1848, 1849, 1850, 1851. Paris: Mallet-Bachelier; Cherbourg: Bedelfontaine et Syffert, Imp., 1860.

"Observations astronomiques et physiques sur la comète découverte à Olinda le 26 février 1860, et éléments de la même comète." *Comptes rendus hebdomadaires des séances de l'Académie des Sciences* 50 (1860): 1089–1093.

"Sur la nouvelle planète annoncée par M. Lescarbault." *Astronomische Nachrichten* 52, no. 1248 (1860): 370–378.

"Détermination de la longitude de Paranagua au moyen d'épreuves photographiques de l'éclipse du 7 septembre 1858." *Comptes rendus hebdomadaires des séances de l'Académie des Sciences* 53 (1861): 29–32.

L'Espace Céleste et la Nature Tropicale; ou Description Physique de l'Univers d'après des observations personnelles faites dans les deux hémisphères. Paris: Garnier Frères, [1865].

Hydrographie du Haut San-Francisco et du Rio das Velhas; ou Résultats au point de vue hydrographique d'un voyage effectué dans la province de Minas-Geraes. Paris: Garnier Frères; Rio de Janeiro: B.L. Garnier, 1865. A first report on the scientific expedition to Minas Gerais, containing an analysis of the navigability of the São Francisco and Das Velhas rivers and fully illustrated with maps.

Traité d'Astronomie appliquée à la Géographie et à la Navigation suivi de la Géodesie pratique. Paris: Garnier Frères, 1867. A handbook on astronomical instruments and the physical theories behind them, published at the Brazilian government's expense.

Climats, Géologie, Faune et Géographie botanique du Brésil. Paris: Garnier Frères, 1872. In this book Liais plunged into the domain of nineteenth-century natural history, putting momentarily aside his alleged preference for astronomy.

Suprématie intellectuelle de la France: Réponse aux allégations germaniques. Paris: Garnier Frères, 1872.

OTHER SOURCES

Ancellin, Jacques. "Un homme de science du XIXème siècle: l'astronome Emmanuel Liais." In *Mémoires de la Société Nationale des Sciences Naturelles et Mathématiques de Cherbourg* 57, edited by M. Maurice Durchon. Coutances, France: Imprimerie OCEP, 1975–1978. The most comprehensive available biography.

Barboza, Christina Helena. "Nice Weather, Meteors at the End of the Day." In *From Beaufort to Bjerknes and Beyond: Critical Perspectives on Observing, Analyzing, and Predicting Weather and Climate,* edited by Stefan Emeis and Cornelia Lüdecke. Augsburg, Germany: Dr. Erwin Rauner Verlag, 2005. An account of Liais's contribution to the creation of a telegraphic meteorological network in France.

Mourão, Ronaldo Rogério de Freitas. "Liais, Emmanuel." In *Dicionário Enciclopédico de Astronomia e Astronáutica.* Rio de Janeiro: Nova Fronteira, 1995. A concise but very accurate biography of Liais.

Pyenson, Lewis. "Functionaries and Seekers in Latin America: Missionary Diffusion of the Exact Sciences, 1850–1930." *Quipu* 2 (1985): 387–420. In one of the few available texts in English devoted to Liais's scientific accomplishments, the author examines particularly his role in the establishment of a French tradition in the National Observatory of Brazil.

Christina Helena Barboza

LIBBY, WILLARD FRANK (*b.* Grand Valley, Colorado, 17 December 1908; *d.* Los Angeles, California, 8 September 1980), *chemistry, nuclear science, radiochemistry, radiochemical dating, paleoarchaeology, paleoanthropology.*

Libby is best known as the developer of the radiocarbon dating technique for determining the age of artifacts based on the radioactive isotope carbon-14 (ordinary carbon is primarily carbon-12), for which he was awarded the Nobel Prize in Chemistry in 1960. He also developed a radioactive dating technique for substances using tritium (hydrogen-3). He served as Atomic Energy commissioner and advocated the use of fallout shelters and other measures to counter the perceived nuclear threat from the Soviet Union. His political stance as a "cold warrior" was controversial.

Early Life and Education. Libby was the son of a farmer, Ora Edward Libby, and his wife, Eva May Libby (née Rivers). In 1913 the family, including Willard and his two brothers and two sisters, moved to an apple ranch north of San Francisco. A tall youth who would eventually grow to six feet, three inches, Willard developed his legendary strength by working on the farm. He attended elementary and high school in Sebastopol, California. In high school he played tackle and was called Wild Bill, a nickname that followed him all his life.

In 1926 Libby graduated from high school, and the following year he entered the University of California at Berkeley. (While there he earned money by building apple boxes, which sometimes brought in as much as one hundred dollars per week.) Libby was interested in English history and literature, but he decided on a more practical career and enrolled as a mining engineer at Berkeley. Because his boardinghouse roommates were chemistry graduate students, he became interested in chemistry in his junior year and enrolled in chemistry, physics, and mathematics courses. After receiving his BS degree in 1931, he continued his university work at Berkeley, studying under physical chemists Gilbert Newton Lewis, dean and chairman of the College of Chemistry, and Wendell M. Latimer.

As a graduate student, Libby built his first Geiger-Müller tube and improved it to detect minute amounts of radioactivity, including elements not theretofore believed to be radioactive, such as the lanthanide element, samarium. Throughout his life he constructed Geiger counters, which he claimed to be more sensitive than those available commercially.

Academic Positions. In 1933 Libby received his PhD degree and joined the Berkeley faculty, becoming first an instructor (1933–1938), then an assistant professor (1938–1945), and subsequently in 1945, an associate professor. In October 1945 he moved to Chicago. On 9 August 1940 he married Leonor Lucinda Hickey. Their twin daughters, Janet Eva and Susan Charlotte, were born in 1945. On 8 December 1941, the day after the Japanese attacked Pearl Harbor, Libby's sabbatical Guggenheim Fellowship at Princeton University was interrupted, and in 1942 he joined the Metallurgical Laboratory at the University of Chicago to work on the top secret Manhattan Project to develop a nuclear bomb, remaining there until 1945. Simultaneously, he also worked under Harold C. Urey, the 1934 Nobel chemistry laureate, at Columbia University to develop methods for separating uranium isotopes by gaseous diffusion for production of the bomb. This led to Libby's interest in nuclear science. In 1946 he showed that cosmic radiation in the upper atmosphere produces traces of tritium, the heaviest isotope of hydrogen (hydrogen-3), which can be used as a tracer for atmospheric water. By measuring tritium concentrations, he developed a method for dating well water and wine as well as for measuring circulation patterns of water and the mixing of ocean waters.

In October 1945 Libby became professor of chemistry in the Department of Chemistry and the Institute for Nuclear Studies (now the Enrico Fermi Institute for Nuclear Studies) at the University of Chicago (1945–1959). At age thirty-six he became the youngest full professor at Chicago, where he carried out the work resulting in his winning the 1960 Nobel Prize in Chemistry "for his method to use carbon-14 for age determination in archaeology, geology, geophysics, and other branches of science."

Atomic Energy Commission. On 1 October 1954 Libby was appointed to the U.S. Atomic Energy Commission (AEC). He continued to mentor graduate students at Chicago but reduced his research activities and concentrated on his AEC duties. Because he was already a member of the AEC General Advisory Committee, which developed the commission's policy, he was familiar with its modus operandi.

Libby soon became deeply involved in the problem of nuclear fallout. In 1953, on the recommendation of the Rand Corporation of Santa Monica, California, he established and directed Project Sunshine to study the worldwide effect of nuclear weapons. He was the first person to measure nuclear fallout in dust, soil, rain, human bone, and other sources, and he wrote articles and testified before the U.S. Congress on this problem. He stated that all human beings are exposed to some fallout of natural radiation from sources such as drinking water and claimed that the combination of the body's natural radioactivity, cosmic radiation, and natural radiation of the earth's surface was more hazardous than the fallout resulting from nuclear testing. Along with most scientists at the time, he believed that the effect of nuclear fallout on human genetics was minimal. It later became known that testing of nuclear weapons resulted in a large global increase in the carbon-14 levels in the atmosphere, which decreased exponentially after the cessation of atmospheric testing in 1963.

As a result of his post on the AEC, Libby became a well-known and controversial figure, and his portrait appeared on the cover of the 15 August 1955 issue of *Time* magazine. Many scientists considered him to be a mere "yes man" for the Eisenhower administration. Libby, however, defended his position and responded to what he considered misguided thinking. For example, he wrote to Albert Schweitzer, who had stated that future generations would probably suffer from fallout, that Schweitzer was unaware of the most recent data and that continued nuclear testing was needed for the defense of the United States and the survival of the free world. On 30 June 1959 he resigned from the AEC to resume scientific research but continued to assert the need for nuclear testing. He suggested that industries use isotopes in factories and farms. He was a member of the international Atoms for Peace project, which supported nuclear energy for nonmilitary purposes. Libby thought that more scientists should assume positions of political power rather than serve as mere advisors, and he was pleased when 1951 Nobel chemistry laureate Glenn T. Seaborg was appointed chairman of the AEC in 1961.

Libby resigned from the AEC largely because his wife wanted to return to California, and in 1959 he became professor of chemistry at the University of California at Los Angeles (UCLA), a position that he retained until his death. Ideologically committed to the Cold War, he joined nuclear physicist Edward Teller in opposing two-time Nobel laureate Linus Pauling's petition that nuclear testing be banned.

To prove that nuclear war was survivable Libby built a fallout shelter at his new home, using sandbags and railroad ties. He assumed that a shelter would provide safety in case of a nuclear attack. In a series of articles for the Associated Press News Service, he argued that every home should have a shelter. After a fire burned Libby's shelter, nuclear physicist and nuclear testing critic Leo Szilard joked, "This proves not only that there is a God but that he has a sense of humor" (Seymour and Fisher, 1988, p. 288).

Nobel Prize. On 10 December 1960, Libby received the Nobel Prize in Chemistry "for his method to use carbon-14 for age determination in archaeology, geology, geophysics, and other branches of science." According to Professor Arne Westgren, chairman of the Nobel Committee for Chemistry of the Royal Swedish Academy of Sciences,

> The idea you had 13 years ago of trying to determine the age of biological materials by measuring their C-14 activity was a brilliant impulse. Thanks to your great experimental skill, acquired during many years devoted to the study of weakly radioactive substances, you have succeeded in developing a method that is indispensible [sic] for research work in many fields and in many institutes throughout the world. Archaeologists, geologists, geophysicists, and other scientists are greatly indebted to you for the valuable support you have given them in their work. (Westgren, 1964, p. 592)

Carbon-14. Libby's radiocarbon dating method is based on carbon-14, the radioisotope discovered by Martin D. Kamen and Samuel Ruben. In 1936 Kamen was working at the Radiation Laboratory of future Nobel physics laureate Ernest O. Lawrence as one of the chemists working among a larger number of physicists. (The Ernest Orlando Lawrence Berkeley National Laboratory [LBNL], formerly the Berkeley Radiation Laboratory and usually shortened to Berkeley Lab or LBL, is a U.S. Department of Energy [DOE] national laboratory conducting unclassified scientific research. Managed and operated by the University of California, the Berkeley Lab holds the distinction of being the oldest of the U.S. Department of Energy's National Laboratories. The word *Radiation* was removed from the title because of the public's fear of radiation.) Kamen performed numerous photosynthetic studies with Ruben, using the short-lived carbon-11, with a half-life of only twenty-one minutes.

Because very few persons could then run a mass spectrometer, which was much more difficult to use than a Geiger counter, Lawrence campaigned to find a long-lived radioisotope of carbon. In the fall of 1939 he assigned Kamen and Ruben the task of finding carbon-14 or any long-lived activity in that part of the periodic table. Harold C. Urey, the 1934 Nobel chemistry laureate, and his group at Columbia University were competing with Lawrence's group in a race to use isotopes as biological tracers.

To help Kamen and Ruben, Lawrence offered both his 37-inch and 60-inch cyclotrons; all the time that they needed; and help from Emilio Segrè, Glenn T. Seaborg, and anyone else at the Radiation Laboratory. In September 1939 Kamen planned a detailed program dealing with every conceivable method for preparing long-lived isotopes of carbon, nitrogen, and oxygen. He was especially interested in reactions that would produce isotopes chemically separable from the target material bombarded in the cyclotron, such as carbon-14 from nitrogen-14 in ammonium nitrate—to prevent dilution of the radioactive isotope by its stable isotope.

During January 1940 Kamen began continuously exposing a graphite probe target to collect stray deuterons in the internal beam of the 37-inch cyclotron as the most likely nuclear reaction to yield carbon-14:

$$12C + 2H \longrightarrow 13C + 1H$$

$$13C + 2H \longrightarrow 14C + 1H$$

By 27 February 1940 Kamen and Ruben had removed the last uncertainty—that the activity of the sample might have resulted from the long-lived sulfur-35 produced by the reaction of deuterons on the sulfur as a possible contaminant of the graphite target. They estimated the half-life of carbon-14 as at least one thousand years and wrote a preliminary account for publication as a short letter, "Radioactive Carbon of Long Half-Life" (1940). Because of Ruben's concern with departmental politics and with obtaining tenure, Kamen allowed him to list his name first on the letter, and thus Kamen's contribution to one of the major discoveries in nuclear science was slighted.

Although this discovery was certainly of Nobel caliber, Ruben's death in 1943 from an accident involving phosgene precluded a joint award to him and Kamen, because Nobel Prizes are not awarded posthumously. In 1960 neither Westgren's presentation speech nor Libby's acceptance lecture for his Nobel Prize mentioned Kamen and Ruben's work—a gross miscarriage of justice in the opinion of some.

Radiocarbon Dating. Libby claimed that he first thought that the "notion of radiocarbon dating" was "beyond reasonable credence," and at the start he decided that he should pursue the project in secret. His second wife, Leona Marshall Libby, noted that he "did not tell anyone of his final goal of proving [that] radiocarbon dating would be able to reveal the history of civilization because he felt that if he talked about such a crazy idea he would be labeled a crackpot and would not be able to get money to fund his research nor [*sic*] students to help him" (Taylor, 2000–2001, p. 38). When asked what the most difficult and critical part of the work was, Libby stated: "Being smart enough to keep it secret until it was in hand.... I don't care who you are. You couldn't get anyone to support it. It's obviously too crazy" (Taylor, 2000–2001, p. 38). Apparently, at first Urey was the only one who knew the goal of the work.

In 1939 cosmic-ray physicist Serge Alexander Korff, whom Libby acknowledged in his Nobel lecture, discovered that cosmic rays create showers of neutrons when they strike atoms in the atmosphere. Because nitrogen, which constitutes about 78 percent of the atmosphere, easily absorbs neutrons and then decays into the radioactive isotope carbon-14,

Libby

$${}_{7}N^{14} + {}_{0}n^{1} \longrightarrow {}_{6}C^{14} + {}_{1}p^{1}$$

Libby believed that traces of carbon-14 should always occur in atmospheric carbon dioxide (CO_2) and that because carbon dioxide is continuously being incorporated into plant tissues during photosynthesis, plants should also contain traces of carbon-14. Because animal life depends on plant life, animals should also contain traces of carbon-14. After an organism died, no additional carbon-14 would be incorporated into its tissues, and that which was already present would begin to decay at a constant rate. Kamen had found its half-life to be 5,730 years—a short time compared to the age of Earth but long enough for an equilibrium to be established between the production and decay of carbon-14. According to Westgren, "it should be possible, by measuring the remaining activity, to determine the time elapsed since death, if this occurred during the period between approximately 500 and 30,000 years ago" (Westgren, 1964, p. 590).

Libby collaborated with Aristide von Grosse of Temple University, who was then working at the Houdry Process Corporation and who had constructed there an apparatus that could concentrate the heavier isotopes of carbon, with Libby's first graduate student at Chicago, Ernest C. Anderson, and with postdoctoral fellow James R. Arnold in successfully separating radiocarbon in nature—in methane (CH_4) produced by the decomposition of organic matter. Because a diffusion column such as von Grosse's was expensive to operate, Libby and Anderson used an inexpensive Geiger counter to construct a device that was very sensitive to the radiation of a given sample. They were able to eliminate 99 percent of the background radiation occurring naturally in the environment with 8-inch-thick iron walls to shield the counter and to use a chemical process to burn the sample into pure lampblack (an electrical conductor), which they then placed on the inner walls of a Geiger counter's sensing tube.

To check the accuracy of his radiocarbon dating technique, Libby applied it to samples of redwood and fir trees whose exact ages had been determined by counting the annual rings and to historical artifacts whose ages were known, such as a piece of timber from Twelfth Dynasty Egyptian pharaoh Sesostris III's funerary boat. (It was actually 3,750 years old and was estimated by Libby's method to be 3,261 years old.) In his University of Chicago doctoral dissertation, Anderson established the fact that there was little variation with latitude by determining the radioactivity of animal and plant material obtained worldwide from the North Pole to the South Pole. By 1947 Libby had perfected his radiocarbon dating technique.

Among the archaeological objects that Libby accurately dated were prehistoric sloth dung from Chile, linen

Willard Frank Libby. *Willard Libby in UCLA lab with carbon-14 process-dated early American objects of great antiquity.* J.R. EYERMAN/TIME LIFE PICTURES/GETTY IMAGES.

wrappings from the Dead Sea Scrolls, bread from a house in Pompeii buried by volcanic ashes in the eruption of Vesuvius in 79 CE, charcoal from a campsite at Stonehenge, and corncobs from a cave in New Mexico. By using wood samples from forests once buried by glaciers, he showed that the last Ice Age in North America ended from ten thousand to eleven thousand years ago, not the twenty-five thousand years previously estimated by geologists. By dating human-made artifacts from North America and Europe, such as a primitive sandal from Oregon and charcoal specimens from various campsites, he repudiated the idea of an Old World and a New World and showed that the oldest dated human settlements around the world began at about the same era. Radiochemical dating was rapidly recognized as a basic technique for determining dates within the last seventy thousand years. Perhaps its most publicized use has been in dating the shroud of Turin.

Radiocarbon dating has been applied to a broad variety of problems. For example, because the generation of cells in the human body has been difficult to study, understanding of cell turnover has been limited. On 15 July 2005 Swedish researchers demonstrated that the level of carbon-14 in genomic DNA closely parallels its levels in the atmosphere and can be used to determine the time point when the DNA was synthesized and cells were born. They used Libby's technique to establish the age of cells in the cortex of the adult human brain and showed that while non-neuronal cells are exchanged, occipital neurons are as old as the individual. This supports the view that postnatal neurogenesis does not occur in this region. Retrospective birth dating has become a generally applicable strategy that can be used to measure cell turnover in humans under physiological and pathological conditions.

Libby concluded his Nobel lecture with this description of his radiocarbon dating technique:

> In general, the samples may have to be inspected with some care under a relatively high-powered glass and then treated with properly chosen chemicals. But all of these things can be done and with techniques that are no more difficult than those used by the average hospital technician, and a sample can be obtained which should give authentic radiocarbon dates. The dating technique is one which requires care, but which can be carried out by adequately trained personnel who are sufficiently serious-minded about it. It is something like the discipline of surgery—cleanliness, care, seriousness, and practice. With these things it is possible to obtain radiocarbon dates which are consistent and which may indeed help roll back the pages of history and reveal to mankind something more about his ancestors, and in this way perhaps about his future. (Libby, 1964, p. 610)

In his presentation speech, Westgren said that Libby's method "is so simple—which is probably not always the case with chemical research distinguished with the Nobel Prize—that everyone should be able to understand the conditions and principles for its execution" (Westgren, 1964, p. 589). A scientist who had nominated Libby for the prize characterized his work as follows: "Seldom has a single discovery in chemistry had such an impact on the thinking in so many fields of human endeavour. Seldom has a single discovery generated such wide public interest" (Westgren, 1964, pp. 591–592).

Honors and Professional Activities. In addition to the Nobel Prize, Libby received a number of other awards. They included the Research Corporation Award for the radiocarbon dating technique (1951), Columbia University's Chandler Medal for outstanding achievement in the field of chemistry (1954), the American Chemical Society Maryland Section's Remsen Memorial Lecture Award (1955), City College of New York's Bicentennial Lecture Award (1956), the American Chemical Society's Glenn T. Seaborg Award for Nuclear Chemistry (1956), the Franklin Institute's Elliott Cresson Medal (1957), the American Chemical Society Chicago Section's Willard

Gibbs Medal (1958), Dickinson College's Priestley Award (1959), the Albert Einstein Society's Albert Einstein Medal (1959), the Geological Society of America's Day Medal (1961), the California Alumnus of the Year Award (1963), and the American Institute of Chemists' Gold Medal (1970).

Libby received honorary degrees from Wesleyan University (1955), Syracuse University (1957), the University of Dublin's Trinity College (1957), Carnegie Institute of Technology (1959), Georgetown University (1962), Manhattan College (1963), University of Newcastle-upon-Tyne (1965), Gustavus Adolphus College (1970), the University of South Florida (1975), and the University of Colorado (1977). He was a scientific and technology consultant for industrial firms, the U.S. Department of Defense, scientific organizations, and universities. The author of numerous articles and several books, he belonged to the editorial boards of several scientific journals and magazines and was a member of numerous learned societies in the United States and abroad.

Last Days. In 1962 Libby became director of the Institute of Geophysics and Planetary Physics at UCLA, a post that he retained for the rest of his life. He considered the new scientific frontier to be outer space and argued that the United States should support a large space exploration program to prevent the Soviet Union from controlling outer space, an outcome which, he believed, would mean domination of the world by the USSR.

In 1966 Libby and his wife Leonor divorced, and in December 1966 he married physicist Leona Marshall (née Woods), a professor of environmental engineering at UCLA and a staff member of the Rand Corporation. Libby retired in 1976 and died at age seventy-one from the complications of pneumonia. He was cremated.

BIBLIOGRAPHY

The Willard F. Libby Papers, c. 1954–1976, collection number 1276, comprising 278 boxes (139 linear ft.), four cartons (four linear ft.), and nine oversize boxes, are preserved at the University of California, Los Angeles Library, Department of Special Collections, Los Angeles, CA 90095-1575. They consist of correspondence, notebooks, research materials, publications, lectures, and memorabilia related to Libby's career as a chemist, and items related to his involvement in the U.S. Atomic Energy Commission, tritium experiments, radiocarbon dating, the Institute of Geophysics and Planetary Physics, and UCLA. Further information is available from http://www.oac.cdlib.org/findaid/ark:/13030/kt9j49q5hh.

WORKS BY LIBBY

With Ernest C. Anderson, Sidney Weinhouse, et al. "Radiocarbon from Cosmic Radiation." *Science* 105 (1947): 576–577.

With Ernest C. Anderson and James R. Arnold. "Age Determinations by Radiocarbon Content: World-Wide Assay of Natural Radiocarbon." *Science* 109 (1949): 227–228.

Sensitive Radiation Detection Techniques for Tritium, Natural Radioactivities, and Gamma Radiation. Dayton, OH: U.S. Air Force, Air Material Command, 1951.

Radiocarbon Dating. 2nd ed. Chicago: University of Chicago Press, 1955.

Technical Report on "Chemical Effects of Radiation." AFOSR No. TN-60-1269. Washington, DC: U.S. Department of Commerce, Office of Technical Services, 1961.

"Radiocarbon Dating." In *Nobel Lectures Including Presentation Speeches and Laureates' Biographies: Chemistry 1942–1962.* Amsterdam and New York: Elsevier, 1964. Also available from http://nobelprize.org/chemistry/laureates/1960/libby-lecture.html.

Marlowe, Greg. *Oral History Interview with Willard Frank Libby, April 12 and 16, 1979.* OH 290. College Park, MD: Center for History of Physics, American Institute of Physics, 1979.

Collected Papers/Willard F. Libby. Edited by Rainer Berger and Leona Marshall Libby. 7 vols.: Vol. 1, *Tritium and radiocarbon;* Vol. 2, *Radiochemistry, Hot Atoms & Physical Chemistry;* Vols. 3–4, *Radioactivity and Particle Physics and Radioactive Fallout and Technology;* Vol. 5, *Solar System Physics and Chemistry;* Vol. 6, *Papers for the Public;* Vol. 7, *Talking to People.* Santa Monica, CA: Geo Science Analytical, 1981.

OTHER SOURCES

Asimov, Isaac. *Asimov's Biographical Encyclopedia of Science and Technology.* New rev. ed. Garden City, NY: Doubleday, 1972.

Farber, Eduard. *Nobel Prize Winners in Chemistry, 1901–1961.* Rev. ed. New York: Abelard-Schuman, 1963.

Kauffman, George B. "Willard Frank Libby (1908–1980)." In *The History of Science in the United States: An Encyclopedia,* edited by Marc Rothenberg. New York and London: Garland Publishing, 2000.

Millar, David, Ian Millar, John Millar, et al. *Chambers Concise Dictionary of Science.* Cambridge, U.K.: Cambridge University Press, 1989.

Porter, Roy, ed. *The Biographical Dictionary of Scientists.* 2nd ed. New York: Oxford University Press, 1994.

Ruben, Samuel, and Michael D. Kamen. "Radioactive Carbon of Long Half-Life." *Physical Review* 57 (1940): 549.

Seymour, Raymond B., and Charles H. Fisher. *Profiles of Eminent American Chemists.* Sydney, Australia: Litarvan Enterprises, 1988.

Taylor, Royal Ervin. "Fifty Years of Radiocarbon Dating." *American Scientist* 88 (2000): 60–67.

———. "Origins of a Nobel Idea: The Conception of Radiocarbon Dating." *Chemical Heritage* 18, no. 4 (2000–2001): 8–9, 36–40.

Wasson, Tyler, ed. *Nobel Prize Winners.* New York: W.H. Wilson, 1987.

Westgren, Arne. "Chemistry 1960." In *Nobel Lectures Including Presentation Speeches and Laureates' Biographies: Chemistry 1942–1962.* Amsterdam and New York: Elsevier, 1964. Also available from http://nobelprize.org/chemistry/laureates/1960/index.html.

"Willard Frank Libby—Biography." In *Nobel Lectures Including Presentation Speeches and Laureates' Biographies: Chemistry 1942–1962.* Amsterdam and New York: Elsevier, 1964. Also available from http://nobelprize.org/nobel_prizes/chemistry/laureates/1960/libby-bio.html.

Wollaston, George F. "Willard Libby 1908–1980." In *Nobel Laureates in Chemistry, 1901–1992,* edited by Laylin K. James. Washington, DC: American Chemical Society, 1993.

Young, Robyn V., ed. *World of Chemistry.* Farmington Hills, MI: Gale Group, 2000.

George B. Kauffman

LICHNÉROWICZ, ANDRÉ (*b.* Bourbon l'Archambault, France, 21 January 1915; *d.* Paris, France, 11 December 1998), *mathematics, differential geometry, general relativity.*

Lichnérowicz, "Lichné" for his friends, was born 21 January 1915, in Bourbon l'Archambault, a small town in the center of France. His parents were highly educated people. Both of them were teachers, his father in the humanities and his mother in mathematics.

Career. Lichnérowicz studied differential geometry under the direction of Élie Cartan at the University of Paris. The thesis he defended under the direction of Georges Darmois in 1939 exemplifies the scientific orientation he took throughout his life, namely, bringing global geometric considerations into general relativity with concepts and formulas that allowed him to draw significant physical consequences. Indeed, he remained an active researcher until the end of his life.

In 1941 Lichnérowicz took the position of "maître de conferences" in mechanics at the Faculté des Sciences in Strasbourg that, during World War II, was operating in Clermont-Ferrand in the middle of France. When the war ended, the university returned to Strasbourg. In 1949 he was elected to a professorial position at the University of Paris, and in 1952 he received the great honor of being named to a chair at the Collège de France in Paris which, in the tradition of that special and elitist institution, bore a name tailored after the holder of the chair, in Lichnérowicz's case, mathematical physics. He remained a professor there until 1986, when he reached the normal age of retirement.

In his lectures at Collège de France, which by the rules of the Collège had to cover a new topic each year, Lichnérowicz embraced a very broad landscape in mathematics and theoretical physics. He took the courses he taught as a natural opportunity to present the results of his research or to test out the content of a new book.

Unlike many of his colleagues, he did not limit his activity to producing new scientific results, and involved himself in changing mentalities in the rather conservative university environment in France. In particular, he played a crucial role in the organization of conferences in Caen and Amiens, in 1956 and 1960 respectively, which were instrumental in the modernization of the French university system. From 1966 to 1973 he also chaired the Commission for the Reform of Mathematics Teaching in France, (Commission pour la réforme de l'enseignement des mathématiques en France) known unofficially as the Lichnérowicz Commission, which had a direct impact on the mathematics instruction. There are diverse opinions about the changes that came out of the commission's work, but it cannot be denied that the panel stirred up communities, and it made it plain that the training of teachers could not be insulated from the new developments in mathematics.

Research Interests. Lichnérowicz's interests covered a wide variety of areas in mathematics and theoretical physics. He was a very important actor in the movement that brought geometry and physics closer together at the end of the twentieth century. He did so through the production of many new results; the training of research students; and the publication of a number of influential books, survey articles, and lectures.

His mathematical contributions belong to modern differential geometry as it developed in the twentieth century, moving from local considerations to global ones. It eventually led to the formal emergence of the central notion of a differential manifold, already envisioned by Bernhard Riemann. The growing importance of the field was significantly furthered by the parallel development of a theory of systems of nonlinear partial differential equations. General relativity exemplifies this revolution, particularly if one considers the much more ambitious challenges relating to the field that in the early 2000s can be tackled; the first fairly general situation where a global solution of the Einstein equation could be found was dealt with in 1990. The formulation of the equation that made this great achievement possible used the so-called "conformal formulation" introduced by Lichnérowicz in the 1980s.

Lichnérowicz's approach to doing mathematics was peculiar in that he was always working on some new calculation, and he retained the urge to do so up to the very end of his life. He had an exceptional ability to see geometric facts through formulas. In some sense he was the perfect antidote to the Bourbaki approach to mathematics: He was continuously motivated by physical considerations and relied heavily on explicit computations as crystallizations of geometric facts.

Another key feature of Lichnérowicz's considerable scientific legacy was his fascinating ability to put himself ahead of fashion. On topics such as holonomy groups, transformation groups, harmonic maps, symplectic geometry, and deformations of algebras of observables, he made substantial contributions precisely when these topics were not considered of central importance or even, in some cases, were viewed as marginal.

Lichnérowicz contributed substantially to a number of key areas. One was global differential geometry. In that field the notion of holonomy group attached to a covariant derivative took some time to emerge. It was introduced by Élie Cartan but it was Armand Borel and Lichnérowicz in 1952, and later Marcel Berger and Jim Simons, who established the key theorems that made it possible for the theory to become important.

In addition, a number of Lichnérowicz's works deal with spaces with transitive transformations groups, the so-called homogeneous spaces. The special category of symmetric spaces, identified by Élie Cartan in the 1920s as providing very important families of models for important mathematical and physical situations that are completely amenable to algebraic computations, appears often in Lichnérowicz's articles.

Lichnérowicz also gave great attention to the development of complex geometry and to the study of holomorphic maps between complex spaces. Motivated by his interest in theoretical mechanics, he pushed forward central concepts regarding symplectic geometry and groups of symplectic automorphisms. He played a key role in the establishment of Poisson geometry, one of symplectic geometry's far-reaching generalizations, as a natural framework for the study of physically relevant models. It is therefore not surprising that he played an important role in improving the understanding of the structure of Kählerian manifolds, a notion that is at a crossroad between complex and symplectic geometries.

The theory of harmonic maps is in the early 2000s an established subject. This nonlinear generalization of harmonic analysis exists in modern theoretical physics under the name of nonlinear σ-models. Very early on, already in the 1970s, Lichnérowicz drew attention to the subject and made an important study of its relation to holomorphic maps when the spaces involved are complex and more specifically Kählerian.

It may be that Lichnérowicz proved himself a pioneer in the most obvious way in his development of a global theory of spinors on Riemannian and Lorentzian manifolds. Very early on, indeed in 1963, just after the proof of the index index theorem was announced, he understood the topological consequences for a compact spin manifold of the existence of a Riemannian metric with positive scalar curvature, by establishing a key formula relating the square of the Dirac operator to the usual Laplace-Beltrami operator on spinor fields. This formula was actually known to Erwin Schrödinger, a fact that has been overlooked for many years. Lichnérowicz was a strong advocate for the development of a self-standing spinorial geometry. The fact that he devoted his last works to Killing spinors, which mathematically underpin the concept of supersymmetry, and related objects is a proof of that opinion.

Lichnérowicz's contributions to physics are also numerous, but general relativity is the topic to which he came back over and over again. As already mentioned, he strengthened the links between this theory and the strictly mathematical facts relevant to it. This led him to the study of a number of special models, especially in relation to the propagation of relativistic waves. He played a crucial role in the development of relativistic magnetohydrodynamics, a theory that ties together in a complicated way a physical analysis of the concepts and the mathematical tools needed to get something out of them.

On several occasions, he came back to the basic question of developing the concepts that allow deforming classical mechanics into quantum mechanics. For example, he explored the interplay between such tools and the structure of the underlying space.

Lichnérowicz published more than 350 research articles (including his lecture notes and contributions to conference proceedings) and seven books, which have been very influential and translated into several foreign languages. He was a dedicated thesis adviser who took great care of his research students. They in return showed great loyalty towards him.

Like his parents, Lichnérowicz was a highly cultivated man, curious about many culture and philosophies.

BIBLIOGRAPHY

WORKS BY LICHNÉROWICZ

With Armand Borel. "Groupes d'holonomie des variétés riemanniennes." *Comptes Rendus Hebdomadaires des Séances de l'Académie des Sciences Paris* 234 (1952): 1835–1837.

"Propagateurs et commutateurs en relativité générale." *Publications mathématiques de l'Institut des hautes études scientifiques* 10 (1961): 293–344.

"Spineurs harmoniques." *Comptes Rendus Hebdomadaires des Séances de l'Académie des Sciences Paris* 257 (1963): 7–9.

"Variétés kählériennes et première classe de Chern." *Journal of Differential Geometry* 1 (1967): 195–223.

Global Theory of Connections and Holonomy Groups. Translated from the French and edited by Michael Cole. Leyden: Noordhoff International Publishing, 1976.

Geometry of Groups of Transformations. Translated from the French and edited by Michael Cole. Leyden: Noordhoff International Publishing, 1977

"Les variétés de Poisson et leurs algèbres de Lie associees." *Journal of Differential Geometry* 12 (1977): 253–300.

Choix de oeuvres mathématiques [Selected mathematical works]. Paris: Hermann 1982.

"Killing Spinors, Twistor-Spinors and Hijazi Inequality." *Journal of Geometry and Physics* 5, no. 1 (1988): 1–18.

With Alexandre Favre, Henri Guitton, Jean Guitton, and Etienne Wolff. *Chaos and Determinism: Turbulence as a Paradigm for Complex Systems Converging toward Final States.* Translated from the 1988 French original by Bertram Eugene Schwarzbach. With a foreword by Julian C. R. Hunt. Baltimore, MD: Johns Hopkins University Press, 1995.

Magnetohydrodynamics: Waves and Shock Waves in Curved Space-Time. Mathematical Physics Studies, 14. Dordrecht: Kluwer Academic Publishers Group, 1994.

With Alain Connes, and Marcel Paul Schützenberger *Triangle of Thoughts.* Translated from the 2000 French original by Jennifer Gage. Providence, RI: American Mathematical Society, 2001.

OTHER SOURCES

Berger, Marcel; Jean-Pierre Bourguignon, Yvonne Choquet-Bruhat, et al. "André Lichnerowicz (1915–1998)." *Notices of the American Mathematical Society* 46 (December 1999): 1387–1396.

Cahen, Michel, and Moshe Flato, eds. *Differential Geometry and Relativity. A Volume in Honour of André Lichnerowicz on His 60th Birthday. Mathematical Physics and Applied Mathematics,* Vol. 3. Dordrecht and Boston: Reidel Publishing Co., 1976.

Jean-Pierre Bourguignon

LICKLIDER, JOSEPH CARL ROBNETT

(*b.* St. Louis, Missouri, 11 March 1915; *d.* Arlington, Massachusetts, 26 June 1990), *acoustics, engineering psychology, computer science.*

J.C.R. Licklider started his academic career as an experimental psychologist. His area of concern broadened from acoustics to engineering psychology and then to human-computer interaction and computerized library systems. From 1962 to 1964 Licklider served as the first director of the Information Processing Techniques Office (IPTO) of the Department of Defense's Advanced Research Projects Agency, the largest federal funding agency of the time for the computing field. His administration of the office galvanized the emerging computer science field into shaping research agendas in time-sharing systems, artificial intelligence, computer graphics, and later, networking. The essential features of modern personal computing can be traced to his inspiration.

Family Background and Education. Licklider was the son of Joseph Parron Licklider (b. 1873) and Margarete Robnett Licklider (b. 1881) and grew up in St. Louis, Missouri. In 1937 Licklider graduated from Washington University in St. Louis, where he majored in physics, mathematics, and psychology. A year later he received a master's degree from Washington University and in 1942 a PhD in Psychology from the University of Rochester. His dissertation topic was "An Electrical Investigation of Frequency-Localization in the Auditory Cortex of the Cat." On 20 January 1945 he married Alberta Louise Carpenter (b. 1919); they would have two childen, Tracy Robnett (b. 1947) and Linda Louise (b. 1949).

As an Academic Psychologist. Licklider taught at Swarthmore College as a research associate in the Department of Psychology for one year while completing his PhD and in 1942 joined the Psycho-Acoustic Laboratory at Harvard University as a research associate. Mobilized during World War II, the laboratory, under the leadership of Professor Stanley Smith Stevens, contributed applied research on acoustics. Licklider concentrated on acoustic intelligibility under combat conditions. The primary area of Licklider's research was the effect upon intelligibility of various distortions of speech; he also studied the effects of high altitude on speech communication and the effects of static on communication by radio receivers.

When the war was over, Licklider, like his colleagues, made good use of the laboratory's accumulated knowledge and research environment; he went on to investigate interaural phase, binaural beats, and, in conjunction with George A. Miller, the intelligibility of interrupted speech. His publication of that research after the war earned peer recognition of his most important contribution to the acoustics area, a new formulation of the problem of pitch perception. Licklider was a guest at the seventh Macy Conference, "Cybernetics: Circular Causal and Feedback Mechanisms in Biological and Social Systems," held on 23–24 March 1950. Licklider gave the talk "The Manner In Which and Extent To Which Speech Can Be Distorted and Remain Intelligible," referring to his recent papers published in the *Journal of the Acoustical Society of America.* In the same year, Licklider received the society's Biennial Award for Outstanding Contributions to Acoustics and moved from Harvard, where he had been lecturer in psychology since 1946, to the Massachusetts Institute of Technology's Department of Electrical Engineering as associate professor of psychology of communications.

In the *Handbook of Experimental Psychology* (1951) edited by Stanley Smith Stevens, which became an essential reference in the field, Licklider contributed chapter 25, "Basic Correlates of the Auditory Stimulus," and, with George A. Miller, chapter 26, "The Perception of Speech." He subsequently served as the president of the Acoustical Society of America from 1958 to 1959.

A Truly SAGE System. With the advent of the Cold War, Licklider participated in various summer studies in which interdisciplinary groups of scientists and engineers sought solutions to military problems. In 1950 Project Hartwell was held at MIT to study overseas transport and undersea warfare; in 1952 Project Charles at MIT studied air defense; in 1956 a group at the University of Michigan discussed battlefield surveillance; and in 1957 and 1958 an Air Force Special Summer Study on general subjects was held at Woods Hole, Massachusetts. Project Charles led in 1954 to a project for a nationwide air defense system, subsequently called SAGE (Semi-Automatic Ground Environment). SAGE was a command and control network for air defense, designed to coordinate all air defense components at individual Direction Centers by a digital computer and a backup at each location. In tandem with Project Charles, Project Lincoln at MIT carried out experimental work on some elements of the air defense system, jointly supported by the U.S. Army, Navy, and Air Force, which led to the establishment of the MIT Lincoln Laboratory. George Miller became the official group leader of the laboratory's psychology department. As a consultant to the group, Licklider closely collaborated with Miller on various projects in the early 1950s.

Licklider was the only psychologist who participated in Project Charles and who was exposed to its knowledge about digital computers. Having the SAGE system as a model in mind, Licklider conceived of the "Truly SAGE system or Toward a Man-Machine System for Thinking" and wrote a proposal for it to the Air Force in 1957. As Licklider wrote on the cover page, the aim of the document was "to suggest a kind of system that seems desirable and reasonable from a psychological point of view and not totally unfeasible from a psychologist's approximation of an engineering point of view." Following the model of SAGE, Licklider proposed a system of networked "information centers." In each center, "there is of course a large-scale digital computer with a very extensive memory," and "the centers for related fields are connected one with another by telecommunication channels" (p. 2).

Although SAGE as an air defense system for bombers was already out of date when it became operational in 1958 because of the intercontinental ballistic missile (ICBM), it had a great impact on military system design, computer research, and industry. Air traffic control systems and online reservation systems such as American Airlines' SABRE were among direct applications of SAGE. As a form of computing, SAGE was an online system that handled many inputs and outputs during a process, a real-time system designed to meet the deadlines for real world demands, and a data communication system using phone lines through modems (devices that modulate-demodulate digital information into analog signals and vice versa).

Indeed, the computing style in each center of the "Truly SAGE System" would be online real-time, as was SAGE. However, it would allow simultaneous access by multiple users to share the information stored in the centers, so that the system would be based on a "time-sharing" technique, which was not yet realized in SAGE: "The computer is operated on a time-sharing basis by a number of people. The arrangements for displaying information to the people are highly developed. They include digital-analogue converters, curve plotters, large-screen cathode-ray tubes, automatic typing or printing machines, and loudspeakers." It is notable that Licklider mentioned "time-sharing by a number of people" here: This 1957 document is one of the earliest written records mentioning "time-sharing" in this sense. It was possible that Licklider had been exposed to the phrase "time-sharing" at the SAGE project, although it was used there to describe the SAGE program's cyclic scheduling process. The phrase "time-sharing" was used as a technical term from the mid-1950s through the 1960s with different meanings. Although the diversity of its usage even became controversial in the mid-1960s, in his 1957 document Licklider's use of the term was clear, and thereafter he pursued the development of time-sharing techniques in his sense.

At Bolt Beranek & Newman. In 1957 Licklider left MIT to become head of the departments of psychoacoustics, engineering psychology, and information systems at Bolt Beranek & Newman Inc. (BBN) in Cambridge, Massachusetts. With little administrative support, Licklider had abandoned the effort to form a psychology section at MIT, although he had already recruited prominent young psychologists. Among them, two assistant professors, David Green and John A. Swets, joined him as part-time workers at BBN. At MIT's Lincoln Laboratory, a group headed by Wesley Clark built TX computers that could support online conversational computing through graphical outputs on a display and through input devices such as a light pen and a keyboard. Licklider witnessed the usability of the system and had a concrete image of advanced human-computer interaction. He persuaded Leo Beranek, one of the founders of BBN, to buy a digital computer, a Royal McBee LPG-30, and learned programming with the help of Edward Fredkin. In 1959 the prototype of the Digital Equipment Corporation's first product, PDP-1, designed after the TX computers, was brought to BBN. In a short time Licklider started working with John McCarthy and Marvin Minsky of MIT, colleagues at BBN, Fredkin, and others on "time-sharing" of the machine. By 1962 the technique was established, and BBN commenced time-sharing service connecting an external memory unit.

During his BBN days Licklider set up a plan for research on future library systems as a prototype of an

Licklider

envisioned network of thinking centers. Though interrupted by his service at the Advanced Research Projects Agency from 1962 to 1964, this research came to fruition in his book *Libraries of the Future*, published in 1965. In it, Licklider set forth a technological vision for a "procognitive system," which was a layered computer network system allowing users to access stored information from consoles.

Human Factors in Man-Machine Systems. The SAGE system was designed to automate a preexistent information system, because manual procedures for transmitting military information through telephone and teletype systems lacked speed and precision. But in fact SAGE had to include human elements, such as computer operators who helped with the procedures and commanders who conducted the decision-making process. As a psychologist whose major area of research was function of the brain and human intelligibility, Licklider was skeptical about early realization of a complete automation of the system that included human decision-making and problem-solving. So he coined the term "man-computer symbiosis" to denote semiautomatic systems that include human factors. In 1958 he wrote the report "Man-Computer Symbiosis: Part of the Oral Report of the 1958 NAS-ARDC Special Study, Presented on Behalf of the Committee on the Roles of Men in Future Air Force Systems, 20–21 November 1958." Licklider argued that it would be many years before "developments in artificial intelligence make it possible for machines alone to do much thinking or problem solving of military significance" (p. 4). Because the system had to include human elements, he stressed the importance of the research area called "man-computer communication," which included the design of programming languages and other related areas that later would be called "human-computer interaction" (HCI).

In the academic sphere of psychology, such human-factors research emerged as engineering psychology. Actually, around 1960 this interdisciplinary area was called by various names, such as "human engineering," "biomechanics," "applied experimental psychology," and "ergonomics." In 1957 the American Psychological Association announced the formation of the new division of the Society of Engineering Psychologists in an effort to form an academic area that would gather research results on the human element in man-machine systems. Licklider was so supportive of the establishment of this academic field that he served on the editorial board of a newly published periodical, *Human Factors*, in late 1950s, published "Man-Computer Symbiosis" in 1960 as the first paper in the first issue of *IRE Transactions on Human Factors in Electronics*, and in 1961–1962 served as the fifth president of the Society of Engineering Psychologists. In addition, he received the society's Franklin V. Taylor Award in 1965.

While the 1958 "Man-Computer Symbiosis" report had been situated within the military context, the 1960 version was generalized and academically shaped as a manifesto of the research area of human-machine communication in computing, though some of the elements already laid out in the 1958 report were reused. In the introduction of the 1960 paper Licklider tried to differentiate man-computer symbiosis from artificial intelligence and suggested that a computer could be an ideal partner for human beings in "formulative thinking," which would be preferable to complete automation of the man-machine system. The paper insisted that online real-time computing would be needed for a symbiotic thinking system, as about 85 percent of "'thinking' time was devoted mainly to activities that were essentially clerical or mechanical: searching, calculating, plotting, transforming, determining the logical or dynamic consequences of a set of assumptions or hypotheses, preparing the way for a decision or an insight" (p. 6), and computers could help such "routinizable, clerical operations that fill the intervals between decisions" (p. 7). Then Licklider itemized five prerequisites for realization of man-computer symbiosis, such as memory hardware and organization, programming languages, and input-output devices. Among these prerequisites, the first item he pointed out was "speed mismatch between men and computers," for which he insisted that time-sharing systems were needed. He continued:

> It seems reasonable to envision, for a time 10 or 15 years hence, a "thinking center" that will incorporate the functions of present-day libraries together with anticipated advances in information storage and retrieval and the symbiotic functions suggested earlier in this paper. The picture readily enlarges itself into a network of such centers, connected to one another by wide-band communication lines and to individual users by leased wire services. In such a system, the speed of the computers would be balanced, and the cost of the gigantic memories and the sophisticated programs would be divided by the number of users. (p. 7)

In this paper, Licklider's idea of a network of thinking centers is not stressed as a priority, which is typical of the way he presented things: In writings where he accentuated the importance of development of online real-time computing, the idea of a network of thinking centers was mentioned as an application. But when he put the priority of description on the network of thinking centers, as in "Truly SAGE system" or *Libraries of the Future*, Licklider explained the network scheme first.

Information Processing Techniques Office. Under the Kennedy administration, command and control systems were to be improved, so the Office of the Director of Defense Research and Engineering assigned a command and control project to the Advanced Research Projects Agency (ARPA) in June 1961. When the Department of Defense set up the Office for Behavioral Sciences and Command & Control Research in the ARPA in 1962, Licklider became its first director and stayed there until 1964. He renamed the office as the Information Processing Techniques Office (IPTO); its budget surpassed the sum of all other federal funding agencies' budgets for the computing field of the time.

Licklider adopted a strategy of dividing the budget among only a handful of organizations, and eight groups of institutions to be funded were related to "time-sharing" in 1963. However, the early time-sharing systems could not sustain the man-computer interaction through graphics, which was indispensable for the IPTO's primary objective, the advancement of command and control systems toward the ultimate goal of its automation. One rationale for support of the development of time-sharing systems was laid out in his paper "Artificial Intelligence, Military Intelligence, and Command and Control" in *Military Information Systems* (1964), based on the First Congress on the Information System Sciences held in November 1962. There Licklider argued that a complex information system required large programming tasks and that a large, fast computer based upon the concept of "sensibly simultaneous time sharing" could be economically used to facilitate the efforts of one programmer. In August 1963 Licklider clearly articulated his interest in time-sharing in his talk "Problems in Man-Computer Communications" at the fourth NATO Symposium on Communication Processes, saying that "the truly important thing is not interaction between one man and a computer, but interaction between several or many men and a computer" (*Proceedings*, p. 260).

At the same time, Licklider also showed his enthusiasm for constructing a network in an internal document sometimes referred to as the "Intergalactic Computer Network" memo. The subject of the memo was "Topics for Discussion at the Forthcoming Meeting," dated 25 April 1963 and sent to the principal investigators he funded, whom he called "Members and Affiliates of the Intergalactic Computer Network." He wrote: "It will possibly turn out, I realize, that only on rare occasions do most or all of the computers in the overall system operate together in an integrated network. It seems to me to be interesting and important, nevertheless, to develop a capability for integrated network operation" (p. 4). He then continued for three of the memo's eight pages with a detailed description of hypothetical computer network usage at a console that includes cathode-ray-tube display, light-pen, and typewriter by an experimental psychologist. Toward the end of the memo, he noted: "The fact is, as I see it, that the military greatly needs solutions to many of most of the problems that will arise if we tried to make good use of the facilities that are coming into existence" (p. 7). Thus, although there was no explanation of the direct military need for time-sharing systems or their networking, Licklider tried to convince researchers that the development of time-sharing systems would benefit the military as a consequence. During his administration, however, networking was not seriously pursued. Instead, the IPTO's strong interest in time-sharing made it one of the main research agendas in computer field.

Other than time-sharing, artificial intelligence received the most funding from the IPTO from 1962 to 1975. Licklider did not expect that it would suffice for full automation of command and control systems, but rather that its techniques would further flexible, interactive computing. Whereas he refrained from allocating a large budget to BBN in order to avoid conflict of interest, he did help the formation of a center of excellence at MIT headed by Robert Fano and called Project MAC, which simultaneously stood for Machine-Aided Cognition as a goal and Multiple-Access Computer as a tool. To advance human-computer interaction research, Licklider recommended Ivan E. Sutherland, whose seminal dissertation "Sketchpad: A Man-Computer Graphical Communication System" was submitted to MIT in 1963, as his successor at the IPTO. There Sutherland set up an academic research agenda and funding scheme for computer graphics as one of the most advanced areas of human-computer communication research. During the 1960s the computer field was institutionalized as "computer science" in academia, and the IPTO funding had an influence on it, both through the selection of funding areas and through support for leading figures in universities such as MIT, Carnegie Mellon, University of California at Berkeley and Los Angeles, University of Utah, and Stanford.

Computer Science at MIT. After his term at the IPTO, Licklider served as a consultant to the director of research of the International Business Machines Corporation (IBM) from 1964 to 1966. Even during this period, Licklider also was an impetus to the establishment of Project Intrex at MIT, beginning with his participation in its Planning Conference on Information Transfer Experiments held from 2 August to 3 September 1965. This project focused on an information network for university libraries and the online usage and community it would bear. Licklider wrote four of the twenty-three papers selected for presentation, including "An On-Line Information Network," "Proposed Experiments in Browsing," "The Nature of the Experiments to be Carried Out by Project Intrex," and "A Technique of Measurement That May Be Useful in Project Intrex Experiments." On 1 December 1966 he returned to MIT as professor of electrical engineering and headed Project MAC from 1968 to 1970. In 1969 Licklider was nominated for membership in the National Academy of Science in the psychology section, though he then switched to the engineering section instead.

In 1968, with the IPTO's third director, Robert W. Taylor, who had initiated the ARPA Networking Project to connect the time-sharing systems it funded, Licklider coauthored a paper, "Computer as a Communication Device." In this joint paper they envisioned a networked society, several months before the launch of the ARPA's network in 1969.

Because Project MAC had two different objectives, in 1970 Marvin Minsky seceded from it to establish the MIT Artificial Intelligence Laboratory, and Project MAC was renamed in 1975 as the MIT Laboratory for Computer Science (LCS). After a short return to the IPTO from 1974 to 1975, without significant new initiatives on this second duty, Licklider came back to LCS and stayed there through his mandatory retirement in 1985 until his death in 1990.

J. C. R. Licklider was a member of the National Academy of Sciences, the American Academy of Arts and Sciences, the New York Academy of Sciences, and the Washington Academy of Sciences.

BIBLIOGRAPHY

The personal papers of J. C. R. Licklider are kept in Manuscript Collections, MC 499, Institute Archives and Special Collection, MIT Libraries, Massachusetts Institute of Technology. Some curricula vitae and bibliographies prepared by Licklider himself exist in the manuscripts.

WORKS BY LICKLIDER

Books and Chapter Articles

"Basic Correlates of the Auditory Stimulus." In *Handbook of Experimental Psychology*, edited by S. S. Stevens. New York: Wiley, 1951.

"The Manner in Which and Extent to Which Speech Can Be Distorted and Remain Intelligible." In *Cybernetics: Circular Causal and Feedback Mechanisms in Biological and Social Systems. Transactions of the Seventh Conference March 23–24, 1950, New York, N. Y.*, edited by Heinz von Foerster. New York: Josiah Macy Jr. Foundation, 1951.

With George A. Miller. "The Perception of Speech." In *Handbook of Experimental Psychology*, edited by S. S. Stevens. New York: Wiley, 1951.

"Artificial Intelligence, Military Intelligence, and Command and Control." In *Military Information Systems: The Design of Computer-Aided Systems for Command*, edited by Edward E. Bennett, James Degan, and Joseph Spiegel. New York: Frederick A. Praeger, 1964.

Libraries of the Future. Cambridge, MA: MIT Press, 1965.

"An On-Line Intellectual Community" (Appendix B), "Proposed Experiments in Browsing" (Appendix I), "In Nature of the 'Experiments' to Be Carried Out by Project Intrex" (Appendix L), and "A Technique of Measurement That May Be Useful in Project Intrex Experiments" (Appendix O). In *Intrex: Report of a Planning Conference on Information Transfer Experiments*, edited by Carl F. J. Overhage and R. Joyce Harman. Cambridge, MA: MIT Press, 1965.

"Problems in Man-Computer Communications." In *Communication Processes, Proceedings of a Symposium Held in Washington, 1963*, edited by Frank A. Geldard. New York: Macmillan, 1965.

"Man-Computer Communication." In *Annual Review of Information Science & Technology*, edited by Carlos A. Cuadra. Chicago: Encyclopaedia Britannica, 1968.

"Communication and Computers." In *Communication, Language, and Meaning*, edited by George A. Miller. New York: Basic Books, 1973.

"Potential of Networking for Science and Education." In *Networks for Research and Education: Sharing Computer and Information Resources Nationwide*, edited by Martin Greenberger et al. Cambridge, MA: MIT Press, 1974.

"Computers and Government." In *The Computer Age: A Twenty-Year View*, edited by Michael L. Dertouzos and Joel Moses. Cambridge, MA: MIT Press, 1979.

Journal or Other Periodical Articles

With J. C. Webster. "The Discriminabilitv of Interaural Phase Relations in Two-Component Tones." *Journal of the Acoustical Society of America* 22 (1950): 191–195.

"The Intelligibility of Amplitude-Dichotomized Time-Quantized Speech Waves." *Journal of the Acoustical Society of America* 22 (1950): 820–823.

With George A. Miller. "The Intelligibility of Interrupted Speech." *Journal of the Acoustical Society of America* 22 (1950): 167–173.

With J. C. Webster and J. M. Hedlun. "On the Frequency Limits of Binaural Beats." *Journal of the Acoustical Society of America* 22 (1950): 468–473.

"A Duplex Theory of Pitch Perception." *Experientia* 7 (1951): 127–134.

"On the Process of Speech Perception." *Journal of the Acoustical Society of America* 24 (1952): 590–594.

"Man-Computer Symbiosis." *IRE Transactions on Human Factors in Electronics* 1 (1960): 4–10. Reprinted in *Perspective on the Computer Revolution*, edited by Zenon W. Pylyshyn, 306–318. Englewood Cliffs, N.J.: Prentice Hall, 1970.

With Welden E. Clark. "On-Line Man-Computer Communication." *Proceedings of Spring Joint Computer Conference* 21 (1962): 113–128.

"Periodicity Pitch and Related Auditory Process Models." *International Audiology* 1 (1962): 11–36.

With John McCarthy, Sheldon Boilen, and Edward Fredkin. "A Time-Sharing Debugging System for a Small Computer." *Proceedings of Spring Joint Computer Conference* 23 (1963): 51–57.

"Man-Computer Partnership." *International Science & Technology* (May 1965): 18–26.

With Daniel G. Bobrow, R. Y. Kain, and Bertram Raphael. "A Computer-Program System to Facilitate the Study of

Technical Documents." *American Documentation* 17 (1966): 186–189.

With Robert W. Taylor and Evan Herbert. "Computer as a Communication Device." *Science & Technology* 76 (1968): 21–31.

With A. Vezza. "Applications of Information Technology." *Proceedings of the IEEE* 66 (1978): 1330–1346.

OTHER SOURCES

Aspray, William, and Arthur L. Norberg. "J. C. R. Licklider." OH 150. 28 October 1988, Cambridge, MA; Charles Babbage Institute, University of Minnesota, Minneapolis. A scholarly oral history record.

"Bolt Beranek and Newman: The First 40 Years." Special issues, *IEEE Annals of the History of Computing* 27, no. 2 (2005); 28, no. 1 (2006). Some of these papers refer to Licklider's contribution in the establishment of BBN's role in the history of computing and networking: Leo Beranek, "BBN's Earliest Days: Founding a Culture of Engineering Creativity" (27, no. 2: 6–14); John A. Swets, "ABC's of BBN: From Acoustics to Behavioral Sciences to Computers" (27, no. 2: 15–29); Sheldon Baron, "Control Systems R&D at BBN" (27, no. 2: 52–64); Wallace Feurzig, "Educational Technology at BBN" (28, no. 1: 18–31); John Makhoul "Speech Processing at BBN" (28, no. 1: 32–45); Ralph Weischedel, "Natural-Language Understanding at BBN" (28, no. 1: 46–55).

Fano, Robert M. "Joseph Carl Robnett Licklider: March 11, 1915–June 26, 1990." *Biographic Memoirs* 75 (1998): 190–213.

Garfinkel, Simon L. *Architects of the Information Society: 35 Years of the Laboratory for Computer Science at MIT.* Edited by Hal Abelson. Cambridge: MIT Press, 1999. As a book on the history of Project MAC and LCS, it refers to Licklider and contains a few images of him. It is based on an MIT-centered history account and stresses the importance of Licklider's idea of the "Intergalactic Computer Network" as the origin of the network development at the ARPA.

Kita, Chigusa Ishikawa. "J. C. R. Licklider's Vision for the IPTO." *IEEE Annals of the History of Computing* 25, no. 3 (2003): 62–77. An extended version of this work is published as a book in Japanese, *J. C. R. Licklider and His Age.* Tokyo: Seido-sha, 2003.

Lee, John A. N., and Robert Rosin. "The Project MAC Interviews." *IEEE Annals of the History of Computing* 14, no. 2 (1992): 14–35. This is a record of group interview (10 October 1988) that includes a biographical sketch of Licklider related to the early history of interactive computing at MIT. Participants are Fernando J. Corbató, Robert M. Fano, Martin Greenberger, Joseph C. R. Licklider, Douglas T. Ross, and Allan L. Scherr, as listed. (This entire *Annals* issue is dedicated to Project MAC.)

Norberg, Arthur L. "Changing Computing: The Computing Community and DARPA." *IEEE Annals of the History of Computing* 18, no. 2 (1996): 40–53.

Norberg, Arthur L., and Judy E. O'Neill, with contributions by Kerry J. Freedman. *Transforming Computer Technology: Information Processing for the Pentagon, 1962–1986.* Baltimore: Johns Hopkins University Press, 1996. Most reliable history book on IPTO based on the historical documents of the agency that include normally inaccessible records.

O'Neill, Judy E. "The Role of ARPA in the Development of the ARPANET, 1961–1972." *IEEE Annals of the History of Computing* 17, no. 4 (1995): 76–81.

Waldrop, M. Mitchell. *The Dream Machine: J. C. R. Licklider and the Revolution That Made Computing Personal.* New York: Viking, 2001. Biographical history book for general readers based on documents and interviews. Author's interviews with Louise Licklider and others result in a lively description of Licklider's personality.

Chigusa Ishikawa Kita

LIE, MARIUS SOPHUS

(*b.* Nordfjordeide, Norway, 17 December 1842; *d.* Christiania [now Oslo], Norway, 18 February 1899), *mathematics.* For the original article on Lie see *DSB,* vol. 8.

Hans Freudenthal's essay in the original *DSB* offers a perceptive account of Lie's mathematical interests, the conflicts he experienced, and those parts of his legacy of greatest importance for the mathematics of the twentieth century. As a leading expert on topological groups and geometric aspects of exceptional Lie groups, Freudenthal had a deep appreciation of modern Lie theory. At the same time, his familiarity with Lie's original ideas enabled him to recognize the yawning gap that separated Lie's grandiose vision from that which he and his disciples were able to realize. Freudenthal was less familiar with Lie's biography (Lie had two daughters and one son), and he relied to some extent on folklore, as in his recounting of Lie's hostility toward Wilhelm Killing. Since then much new documentary evidence has become available that helps clarify important episodes in Lie's career.

Background to Conflicts. During his lifetime Lie was a highly controversial figure, and his legend lives on in Norway even in the early twenty-first century. As new facets of his life and work have been brought to light, a picture emerges of a brilliant but troubled man whose career was filled with inner and outer conflicts. His long-forgotten early work with Felix Klein has been reexamined, leading to new assessments of their partnership and its significance for Lie's gradual immersion in the theory of continuous groups. Lie's work in this field eventually spawned what became modern Lie theory, a field of central importance for quantum mechanics. Yet while nearly every theoretical physicist knows about Lie groups, certainly very few have ever read a word of his work. In his pioneering studies, Thomas Hawkins helps remedy that problem. Hawkins not only uncovers the main sources of Lie's

Sophus Lie. SPL / PHOTO RESEARCHERS, INC.

inspiration but he also lays bare the thorny paths followed afterward by numerous others—Killing, Georg Frobenius, Issai Schur, Élie Cartan, Hermann Weyl, and others—whose work created the modern theory of Lie groups. None of these figures, to be sure, was allied with Lie's Leipzig school; indeed, Killing and Frobenius, both trained in Berlin, actively opposed Lie's claims to authority.

Lie drew on two main sources of inspiration in developing his ideas for a theory of continuous groups. The first involved a wide range of geometrical problems that culminated with his discovery of the line-to-sphere transformation in 1870, a breakthrough that opened the way to his investigations on general contact transformations. Much of this work was undertaken in collaboration with Klein, whose "Erlangen Programm" of 1872 (*Vergleichende Betrachtungen über neuere geometrische Forsuchungen*) strongly reflects the impact of Lie's ideas. Soon afterward Lie found a second major source of inspiration in Carl Gustav Jacob's analytic methods in the theory of differential equations. In this he was aided by the Leipzig analyst Adolf Mayer, who encouraged Lie to translate his geometric ideas into the language of Jacobian analysis. By 1874 Mayer had become Lie's most important mathematical resource.

Part of the tragedy surrounding Sophus Lie's life stemmed from his involvement in clashes between prominent mathematicians, many of whom were associated with leading mathematical schools in Germany. Avoiding such entanglements would have been virtually impossible because of his close association with Klein, Leipzig's controversial professor of geometry during the early 1880s. Against strong opposition, both within the Leipzig faculty and in Berlin, Klein managed to orchestrate Lie's appointment as his successor in 1886. From the moment the Norwegian arrived, Leipzig's senior mathematician, Carl Neumann, sought to undermine his position by offering courses and seminars on geometrical topics. Nevertheless, during the course of his twelve-year tenure there, Lie managed to build up an important school whose members specialized in one facet or another of the master's vast research program. Still, he paid a heavy personal price in exchanging the calm tranquillity of Christiania, where he held a parliamentary professorship since 1872, for the dreary urban life he encountered in Leipzig. He found his teaching responsibilities time-consuming, particularly because of difficulties with the German language, and he worried about his wife's health after a tumor was detected in one of her breasts. On top of these daily pressures, he became concerned about a new competitor who suddenly appeared on the horizon: Wilhelm Killing.

Lie had always been suspicions of potential rivals—the French geometers Gaston Darboux and Georges Halphen being two notable cases—but these feelings intensified and spread once he arrived in Leipzig, a far more competitive environment than Christiania. By 1888 he was deeply convinced that his principal disciple, Friedrich Engel, had betrayed his trust. Thus began a long, painful period during which Lie gradually broke off relations with nearly all his friends and supporters in Germany. It was this factor—betrayal, whether real or imagined—that played a major role during the last decade of Lie's ultimately tragic life.

Illness. Initially no apparent signs of conflict arose when Killing met with Lie and Engel in the summer of 1886. Lie presumably knew all along that Engel had been writing to Killing and hoped that the latter's work would enhance the stature of his theory. He changed his mind, however, in early 1888 when he saw the first installment of Killing's four-part study in *Mathematische Annalen.* Lie wrote to Klein: "Mr. Killing's work … is a gross outrage against me, and I hold Engel responsible. He has certainly also worked on the proof corrections" (Rowe, 1988, p. 41). Lie concluded that too many of his ideas had been communicated to Killing by Engel, ideas Lie regarded as his exclusive intellectual property. His relationship with Engel never fully recovered from this bitter episode.

The following year Lie had to be placed in a psychological clinic as he could no longer sleep at night. His wife brought him home in the summer of 1890, but his condition did not improve until long afterward. This dark interlude strongly colored the last decade of Lie's life. Whether or not it affected Lie's personality, as Freudenthal wrote based on Engel's original claims, it undoubtedly affected the way he saw the world and especially his relationships within the German mathematical community.

Conflict with Klein. During the period 1889–1892, when Lie was severely depressed, Klein was returning to several topics in geometry that he had pursued twenty years earlier, the period when he had collaborated closely with Lie. He was also approached by the algebraic geometer Corrado Segre, whose student, Gino Fano, prepared an Italian translation of Klein's "Erlangen Programm" from 1872. This famous survey underscored the role of transformation groups and their invariants in geometry; indeed, it proclaimed that all other aspects (even the dimension of the manifold in question) were of secondary significance for geometrical studies. Soon afterward, the Erlangen program appeared in French and English translations, and Klein wanted to republish it in German too, along with several of Lie's earlier works.

By calling attention to this earlier work, Klein hoped to draw the lines between the intuitive geometric style of mathematics he favored and the dominant research ethos of the period, typified by the trend toward "arithmetization" as practiced in Berlin by Karl Weierstrass and Leopold Kronecker. Lie had become very troubled by Klein's sudden interest in resurrecting their earlier work, and he became increasingly distrustful of the Göttingen mathematician's schemes. Yet he failed to signal these concerns to Klein, who continued to view Lie as his principal ally in an ongoing battle with the Berlin mathematicians. Klein hoped their alliance was still intact in the late summer of 1893 when he delivered his Evanston Colloquium Lectures, two of which gave a highly personal synopsis of Lie's mathematics in which he emphasized the geometrical inspiration behind Lie's work on continuous groups as well as differential equations.

These circumstances loomed in the background when Klein began pressuring Lie regarding his plan to republish their earlier work in *Mathematische Annalen.* Klein even wrote two drafts for an introductory essay on their collaboration during the period 1869–1872 only to learn that Lie profoundly disagreed with his portrayal of these events. Lie rightly noted that his own subsequent research program had little to do with Klein's Erlangen program. Had he confined his critical remarks to their private correspondence, few probably would have known that his relationship with Klein had by this time soured completely. Instead, however, he chose to "set the record straight" in the introduction to the third volume of his treatise on transformation groups (all three were largely written by Engel) by proclaiming: "I am no pupil of Klein's. Nor is the reverse the case, even though it perhaps comes closer to the truth. I value Klein's talent highly and will never forget the sympathetic interest with which he has always followed my scientific endeavors. But I do not feel that he has a satisfactory understanding of the difference between induction and proof, or between a concept and its application" (Lie, 1893, p. xvii). These remarks, not surprisingly, scandalized many within Klein's extensive network, but several others were also criticized by name, including Hermann von Helmholtz, Joseph-Marie de Tilly, Ferdinand von Lindemann, and Killing.

Although prone to outbursts, Lie was tenaciously firm when it came to protecting what he regarded as his intellectual property rights. During the years following his estrangement from Engel, he acquired the services of a new assistant, Georg Scheffers, who edited several of Lie's lecture courses for publication. Reacting to the volume on Lie's theory of contact transformations prepared by Scheffers, Klein privately expressed these revealing remarks:

> That is the true Lie, as he was from 1869–1872, supplemented and completed by careful historical and comparative studies along with excellent drawings by Scheffers. But he breaks off everywhere where my complementary investigations or our collaborative work begins. Why? That's the spirit of latent jealousy. The impression could otherwise possibly arise that I had some kind of share in the ideas that Lie regards as his exclusive property. (Niedersächsische Staats- und Universitätsbibliothek Göttingen, Cod. Ms. F. Klein, 22f)

The Turn to France. Much to Klein's chagrin, Lie lost all interest in the German domestic scene and turned toward France, where the younger generation showed a keen interest in his group-theoretic approach to differential equations. Lie's interest in the reactions of the French community went hand in hand with growing disillusionment with the reception of his work in the German mathematical world. Craving recognition for his theory, he was not content with the kind of support he got from the likes of Engel and Eduard Study, whom he regarded as marginal figures in the German mathematical community. Darboux had shown an early interest in Lie's work, and in 1888 he encouraged two graduates of the École Normale, Vladimir de Tannenberg and Ernest Vessiot, to study with

Lie in Leipzig. Vessiot, following the lead of Émile Picard, took up Lie's original vision, namely to develop a Galois theory of differential equations. Nearly all the French mathematicians were primarily interested in applications of Lie's theory, not in the structure theory itself; even Cartan shared this viewpoint to some extent.

This open-minded attitude of the Parisian community to Lie's theory contrasted sharply with the rejection voiced by Frobenius, who became Berlin's leading mathematician after Weierstrass retired in 1892. The latter considered Lie's work—presumably in the form presented by Engel in *Theorie der Transformationsgruppen*—so wobbly that it would have to be reworked from the ground up. Frobenius went even further, claiming that even if it could be made into a rigorous theory, Lie's approach to differential equations represented a retrograde step compared with the more natural and elegant techniques for solving differential equations developed by Leonhard Euler and Joseph-Louis Lagrange. Needless to say, the leading French mathematicians felt otherwise. Among the younger generation, Cartan, whose work was directly linked to Killing's, showed the strongest affinity for the abstract problems associated with Lie's theory.

In the original *DSB* article, Freudenthal suggests that Lie tried "to adapt and express in a host of formulas, ideas which would have been better without them.... [For] by yielding to this urge, he rendered his theories obscure to the geometricians and failed to convince the analysts" (p. 325). Leaving aside the issue of whether or not Lie himself felt any urge to dress up his theory for analysts, there can be no doubt that he sought their recognition. Lie had long bemoaned his isolation in Norway, and he felt frustrated over the difficulties he encountered in trying to gain an audience for his work. His two most trusted allies in Germany, Klein and his Leipzig colleague Mayer, were well aware of these circumstances. Presumably both reached the conclusion that Lie's mathematics had to be made more palatable for analysts—particularly those closely associated with Weierstrass's school in Berlin—and together they counseled young Engel to carry out this plan.

As the "ghostwriter" of Lie's three volumes on the theory of transformation groups, Engel clearly played a major role in this endeavor. Whether or not Lie valued this effort, he apparently never felt quite at home with the results. According to his student Gerhard Kowalewski, when discussing his work Lie never referred to the three volumes written by Engel, with their "function-theoretic touch," but rather always cited his own papers. This suggests that the "true Lie"—to take up Klein's image—should not be sought in the volumes produced with Engel's assistance but rather in his own earlier papers.

Kowalewski, Klein, and Engel were fascinated by Lie's powerful, Nordic mathematical persona; all three left lively recollections of their encounters with him. Numerous others, including his many students, bore witness to his brilliant originality. Yet despite his numerous achievements, the recognition he received from his many pupils and admirers, and the honors and accolades accorded him by distinguished societies, he spent the last years of his life trying to frame his place in the history of mathematics as Évariste Galois's true successor and Norway's "second [Niels Henrik] Abel." After Lie's death, Engel devoted the last twenty years of his life to preparing the publication of Lie's collected works in six volumes. The seventh volume appeared only many years afterward in 1960, but the editors chose to omit Engel's essay on the conflict between Klein and Lie.

SUPPLEMENTARY BIBLIOGRAPHY

WORK BY LIE

Theorie der Transformationsgruppen. Bd. 3. Leipzig, Germany: Teubner, 1893.

OTHER SOURCES

Hawkins, Thomas. *Emergence of the Theory of Lie Groups: An Essay in the History of Mathematics, 1869–1926.* Berlin, Heidelberg, and New York: Springer-Verlag, 2000. Surveys the broader development of Lie theory.

Rowe, David E. "Der Briefwechsel Sophus Lie–Felix Klein, eine Einsicht in ihre persönlichen und wissenschaftlichen Beziehungen." *NTM* 25, no. 11 (1988): 37–47. Discusses Lie's conflicts with Killing and Klein.

Stubhaug, Arild. *The Mathematician Sophus Lie: It Was the Audacity of My Thinking.* Translated by Richard H. Daly. Berlin, Heidelberg, and New York: Springer-Verlag, 2002. Contains many heretofore unknown aspects of Lie's life.

David E. Rowe

LIEBIG, JUSTUS VON (*b.* Darmstadt, Grand Duchy of Hesse-Darmstadt, 12 May 1803; *d.* Munich, Germany, 18 April 1873), *chemistry.* For the original article on Liebig see *DSB,* vol. 8.

Liebig's life encompassed innovation in teaching, important contributions to organic chemistry and, above all, the significant application of chemistry to agriculture, physiology, medicine, nutrition, and industry, as well as to the popularization of chemistry. He has attracted considerable attention since Frederic L. Holmes's fine article was published in 1973. Historical interest has been concentrated on the publication of critical editions of Liebig's extensive correspondence with other chemists and pharmacists, his publishers, and the chancellor of the University of Giessen; the development of a deeper understanding of his role in organic chemistry, especially through his improvement of analytical techniques and the use of chemical formulas; the enrichment of knowledge concerning Liebig's methods for training chemists and the influence of the Giessen school on the international development of laboratories and research schools; the investigation of the worldwide impact of the publication of his writings on agricultural and physiological chemistry; and the role of his *Chemische Briefe* in the popularization of chemistry. While Jakob Volhard's two-volume German biography of Liebig (1909) remains the essential introduction, there is now a substantial English and German biography by William H. Brock (1997) that examines the controversial nature of Liebig's opinions and character, portraying him as a gatekeeper who helped to transform public and governmental awareness of chemistry and its essential role in a modern society.

Liebig became a world celebrity during his lifetime and was one of the most significant nineteenth-century scientists in shaping an international vision of science. For Liebig, science was a body of knowledge that ignored and transcended national boundaries. What could be said of German or British agriculture applied equally to that in Italy, America, or Japan. He wanted to universalize the localized nature of knowledge and practice and did this by forging educational tools, the exchange of information and research, and the popularization of chemical knowledge. One vital aspect of his determination to communicate the central significance of chemistry was his use of Roman instead of Gothic (Fraktur) type in his monthly periodical *Annalen der Chemie* and in the German editions of his books, knowing that this typography made it easier for foreign readers (for whom Fraktur was as unintelligible as Greek or Cyrillic print). Translation of his work was another important means of communication, and he did everything possible to encourage it. For example, *Die organische Chemie in ihrer Anwendung auf Agricultur und Physiologie* (1840) appeared in at least nineteen editions in about nine different languages. Such internationalism reached its long-term fulfillment in the twentieth century when *Liebigs Annalen der Chemie* was amalgamated into the *European Journal of Organic Chemistry* in 1998.

Liebig's fame was not so much that he made a startling new chemical discovery. It was largely due to his demonstration with Friedrich Wöhler that it was possible to use the paper tools of Berzelian chemical symbols to make sense of analytical results by inspecting and juggling with the compositions of reactants and products. As Ursula Klein (2003) has shown, the 1830s produced novel ways of individuating, identifying, and classifying organic compounds, chief of which was the exploitation of Jacob Berzelius's chemical formulas and their manipulation on paper in an attempt to understand the composition of the dazzling parade of new derivatives that were totally

Justus von Liebig. © AUSTRIAN ARCHIVES/CORBIS.

unknown in nature. Liebig, Wöhler, and Jean-Baptiste Dumas excelled at this practice and were considerably helped by the sophisticated method of organic analysis with the so-called Kaliapparat that Liebig developed in 1830 when attempting to understand the composition of plant alkaloids. Replication of these gravimetric experiments by Melvyn C. Usselman and others (2005) has shown historians how accurate Liebig's results were for carbon, hydrogen, and oxygen content (though nitrogen content remained an acute problem). This development of a rapid and accurate method of gravimetric organic analysis using the Kaliapparat acted as a trigger for the explosion of organic (as opposed to inorganic) chemistry. These two techniques, paper chemistry and accurate compositional analysis, forged a new ontology of carbon chemistry, as opposed to the tradition of vegetable and animal chemistry, and enabled chemists to classify and interpret analyses in terms of common groups or radicals and, later, in terms of "chemical types."

The "Giessen Model." However, Liebig's contribution to the perfection of inorganic analysis and its dissemination must not be underestimated. Building upon the long historical tradition of tests for mineralogical composition, at Giessen he taught systematic methods of inorganic analysis, though he left it to pupils and assistants such as Carl Fresenius and Heinrich Will to publish these methods in the 1830s. These systematic group separation methods (wet qualitative and quantitative analysis) were taught to every student of practical chemistry into the 1950s.

The fame and celebrity status that Liebig sought as a young man came about through teaching these systematic methods of inorganic and organic analysis. Beginning with a majority of pharmacy students, he successfully attracted an international body of chemistry students to Giessen where, from 1835 until he left for Munich in 1852, he engaged in line-production research investigating the chemistry of living systems of plants and animals. Whether Giessen was the model for future research schools has been the subject of great historical interest since the publication of Jack Morrell's heuristic model in 1972. Joseph S. Fruton (1990) has contrasted Liebig's style of research leadership with other chemists and biochemists and has also provided a valuable list of the majority of the active researchers who studied with him at Giessen or Munich. Alan J. Rocke (2003) suggests that Liebig was able to establish something new and unique in chemical education by exploiting isomerism, the use of formulas to understand composition, and the Kaliapparat to ensure accurate quantitative compositions. This may explain why and how Liebig's "Giessen model" spread rapidly far and wide. Liebig himself bombastically advertised the Giessen method in his polemic against the Prussian government in 1840. His letters to his publisher Friedrich Vieweg demonstrate his control of the monthly *Annalen der Pharmacie* and the ambitious nature of his own book publication program, as well as his ambition to publish German translations of important English cultural works by Charles Darwin and John Stuart Mill. Other letters to Justin von Linde, the Catholic chancellor of the University of Giessen, are revelatory in two respects. On the one hand, they demonstrate Liebig's inexhaustible dedication to chemistry, and on the other, his determination to promote his university as a leading European institution of scientific learning. Finally, his correspondence with Georg von Cotta, the aristocratic owner of the *Augsberger Allgemeinen Zeiting,* reveals Liebig's ambitions as a popular writer and his determination to promote chemistry as the central science for economic prosperity through his *Chemische Briefe* from 1844 onward.

BIBLIOGRAPHY

Carlo Paoloni, Justus von Liebig. Eine Bibliographie sämtlicher Veröffentlichungen *(Heidelberg, Germany: Carl Winter, 1968), the standard bibliography cited by Holmes, is not entirely reliable. A good guide to the literature up to 1996 is found in William H. Brock,* Justus von Liebig: The Chemical Gatekeeper *(Cambridge, U.K. and New York: Cambridge University Press, 1997; paperback, 2002); German translation* Justus von Liebig: Eine Biographie des großen Wissenschaftlers und Europäers *(Braunschweig and Wiesbaden, Germany: Vieweg, 1999). The following bibliography lists primary sources published since the 1973 entry, but, with some exceptions, only secondary works published since 1996.*

WORKS BY LIEBIG

Animal Chemistry (New York, 1842). Edited by Frederic L. Holmes. New York: Johnson Reprint, 1964. Holmes's introduction, pp. vii–cxvi. For a German translation of Holmes's essay, see Büttner and Lewicki (2001), cited below, pp. 1–107.

Die Organische Chemie in ihrer Anwendung auf Physiologie und Pathologie (Braunschweig, 1842). Facsimile edition with appendix containing reprints of essays on the history of physiological chemistry, edited by Wilhelm Lewicki. Pinneberg, Germany: AgriMedia Verlag Alfred Strothe, 1992.

Die Chemie in ihrer Anwendung auf Agricultur und Physiologie, 9th posthumous edition by Philipp Zöller (1876), facsimile in 2 vols. Holm, Germany: Agrimedia, 1995. Issued with a supplementary volume, edited by Wilhelm Lewicki, containing reprints of essays on the history of agricultural chemistry.

Aus Justus Liebig's und Friedrich Wöhler's Briefwechsel in den Jahren 1829–1873. 2 vols. Edited by August Wilhelm Hofmann. Braunschweig, Germany: Friedrich Vieweg, 1888. Reprint edited by Wilhelm Lewicki. Göttingen, Germany: Jürgen Cromm, 1982. Hofmann's edition is an uncritical and massively expurgated extracts of the whole correspondence. A complete critical edition of the surviving 1,700 letters is in production at the University of Regensburg, edited by Christoph Meinel and Thomas Steinhauser.

Berzelius und Liebig: Ihre Briefe von 1831–1845. Edited by Justus Carrière. Munich, Germany: J.F. Lehmann, 1893; 2nd ed., 1898. Reprint edited by Till Reschke. Göttingen, Germany: Jürgen Cromm, 1978.

Justus von Liebig "Hochwohlgeborner Freyherr": Die Briefe an Georg von Cotta und die anonymen Beiträge zur Augsburger Allgemeinen Zeitung. Edited by Andreas Kleinert. Mannheim, Germany: Bionomica-Verlag, 1979.

Liebigs Experimentalvorlesung. Vorlesungsbuch und Kekulés Mitschrift. Edited by Otto Paul Krätz and Claus Priesner. Weinheim, Germany: Verlag Chemie, 1983. Facsimiles and transcriptions of Liebig's Giessen lectures on organic chemistry, together with Kekulé's student notes of 1848 and valuable commentaries.

Justus von Liebig und August Wilhelm Hofmann in ihren Briefen (1841–1873). Edited by William Hodson Brock. Weinheim, Germany: Verlag Chemie, 1984. With English abstracts; note supplement by Heuser and Zott below (1988).

Justus von Liebig. Briefe an Vieweg. Edited by Margarete Schneider and Wolfgang Schneider. Braunschweig and Wiesbaden, Germany: Vieweg & Sohn, 1986. Liebig's letters to his publisher, 1823–1872.

The Letters from Gerrit Jan Mulder to Justus Liebig (1838–1846). Edited by Harry A. M. Snelders. Amsterdam: Rodopi, 1986.

Justus Liebig und Julius Eugen Schlossberger in ihren Briefen von 1844–1860. Edited by Fritz Heße and Emil Heuser, eds. Mannheim, Germany: Bionomica-Verlag, 1988.

Justus von Liebig und August Wilhelm Hofmann in ihren Briefen, Nachträge 1845–1869; Justus von Liebig und Emil Erlenmeyer in ihren Briefen von 1861–1872. Edited by Emil Heuser and Regine Zott. Mannheim, Germany: Bionomica-Verlag, 1988. The Hofmann-Liebig letters supplement the edition by Brock.

Justus von Liebig und der Pharmazeut Friedrich Julius Otto in ihren Briefen von 1838–1840 und 1856–1867. Edited by Emil Heuser. Mannheim, Germany: Bionomica-Verlag, 1989.

Die Nachlässe von Martius, Liebig und den Brüdern Schlagintweit in der Bayerischen Staatsbibliothek. Edited by Anne Büchler and Rolf Schumacher. Wiesbaden, Germany: Otto Harrassowitz, 1990. Lists Liebig's archives in Munich.

Universität und Ministerium im Vormärz: Justus Liebigs Briefwechsel mit Justin von Linde. Edited by Eva-Marie Felschow and Emil Heuser. Giessen, Germany: Verlag der Ferber'schen Universitäts-Buchhandlung Giessen, 1992.

Die streitbaren Gelehrten: Justus Liebig und die preußischen Universitäten. Edited by Regine Zott and Emil Heuser. Berlin: ERS-Verlag, 1992. Documents the controversy arising from Liebig's polemic concerning the state of science teaching in Prussia.

Justus von Leibig and Hermann Kolbe in ihren Briefen, 1846–1873. Edited by Alan J. Rocke and Emil Heuser. Mannheim, Germany: Bionomica Verlag, 1994.

Kleine Schriften. Edited by Hans-Werner Schütt. Hildesheim, Germany, Zürich, and New York: Olms-Weidmann, 2000. A reprint of Liebig's papers from *Annalen der Chemie und Pharmacie.*

Justus Liebig in Grossbritannien: Justus Liebigs Briefe aus Grossbritannien an seine Frau Henriette. Edited by Günther Klaus Judel. Giessen: Liebig-Gesellschaft, 2003.

OTHER SOURCES

Berichte der Justus Liebig-Gesellschaft zu Giessen, vols. 1 (1990) to date. Giessen, Germany: Justus Liebig-Gesellschaft.

Billig, Christine. *Pharmazie und Pharmaziestudium an der Universität Giessen.* Stuttgart, Germany: Wissenschafttliche Verlagsgesellschaft, 1994. Liebig's pharmaceutical teaching.

Brock, William H. "Breeding Chemists in Giessen." *Ambix* 50 (2003): 25–70.

Büttner, Johannes, and Wilhelm Lewicki, eds. *Stoffwechsel im tierischen Organismus: Historische Studien zu Liebigs, "Thier-Chemie."* Seesen, Germany: HisChymia Buchverlag, 2001.

Finlay, Mark R. "Justus von Liebig and the Internationalization of Science." *Berichte der Justus Liebig-Gesellschaft* 4 (1998): 57–76.

Fruton, Joseph S. *Contrasts in Scientific Style: Research Groups in the Chemical and Biochemical Sciences.* Philadelphia: American Philosophical Society, 1990. Pages 277–307 contain comprehensive lists of Liebig's students. Reprint in Büttner & Lewicki (2001), pp. 373–412.

Heilenz, Siegfried. *Eine Führung durch das Liebig-Museum in Giessen.* Giessen: Verlag Liebig-Gesellschaft, 1994. An illustrated guide to Liebig's laboratory.

Holmes, Frederic L. "The Complementarity of Teaching and Research in Liebig's Laboratory." *Osiris* 5 (1989): 121–164.

———. "Justus Liebig and the Construction of Organic Chemistry." In *Chemical Sciences in the Modern World,* edited by Seymour H. Mauskopf, 119–134. Philadelphia: University of Pennsylvania Press, 1993.

Hormuth, Stefan, ed. *Justus Liebig: Seine Zeit und Unsere Zeit; Der streitbare Gelehrte; Die "Chemischen Briefe."* 3 vols. Giessen: Justus Liebig-Universität, 2003. Three scholarly illustrated exhibition catalogs celebrating the 200th anniversary of Liebig's birth.

Jaschke, Brigitte. *Ideen und Naturwissenschaf:. Wechselwirkungen zwischen Chemie und Philosophie am Beispiel des Justus von Liebig und Moriz Carrière.* Stuttgart, Germany: Chemisches Institut der Universität Stuttgart, 1996. Carrière was Liebig's son-in-law.

Kirschke, Martin. *Liebigs Lehrer Karl W. G. Kastner (1783–1857): Eine Professorenkarriere in Zeiten naturwissenschaftlichen Umbruchs.* Berlin-Diepholz: GNT Verlag, 2001.

Klein, Ursula. *Experiments, Models, Paper Tools: Cultures of Organic Chemistry in the Nineteenth Century.* Stanford, CA: Stanford University Press, 2003. For Liebig's exploitation of chemical formulas.

———. "Contexts and Limits of Lavoisier's Analytical Plant Chemistry: Plant Materials and Their Classification." *Ambix* 52 (2005): 107–158. Discusses the assimilation of vegetable chemistry into organic chemistry in the 1830s.

Meimberg, Paul, ed. *Justus v. Liebig 1803–1873,* special issue of *Gießener Universitätsblätter,* Jahrgang 6 (April 1973). Contains valuable essays and list of archives in Liebig Museum.

Munday, Pat. "'Politics by Other Means': Liebig and Mill." *British Journal for the History of Science* 31 (1998): 403–418. On Liebig's promotion of a translation of J. S. Mill's *Logic.*

Rocke, Alan J. "Origins and Spread of the 'Giessen Model' in University Science." *Ambix* 50 (2003): 90–115.

Schwedt, Georg. *Liebig und seiner Schüler: Die neue Schule der Chemie.* Berlin: Springer, 2002. A popular illustrated study.

Strube, Wilhelm. *Justus Liebig: Eine Biographie.* Beucha, Germany: Sax-Verlag, 1998. A short biography.

Usselman, Melvyn C. "Liebig's Alkaloid Analyses: The Uncertain Route from Elemental Content to Molecular Formulae." *Ambix* 50 (2003): 71–89.

Usselman, Melvyn C., Alan J. Rocke, Christine Reinhart, et al. "Restaging Liebig: A Study in the Replication of Experiments." *Annals of Science* 62 (2005): 1–55. A study of the Kaliapparat.

Werner, Petra, and Frederick L. Holmes. "Liebig and the Plant Physiologists." *Journal of the History of Biology* 35 (2002): 421–441.

W. H. Brock

LINNAEUS, CARL (*b.* Södra Råshult, Småland, Sweden, 23 May 1707; *d.* Uppsala, Sweden, 10 January 1778), *botany, zoology, mineralogy, geology, medicine, economy.* For the original article on Linnaeus see *DSB*, vol. 8.

As Sten Lindroth emphasizes right from the start of his *DSB* article, Linnaeus had few, if any, equals in reshaping the ways in which natural history was practiced in the eighteenth century. His literary output was prolific, and eighteenth-century reeditions (both authorized and pirated), translations, and popular adaptations of his main works—such as the *Systema Naturae*, the *Genera Plantarum*, and the *Philosophia Botanica*—number in the tens and hundreds. By the end of the century it was effectively impossible to be taken seriously as a naturalist without complying with the conventions of Linnaean taxonomy and nomenclature. Explaining this astonishing impact remains one of the main challenges for Linnaean scholarship, especially because Linnaeus's work presents an amalgam of neo-Stoicism, scholastic reasoning, Cartesian iatromechanism, Baconian empiricism, Paracelsian natural philosophy, protestant physicotheology, and Scandinavian folklore that was surely hard to digest for his contemporaries. As Lindroth's article amply demonstrates, Linnaeus was a highly idiosyncratic and syncretistic thinker, but also marginal by geographic origin and social standing, especially if compared with his life-long rival Georges-Louis Leclerc, Comte de Buffon, superintendent of the *Jardin du Roi* in Paris and initiator and main author of the voluminous *Histoire naturelle, générale et particulière* (44 vols., 1749–1804).

The reasons for Linnaeus's success must therefore be sought in the wider cultural context of his time rather than in the force and authority of the ideas and arguments he put forward. Scholarship since the 1980s has done a lot to explore these contexts, focusing on three interrelated themes: Linnaeus's role in transforming the culture of collecting that supported natural history; the economic agendas that motivated his reform of natural history; and the way in which his work resonated with eighteenth-century concerns with sexuality and human difference.

Collections and the Natural System. Linnaean scholarship has long suffered from a kind of inverse anachronism. Eager to reject the image of botany and zoology as mere "stamp collecting," plant physiologists such as Julius von Sachs in the nineteenth century, as well as evolutionary biologists such as Arthur J. Cain and Ernst Mayr in the twentieth, made Linnaeus into a representative of a disciplinary past that was dominated by a mistaken "essentialist" methodology. However, factual historical evidence supporting the view that Linnaeus "excelled in [scholastic] logic"—as Mayr put it in his influential *The Growth of Biological Thought* (1982, p. 173)—is scant, to say the

Carl Linnaeus. *Carl Linnaeus wearing the typical dress of a Laplander, circa 1740.* HULTON ARCHIVE/GETTY IMAGES.

least. His school and university teachers shared the same eclecticism that Linnaeus himself later espoused, and the only authorities he ever quoted on methodological and epistemological issues were Francis Bacon and Herman Boerhaave. It was Linnaeus himself, moreover, who was the first to denounce classifications reached by division *per genus et differentiam*, including his own so-called sexual system, as artificial.

The aim of natural history, according to Linnaeus, was not to provide diagnostic tools but to uncover what he called the "natural system," and he was clearly aware of the fact that this aim could not to be reached by a priori reasoning, but by empirical research alone. In natural history, such research included the acquisition of specimens through fieldwork or exchange with other naturalists, their accumulation in specialized institutions (botanical gardens and museums), and careful comparative work, as well as meticulous recording of data in catalogs, synoptic tables, and descriptions, cross-referenced by indices and stable names. Linnaeus's key to success was the integration of such technologies with a system of social recognition that favored individualized contributions to a common agenda over grand theoretical designs. Acutely aware of the importance of collectors' networks for the practice of natural history, Linnaeus specified rules of taxonomy and nomenclature—first in his *Fundamenta Botanica* (1736), and later in the *Philosophia Botanica* (1751), which amounted to a manual of do-it-yourself botany—that were explicit enough to allow every person capable of reading and writing to contribute to natural history (e.g., by publishing the description of a "new" species), yet afforded the authority and expertise of central institutions to synthesize the amassing data and clear competing claims of discovery and priority. Linnaeus himself disposed of a vast network of correspondents, including both metropolitan naturalists and peripheral amateurs, and for a time it became virtually impossible to lay claim on the discovery of a "new" species without having it sanctioned by Linnaeus.

The "natural system" has remained the focal research object of botanical and zoological taxonomy since Linnaeus. Not surprisingly, however, its meaning has changed considerably over time. For Linnaeus, "natural affinities" among organisms became manifest in two ways: first, by the constancy with which organisms reproduced their specific characters from one generation to the next; second, by the correlation of traits that allowed the naturalist to describe not only species, but also genera and units of higher taxonomic rank such as orders and classes. Both of these aspects were closely intertwined with the practical basis of Linnaean natural history. Linnaeus's species concept shifted the emphasis from the observation of similarities to the observation of "laws of generation," according to which certain characters proved "constant" in the transplantations specimens were undergoing when exchanged among naturalists. And Linnaeus's concept of "natural" genera and orders established higher taxonomic units as research subjects, to be described and presented on their own, independent of any universally accepted method or system. The hybridization theory that Linnaeus developed in later years to account for the origin of species in time tried to unite both of these aspects. Species transformations and hybridizations between widely different species had been assumed possible since antiquity, but with Linnaeus they became the subject of an organized, collective research activity. In 1759 Linnaeus performed a hybridization experiment with *Tragopogon pratensis* (goatsbeard), which initiated a research tradition leading right up to Gregor Mendel's famous experiments with peas.

Economy and Natural History. Linnaeus promoted reform not only in the international community of naturalists, to which he liked to refer as a "free republic," but in his own home country as well, developing a distinct politico-economic agenda that built on a close connection between science, rational governance, and economic well-being. The Swedish Royal Academy of Sciences in Stockholm, which he cofounded in 1739, was initially called the Economic Society of Science, and the first volume of its *Transactions* contained a contribution by Linnaeus, in which he argued that natural history provided knowledge of natural resources, and thus one of the bases for the economic well-being of a nation. In a similar vein, the inaugural lecture Linnaeus gave in 1741, when assuming his professorship at Uppsala University, propagated "the necessity of traveling in one's own country" (*Oratio qua peregrinationum intra patriam asseritur necessitas,* 1741), and Linnaeus's surveys of various provinces of Sweden in the 1740s were carried out on behalf of the Manufacture and Trade Deputation of the National Estates. Linnaeus was also among those who called for obligatory education in economics for obtaining any university degree, and who promoted the installation of professorships in economics at Swedish universities during the 1740s. One of his own students, Pehr Kalm, was the first to be appointed to such a professorship at the University of Åbo (Turku, Finland) in 1747, shortly before he embarked on his three-year journey through North America.

Linnaeus's engagement with economics was motivated by a peculiar mix of cameralist and mercantilist notions of economy. Linnaeus had close connections to members of the mercantilist, so-called Hat Party, which dominated Sweden during its so-called era of freedom (1718–1772), and he subscribed to the mercantilist view that a negative trade balance was detrimental to national economies. Import of exotic luxury goods such as tea, sugar, spices, or printed cotton led to a loss of bullion, and thus to a country's economic decline. For Linnaeus, however, the solution to this problem did not lie in counterbalancing imports by exports, but in striving for the cameralist ideal of a rationalistically governed autarky, or economic self-sufficiency. "The idea was," as Lisbet Koerner put it succinctly in her 1999 book *Linnaeus: Nature and Nation,* "to create a miniaturized mercantilist empire within the borders of the European state" (p. 188), either by identifying domestic natural resources that could supplant foreign imports, or by importing and "acclimatizing" exotic plants and animals for domestic use.

Natural history had an important role in this, as it provided an inventory of both domestic and exotic natural resources. University training in economics—combining natural history with the teaching of known uses of minerals, plants, and animals—could thus provide the state with a broad, disciplined basis of administrators, parsons, physicians, and engineers that furthered national prosperity through systematic allocation and exploitation of natural resources. Linnaeus's own travels through his home country, and the questionnaires he published to obtain information, for example, on the calendar dates at which different species of trees came to leaf in different geographic regions in Sweden (*Vernatio Arborum,* 1753), betray this proto-statistical ambition. And also the

Linnaeus

worldwide travels that some of his students undertook, often with active support from the Royal Academy of Sciences and the Swedish East India Company, have to be seen in this context. Throughout his entire career Linnaeus tried desperately to get hold of exotic animals and plants, such as silkworms, rice, and tea, to cultivate them in Sweden, and developed grand designs to forest Lapland. Most of these projects failed, with two exceptions: Linnaeus introduced rhubarb to Sweden, and successfully seeded domestic mussels with pearls, for which he received, in 1761, enoblement, three thousand silver thalers, and the right to choose his own successor.

Sexuality, Anthropology, and the Politics of Nature. A key element of Linnaeus's popular fame was without doubt his sexual system, first published in the *Systema Naturae* (1735). It was not only a handy, diagnostic tool for the identification of plants, but also made use of anthropomorphic metaphors in describing the sexual organs of plants. The result was a kind of ethnography of human sexuality, including adultery, polygamy, prostitution, incest, and homosexuality. Using these metaphors was, in part, surely meant to attract public attention, but also reflects the degree to which Linnaeus's whole thinking was preoccupied with sexuality. The two substances of which all living beings consisted according to Linnaeus's physiological teaching, "marrow" (*medulla*) and "bark" (*cortex*), were strongly gendered. The "marrow" was a substance passed on maternally, occupying the interior of organisms, and endowed with a tendency to grow and multiply. The "bark," on the contrary, was a substance passed on paternally, enclosing and restraining the medulla, and endowed with the ability to protect and nourish it. In contrast to most of his contemporaries, Linnaeus was not a preformationist, but believed that organisms developed epigenetically through the antagonistic interaction of *medulla* and *cortex*. In many ways, this theory reflected Linnaeus's own, strained relationship with his wife Sara Elisabeth Moraea, who bore him four daughters and a son and managed the "household," which, at the height of Linnaeus's career, consisted of three large rural estates.

Linnaeus's tendency to naturalize social relations is prominent also in his anthropological work. The *Systema Naturae* of 1735 classified humans together with the apes and the sloth as *Anthropomorpha*, and subdivided humanity into four "varieties," "white Europeans," "red Americans," "yellow Asians," and "black Africans." In later editions of the *Systema Naturae*, Linnaeus would supplement this classification by skin color with psychological criteria borrowed from the medical doctrine of the four temperaments. By contrast, Linnaeus portrayed the native inhabitants of Lapland, the Sami, as "noble savages" in his *Flora Lapponica* (1737). Anticipating Jean-Jacques Rousseau (1712–1778), he described their way of life as a model of virtue and health, free of the corruptions that civilization, especially at courts and in cities, brought with it. This may in part explain the fact that postrevolutionary France indulged in a downright cult of Linnaeus and his science, resulting in the foundation of a great number of provincial *Sociétés linnéennes* in the early nineteenth century.

Both sexuality and human difference would become prominent themes in the Enlightenment. Their pronounced presence in Linnaeus's work betrays the fact that he did not belong to an outdated tradition. He rather broke in many ways with traditional thought, especially with the idea of nature forming a scale of perfection on which each being was assigned its preordained place. Humankind's place in nature, in Linnaeus's eyes, was not at the top of the *scala naturae*, but rather in the center of a web of relations that interconnected all beings and was open to progress. The two Linnaean essays that propounded this view of nature, *Oeconomia Naturae* (1749) and *Politia Naturae* (1761), were then the ones also, whose legacy lasted well into the nineteenth century.

BIBLIOGRAPHY

Editions of Linnaeus's original publications abound, and there is no modern scholarly edition of his work. The best access to his work is still provided by B. H. Soulsby's 1933 bibliography of Linnaeana in the holdings of the British Library (see Lindroth's article for bibliographic detail). Wilfrid Blunt's The Compleat Naturalist *(see below) contains a useful overview of Linnaeus's main works. The bibliographies by Sven-Erik Sandermann Olsen and R. W. Kiger et al. cover the two hundred dissertations completed under Linnaeus and present important biobibliographic information on his students. Linnaeus published in Latin and Swedish, but there exist translations into many different languages. The following list compiles English translations of his major works only, for other translations see Soulsby (1933).*

WORKS BY LINNAEUS

Linnaeus's Öland and Gotland Journey, 1741. Edited by William T. Stearn. Translated by Marie Åsberg. London: Academic Press for the Linnean Society of London, 1973.

Miscellaneous Tracts relating to Natural History, Husbandry, and Physick. Translated by Benjamin Stillingfleet. London: R. & J. Dodsley, 1759. A translation of select dissertations from the collection of theses and orations that Linnaeus had published under the title *Ammoenitates academicae* (Stockholm 1749–1769). Contains translations of *Oratio qua peregrinationum intra patriam asseritur necessitas* (1741), *Oeconomia naturae* (1749), and *Vernatio arborum* (1753).

The Families of Plants, with their Natural Characters, according to the Number, Figure, Situation, and proportion of all the parts of the Fructification. Translated by a Botanical Society at Lichfield. Lichfield: John Jackson, 1787. A translation of the

Linnaeus

posthumous seventh edition of the *Genera plantarum* (edited by J. J. Reichard, Frankfurt/M. 1778).

Select Dissertations from the Amoenitates Academicae. *A Supplement to Mr. Stillingfleets's Tracts relating to Natural History.* Translated by F. J. Brand. London: G. Robinson & J. Robson, 1781. Contains translations of *Curiositatis naturalis* (1748), *Telluris habitabilis incremento* (1744), and *Politia naturae* (1761).

A General System of Nature, through the three grand kingdoms of Animals, Vegetables, and Minerals. 7 vols., translated by W. Turton. London: Lackington, Allen & Co., 1802–1806. A translation of the posthumous, thirteenth edition of the *Systema naturae* (edited by J. F. Gmelin, Leipzig, 1788–1793).

Nemesis Divina. Edited by Michael John Petry. Dordrecht: Kluwer Academic Publishers, 1996. Translation of a secret diary Linnaeus kept to compile evidence for divine retaliation.

Lachesis Lapponica, or a Tour in Lapland. Edited by James Edward Smith, translated by Carl Troilius. London: White & Cochrane, 1811. This is the first edition of Linnaeus Lapland journal, which he himself never came to publish.

Linnaeus' Philosophia botanica. Translated by Stephen Freer. Oxford, U.K.: Oxford University Press, 2003.

Musa Cliffortiana. Introduction by S. Müller-Wille; translated by Steven Freer. Koenigstein: Koeltz Scientific Publishers for International Association for Plant Taxonomy. In press.

OTHER SOURCES

Beretta, Marco. "The Société Linnéenne de Paris (1787–1827)." *Svenska Linnésällskapets Årsskrift* (1991): 151–175.

Blunt, Wilfrid. *The Compleat Naturalist: A Life of Linnaeus.* 2nd ed. London: Lincoln, 2001. Still the best biography of Linnaeus, with an updated bibliography.

Broberg, Gunnar, Allan Ellenius, and Bengt Jonsell. *Linnaeus and His Garden.* Uppsala: Swedish Linnaeus Society, 1983.

Drouin, Jean-Marc. "Linné et l'économie de la nature." In *Science, Techniques & Encyclopédies,* edited by Denis Hue. Paris: Association Diderot, l'Éncyclopédisme & autres, 1991.

Duris, Pascal. *Linné et la France, 1780–1850.* Geneva: Droz, 1995.

Frängsmyr, Tore, ed. *Linnaeus: The Man and His Work.* Berkeley: University of California Press, 1983. Contains translations of important contributions by Swedish scholars on Linnaeus: Sten Lindroth's classic "The Two Faces of Linnaeus," Gunnar Eriksson's "Linnaeus the Botanist," Frängsmyr's "Linnaeus the Geologist," and Gunnar Broberg's "Homo Sapiens: Linnaeus's Classification of Man."

Gardiner, Brian G. "Linnaeus's Medical Career." *The Linnean* 1 (1984): 11–17.

———. "Linnaeus and Tobacco." *The Linnean* 6 (1990): 15–20.

Heller, John Lewis. *Studies in Linnaean Method and Nomenclature.* Frankfurt: Lang, 1983. A collection of essays providing, among other things, a translation of the dedication to the *Hortus Cliffortianus* (1737), an interesting essay on the origins of binomial nomenclature in bibliographic practices, and expansions of the abbreviations Linnaeus used to refer to older taxonomic literature.

Hocquet, Thierry, ed. *Les fondements de la botanique: Linné et la classification des plantes.* Paris: Vuibert, 2005. A collection of essays by Giulio Barsanti, Pietro Corsi, Jean-Marc Drouin, Pascal Duris, and Staffan Müller-Wille. Also contains translations of the *Fundamenta Botanica* (1736) and the introduction to the *Genera Plantarum* (1737).

Hövel, Gerlinde. *"Qualitates vegetabilium," "vires medicamentorum," und "oeconomicus usus plantarum" bei Carl von Linné (1707–1778): Erste Versuche einer zielgerichteten Forschung nach Arznei- und Nutzpflanzen auf wissenschaftlicher Grundlage.* Stuttgart: Deutscher Apotheker-Verlag, 1999. An important study of Linnaeus's pharmacological doctrine.

Jarvis, Charlie. *Order out of Chaos: Linnaean Plant Names and Their Types.* London: Linnean Society, 2007.

Kiger, Robert W., et al. *Index to Scientific Names of Organisms Cited in the Linneaen Dissertations: Together with a Synoptic Bibliography of the Dissertations and a Concordance for Selected Editions.* Pittsburgh: Hunt Institute for Botanical Documentation, 1999.

Koerner, Lisbet (née Rausing). *Linnaeus: Nature and Nation.* Cambridge, MA: Harvard University Press, 1999. A detailed study of Linnaeus's economic thought, his relation to Lapland, and the acclimatization projects he and his students undertook.

Lafuente, Antonio, and Nuria Valverde. "Linnaean Botany and Spanish Imperial Biopolitics." In *Colonial Botany: Science, Commerce, and Politics in the Early Modern World,* edited by Londa Schiebinger and Claudia Swan. Philadelphia: University of Pennsylvania Press, 2005.

Larson, James L. *Interpreting Nature: The Science of Living Form from Linnaeus to Kant.* Baltimore: Johns Hopkins University Press, 1994. An important study of the late-eighteenth-century legacy of Linnaeus, Buffon, and Albrecht von Haller.

Müller-Wille, Staffan. "Gardens of Paradise." *Endeavour* 25, no. 2 (2001): 49–54.

———. "Joining Lapland and the Topinambes in Flourishing Holland: Center and Periphery in Linnaean Botany." *Science in Context* 16 (2003): 461–488.

———. "Nature as a Marketplace: The Political Economy of Linnaean Botany." In *Oeconomies in the Age of Newton,* edited by Neil De Marchi and Margaret Schabas. *History of Political Economy* annual supplement, Vol. 35. Durham, NC: Duke University Press, 2003.

———. "Walnuts at Hudson Bay, Coral Reefs in Gotland: The Colonialism of Linnaean Botany." In *Colonial Botany: Science, Commerce, and Politics in the Early Modern World,* edited by Londa Schiebinger and Claudia Swan. Philadelphia: University of Pennsylvania Press, 2005.

———. "Linnaeus' Herbarium Cabinet: A Piece of Furniture and Its Function." *Endeavour* 30, no. 2 (2006): 60–64.

———. "Collection and Collation: Theory and Practice of Linnaean Botany." In *Studies in History and Philosophy of the Biological and Biomedical Sciences* 38, no. 3 (2007). Provides an analysis and English translation (coauthored with K. Reeds) of the introduction to the *Genera Plantarum* (1737).

Natural History Museum (London). "Linnaeus Link." Available from http://www.nhm.ac.uk/research-curation/projects/linnaeus-link/index.html. Collection of information on holdings of Linnaeana in museums worldwide.

Rausing, Lisbet. "Underwriting the Oeconomy: Linnaeus on Nature and Mind." In *Oeconomies in the Age of Newton,* edited by Neil De Marchi and Margaret Schabas. *History of Political Economy* annual supplement, vol. 35. Durham, NC: Duke University Press, 2003.

Sandermann Olsen, Sven-Erik. *Bibliographia discipuli Linnaei: Bibliographies of the 331 Pupils of Linnaeus.* Copenhagen: Bibliotheca Linnaeana Danica, 1997.

Schiebinger, Londa. "The Private Life of Plants: Sexual Politics in Carl Linnaeus and Erasmus Darwin." In *Science and Sensibility: Gender and Scientific Enquiry, 1780–1945,* edited by Marina Benjamin. Oxford, U.K.: Blackwell, 1991.

———. "Why Mammals Are Called Mammals: Gender Politics in 18th-Century Natural History." *American Historical Review* 98 (1993): 382–411.

Sloan, Phillip R. "The Buffon-Linnaeus Controversy." *Isis* 67 (1976): 356–375.

———. "The Gaze of Natural History." In *Inventing Human Science: Eighteenth-Century Domains,* edited by Christopher Fox, Roy Porter, and Robert Wokler. Berkeley: University of California Press, 1995. An important study of Linnaeus's and Buffon's anthropology.

Sörlin, Sverker. "Scientific Travel: The Linnaean Tradition." In *Science in Sweden: The Royal Swedish Academy of Sciences, 1739–1989,* edited by Tore Frängsmyr. Canton, MA: Science History Publications, 1989.

———. "Ordering the World for Europe: Science as Intelligence and Information as Seen from the Northern Periphery." In *Nature and Empire: Science and the Colonial Enterprise,* edited by Roy MacLeod. Chicago: University of Chicago Press, 2000. An interesting account of Linnaeus's position in the world of eighteenth-century natural history.

Stevens, Peter F., and Steven P. Cullen. "Linnaeus, the Cortex-Medulla Theory, and the Key to His Understanding of Plant Form and Natural Relationships." *Journal of the Arnold Arboretum* 71 (1990): 179–220.

Swedish Linnaean Society, the Royal Swedish Academy of Sciences, Uppsala University and its library, and the Linnean Society of London, with the collaboration of the Centre International d'Étude du XVIIIe Siècle. "The Linnaean Correspondence." Available from http://linnaeus.c18.net/. An online edition of all preserved letters to and from Linnaeus, with explanatory notes and English summaries.

Weinstock, John, ed. *Contemporary Perspectives on Linnaeus.* Lanham, MD: University Press of America, 1985. An essay collection with interesting contributions on Linnaeus and folk taxonomy, his classificatory practices, and his background in Swedish culture.

Staffan Müller-Wille

LIONS, JACQUES-LOUIS

(*b.* Grasse, France, 2 May 1928; *d.* Paris, France, 17 May 2001), *mathematics and applications of mathematics, analysis, applied analysis, scientific computing, control theory.*

Lions was a scientist of remarkable prescience and immense energy. His vision extended to the development of entire areas of mathematical science. He understood that mathematics could make a great contribution to science and he worked to see this realized. While Lions's name is rarely attached to the specific results of his researches, his mathematical legacy is extremely important. He authored or co-authored twenty books and nearly 600 articles, and had a considerable influence on the development of the French school of applied mathematics, and on many mathematicians and many mathematical institutions worldwide. He was also active in other areas of scientific research, in industry, and in politics.

Jacques-Louis Lions was the only son of Honoré Lions and Annette Muller. For more than thirty years, his father was the mayor of the old and charming town of Grasse located in the hills above the Mediterranean seashore, in a region producing many of the flowers used by the French perfume industry. Lions grew up in Grasse and he was very attached to the town and to this part of France. In 1950 he married his lifelong companion, Andrée Olivier, whom he had met during World War II in the Resistance. Their only son, Pierre-Louis, was born in Grasse in 1956. Pierre-Louis, like his father, became a famous mathematician, and he was awarded one of the Fields Medals at the International Congress of Mathematicians in Zurich in 1994.

After high school Lions studied at a small one-year college that would become the Université de Nice. His potential was detected by an examiner who advised him to prepare for the competitive École Normale Supérieure in Paris, which Lions entered in 1947 and attended until 1950. After the Ecole Normale Supérieure, instead of becoming a schoolteacher (as did many of his fellow students), Lions chose to become a university teacher. He obtained a fellowship from the Centre national de la recherche scientifique (CNRS) and started research under the direction of Laurent Schwartz at the Université de Nancy. Also in 1950, Schwartz was awarded the Fields Medal at the International Congress of Mathematicians at Harvard University in Cambridge, Massachusetts, for his work on the theory of distributions. Schwartz thought that the theory of partial differential equations should be completely recast in the context of distribution theory. Lions was one of several students whom Schwartz directed to take this new approach, and Lions's doctoral thesis developed what has become the standard variational theory of linear elliptic and evolution equations. From then on Lions developed his own research, working hard and without any interruption to the end, even when he had other important and time-consuming responsibilities.

Lions received a doctor of science degree in 1954 and was appointed maître de conférences (associate professor) at the Faculté des sciences of Nancy. Later he became professor and, in 1963, he was appointed professor at the Faculté des sciences at the University of Paris. In 1973 he became Professor at the famous Collège de France in Paris, holding the chair that was named, by his choice, "Analyse mathématique des systèmes et de leur contrôle" ("Mathematical analysis of systems and their control"). He retired from the Collège de France in 1998.

The mathematical work of Lions is at the same time very diverse and well unified. Part but not all of his work is contained in the three-volume collected works published in 2003. This article will describe his work following the titles of these three volumes.

Partial Differential Equations (PDEs) and Interpolation. This work started in the early 1950s; continuing his thesis work, it contains many of the building blocks of what would be Lions's "Analyse des Systèmes." He first addressed, alone or in collaboration, many issues in linear PDE theory and in distribution theory. In the late 1950s Lions started his first three major and lasting works. Under the influence of Jean Leray, Lions became interested in nonlinear partial differential equations, in particular the incompressible Navier-Stokes equations. The mathematical analysis of the Navier-Stokes equations, which had been inactive since the pioneering work of Leray in the 1930s, came back to life in 1951, when Eberhard Hopf established the long-time existence of weak solutions for bounded domains in three dimensions. The contributions of Lions to the subject are twofold. He and Giovanni Prodi proved independently the uniqueness of weak solutions in two space dimensions, publishing the result together in 1959. Also, as part of his better understanding of evolution equations, Lions was able to shorten and make considerably more accessible the results of Hopf; he thus contributed, with Olga Aleksandrovna Ladyzhenskaya, James Serrin and others, to the beginning of the modern theory of mathematical fluid dynamics.

Also around this time, Lions began his work with Enrico Magenes on nonhomogeneous boundary value problems, which led eventually to the publication of a three-volume book in 1968. A wealth of results were developed as part of this work that necessitated a better understanding and many different characterizations of the Sobolev spaces (introduced by Sergei L. Sobolev in the 1930s), the systematic study of Sobolev spaces with fractional exponents, and the theory of linear elliptic and parabolic equations in such spaces.

The third major work started in that period—parallel to, and necessary for Lions's two other projects—was the theory of interpolation between Hilbert or Banach spaces; that is, constructing a space intermediate in the sense of topology and set inclusion between two given spaces. Lions made substantial contributions to interpolation and he initiated the interpolation between Hilbert spaces in 1953. Although "linear," the last two areas of research eventually proved to be very important for nonlinear problems.

Following an idea of George Minty and Felix Browder, Lions was in his subsequent work involved in the development of the theory of strongly nonlinear equations that are monotone in their highest arguments. In his only work with Jean Leray, he published in 1965 one of the most general results in that direction, extending and simplifying an earlier result of Mark Vishik. Also, together with Guido Stampacchia, he published in 1965 and 1967 two articles that laid the basis of the theory of variational inequalities. Subsequently Lions continued to develop this theory alone, and with Haim Brezis, and, later, with Georges Duvaut in a book devoted to the application of variational inequalities to many concrete and specific problems in continuum mechanics and physics (1972). Lions's theoretical work on nonlinear PDEs is included in a book published in 1969 that was very exhaustive at the time.

Numerical Analysis, Scientific Computing, and Applications. Lions became interested in this subject in the early 1960s. At that point he was influenced by another of his intellectual mentors, John von Neumann, who had designed the first computers and started to use numerical methods for the solution of partial differential equations from fluid mechanics and meteorology. Lions dreamed, almost alone in France, that there was an important future for mathematics in that direction, and he threw himself into this subject. At that time the French mathematical school was almost exclusively engaged in the development of the Bourbaki program in pure mathematics, and it is one of Lions's major achievements and vision to have predicted this development and to have made applied mathematics accepted in the French mathematical community. He did not publish much in the area at first, but he started the French school of numerical analysis. He taught numerical analysis at the Institute Blaise Pascal in Paris and at Ecole Polytechnique. His students of that time and their academic descendants represent a very significant part of the French numerical analysis community.

In his courses, Lions used from the beginning the variational theory of boundary value problems that he developed himself in his thesis and subsequently. This point of view was further developed in the first theses that he directed in numerical analysis, thus producing the appropriate framework for the development of the finite element methods and for many other important subsequent and contemporary developments in numerical analysis. This includes the thesis and work of Jean Céa on the use of variational methods in numerical analysis, and the works of Philippe Ciarlet and Pierre-Arnaud Raviart on the numerical analysis of the finite elements methods. In this way Lions also played an important indirect role in the research on the numerical analysis of PDEs. Lions himself subsequently published in the area of numerical analysis and scientific computing, including works with Roland Glowinski on the numerical analysis of variational inequalities and control theory (1981), and with Olivier Pironneau on the domain decomposition method.

Among Lions's many works related to applications, in the 1990s, while he was President of the Centre national d'études spatiales (CNES) and President of the Scientific Council of the National Meteorological Office in France, he developed an interest in the mathematical problems of the ocean, atmosphere, and environment. In one of his books on control (1992), he introduced and studied the concept of sentinels for the control and detection of pollution. With two collaborators, Lions wrote a series of eleven articles and a monograph on the mathematical problems raised by the primitive equations of the atmosphere, of the ocean, and of the coupled atmosphere and ocean, and by related asymptotic and numerical issues.

Another massive work in this area is the nine-volume series that he published and edited with Robert Dautray in 1988, *Mathematical Analysis and Numerical Methods for Science and Technology*, which addresses the problems discussed in the classical book of Richard Courant and David Hilbert (*Methods of Mathematical Physics*, vol. 1 and 2, 1953, 1962) in light of the modern methods. This series was translated into English in a six-volume series. Lions also started to publish and edit with Ciarlet the *Handbook for Numerical Analysis* series, which Ciarlet continued to publish by himself after the death of Lions (1990–2002); together they also edited two series of books in applied mathematics.

Control and Homogenization. Lions began new directions of research, "System theory" and optimal control, in the late 1960s when he was scientific director at the newly created Institut de Recherche en Informatique et Automatique (Research Institute for Informatics and Automated Systems, or IRIA). He then worked on control theory corresponding to the second part of the title of his chair at the Collège de France. In its general and industrial sense, control theory refers to the procedure by which one determines the best input in a system—e.g., an industrial plant—to obtain the best output: the closest to the desired result in terms of production cost or quality of the product. It can also refer to the description of the states that a given system can or cannot reach. Depending on the nature of the equations governing the system this can result in substantially difficult mathematical problems. Instead of publishing articles, Lions directly published a research monograph on the optimal control of systems governed by partial differential equations (1971). This unique book, originally published in 1968, became the reference book on the subject; like others of his books, it was translated into English, Russian, Japanese, and Chinese. Subsequently Lions considerably developed the subject, writing nine books partly or totally devoted to control theory, which were published between 1968 and 1992. Two of them, in 1978 and 1982, were written with Alain Bensoussan; one was devoted to the applications of variational inequalities to stochastic control, and the other one to impulse control and variational inequalities. In the 1980s Lions was interested in controllability and he introduced the Hilbert Uniqueness Method (HUM), which he developed in a book published in 1988; this was also the topic of his John von Neumann lecture at the Society for Industrial and Applied Mathematics (SIAM) meeting in Boston in 1986.

Another direction of research in the late 1970s and through the 1980s was homogenization. The purpose of this research is the macroscopic description of materials with a complex microscopic structure; PDEs, asymptotics and stochastic analysis are the tools needed here. His first major work appeared in 1978 in a book with Bensoussan and George Papanicolaou. Lions also followed very closely the related work on G and Gamma Convergence of the Italian school around Ennio De Giorgi.

Lions worked in many areas of mathematics, started and developed numerous ideas, theories, and concepts. Relatively few mathematical objects or results bear his name. In fact he did not want mathematical objects or results to be named after him and he often sent signals to that effect to his close collaborators. Instead he was very good at coining appealing names for new mathematical concepts or objects that he introduced or contributed to.

Scientific Responsibilities and Other Activities. The above description of Lions's scientific work does not give a proper idea of the considerable impact of his work, or the tremendous activity behind it: the original courses and lectures he gave, the plenary lectures at major international congresses, or the seminars in small departments (often in developing countries), his frequent travels to distant destinations, the hundreds of pages of faxes that he exchanged weekly with his collaborators (at a time predating e-mail). Lions was also extremely influential with his students. Lions attracted many young people around him, both French and foreign. A very partial list of his graduate students and scientific descendants appears in the list of the Mathematics Genealogy Project of the American

Mathematical Society. In 2007, twenty-four students and 709 descendants were mentioned in their database. Lions had at least fifty students for PhD theses, Thèses d'Etat, or Habilitations corresponding to the postdoctoral level. Many of his students became well-established mathematicians; at the end of his life he had scientific descendants of the sixth generation.

Lions also had regular scientific contacts with many high-level scientists worldwide, whom he visited regularly or who visited him in Paris. The regular visitors included Browder, Peter Lax, and Louis Nirenberg from North America; de Giorgi, Magenes, and Guido Stampacchia from Italy; and Sobolev and Vishik from Russia. Magenes recalls that, among Lions's countless mathematical initiatives, at the end of World War II, he was the first French mathematician (with Schwartz) to reestablish contact with the Italian mathematical community and to visit Italy. This led to the lasting and very active interaction and collaboration with de Giorgi, Magenes, Prodi, and Stampacchia. Lions also contributed to the development of applied mathematics in Spain and India (Bangalore), and was always very generous of his time with young people for correspondence, advising, and visits.

The scientific research of Lions was only part of his work; the other part was his role as manager and consultant, his responsibilities in governmental organizations, and later his role in high-level industrial companies. He seems to be one of very few mathematicians in modern history to have had at the same time an important research activity and important positions in governmental and industrial organizations.

In 1980 the Research Institute for Informatics and Automated Systems (IRIA) became the Institut National de recherche en informatique et automatique (INRIA) and Lions became its first president, a position which he held until 1984. At INRIA, Lions was both the manager and the scientific head of this new institute, which he literally molded. INRIA has played and still plays an important role in the development of computer sciences in France. Lions got involved as much as possible in all of its scientific and organizational aspects.

In 1984 Lions became president of the Centre National d'Etudes Spatiales (CNES), the French space agency. The previous president of CNES, Hubert Curien, himself a physicist who would become the Minister of Research, foresaw the important role that mathematics would play in space research; he asked Lions to accept this responsibility. In this new position, Lions was confronted with new challenges: besides the scientific ones (to supervise works on mathematics, physics, chemistry, and engineering), he went from directing INRIA—a new institute that he fully shaped—to presiding over a large, active, and well-established institution. Furthermore, he was the first mathematician to hold this position.

A new four-year appointment was proposed to him in 1992, but instead Lions retired from CNES, deciding once more to confront new challenges: the industrial world. For many years he had been working on mathematical problems originating from industry; as president of INRIA and CNES he had many contacts with industry. He decided to enter the industry establishment itself, and he became a member of the scientific council or of the board of directors of large industrial groups. He was president of the scientific committees of Pechiney, Gaz de France, Electricité de France, and France Telecom; high-level scientific consultant at Dassault-Aviation and Elf; and he belonged to the board of directors of Dassault-Systems, Pechiney, Compagnie Saint-Gobain, and Thomson Multimedia.

Lions was president of the French Academy of Sciences from 1997 to 1999. He was secretary (1978–1991) and then president (1991–1994) of the International Mathematical Union. He was also member, secretary, or chairman of countless committees and institutions related to research. He was always dedicated in his efforts to help individuals or young groups in isolated places, especially in developing countries.

Lions received many awards and distinctions for all his activity. He was member or foreign member of about twenty academies, including the French Academy of Sciences, the (U.S.) National Academy of Sciences, the American Academy of Arts and Sciences, the Russian Academy of Sciences, and the Third World Academy of Sciences. He received about twenty honorary degrees. Lions was awarded the John von Neumann Prize in 1986, the Japan Prize and the Harvey Prize in 1991, and the Lagrange Prize in 1999, among others. In France, he was named Commandeur de la Légion d'Honneur and Grand Officier dans l'Ordre National du Mérite.

Jacques-Louis Lions was an exceptional person in many respects. He was a charismatic man, generous, very open and accessible, adept at avoiding conflict. One of the most striking aspects of his personality was his long-term vision; he was able to see and get involved in things that only came to fruition five, ten, or twenty years later. He had many good ideas and he had the mathematical talent, the physical strength, and the human abilities to implement them (Temam, 2001).

BIBLIOGRAPHY

WORKS BY LIONS

Quelques méthodes de résolution des problèmes aux limites non linéaires. Paris: Dunod, 1969.

Optimal Control of Systems Governed by Partial Differential Equations. Translated by S. K. Mitter. Berlin: Springer-Verlag, 1971.

With Alain Bensoussan and George Papanicolaou. *Asymptotic Analysis for Periodic Structures.* Amsterdam: North-Holland, 1978.

With Roland Glowinski and Raymond Trémolières. *Numerical analysis of variational inequalities.* Amsterdam: North-Holland, 1981.

With Robert Dautray. *Mathematical Analysis and Numerical Methods for Science and Technology.* 6 vols. Translated by Ian N. Sneddon. Berlin: Springer-Verlag, 1988–1993.

As editor, with Phillipe G. Ciarlet. *Handbook of Numerical Analysis,* 11 vols. Amsterdam: North-Holland, 1990–2002.

Sentinelles pour les systèmes distribués à données incomplètes [Sentinels for distributed systems with incomplete data]. *Recherches en Mathématiques Appliquées,* 21. Paris: Masson, 1992.

Oeuvres choisies de Jacques-Louis Lions, Vol. I, *Équations aux dérivées partielles. Interpolation;* Vol. II, *Contrôle, Homogénéisation;* Vol. III, *Analyse numérique, Calcul scientifique, Applications.* Edited by Alain Bensoussan, Philippe G. Ciarlet, Roland Glowinski, Roger Temam, François Murat, and Jean-Pierre Puel. Paris: EDP Sciences, 2003.

OTHER SOURCES

Ciarlet, Philippe G. "Jacques-Louis Lions, 1928–2001." *Matapli* 55 (2001): 5–16.

Temam, Roger. "Jacques-Louis Lions, 1928–2001." *SIAM News* 34, no. 6 (2001).

Roger Temam

LIPMANN, FRITZ ALBERT (*b.* Königsberg, East Prussia [now Kaliningrad Oblast, Russia], 12 June 1899; *d.* Poughkeepsie, New York, 24 July 1986), *biochemistry, role of organic phosphates in metabolism, "energy-rich" bonds, coenzyme A, basic mechanisms of protein biosynthesis.*

> *"There is no biochemist like Fritz Lipmann. He has always been at the heart of biochemistry ... [and] occupies a pivotal position, his lab playing a central role in understanding how central metabolic processes ... [work] ... together."* (Organizer, 1969, p. vii)

Along with Otto Heinrich Warburg and Otto Meyerhof, Lipmann was the deepest thinking, most clear-sighted, and creative of the architects of biochemistry in its construction phase, laying the foundations of metabolic enzymology. Trained in both medicine and chemistry, he combined these two areas of bioscientific research, both by experimental skill and artistic intuition. He formed a great international school of pupils and disciples, whom he led into new fields and left to develop as he turned to plow new ground. His primary interest lay in the physiological use of organic bound phosphate and the turnover of metabolic energy by its utilization (as "energy rich" or "squiggle" phosphate: ~P) in carbohydrate, fatty acid, amino acid, and nucleic acid syntheses as well as degradations. In this course he discovered the coenzyme of enzymatic transacetylations, coenzyme A, acronymized "CoA" or "CoA-SH," containing the transfer catalyzing thiol(SH)-function of its pantetheine moiety, a vitamin composed of pantothenic acid and cysteamine. He also initiated research on the nonribosomal synthesis of linear and cyclic (antibiotic) peptides by specifically arranged multienzyme complexes with pantetheine at its active center and on the milieu dependence of the chemical phosphate potential of phosphorylated amino acids in phosphoproteins.

For the discovery of coenzyme A as cocatalyst of the metabolic transfer of activated acetate-groups he received the Nobel Prize for Physiology or Medicine in 1953, jointly with Hans Adolf Krebs, who was laureated for untangling the energy-yielding tricarboxylic acid (or "Krebs") cycle, which also (re)utilizes CoA-activated acetate, derived from glucose-borne pyruvate.

Curriculum Vitae and Personality. Lipmann was born in Königsberg, East Prussia, on 12 June 1899, the second son of the lawyer Leopold Lipmann and Gertrud Lachmanski. His brother was Heinz Erich Lipmann, playwright, actor, and theater manager at Leopold Jessner's Berlin Staatstheater. In 1931 Lipmann married Elfreda (Freda) M. Hall, a distinguished artist, photographer, and fashion designer. Their son, Steven Hall Lipmann, became a professor of comparative literature in Boston, Massachusetts.

Lipmann came from a typical liberal German-Jewish middle-class family in which general education, arts, and social engagement were de rigueur, while religion was less important. Besides life sciences, he also loved life per se and beauty. Though a serious student, he was broadly interested, though not active, in theatrical and modern art. He interrupted his medical studies in Königsberg in 1921 and spent one semester in Munich's Schwabing district with his brother, and a second semester in Berlin to enjoy the postwar Roaring Twenties with its masqued balls, world-renowned stage performances, and theaters. In Berlin he met his future wife, a practicing artist, and he soon used the pretext of learning new scientific methods to return from Heidelberg again to Berlin.

Lipmann's research combined originality with critical strictness and an impressive ability to extrapolate into the "right," experimentally approachable, chemical and biological direction to uncover metabolic pathways. He had an almost intuitive ability to see the "possible." The leitmotif of his work was the enzymatic utilization of chemical-bond-energy from degradation of foodstuffs (essentially, glucose from glycolysis) for the biosynthetic processes involving phosphates and vitamin-derived coenzymes, leading to condensed biomacromolecules such as proteins, nucleic acids, fats, and their conjugates.

Lipmann's experiments were simple, yet well conceived and to the point. His methods remained those of his training: Warburg respirometry; test-tube incubations, fractionations, and separations by all current methods; (spectro)photometry; and slide-rule. He saw technical progress in methodology and data collecting as necessary for accumulating verifiable and falsifiable evidence, not as means *per se.* He dared extrapolations but rarely overstressed the results, using them instead to suggest further investigations.

After his training at Berlin and Heidelberg in the collegial team of independent junior research workers under Otto Meyerhof's steering synthetic and Karl Lohmann's bridling analytic supervision, he started on his own at Boston's Massachusetts General Hospital, where he quickly developed his intrinsic leadership qualities. He was brains but not boss. Private frank discussions in small circles on writing reports and papers, common seminars, and coffee hours were the fruitful ground for continuing branching and diversification, even when international fame increased the momentum.

After receiving a classical education in the local humanistic gymnasium, Lipmann studied medicine (1917–1923), influenced by his Lachmanski uncle, an admired pediatrician. He also studied chemistry with the inspiring Hans Leberecht Meerwein, first at the University of Königsberg, then at Munich and Berlin. His studies were interrupted by military service in the medical corps at the Marne front and during the influenza epidemic of 1918–1919, and by a research fellowship with Ernst Laqueur of the University of Amsterdam in 1923. He received an MD in 1924 from Berlin University with Peter Rona, who trained many of the later specialists in enzymology, and in 1928 a PhD in chemistry with Meyerhof, under the formal aegis of the well-known biochemist Carl Neuberg from Technische Hochschule Berlin, investigator of yeast fermentation.

Between 1931 and 1939, Lipmann became research assistant to Meyerhof, at the Kaiser Wilhelm Institute (KWI) for Medical Research, Berlin and Heidelberg; to Albert Fischer, at the KWI for Biology, Berlin, cell culture division; as Rockefeller Fellow to Phoebus A. T. Levene at the Rockefeller Institute for Medical Research, New York, New York; and again to Fischer, at the Biological Institute

Fritz Albert Lipmann. THE LIBRARY OF CONGRESS.

of the Carlsberg Foundation, Copenhagen. In 1933, when Adolf Hitler seized power in Germany and his government introduced the anti-Semitic race laws, Lipmann remained in Denmark, now supported by the Rockefeller Foundation, to find a place. He emigrated in 1939 to the United States, where he was naturalized in 1944. He served as research associate with Vincent du Vigneaud at Cornell University School of Medicine in New York City (1939–1941), then as a member of the staff at Massachusetts General Hospital, Harvard School of Medicine, in Boston, Massachusetts, first as a research fellow, later as an associate and professor in biological chemistry (1941–1957). In 1957 Lipmann was appointed (emeritus) professor of biochemistry at the Rockefeller Institute of Medical Research, Rockefeller University, New York City. He died in Poughkeepsie (not far from his country home in Rhinebeck, New York) on 24 July 1986.

Lipmann was multiply recognized by honorary academic degrees: MD, University of Marseilles (1942) and University of Copenhagen (1953); MA, Harvard University (1949); DSc, University of Chicago (1953), University of Paris (1966), Harvard University (1967), and Rockefeller University (1971); LHD, Brandeis University

Waltham (1954) and Yeshiva University (1964). In 1953 he received the Nobel Prize in Physiology or Medicine; in 1948 both the Mead Johnson Award and the Carl Neuberg Medal; in 1966 the National Medal of Science; and in 1975 the Pour le Mérite (German order of merit).

Lipmann was an elected or honorary member of almost all major academies of science worldwide, including the Danish Royal Academy of Sciences and the New York Academy of Sciences (1949); the U.S. National Academy of Sciences (1950); the American Philosophical Society (1959); the Royal Society of London (1962); and the German Academy of Sciences Leopoldina (1969); as well as many bioscientific societies.

Lipmann stands in the hub of the wheel of evolving metabolic molecular biology. His peers were school-forming colleagues of the Meyerhof-Lohmann team, among others: Severo Ochoa, David Nachmansohn, Karl Mayer, and Dean Burk; his pupils came from all over the world and spread internationally, forming the nuclei of new schools of experiment and thought in the chemical and medical biosciences.

Scientific Achievements. During his medical and post-clinical training on the ward and at the section table, Lipmann developed scruples on the material aspects of the vocation but also on his avocation as a general or specialized practitioner. He wanted a more scientific understanding of the processes he observed. Having understood early on how essentially interwoven chemical and life processes are, he completed, parallel to his studies in medicine, additional time-consuming courses at the teacher's level in general and analytical chemistry and also passed the *Staatsexamen* (teacher's examination), which would allow him to work in the Prussian school service if need arose.

Energy Cycles Involving Phosphocreatine. Luckily, he followed the scientific career instead. Part of his medical practical year in Berlin he spent with Ludwig Pick in pathology, the other part on the famous three-months training course in quantitative physical methods of then-modern colloid chemistry at Leonor Michaelis's and Peter Rona's research laboratory. There he met some of his later peers in the nascent biochemical sciences. The quarter of a year studying the ion dependency of colloidal ferric hydroxyde sols as model of proteins sufficed for obtaining the MD from inflation-stricken Berlin University, followed by six months of laboratory practice in affluent Amsterdam. Thus trained practically and prepared mentally for research in enzymology and intermediary metabolism, Lipmann chose to study biochemistry at one of the foremost places in Germany: Meyerhof's laboratory at Warburg's section of the KWI for Biology in Berlin-Dahlem.

The focus of interest of biochemists at that time when cell-free systems opened for detailed enzyme studies were the glycolytic processes in yeast during alcoholic fermentation (glucose yielding 2 moles each of ethanol and CO_2 *plus* energy) and in muscle during work (glucose yielding 2 moles of lactate *plus* energy), in both of which phosphate is involved. Meyerhof and Archibald V. Hill had just received the Nobel Prize in Physiology or Medicine (1922) for their studies on the energetic stoichiometry in muscle contraction. Lohmann and Meyerhof had demonstrated that the two processes run parallel except for the last step, when pyruvate is either reduced (in muscle) to lactic acid or split (in yeast) to CO_2 and acetaldehyde followed by reduction to ethanol. Both reductions use pyridine nucleotide (NAD^+) as a hydrogen carrier. It was postulated that lactic acid causes muscular contraction. It became clear that energy production and intermediary metabolism are mechanistically, that is, chemically, connected.

This new concept struck Lipmann, and he applied for and was accepted into Meyerhof's select research group to work on his PhD thesis in chemistry. This resulted in important studies: (1) on creatine phosphate as energy store, and (2) on interconnected metabolic effects of fluoride on glycolysis, respiration, and oxidized hemoglobin. Despite being rather preliminary, they formed the basis for later fundamental insights. When Einar Lundsgaard showed in 1930 that iodoacetate (or fluoride) poisoning of muscle does not block contraction, which goes on until creatine phosphate stores are exhausted, it became clear that the latter, not the acidification, is the prime energy donor in working muscle. By manometrically following the primary breakdown of creatine phosphate and the secondary formation of lactate during electrostimulation of (frog) muscle, Lipmann and Meyerhof proved this definitely. Lohmann showed calorimetrically the energy equivalency of creatine phosphate and the "energy rich" end-(or "seven minute")-phosphate of his newly discovered adenosine triphosphate (ATP). Gustav Embden stated that hexose-diphosphates, the initiating phosphorylated intermediates in glycolysis, are split between carbon #3 and #4 in the middle of the six-carbon chain of the hexose sugar to two molecules of acid-stable triosephosphate *via* instable phosphorylated derivatives, and set the keystone on the arch of the "Embden-Meyerhof" glycolytic cycle.

Now everything fell into place: "Energy rich" phosphate intermediates of glycolysis, phosphoglycerylphosphate (PGA) and phosphoenolpyruvate (PEP), transfer enzymatically the active phosphate group to adenosine diphosphate (ADP) to form ATP, which then is either used directly for all energy-driven cellular processes or intermediarily phosphorylates creatinine to creatine phosphate as energy buffer and store while regenerating ADP

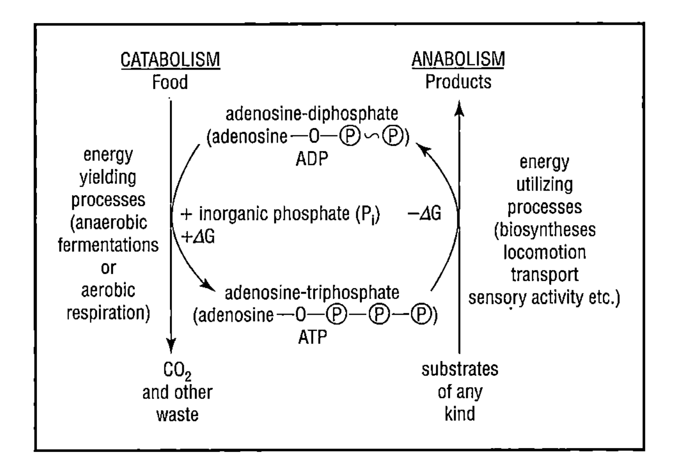

***Figure 1.** Role of ATP in Metabolism.*

for further transphosphorylation cycles. ATP, formed by electron withdrawing (oxidative) enzymatic processes, is used as the universal energy source to drive chemical work in metabolism as well as mechanical work in muscle, as schematized in Figure 1.

Concept of the "Energy Rich Bond" in Biochemical Systems. This became the origin of the concept of "energy rich" or "squiggle" (~) bonds as the site of the event in group transfer reactions. Biochemists took up the concept enthusiastically, as it gave them a feeling for the potential or pressure inherent in the event, while physicochemists remained skeptical about the "loose speak" of the non-initiated. Nevertheless, it is a useful and suggestive symbol for a molecule in a special quantum chemical conformation of its binding system. Lipmann introduced the concept into scientific discussion in his groundbreaking 1941 review article, "Metabolic Generation and Utilization of Phosphate Bond Energy," in the first volume of *Advances in Enzymology and Related Areas of Molecular Biology.*

The ~ symbolizes the chemical bond between carrier (C) and donor (D) moieties from electron (e⁻) or proton (H^+) to groups (for example, phosphate, P) and larger functional portions of a substrate in a transfer system to an acceptor (A): C~D + A ⇌ A-D + C + [-(G°]. G° is the Gibbs Free Energy (reversible work [enthalpy] *plus* entropy), measured in energy units (kcal or kJ). The value of –(G° denotes the energy gradient between the left and right side of the equation. The *minus* (–) sign in biochemistry is a signature for a downhill reaction. Conventionally, values > –7 kcal (c. 28 kJ) are termed "energy rich." In a freely reversible (equilibrium) system the (G°s of both sides cancel each other.

When new techniques became available—such as working with tissue culture of whole cells, which he learned from and with Fischer in Berlin and

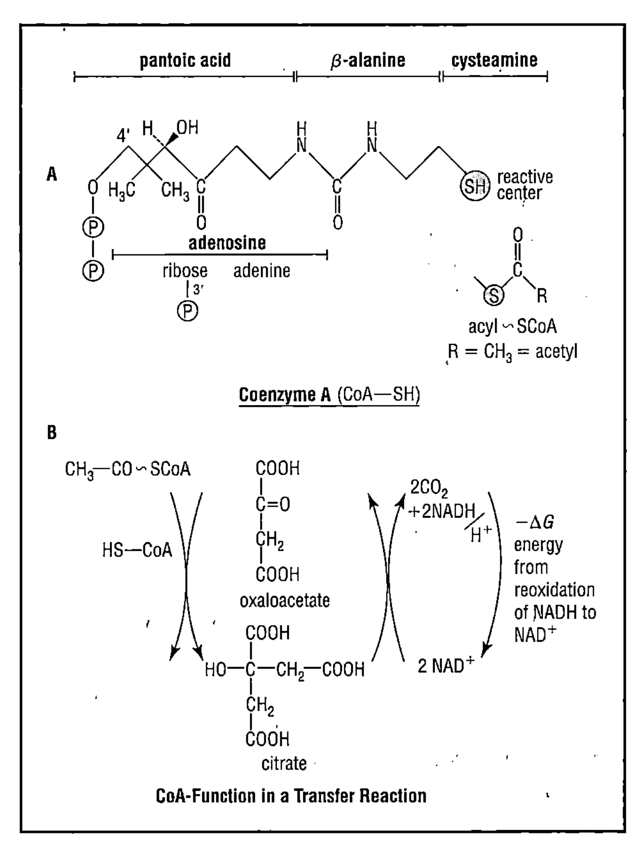

***Figure 2.** Coenzyme A (CoA) and its Function/CoA - Function in a Transfer Reaction.*

Copenhagen—Lipmann turned to the study of aerobic glycolysis in chicken heart fibroblasts and other blasts and in embryonic cells, which yielded much möre energy than the anaerobic process. He combined it with the manometry of Warburg, who had discovered the large aerobic metabolism in malignant tissue. He replaced cell counting by quantitative metabolic parameters such as respiratory O_2 uptake and its dependence on phosphate binding. He measured the aerobic repression of the wasteful energy supply by glycolysis through the highly economical respiratory energy production, what Warburg termed the "Pasteur effect," and found it similar in normal cultured cells. It was surmised to depend on the redox potential, measurable by dye-indicators: Titratable sulfhydryl (SH) groups of glutathione or SH-enzymes (actually, glyceraldehydephosphate-dehydrogenase, GAPDH) disappeared coincident with glycolysis. This was an early hint to the metabolic formation of "energy-rich phosphate (~P)" bonds that finally lead to the high-energy intermediates of glycolysis by the GAPDH twin-system and by PEP-kinase. We now know that the true Pasteur effect is attributable to a feedback loop by high levels of ATP on phosphofructokinase, phosphorylating fructose-6-phosphate to fructose-1, 6-bisphosphate. But it was at that time that Lipmann became interested in the role of organic phosphates in metabolism.

Pyruvic Acid Oxidation, Activated Phosphate and Acetate, Coenzyme A. Cell-free extracts of *Lactobacillus delbrueckii* require inorganic phosphate (P_i) for oxidation of the glycolysis intermediate pyruvic acid and yield ATP with added adenylic acid (AMP). As Lipmann and L. Constance Tuttle showed in 1944, the phosphoroclastic reaction of pyruvate is reversible; pyruvate and P_i can be replaced by acetyl phosphate, or, as observed soon with a fractionated enzyme system, by pyruvate decarboxylase, the electron carrier nicotine adenine dinucleotide (NAD^+) and transacetylase *plus* a heat stable coenzyme—also found by David Nachmansohn in New York, and by Wilhelm Feldberg in Cambridge, England, as cofactor in the acetylation of choline for the synthesis of the neurotransmitter acetyl choline—that Lipmann named coenzyme A (CoA) and identified as a general carrier in transacetylations.

In combined work with Esmond E. Snell and other biochemists, it was found that CoA is an unusually complex molecule, consisting of phosphorylated adenylic acid and pantetheine phosphate (phosphate ester of the vitamin pantothenic acid peptidically bound to mercaptoethylamine) joined by a diphosphate bridge (shown in Figure 2). Soon after, Feodor Lynen and Ernestine Reichert found that the acetic acid in acetyl-CoA is bound as thermodynamically "energy rich" thioester or acylthiol to the activating acidic SH-terminus, thus S-acetyl-CoA = $CoA\text{-}S\text{-}COCH_3$.

As a kinetically stable, thermodynamically labile, transport metabolite, acetyl-CoA has multiple functions in energy transfer in all known organisms. In the metabolism of carbohydrates, fats, sterols, and proteins it is an intermediary of enzymatic cellular condensing reactions, for example, in the Krebs citric acid cycle and in the Lynen fatty acid forming and degrading spirals. S-acetyl-CoA is energetically similar and mechanistically comparable to acetylchloride used in organic syntheses, also an acid anhydride. The principle of group activation was verified in carbamoyl phosphate generation (in urea and pyrimidine syntheses) and sulfate activations (to chondroitins, cerebrosides, or tyrosines in proteins).

Protein Biosynthesis and Molecular Evolution. Group activation is also the basis of peptide bond formation in the biosynthesis of proteins and, in general, of polycondensations (of sugars, nucleic acids, lipids, and esters). Amino acids are ATP-activated by phosphorylation of their carboxyl group, then, as postulated early (1950) by the cytobiologist Daniel Mazia, transferred to RNA-carriers with amino acid–specific triplet marks and

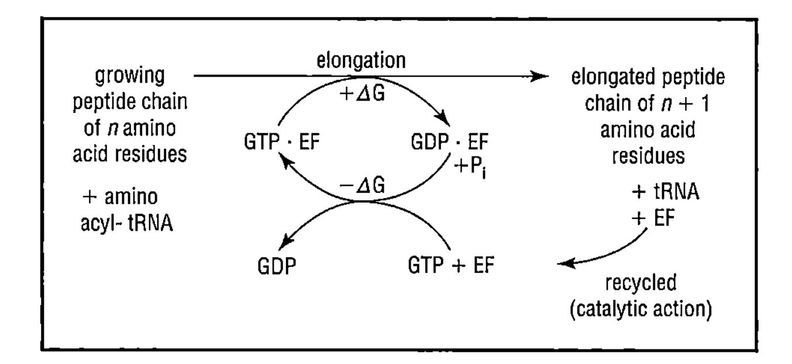

Figure 3. *Function of Elongation Factors (EF) in Pepide Synthesis.*

brought into the proper sequence by means of the anticodons of messengers that are blots of informing DNAs of the genome. The machinery is in the ribosomes, as first shown in 1955 by Philip Siekevitz in the team of Paul C. Zamècnik and Mahlon B. Hoagland at Harvard University working with microsomal cell fractions of liver and bacteria. Lipmann used the latter system. The mechanism of bacterial polypeptide synthesis runs in three phases, as shown in Figure 3: (1) initiation; (2) polymerization; and (3) termination. The elongation factors (unstable T_u, stable T_s, and G) were separated and their function at the acceptor (A) site and donor (D) site of the ribosome complex in the ribosomal peptide bond established. This forms the basis of the present understanding of the whole process of protein biosynthesis, which is mechanistically analogous, yet not materially homologous, in eukaryotes.

Phosphoproteins and Site-Dependent Increase of Phosphorylation Potential of Serine- and Tyrosine-O-Phosphates. Phosphoproteins as nutritional phosphate carriers are found in milk casein (3% P) and in egg yolk vitellinic acid (10% P). P is O-bound to serine and/or tyrosine residues of the polypeptides, and most of it is relatively acid-stable. Phosphorylated proteins had interested Lipmann during his work at Phoebus Levene's laboratory in 1932 and remained in his subconscious for more than thirty years. Then the transphosphorylations from and to proteins came into focus, and he again turned to them. He showed that the phosphate groups of serine and tyrosine sites have different thermodynamic potential and are turned over from and to ATP by specific protein kinases at different speed. The ATP-dependent protein kinases act partially reversible (as shown with phosvitin or immunoglobulins), so that ADP may be up-phosphorylated to ATP by certain clustered protein phosphate groups within the sequence if a strong pull is applied. This means that in such serine or tyrosine clusters the otherwise very stable phosphate becomes energy rich and may be transferred between donors and acceptors by protein transphosphorylations in receptor functions and cell-regulatory signal chains. This is essentially the basis for the inter- and intracellular communication by protein factors and chemical messages.

Polypeptide Antibiotics on Nonribosomal Templates by Thiol-Linked Peptide Activation and Polymerization. In addition to the peptide bond formation by the ribosomal apparatus, peptides are also formed in special cases and for special vital purposes by nonribosomal multienzyme complexes, which may have evolved earlier from the chain of events in fatty acid ana- and catabolism. Early on, Lipmann became aware of such energy-utilizing processes as exemplified by glutathione and gramicidin biosyntheses.

Bacilli synthesize straight chain and cyclic antibiotic peptides, often containing a D-amino acid or other fitting molecules. The catalyzing binary enzyme complexes resemble in certain aspects the crank-wheel mechanism of fatty acid syntheses, with pantetheine in the activating, transporting, and assembling center. The first step is the ATP-activation of L-phenylalanine by the small aminoacyl carrier protein; followed by the transfer of the activated amino acid to the central SH-group of the big polyenzyme, which also contains a racemase forming the D-stereomer of the initiating amino acid; and the polycondensation of more amino acids according to the given pattern of the subunits with or without head-to-tail cyclization to form the loop or cycle, respectively. This basic type can be experimentally, combinatorially, and genetically varied to great applicable extent.

BIBLIOGRAPHY

The volume edited by Horst Kleinkauf, Hans von Döhren, and Lothar Jaenicke, The Roots of Modern Biochemistry, *contains Lipmann's complete bibliography.*

WORKS BY LIPMANN

"Metabolic Generation and Utilization of Phosphate Bond Energy." *Advanced Enzymology* 1 (1941): 99–162.

Wanderings of a Biochemist. New York: Wiley-Interscience, 1971.

"A Long Life in Times of Great Upheaval." *Annual Review of Biochemistry* 53 (1984): 1–33.

OTHER SOURCES

Duve, Christian de. "Fritz Lipmann: In Memoriam." *FASEB Journal* 1 (1987): 1–3.

Jaenicke, Lothar. "Zum Goldenen Nobel Jubiläum." *BIOspektrum* 10 (2004): 170–173.

Kalckar, Herman M. "Lipmann and the 'Squiggle.'" In *Current Aspects of Biochemical Energetics: Fritz Lipmann Dedicatory Volume,* edited by Nathan Oram Kaplan and Eugene P. Kennedy, 1–9. New York: Academic Press, 1966.

Kleinkauf, Horst, Hans von Döhren, and Lothar Jaenicke, eds. *The Roots of Modern Biochemistry: Fritz Lipmann's Squiggle and Its Consequences.* Berlin: Walter de Gruyter, 1988. Also contains Lipmann's complete bibliography.

Levitan, Thomas N. *Laureates: Jewish Winners of the Nobel Prize.* 1960.

Medawar, Jean, and David Pyke. *Hitler's Gift: The True Story of the Scientists Expelled by the Nazi Regime.* New York: Arcade, 2001.

Nachmansohn, David. *German-Jewish Pioneers in Science, 1900–1933: Highlights in Atomic Physics, Chemistry, and Biochemistry.* Berlin: Springer-Verlag, 1979.

Organizer. "The Mechanisms of Protein Synthesis." *CSH Symposia on Quantitative Biology* 34 (1969): vii.

Lothar Jaenicke

LIVINGSTON, MILTON STANLEY

(*b.* Broadhead, Wisconsin, 25 May 1905; *d.* Santa Fe, New Mexico, 25 August 1986), *nuclear physics, accelerator design and construction, science administration.*

Livingston pioneered the development of particle accelerators. He was the principal collaborator of Ernest O. Lawrence, inventor of the cyclotron, during the first four years of its development, and a codiscoverer of the strong focusing principle essential to synchrotrons. Livingston was also an administrator for several key accelerator programs: first head of the Accelerator Project at Brookhaven National Laboratory, director of the Cambridge Electron Accelerator, and associate director of the National Accelerator Laboratory (later known as Fermilab).

Early Years. M. Stanley Livingston was born in 1905, the son of the minister of a small church in Broadhead, Wisconsin. When he was five, his father moved the family to southern California to take up the more promising career of schoolteacher, and also bought a ranch with an orange grove. Livingston later attributed his mechanical interests and abilities to the fact that it was his job to repair and maintain the farm tools and instruments, including tractors and trucks. His mother died when he was twelve, and his father remarried.

Livingston graduated from high school in 1921, having studied chemistry but not physics. He attended the nearby Pomona College and initially majored in chemistry—for, as he once explained in a videotaped interview in 1982, World War I had made chemistry *the* science to study just as World War II made physics *the* science. When Livingston left the farm to live on campus, his roommate was a physics major named Victor Neher. Livingston was losing interest in chemistry, and Neher fascinated Livingston with his tales as a lab assistant. In his senior year, Livingston took course overloads to allow him

M. Stanley Livingston. Livingston shown with cyclotron. **LAWRENCE BERKELEY LABORATORY/SCIENCE PHOTO LIBRARY.**

to major in physics and chemistry. After his graduation in 1926, his physics professor, Roland R. Tileston, secured him a position as an instructor at Dartmouth College, where Tileston had connections. Livingston received an MA from Dartmouth in 1928, and stayed for another year as an instructor.

Lawrence and Cyclotrons. In 1929 Livingston obtained a fellowship to enter the University of California at Berkeley as a graduate student in physics. His MA and his own inventiveness gave him a head start over his classmates, and by the end of his first year—summer 1930—he had completed his course work, passed his qualifying exams, and asked several faculty members for ideas regarding thesis topics. The most intriguing suggestion came from Lawrence, Livingston's instructor in the electricity and magnetism course: to look for the resonance of hydrogen ions, moving in a magnetic field, excited by an alternating electric current operating at radio frequencies. Not only was the topic exciting but so was Lawrence, an intense and driven experimenter who exuded confidence and ambition, Livingston once said, "gave everyone a feeling of enthusiasm and confidence."

Lawrence, who had arrived at Berkeley in 1928, had seen a paper by Rolf Wideröe outlining a way to accelerate charged particles, and decided to fashion his own version. Lawrence had a graduating student, Niels Edlefsen, construct a primitive device, but the results were inconclusive. When Edlefsen received his degree in 1930 and left the laboratory, Lawrence sought another grad student—and along came Livingston. Livingston discovered that other faculty members had serious doubts about the idea, pointing out that the slightest perturbation would send the particles spiraling out of a flat plane and thus be lost. But Lawrence rekindled his enthusiasm, and Livingston took on the project.

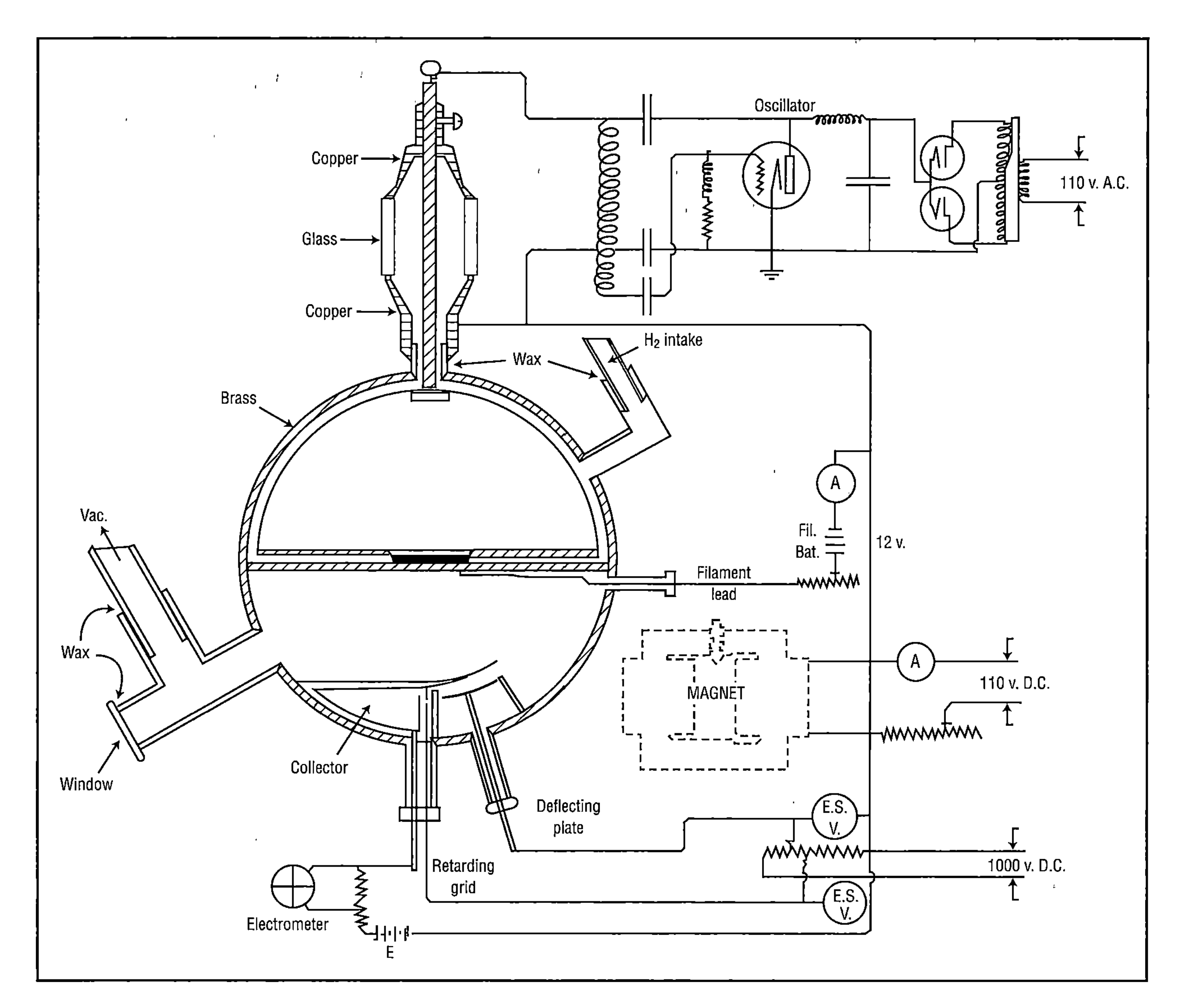

Figure 1. *Diagram of Livingston's 4-inch device, on which he detected resonance for what was surely the first time, from his doctoral thesis.*

Livingston rebuilt the device virtually from scratch, taking advantage of the close fraternity of other students working on their thesis projects and with the supervision of Lawrence, who dropped in almost every day to check on progress. The device might be described as in the form of a circular metal sandwich. The middle layer consisted of a vacuum chamber, and inside Livingston put a hollow D-shaped electrode. This chamber was placed between the cylindrical poles of a 4-inch (10.16-centimeter) electromagnet. The magnet would cause charged particles released in the center of the chamber to run in circular orbits, entering and leaving the D on each traversal of the circle. The D would be charged by an alternating radio-frequency excitation; inside the D was a field-free region. The particles would be alternately attracted and repelled by the D, gaining a slight amount of energy each time they entered and left it. The radius of the particle orbits would slowly increase, and eventually the particles would leave the chamber via a beam port. Such devices came to be called cyclotrons (see Figure 1).

Lawrence had realized the significance of resonance: that the excited particles would stay in step at the same frequency regardless of how fast they were traveling and thus how big their orbits grew—and therefore that repeated application of a small voltage V would result in a large total energy E ($E = NeV$, where N is the number of times the particle entered and left the D). Particles of mass M and charge e would revolve with frequency ω that was independent of the energy of the particles as they increased in speed, meaning that a constant frequency

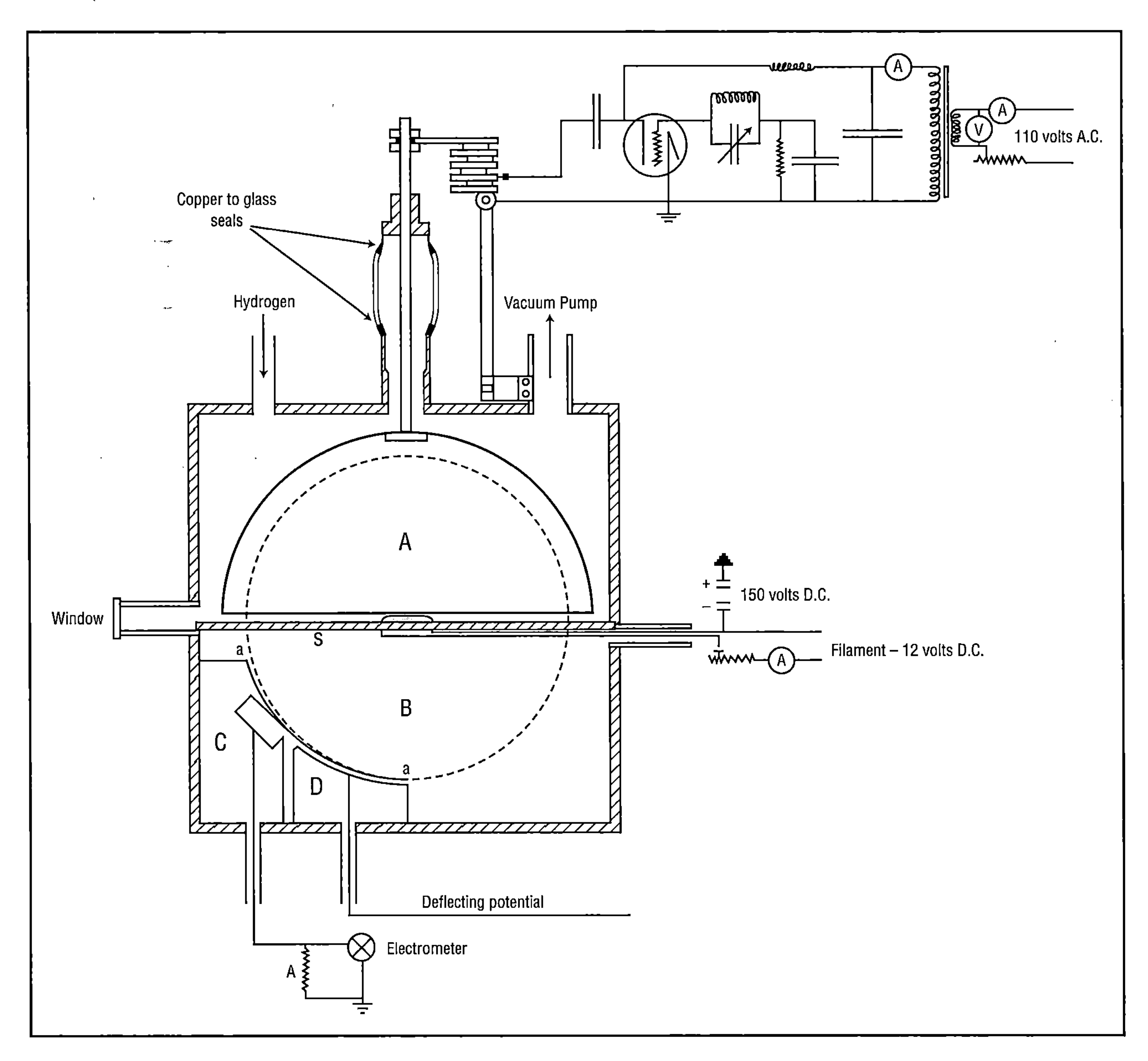

__Figure 2.__ Diagram of the Livingston-Lawrence 11-inch cyclotron, the first accelerator to reach 1 million electron volts.

applied to the D would continue to accelerate the particles ($\omega = eB/Mc$, where B is the magnetic field and c is the velocity of light).

Livingston began work in summer 1930, and shortly discovered that Edlefsen had not detected resonance but another kind of acceleration effect. In November Livingston detected resonance for what was almost certainly the first time. Livingston then borrowed a bigger magnet, and in January 1931 used it and a thousand electron volt (KeV) voltage to accelerate hydrogen ions (H_2^+) to 80 KeV. That result excited Lawrence and sent him, Livingston once recalled, "off to the races," on a decade-long quest, successful even during the Depression, for ever-larger funding and support to build ever-bigger cyclotrons.

Then Lawrence forced Livingston to interrupt his work and finish his PhD requirements. Lawrence needed Livingston to have a PhD, for he had secured Livingston an instructorship, contingent on the PhD, which would enable Livingston to remain at Berkeley and continue constructing cyclotrons. Livingston wrote up and submitted his work in two weeks, but Lawrence forbade him to study for the orals, saying there was too much to do and he could brush up on nuclear physics later. In April 1931, despite being less prepared on nuclear physics—his forte lay in his mechanical abilities—Livingston took and passed the exam.

By this time, Lawrence had received a $1,000 grant from the National Research Council to build a cyclotron with an 11-inch (28-centimeter) magnet. Livingston set to work, this time using a pair of Ds with a small gap between them (see Figure 2). On 9 January 1932, the device passed an important milestone, becoming the first accelerator to exceed a million electron volts (MeV). Years later Livingston recalled: "As the galvanometer spot swung across the scale, indicating that protons of 1-MeV energy were reaching the collector, Lawrence literally danced around the room with glee. The news quickly spread through the Berkeley laboratory, and we were busy all that day demonstrating million-volt protons to eager viewers" (1969, p. 29). On 20 February 1932, they mailed their first publication to *Physical Review*, announcing that they had reached 1.2 million volts. In the early twenty-first century the machine is in the Science Museum in South Kensington, London.

A few months later, John Cockcroft and Ernest Walton of the Cavendish Laboratory in England announced that that they had used a different type of high-voltage acceleration device they had been developing to disintegrate nuclei. That reminded Lawrence and Livingston that they had rivals as accelerator builders, and of the need to pay attention to nuclear physics applications. Lawrence began to push his assistant even harder, with Livingston working until midnight most nights, including weekends and holidays. Lawrence and Livingston built ever-larger cyclotrons, the next being a device with 27.5-inch (69.85-centimeter) pole faces in the summer of 1932. It reached 5 MeV. They began nuclear experimentation with the aid of their colleague Gilbert Newton Lewis, who supplied deuterium (a hydrogen isotope that contains both a proton and a neutron), an effective projectile.

Somehow Livingston found time to meet and marry Lois Robinson, with whom he had two children, Diane (b. 1935) and Stephen (b. 1943).

During his four years at Berkeley, Livingston was the "mechanically minded person who could make these things happen," he once said (AIP interview, 1967, p. 22), while Lawrence provided the funding, support, contacts, and drive. "Lawrence made waves in science," Livingston recalled. "He made such big waves that we all got thrown up to the top. I coasted down the front side of that wave for the rest of my career, using what I had acquired in the way of a reputation, experience, and contacts" (Video interview, 1982, cited in Crease, 1999).

Yet it was Livingston who was responsible for two improvements essential to the success of cyclotrons, one having to do with the electric focusing, the other with the magnetic focusing. The first cyclotrons had been built with wires strung across the opening of the D to create an electric plane thought essential to focus the particles; Livingston discovered that the wires were not only unnecessary, because of an unsuspected focusing effect, but that their removal boosted the intensity. Livingston's second improvement was to use shims to create a "fringing field," or slightly bowed shape to the magnetic field, which focused the particles in midplane.

The Berkeley cyclotron team, however, made a slow start on nuclear physics experimentation, and became involved in an embarrassing episode involving the disintegration of deuterium in which target contamination led them to misjudge the mass of the neutron. And one day in 1934, Lawrence came running into the lab waving a copy of the French journal *Comptes Rendus*, containing an article by Frédéric Joliot-Curie and Irène Joliot-Curie announcing the discovery of induced radioactivity, a phenomenon that the Berkeley crew easily could have observed earlier. Half an hour later, they were observing it as well.

But the cyclotron soon put Berkeley on the map as a center for nuclear physics. It was the beginning of the era when accelerators were the key to nuclear physics—except in slow neutron physics of the sort Enrico Fermi was doing—and the ability to make discoveries was a function of the size of the accelerator. "Nuclear physics just blew wide open," Livingston once said. "Here was this machine, where essentially everything you touched made a publishable paper. … The papers just poured out of that lab" (Video interview, 1982, cited in Crease, 1999). The historians John Heilbron and Robert Seidel have written, "The technical achievement was mainly Livingston's; the inspiration, push, and above all, the vision of future greatness, were Lawrence's."

Cornell and MIT. But Livingston was increasingly unhappy at Berkeley. He felt more and more in Lawrence's shadow, and chafed at the way Lawrence somehow retained all the credit. Lawrence's 1932 patent application for the cyclotron did not mention Livingston (though this seems to have been thoughtless rather than intentional, as Lawrence was in a rush to beat another applicant) and when Lawrence was awarded the Nobel Prize, in 1939, it was "for the invention and development of the cyclotron." In 1933 Livingston had been offered a job elsewhere, but Lawrence persuaded him to turn it down. In 1934 Livingston received an offer from Cornell, and this time, he accepted.

Cornell was an increasingly influential center for nuclear research thanks to the presence of physicists such as Robert Bacher and Hans Bethe. Livingston was attracted by the prospect of being able to build cyclotrons

Livingston

himself, and by the opportunity to catch up on nuclear physics. At Cornell, with $800 and half a dozen graduate students, he built the first cyclotron outside Berkeley, a 16-inch (41 centimeter) 2 MeV proton machine. He also collaborated with Bethe on one of Bethe's three seminal articles in *Reviews of Modern Physics* (published in 1936 and 1937) that surveyed the field of nuclear physics and that scientists in the field affectionately referred to as "Bethe's Bible."

In 1938 Livingston moved to the Massachusetts Institute of Technology (MIT). This move was motivated not by unhappiness but by opportunity. The physicist Robley Evans was about to build a large cyclotron financed by the Markle Medical Foundation, and used the project to lure Livingston. There he built, for $60,000, a 42-inch (107-centimeter) 12-MeV cyclotron, which came into operation in 1940.

During World War II, when some of Livingston's colleagues moved to the MIT Radiation Laboratory—the RadLab—to work on radar technology, and others went to various outposts of the Manhattan Project, Livingston remained for the first part of the war at his cyclotron, which was performing war duty by producing radioisotopes. In 1944 Livingston was recruited by his MIT colleague Philip Morse to work at the Office of Scientific Research and Development (OSRD) at Morse's underwater sound laboratory in Washington, D.C.

Brookhaven and Synchrotrons. After the war in mid-1946, Morse was chosen by a consortium of nine northeast universities—MIT among them—to become head of Brookhaven National Laboratory (BNL), one of the first U.S. national laboratories. Morse convinced Livingston to take leave of MIT to head BNL's accelerator program. Livingston was the obvious choice for his experience, ambition, and connections. He, in turn, saw it as the chance—finally—to create a wave of his own, thanks to the substantial government backing of the new lab. The initial plan was to build four accelerators: a van de Graaff for the low energies, a cyclotron for intermediate energies, a 240-inch (610-centimeter) offspring of the cyclotron called a synchrocyclotron for the high end, and a yet-to-be-designed accelerator for the future.

Synchrocyclotrons were a recent discovery, designed to overcome a limitation on cyclotrons. For there was a limit to the resonance principle, created by the fact that particles increase in mass as they approach the speed of light. This means that they take longer to complete an orbit, and thus they get out of phase with respect to the frequency of the voltage changes. In 1945 a way around this problem was announced by the Soviet physicist Vladimir Veksler and independently by the Berkeley physicist Ed McMillan. The idea involved decreasing the frequency of the voltage changes with the rise in energy of the particles to compensate for their increased energy and to keep the particles synchronous—hence the name "synchrocyclotron." This idea not only extended the lifetime of cyclotrons, but also suggested a new kind of accelerator (called a synchrotron), in which the magnets surrounded the particle path like beads on a string.

Livingston was particularly interested in the synchrocyclotron, which would be able to reach 600–1000 MeV. This was an important region because pions, whose study was now important to nuclear physics, would be produced easily. It would not only be the world's most powerful accelerator, but it would be the last hurrah of cyclotrons before they gave way to synchrotrons. But support for the project eroded when other BNL scientists embraced the idea of putting the lab's resources into a synchrotron instead. Livingston was reluctant to abandon the synchrocyclotron and did not like the thought of placing all the lab's high-energy funds in one project, gambling on an entirely new type of machine. But in April 1947 the laboratory decided to abandon the synchrocyclotron project to focus on a synchrotron. Though Livingston was initially furious, he gracefully overcame his objections and contributed to the design of what would become known as the Cosmotron. In fall 1948 he returned to MIT but continued to return to BNL as a summer visitor. And in that capacity he became involved in one of the most momentous breakthroughs in accelerator physics, the discovery of strong focusing.

The Cosmotron was completed in May 1952 and was the first accelerator to exceed a billion electron volts (1 GeV). Meanwhile, a new European accelerator laboratory was being set up outside Geneva, then known as the Conseil Europėen pour la Recherche Nucléaire, now known as CERN. The European scientists planned to build a synchrotron, and a trio of CERN accelerator scientists—including Wideröe—was sent to BNL to examine the Cosmotron. In advance of their visit, Livingston organized a study group to brainstorm for ideas to give to the Europeans. One major drawback to the Cosmotron, for instance, was the fact that its magnets were C-shaped, meaning that the beam could be accessed from only one side. What, Livingston asked BNL accelerator designer Ernest Courant, if the magnets alternated in sectors, with some pointing in and some out?

Courant's immediate response was to cite the conventional wisdom that this would disrupt orbital stability, but promised to look into it. When he did, he was astonished to arrive at the opposite conclusion—that alternating the magnets would improve focusing and orbital stability. The method turned out to have an antecedent, called the butterfly cyclotron, and it had also been discovered previously by a Greek engineer, Nicholas Christofilos, whose work had been overlooked. But by the time the CERN team arrived at BNL in August, Livingston, Courant, and the third member of the group, Hartland Snyder, realized that they had stumbled onto a powerful new type of accelerator focusing method. The method was used to build BNL's next accelerator, the Alternating Gradient Synchrotron (AGS) as well as CERN's first major accelerator, the Proton Synchrotron (PS). The PS came on line in 1959, and the AGS in 1960.

Scientists at Harvard and MIT then came up with a proposal for the Cambridge Electron Accelerator (CEA), and Livingston became the new laboratory's director. Livingston performed well during his tenure, and oversaw the rebuilding of the laboratory after an explosion of a hydrogen bubble chamber destroyed much of the laboratory and killed one person. He also supervised the construction of a beam bypass at the CEA, transforming the strong-focusing electron synchrotron into an electron-positron collider.

Livingston had divorced Lois in 1949, and married Margaret (Peggy) Hughes in 1952, but after the latter's death in 1959, he remarried Lois that same year. In 1967 he became associate director of yet another laboratory, the National Accelerator Laboratory (later renamed Fermilab), outside Chicago. He served as chairman of the Federation of American Scientists in 1954 and in 1959. He retired from Fermilab in 1970 and with Lois moved to Santa Fe, New Mexico, where he continued working at Los Alamos. In 1986 he had a prostate cancer operation from which he never fully recovered, and he died on 25 August 1986.

Livingston knew the business of cyclotrons as well as anyone, with the possible exception of Lawrence. He was one of the most important accelerator builders during the age when they were the central tool of nuclear and high-energy physics, a time which may well go down in history as the golden age of U.S. physics.

BIBLIOGRAPHY

Livingston's papers from 1928 to1986 can be accessed at the Massachusetts Institute of Technology, Archives and Special Collections. Cambridge, Massachusetts. The records of the Cambridge Electron Accelerator (1952–1974) can be accessed at the Harvard University Archives, Pusey Library, Cambridge Massachusetts, and these contain Livingston's director's files, including extensive correspondence and machine design, from 1962 to 1974.

WORKS BY LIVINGSTON

With Ernest O. Lawrence. "The Production of High-Speed Protons without the Use of High Voltages." *Physical Review* 40(1932): 19–35.

With Erenst D. Courant and Hartland S. Snyder. "The Strong-Focusing Synchrotron: A New High Energy Accelerator." *Physical Review* 88 (1952): 1190–1196.

With John P. Blewett. *Particle Accelerators.* New York: McGraw-Hill, 1962.

Interview by Charles Weiner and Neil Goldman. 21 August 1967, Harvard University. AIP oral history interview, deposited at the Center for History of Physics of the American Institute of Physics.

Particle Accelerators: A Brief History. Cambridge, MA: Harvard University Press, 1969.

OTHER SOURCES

Courant, Ernest D. "Milton Stanley Livingston 1905–1986." In *Biographical Memoirs* 72. Washington, DC: National Academy Press, 1997.

Crease, Robert P. *Making Physics: A Biography of Brookhaven National Laboratory, 1946–1972.* Chicago: University of Chicago Press, 1999. Especially chapter 7.

Heilbron, John L., and Robert W. Seidel. *Lawrence and His Laboratory: A History of the Lawrence Berkeley Laboratory.* Berkeley: University of California Press, 1989.

Paris, Elizabeth. "Lords of the Ring: The Fight to Build the First U.S. Electron-Positron Collider." *Historical Studies in the Physical and Biological Sciences* 31, no. 2 (2001): 355–380.

Robert P. Crease

LORENTZ, HENDRIK ANTOON

(*b.* Arnhem, Netherlands, 18 July 1853; *d.* Haarlem, Netherlands, 4 February 1928), *theoretical physics.* For the original article on Lorentz see *DSB,* vol. 8.

This supplementary note focuses on Lorentz's attitude toward the new developments of the early twentieth century, special relativity and quantum physics.

Treatment of Moving Bodies. Central to any comparison of Lorentz's approach to the electrodynamics of moving bodies to Einstein's is Lorentz's "theorem of corresponding states." Lorentz first formulated this theorem in the *Versuch* of 1895 to deal with problems in optics in frames of reference moving through the ether. He improved and extended it beyond optics in the "Théorie Simplifiée …" of 1899 and again in "Electromagnetic Phenomena in a System Moving with Any Velocity Smaller than That of Light" of 1904. It received its final form in the second edition of *The Theory of Electrons* of 1915. In modern terms the theorem asserts that James Clerk Maxwell's equations are invariant under what Henri Poincaré proposed to call Lorentz transformations. In the earliest version of the theorem, this was true only for the macroscopic equations governing the propagation of light waves and only to first

Lorentz

order in v/c, the ratio of the velocity of the moving frame to the velocity of light. Eventually, Lorentz was able to prove an exact version of the theorem for the fundamental microscopic equations governing all of electrodynamics.

Lorentz's starting point in 1895 was the free Maxwell equations transformed under what would eventually be called a Galilean transformation from a frame at rest in the ether to a moving frame. He replaced the time variable and the electric and magnetic fields in the moving frame with velocity-dependent auxiliary quantities such that the auxiliary fields as functions of position in the moving frame and of the new time variable—called "local time" because of its dependence on position—would satisfy the free Maxwell equations to first order in v/c. Hence, for any field configuration in a frame at rest in the ether allowed by the free Maxwell equations there is another field configuration that to a good approximation is allowed in the moving frame. Lorentz used this first version of the theorem of corresponding states to explain why no experiment accurate to order v/c had ever detected any ether drift, in other words, any motion of the ether with respect to the Earth. Such experiments, Lorentz pointed out, all boiled down to the observation of patterns of brightness and darkness, and such patterns would be the same in field configurations related to one another through his theorem.

In 1899 Lorentz changed the auxiliary quantities in the moving frame somewhat so that the auxiliary fields as functions of local time and new auxiliary coordinates satisfied the free Maxwell equations exactly. The patterns of brightness and darkness described by two field equations related by this new theorem of corresponding states are now no longer identical. If an arbitrary overall scaling factor is set equal to one (as Lorentz did in 1904), the pattern in the moving frame is obtained from the pattern in the frame at rest through contraction of the latter in the direction of motion by a factor familiar from the Lorentz-FitzGerald contraction hypothesis. To predict negative results for ether drift experiments, no matter how accurate, that eventually turn on the observation of patterns of brightness and darkness (such as the variant on the Michelson-Morley experiment suggested by Alfred Liénard that partly inspired Lorentz's 1899 paper), Lorentz needed to add a physical assumption to the purely mathematical theorem of corresponding states. He needed to assume that the material components in such experiments experience the same contraction as the electromagnetic field configurations do according to Maxwell's equations and the theorem of corresponding states. More precisely, if a material system producing a certain electromagnetic field configuration in a frame at rest in the ether is set in motion, it turns into the material system producing the corresponding field configuration in the moving frame. This assumption, which can be called the generalized contraction hypothesis, was stated explicitly by Lorentz in 1899. In modern terms, it is the assumption that the laws governing material systems, such as those governing the fields with which they interact, are Lorentz invariant. Stating the assumption in this way, one recognizes that it is, in fact, also needed to explain the null results of first-order ether drift experiments. Most of the differences between the system at rest and the system in motion, however, are of second order in v/c. In addition to the length contraction effect, Lorentz already found the relativistic results for time dilation and the velocity dependence of forces and masses in his 1899 paper. Eventually, rather than the basic assumption that systems at rest turn into their corresponding states when set into motion, Lorentz adopted a number of more specific hypotheses about the effect of motion through the ether on forces and masses from which this basic assumption could be derived.

By 1900 Henri Poincaré had already realized that Lorentz's local time would be the time registered by co-moving observers synchronizing their clocks with the help of light signals as if they were at rest in the ether. It was not until after Einstein's work of 1905, however, that Lorentz realized that the auxiliary quantities of his theorem of corresponding states are the measured quantities for the moving observer. This immediately made it clear that motion through the ether could never be detected. It no longer required any special argument of the kind needed in earlier versions of Lorentz's theory that corresponding states agree in such observable consequences as the interference pattern detected on a screen. Lorentz nonetheless continued to treat Lorentz invariance not as reflecting the symmetries of a new relativistic space-time but as a feature accidentally shared by all laws governing systems in Newton's space and time.

Reaction to Quantum Theory. As becomes clear from his correspondence (in particular with Max Planck and Einstein), as well as from his publications, Lorentz continued to try to find a classical interpretation for the quantum hypothesis. He rejected Einstein's light quantum hypothesis on the grounds that many well-established phenomena, such as interference and diffraction, were impossible to reconcile with a particulate nature of light. He also pointed out that attempts to estimate the size of a light quantum led to absurd results.

A key moment in his thinking about the quantum was the lecture he gave at the Fourth International Congress of Mathematicians in Rome in 1908 (Lorentz, 1908). There he presented a very general classical theory of charged particles and radiation, employing the still relatively unknown technique of Gibbsian statistical mechanics. He came to the conclusion that the Rayleigh-Jeans radiation law was the only law compatible with classical molecular mechanics and that for the system considered by him the law of equipartition of energy had to be valid. This destroyed his hope that the experimental results that seemed to favor Planck's law over the Rayleigh-Jeans law could be explained by a breakdown of the equipartition law. Instead, he adopted James Jeans's suggestion that the experimental discrepancy was due to the fact that in the experiments there was no true equilibrium between ether (or radiation) and matter, and that in fact it might take extremely long for equilibrium to set in. Immediately after his lecture it was pointed out to him, by Wilhelm Wien among others, that his conclusion was untenable on experimental grounds. The argument that convinced him was that the Rayleigh-Jeans law would imply that even at temperatures as low as room temperature a bar of silver would be clearly visible in the dark.

Lorentz quickly and publicly retracted the conclusion drawn in the Rome lecture and admitted the experimental validity of Planck's law. The theoretical considerations presented in the lecture, however, remained valid and served as an important clarification of the problems of radiation theory.

In spite of his acceptance of Planck's law, Lorentz remained skeptical about quantum theory; as late as 1923, for instance, he lamented that we "do not understand" quantum theory or Planck's hypothesis (Lorentz, 1923).

Hendrik Antoon Lorentz. SPL / PHOTO RESEARCHERS, INC.

SUPPLEMENTARY BIBLIOGRAPHY

WORKS BY LORENTZ

"Le partage de l'énergie entre la matière pondérable et l'éther." *Nuovo Cimento* 16 (1908): 5–34. Version with additional note in *Atti del IV Congresso Internazionale dei Matematici (Roma, 6–11 Aprile 1908)* edited by G. Castelnuovo. Vol. 1. Rome: R. Accademia dei Lincei, 1909.

"l'Ancienne et la nouvelle mécanique." (Lecture, Paris, 10 December 1923.) In *Le livre du cinquantenaire de la Société française de Physique.* Paris: Éditions de la Revue d'Optique Théorique et Instrumentale, 1925.

Hendrik Antoon Lorentz: Selected Scientific Correspondence, edited by A. J. Kox. Vol. 1. New York: Springer, 2008. This volume includes a full bibliography of Lorentz's work. A second volume is foreseen for 2010.

OTHER SOURCES

Arabatzis, Theodore. "The Zeeman Effect and the Discovery of the Electron." In *Histories of the Electron: The Birth of Microphysics,* edited by Jed Z. Buchwald and Andrew Warwick. Cambridge, MA: MIT Press, 2001. On Lorentz and the Zeeman effect.

Darrigol, Olivier. "The Electron Theories of Larmor and Lorentz: A Comparative Study." *Historical Studies in the Physical and Biological Sciences* 24 (1994): 265–336.

———. *Electrodynamics from Ampère to Einstein.* Oxford: Oxford University Press, 2000.

Frisch, Matthias. "Mechanisms, Principles, and Lorentz's Cautious Realism." *Studies in History and Philosophy of Modern Physics* 36 (2005): 659–679. On Lorentz's methodology.

Janssen, Michel. "H. A. Lorentz's Attempt to Give a Coordinate-Free Formulation of the General Theory of Relativity." In *Studies in the History of General Relativity,* edited by Jean Eisenstaedt and A. J. Kox. Boston: Birkhäuser, 1992. On Lorentz and general relativity.

———. *A Comparison between Lorentz's Ether Theory and Einstein's Special Theory of Relativity in the Light of the Experiments of Trouton and Noble.* PhD diss., University of Pittsburgh, 1995.

———. "Reconsidering a Scientific Revolution: The Case of Lorentz versus Einstein." *Physics in Perspective* 4 (2002): 421–446.

———, and Matthew Mecklenburg. "From Classical to Relativistic Mechanics: Electromagnetic Models of the Electron." In *Interactions: Mathematics, Physics and Philosophy, 1860–1930,* edited by Vincent F. Hendricks, et al. Dordrecht, Netherlands: Springer, 2006.

———, and John Stachel. "Optics and Electrodynamics in Moving Bodies." In *Going Critical,* edited by John Stachel. Berlin: Springer, in preparation.

Kox, A. J. "Hendrik Antoon Lorentz, the Ether, and the General Theory of Relativity." *Archive for History of Exact Sciences* 38 (1988): 67–78. On Lorentz and general relativity.

———. "The Discovery of the Electron: II The Zeeman Effect." *European Journal of Physics* 18 (1997): 139–144. On Lorentz and the Zeeman effect.

Lorenz

Kuhn, Thomas S. *Black-Body Theory and the Quantum Discontinuity, 1894–1912.* Oxford and New York: Oxford University Press, 1978; rev. ed.: Chicago: University of Chicago Press, 1987. On Lorentz and early quantum physics.

Miller, Arthur I. *Albert Einstein's Special Theory of Relativity. Emergence (1905) and Early Interpretation (1905–1911).* Reading, MA: Addison-Wesley, 1981; Rep.: New York: Springer, 1998.

Michel Janssen
A. J. Kox

LORENZ, KONRAD ZACHARIAS (*b.* Vienna, Austria, 7 November 1903; *d.* Vienna, 27 February 1989), *ethology, animal behavior.*

Lorenz was the primary founder of the science of ethology, the biological study of behavior. He was a bold theorist, a charismatic teacher, a successful popularizer, and one of the most colorful and controversial figures of twentieth-century biology. For his contributions to the study of animal behavior he was awarded in 1973 the Nobel Prize in Physiology or Medicine, which he shared with Karl von Frisch and Nikolaas Tinbergen.

Life and Career. Konrad Lorenz was the son of Dr. Adolf Lorenz (1854–1946), an internationally famous professor of orthopedic surgery at the University of Vienna, and Emma Lorenz, née Lecher (1861–1938). He had but one sibling, his brother Albert, who was already eighteen years old at the time Konrad was born.

As a child, Lorenz had a passion for raising pets, an activity in which his parents indulged him. At the family home in the village of Altenberg, a short commute from the city of Vienna, the young boy surrounded himself with ducks, geese, and numerous other birds and animals. Years later, as a mature scientist, he would insist that his scientific practices were continuous with practices he had developed in his youth as an animal lover and raiser. He claimed that being an animal lover was a prerequisite to being a good observer of animal behavior, for without a love of animals in the first place, it was inconceivable that anyone would ever be patient enough to watch an animal over the length of time necessary to come to know its entire behavioral repertoire.

Lorenz attended the elite Schottengymnasium in Vienna, graduating from there in 1922 at the age of eighteen. Although his primary interests at that point were in zoology and especially paleontology, his father insisted that he study medicine instead and sent him to the United States for premedical studies at Columbia University. Lorenz remained there for only one semester. He returned home and enrolled in 1923 as a medical student in the Second Anatomical Institute of the Medical Faculty at the University of Vienna. In 1927 he married his childhood sweetheart, Margarethe Gebhardt (1900–1986). They lived at the Lorenz family home in Altenberg. They had three children: Thomas, born in 1928; Agnes, born in 1930; and Dagmar, born in 1941.

Lorenz earned his doctorate in medicine in 1928. He thereupon enrolled in the university's Zoological Institute, from which he earned a PhD in 1933 for a comparative study of the mechanisms of flight and the adaptations of wing form in birds. From 1931 to 1935 he served as a paid assistant at the Second Anatomical Institute, directed by the comparative anatomist Ferdinand Hochstetter. Hochstetter allowed Lorenz to pursue his own comparative studies of bird behavior and to begin preparing for the habilitation exam that would entitle him to be a privatdozent (lecturer) in zoology and animal psychology. Upon Hochstetter's retirement in 1935, Lorenz moved to another institute at the university, the Psychological Institute directed by Karl Bühler. He passed his exam in 1936.

In the 1930s Lorenz rose quickly to prominence in the fields of ornithology and animal psychology as the result of his pathbreaking studies on instinctive behavior in birds. In addition to the support he received at the University of Vienna from Hochstetter and Bühler, he was greatly aided by the encouragement and mentorship of the leading German ornithologists of his day, Oskar Heinroth and Erwin Stresemann. His growing recognition as an important contributor to science was not immediately accompanied, however, by remunerative employment. As of 1937 his only post was that of an unpaid *Dozent* in Bühler's Psychological Institute. He believed his chances for employment were hindered by the fact that he was a Protestant in a country ruled by Catholics and because his ideas about the biological foundations of human social behavior ran contrary to the views of the Catholic educational establishment. He began to seek research funding from Germany. His most cherished hope was that Germany's primary institution for the support of scientific research, the Kaiser Wilhelm Gesellschaft (KWG), would establish for him in Altenberg an institute devoted to comparative behavior studies.

Lorenz's expectations were raised in March 1938 when the *Anschluss* incorporated Austria into Germany. He welcomed the *Anschluss* with great enthusiasm, as did many Austrians. He did so in part because his family regarded itself as culturally German, but he also clearly believed that the incorporation of Austria into Germany would improve his career opportunities. In May 1938 the KWG Senate reviewed favorably the idea of an institute for Lorenz. In June Lorenz applied for membership in the Nazi Party. His application was accepted. At scientific meetings and in scientific journals he began presenting papers in which he claimed that evidence from the study of animal behavior had a bearing on questions about racial degeneration in humans.

To Lorenz's disappointment, the funds that the KWG needed to establish an institute for him failed to materialize. However, he was eventually awarded a new *Dozent's* position in zoology, with stipend, at the University of Vienna. Then in 1940 he was chosen for the chair of psychology at the University of Königsberg. On 1 January 1941 he was officially named professor of psychology and director of a new institute for comparative psychology at the university.

Lorenz's tenure as a professor at Königsberg was brief. He was drafted for military service in October 1941. He was sent to Posen (now Poznań), Poland, where he served as a military psychologist until May 1942 and then as a neurologist and psychiatrist treating hysteria and other neuroses. In April 1944 he was made a troop physician and transferred to the eastern front. He was taken prisoner by Russian forces on 26 June 1944 in Vitebsk (now Vitsyebsk, Belarus). For the next three and a half years he served as a camp doctor in a series of Soviet prisoner-of-war camps. He did not return home to Austria until February 1948.

Upon his return, Lorenz once again experienced difficulties finding an academic position or research support. In 1949, his private research station at Altenberg was officially designated an institute for "Comparative Behavior Study under the Direction of the Austrian Academy of Sciences," but the financial support that came with this impressive title was minimal. He wrote popular books in an attempt to make ends meet. When, in 1950, political and ideological considerations combined to undermine his candidacy for the professorship of zoology at the University of Graz, he decided the time had come for him to find a job outside Austria. He asked friends in Great Britain if they could line up a position for him there. The danger of losing another top German scientist to a foreign country inspired the new Max Planck Gesellschaft (MPG) to do what its predecessor, the KWG, had never accomplished: it founded an institute for Lorenz. The institute was set up in Buldern, Westphalia, under the auspices of Erich von Holst's Max Planck Institute for Marine Biology in Wilhelmshaven.

Lorenz worked at Buldern until 1956, when he moved to a new Max Planck Institute for Behavioral Physiology that the MPG established for him and Holst at Seewiesen, near Starnberg, in Bavaria. Holst was named director of the institute and Lorenz was named deputy director. Lorenz became director in 1961 when Holst relinquished that post due to illness. Lorenz remained at Seewiesen until his retirement in 1973, after which he returned home to Altenberg. There he received institutional support once again from the Austrian Academy of Sciences. In the 1970s and 1980s he joined forces with the new Austrian Green Party in opposing the construction of nuclear power plants in Austria.

Over the course of his career Lorenz received a great many awards, including numerous honorary doctoral degrees, and in 1973, the Nobel Prize in Physiology or Medicine. He died in Vienna on 27 February 1989. He was preceded in death by his wife and his son.

Early Scientific Development and Professional Contacts. Lorenz was already a firm believer in evolution when he entered the University of Vienna as a student. It was there, however, that he learned from his professor of comparative anatomy, Hochstetter, how to study evolution through the comparative method. Comparative anatomists seek to identify homologous structures, that is, parts whose position or relation to a general pattern are indicative of common ancestry. The horse's foreleg, the bird's wing, and the whale's flipper, for example, are understood by comparative anatomists to be homologous structures. With Hochstetter's encouragement, Lorenz concluded that the methods of comparative anatomy could be applied to the instinctive behavior patterns of animals just as instructively as they could be applied to animal structures. In other words, behavior patterns could be used just like organs to reconstruct phylogenies (evolutionary lineages). Charles Otis Whitman in the United States and Heinroth in Germany had already promoted this idea. Lorenz would later characterize this discovery as the "Archimedean point" from which modern ethology took its origin.

Lorenz continued to collect and raise animals even while he was a medical student. An acquisition that proved of special importance to him was his purchase in the summer of 1926 of a young jackdaw (*Corvus monedula*). After a few days, the jackdaw had developed such an attraction for him that he could allow the bird to fly freely after him when he went outdoors. Constrained neither by fear nor the limits of a cage, the bird performed in Lorenz's presence many of the natural behavior patterns of the species. The bird's behavior became the subject of Lorenz's first published scientific paper, the production of which brought him into contact with the ornithologists Heinroth and Stresemann.

Lorenz found watching the behavior of a single jackdaw to be highly instructive, but he also appreciated that one bird's actions were not sufficient to illuminate how a normal jackdaw colony functioned. He supposed that certain of the bird's seemingly instinctive and "ceremonial" behavior patterns were actions that in the normal life of a jackdaw would elicit appropriate reactions from other jackdaws. In order to study this, he converted the attic of the family home in 1927 into an aviary for fourteen young, hand-reared jackdaws. He marked the birds for identification and proceeded to observe their behavior closely over the course of the next several years. His successes with jackdaws led him in 1931 to undertake a study of the social behavior of the night heron (*Nycticorax nycticorax*). The following year he began raising greylag geese (*Anser anser*), the species with which his name is today most frequently associated. His studies of tame, free-flying birds, conducted at his home research station in Altenberg, provided the empirical basis for a good deal of his subsequent theorizing.

Lorenz received valuable encouragement in these early studies from Heinroth and Stresemann. Although Heinroth held the post of director of the aquarium at the Berlin Zoo, birds and the "finer details" of their behavior were the primary focus of his research. As early as 1910 he had promoted the view that "species-specific instinctive actions" could be used like anatomical structures to determine the genetic affinities of species, genera, and subfamilies of birds. He also studied the way that instinctive behavior patterns function in the family lives of the ducks, geese, and swans (the birds that were his favorite subjects). As of 1930, by which time Lorenz was actively corresponding with him, Heinroth and his wife Magdalena had published the first three volumes of their elaborate, four-volume study of the behavior of the birds of central Europe. Their research featured raising baby birds in isolation from other birds and then studying their characteristic behavior patterns. Their aim was to distinguish species-specific actions that were innate from those that were acquired.

Lorenz was thrilled to find in Heinroth an authority whose views and practices were very much like his own. He praised Heinroth for undertaking to establish animal psychology as a branch of biology. He was also excited by remarks Heinroth had made about the continuity between the social instincts of animals and humans. Heinroth himself, however, was temperamentally indisposed to constructing broad theories or building a new discipline. As Lorenz's work became increasingly theoretical, he was inclined to try out his writings first on Stresemann, the editor of the *Journal für Ornithologie,* and only show them to Heinroth after he could claim Stresemann's support. Heinroth in effect left it to Lorenz to take the lead in developing the study of animal behavior in an energetic, programmatic way.

In order to establish animal psychology as a branch of biology, Lorenz believed, there was a proper order and a proper method to follow in doing so. Fundamental to his thinking throughout his entire career was the notion that instinctive and learned behavior were fundamentally distinct from one another. They represented discrete units, he insisted, even when they were "intercalated" in complex behavior chains. In his view, the proper way to study the behavior of any species of animal was to raise individuals of the species under conditions of semifreedom where one could observe the full range of their natural behavior patterns. The investigator's first task was to recognize which of the animal's behavior patterns were instinctive. Lorenz did not deny the importance of learning in animal life, but he insisted that one could never really gauge an animal's abilities to learn without a prior knowledge of what it did instinctively.

Lorenz felt that anyone with extensive experience raising animals would be able to recognize intuitively which behavior patterns are instinctive and which are not. Nonetheless, he did have a list of criteria that could be used to make the distinction. He was disposed to judge that a behavior pattern was instinctive, he wrote in 1932, if it satisfied one or more of the following five conditions:

1. if a young animal reared in isolation from members of its own species displayed the species-specific behavior pattern in question without having had any models from which to learn;
2. if all the individuals of a species performed the same behavior pattern in the same, stereotyped way;
3. if there was a conspicuous incongruity between the normal intelligent abilities of an animal (as seen in other situations) and the abilities that would be necessary for the animal to perform the behavior in question by insight;
4. if the behavior pattern was performed incompletely or in a situation when the appropriate biological goal was not present, thus making it clear that the animal was not conscious of the biological purpose of the action;
5. if the rigidity of the behavior pattern and its resistance to environmental influences continued to be displayed under conditions far different from those under which the pattern originally evolved.

Lorenz's reputation as a pathbreaking researcher and theoretician was firmly established by his monumental work of 1935, "Der Kumpan in der Umwelt des Vogels: Der Artgenosse als auslösendes Moment sozialer Verhaltungsweisen" (The companion in the bird's world: Fellow members of the species as releasers of social behavior). It represented his attempt to organize into a coherent framework the knowledge and insights he had gleaned from raising and observing as many as 350 individual birds representing nearly thirty different species. His goal in this case was to elaborate concepts on which a continuing program of research could be built. He dedicated the study to the German biologist Jakob von Uexküll, who

Konrad Lorenz. *Konrad Lorenz being followed by a group of ducklings.* SPL / PHOTO RESEARCHERS, INC.

had developed the *Umwelt* concept, identifying an animal's effective environment as the sum total of factors that the animal perceives and to which it responds. Uexküll had also introduced the notions of *releasers* and *companions* in animal life. These were the key concepts of Lorenz's 1935 monograph, exemplified and developed through an extended discussion of the social life of birds.

Following Uexküll, Lorenz explained that humans tend to recognize objects in their environment as things, on the basis of a compilation of multiple stimuli impinging upon their sense organs. It is the successful integration of stimuli emanating from objects in the environment that allows humans to build up a knowledge of the causal relationships of things in their environment, which in turn allows them to survive in the environment. In contrast, lower animals such as birds are adapted to their environments primarily through highly differentiated instinctive behavior patterns. These animals act not according to insight but according to instinctively determined responses that have been built up through evolution by virtue of their survival value. To be effective, they need only be elicited or released by one or at most a very few of the stimuli emanating from the objects in their environment. As Lorenz envisioned it, the proper combination of stimuli triggers an "innate schema" (this would later be called an "innate releasing mechanism"), thereby allowing the associated instinctive action to be performed. For this to be of service to the organism the stimuli triggering the innate schema needed to characterize the object sufficiently closely so that stimuli from other, inappropriate, objects did not accidentally elicit the animal's instinctive response.

Lorenz explained that the fine-tuning of this relation by adaptive evolution could only go so far when it involved foreign objects in the animal's environment. On the other hand, it could continue almost indefinitely when the object of the instinctive reaction was a fellow member of the same species. In that case, both the receiving and the issuing of stimuli were subject to evolutionary modification, leading to combinations of such general improbability as to make certain that the instinctive reactions were not elicited by stimuli from the "wrong" sources. Lorenz used the term *releasers* to designate the stimuli that serve within the species to activate the innate releasing mechanisms of fellow members of the species, thereby bringing forth the performance of appropriate chains of instinctive behavior patterns. Releasers, Lorenz explained, could be either morphological structures or conspicuous behavior patterns or a combination of both.

Lorenz

The overall claim of Lorenz's "Kumpan" monograph was that the complex, intraspecific social relations of birds depend largely on releasers. The highly organized social life of jackdaws, for example, is built upon a remarkably small number of simple, innate responses to the releasers provided by conspecifics. According to Lorenz, each jackdaw has a number of different social drives with respect to which other jackdaws play the role of "companion." These companions provide the individual with the particular stimuli necessary to release the instinctive behavior patterns appropriate to its respective drives.

One of the most intriguing phenomena Lorenz described in his "Kumpan" monograph was that which he called imprinting (*Prägung*). He was not the first to observe the phenomenon. It had been well known to Whitman. Heinroth for his part had reported that newly hatched greylag goslings do not instinctively recognize adult greylag geese as conspecifics. If exposed to a human being before they are exposed to a mother goose, goslings will follow that human as if he or she were their parent and will direct toward this foster parent the innate responses that they would under normal circumstances direct toward their own kind. Lorenz had encountered imprinting with his jackdaws, his geese, and most of his other hand-reared birds. As he represented the process, a young bird, at a very brief stage of its early development, has irreversibly imprinted upon it the object that will then serve to release certain of its instinctive behavior patterns, and this process, furthermore, is irreversible. Imprinting, he insisted, is quite distinct from true learning. To him it appeared more like the process of induction in embryological development.

Comprehensive Theory of Instinctive Behavior. In his "Kumpan" monograph Lorenz set forth his general understanding of how instinctive behavior patterns function in avian social life. He had not yet arrived, however, at a general physiological explanation of how instincts work. Up through the publication of his 1935 monograph, he was inclined to think of instinctive actions as chains of reflexes. This view appealed to him because of its nonvitalistic character and because it fit most of the major facts of animal behavior as he knew them: in particular, the stereotyped, species-typical nature of instinctive behavior patterns; the automatic way that an animal performs its different instinctive behavior patterns upon being presented with the appropriate stimuli; and the animal's apparent unawareness of the biological purposes of its instinctive acts. However, in the period 1935–1937, profiting especially from interactions with the American psychologist Wallace Craig and the German physiologist Erich von Holst, Lorenz became convinced that the chain-reflex theory of instinct was inadequate and needed to be replaced.

Craig, in a paper of 1918, had distinguished between the "appetitive behavior" that begins an instinctive behavior cycle and the "consummatory act" that brings the cycle to an end. Lorenz valued this distinction. At the same time, he found it inconsistent with the way Craig identified the whole behavior cycle as instinctive. As Lorenz saw it, each behavior cycle needed to be broken down into its component parts. Some of these parts were purposive and modifiable. Others, those that were truly instinctive, were nonpurposive and nonmodifiable. Benefiting from an extended conversation on the subject with Craig, Lorenz wrote a paper arguing that the failure to separate these distinct components of behavior was what had led earlier writers such as Herbert Spencer, Conwy Lloyd Morgan, and William McDougall to think that animals had insight into their instinctive behavior and that animal instincts were modifiable by experience. He delivered the paper, titled "The Formation of the Instinct Concept," in Berlin in 1936 at Harnackhaus, the special new conference site of the KWG.

Lorenz at this point continued to endorse the chain-reflex theory of instincts, but he acknowledged there was a significant difference between instincts and reflexes: animals strive to secure the release of their instinctive behavior patterns, they do not strive to secure the performance of their simple reflexes. For this reason, he said, the instinctive behavior pattern should be regarded as a special category of reflex, namely, "a striven-after reflex action."

Years later Lorenz liked to tell how one of the auditors of his 1936 lecture was the young physiologist Erich von Holst. By Lorenz's account, it took Holst only a matter of minutes after the lecture to persuade Lorenz that instinctive behavior patterns were better interpreted as the result of internally generated and coordinated impulses than as chains of reflexes set in motion by external stimuli. Lorenz's recollection in this case was inaccurate. His correspondence shows that the conversion in question did not take place until the following spring. Nevertheless, the essence of the conversion was as Lorenz reported it. It shifted his attention away from external stimuli to the animal's internal state. The newer view made sense of appetitive behavior, where an animal, evidently motivated by its own internal condition, seeks out particular external stimuli. It also made sense of two apparently related sets of phenomena that the chain-reflex theory had not been able to handle satisfactorily. Lorenz referred to these as *threshold lowering* and *vacuum activities* ("*Leerlaufreaktion*").

In threshold lowering, an instinctive motor pattern became easier to release the longer it had been since the pattern was last performed. A vacuum activity was when the threshold for eliciting a behavior pattern was reduced to such a point that the behavior pattern "went off" without serving its proper function, like an engine running in neutral. Threshold lowering and vacuum activities suggested to Lorenz that inner stimuli increased in intensity in an animal in proportion to the amount of time since an instinctive action had last been released. This led him to envision a kind of "damming up" of a "reaction-specific energy." In his Berlin lecture of 1936 he compared this to a gas being continually pumped into a container, causing an increase in pressure until under very specific conditions it was discharged. What eventually excited Lorenz about Holst's work was that Holst's findings on the endogenous generation and central coordination of nervous impulses appeared to support the idea of some kind of "action-specific energy" building up internally in an organism to the point that the corresponding instinctive action would erupt spontaneously if it were not released.

Lorenz and Tinbergen. In November 1936 Lorenz traveled to the Netherlands to the University of Leiden to participate in a conference on instinct. There he met the young Dutch zoologist Niko Tinbergen. Their collaboration proved critical for ethology's development over the next three or more decades. The two men quickly recognized that their talents were complementary. Lorenz was the bold, intuitive theorist. Tinbergen was the more critical and careful analyst who would demand and devise observational and experimental methods to test Lorenz's theorizing. Whereas Lorenz was primarily an animal raiser, Tinbergen was primarily a field naturalist who preferred to watch animals in their natural settings. But Tinbergen had also conducted laboratory experiments on animal behavior. Lorenz was delighted to learn of the experiments Tinbergen and his students at Leiden had conducted on the reproductive behavior of a fish, the three-spined stickleback. Lorenz felt that this work was just what was needed to advance his emerging science of animal behavior. Tinbergen arranged a leave from the University of Leiden to travel to Altenberg in March 1937. He spent the next three-and-a-half months working with Lorenz. It was there that they conducted among other experiments their classic study of the egg-rolling behavior of the greylag goose.

The experiments on the egg-rolling behavior of the greylag showed that the motor sequence employed by a goose in returning a stray egg to its nest involves two separate components: an instinctive motor pattern and a taxis. The instinctive component of the motor sequence was released by the visual stimulus of an egg or egglike object outside the nest. It involved a bending downward of the goose's head and neck, so that the egg rested against the lower side of the bird's beak, and then a pushing of the egg toward the nest. The taxis component of the motor sequence, in contrast, consisted of lateral balancing movements that kept the egg rolling in the direction of the nest.

Konrad Zacharias Lorenz. © BETTMANN/CORBIS.

Together, the two components combined to produce a unified and adaptive behavioral sequence.

The form of the instinctive motor pattern proved invariable. It was not influenced either by the shape of the object that was rolled or the pathway on which the object was rolled. What is more, this motor pattern, once begun, would continue all the way to the nest even if the egg were removed along the way. The ethologists interpreted these results as being consistent with the hypothesis that an instinctive behavior pattern, once released, maintains its form independently of external stimuli because it depends upon internal processes in the central nervous system coordinating the impulses sent to the muscles.

The value of Tinbergen's stay with Lorenz in Altenberg amounted to much more than the generation of experimental results. Tinbergen had found in Lorenz a man with an immense knowledge of animal behavior and the theoretical talents to match. Lorenz had found in Tinbergen a fellow scientist who was prepared to devote himself to helping develop a new, self-consciously "objectivistic" approach to the study of animal instincts and who had experimental skills that Lorenz lacked. The two men established a lifelong friendship. In the years that followed, Tinbergen would be the greatest champion of

Lorenz's accomplishment as the founder of the new science of ethology. This was so even despite the way Tinbergen and his country suffered at the hands of the Germans and Austrians during World War II, and despite Tinbergen's awareness that Lorenz had at least for a time developed an enthusiasm for National Socialism.

Linking "Domestication" Phenomena and Racial Degeneration. As indicated above, Lorenz welcomed the *Anschluss,* the incorporation of Austria into the Third Reich in March 1938. Soon thereafter, his writings began to reflect his new political surroundings. Earlier, in his "Kumpan" paper of 1935, he had noted that domesticated animals are not good subjects for the study of innate behavior patterns because their innate behavior patterns, like those of animals in poor health, showed pathological breakdown. The conditions of domestication, he said, lead to the disintegration of the complex, well-integrated behavior patterns that allow animals to survive in the wild. After the *Anschluss,* and after he joined the Nazi Party, he began writing about the parallels between the deleterious effects domestication had on the behavior of animals and the degenerative effects civilization had on the behavior of humans.

At the 1938 meeting of the German Society for Psychology, Lorenz delivered an address titled "Über Ausfallserscheinungen im Instinktverhalten von Haustieren und ihre sozialpsychologische Bedeutung" (Breakdowns in the instinctive behavior of domestic animals and their social-psychological meaning). There he argued that the breakdowns in the instinctive behavior of domestic animals are strictly analogous to the "signs of decay" in the behavior of human beings in civilization. The cause in both cases, he asserted, was the relaxation of selection. He went on to claim that the essence of racial health was to be found in the value placed on innate social behavior patterns. The nondegenerate individual, he explained, has an instinctive, intuitive ability to recognize the good or bad ethical (and genetic) character of the social behavior of others, and this instinctive response is truer (and more important for the future of the race) than any reasoned response.

The great danger, he warned, lay in the undesirable types that proliferated under the conditions of civilization and spread through the race like cancer cells in a body. Degeneration within the race led to the additional deleterious problem of race mixing, because the racially degenerate organism (whether a barnyard animal or a human) was inclined to be insufficiently discriminating in choosing a mate. In this paper and others Lorenz argued that dangerous defectives had to be rigorously weeded out of the population if the cascading consequences of degeneration were to be held in check.

Lorenz's support of Nazi racial purity laws did not go unnoticed by some of his critics in the postwar period. Another wartime publication that proved embarrassing to him after the war was a piece he published in 1940 in the Nazi biology teachers' journal, *Der Biologe* (The biologist). There he challenged the views of two Nazi authorities who were critical of evolutionary theory. His approach in this case was not to cite the wealth of evidence in evolution's favor but instead to argue that Darwinism and Nazism were fundamentally compatible. He claimed that as soon as one recognized that it was not the whole of humanity but rather the race that was the essential biological unit, it became obvious that Darwinism led not to communism or socialism but to national socialism.

Interpretations of Lorenz's behavior as a scientist under the Third Reich have varied greatly. The subject is complicated by the fact that National Socialist biology was not a monolithic entity with one, single ideological purpose. Lorenz evidently believed he could advance his career by emphasizing the ways in which his ideas ran in parallel with those who believed that the state should be run according to biological principles. It bears noting that his papers on genetic degeneration never targeted Jews or any other race. Furthermore, after the war, when he disavowed ever having had any genuine Nazi sympathies, he continued to write about the genetic and moral dangers of civilization and domestication.

On the other hand, if he separated himself in his own mind from Nazi ideologues, his use of the rhetoric of "elimination" as a scientist in the Third Reich has made it difficult for subsequent observers to discern this separation so clearly or to feel that Lorenz did not in some measure compromise himself in this period. At the time of his receipt of the Nobel Prize in 1973, in response to questions and criticisms concerning his wartime writings, he expressed his regret for his wartime rhetoric and his naïveté about the Nazis' aims. He stopped short, however, of recognizing or allowing it was possible that by lending his authority to contemporary discussions of racial hygiene he might have contributed, even in a small, indirect way, to an enterprise that resulted in genocide.

Instinctive Behavior Patterns, Evolution, and Epistemology. Not all of the papers that Lorenz published as a scientist under the Third Reich evidenced political or ideological dimensions. In an important monograph in the *Journal für Ornithologie* in 1941 he made good on his long-standing claims about the value of studying instinctive behavior patterns for the purpose of determining genealogical affinities. He described in detail the instinctive motor patterns of eighteen different species of ducks. Identifying thirty-three different instinctive motor patterns such as "introductory body-shaking," the "down-up movement," "chin-raising," and "nod-swimming by the female," he used these together with fifteen different morphological characters to construct a diagram of the phylogenetic relations among the species in question, adding two goose genera for further comparison. With this study, he felt he had conclusively established that the comparative method and the concept of homologies could be successfully applied to innate behavior patterns.

A newer project for Lorenz in the same period was an exploration of the epistemological implications of biological evolution. When he assumed the professorship of psychology at the University of Königsberg, he took up a position that descended directly from the chair of philosophy occupied in the eighteenth century by Immanuel Kant. Lorenz had developed an interest in Kantian epistemology prior to the appointment. In his new position he pursued the relation between Kant's teaching and modern science. In a paper of 1941 titled "Kants Lehre vom Apriorischen im Lichte gegenwärtiger Biologie" (Kant's doctrine of the a priori in the light of contemporary biology), he argued that human reason, just like the human brain itself, has evolved through a continuous interaction with nature—and according to the laws of nature. The categories and forms of intuition of the human mind are nothing less than hereditary differentiations of the central nervous system, characteristic of the species. They have been developed over the course of evolutionary history because of their survival value.

Where Kant had viewed the a priori categories of human thought were part of a God-given, immutable system, Lorenz offered instead a strictly naturalistic interpretation of human reason. He maintained that the apparatus through which the human mind apprehends the external world is the consequence of the long-term operation of heredity and evolution.

The idea of inborn forms of ethical and aesthetic judgment, developed over time as a result of their survival value, was also a central theme of Lorenz's large monograph of 1943 titled "Die angeborenen Formen möglicher Erfahrung" (The inborn forms of possible experience). There Lorenz offered a synthesis of his ideas on instinctive behavior, race hygiene, the dangers of domestication, the categorical imperative, humanity as the architect of its own destiny, and more. He pursued questions of evolutionary epistemology further in a large manuscript he wrote while imprisoned by the Russians. Although this manuscript was not published in his lifetime, it served as the basis for his book, *Behind the Mirror* (1977), first published in 1973 as *Die Rückseite des Speigels.*

Lorenz and Postwar Development of Ethology. Lorenz made his most brilliant and influential contributions to the study of animal behavior in the 1930s and 1940s. In

Lorenz

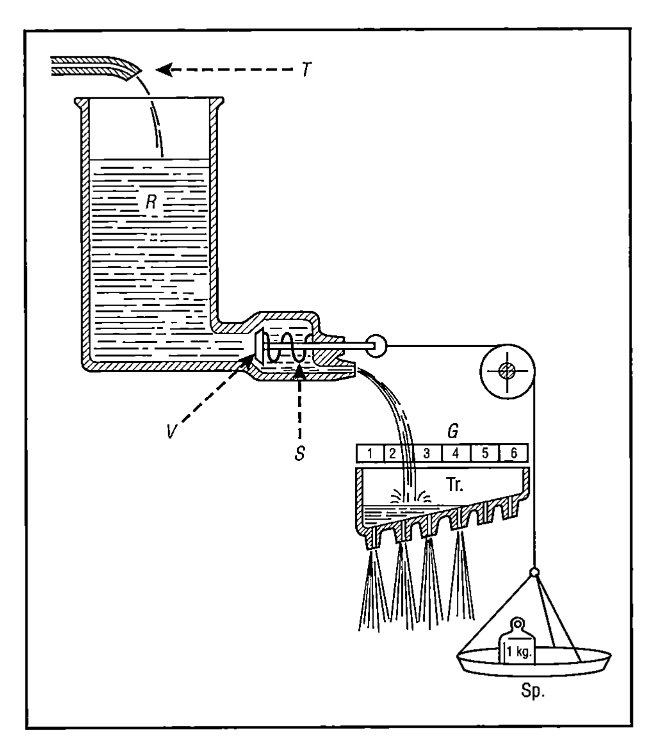

Lorenz's "hydro-mechanical" or "psycho-hydraulic" model of instinctive action.

1949 he rehearsed his major claims at a special symposium on physiological mechanisms in animal behavior, held in Cambridge, England. While insisting on the inductive foundations of his work, he also set forth a qualitative, "hydro-mechanical" or "psycho-hydraulic" model to represent the most salient features of how instincts work. In the diagram he presented, the liquid filling the reservoir corresponds to the hypothesized internal buildup of "action-specific energy," the spring in the valve and the scale pan attached to it by the string over the pulley correspond to the innate releasing mechanism, the weight on the scale represents the impinging stimuli that serve to trigger the releasing mechanism, and the jet of liquid that comes through the valve represents the instinctive motor reaction itself, with different results produced according to the strength of the reaction.

Lorenz's psycho-hydraulic model, together with a diagram Tinbergen offered of the hierarchical organization of drives, embodied the ethologists' basic assessment, as of 1950, of the physiological foundations of instinctive action. Representing what is now referred to as classical ethology, these models had considerable heuristic value. They served to organize a wide range of evidence. At the same time, they became subjects of critical scrutiny.

The ethologists' models and assumptions came under question both from within the ethological community (from new recruits to the field) and from outside of it. The American psychologist Daniel Lehrman launched the most severe attack. Lehrman's primary objection was that Lorenz's dichotomy between innate and learned behavior precluded any serious attention to the problem of how behavior develops in the individual. As Lorenz had originally defined it, "innate" behavior was behavior that developed without any environmental influence. Lehrman was willing to credit Lorenz with the taxonomic use he had made of behavioral characteristics, but he insisted that just because a characteristic was of taxonomic value did not mean that it necessarily developed free from any environmental input. Several ethologists, including Tinbergen and the Cambridge biologist Robert Hinde, came to agree with Lehrman that ethologists had used the word *innate* too uncritically and had not paid sufficient attention to the process of behavioral development.

Lorenz ultimately acknowledged that the word *innate* should not have been defined as the opposite of the word *learned.* He argued, however, that instead of giving up the word *innate,* one should use it to designate behavior acquired over the course of evolution. There are only two sources, he insisted, of adaptive behavior: It can be acquired phylogenetically, over the course of evolution, or it can be acquired individually, in the course of the individual's development. He maintained it was legitimate to use the term *innate* to designate those behavior patterns that had been acquired over the course of evolution through natural selection. Then, as a way of turning the tables on those of his critics who emphasized the importance of learning, he noted that learned behavior itself must have an innate basis, as demonstrated by the fact that animals in their natural environments are genetically predisposed to learn the kinds of information that help them to survive. He jokingly referred to this genetic disposition to learn the right thing as "the innate schoolmarm."

Lorenz also took exception to the way Hinde had questioned the models of drive that served as the motive force of Lorenz's and Tinbergen's respective models of the physiological causation of instinctive behavior. Hinde argued that there were no known mechanisms in the nervous system for the gradual buildup and discharge of action-specific energy that Lorenz's theory called for, and likewise that experiments failed to show the kind of threshold reduction over time that Lorenz's theory predicted. Lorenz continued to believe that certain behavioral phenomena like vacuum activities and threshold lowering indicated the existence of something that was very real, that was endogenously produced, and that was consumed or eliminated when the instinctive motor pattern was performed. In the general survey of the field that he published in the late 1970s, *The Foundations of Ethology,* he made some modifications to his old psycho-hydraulic model but retained the basic idea of a fluid that built up in a reservoir and eventually needed release.

Lorenz was unquestionably better at generating ideas than he was at testing them experimentally. Through the 1950s and 1960s he continued to be a charismatic leader of ethology, attracting new recruits to the field. However, many of the specific elements of his theorizing—action-specific energy, innate releasing mechanisms, and the psycho-hydraulic model—did not hold up to experimental examination. By the 1970s, furthermore, students of animal behavior were paying increasing attention to issues of behavioral ecology, an area of study in which the field naturalist Tinbergen had much more to offer than did the animal raiser Lorenz. Lorenz, who had been content to write about natural selection acting for "the good of the species," was criticized by biologists who insisted that selection acts at the level of the individual or the gene and not for the good of the species as a whole. Nonetheless, Lorenz's broad influence in the postwar period should not be undervalued. His insistence on the species-specific nature of much of animal behavior and on the methods by which animal behavior should be studied worked a fundamental transformation in the way animal behavior was studied and understood.

Overall Significance and Influence. Lorenz's primary and most lasting achievement as a biologist was to bring biological perspectives to bear on the study of animal behavior. He insisted that animal psychology needed to be studied as a biological science and that it needed to embrace a certain methodology and to address certain topics before it went on to others. Opposing the methodological orientation of the behaviorist psychologists, he maintained that an extensive period of general observation of an animal was absolutely necessary before attempting experiments on it, and that a thorough knowledge of an animal's instinctive behavior patterns was crucial before one studied behavior patterns that involved learning.

He based his claims to scientific authority on the unique character of his practice: The animal raiser had special advantages, he argued, not only over the laboratory biologist or psychologist but also over the field naturalist with regard to what daily experiences allowed him to see. He raised animals under conditions that enabled him to inventory and analyze their behavior patterns and to witness instructive occasions when these "misfired." While these practices profoundly shaped his understanding of animal behavior, they also had their limitations. Lorenz's friend Tinbergen, as a field naturalist, proved to be in a better position to develop studies of behavioral function and to examine how selection works in the natural world.

Lorenz promoted the idea that there are innate species differences in behavior that can be identified and examined analytically and that can be used for taxonomic purposes. He likewise insisted that an animal's genetic composition provides "structural restraints" for its behavioral abilities. Beyond this, he argued that once one recognizes that instinctive behavior patterns are comparable to organs, one can inquire about the similarities of animal and human experience in cases where instinctive behavior patterns in animals have phylogenetic "homologues" in human beings.

Lorenz was a bold, intuitive thinker and charismatic individual. He was the key figure in bringing a conceptual framework to comparative behavior studies, and he attracted many others to the new field that he founded. He was also a pioneer in the area of evolutionary epistemology. His popular books, including *King Solomon's Ring* (1952; originally published in 1949 as *Er redet mit dem Vieh, den Vögeln und den Fischen*), *On Aggression* (1966; first published in 1963 as *Der sogenannte Böse*), and *Civilized Man's Eight Deadly Sins* (1974; appearing initially in 1973 as *Die acht Todsünden der zivilisierten Menschheit*), offered the provocative view that developments in modern, civilized society have created settings that are increasingly at odds with the biological endowment of the human species. Though his ideas on the biological basis of human behavior and of aggression in particular have been widely criticized, his general claim that human behavior needs to be understood in the context of biological evolution remains of fundamental importance.

BIBLIOGRAPHY

The most significant collection of Lorenz's manuscript papers and correspondence is held by the Konrad-Lorenz-Institut für Evolutions- und Kognitionsforschung in Altenberg, Austria. Lorenz's correspondence with the ornithologists Oskar Heinroth and Erwin Stresemann is to be found in the Oskar Heinroth papers and the Erwin Stresemann papers at the Staatsbibliothek Berlin (West), Preußisher kulturBesitz. Additional archival sources for Lorenz papers are identified in Taschwer and Föger, Konrad Lorenz: Biographie, *below. For an extended bibliography of Lorenz's published writings, see Krebs and Sjölander, below.*

WORKS BY LORENZ

"Beobachtungen an Dohlen." *Journal für Ornithologie* 75 (1927): 511–519.

"Betrachtungen über das Erkennen der arteigenen Triebhandlungen der Vögeln." *Journal für Ornithologie* 80 (1932): 50–98.

"Beobachtetes über das Fliegen der Vögel und über die Beziehungen der Flügel- und Steuerform zur Art des Fluges." *Journal für Ornithologie* 81 (1933): 107–236.

"Der Kumpan in der Umwelt des Vogels: Der Artgenosse als auslösendes Moment sozialer Verhaltungsweisen" [The companion in the bird's world: Fellow members of the species as releasers of social behavior]. *Journal für Ornithologie* 83 (1935): 137–215, 289–413. The monograph that established Lorenz's reputation as the leading theorist of avian social behavior.

"Über den Begriff der Instinkthandlung." *Folia Biotheoretica,* series B, 2 (1937): 17–50.

"Über die Bildung des Instinktbegriffes" [On the formation of the instinct concept]. *Die Naturwissenschaften* 25 (1937): 289–300, 307–318, 324–331.

"Über Ausfallserscheinungen im Instinktverhalten von Haustieren und ihre sozialpsychologische Bedeutung." In *16. Kongress der Deutschen Gesellschaft für Psychologie in Bayreuth, 1938.* Leipzig: Johan Ambrosius Barth, 1938. Lorenz's first attempt to show the significance of his work for questions of racial hygiene.

With Nikolaas Tinbergen. "Taxis und Instinkthandlung in der Eirollbewegung der Graugans, I." *Zeitschrift für Tierpsychologie* 2 (1938): 1–29.

"Durch Domestikation verursachte Störungen arteigenen Verhaltens." *Zeitschrift für angewandte Psychologie und Charackterkunde* 59 (1940): 1–81.

"Nochmals: Systematik und Entwicklungsgedanke im Unterricht." *Der Biologe* 9 (1940): 24–36.

"Kants Lehre vom Apriorischen im Lichte gegenwärtiger Biologie" [Kant's doctrine of the a priori in the light of contemporary biology]. *Blätter für Deutsche Philosophie* 15 (1941): 94–125.

"Vergleichende Bewegungsstudien an Anatiden." *Journal für Ornithologie* 89, suppl. 3 (1941): 194–293.

"Die angeborenen Formen möglicher Erfahrung." *Zeitschrift für Tierpsychologie* 5 (1943): 235–409. Lorenz's major wartime synthesis of his thoughts on instinctive behavior, racial hygiene, and evolutionary epistemology.

"The Comparative Method in Studying Innate Behaviour Patterns." *Symposia of the Society for Experimental Biology* 4 (1950): 221–268.

So kam der Mensch auf den Hund. Vienna: Verlag Dr. G. Borotha-Schoeler, 1950.

King Solomon's Ring. London: Methuen, 1952. English translation of *Er redet mit dem Vieh, den Vögeln und den Fischen.* Vienna: Verlag Dr. G. Borotha-Schoeler, 1949. A charming, popular account of Lorenz's experiences as an animal raiser.

"Phylogenetische Anpassung und adaptive Modifikation des Verhaltens." *Zeitschrift für Tierpsychologie* 18 (1961): 139–187.

Evolution and Modification of Behavior. Chicago: University of Chicago Press, 1965. An expanded, English version of the preceding paper, representing Lorenz's response to the critics of his use of the term *innate.*

On Aggression. New York: Harcourt Brace and World, 1966. English translation of *Der sogenannte Böse.* Vienna: Verlag Dr. G. Borotha-Schoeler, 1963. Lorenz's popular and controversial portrayal of aggression as a valuable instinct that builds up internally in the body and in modern society needs to be directed into nondestructive or even productive channels.

Studies in Animal and Human Behaviour. Translated by Robert Martin. 2 vols. Cambridge, MA: Harvard University Press,

1970–1971. A helpful collection of some of Lorenz's most important papers.

Civilized Man's Eight Deadly Sins. New York: Harcourt Brace, 1974. Translation of *Die acht Todsünden der zivilisierten Menschheit.* Munich: Piper, 1973.

"Konrad Lorenz." *Les Prix Nobel en 1973.* 1974. See also http://nobelprize.org/nobel_prizes/medicine/laureates/1973/index.html for portrait, autobiography, Nobel lecture, and other information.

Behind the Mirror: A Search for a Natural History of Human Knowledge. London: Methuen, 1977. Translation of *Die Rückseite des Speigels.* Munich: Piper, 1973.

The Foundations of Ethology. New York: Springer-Verlag, 1981. Translation of *Vergleichende Verhaltensforschung: Grundlagen der Ethologie.* Vienna: Springer-Verlag, 1978.

"My Family and Other Animals." In *Studying Animal Behavior: Autobiographies of the Founders,* edited by Donald A. Dewsbury. Chicago: University of Chicago Press, 1989. The most detailed of Lorenz's autobiographical recollections.

The Natural Science of the Human Species: An Introduction to Comparative Behavioral Research: The "Russian Manuscript" (1944–1948). Cambridge, MA: MIT Press, 1996. Translation of *Die Naturwissenschaft vom Menschen: Eine Einführung in die vergleichende Verhaltensforschung,* edited by A. von Cranach. Munich, Germany: Piper, 1992.

OTHER SOURCES

Burkhardt, Richard W., Jr. *Patterns of Behavior: Konrad Lorenz, Niko Tinbergen, and the Founding of Ethology.* Chicago: University of Chicago Press, 2005. The most extended discussion of Lorenz's life and work in the context of the history of ethology.

Craig, Wallace. "Appetites and Aversions as Constituents of Instincts." *Biological Bulletin of the Marine Biological Laboratory* 34 (1918): 91–107.

Deichmann, Ute. *Biologists under Hitler.* Translated by Thomas Dunlap. Cambridge, MA: Harvard University Press, 1996. Translation of *Biologen unter Hitler: Vertreibung, Karrieren, Forschungsförderung.* Frankfurt, Germany: Campus, 1992.

Föger, Benedikt, and Klaus Taschwer. *Die andere Seite des Speigels: Konrad Lorenz und der Nationalsozialismus.* Vienna: Czernin Verlag, 2001. The most detailed evaluation to date of Lorenz's activities as a scientist in the Third Reich.

Heinroth, Oskar. "Beiträge zur Biologie: namentlich Ethologie und Psychologie der Anatiden." In *Verhandlungen des 5. Internationalen Ornithologen-Kongresses in Berlin, 30 Mai bis 4 Juni 1910,* edited by Herman Schalow, pp. 589–702.

Kalikow, Theodora J. "Konrad Lorenz's Ethological Theory: Explanation and Ideology, 1938–1943." *Journal of the History of Biology* 16 (1983): 39–73.

Krebs, John R., and S. Sjölander. "Konrad Zacharias Lorenz." *Biographical Memoirs of Fellows of the Royal Society* 38 (1992): 210–228.

Lehrman, D. S. "A Critique of Konrad Lorenz's Theory of Instinctive Behavior." *Quarterly Review of Biology* 28 (1953): 337–363.

Nisbett, Alec. *Konrad Lorenz.* New York: Harcourt Brace Jovanovich, 1976. The first full-length biography of Lorenz.

Taschwer, Klaus, and Benedikt Föger. *Konrad Lorenz: Biographie.* Vienna: Paul Zolnay Verlag, 2003. The most recent biography of Lorenz, based on an extensive reading of Lorenz's manuscript correspondence.

Richard W. Burkhardt Jr.

LOVELACE, ADA

SEE **King, Ada Augusta, the Countess of Lovelace.**

LULL, RAMON (also known as Ramon Llull) (*b.* Ciutat de Mallorques [now Palma de Mallorca], c. 1232; *d.* Ciutat de Mallorques [?], January/March [?] 1315), *polymathy, alchemy.* For the original article on Lull see *DSB*, vol. 8.

The Catalan Lull left a magisterial work in both the philosophic and spiritual domains. His aura was such that a large number of apocryphal texts appeared after his death. In the field of alchemy these texts constitute one of the most important corpora of this medieval science so often disdained and misunderstood. The pseudo-Lullian texts, exhumed and studied by Michela Pereira and others, have illuminated fourteenth-century alchemy whose influence in its domain and in others was decisive.

Even though Ramon Lull in his *De ente reali* stigmatized the foolishness of alchemists who thought they could create gold from silver and silver from mercury, a large number of alchemical texts were attributed to him in the fourteenth and fifteenth centuries. This pseudepigraphical vein dates from about 1370: The alchemist Sedacer cited Lull among the authorities of this art in his *Sedacina totius artis alkimiæ* (circa 1378). Such a profusion (one-hundred forty-three texts) can be explained as much by the affinity that certain of these texts manifest with those of the historic Lull (alphabets, diagrams) as by the transmission of similar alchemical doctrines and techniques, or quite simply by an effect of contagion, an isolated text in a mainly pseudo-Lullian manuscript being sometimes then recopied with this attribution. There are sixty-five pseudo-Lullian treaties, forty-nine occasionally attributed to Ramon Lull, and twenty-eight that should be excluded from the corpus, attributed to Lull in only a single manuscript or in no manuscript at all, only printed materials. The tradition begins with the *Testamentum,* the first alchemical text accepted as pseudo-Lullian.

The *Testamentum*. Between 1330 and 1332 a Catalan alchemist wrote in Latin, then translated into his own language (Catalan) a summa divided into three parts (Theory, Practice, and *Liber mercuriorum*) to which was attached a poem defining alchemy as a virtuous and eminently Christian science (*Cantilena*). In 1443 John Kirkeby translated the book into Latin.

Although it is the work of an anonymous master, the *Testamentum* contains enough Lullian elements to initiate a tradition of alchemical texts called pseudo-Lullian. In addition to citing two authentic works of Ramon Lull, it uses the figures of the Lullian *Ars combinatoria* and alphabets as ways to implant the alchemical process in the memory. Lull is not the only authority to whom this important work refers; the physician Arnald of Villanova is also one of its major sources, in particular his determinative book of medieval medicine, the *Aphorismi de gradibus.*

The *Testamentum,* following the example of the *Rosarius philosophorum* of the pseudo-Arnald of Villanova, is as much a book of alchemy as it is a medical book. It proposes an efficacious way, founded on a sound philosophic doctrine, to produce an elixir as effective in multiplying gold or in creating precious stones as it is in treating the human body. This is no miracle, the pseudo-Lull was careful to state, nor is it a magical act, but the simple effect of natural heat "concentrated [he says] in its humid radical" ("Istud vero, fili, non est nisi calor naturalis infixus in suo humido radicale," Pereira and Spaggiari, eds, *Testamentum,* II, §30, 31, 378). More precisely, the plan of the *Testamentum* is rooted as deeply in the tradition of Roger Bacon seeking to "prolong life" by alchemy as it is in that of the medicine taught at Montpellier with the ideal of creating a universal "medicina," the sign of the perfection of matter. In this respect, it marks an important date in the history of alchemy, for it places the elixir of metals, obtained through the science of the pseudo-Geber, in a principally therapeutic perspective.

The success of the *Testamentum* inspired the writing of other alchemical texts. Among these, the most important are the *Liber de intentione alchimistarum* (also linked to Arnald of Villanova), the *Epistola accurtationis* (dedicated to King Robert of Naples), the *Clavicula,* the *Testamentum ultimum* (a commentary to the *Testamentum*), and the *Codicillus.* Others that might be mentioned are the treatises on the *ars lapidifica,* devoted to the making of precious stones, an original aspect of this corpus. All these texts have in common the presentation of an alchemical theory that is relatively complex from the point of view of its ideas. For example, in the case of the *ars lapidifica* accomplished by using human urine, it can appear as an attempt to integrate *via* the concept of man as a microcosm the basest material reality into thoughts of the most sublime philosophy. Moreover, because these treatises developed in the South of France and in Catalonia, they came into direct contact with other alchemical works such as those of the pseudo-Arnald of Villanova and of Jean de Roquetaillade, thus bringing to prominence a group of related texts. In this regard one might even speak of a veritable alchemical school.

The *Liber de secretis* ... The production of pseudo-Lullian alchemical texts culminated at the end of the fourteenth century with an important work, the *Liber de secretis naturæ sive de quinta essentia.* At that time the formation of this corpus of texts entered a second stage. In the *Liber de secretis naturæ sive de quinta essentia* the alchemical practice of the *Testamentum* becomes linked to the fifth essence of wine, a distillation technique popularized in by Jean de Roquetaillade in 1350. Moreover, its author said on several occasions that he relied on the *Testamentum* and other alchemical texts, thus recognizing Lull as an alchemist. If the *Liber de secretis naturæ sive de quinta essentia* seems to be a medical book guided by the thought and the style of Lull, it is also notable for its author's interest in turning matter into gold, unlike John Roquetaillade who for religious reasons was not mainly interested in such transmutation. It begins with a prologue consisting of a conversation between Lull and a monk, then come the two books paraphrasing Roquetaillade's *De quinta essentia.* It ends with a *Tertia distinctio* devoted to an alchemical application of the Lullian method (alphabets and trees). Even if the *Liber de secretis naturæ sive de quinta essentia* suffered, like a number of alchemical works, from a very unreliable textual tradition in both manuscript and printed form, it enjoyed great success in the sixteenth century.

The Legend of Lull as Alchemist. Numerous legends proliferated around Lull, whose life and works were far from ordinary. One of the most persistent is that Ramon Lull devoted himself to alchemy. In some works of the corpus, it is reported that Lull was introduced to this science by Arnald of Villanova who had met Lull through King Robert of Naples. The most prominent legend—one even mentioned in Peter Acroyd's book on London (*London the Biography,* 2000)—pertains to a trip to England. There Ramon Lull supposedly undertook some transmutations into gold on behalf of Edward III. Troubled by the inappropriate uses that the King made of the alchemical gold, Lull is reputed to have protested and been thrown into prison. This legend was nourished on by the mention of King Edward and the city of London in the colophon of the *Testamentum,* by the historic interest that English princes showed in alchemy, and by their disappointment in it, which they manifested by incarcerating alchemists. It should be noted that if the first legend shows an Arnald of Villanova bequeathing to Lull an alchemy of distillation whose purpose is the quest of the elixir, the English legend emphasizes the transmutatory character of his alchemy. In the fifteenth century Guillaume Fabri de Dye, in his *Liber* *de lapide philosophorum* (Chiara Crisciani edition, 2002), combined the elements of the two legends into a single one, grouping Arnald of Villanova, Ramon Lull, and John Dastin together as three alchemists working under the protection of good King Edward for the fulfillment of the opus and writing books.

The Success of the Pseudo-Lull. The pseudo-Lullian alchemy arose initially in Catalonia and in the South of France. It achieved a remarkable breakthrough in England as evidenced by the translations of John Kirkeby, by the works of the pseudo-Lullian George Ripley (fl. 1476), and by the fact that it is in England that one finds the oldest manuscript copies. From there, the movement reached the continent, Italy (Venice, Florence, Naples), Germany, and France where it attracted interest in alchemical and medical circles. In the Renaissance the pseudo-Lullian alchemy was regarded as highly as was the *Ars combinatoria* of Lull, which was particularly in vogue among Italian humanists. The Venetian Pantheus then produced a work, the *Voarchadumia* (1530), in which the pseudo-Lull's combinatory art in the service of alchemy mixes with the cabala, a prelude to the encyclopedic works of the mathematician and astrologer John Dee. Moreover, the alchemical legend attached to Lull's name, which became in the course of centuries increasingly complex and embellished, contributed greatly to the success of the pseudo-Lullian corpus, which continued to grow until the eighteenth century.

Despite the critics who, beginning in the middle of the sixteenth century, doubted the authenticity of the corpus, the success of the published editions of the pseudo-Lullian alchemical corpus, especially that of the *Testamentum* and the *Liber de secretis naturæ sive de quinta essentia*, and the interest accorded to the treatises on distillation were influential in preparing minds for the alchemical medicine of Paracelsus. In the eighteenth century, Ivo Salzinger, the inventor of a new system of musical harmony and the publisher of the *Lullian Opera omnia* (Mainz, 1721–1742), placed alchemy at the heart of the Lullian system, sweeping away all objections and demonstrating that the inspired doctor knew, practiced, and taught in his books the great work. Even Newton possessed in his library six alchemical works attributed to Ramon Lull.

SUPPLEMENTARY BIBLIOGRAPHY

Besides the editions of the Renaissance already cited and accessible on the Internet, see also the Ramon Lull Databse (Base de Dades Ramon Llull) at the Universitat de Barcelona, available online from http://orbita.bib.ub.es/llull/. It contains a bibliographic list of all the works in the apocryphal alchemical corpus, the principal manuscripts and editions (both ancient and modern) that have transmitted them, as well as catalogues, articles, and books where Lull's work is mentioned. This is a complete scholarly tool that is extremely useful.

WORKS BY LULL

Apertorium, Compendium animæ transmutationis metallorum, etc. Johann Petreius gathered these into a first collection in 1546: *Raimundi Lullii Majoricani, De Alchimia Opuscula.* Nuremberg, 1546.

Apertorium, Ars intellectiva, Liber de intentione alchimistarum ...: Veræ alchemiæ artisque metallicæ. Basel, G. Grataroli, 1561. Available from http://www.bium.univ-paris5.fr/histmed/medica.htm, Dióscorides: http://www.ucm.es/BUCM/foa/presentacion.htm. The great collections of the sixteenth and seventeenth centuries were obviously fond of pseudo-Lullian texts.

Raymundi Lullii, Libelli aliquot Chemici. Basel, Michael Toxites, 1572. This contains *Cantilena, Codicillum, Elucidatio Testamenti, Epistola accurtationis.*

Artis auriferae quam chemiam vocant, 3 vols. Basel, 1572, 1593, 1601, 1610.

Testamentum, De secretis naturæ sive de quinta essentia. Reprinted in the collections of Zetzner (*Theatrum chemicum,* I-V, Strasbourg, 1659–1661, repr. Turin: La Bottega d'Erasmo, 1981).

Bibliotheca Chemica Curiosa, I. Edited by Jean-Jacques Manget. Geneva, 1702 (repr. Bologna, 1976).

Catalogue of Latin and Vernacular Alchemical Manuscripts in Great Britain and Ireland Dating from before the XVI Century. Edited by Dorothea Waley Singer. Brussels: Union Académique Internationale, 1928–1931.

Catalogue des manuscrits alchimiques latins, manuscrits des bibliothèques publiques de Paris, antérieurs au XVIIe siècle (*Catalogue of Latin Alchemical Manuscripts, Pre-Seventeenth Century Manuscripts in the Public Libraries of Paris*). Edited by James Corbett. Brussels: Union académique Internationale, 1939; James Corbett, Catalogue des manuscrits alchimiques latins, Manuscrits des bibliothèques publiques des départements français antérieurs au XVIIe siècle, (*Catalogue of Latin Alchemical Manuscripts, Pre-Seventeenth Century Manuscripts in the Public Libraries of French Departments*). Brussels, 1951.

Pereira, Michela. "Un lapidario alchemico, il 'Liber de investigatione secreti occulti' attribuito a Raimondo Lullo. Studio introduttivo ed edizione" ("An Alchemical Lapidary, the 'Liber de investigatione secreti occulti' attributed to Ramon Lull: An Introductory Study and Edition"). *Documenti e Studi per la Tradizione Filosofica Medievale* (Documents and Studies for the Medieval Philosophical Tradition) 1 (1990): 549–603.

———, and Barbara Spaggiari. *Il "Testamentum" Alchemico attribuito a Raimondo Lullo* (The Alchemical Testamento Attributed to Ramon Lull). Florence: SISMEL Edizioni del Galluzo, 1999. An edition of the Latin *Testamentum* with the parallel Catalan text on the facing page.

OTHER SOURCES

Crisciani, Chiara. *Il Papa e l'Alchimia: Felice V, Guglielmo Fabri e l'Elixir* (The Pope and Alchemy: Felix V, Guillaume Fabri and the Elixir). Rome: Viella, 2002.

Pereira, Michela. *The Alchemical Corpus Attributed to Raymond Lull.* London: The Warburg Institute, 1989. This is the most complete study to date on the pseudo-Lull alchemist.

———. *L'oro dei filosofi. Saggio sulle idee di un alchemista del Trecento* (The Philosophers' Gold: Essay on the Ideas of a Fourteenth-Century Alchemist). Spoleto, 1992.

———. "Maestro di segreti o caposcuola contestato? Presenza di Arnaldo da Villanova e di temi della medicina arnaldiana in alcuni testi alchemici pseudo-lulliani" ("A Master of Mysteries or the Controversial Leader of a School? The Presence of Arnald of Villanova and of Themes of Arnaldian Medicine in Some Pseudo-Lullian Alchemical Texts"). *Arxiu de Textos Catalans Antics,* 23/24: II Trobada Internacional d'Estudis sobre Arnau de Vilanova (*Archives for Ancient Catalan Texts*: II International Congress for Studies on Arnald of Villanova) (2004–2005): 381–412.

Szulakowska, Urszula. "Thirteenth Century Material Pantheism in the Pseudo-Lullian 'S'-Circle of the Powers of the Soul." *Ambix* 35, no. 3 (1988): 127–154.

Thorndike, Lynn. *A History of Magic and Experimental Science.* New York: Columbia University Press, IV, 1934: 3–64, 619–652.

Waley Singer, Dorothea "The Alchemical 'Testamentum' Attributed to Raymond Lull." *Archeion* 9 (1928–1929).

Antoine Calvet

LURIA, ALEXANDER ROMANOVICH

(*b.* Kazan, Russia, 16 July 1902; *d.* Moscow, Russia, 14 August 1977), *neuropsychology, neurolinguistics, cultural psychology, classical science, romantic science, neuroanatomy, neurophysiology.*

Luria's major scientific accomplishments were in the broad area of neuropsychology, with major contributions to the study of brain damage in adults, with resulting motor-sensory disruption, memory disorders, and language loss (aphasia). He worked early in his career on the cultural aspects of cognition, but later studied neurology in order to enter the medical profession and work directly with diseases of the nervous system. He is most widely known for his prodigious research and publication on the adult language disorders of aphasia, concentrating on frontal lobe functions, but extending his model of brain and language to many other regions of the cerebral cortex as well as subcortical regions. His detailed case studies of perception and memory loss are known throughout the world, and all neurolinguists have read and studied his major publications in the area. Evident in all his work was his balanced view of the mental and the physical. His psychophysical model of brain and behavior provided him with a dualism that allowed him to steer through the hazardous scientific milieu of the most repressive years of Communist Russia.

Biographical Overview. Alexander Romanovich Luria was born on 16 July 1902 in the city of Kazan, the present-day capital of the Republic of Tatarstan, in the heart of Russia, about 500 miles (800 km) east of Moscow in a region between the Volga River and the Ural Mountains. The language of Luria's birthplace is Tatar, a Turkic language of the Altaic family. Little is said regarding whether Luria knew or spoke much Tatar; he spoke Russian. Although there was a degree of bilingualism in the region, the official language was Russian. Luria's father was a gastroenterologist who taught medicine at the University of Kazan. After the Revolution of 1917, he founded and became chief of the Kazan Institute of Advanced Medical Education. Luria's mother was a dentist, and his sister eventually became a psychiatrist—no easy feat in Communist Russia.

At the age of seven, little Alexander was considered a genius; he started the *gymnasium* at that age. Eugenia D. Homskaya (2001) writes that Luria enjoyed literature, history, and philosophy. He took well to the study of languages, especially Latin, German, French, and English. When he was fifteen, the Communist revolution took place, and his *gymnasium* was closed that year. In 1921 he left Kazan for Moscow, where he worked until his death in 1977.

At the age of twenty-one, Luria married Vera Nikolayevna Blagovidova—a marriage that lasted six years and produced no children. In 1933 Luria met and married Lena Pimenovna Linchina. This marriage lasted until Luria's death in 1977; Lena died a year later. They had one child—a daughter, Elena, who was born on 21 June 1938. She eventually moved to the United States, where she practiced psychiatry in New York City for many years until her death on 20 January 1992.

After several years of cardiovascular signs of disease, Luria suffered a fatal heart attack on 14 August 1977 in Moscow, shortly after turning seventy-five. He was active up until the very end of his life. Homskaya relates that "In the few minutes before his fatal heart attack, he was writing an article on the pathologies of memory" (2001, p. 117). Even though this paper was unfinished, Luria's colleagues saw to it that it was published, first in the original Russian, and later, as was usually the case, in an English-language version under the title "Paradoxes of Memory" (1982).

Luria and the Soviet Union. It is unwise to discuss Luria without reference to the fact that he lived and worked in the Soviet Union, and that he was Jewish. Luria had read Sigmund Freud's works on the psychoanalytical approach and attempted to incorporate them into his first two monographs, the first titled: "The Principles of True Psychology," followed by a second monograph on

"Psychoanalysis in Light of the Principal Tendencies of Contemporary Psychology." These were written in Kazan by a very young Luria, shortly before he left for Moscow.

In Moscow, Luria started working in education, and at the age of twenty-one became head and chair of psychology at the Academy of Communist Upbringing (Education) of Krupskaya. In 1932 Luria published a summary of his early Freudian monographs in a book titled *The Nature of Human Conflicts*. Only thirty years old, Luria was brash enough to send a copy of his summary to Ivan Pavlov. Pavlov blazed into Luria's office the next day, pulled out the monograph, tore it in half, tossed it on the floor, and upbraided Luria for describing behavior "as a whole" with high-level generalizations. Pavlov's claim was that science proceeds from low-level units and progresses upward. Thus began the official Soviet condemnation of Luria's work as "un-Soviet," which at that time meant "un-Pavlovian." From that point, Luria was officially prohibited from teaching, conducting research, or publishing anything that assumed a psychoanalytic stance. The ironic aspect of much of this was that, in actuality, there were points at which Luria went beyond strict Freudian mind-brain theory to search for outward manifestations of the mental, focusing upon the lower Pavlovian phenomena of motor and visceral response correlations.

Earlier, in 1924, Luria and Lev Semyonovich Vygotsky met and combined their admiration for Freud's psychology into some of the first studies of the cognitive aspects of human sociocultural history, which they scaffolded across neurocognitive systems. Luria immediately took to Vygotsky's claims that in order to understand human cognition, one had to consider the historical sociocultural background of the individual. It was not long until Luria, with the blessing of Vygotsky, was off to Uzbekistan in central Asia to study the effects of literacy and social change on inference, memory, and perceptual categorization. Luria, along with a colleague, F. N. Shemiakin, had success in demonstrating that in the small villages of the region, illiterate and educated Uzbeks behaved differently in the perceptual processing of photographs and drawings. The illiterate population was not able to perceive depth in logical terms, but rather only in terms of situations depicted in the visual material. Luria set out again on another expedition to Uzbekistan, this time learning some of the Uzbek language, along with further studies of deductive reasoning; of how metaphors, symbols and logic were processed; of the perception of shapes, colors, and optical geometric illusions; and of drawing, as well as how the Uzbeks calculated and remembered (Homskaya, 2001). This significant work was blocked from publication, and the Soviet regime prohibited any sociocultural work by Luria, Vygotsky, or anyone else. The work would not see the light of day until forty years later. Michael Cole, later as a young student with Luria, inspired his mentor to continue these sociocultural and historical studies of human cognition, and of late it has been Cole who has brought this research into clearer relief.

One fascinating aspect of Luria's scientific life in Russia is that he managed to keep as busy as he was, working and writing almost daily, drawing on his patience and self-protective skills to work under the dictates of totalitarianism. In Moscow, Luria was working on medical genetic investigations at the State Institute of Experimental Medicine. In 1936 genetic research was suddenly proclaimed illegal and the institute was closed. Luria, evidently, had sensed that the work at the institute was doomed, and had left a month earlier to pursue the full-time study of medicine, at the First Medical Institute of Moscow; he became a medical doctor. Despite his move to the medical sciences, he and many of his colleagues were known for their tendency toward liberal idealism in such spheres as art, science, and literature. Luria was among the suspected "subversives," and in 1951 his laboratory at the Institute of Neurosurgery of Burdenko was closed. He quietly and deferentially transferred to the Institute of Defectology of the Academy of Pedagogical Sciences of the Russian Federation. Not longer than a year later, in 1952, Jewish doctors fell prey to a large anti-Semitic campaign. Many Jewish physicians were fired during this time, but Luria remained relatively unscathed—likely due to the extraordinary work and responsibilities he had at the Institute of Defectology.

Case Histories and Psycholinguistic Models. Luria is well known for his book-length stories on people such as Zasetsky, who lived only in the present, and Solomon Shereshevsky, whose memory and perception were extraordinarily fixed to particulars, to the extent that he could not categorize. These books represent the best of romantic science; they are long and detailed case histories (anemnesis) and recollections from the patient. The story-telling schemata for patient behavior are often referred to as psycho-biography, or as "portraits."

The romantic science of Luria and the tale of Shereshevsky have direct relevance to the psycholinguistic models of discourse analysis referred to as "mental models." To comprehend, interpret, and compute inferences in conversation properly, listeners construct an on-line view in their minds of the scenes and events being involved in conversation. Normal discourse allows for this. Theoretically, however, it has always been the case that there are no inherent restrictions or architectural constraints as to thedetail needed for hearers when constructing a mental model as they listen and comprehend. The mental model claim is that hearers need to get the "gist" of what is said. Luria's tale of the mnemonist he called "S" presents instances in which the pathological press for particulars and extraordinary detail renders getting the gist of the story impossible because the model so constructed is cluttered with ornate detail. "S" was oversaturated by the particular to the point of clouding comprehension. The mnemonist described in Luria's tales cannot appreciate generalities or categories, so he cannot "get the gist."

Luria's Brain Science: Pavlov's Influence. Luria entered medical school in the early 1930s, specializing in neurology. This move enabled him to fasten the physical to his previous training and research in psychology, approaching more closely a true neuropsychology. It also provided him with added scientific armor to withstand the pressures of the materialism of the official party line. Although rebuked in his youth by Pavlov, Luria continued to utilize many of the physiological concepts for cortical neurodynamics. Spread of activation, inhibition, post-activation rebound, strength or weakness of each, threshold levels, and decay to steady states were the jargon of cortical processing. Luria's "normal rules of force" operate where strong or important stimuli evoke strong reactions, and weak and unimportant stimuli evoke weak reactions. When the rules of force break down, the physiology changes. In the so-called disrupted inhibitory phase, both strong and weak stimuli evoke reactions of the same strength—a so-called phase of equalization. Paradoxically, a physiologically altered phase may come about in which weak or insignificant stimuli will actually evoke stronger reactions than strong stimuli.

The frontal lobes were the anatomical structures that received most of Luria's attention, in that they appeared to him to be the central control and regulation systems for planning actions and carrying them out. In a real sense, in Luria's scheme, the frontal lobes were the sine qua non of personal humanity, because they housed systems for goal-oriented intentional behavior, evoking the movements, guiding them to their ultimate target shapes. Major execution systems reside in the frontal lobes, and many important neuropharmacological neuronal networks course through them—especially dopaminergic systems in mesolimbic zones of the frontal lobes.

Luria wrote at length on motor perseveration, where activated movements tend to be erroneously reactivated at subsequent points; he distinguished two forms of frontal lobe motor perseveration. One resulted from an efferent pathological inertia, in which lower-level motor sequencing was disrupted by lesions in anterior regions of the frontal lobe. A major responsibility of the anterior motor system in Luria's model is to smoothly sequence the surface order chaining of motor gestures. Disrupted sequencing often resulted in the carryover of prior action, and in speech, articulatory gestures would be erroneously repeated. A second type of perseveration affected slightly higher-level programs of action, in which raw articulatory units were not so much affected as the units at the level of the phonological plan. There has been an upsurge in the study of perseveration in the early twenty-first century, but Luria laid much of the foundation in the 1960s. His 1973 publication, *The Working Brain*, contains an excellent survey of his "classical" and nomothetic neuropsychology.

Throughout his neuropsychological studies of language deficits secondary to brain damage, Luria maintained a Freudian strategy of eschewing any overly localistic and physical mapping from form to function. As Freud, and John Hughlings Jackson before him, Luria resisted strict localizationist interpretations of the classical aphasiologists (Pierre Paul Broca, Carl Wernicke, and Ludwig Lichtheim), opting for a greater deal of computational simultaneity of related but noncontiguous regions. His constant awareness of the effects of culture on cognitive patterns kept him even further away from innocent localization of function in the brain. Pavlov's physiology tended to reduce to the physical all too easily. Behaviorist reductionism was to be countered at all times, replaced by Luria with some form of psychophysical identity or parallelism.

Luria's Linguistic Aphasiology. Luria never supported classical aphasia localization, and therefore was a constant critic of the neolocalizationists who followed Norman Geschwind. It is likely that much of Luria's nonclassical localizationist aphasia model was due in large part to the overwhelming bias in his data corpus of World War II penetrating missile wounds and war-related traumatic brain injuries (closed head injury) in young military patients. Little wonder that stroke etiology in older patients played a minor role in Luria's observations of aphasic disorders in the war-torn late 1930s and 1940s. Etiology of aphasia often determines the nature of its symptomatology.

Central to Luria's breakdown of cortical function is Pavlov's tripartite division into three types of complex neural circuitry, called analyzers. There are analyzers for the input-output systems that most directly connect the body with the outside world (primary systems), analyzers for internal proprioception (secondary systems), and a third set of analyzers that interconnect the other analyzers in many ways (tertiary systems). Human language capacity results from massive interaction of the separate analyzers. For example, a highly complex analyzer for speech output would involve tight interaction among phonological plans, phonetic detail of allophones, and subcortical/cranial nerve function, ultimately synapsing upon muscular structures in the articulatory periphery. Acoustic analyzers are obviously set to underlie speech perception as

Luria

part of the gateway to comprehension. Since evidence points to sensory guidance for muscular movement systems, human articulatory function must integrate auditory and speech movement programs for a proper output. Phonemes are multifaceted abstract units of both sensory and motor interaction. The phoneme is considered to be a set or "bundle" of features—mostly articulatory, but some acoustic. Words, concepts, and conceptual units are complex elements, which can be dissociated as a consequence of damage to the nervous system; the system is highly interactive.

Luria's Aphasic Syndromes. Luria's early notions of frontal lobe function led him to the conclusion that the aphasias resulting from brain damage there mostly left the patient with disruptions in the sequencing of elements in the volitional production of language, that is, serial ordering. The temporo-parietal regions of the dominant hemisphere, however, mediated language codes primarily through the selection of elements, based on similarity of form function in hierarchical systems of associative relations. Sequencing, contiguity, and syntactic ordering are not considered typical functions of posterior cortical language regions. This strict dichotomy between frontal lobe sequential ordering and temporo-parietal lobe selection has been criticized as overly simplistic.

For Luria, there are essentially five aphasia syndromes, and he outlined these in a presentation at the Ciba Foundation on Disorders of Language in the early 1960s. The first aphasic syndrome is referred to as sensory aphasia. Luria defined many types of sensory aphasia, but all involve some aspect of the auditory analyzer, and for the most part, the brain damage is in the dominant temporal lobe in the posterior third of the superior convolution. Compromised are the comprehension of spoken language and the inability to repeat words and name objects. Writing and spontaneous speaking are compromised. Much of the problem with sensory aphasia for Luria involved a disturbance in the utilization the distinctive features of phonemes—especially the acoustic features, sets of which constitute the phonemic architecture. The phonology of the linguist Roman Jakobson is readily apparent in most of Luria's linguistic descriptions of phonemic hearing. The syndrome of acoustic-amnestic aphasia in Luria's scheme is characterized by a short-term verbal buffer memory breakdown, whereby the patient cannot "hold onto" the acoustic impression long enough to operate on it for its production. The disruption here is not so much representation retrieval, but rather that the sensory acoustic activation decays too quickly.

Motor Aphasia: Two Types. For Luria, there is not just one motor aphasia, classically labeled as Broca's aphasia. Luria postulates two motor aphasias; one directly invades the efferent/kinetic motor articulatory programming. The creation of smooth, serially ordered sequences of sound is impeded, and the speech is slow and laborious. This Luria labels "efferent motor aphasia" and locates the responsible brain damage in Broca's area in the posterior third of the inferior frontal lobe convolution. Afferent motor aphasia arises as a consequence of damage to the sensory strip region across the Rolandic fissure from the primary motor cortex for the speech articulators. Without internal sensation of touch for the articulators, the speaker cannot maintain guided and targeted movement. Due to the loss of internal sensation, for example, the tongue will have difficulty finding its way through the oral cavity, and will be prevented from making the fine articulatory gestures with respect to place of articulation, manner of articulation, and quite likely, proper manipulation of the vocal cords to turn glottal pulsation on and off. This is essentially an internal kinesthetic disruption.

Semantic Aphasia. Luria's characterization of semantic aphasia is eclectic and pulls from different theories, especially the work of Henry Head. The region most vulnerable to lesions causing this form of aphasia is in a sort of way-station area, where the regions are at the interface of vision, audition, and tactile. Accordingly, Luria locates the general cortical zone for this type of aphasia in the tertiary parieto-temporal-occiptal cortex, where there are overlapping functions. When the semantic systems are functioning normally, there is a great degree of simultaneous synthesis of lexical, grammatical, logical and relational computations. Many of the lexical relations are fixed by a certain similarity of function, which then form into so-called associative fields. The fields are groupings of words based on a similarity of sound or on a similarity of function or on levels of cohyponyms. Semantic breakdowns will occasion word substitutions among these types of word associates. Difficulties arise in the computation and comprehension of sentences with inversions and with complex intersentential referential relations. Spoken human language output is extremely fast, and many semantic and grammatical processes are computed cooperatively and with a speed that almost approaches simultaneity. At times, semantic aphasia will involve extreme difficulty with the retrieval of words from the patient's mental store of words. As expected from lesions so far from the motor centers, the semantic aphasic has no paralysis, ambulates with no effort, and articulates normally.

Anomia. Early on, Luria (1964) embedded an amnestic aphasia under sensory aphasia. Others have called this amnestic aphasia an anomia, in which the overriding disturbance is one of accessing words—most often nouns. As sensory aphasia, fluent anomias are considered to arise from lesions to posterior temporo-parietal in the dominant hemisphere. For Luria, there are three types of amnestic syndromes. The first is an access disruption in the visual modality specifically. *Optic aphasia* is a term that many use in present-day aphasiology, and it usually follows as a consequence of damage to the occipito-parietal regions of the left hemisphere. The second type of anomia for Luria is more of a phonological execution breakdown of the phonological structure of the word. Here, a portion of a word or its metrical structure is retrieved, but one or more of the segments are altered—often with phonemes that share perhaps two of three features. For example, the word *dinner* may be produced as "tinner." The "d" and the "t" share manner (oral stops) and place (alveolar), but differ in the feature (voice). The third type of anomia in Luria's scheme produces more exclusively semantic associates, in which the substituted word is within the same semantic sphere as the word sought after. Luria's "rules of force" of cortical neurodynamics are altered here, with the result being a leveling of activation strength such that semantic sphere associates of the target have close to an equal chance of being retrieved.

Conduction Aphasia. Luria always resisted allocating syndrome status for conduction aphasia. The sine qua non of this type of syndrome is a failure to repeat a heard verbal stimulus successfully. Repetition failure is the major response indicator of acoustic-amnestic aphasia, and so in a sense conduction aphasia, as anomia, for Luria has been embedded in an overarching syndrome category. In conduction aphasia, responses are typically replete with phoneme errors, such as substitutions, deletions, or incorrect serial ordering—either anticipatory or carryover. Phonemic paraphasias are also a marker of this aphasia type. What establishes conduction aphasia as truly autonomous is the fact that many of the repletion errors are not caused by short-term verbal auditory memory. Rather, the repetition errors seem to adumbrate faulty manipulation of phoneme features rather than a fast fade of the auditory stimulus. Often, these patients will take a rather long time to provide the repetition, but they nevertheless stay on target as they approach the correct production. This mechanism would argue for conduction aphasia being more than the result of a working memory or operating buffer breakdown. Furthermore, Luria implicates an element of afferent motor aphasia, because he claims that faulty proprioception (kinesthesia) within the oral cavity would weaken correct specification for features such as place or manner, since without tactile sensation, the articulators would lack crucial knowledge for proper gestural achievement. Luria, further, accepts a suggestion from a contemporary aphasiologist, Kurt Goldstein, that the mere act of repeating a heard stimulus upon command by the examiner is far removed from anything natural regarding language production. The patient must attain a highly abstract cognitive stance for this special form of conscious activity. It is that ability that is possibly disrupted in conduction aphasia. Lastly, Luria has always been suspicious of disconnection accounts of this syndrome. Classical aphasia models specified a lesion that disrupts neither Broca's area nor Wernicke's area, but rather the myelinated white matter fiber tract that connects the two (the arcuate fasciculus).

Dynamic Aphasia. The category of aphasia in Luria's scheme that he labeled *dynamic aphasia* implicates several frontal lobe systems, and also closely relates to one other classical aphasia type: transcortical motor aphasia. If one considers why Luria selected the adjective "dynamic" for this syndrome, one can approach the cognitive mechanism he is suggesting. A crucial role of the frontal lobes is not only executing the dynamics of communicative production, but of first planning and calculating the intention to act. There must be a way of transferring these intentions into sequenced movements, and where language is involved, the sequences involve the construction of phrases and sentences. Volitional evocation of action matrices is seriously distrurbed; sentences are slow to come. The spontaneity of speech production for planned communicative narrative is difficult, at best. This breakdown is easily disassociated from much else of the language code, for the simple reason that these subjects can understand heard speech and can perform curiously well in naming (single objects or other elements); they have fluent speech (when the speech is less volitional and more automatic), their audition is unaffected, and they can repeat single or automated short stretches of speech. Without any intentional processing, these patients often "echo" what they hear from speakers—what is referred to as *echolalia.* Finally, patients with dynamic aphasia have much more difficulty with verb access than they do with nouns; the opposite accrues for sensory and semantic aphasia; there, the breakdown is more severe for nouns.

Transcortical Motor Aphasia. Later in his life, Luria integrated the classical syndrome known as transcortical motor aphasia, because that aphasia type has been in models of language breakdown since Ludwig Lichtheim's (1885) classic study and because its behavioral and neuroanatomical correlates have nearly matched Luria's dynamic aphasia. Luria's (1977) last assessment of these two anterior frontal syndromes is that transcortical motor aphasics subdivide upon close examination into one group whose volitional sequencing errors are replete with perseverations and another group who do not perseverate. Otherwise, the patient populations share much in common—neither group initiating much novel language production on their own. Both at least seem to be a

Alexander Luria. RIA NOVOSTI/SCIENCE PHOTO LIBRARY.

consequence of lesions anterior to Broca's area in prefrontal regions. Obviously, as Luria writes, much further careful neuropsychological and neurolinguistic research is needed with these dominant hemisphere frontal lobe syndromes.

Luria's International Influence. Alexander Luria put forth a herculean effort to penetrate the West and influence its scientific thought from the closed society that was the Soviet Union. Very few Soviet scientists had the success of Luria in projecting ideas from inside the Kremlin during most of the twentieth century. He paid early visits to Germany in 1925, where he met Kurt Goldstein, and to the United States in 1929, where he met the Czech linguist Roman Jakobson, who eventually taught at Harvard University. From 1930 to 1960, Luria had great difficulty in maintaining steady research laboratory investigative work, and during this period only a relatively small number of his larger works and some scattered papers in English reached beyond Russia's border.

Homskaya (2001, chap. 6) defines the 1960s as the decade of increased intensive research and active international outreach—largely through numerous visits to the Soviet Union by American and other renowned Western neuroscientists, such as Karl Pribrum. Luria took advantage of an increasing number of international congresses—especially the meetings of the International Congress of Psychology. In 1966 Luria organized and presided over the Eighteenth International Psychological Congress, which conveniently took place in Moscow. At this meeting, he also actively participated in other seminars and colloquia that addressed pathopsychology, electrophysiology, and biological bases of memory traces. That Eighteenth International Psychological Congress turned out to be one of the most influential meetings, with a swath of nations represented by the participants, as well as the scale of neuroscientific issues addressed. Participants in this Eighteenth Congress included Karl Pribrum and Hans-Lukas Teuber (U.S.), Brenda Milner (Canada), Oliver Zangwill (U.K.), and Henri Hecaen (France). Also in 1966, Luria was elected vice president of the International Association of Scientific Psychology, became an honorable foreign member of the American Academy of Arts and Sciences, and was elected as an honorary member in a number of national psychological societies throughout Europe.

Luria's works were eventually translated into English, German, French, Spanish, and some other languages. Mouton and Basic Books were high-profile publishing outlets for Luria's earlier works on traumatic aphasia, the working brain, and the novelesque works on two famous patients with disrupted memory and perception. Several of his books were later published by Harvard University Press. He read and wrote English quite well and worked judiciously with his English-language galleys—often helping with the original translations. He was an active member and manuscript reviewer on editorial boards for a number of international journals in the neurosciences: *Neuropsychologia, Cortex, Cognition, and Brain and Language*—to mention just a few.

During the Cold War years, many young Latin American students went to Russia to do graduate work in neuropsychology with Luria, all of whom brought back his theories and models to their native countries. Luria's work was in turn shared at the various conferences on neuropsychology in Latin America—especially in Mexico, Colombia, and Argentina. One of the most prominent and prodigious scholars from Luria's lab is Alfredo Ardila, a Colombian who teaches at Florida International University in Miami. Alexander Luria remains one of the most renowned and influential investigators of the neuropsychological sciences of brain and language, despite the fact that his research and publication trajectory emerged from the sealed society of the Soviet Union.

BIBLIOGRAPHY

The scientific biography by Eugenia D. Homskaya, cited below, contains the most exhaustive bibliography of works by and about Luria that has been published to date. Luria's publications are subdivided into Russian, English, and other languages. In the Spanish language listings, many works are edited and translated by Alfredo Ardila, Juan E. Azcoaga, and Jordi Pena-Casanova. The citations of Luria's works run from pages 127 to 169.

WORKS BY LURIA

Traumatic Aphasia. The Hague: Mouton, 1947.

Restoration of Functions after Brain Injury. New York: Macmillan, 1963.

"Factors and Forms of Aphasia." In *Disorders of Language,* edited by Anthony V. S. de Reuck and Maeve O'Connor. Proceedings from the Ciba Foundation Symposium on Disorders of Language. London: Churchill Livingstone, 1964. To be read along with Jakobson's paper from the same conference.

Higher Cortical Functions in Man. New York: Basic Books, 1966.

The Mind of a Mnemonist: A Little Book about a Vast Memory. New York: Basic Books, 1968. Reissued, Cambridge, MA: Harvard University Press, 1987. A major work of Luria's "romantic" science.

The Man with a Shattered World. New York: Basic Books, 1972. Reissued, Cambridge, MA: Harvard University Press, 1987. Another major work of Luria's "romantic" science.

The Working Brain. New York: Basic Books, 1973.

Basic Problems in Neurolinguistics. The Hague: Mouton, 1976.

Cognitive Development. Cambridge, MA: Harvard University Press, 1976.

Fundamentals of Neurolinguistics. New York: Basic Books, 1976.

Neuropsychological Studies in Aphasia, edited by Richard Hoops and Yvan Lebrun. Amsterdam: Swets and Zeitlinger, 1977. Some of Luria's major papers on aphasia—some not readily available elsewhere.

The Making of Mind. Cambridge, MA: Harvard University Press, 1979.

"Paradoxes of Memory." *Soviet Neurology and Psychiatry* 14 (1982): 3–13.

OTHER SOURCES

Ardila, Alfredo. "Spanish Application of Luria's Assessment Methods." *Neuropsychology Review* 9 (1999): 63–69. This work has an extensive bibliography of the Spanish-language publications on, about, or influenced strongly by Luria.

Cole, Michael. *Cultural Psychology: A Once and Future Discipline.* Cambridge, MA: Belknap/Harvard University Press, 1996.

Cole, Michael, and J. Wertsch. *Contemporary Implications of Vygotsky and Luria.* Heinz Werner Lecture Series, Vol. XXI. Worcester, MA: Clark University Press, 1976.

Goldberg, Elkhonon, ed. *Contemporary Neuropsychology and the Legacy of Luria.* Hillsdale, NJ: Lawrence Erlbaum, 1990.

Homskaya, Eugenia D. *Alexander Romanovich Luria: A Scientific Biography.* Edited, with a foreword, by David E. Tupper. Translated from the Russian by Daria Krotova. New York: Kluwer Academic/Plenum Publishers, 2001. This book is a major biography by one of Luria's students; it contains heretofore unavailable information on Luria from birth to death and includes an exhaustive bibliography of his works.

Kaczmarek, Bozydar L. J., ed. Special Issue for A. R. Luria. *Aphasiology* 9, no. 2 (1995): 97–206.

Pena-Casanova, Jordi, ed. Special Issue: A. R. Luria. *Journal of Neurolinguistics* 4, no. 1 (1989): 1–178.

Hugh Buckingham

LURIA, SALVADOR EDWARD (SALVATORE)

(*b.* Turin, Italy, 13 August 1912; *d.* Lexington, Massachusetts, 6 February 1991), *virology, bacterial genetics, molecular biology, cancer research.*

Luria was one of the central figures in the development of twentieth-century life sciences. His 1940s research on bacteriophage and its bacterial hosts laid the groundwork for the emergence of bacterial genetics and virology as independent disciplines. He shared the 1969 Nobel Prize in Physiology or Medicine with his longtime collaborators Max Delbrück and Alfred Hershey for "discoveries concerning virus replication and genetics and … the importance of your contributions to the biological and medical sciences." Luria also played a role in the establishment of molecular biology, as his first graduate student was James D. Watson. His career trajectory from basic research on microorganisms to cancer research parallels the development of the life sciences in the postwar era. An immigrant to the United States, Luria was a passionate participant in American politics who publicly articulated his scientific and political ideals in the context of the Cold War.

Origins and Early Scientific Training. Salvatore Luria was born in the Northern Italian city of Turin, the second son of a lower-middle-class Jewish family. He was educated at Turin's elite institutions, the Liceo Massimo d'Azeglio and the medical school of the University of Turin, at a time when the city was a seat of antifascist activity. At the Liceo, he studied philosophy and literature with Augusto Monti, a famous antifascist intellectual. In medical school, Luria worked for several years in the histology laboratory of the talented researcher Giuseppe Levi, who was also a committed antifascist. Although he was not politically active as a youth, these mentors impressed him with their dedication to democratic and rational ideals. Luria graduated from medical school with highest honors in 1935.

Luria was not interested in practicing medicine. During medical school, he became interested in the ways in which modern physics could be applied to problems in biology and genetics. After completing his required year of army service as a medical officer, he arranged to spend 1937–1938 studying with the physicist Enrico Fermi in Rome. During his year in Rome, he first encountered the work of physicist-turned-biologist Max Delbrück and began experimenting with bacteriophage, the viruses that attack bacteria. When Benito Mussolini passed a set of race laws that restricted the participation of Jews in public life in July 1938, Luria was prevented from pursuing a career in science in Italy. Luria took the opportunity to leave both Italy and medicine behind, and he moved to France to pursue virus research at the Institut Pasteur. He collaborated with Fernand Holweck and Eugène Wollman, and learned their statistical techniques for analyzing virus growth. When the Nazis invaded France in the spring of 1940, Luria fled again, this time to the United States via Portugal. He arrived in New York City on 12 September 1940.

Bacteriophage Research. As soon as he reached the United States, Luria began the process of becoming an American scientist. He secured funding from the Rockefeller Foundation and the Dazian Medical Foundation for Medical Research and was appointed to a temporary position in the Bacteriology Department at Columbia University. On his application for citizenship, Luria changed his first name from the Italian "Salvatore" to the Spanish spelling "Salvador" and added the English middle name "Edward." He first met Delbrück in December 1940, and the two immediately began to collaborate.

Despite their geographic separation, Luria and Delbrück had an instant intellectual closeness. They arranged to spend the summer of 1941 at Cold Spring Harbor Laboratories, in New York, investigating the genetic and biochemical properties of bacteriophage and its host *Escherichia coli.* Their first joint publication described the curious way that two different strains of virus interfere with each other's growth when both infect a colony of bacteria. They soon began to consider the implication of virus research for questions of resistance and evolutionary adaptation. Luria and Delbrück also collaborated with Thomas Anderson at the RCA Laboratories, taking some of the first electron micrograph images of bacteriophage.

During that summer at Cold Spring Harbor, Luria first met Milislav Demerec and Hermann Muller, both of whom were instrumental in helping him win a Guggenheim fellowship in 1942 and eventually to secure a permanent academic position. In January 1943, Luria moved to Bloomington, Indiana, to begin teaching in the Botany and Bacteriology Department at Indiana University. Luria and Delbrück returned to Cold Spring Harbor nearly every summer during the 1940s and 1950s to collaborate and to teach.

Luria flourished at Indiana professionally and personally. He formed close friendships with his colleagues including Leland McClung, Muller, and Tracy Sonneborn and began to train graduate students and postdoctoral researchers. His first advisee was James Dewey Watson, who soon became part of the larger community of phage researchers. In 1947 Luria invited his Italian medical school colleague Renato Dulbecco to work in his laboratory. Later, he was joined by Giuseppe Bertani, who worked on lysogeny and developed the standard culture for growing bacteria. Bertani referred to his "lysogeny broth" medium as "LB" in a 1951 paper, and the abbreviation soon became known as "Luria broth." In 1945, Luria married Zella Hurwitz, the only female graduate student in the Department of Psychology. Their son Daniel was born in 1948.

Bacterial Genetics. A few weeks after he arrived in Bloomington, Luria got an idea for an experiment one Saturday night while watching a colleague play a slot machine at a faculty dance. Because he and Delbrück had first worked with bacterial resistance to phage, Luria had considered the problem of how that resistance arose. He considered two possible hypotheses: that the resistance was somehow triggered by the presence of bacteriophage, or that some of the bacteria were already resistant as a result of random genetic mutations. However, it was unclear how to test these possibilities experimentally.

As he teased his friend about gambling, Luria realized that mutations in bacteria could be considered analogous to jackpots. He saw that jackpots in an unprogrammed slot machine are a series of random events that could be described with a Poisson distribution of rare independent events. Luria hypothesized that if bacterial resistance to bacteriophage was the result of spontaneous random mutations, then it would appear in a random distribution of "jackpots" in a series of cultures. Some cultures could have many resistant colonies while others could have none. If, on the other hand, resistance was acquired as a result of contact with phage, resistant colonies would be evenly distributed across all cultures. In either case, the average number of resistant colonies across all cultures could be the same, but the distribution would indicate whether the cause was spontaneous or not.

Luria ran the experiment with α, a virulent strain of phage, and as he anticipated, he found several jackpots randomly distributed among the plates. Luria was fortunate to have chosen virulent phage. If he had used a temperate strain, he would have seen that the presence of phage did induce resistance, and he would have concluded that it was acquired rather than hereditary. Excited by his results, Luria consulted with Delbrück, who noted that the experiment also provided a way to determine mutation rates with greater precision than had been possible before.

This experiment, known as the fluctuation test, was published in *Genetics* in 1943. The implications of Luria and Delbrück's work went far beyond the handful of researchers interested in phage. As Luria and others have pointed out, the 1943 paper dealt a blow to the neo-Lamarckian view that viruses somehow induced mutations in bacteria. In a 1947 review, Luria noted that bacteriology had been "one of the last strongholds of Lamarckism," because of the difficulty in providing direct evidence for the existence of both Mendelian traits and the characteristic Darwinian criteria of random change (p. 1). The fluctuation test marked the beginning of Luria's appreciation for the evolutionary implications of his genetics research, and his interest in larger biological questions was evident in his publications through the 1950s. Gunther Stent's classic textbook *Molecular Genetics* equates this publication with Gregor Mendel's 1865 paper on "Versuche über Pflanzenhybriden" (Experiments on plant hybrids), and historian Thomas Brock identifies it as the founding document of modern bacterial genetics.

Luria had a small role in the Western scientific campaign to discredit Trofim Lysenko's Lamarckian genetics program in the Soviet Union. In 1946, Luria drafted a letter to the British biologist J. B. S. (John Burdon Sanderson) Haldane asking him to publish a critique of Lysenko's book *Heredity and Its Variability*. Muller, Theodosius Dobzhansky, Leslie Dunn, Curt Stern, and Luria signed the letter, which enraged Haldane, a committed Communist, so much that he not only refused to write the review, but returned the original letter to Muller. Later in the 1940s and early 1950s, Luria served as a member of the Genetics Society of America's Committee to Aid Geneticists Abroad and he ran unsuccessfully for a spot on the society's Committee to Combat Anti-genetic Propaganda.

Luria's early experiments with bacteriophage earned him recognition from the biological community and helped establish viruses as useful research organisms. In 1945 he won a grant from the American Cancer Society's Committee on Growth, which was renewed for over twenty years. Luria delivered the first Slotin Memorial Lecture in 1948 and gave the 1950 Jesup Lectures at Columbia University on "The Reproduction of Viruses." Several top biology and bacteriology departments tried to recruit Luria and in 1950 he accepted an offer from the University of Illinois at Urbana-Champaign to join their Department of Bacteriology.

Restriction-Modification. Luria continued his innovative virus research in Illinois. His second key finding—host-induced modifications of viruses—was the result of a laboratory accident rather than a planned experiment. In the spring of 1952, Luria and his graduate student Mary Human reported a "novel situation" that they had observed. A test tube full of phage-sensitive *E. coli* culture broke, so Luria substituted a different bacterium, *Shigella dysenteriae,* and observed a surprising phenomenon. After exposure to *Shigella,* the viruses would not grow in *E. coli.* Some bacterial mutants had temporarily modified their viral invaders so that they could not reproduce in certain bacterial hosts.

In a series of experiments with different phages and a range of *E. coli* and *Shigella* hosts, some of the phages seemed to disappear, only to reappear when cultured with different bacteria. Luria and Human observed that one generation later, the phages once again reproduced in the original hosts. They concluded that whatever had caused the modification in the viruses was not a mutation, which would have caused permanent genetic change, or even evidence of the "peculiar plasticity of virus heredity" (Luria, 1953a, p. 237). Rather, they had found a new type of genetic variation, which they labeled "host-induced" (Luria and Human, 1952, p. 557) and later became known as the restriction-modification phenomenon.

For several years, restriction-modification was a laboratory curiosity for virologists and bacterial geneticists. In the 1960s, however, other researchers discovered that the temporary changes were the result of bacterial restriction enzymes that degraded viral DNA as a defense mechanism. Restriction enzymes, which recognize and target short strands of DNA, were key for the development of recombinant DNA technology in the late 1960s and early 1970s, which in turn is the basis for genetic engineering and other types of genetic manipulation. Luria later acknowledged that his role in the history of recombinant DNA technology was "accidental" and "serendipitous" (Luria, 1983, p. 57); this episode provides further evidence of his far-reaching influence on modern biology.

Virology. The 1950s were a critical decade in the history of virology. Luria was a strong force in the establishment of virology as an independent discipline and he published several important pieces on the utility of viruses as a fundamental biological unit. In "Bacteriophage: An Essay on Virus Reproduction" published in May 1950 in *Science,* he made a clear argument for virology as the key discipline that would help biologists reach their "ultimate goal, the identification of the elementary 'replicating units' of biological material and the clarification of their mode of reproduction" (p. 511). This essay, also published as a core reading for a California Institute of Technology conference on viruses in March 1950, gives an operational definition of viruses that emphasizes their research functions rather than their disease-causing features. A virus is "an exogenous submicroscopic unit capable of

Salvador Edward Luria. © BETTMANN/CORBIS.

multiplication only inside specific living cells" (p. 507). It describes the relationship between viruses and hosts as "parasitism at the genetic level," and argues that understanding the biochemical and genetic features of that relationship would lead directly to a detailed description of all biological replication (pp. 510–511).

Nearly three years later, in April 1953, Luria's student James Watson and his Cambridge colleague Francis Crick announced that they had discovered the double helical structure of DNA, the molecule that constitutes genetic material. Watson, Stent, and John Cairns bound the early history of molecular biology with bacteriophage in their festschrift for Delbrück's sixtieth birthday, the classic *Phage and the Origins of Molecular Biology.* Although historians have demonstrated that many disciplinary strands came together to form molecular biology, the bacteriophage experiments performed by Seymour Cohen, Alfred Hershey, and Martha Chase provided critical evidence that DNA is the physical location of genetic material.

General Virology. The year 1953 was a turning point in the independent disciplinary history of virology as well. That year, Luria published *General Virology,* the first comprehensive textbook in the field. The text emerged from Luria's teaching experience in Indiana. In the preface, he recalled how in 1946 he began planning a new course in virology, one that was not "a watered-down course in virus diseases," but rather "a new type of course, in which virology would be presented as a biological science, like botany, zoology, or general bacteriology" (p. ix). His approach for the course and for the textbook accepted the ambiguities inherent in nonmedical virus research, and made it "a central concept, that of the dual nature of viruses as inert particles on the one hand, and as operating constituents of functional cells on the other hand" (p. x). Physics, chemistry, biochemistry, and cell physiology were thus integrated into the study of virus properties and behavior, while pathology was relegated to a supporting role.

In the textbook, Luria argued that virology should be considered an integrative and interpretive science, rather than a taxonomic one. By defining virology as the science concerned with the genetic and chemical properties and functions of viruses, he emphasized methodology over taxonomy. He was unequivocal about the relevance of virus research to fundamental biological questions. Some biologists felt that viruses should not be considered living, since they could not reproduce independently. However, Luria felt that the genetic dependence of viruses "makes them invaluable to the biologist, whom they present with the unique opportunity of observing in isolation the active determinants of biological specificity, which are truly the stuff of which all life is made" (p. 363). Luria took pains to establish virology as a basic science, but from the very start, his research had implications for the applied field of cancer research.

Viruses and Cancer. In 1959 the American Cancer Society organized a three-day meeting to discuss "The Possible Role of Viruses in Cancer" and to assess the state of the field. The conference brought together a diverse group of the most prominent researchers in the areas of animal, plant, and bacterial viruses, including Peyton Rous, Wendell Stanley, Renato Dulbecco, François Jacob, and André Lwoff. Luria was invited to give the keynote orienting address for the meeting and the papers were published in *Cancer Research* in 1960.

In "Viruses, Cancer Cells, and the Genetic Concept of Virus Infection," Luria reviewed basic virology findings that were relevant to cellular function, which is a crucial element in the understanding of cancer as the result of abnormal cell growth and control. He noted that viruses could be implicated in the development of cancer either as the cause of somatic mutations or as a direct infection of a tumor-causing agent. In both cases, viruses could be seen as causing a type of cellular mutation since the "entry of the viral genome is a genetic change" (p. 679) and viral tumors thus would be examples of "infective heredity at the cellular level" (p. 680). While this view of the relationship between cancer and viruses was not universally accepted, it was a strong approach taken by many researchers for the next several decades.

1950s: Early Political Activity. Because of his left-leaning political affiliations, Luria attracted the attention of the Federal Bureau of Investigation (FBI) in 1950. A two-year investigation yielded no evidence of Communist or other subversive activity, but Luria was nevertheless denied a passport in 1952. He was scheduled to give a lecture at the Society for General Microbiology meeting in Oxford, England, and to visit his mother in Italy, but he was told by Ruth Shipley, chief of the passport division of the State Department, that his "proposed travel would not be in the best interest of the United States" (letter from Shipley to S.E.L., 25 January 1952, State Department correspondence folder, Luria papers, American Philosophical Society). Despite appeals by Luria and the president of the University of Illinois, Luria was not granted permission to leave the United States. James Watson reported on the latest phage research in Luria's stead, and Luria was not granted a passport until 1959.

This treatment did not deter him, however, and Luria increased his political activities during the 1950s. He helped organize protests against Illinois state laws requiring university professors and other employees to sign loyalty oaths, and was active in local labor and desegregation campaigns. In his capacity as a vice president of the American Association for the Advancement of Science, he participated in a lively debate about the appropriateness of holding a national scientific meeting in racially segregated Atlanta. Luria was one of the first signers of Linus Pauling's 1957 "Appeal by American Scientists to the Government and Peoples of the World" to ban nuclear weapons.

Move to Boston. In 1959 the Luria family moved to Boston. Because rules against nepotism barred two members of the same family from teaching at the University of Illinois, Zella Luria could not find work as a psychology professor in Urbana. She was offered a position at Tufts University and Salvador accepted an invitation from the Massachusetts Institute of Technology (MIT) to join their Department of Biology. With the exception of the time he spent at the Pasteur Institute in 1963 as a Guggenheim Fellow, Luria remained at MIT for the rest of his career.

At MIT, Luria was instrumental in transforming their Department of Biology into a world-class research community. Along with fellow faculty members Boris Magasanik, Alexander Rich, and Irwin Sizer, he recruited top researchers and graduate students, including David Baltimore. He helped shift the department's pedagogical focus toward genetics and microbiology and in the process increased undergraduate enrollment significantly. Luria was elected to the National Academy of Sciences in 1960. In 1966, he was named the first William Thompson Sedgwick Professor of Biology and in 1970 he became an institute professor. Luria's status as an authority on viruses and cancer helped MIT win one of the first federal grants for basic cancer research as part of the "war on cancer." Luria served as the director of MIT's Center for Cancer Research from 1972 until he retired in 1985.

Protesting the Vietnam War. The height of Luria's political involvement came in the 1960s and 1970s. He was one of the founding members of the Boston Area Faculty Group on Public Issues, a cohort of academics who sponsored advertisements and wrote editorials on political issues. Their main focus was on protesting the Vietnam War in a series of advertisements in the *New York Times* and other national publications. The most dramatic advertisement appeared in January 1967. Surrounding the words "Mr. President, Stop the Bombing" were the names of more than two thousand academics from around the United States.

Luria's leadership role in the scientific community gave him the opportunity to mobilize biologists against the Vietnam War. As the president of the American Society for Microbiology in 1968, he announced that the society would terminate its advisory relationship with the U.S. Army's Biological Laboratory at Fort Detrick, Maryland. In a widely publicized address on "The Microbiologist and His Times," Luria articulated a vision of "a society in which science will flourish, both as a liberating intellectual activity and as the source of a beneficial technology" (1968, p. 403).

Public Figure. Luria's public roles converged in one dramatic weekend in October 1969. He helped organize the Vietnam War Moratorium in Boston on 15 October. The next day, his Nobel Prize was announced, and Luria declared that he would donate part of his prize money to the peace movement. On 20 October, Luria again made headlines when his name was found on a list of scientists who had been blacklisted from serving on advisory committees at the National Institutes of Health, presumably because of their political views.

Late in his career, Luria had many opportunities to argue for the value of science in a democratic society. He testified before Senate and Congressional committees on science policy, and was one of the first fellows of the Salk

Institute in La Jolla, California. He wrote articles, textbooks, curricula, and a popular science book. In *Life: The Unfinished Experiment* (1973), he explained evolution, genetics, and molecular biology to a general audience while critiquing sociobiology and other examples of biological determinism. This book was a critical success and won the National Book Award in 1974. Luria continued to publish essays and op-eds on scientific and political issues until his death from cancer on 6 February 1991.

BIBLIOGRAPHY

A complete bibliography is available in Luria's papers, deposited at the American Philosophical Society Library in Philadelphia. A list of many of his scientific publications since 1955 can be retrieved through the Web of Science database. Other archival sources include the collections at the American Society for Microbiology, the Cold Spring Harbor Laboratory, Indiana University, the Massachusetts Institute of Technology, the Rockefeller Archive Center, and the University of Illinois, Urbana-Champaign. Luria's FBI files were obtained by this author through a Freedom of Information Act request in 1999.

WORKS BY LURIA

With Thomas F. Anderson. "The Identification and Characterization of Bacteriophages with the Electron Microscope." *Proceedings of the National Academy of Sciences of the United States of America* 28 (1942): 127–130.

With Max Delbrück. "Interference between Bacterial Viruses: I. Interference between Two Bacterial Viruses Acting upon the Same Host, and the Mechanism of Virus Growth." *Archives of Biochemistry* 1 (1942): 111–141.

With Max Delbrück. "Interference between Inactivated Bacterial Virus and Active Virus of the Same Strain and of a Different Strain." *Archives of Biochemistry* 1 (1942): 207–218.

With Max Delbrück. "Mutations of Bacteria from Virus Sensitivity to Virus Resistance." *Genetics* 28 (1943): 491–511.

With Max Delbrück and Thomas F. Anderson. "Electron Microscope Studies of Bacterial Viruses." *Journal of Bacteriology* 46 (1943): 57–76.

"Recent Advances in Bacterial Genetics." *Bacteriological Reviews* 11 (1947): 1–40.

With Renato Dulbecco. "Genetic Recombinations Leading to Production of Active Bacteriophage from Ultraviolet Inactivated Bacteriophage Particles." *Genetics* 34 (1949): 93–125.

"Bacteriophage: An Essay on Virus Reproduction." *Science* 111 (12 May 1950): 507–511. Reprinted in *Viruses 1950,* edited by Max Delbrück. Pasadena: Division of Biology of the California Institute of Technology, 1950.

With Mary L. Human. "A Nonhereditary, Host-Induced Variation of Bacterial Viruses." *Journal of Bacteriology* 64 (1952): 557–569.

General Virology. New York: John Wiley and Sons, 1953a.

"Host-Induced Modifications of Viruses." In *Viruses.* Cold Spring Harbor Symposia on Quantitative Biology 18. Cold Spring Harbor, NY: Cold Spring Harbor Press, 1953b.

"Viruses, Cancer Cells, and the Genetic Concept of Virus Infection." *Cancer Research* 20 (June 1960): 677–688.

"Mutations of Bacteria and of Bacteriophage." In *Phage and the Origins of Molecular Biology,* edited by John Cairns, Gunther Stent, and James D. Watson. Cold Spring Harbor, NY: Cold Spring Harbor Laboratory Press, 1966. Revised and expanded ed., 1992.

With James E. Darnell Jr. *General Virology.* 2nd ed. New York: John Wiley and Sons, 1967.

"The Microbiologist and His Times." *Bacteriological Reviews* 32 (December 1968): 401–403.

Life: The Unfinished Experiment. New York: Charles Scribner's Sons, 1973.

36 Lectures in Biology. Cambridge, MA: MIT Press, 1975.

"Phage; Colicins and Macroregulatory Phenomena." In *Nobel Lectures in Molecular Biology, 1933–1975,* edited by David Baltimore. New York: Elsevier, 1977.

With James E. Darnell Jr., David Baltimore, and Allan Campbell. *General Virology.* 3rd ed. New York: John Wiley and Sons, 1978.

With Stephen Jay Gould and Sam Singer. *A View of Life.* Menlo Park, CA: Benjamin/Cummings, 1981.

"Ethical and Institutional Aspects of Recombinant DNA Technology." In *Recombinant DNA Research and the Human Prospect,* edited by Earl D. Hanson. Washington, DC: American Chemical Society, 1983.

A Slot Machine, a Broken Test Tube: An Autobiography. New York: Harper and Row, 1984.

OTHER SOURCES

Abir-Am, Pnina G. "Entre mémoire collective et histoire en biologie moléculaire: les premiers rites commémoratifs pour les groupes fondateurs." In *La mise en mémoire de la science: Pour une ethnographie historique des rites commémoratifs,* edited by Pnina G. Abir-Am. Amsterdam: Overseas Publishers Association, 1998.

Bertani, Giuseppe. "Lysogeny at Mid-Twentieth Century: P1, P2, and Other Experimental Systems." *Journal of Bacteriology* 186 (2004): 595–600.

Brock, Thomas D. *The Emergence of Bacterial Genetics.* Cold Spring Harbor, NY: Cold Spring Harbor Laboratory Press, 1990.

Cairns, John, Gunther Stent, and James D. Watson, eds. *Phage and the Origins of Molecular Biology.* Cold Spring Harbor, NY: Cold Spring Harbor Laboratory Press, 1966. Revised and expanded ed., 1992. This volume, published in honor of Max Delbrück's sixtieth birthday, was the first attempt by molecular biologists to write their own history.

Fischer, Ernst Peter, and Carol Lipson. *Thinking about Science: Max Delbrück and the Origins of Molecular Biology.* New York: W. W. Norton, 1988.

Judson, Horace Freeland. *The Eighth Day of Creation: Makers of the Revolution in Biology.* New York: Simon & Schuster, 1979. One of the standard historical accounts of the emergence of molecular biology, based on extensive interviews with the participants.

Kay, Lily E. "Conceptual Models and Analytical Tools: The Biology of Physicist Max Delbruck." *Journal of the History of Biology* 18, no. 2 (Summer 1985): 207–247.

———. *The Molecular Vision of Life: Caltech, the Rockefeller Foundation and the Rise of the New Biology.* Oxford: Oxford University Press, 1993.

———. *Who Wrote the Book of Life: A History of the Genetic Code.* Stanford, CA: Stanford University Press, 2000.

Krementsov, Nikolai. "A 'Second Front' in Soviet Genetics: The International Dimension of the Lysenko Controversy, 1944–47." *Journal of the History of Biology* 29 (1996): 229–250.

Lysenko, Trofim Denisovich. *Heredity and Its Variability.* Translated from the Russian by Theodosius Dobzhansky. New York: King's Crown Press, 1945.

Mendel, Gregor. "Versuche über Pflanzenhybriden." *Verhandlungen des Naturforschenden Vereins in Brünn* 4 (1866): 3–57. Read at 1865 meeting.

Morange, Michel. *A History of Molecular Biology.* Cambridge, MA: Harvard University Press, 1998.

"Nobel Prize in Physiology or Medicine, 1969." Nobel Prize Web site. Available from http://nobelprize.org/nobel_prizes/medicine/.

Olby, Robert. *The Path to the Double Helix: The Discovery of DNA.* Seattle: University of Washington Press, 1974. 2nd ed., 1994. Another standard historical account, focusing more on the technological and experimental roots of the discovery of DNA.

Rasmussen, Nicolas. *Picture Control: The Electron Microscope and the Transformation of Biology in America, 1940–1960.* Stanford, CA: Stanford University Press, 1997.

Selya, Rena. "Salvador Luria's Unfinished Experiment: The Public Life of a Biologist in Cold War America." PhD diss., Harvard University, 2002.

Stent, Gunther S. *Molecular Genetics: An Introductory Narrative.* San Francisco: W. H. Freeman, 1970.

Summers, William C. "How Bacteriophage Came to Be Used by the Phage Group." *Journal of the History of Biology* 26 (1993): 255–267.

Watson, James D. "Growing up in the Phage Group." In *Phage and the Origins of Molecular Biology,* edited by John Cairns, Gunther Stent, and James D. Watson. Cold Spring Harbor, NY: Cold Spring Harbor Laboratory Press, 1966. Revised and expanded ed., 1992.

———. *The Double Helix: A Personal Account of the Discovery of the Structure of DNA.* Norton Critical Edition, edited by Gunther Stent. New York: W. W. Norton, 1980. James Watson was Luria's first graduate student. These two works reflect Watson's intense interest in shaping how the history of molecular biology is written and presented to the public.

Rena Selya

LUX

SEE **Royer, Clémence.**